COLLEGE ALGEBRA
VERSION $\lfloor \pi \rfloor$
CORRECTED EDITION

BY

Carl Stitz, Ph.D.   Jeff Zeager, Ph.D.
Lakeland Community College   Lorain County Community College

JULY 4, 2013

ACKNOWLEDGEMENTS

While the cover of this textbook lists only two names, the book as it stands today would simply not exist if not for the tireless work and dedication of several people. First and foremost, we wish to thank our families for their patience and support during the creative process. We would also like to thank our students - the sole inspiration for the work. Among our colleagues, we wish to thank Rich Basich, Bill Previts, and Irina Lomonosov, who not only were early adopters of the textbook, but also contributed materials to the project. Special thanks go to Katie Cimperman, Terry Dykstra, Frank LeMay, and Rich Hagen who provided valuable feedback from the classroom. Thanks also to David Stumpf, Ivana Gorgievska, Jorge Gerszonowicz, Kathryn Arocho, Heather Bubnick, and Florin Muscutariu for their unwaivering support (and sometimes defense) of the book. From outside the classroom, we wish to thank Don Anthan and Ken White, who designed the electric circuit applications used in the text, as well as Drs. Wendy Marley and Marcia Ballinger for the Lorain CCC enrollment data used in the text. The authors are also indebted to the good folks at our schools' bookstores, Gwen Sevtis (Lakeland CC) and Chris Callahan (Lorain CCC), for working with us to get printed copies to the students as inexpensively as possible. We would also like to thank Lakeland folks Jeri Dickinson, Mary Ann Blakeley, Jessica Novak, and Corrie Bergeron for their enthusiasm and promotion of the project. The administrations at both schools have also been very supportive of the project, so from Lakeland, we wish to thank Dr. Morris W. Beverage, Jr., President, Dr. Fred Law, Provost, Deans Don Anthan and Dr. Steve Oluic, and the Board of Trustees. From Lorain County Community College, we wish to thank Dr. Roy A. Church, Dr. Karen Wells, and the Board of Trustees. From the Ohio Board of Regents, we wish to thank former Chancellor Eric Fingerhut, Darlene McCoy, Associate Vice Chancellor of Affordability and Efficiency, and Kelly Bernard. From OhioLINK, we wish to thank Steve Acker, John Magill, and Stacy Brannan. We also wish to thank the good folks at WebAssign, most notably Chris Hall, COO, and Joel Hollenbeck (former VP of Sales.) Last, but certainly not least, we wish to thank all the folks who have contacted us over the interwebs, most notably Dimitri Moonen and Joel Wordsworth, who gave us great feedback, and Antonio Olivares who helped debug the source code.

TABLE OF CONTENTS

Preface   ix

1 Relations and Functions   1
    1.1 Sets of Real Numbers and the Cartesian Coordinate Plane 1
       1.1.1 Sets of Numbers 1
       1.1.2 The Cartesian Coordinate Plane 6
       1.1.3 Distance in the Plane 10
       1.1.4 Exercises 14
       1.1.5 Answers 17
    1.2 Relations 20
       1.2.1 Graphs of Equations 23
       1.2.2 Exercises 29
       1.2.3 Answers 33
    1.3 Introduction to Functions 43
       1.3.1 Exercises 49
       1.3.2 Answers 53
    1.4 Function Notation 55
       1.4.1 Modeling with Functions 60
       1.4.2 Exercises 63
       1.4.3 Answers 69
    1.5 Function Arithmetic 76
       1.5.1 Exercises 84
       1.5.2 Answers 87
    1.6 Graphs of Functions 93
       1.6.1 General Function Behavior 100
       1.6.2 Exercises 107
       1.6.3 Answers 114
    1.7 Transformations 120
       1.7.1 Exercises 140
       1.7.2 Answers 144

2 Linear and Quadratic Functions — 151

- 2.1 Linear Functions 151
  - 2.1.1 Exercises 163
  - 2.1.2 Answers 169
- 2.2 Absolute Value Functions 173
  - 2.2.1 Exercises 183
  - 2.2.2 Answers 184
- 2.3 Quadratic Functions 188
  - 2.3.1 Exercises 200
  - 2.3.2 Answers 203
- 2.4 Inequalities with Absolute Value and Quadratic Functions 208
  - 2.4.1 Exercises 220
  - 2.4.2 Answers 222
- 2.5 Regression 225
  - 2.5.1 Exercises 230
  - 2.5.2 Answers 233

3 Polynomial Functions — 235

- 3.1 Graphs of Polynomials 235
  - 3.1.1 Exercises 246
  - 3.1.2 Answers 250
- 3.2 The Factor Theorem and the Remainder Theorem 257
  - 3.2.1 Exercises 265
  - 3.2.2 Answers 267
- 3.3 Real Zeros of Polynomials 269
  - 3.3.1 For Those Wishing to use a Graphing Calculator 270
  - 3.3.2 For Those Wishing NOT to use a Graphing Calculator 273
  - 3.3.3 Exercises 280
  - 3.3.4 Answers 283
- 3.4 Complex Zeros and the Fundamental Theorem of Algebra 287
  - 3.4.1 Exercises 295
  - 3.4.2 Answers 297

4 Rational Functions — 301

- 4.1 Introduction to Rational Functions 301
  - 4.1.1 Exercises 314
  - 4.1.2 Answers 316
- 4.2 Graphs of Rational Functions 320
  - 4.2.1 Exercises 333
  - 4.2.2 Answers 335
- 4.3 Rational Inequalities and Applications 342
  - 4.3.1 Variation 350
  - 4.3.2 Exercises 353

| | | 4.3.3 | Answers | 356 |

5 Further Topics in Functions — 359
- 5.1 Function Composition — 359
  - 5.1.1 Exercises — 369
  - 5.1.2 Answers — 372
- 5.2 Inverse Functions — 378
  - 5.2.1 Exercises — 394
  - 5.2.2 Answers — 396
- 5.3 Other Algebraic Functions — 397
  - 5.3.1 Exercises — 407
  - 5.3.2 Answers — 411

6 Exponential and Logarithmic Functions — 417
- 6.1 Introduction to Exponential and Logarithmic Functions — 417
  - 6.1.1 Exercises — 429
  - 6.1.2 Answers — 433
- 6.2 Properties of Logarithms — 437
  - 6.2.1 Exercises — 445
  - 6.2.2 Answers — 447
- 6.3 Exponential Equations and Inequalities — 448
  - 6.3.1 Exercises — 456
  - 6.3.2 Answers — 458
- 6.4 Logarithmic Equations and Inequalities — 459
  - 6.4.1 Exercises — 466
  - 6.4.2 Answers — 468
- 6.5 Applications of Exponential and Logarithmic Functions — 469
  - 6.5.1 Applications of Exponential Functions — 469
  - 6.5.2 Applications of Logarithms — 477
  - 6.5.3 Exercises — 482
  - 6.5.4 Answers — 490

7 Hooked on Conics — 495
- 7.1 Introduction to Conics — 495
- 7.2 Circles — 498
  - 7.2.1 Exercises — 502
  - 7.2.2 Answers — 503
- 7.3 Parabolas — 505
  - 7.3.1 Exercises — 512
  - 7.3.2 Answers — 513
- 7.4 Ellipses — 516
  - 7.4.1 Exercises — 525
  - 7.4.2 Answers — 527

	7.5	Hyperbolas	531
		7.5.1 Exercises	541
		7.5.2 Answers	544

8 Systems of Equations and Matrices — 549

- 8.1 Systems of Linear Equations: Gaussian Elimination . . . 549
  - 8.1.1 Exercises . . . 562
  - 8.1.2 Answers . . . 564
- 8.2 Systems of Linear Equations: Augmented Matrices . . . 567
  - 8.2.1 Exercises . . . 574
  - 8.2.2 Answers . . . 576
- 8.3 Matrix Arithmetic . . . 578
  - 8.3.1 Exercises . . . 591
  - 8.3.2 Answers . . . 595
- 8.4 Systems of Linear Equations: Matrix Inverses . . . 598
  - 8.4.1 Exercises . . . 609
  - 8.4.2 Answers . . . 612
- 8.5 Determinants and Cramer's Rule . . . 614
  - 8.5.1 Definition and Properties of the Determinant . . . 614
  - 8.5.2 Cramer's Rule and Matrix Adjoints . . . 618
  - 8.5.3 Exercises . . . 623
  - 8.5.4 Answers . . . 627
- 8.6 Partial Fraction Decomposition . . . 628
  - 8.6.1 Exercises . . . 635
  - 8.6.2 Answers . . . 636
- 8.7 Systems of Non-Linear Equations and Inequalities . . . 637
  - 8.7.1 Exercises . . . 646
  - 8.7.2 Answers . . . 648

9 Sequences and the Binomial Theorem — 651

- 9.1 Sequences . . . 651
  - 9.1.1 Exercises . . . 658
  - 9.1.2 Answers . . . 660
- 9.2 Summation Notation . . . 661
  - 9.2.1 Exercises . . . 670
  - 9.2.2 Answers . . . 672
- 9.3 Mathematical Induction . . . 673
  - 9.3.1 Exercises . . . 678
  - 9.3.2 Selected Answers . . . 679
- 9.4 The Binomial Theorem . . . 681
  - 9.4.1 Exercises . . . 691
  - 9.4.2 Answers . . . 692

PREFACE

Thank you for your interest in our book, but more importantly, thank you for taking the time to read the Preface. I always read the Prefaces of the textbooks which I use in my classes because I believe it is in the Preface where I begin to understand the authors - who they are, what their motivation for writing the book was, and what they hope the reader will get out of reading the text. Pedagogical issues such as content organization and how professors and students should best use a book can usually be gleaned out of its Table of Contents, but the reasons behind the choices authors make should be shared in the Preface. Also, I feel that the Preface of a textbook should demonstrate the authors' love of their discipline and passion for teaching, so that I come away believing that they really want to help students and not just make money. Thus, I thank my fellow Preface-readers again for giving me the opportunity to share with you the need and vision which guided the creation of this book and passion which both Carl and I hold for Mathematics and the teaching of it.

Carl and I are natives of Northeast Ohio. We met in graduate school at Kent State University in 1997. I finished my Ph.D in Pure Mathematics in August 1998 and started teaching at Lorain County Community College in Elyria, Ohio just two days after graduation. Carl earned his Ph.D in Pure Mathematics in August 2000 and started teaching at Lakeland Community College in Kirtland, Ohio that same month. Our schools are fairly similar in size and mission and each serves a similar population of students. The students range in age from about 16 (Ohio has a Post-Secondary Enrollment Option program which allows high school students to take college courses for free while still in high school.) to over 65. Many of the "non-traditional" students are returning to school in order to change careers. A majority of the students at both schools receive some sort of financial aid, be it scholarships from the schools' foundations, state-funded grants or federal financial aid like student loans, and many of them have lives busied by family and job demands. Some will be taking their Associate degrees and entering (or re-entering) the workforce while others will be continuing on to a four-year college or university. Despite their many differences, our students share one common attribute: they do not want to spend $200 on a College Algebra book.

The challenge of reducing the cost of textbooks is one that many states, including Ohio, are taking quite seriously. Indeed, state-level leaders have started to work with faculty from several of the colleges and universities in Ohio and with the major publishers as well. That process will take considerable time so Carl and I came up with a plan of our own. We decided that the best way to help our students right now was to write our own College Algebra book and give it away electronically for free. We were granted sabbaticals from our respective institutions for the Spring semester of 2009 and actually began writing the textbook on December 16, 2008. Using an open-source text editor called TexNicCenter and an open-source distribution of LaTeX called MikTex 2.7, Carl and I wrote and edited all of the text, exercises and answers and created all of the graphs (using Metapost within LaTeX) for Version $0.\overline{9}$ in about eight months. (We choose to create a text in only black and white to keep printing costs to a minimum for those students who prefer a printed edition. This somewhat Spartan page layout stands in sharp relief to the explosion of colors found in most other College Algebra texts, but neither Carl nor I believe the four-color print adds anything of value.) I used the book in three sections of College Algebra at Lorain County Community College in the Fall of 2009 and Carl's colleague, Dr. Bill Previts, taught a section of College Algebra at Lakeland with the book that semester as well. Students had the option of downloading the book as a .pdf file from our website www.stitz-zeager.com or buying a low-cost printed version from our colleges' respective bookstores. (By giving this book away for free electronically, we end the cycle of new editions appearing every 18 months to curtail the used book market.) During Thanksgiving break in November 2009, many additional exercises written by Dr. Previts were added and the typographical errors found by our students and others were corrected. On December 10, 2009, Version $\sqrt{2}$ was released. The book remains free for download at our website and by using Lulu.com as an on-demand printing service, our bookstores are now able to provide a printed edition for just under \$19. Neither Carl nor I have, or will ever, receive any royalties from the printed editions. As a contribution back to the open-source community, all of the LaTeX files used to compile the book are available for free under a Creative Commons License on our website as well. That way, anyone who would like to rearrange or edit the content for their classes can do so as long as it remains free.

The only disadvantage to not working for a publisher is that we don't have a paid editorial staff. What we have instead, beyond ourselves, is friends, colleagues and unknown people in the open-source community who alert us to errors they find as they read the textbook. What we gain in not having to report to a publisher so dramatically outweighs the lack of the paid staff that we have turned down every offer to publish our book. (As of the writing of this Preface, we've had three offers.) By maintaining this book by ourselves, Carl and I retain all creative control and keep the book our own. We control the organization, depth and rigor of the content which means we can resist the pressure to diminish the rigor and homogenize the content so as to appeal to a mass market. A casual glance through the Table of Contents of most of the major publishers' College Algebra books reveals nearly isomorphic content in both order and depth. Our Table of Contents shows a different approach, one that might be labeled "Functions First." To truly use The Rule of Four, that is, in order to discuss each new concept algebraically, graphically, numerically and verbally, it seems completely obvious to us that one would need to introduce functions first. (Take a moment and compare our ordering to the classic "equations first, then the Cartesian Plane and THEN functions" approach seen in most of the major players.) We then introduce a class of functions and discuss the equations, inequalities (with a heavy emphasis on sign diagrams) and applications which involve functions in that class. The material is presented at a level that definitely prepares a student for Calculus while giving them relevant Mathematics which can be used in other classes as well. Graphing calculators are used sparingly and only as a tool to enhance the Mathematics, not to replace it. The answers to nearly all of the computational homework exercises are given in the text and we have gone to great lengths to write some very thought provoking discussion questions whose answers are not given. One will notice that our exercise sets are much shorter than the traditional sets of nearly 100 "drill and kill" questions which build skill devoid of understanding. Our experience has been that students can do about 15-20 homework exercises a night so we very carefully chose smaller sets of questions which cover all of the necessary skills and get the students thinking more deeply about the Mathematics involved.

Critics of the Open Educational Resource movement might quip that "open-source is where bad content goes to die," to which I say this: take a serious look at what we offer our students. Look through a few sections to see if what we've written is bad content in your opinion. I see this open-source book not as something which is "free and worth every penny", but rather, as a high quality alternative to the business as usual of the textbook industry and I hope that you agree. If you have any comments, questions or concerns please feel free to contact me at jeff@stitz-zeager.com or Carl at carl@stitz-zeager.com.

Jeff Zeager
Lorain County Community College
January 25, 2010

Chapter 1

Relations and Functions

1.1 Sets of Real Numbers and the Cartesian Coordinate Plane

1.1.1 Sets of Numbers

While the authors would like nothing more than to delve quickly and deeply into the sheer excitement that is *Precalculus*, experience[1] has taught us that a brief refresher on some basic notions is welcome, if not completely necessary, at this stage. To that end, we present a brief summary of 'set theory' and some of the associated vocabulary and notations we use in the text. Like all good Math books, we begin with a definition.

> **Definition 1.1.** A **set** is a well-defined collection of objects which are called the 'elements' of the set. Here, 'well-defined' means that it is possible to determine if something belongs to the collection or not, without prejudice.

For example, the collection of letters that make up the word "smolko" is well-defined and is a set, but the collection of the worst math teachers in the world is **not** well-defined, and so is **not** a set.[2] In general, there are three ways to describe sets. They are

> **Ways to Describe Sets**
>
> 1. **The Verbal Method:** Use a sentence to define a set.
>
> 2. **The Roster Method:** Begin with a left brace '{', list each element of the set *only once* and then end with a right brace '}'.
>
> 3. **The Set-Builder Method:** A combination of the verbal and roster methods using a "dummy variable" such as x.

For example, let S be the set described *verbally* as the set of letters that make up the word "smolko". A **roster** description of S would be $\{s, m, o, l, k\}$. Note that we listed 'o' only once, even though it

---

[1] ...to be read as 'good, solid feedback from colleagues'...
[2] For a more thought-provoking example, consider the collection of all things that do not contain themselves - this leads to the famous Russell's Paradox.

appears twice in "smolko." Also, the *order* of the elements doesn't matter, so $\{k, l, m, o, s\}$ is also a roster description of S. A **set-builder** description of S is:

$$\{x \mid x \text{ is a letter in the word "smolko".}\}$$

The way to read this is: 'The set of elements x such that x is a letter in the word "smolko."' In each of the above cases, we may use the familiar equals sign '=' and write $S = \{s, m, o, l, k\}$ or $S = \{x \mid x \text{ is a letter in the word "smolko".}\}$. Clearly m is in S and q is not in S. We express these sentiments mathematically by writing $m \in S$ and $q \notin S$. Throughout your mathematical upbringing, you have encountered several famous sets of numbers. They are listed below.

Sets of Numbers

1. The **Empty Set**: $\emptyset = \{\} = \{x \mid x \neq x\}$. This is the set with no elements. Like the number '0,' it plays a vital role in mathematics.[a]

2. The **Natural Numbers**: $\mathbb{N} = \{1, 2, 3, \ldots\}$ The periods of ellipsis here indicate that the natural numbers contain 1, 2, 3, 'and so forth'.

3. The **Whole Numbers**: $\mathbb{W} = \{0, 1, 2, \ldots\}$

4. The **Integers**: $\mathbb{Z} = \{\ldots, -3, -2, -1, 0, 1, 2, 3, \ldots\}$

5. The **Rational Numbers**: $\mathbb{Q} = \{\frac{a}{b} \mid a \in \mathbb{Z} \text{ and } b \in \mathbb{Z}\}$. Rational numbers are the ratios of integers (provided the denominator is not zero!) It turns out that another way to describe the rational numbers[b] is:

   $$\mathbb{Q} = \{x \mid x \text{ possesses a repeating or terminating decimal representation.}\}$$

6. The **Real Numbers**: $\mathbb{R} = \{x \mid x \text{ possesses a decimal representation.}\}$

7. The **Irrational Numbers**: $\mathbb{P} = \{x \mid x \text{ is a non-rational real number.}\}$ Said another way, an irrational number is a decimal which neither repeats nor terminates.[c]

8. The **Complex Numbers**: $\mathbb{C} = \{a + bi \mid a, b \in \mathbb{R} \text{ and } i = \sqrt{-1}\}$ Despite their importance, the complex numbers play only a minor role in the text.[d]

---

[a] ...which, sadly, we will not explore in this text.
[b] See Section 9.2.
[c] The classic example is the number π (See Section 10.1), but numbers like $\sqrt{2}$ and $0.101001000100001\ldots$ are other fine representatives.
[d] They first appear in Section 3.4 and return in Section 11.7.

It is important to note that every natural number is a whole number, which, in turn, is an integer. Each integer is a rational number (take $b = 1$ in the above definition for $\mathbb{Q}$) and the rational numbers are all real numbers, since they possess decimal representations.[3] If we take $b = 0$ in the

---

[3] Long division, anyone?

1.1 Sets of Real Numbers and the Cartesian Coordinate Plane

above definition of $\mathbb{C}$, we see that every real number is a complex number. In this sense, the sets $\mathbb{N}$, $\mathbb{W}$, $\mathbb{Z}$, $\mathbb{Q}$, $\mathbb{R}$, and $\mathbb{C}$ are 'nested' like Matryoshka dolls.

For the most part, this textbook focuses on sets whose elements come from the real numbers $\mathbb{R}$. Recall that we may visualize $\mathbb{R}$ as a line. Segments of this line are called **intervals** of numbers. Below is a summary of the so-called **interval notation** associated with given sets of numbers. For intervals with finite endpoints, we list the left endpoint, then the right endpoint. We use square brackets, '[' or ']', if the endpoint is included in the interval and use a filled-in or 'closed' dot to indicate membership in the interval. Otherwise, we use parentheses, '(' or ')' and an 'open' circle to indicate that the endpoint is not part of the set. If the interval does not have finite endpoints, we use the symbols $-\infty$ to indicate that the interval extends indefinitely to the left and ∞ to indicate that the interval extends indefinitely to the right. Since infinity is a concept, and not a number, we always use parentheses when using these symbols in interval notation, and use an appropriate arrow to indicate that the interval extends indefinitely in one (or both) directions.

Interval Notation

Let a and b be real numbers with $a < b$.

Set of Real Numbers	Interval Notation	Region on the Real Number Line
$\{x \mid a < x < b\}$	(a, b)	
$\{x \mid a \leq x < b\}$	$[a, b)$	
$\{x \mid a < x \leq b\}$	$(a, b]$	
$\{x \mid a \leq x \leq b\}$	$[a, b]$	
$\{x \mid x < b\}$	$(-\infty, b)$	
$\{x \mid x \leq b\}$	$(-\infty, b]$	
$\{x \mid x > a\}$	(a, ∞)	
$\{x \mid x \geq a\}$	$[a, \infty)$	
$\mathbb{R}$	$(-\infty, \infty)$	

For an example, consider the sets of real numbers described below.

Set of Real Numbers	Interval Notation	Region on the Real Number Line
$\{x \mid 1 \leq x < 3\}$	$[1, 3)$	
$\{x \mid -1 \leq x \leq 4\}$	$[-1, 4]$	
$\{x \mid x \leq 5\}$	$(-\infty, 5]$	
$\{x \mid x > -2\}$	$(-2, \infty)$	

We will often have occasion to combine sets. There are two basic ways to combine sets: **intersection** and **union**. We define both of these concepts below.

Definition 1.2. Suppose A and B are two sets.

- The **intersection** of A and B: $A \cap B = \{x \mid x \in A \text{ and } x \in B\}$
- The **union** of A and B: $A \cup B = \{x \mid x \in A \text{ or } x \in B \text{ (or both)}\}$

Said differently, the intersection of two sets is the overlap of the two sets – the elements which the sets have in common. The union of two sets consists of the totality of the elements in each of the sets, collected together.[4] For example, if $A = \{1, 2, 3\}$ and $B = \{2, 4, 6\}$, then $A \cap B = \{2\}$ and $A \cup B = \{1, 2, 3, 4, 6\}$. If $A = [-5, 3)$ and $B = (1, \infty)$, then we can find $A \cap B$ and $A \cup B$ graphically. To find $A \cap B$, we shade the overlap of the two and obtain $A \cap B = (1, 3)$. To find $A \cup B$, we shade each of A and B and describe the resulting shaded region to find $A \cup B = [-5, \infty)$.

$A = [-5, 3)$, $B = (1, \infty)$ \qquad $A \cap B = (1, 3)$ \qquad $A \cup B = [-5, \infty)$

While both intersection and union are important, we have more occasion to use union in this text than intersection, simply because most of the sets of real numbers we will be working with are either intervals or are unions of intervals, as the following example illustrates.

---

[4]The reader is encouraged to research **Venn Diagrams** for a nice geometric interpretation of these concepts.

1.1 Sets of Real Numbers and the Cartesian Coordinate Plane

Example 1.1.1. Express the following sets of numbers using interval notation.

1. $\{x \,|\, x \leq -2 \text{ or } x \geq 2\}$
2. $\{x \,|\, x \neq 3\}$
3. $\{x \,|\, x \neq \pm 3\}$
4. $\{x \,|\, -1 < x \leq 3 \text{ or } x = 5\}$

Solution.

1. The best way to proceed here is to graph the set of numbers on the number line and glean the answer from it. The inequality $x \leq -2$ corresponds to the interval $(-\infty, -2]$ and the inequality $x \geq 2$ corresponds to the interval $[2, \infty)$. Since we are looking to describe the real numbers x in one of these *or* the other, we have $\{x \,|\, x \leq -2 \text{ or } x \geq 2\} = (-\infty, -2] \cup [2, \infty)$.

    $$-2 \qquad 2$$
    $$(-\infty, -2] \cup [2, \infty)$$

2. For the set $\{x \,|\, x \neq 3\}$, we shade the entire real number line except $x = 3$, where we leave an open circle. This divides the real number line into two intervals, $(-\infty, 3)$ and $(3, \infty)$. Since the values of x could be in either one of these intervals *or* the other, we have that $\{x \,|\, x \neq 3\} = (-\infty, 3) \cup (3, \infty)$

    $$3$$
    $$(-\infty, 3) \cup (3, \infty)$$

3. For the set $\{x \,|\, x \neq \pm 3\}$, we proceed as before and exclude both $x = 3$ and $x = -3$ from our set. This breaks the number line into *three* intervals, $(-\infty, -3)$, $(-3, 3)$ and $(3, \infty)$. Since the set describes real numbers which come from the first, second *or* third interval, we have $\{x \,|\, x \neq \pm 3\} = (-\infty, -3) \cup (-3, 3) \cup (3, \infty)$.

    $$-3 \qquad 3$$
    $$(-\infty, -3) \cup (-3, 3) \cup (3, \infty)$$

4. Graphing the set $\{x \,|\, -1 < x \leq 3 \text{ or } x = 5\}$, we get one interval, $(-1, 3]$ along with a single number, or point, $\{5\}$. While we *could* express the latter as $[5, 5]$ (Can you see why?), we choose to write our answer as $\{x \,|\, -1 < x \leq 3 \text{ or } x = 5\} = (-1, 3] \cup \{5\}$.

    $$-1 \qquad 3 \quad 5$$
    $$(-1, 3] \cup \{5\}$$

    $\square$

1.1.2 The Cartesian Coordinate Plane

In order to visualize the pure excitement that is Precalculus, we need to unite Algebra and Geometry. Simply put, we must find a way to draw algebraic things. Let's start with possibly the greatest mathematical achievement of all time: the **Cartesian Coordinate Plane**.[5] Imagine two real number lines crossing at a right angle at 0 as drawn below.

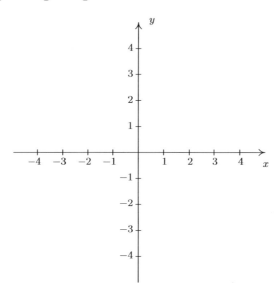

The horizontal number line is usually called the x-**axis** while the vertical number line is usually called the y-**axis**.[6] As with the usual number line, we imagine these axes extending off indefinitely in both directions.[7] Having two number lines allows us to locate the positions of points off of the number lines as well as points on the lines themselves.

For example, consider the point P on the next page. To use the numbers on the axes to label this point, we imagine dropping a vertical line from the x-axis to P and extending a horizontal line from the y-axis to P. This process is sometimes called 'projecting' the point P to the x- (respectively y-) axis. We then describe the point P using the **ordered pair** $(2, -4)$. The first number in the ordered pair is called the **abscissa** or x-**coordinate** and the second is called the **ordinate** or y-**coordinate**.[8] Taken together, the ordered pair $(2, -4)$ comprise the **Cartesian coordinates**[9] of the point P. In practice, the distinction between a point and its coordinates is blurred; for example, we often speak of 'the point $(2, -4)$.' We can think of $(2, -4)$ as instructions on how to

---
[5]So named in honor of René Descartes.

[6]The labels can vary depending on the context of application.

[7]Usually extending off towards infinity is indicated by arrows, but here, the arrows are used to indicate the *direction* of increasing values of x and y.

[8]Again, the names of the coordinates can vary depending on the context of the application. If, for example, the horizontal axis represented time we might choose to call it the t-axis. The first number in the ordered pair would then be the t-coordinate.

[9]Also called the 'rectangular coordinates' of P – see Section 11.4 for more details.

1.1 Sets of Real Numbers and the Cartesian Coordinate Plane

reach P from the **origin** $(0,0)$ by moving 2 units to the right and 4 units downwards. Notice that the order in the ordered pair is important − if we wish to plot the point $(-4, 2)$, we would move to the left 4 units from the origin and then move upwards 2 units, as below on the right.

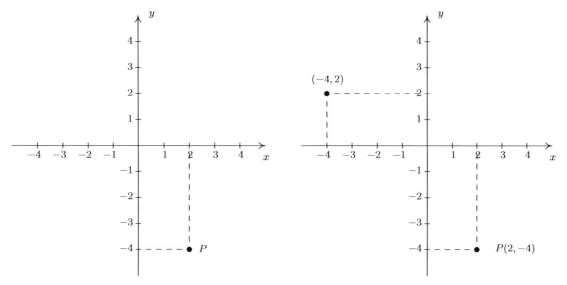

When we speak of the Cartesian Coordinate Plane, we mean the set of all possible ordered pairs (x, y) as x and y take values from the real numbers. Below is a summary of important facts about Cartesian coordinates.

Important Facts about the Cartesian Coordinate Plane

- (a, b) and (c, d) represent the same point in the plane if and only if $a = c$ and $b = d$.
- (x, y) lies on the x-axis if and only if $y = 0$.
- (x, y) lies on the y-axis if and only if $x = 0$.
- The origin is the point $(0, 0)$. It is the only point common to both axes.

Example 1.1.2. Plot the following points: $A(5, 8)$, $B\left(-\frac{5}{2}, 3\right)$, $C(-5.8, -3)$, $D(4.5, -1)$, $E(5, 0)$, $F(0, 5)$, $G(-7, 0)$, $H(0, -9)$, $O(0, 0)$.[10]

Solution. To plot these points, we start at the origin and move to the right if the x-coordinate is positive; to the left if it is negative. Next, we move up if the y-coordinate is positive or down if it is negative. If the x-coordinate is 0, we start at the origin and move along the y-axis only. If the y-coordinate is 0 we move along the x-axis only.

---

[10] The letter O is almost always reserved for the origin.

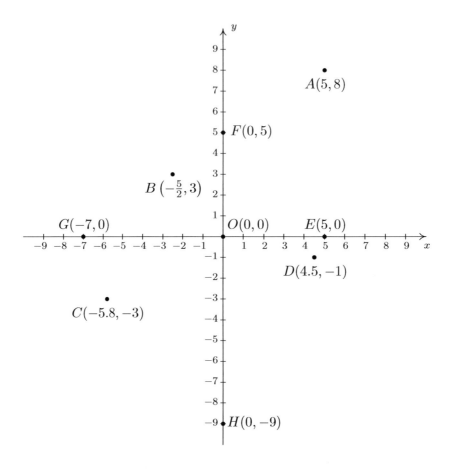

The axes divide the plane into four regions called **quadrants**. They are labeled with Roman numerals and proceed counterclockwise around the plane:

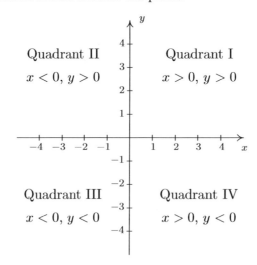

1.1 Sets of Real Numbers and the Cartesian Coordinate Plane

For example, $(1,2)$ lies in Quadrant I, $(-1,2)$ in Quadrant II, $(-1,-2)$ in Quadrant III and $(1,-2)$ in Quadrant IV. If a point other than the origin happens to lie on the axes, we typically refer to that point as lying on the positive or negative x-axis (if $y = 0$) or on the positive or negative y-axis (if $x = 0$). For example, $(0,4)$ lies on the positive y-axis whereas $(-117,0)$ lies on the negative x-axis. Such points do not belong to any of the four quadrants.

One of the most important concepts in all of Mathematics is **symmetry**.[11] There are many types of symmetry in Mathematics, but three of them can be discussed easily using Cartesian Coordinates.

Definition 1.3. Two points (a,b) and (c,d) in the plane are said to be

- **symmetric about the x-axis** if $a = c$ and $b = -d$
- **symmetric about the y-axis** if $a = -c$ and $b = d$
- **symmetric about the origin** if $a = -c$ and $b = -d$

Schematically,

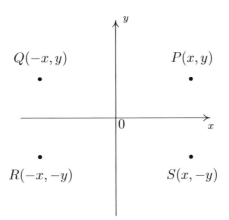

In the above figure, P and S are symmetric about the x-axis, as are Q and R; P and Q are symmetric about the y-axis, as are R and S; and P and R are symmetric about the origin, as are Q and S.

Example 1.1.3. Let P be the point $(-2,3)$. Find the points which are symmetric to P about the:

1. x-axis
2. y-axis
3. origin

Check your answer by plotting the points.

Solution. The figure after Definition 1.3 gives us a good way to think about finding symmetric points in terms of taking the opposites of the x- and/or y-coordinates of $P(-2,3)$.

---

[11] According to Carl. Jeff thinks symmetry is overrated.

1. To find the point symmetric about the x-axis, we replace the y-coordinate with its opposite to get $(-2, -3)$.

2. To find the point symmetric about the y-axis, we replace the x-coordinate with its opposite to get $(2, 3)$.

3. To find the point symmetric about the origin, we replace the x- and y-coordinates with their opposites to get $(2, -3)$.

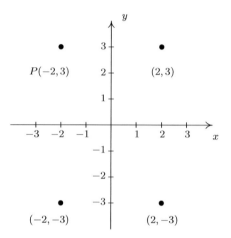

One way to visualize the processes in the previous example is with the concept of a **reflection**. If we start with our point $(-2, 3)$ and pretend that the x-axis is a mirror, then the reflection of $(-2, 3)$ across the x-axis would lie at $(-2, -3)$. If we pretend that the y-axis is a mirror, the reflection of $(-2, 3)$ across that axis would be $(2, 3)$. If we reflect across the x-axis and then the y-axis, we would go from $(-2, 3)$ to $(-2, -3)$ then to $(2, -3)$, and so we would end up at the point symmetric to $(-2, 3)$ about the origin. We summarize and generalize this process below.

Reflections

To reflect a point (x, y) about the:

- x-axis, replace y with $-y$.

- y-axis, replace x with $-x$.

- origin, replace x with $-x$ and y with $-y$.

1.1.3 Distance in the Plane

Another important concept in Geometry is the notion of length. If we are going to unite Algebra and Geometry using the Cartesian Plane, then we need to develop an algebraic understanding of what distance in the plane means. Suppose we have two points, $P(x_0, y_0)$ and $Q(x_1, y_1)$, in the plane. By the **distance** d between P and Q, we mean the length of the line segment joining P with Q. (Remember, given any two distinct points in the plane, there is a unique line containing both

1.1 Sets of Real Numbers and the Cartesian Coordinate Plane

points.) Our goal now is to create an algebraic formula to compute the distance between these two points. Consider the generic situation below on the left.

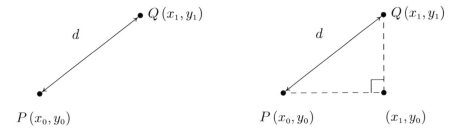

With a little more imagination, we can envision a right triangle whose hypotenuse has length d as drawn above on the right. From the latter figure, we see that the lengths of the legs of the triangle are $|x_1 - x_0|$ and $|y_1 - y_0|$ so the Pythagorean Theorem gives us

$$|x_1 - x_0|^2 + |y_1 - y_0|^2 = d^2$$
$$(x_1 - x_0)^2 + (y_1 - y_0)^2 = d^2$$

(Do you remember why we can replace the absolute value notation with parentheses?) By extracting the square root of both sides of the second equation and using the fact that distance is never negative, we get

> **Equation 1.1. The Distance Formula:** The distance d between the points $P(x_0, y_0)$ and $Q(x_1, y_1)$ is:
> $$d = \sqrt{(x_1 - x_0)^2 + (y_1 - y_0)^2}$$

It is not always the case that the points P and Q lend themselves to constructing such a triangle. If the points P and Q are arranged vertically or horizontally, or describe the exact same point, we cannot use the above geometric argument to derive the distance formula. It is left to the reader in Exercise 35 to verify Equation 1.1 for these cases.

Example 1.1.4. Find and simplify the distance between $P(-2, 3)$ and $Q(1, -3)$.

Solution.

$$\begin{aligned} d &= \sqrt{(x_1 - x_0)^2 + (y_1 - y_0)^2} \\ &= \sqrt{(1 - (-2))^2 + (-3 - 3)^2} \\ &= \sqrt{9 + 36} \\ &= 3\sqrt{5} \end{aligned}$$

So the distance is $3\sqrt{5}$. □

Example 1.1.5. Find all of the points with x-coordinate 1 which are 4 units from the point $(3, 2)$.

Solution. We shall soon see that the points we wish to find are on the line $x = 1$, but for now we'll just view them as points of the form $(1, y)$. Visually,

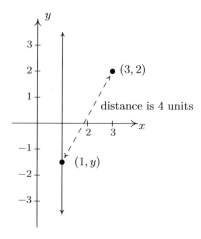

We require that the distance from $(3, 2)$ to $(1, y)$ be 4. The Distance Formula, Equation 1.1, yields

$$
\begin{aligned}
d &= \sqrt{(x_1 - x_0)^2 + (y_1 - y_0)^2} \\
4 &= \sqrt{(1 - 3)^2 + (y - 2)^2} \\
4 &= \sqrt{4 + (y - 2)^2} \\
4^2 &= \left(\sqrt{4 + (y - 2)^2}\right)^2 \qquad \text{squaring both sides} \\
16 &= 4 + (y - 2)^2 \\
12 &= (y - 2)^2 \\
(y - 2)^2 &= 12 \\
y - 2 &= \pm\sqrt{12} \qquad \text{extracting the square root} \\
y - 2 &= \pm 2\sqrt{3} \\
y &= 2 \pm 2\sqrt{3}
\end{aligned}
$$

We obtain two answers: $(1, 2 + 2\sqrt{3})$ and $(1, 2 - 2\sqrt{3})$. The reader is encouraged to think about why there are two answers. □

Related to finding the distance between two points is the problem of finding the **midpoint** of the line segment connecting two points. Given two points, $P(x_0, y_0)$ and $Q(x_1, y_1)$, the **midpoint** M of P and Q is defined to be the point on the line segment connecting P and Q whose distance from P is equal to its distance from Q.

1.1 Sets of Real Numbers and the Cartesian Coordinate Plane

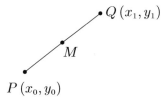

If we think of reaching M by going 'halfway over' and 'halfway up' we get the following formula.

Equation 1.2. The Midpoint Formula: The midpoint M of the line segment connecting $P(x_0, y_0)$ and $Q(x_1, y_1)$ is:
$$M = \left(\frac{x_0 + x_1}{2}, \frac{y_0 + y_1}{2} \right).$$

If we let d denote the distance between P and Q, we leave it as Exercise 36 to show that the distance between P and M is $d/2$ which is the same as the distance between M and Q. This suffices to show that Equation 1.2 gives the coordinates of the midpoint.

Example 1.1.6. Find the midpoint of the line segment connecting $P(-2, 3)$ and $Q(1, -3)$.

Solution.

$$\begin{aligned} M &= \left(\frac{x_0 + x_1}{2}, \frac{y_0 + y_1}{2} \right) \\ &= \left(\frac{(-2) + 1}{2}, \frac{3 + (-3)}{2} \right) = \left(-\frac{1}{2}, \frac{0}{2} \right) \\ &= \left(-\frac{1}{2}, 0 \right) \end{aligned}$$

The midpoint is $\left(-\frac{1}{2}, 0 \right)$. □

We close with a more abstract application of the Midpoint Formula. We will revisit the following example in Exercise 72 in Section 2.1.

Example 1.1.7. If $a \neq b$, prove that the line $y = x$ equally divides the line segment with endpoints (a, b) and (b, a).

Solution. To prove the claim, we use Equation 1.2 to find the midpoint

$$\begin{aligned} M &= \left(\frac{a + b}{2}, \frac{b + a}{2} \right) \\ &= \left(\frac{a + b}{2}, \frac{a + b}{2} \right) \end{aligned}$$

Since the x and y coordinates of this point are the same, we find that the midpoint lies on the line $y = x$, as required. □

1.1.4 Exercises

1. Fill in the chart below:

Set of Real Numbers	Interval Notation	Region on the Real Number Line
$\{x \mid -1 \leq x < 5\}$		
	$[0, 3)$	
		 ∘———• 2 7
$\{x \mid -5 < x \leq 0\}$		
	$(-3, 3)$	
		 •———• 5 7
$\{x \mid x \leq 3\}$		
	$(-\infty, 9)$	
		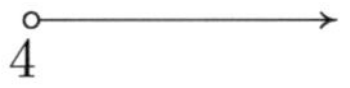 ∘——→ 4
$\{x \mid x \geq -3\}$		

In Exercises 2 - 7, find the indicated intersection or union and simplify if possible. Express your answers in interval notation.

2. $(-1, 5] \cap [0, 8)$ 3. $(-1, 1) \cup [0, 6]$ 4. $(-\infty, 4] \cap (0, \infty)$

5. $(-\infty, 0) \cap [1, 5]$ 6. $(-\infty, 0) \cup [1, 5]$ 7. $(-\infty, 5] \cap [5, 8)$

In Exercises 8 - 19, write the set using interval notation.

8. $\{x \mid x \neq 5\}$ 9. $\{x \mid x \neq -1\}$ 10. $\{x \mid x \neq -3, 4\}$

11. $\{x \,|\, x \neq 0, 2\}$

12. $\{x \,|\, x \neq 2, -2\}$

13. $\{x \,|\, x \neq 0, \pm 4\}$

14. $\{x \,|\, x \leq -1 \text{ or } x \geq 1\}$

15. $\{x \,|\, x < 3 \text{ or } x \geq 2\}$

16. $\{x \,|\, x \leq -3 \text{ or } x > 0\}$

17. $\{x \,|\, x \leq 5 \text{ or } x = 6\}$

18. $\{x \,|\, x > 2 \text{ or } x = \pm 1\}$

19. $\{x \,|\, -3 < x < 3 \text{ or } x = 4\}$

20. Plot and label the points $A(-3, -7)$, $B(1.3, -2)$, $C(\pi, \sqrt{10})$, $D(0, 8)$, $E(-5.5, 0)$, $F(-8, 4)$, $G(9.2, -7.8)$ and $H(7, 5)$ in the Cartesian Coordinate Plane given below.

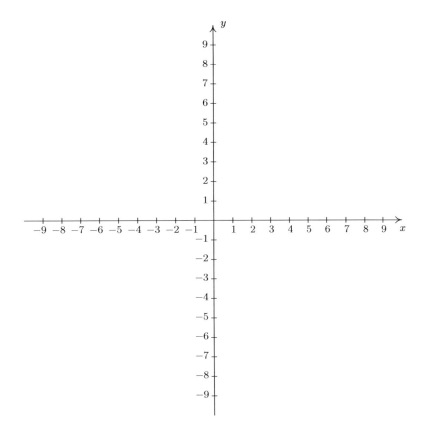

21. For each point given in Exercise 20 above

    - Identify the quadrant or axis in/on which the point lies.
    - Find the point symmetric to the given point about the x-axis.
    - Find the point symmetric to the given point about the y-axis.
    - Find the point symmetric to the given point about the origin.

In Exercises 22 - 29, find the distance d between the points and the midpoint M of the line segment which connects them.

22. $(1, 2)$, $(-3, 5)$

23. $(3, -10)$, $(-1, 2)$

24. $\left(\dfrac{1}{2}, 4\right)$, $\left(\dfrac{3}{2}, -1\right)$

25. $\left(-\dfrac{2}{3}, \dfrac{3}{2}\right)$, $\left(\dfrac{7}{3}, 2\right)$

26. $\left(\dfrac{24}{5}, \dfrac{6}{5}\right)$, $\left(-\dfrac{11}{5}, -\dfrac{19}{5}\right)$.

27. $(\sqrt{2}, \sqrt{3})$, $(-\sqrt{8}, -\sqrt{12})$

28. $(2\sqrt{45}, \sqrt{12})$, $(\sqrt{20}, \sqrt{27})$.

29. $(0, 0)$, (x, y)

30. Find all of the points of the form $(x, -1)$ which are 4 units from the point $(3, 2)$.

31. Find all of the points on the y-axis which are 5 units from the point $(-5, 3)$.

32. Find all of the points on the x-axis which are 2 units from the point $(-1, 1)$.

33. Find all of the points of the form $(x, -x)$ which are 1 unit from the origin.

34. Let's assume for a moment that we are standing at the origin and the positive y-axis points due North while the positive x-axis points due East. Our Sasquatch-o-meter tells us that Sasquatch is 3 miles West and 4 miles South of our current position. What are the coordinates of his position? How far away is he from us? If he runs 7 miles due East what would his new position be?

35. Verify the Distance Formula 1.1 for the cases when:

    (a) The points are arranged vertically. (Hint: Use $P(a, y_0)$ and $Q(a, y_1)$.)
    (b) The points are arranged horizontally. (Hint: Use $P(x_0, b)$ and $Q(x_1, b)$.)
    (c) The points are actually the same point. (You shouldn't need a hint for this one.)

36. Verify the Midpoint Formula by showing the distance between $P(x_1, y_1)$ and M and the distance between M and $Q(x_2, y_2)$ are both half of the distance between P and Q.

37. Show that the points A, B and C below are the vertices of a right triangle.

    (a) $A(-3, 2)$, $B(-6, 4)$, and $C(1, 8)$
    (b) $A(-3, 1)$, $B(4, 0)$ and $C(0, -3)$

38. Find a point $D(x, y)$ such that the points $A(-3, 1)$, $B(4, 0)$, $C(0, -3)$ and D are the corners of a square. Justify your answer.

39. Discuss with your classmates how many numbers are in the interval $(0, 1)$.

40. The world is not flat.[12] Thus the Cartesian Plane cannot possibly be the end of the story. Discuss with your classmates how you would extend Cartesian Coordinates to represent the three dimensional world. What would the Distance and Midpoint formulas look like, assuming those concepts make sense at all?

---

[12]There are those who disagree with this statement. Look them up on the Internet some time when you're bored.

1.1.5 Answers

1.

Set of Real Numbers	Interval Notation	Region on the Real Number Line
$\{x \mid -1 \leq x < 5\}$	$[-1, 5)$	•—————∘ -1 5
$\{x \mid 0 \leq x < 3\}$	$[0, 3)$	•—————∘ 0 3
$\{x \mid 2 < x \leq 7\}$	$(2, 7]$	∘—————• 2 7
$\{x \mid -5 < x \leq 0\}$	$(-5, 0]$	∘—————• -5 0
$\{x \mid -3 < x < 3\}$	$(-3, 3)$	∘—————∘ -3 3
$\{x \mid 5 \leq x \leq 7\}$	$[5, 7]$	•—————• 5 7
$\{x \mid x \leq 3\}$	$(-\infty, 3]$	←—————• 3
$\{x \mid x < 9\}$	$(-\infty, 9)$	←—————∘ 9
$\{x \mid x > 4\}$	$(4, \infty)$	∘—————→ 4
$\{x \mid x \geq -3\}$	$[-3, \infty)$	•—————→ -3

2. $(-1, 5] \cap [0, 8) = [0, 5]$

3. $(-1, 1) \cup [0, 6] = (-1, 6]$

4. $(-\infty, 4] \cap (0, \infty) = (0, 4]$

5. $(-\infty, 0) \cap [1, 5] = \emptyset$

6. $(-\infty, 0) \cup [1, 5] = (-\infty, 0) \cup [1, 5]$

7. $(-\infty, 5] \cap [5, 8) = \{5\}$

8. $(-\infty, 5) \cup (5, \infty)$

9. $(-\infty, -1) \cup (-1, \infty)$

10. $(-\infty, -3) \cup (-3, 4) \cup (4, \infty)$

11. $(-\infty, 0) \cup (0, 2) \cup (2, \infty)$

12. $(-\infty, -2) \cup (-2, 2) \cup (2, \infty)$

13. $(-\infty, -4) \cup (-4, 0) \cup (0, 4) \cup (4, \infty)$

14. $(-\infty, -1] \cup [1, \infty)$

15. $(-\infty, \infty)$

16. $(-\infty, -3] \cup (0, \infty)$

17. $(-\infty, 5] \cup \{6\}$

18. $\{-1\} \cup \{1\} \cup (2, \infty)$

19. $(-3, 3) \cup \{4\}$

20. The required points $A(-3,-7)$, $B(1.3,-2)$, $C(\pi, \sqrt{10})$, $D(0,8)$, $E(-5.5, 0)$, $F(-8, 4)$, $G(9.2, -7.8)$, and $H(7,5)$ are plotted in the Cartesian Coordinate Plane below.

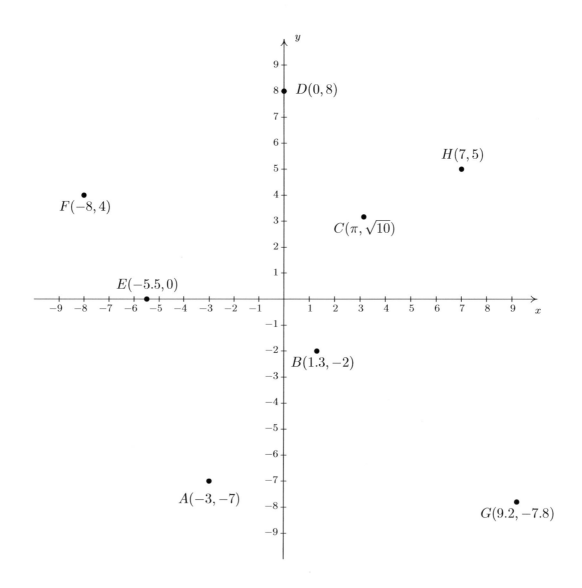

1.1 Sets of Real Numbers and the Cartesian Coordinate Plane

21. (a) The point $A(-3,-7)$ is
    - in Quadrant III
    - symmetric about x-axis with $(-3,7)$
    - symmetric about y-axis with $(3,-7)$
    - symmetric about origin with $(3,7)$

    (b) The point $B(1.3,-2)$ is
    - in Quadrant IV
    - symmetric about x-axis with $(1.3,2)$
    - symmetric about y-axis with $(-1.3,-2)$
    - symmetric about origin with $(-1.3,2)$

    (c) The point $C(\pi,\sqrt{10})$ is
    - in Quadrant I
    - symmetric about x-axis with $(\pi,-\sqrt{10})$
    - symmetric about y-axis with $(-\pi,\sqrt{10})$
    - symmetric about origin with $(-\pi,-\sqrt{10})$

    (d) The point $D(0,8)$ is
    - on the positive y-axis
    - symmetric about x-axis with $(0,-8)$
    - symmetric about y-axis with $(0,8)$
    - symmetric about origin with $(0,-8)$

    (e) The point $E(-5.5,0)$ is
    - on the negative x-axis
    - symmetric about x-axis with $(-5.5,0)$
    - symmetric about y-axis with $(5.5,0)$
    - symmetric about origin with $(5.5,0)$

    (f) The point $F(-8,4)$ is
    - in Quadrant II
    - symmetric about x-axis with $(-8,-4)$
    - symmetric about y-axis with $(8,4)$
    - symmetric about origin with $(8,-4)$

    (g) The point $G(9.2,-7.8)$ is
    - in Quadrant IV
    - symmetric about x-axis with $(9.2,7.8)$
    - symmetric about y-axis with $(-9.2,-7.8)$
    - symmetric about origin with $(-9.2,7.8)$

    (h) The point $H(7,5)$ is
    - in Quadrant I
    - symmetric about x-axis with $(7,-5)$
    - symmetric about y-axis with $(-7,5)$
    - symmetric about origin with $(-7,-5)$

22. $d=5$, $M=\left(-1,\frac{7}{2}\right)$

23. $d=4\sqrt{10}$, $M=(1,-4)$

24. $d=\sqrt{26}$, $M=\left(1,\frac{3}{2}\right)$

25. $d=\frac{\sqrt{37}}{2}$, $M=\left(\frac{5}{6},\frac{7}{4}\right)$

26. $d=\sqrt{74}$, $M=\left(\frac{13}{10},-\frac{13}{10}\right)$

27. $d=3\sqrt{5}$, $M=\left(-\frac{\sqrt{2}}{2},-\frac{\sqrt{3}}{2}\right)$

28. $d=\sqrt{83}$, $M=\left(4\sqrt{5},\frac{5\sqrt{3}}{2}\right)$

29. $d=\sqrt{x^2+y^2}$, $M=\left(\frac{x}{2},\frac{y}{2}\right)$

30. $(3+\sqrt{7},-1)$, $(3-\sqrt{7},-1)$

31. $(0,3)$

32. $(-1+\sqrt{3},0)$, $(-1-\sqrt{3},0)$

33. $\left(\frac{\sqrt{2}}{2},-\frac{\sqrt{2}}{2}\right)$, $\left(-\frac{\sqrt{2}}{2},\frac{\sqrt{2}}{2}\right)$

34. $(-3,-4)$, 5 miles, $(4,-4)$

37. (a) The distance from A to B is $|AB|=\sqrt{13}$, the distance from A to C is $|AC|=\sqrt{52}$, and the distance from B to C is $|BC|=\sqrt{65}$. Since $\left(\sqrt{13}\right)^2+\left(\sqrt{52}\right)^2=\left(\sqrt{65}\right)^2$, we are guaranteed by the converse of the Pythagorean Theorem that the triangle is a right triangle.

    (b) Show that $|AC|^2+|BC|^2=|AB|^2$

1.2 Relations

From one point of view,[1] all of Precalculus can be thought of as studying sets of points in the plane. With the Cartesian Plane now fresh in our memory we can discuss those sets in more detail and as usual, we begin with a definition.

Definition 1.4. A **relation** is a set of points in the plane.

Since relations are sets, we can describe them using the techniques presented in Section 1.1.1. That is, we can describe a relation verbally, using the roster method, or using set-builder notation. Since the elements in a relation are points in the plane, we often try to describe the relation graphically or algebraically as well. Depending on the situation, one method may be easier or more convenient to use than another. As an example, consider the relation $R = \{(-2, 1), (4, 3), (0, -3)\}$. As written, R is described using the roster method. Since R consists of points in the plane, we follow our instinct and plot the points. Doing so produces the **graph** of R.

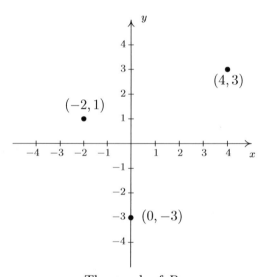

The graph of R.

In the following example, we graph a variety of relations.

Example 1.2.1. Graph the following relations.

1. $A = \{(0,0), (-3,1), (4,2), (-3,2)\}$
2. $HLS_1 = \{(x, 3) \,|\, -2 \leq x \leq 4\}$
3. $HLS_2 = \{(x, 3) \,|\, -2 \leq x < 4\}$
4. $V = \{(3, y) \,|\, y \text{ is a real number}\}$
5. $H = \{(x, y) \,|\, y = -2\}$
6. $R = \{(x, y) \,|\, 1 < y \leq 3\}$

---

[1] Carl's, of course.

1.2 RELATIONS

Solution.

1. To graph A, we simply plot all of the points which belong to A, as shown below on the left.

2. Don't let the notation in this part fool you. The name of this relation is HLS_1, just like the name of the relation in number 1 was A. The letters and numbers are just part of its name, just like the numbers and letters of the phrase 'King George III' were part of George's name. In words, $\{(x,3) \mid -2 \leq x \leq 4\}$ reads 'the set of points $(x,3)$ such that $-2 \leq x \leq 4$.' All of these points have the same y-coordinate, 3, but the x-coordinate is allowed to vary between -2 and 4, inclusive. Some of the points which belong to HLS_1 include some friendly points like: $(-2,3)$, $(-1,3)$, $(0,3)$, $(1,3)$, $(2,3)$, $(3,3)$, and $(4,3)$. However, HLS_1 also contains the points $(0.829, 3)$, $\left(-\frac{5}{6}, 3\right)$, $(\sqrt{\pi}, 3)$, and so on. It is impossible[2] to list all of these points, which is why the variable x is used. Plotting several friendly representative points should convince you that HLS_1 describes the horizontal line segment from the point $(-2,3)$ up to and including the point $(4,3)$.

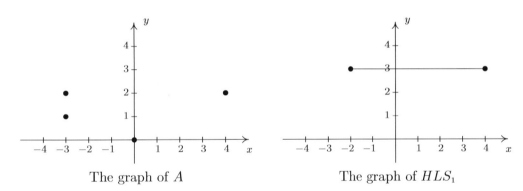

The graph of A The graph of HLS_1

3. HLS_2 is hauntingly similar to HLS_1. In fact, the only difference between the two is that instead of '$-2 \leq x \leq 4$' we have '$-2 \leq x < 4$'. This means that we still get a horizontal line segment which includes $(-2,3)$ and extends to $(4,3)$, but we do *not* include $(4,3)$ because of the strict inequality $x < 4$. How do we denote this on our graph? It is a common mistake to make the graph start at $(-2,3)$ end at $(3,3)$ as pictured below on the left. The problem with this graph is that we are forgetting about the points like $(3.1,3)$, $(3.5,3)$, $(3.9,3)$, $(3.99,3)$, and so forth. There is no real number that comes 'immediately before' 4, so to describe the set of points we want, we draw the horizontal line segment starting at $(-2,3)$ and draw an open circle at $(4,3)$ as depicted below on the right.

---

[2] Really impossible. The interested reader is encouraged to research <u>countable</u> versus <u>uncountable</u> sets.

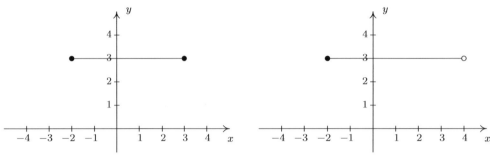

This is NOT the correct graph of HLS_2 The graph of HLS_2

4. Next, we come to the relation V, described as the set of points $(3, y)$ such that y is a real number. All of these points have an x-coordinate of 3, but the y-coordinate is free to be whatever it wants to be, without restriction.[3] Plotting a few 'friendly' points of V should convince you that all the points of V lie on the vertical line[4] $x = 3$. Since there is no restriction on the y-coordinate, we put arrows on the end of the portion of the line we draw to indicate it extends indefinitely in both directions. The graph of V is below on the left.

5. Though written slightly differently, the relation $H = \{(x, y) \,|\, y = -2\}$ is similar to the relation V above in that only one of the coordinates, in this case the y-coordinate, is specified, leaving x to be 'free'. Plotting some representative points gives us the horizontal line $y = -2$.

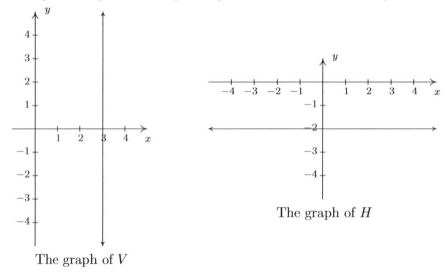

The graph of V The graph of H

6. For our last example, we turn to $R = \{(x, y) \,|\, 1 < y \leq 3\}$. As in the previous example, x is free to be whatever it likes. The value of y, on the other hand, while not completely free, is permitted to roam between 1 and 3 excluding 1, but including 3. After plotting some[5] friendly elements of R, it should become clear that R consists of the region between the horizontal

---

[3] We'll revisit the concept of a 'free variable' in Section 8.1.
[4] Don't worry, we'll be refreshing your memory about vertical and horizontal lines in just a moment!
[5] The word 'some' is a relative term. It may take 5, 10, or 50 points until you see the pattern.

lines $y = 1$ and $y = 3$. Since R requires that the y-coordinates be greater than 1, but not equal to 1, we dash the line $y = 1$ to indicate that those points do not belong to R.

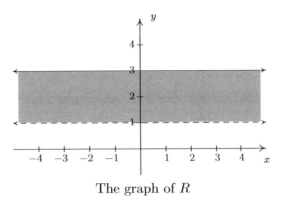

The graph of R □

The relations V and H in the previous example lead us to our final way to describe relations: **algebraically**. We can more succinctly describe the points in V as those points which satisfy the equation '$x = 3$'. Most likely, you have seen equations like this before. Depending on the context, '$x = 3$' could mean we have solved an equation for x and arrived at the solution $x = 3$. In this case, however, '$x = 3$' describes a set of points in the plane whose x-coordinate is 3. Similarly, the relation H above can be described by the equation '$y = -2$'. At some point in your mathematical upbringing, you probably learned the following.

Equations of Vertical and Horizontal Lines

- The graph of the equation $x = a$ is a **vertical line** through $(a, 0)$.
- The graph of the equation $y = b$ is a **horizontal line** through $(0, b)$.

Given that the very simple equations $x = a$ and $y = b$ produced lines, it's natural to wonder what shapes other equations might yield. Thus our next objective is to study the graphs of equations in a more general setting as we continue to unite Algebra and Geometry.

1.2.1 Graphs of Equations

In this section, we delve more deeply into the connection between Algebra and Geometry by focusing on graphing relations described by equations. The main idea of this section is the following.

The Fundamental Graphing Principle
The graph of an equation is the set of points which satisfy the equation. That is, a point (x, y) is on the graph of an equation if and only if x and y satisfy the equation.

Here, 'x and y satisfy the equation' means 'x and y make the equation true'. It is at this point that we gain some insight into the word 'relation'. If the equation to be graphed contains both x and y, then the equation itself is what is relating the two variables. More specifically, in the next two examples, we consider the graph of the equation $x^2 + y^3 = 1$. Even though it is not specifically spelled out, what we are doing is graphing the relation $R = \{(x, y) \mid x^2 + y^3 = 1\}$. The points (x, y) we graph belong to the *relation* R and are necessarily *related* by the equation $x^2 + y^3 = 1$, since it is those pairs of x and y which make the equation true.

Example 1.2.2. Determine whether or not $(2, -1)$ is on the graph of $x^2 + y^3 = 1$.

Solution. We substitute $x = 2$ and $y = -1$ into the equation to see if the equation is satisfied.

$$(2)^2 + (-1)^3 \stackrel{?}{=} 1$$
$$3 \neq 1$$

Hence, $(2, -1)$ is **not** on the graph of $x^2 + y^3 = 1$. $\square$

We could spend hours randomly guessing and checking to see if points are on the graph of the equation. A more systematic approach is outlined in the following example.

Example 1.2.3. Graph $x^2 + y^3 = 1$.

Solution. To efficiently generate points on the graph of this equation, we first solve for y

$$\begin{aligned} x^2 + y^3 &= 1 \\ y^3 &= 1 - x^2 \\ \sqrt[3]{y^3} &= \sqrt[3]{1 - x^2} \\ y &= \sqrt[3]{1 - x^2} \end{aligned}$$

We now substitute a value in for x, determine the corresponding value y, and plot the resulting point (x, y). For example, substituting $x = -3$ into the equation yields

$$y = \sqrt[3]{1 - x^2} = \sqrt[3]{1 - (-3)^2} = \sqrt[3]{-8} = -2,$$

so the point $(-3, -2)$ is on the graph. Continuing in this manner, we generate a table of points which are on the graph of the equation. These points are then plotted in the plane as shown below.

x	y	(x, y)
-3	-2	$(-3, -2)$
-2	$-\sqrt[3]{3}$	$(-2, -\sqrt[3]{3})$
-1	0	$(-1, 0)$
0	1	$(0, 1)$
1	0	$(1, 0)$
2	$-\sqrt[3]{3}$	$(2, -\sqrt[3]{3})$
3	-2	$(3, -2)$

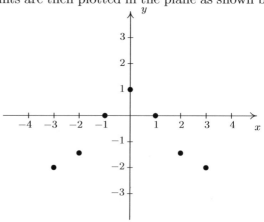

Remember, these points constitute only a small sampling of the points on the graph of this equation. To get a better idea of the shape of the graph, we could plot more points until we feel comfortable

1.2 RELATIONS

'connecting the dots'. Doing so would result in a curve similar to the one pictured below on the far left.

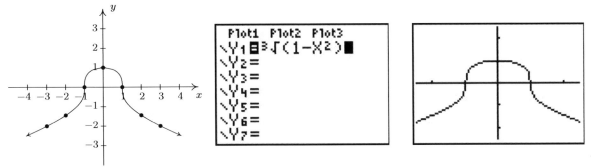

Don't worry if you don't get all of the little bends and curves just right – Calculus is where the art of precise graphing takes center stage. For now, we will settle with our naive 'plug and plot' approach to graphing. If you feel like all of this tedious computation and plotting is beneath you, then you can reach for a graphing calculator, input the formula as shown above, and graph. □

Of all of the points on the graph of an equation, the places where the graph crosses or touches the axes hold special significance. These are called the **intercepts** of the graph. Intercepts come in two distinct varieties: x-intercepts and y-intercepts. They are defined below.

Definition 1.5. Suppose the graph of an equation is given.

- A point on a graph which is also on the x-axis is called an **x-intercept** of the graph.

- A point on a graph which is also on the y-axis is called an **y-intercept** of the graph.

In our previous example the graph had two x-intercepts, $(-1, 0)$ and $(1, 0)$, and one y-intercept, $(0, 1)$. The graph of an equation can have any number of intercepts, including none at all! Since x-intercepts lie on the x-axis, we can find them by setting $y = 0$ in the equation. Similarly, since y-intercepts lie on the y-axis, we can find them by setting $x = 0$ in the equation. Keep in mind, intercepts are *points* and therefore must be written as ordered pairs. To summarize,

Finding the Intercepts of the Graph of an Equation

Given an equation involving x and y, we find the intercepts of the graph as follows:

- x-intercepts have the form $(x, 0)$; set $y = 0$ in the equation and solve for x.

- y-intercepts have the form $(0, y)$; set $x = 0$ in the equation and solve for y.

Another fact which you may have noticed about the graph in the previous example is that it seems to be symmetric about the y-axis. To actually prove this analytically, we assume (x, y) is a generic point on the graph of the equation. That is, we assume $x^2 + y^3 = 1$ is true. As we learned in Section 1.1, the point symmetric to (x, y) about the y-axis is $(-x, y)$. To show that the graph is symmetric about the y-axis, we need to show that $(-x, y)$ satisfies the equation $x^2 + y^3 = 1$, too. Substituting $(-x, y)$ into the equation gives

$$(-x)^2 + (y)^3 \stackrel{?}{=} 1$$
$$x^2 + y^3 \stackrel{\checkmark}{=} 1$$

Since we are assuming the original equation $x^2 + y^3 = 1$ is true, we have shown that $(-x, y)$ satisfies the equation (since it leads to a true result) and hence is on the graph. In this way, we can check whether the graph of a given equation possesses any of the symmetries discussed in Section 1.1. We summarize the procedure in the following result.

Testing the Graph of an Equation for Symmetry

To test the graph of an equation for symmetry

- about the y-axis – substitute $(-x, y)$ into the equation and simplify. If the result is equivalent to the original equation, the graph is symmetric about the y-axis.

- about the x-axis – substitute $(x, -y)$ into the equation and simplify. If the result is equivalent to the original equation, the graph is symmetric about the x-axis.

- about the origin - substitute $(-x, -y)$ into the equation and simplify. If the result is equivalent to the original equation, the graph is symmetric about the origin.

Intercepts and symmetry are two tools which can help us sketch the graph of an equation analytically, as demonstrated in the next example.

Example 1.2.4. Find the x- and y-intercepts (if any) of the graph of $(x-2)^2 + y^2 = 1$. Test for symmetry. Plot additional points as needed to complete the graph.

Solution. To look for x-intercepts, we set $y = 0$ and solve

$$\begin{aligned}
(x-2)^2 + y^2 &= 1 \\
(x-2)^2 + 0^2 &= 1 \\
(x-2)^2 &= 1 \\
\sqrt{(x-2)^2} &= \sqrt{1} \quad \text{extract square roots} \\
x - 2 &= \pm 1 \\
x &= 2 \pm 1 \\
x &= 3, 1
\end{aligned}$$

We get two answers for x which correspond to two x-intercepts: $(1, 0)$ and $(3, 0)$. Turning our attention to y-intercepts, we set $x = 0$ and solve

1.2 Relations

$$(x-2)^2 + y^2 = 1$$
$$(0-2)^2 + y^2 = 1$$
$$4 + y^2 = 1$$
$$y^2 = -3$$

Since there is no real number which squares to a negative number (Do you remember why?), we are forced to conclude that the graph has no y-intercepts.

Plotting the data we have so far, we get

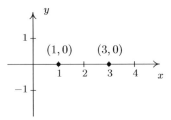

Moving along to symmetry, we can immediately dismiss the possibility that the graph is symmetric about the y-axis or the origin. If the graph possessed either of these symmetries, then the fact that $(1,0)$ is on the graph would mean $(-1,0)$ would have to be on the graph. (Why?) Since $(-1,0)$ would be another x-intercept (and we've found all of these), the graph can't have y-axis or origin symmetry. The only symmetry left to test is symmetry about the x-axis. To that end, we substitute $(x, -y)$ into the equation and simplify

$$(x-2)^2 + y^2 = 1$$
$$(x-2)^2 + (-y)^2 \stackrel{?}{=} 1$$
$$(x-2)^2 + y^2 \stackrel{\checkmark}{=} 1$$

Since we have obtained our original equation, we know the graph is symmetric about the x-axis. This means we can cut our 'plug and plot' time in half: whatever happens below the x-axis is reflected above the x-axis, and vice-versa. Proceeding as we did in the previous example, we obtain

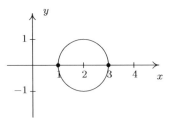

□

A couple of remarks are in order. First, it is entirely possible to choose a value for x which does not correspond to a point on the graph. For example, in the previous example, if we solve for y as is our custom, we get
$$y = \pm\sqrt{1-(x-2)^2}.$$
Upon substituting $x = 0$ into the equation, we would obtain
$$y = \pm\sqrt{1-(0-2)^2} = \pm\sqrt{1-4} = \pm\sqrt{-3},$$
which is not a real number. This means there are no points on the graph with an x-coordinate of 0. When this happens, we move on and try another point. This is another drawback of the 'plug-and-plot' approach to graphing equations. Luckily, we will devote much of the remainder of this book to developing techniques which allow us to graph entire families of equations quickly.[6] Second, it is instructive to show what would have happened had we tested the equation in the last example for symmetry about the y-axis. Substituting $(-x, y)$ into the equation yields

$$\begin{aligned}(x-2)^2 + y^2 &= 1 \\ (-x-2)^2 + y^2 &\stackrel{?}{=} 1 \\ ((-1)(x+2))^2 + y^2 &\stackrel{?}{=} 1 \\ (x+2)^2 + y^2 &\stackrel{?}{=} 1.\end{aligned}$$

This last equation does not *appear* to be equivalent to our original equation. However, to actually prove that the graph is not symmetric about the y-axis, we need to find a point (x, y) on the graph whose reflection $(-x, y)$ is not. Our x-intercept $(1, 0)$ fits this bill nicely, since if we substitute $(-1, 0)$ into the equation we get

$$\begin{aligned}(x-2)^2 + y^2 &\stackrel{?}{=} 1 \\ (-1-2)^2 + 0^2 &\neq 1 \\ 9 &\neq 1.\end{aligned}$$

This proves that $(-1, 0)$ is not on the graph.

---

[6] Without the use of a calculator, if you can believe it!

1.2 Relations

1.2.2 Exercises

In Exercises 1 - 20, graph the given relation.

1. $\{(-3,9),\ (-2,4),\ (-1,1),\ (0,0),\ (1,1),\ (2,4),\ (3,9)\}$

2. $\{(-2,0),\ (-1,1),\ (-1,-1),\ (0,2),\ (0,-2),\ (1,3),\ (1,-3)\}$

3. $\{(m, 2m) \mid m = 0, \pm 1, \pm 2\}$

4. $\{\left(\frac{6}{k}, k\right) \mid k = \pm 1, \pm 2, \pm 3, \pm 4, \pm 5, \pm 6\}$

5. $\{(n, 4 - n^2) \mid n = 0, \pm 1, \pm 2\}$

6. $\{(\sqrt{j}, j) \mid j = 0, 1, 4, 9\}$

7. $\{(x, -2) \mid x > -4\}$

8. $\{(x, 3) \mid x \leq 4\}$

9. $\{(-1, y) \mid y > 1\}$

10. $\{(2, y) \mid y \leq 5\}$

11. $\{(-2, y) \mid -3 < y \leq 4\}$

12. $\{(3, y) \mid -4 \leq y < 3\}$

13. $\{(x, 2) \mid -2 \leq x < 3\}$

14. $\{(x, -3) \mid -4 < x \leq 4\}$

15. $\{(x, y) \mid x > -2\}$

16. $\{(x, y) \mid x \leq 3\}$

17. $\{(x, y) \mid y < 4\}$

18. $\{(x, y) \mid x \leq 3,\ y < 2\}$

19. $\{(x, y) \mid x > 0,\ y < 4\}$

20. $\{(x, y) \mid -\sqrt{2} \leq x \leq \frac{2}{3},\ \pi < y \leq \frac{9}{2}\}$

In Exercises 21 - 30, describe the given relation using either the roster or set-builder method.

21.

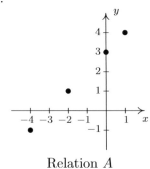

Relation A

22.

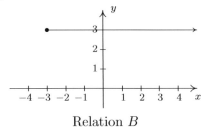

Relation B

23.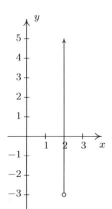

Relation C

24.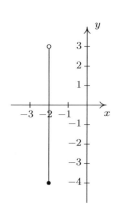

Relation D

25.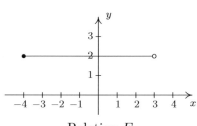

Relation E

26.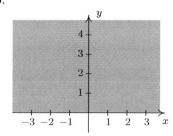

Relation F

27.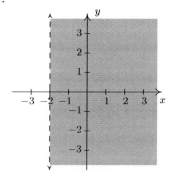

Relation G

28.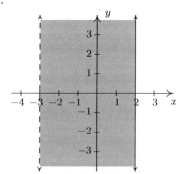

Relation H

1.2 RELATIONS

29.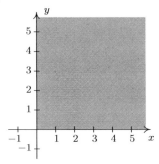

Relation I

30.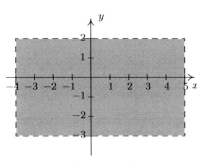

Relation J

In Exercises 31 - 36, graph the given line.

31. $x = -2$

32. $x = 3$

33. $y = 3$

34. $y = -2$

35. $x = 0$

36. $y = 0$

Some relations are fairly easy to describe in words or with the roster method but are rather difficult, if not impossible, to graph. Discuss with your classmates how you might graph the relations given in Exercises 37 - 40. Please note that in the notation below we are using the ellipsis, ..., to denote that the list does not end, but rather, continues to follow the established pattern indefinitely. For the relations in Exercises 37 and 38, give two examples of points which belong to the relation and two points which do not belong to the relation.

37. $\{(x, y) \,|\, x \text{ is an odd integer, and } y \text{ is an even integer.}\}$

38. $\{(x, 1) \,|\, x \text{ is an irrational number }\}$

39. $\{(1, 0), (2, 1), (4, 2), (8, 3), (16, 4), (32, 5), \ldots\}$

40. $\{\ldots, (-3, 9), (-2, 4), (-1, 1), (0, 0), (1, 1), (2, 4), (3, 9), \ldots\}$

For each equation given in Exercises 41 - 52:

- Find the x- and y-intercept(s) of the graph, if any exist.
- Follow the procedure in Example 1.2.3 to create a table of sample points on the graph of the equation.
- Plot the sample points and create a rough sketch of the graph of the equation.
- Test for symmetry. If the equation appears to fail any of the symmetry tests, find a point on the graph of the equation whose reflection fails to be on the graph as was done at the end of Example 1.2.4

41. $y = x^2 + 1$

42. $y = x^2 - 2x - 8$

43. $y = x^3 - x$

44. $y = \frac{x^3}{4} - 3x$

45. $y = \sqrt{x - 2}$

46. $y = 2\sqrt{x + 4} - 2$

47. $3x - y = 7$

48. $3x - 2y = 10$

49. $(x + 2)^2 + y^2 = 16$

50. $x^2 - y^2 = 1$

51. $4y^2 - 9x^2 = 36$

52. $x^3 y = -4$

The procedures which we have outlined in the Examples of this section and used in Exercises 41 - 52 all rely on the fact that the equations were "well-behaved". Not everything in Mathematics is quite so tame, as the following equations will show you. Discuss with your classmates how you might approach graphing the equations given in Exercises 53 - 56. What difficulties arise when trying to apply the various tests and procedures given in this section? For more information, including pictures of the curves, each curve name is a link to its page at www.wikipedia.org. For a much longer list of fascinating curves, click here.

53. $x^3 + y^3 - 3xy = 0$ Folium of Descartes

54. $x^4 = x^2 + y^2$ Kampyle of Eudoxus

55. $y^2 = x^3 + 3x^2$ Tschirnhausen cubic

56. $(x^2 + y^2)^2 = x^3 + y^3$ Crooked egg

57. With the help of your classmates, find examples of equations whose graphs possess

- symmetry about the x-axis only
- symmetry about the y-axis only
- symmetry about the origin only
- symmetry about the x-axis, y-axis, and origin

Can you find an example of an equation whose graph possesses exactly *two* of the symmetries listed above? Why or why not?

1.2 Relations

1.2.3 Answers

1.

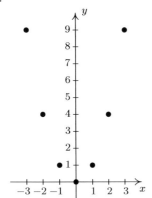

2.

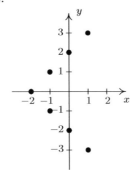

3.

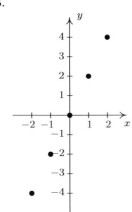

4.

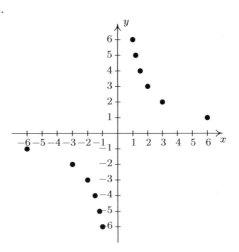

5.

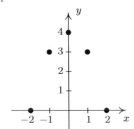

6.

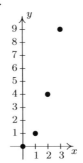

7.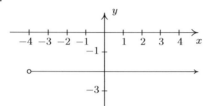

8.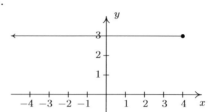

9.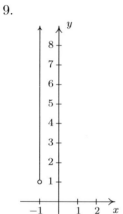

10.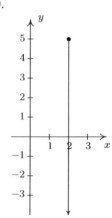

11.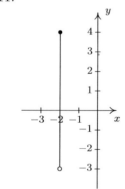

12.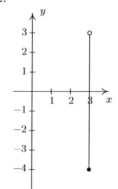

1.2 Relations

13.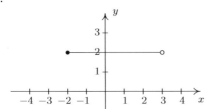

14.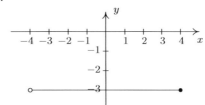

15.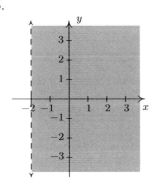

16.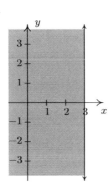

17.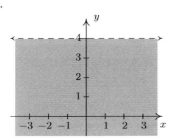

18.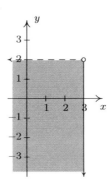

19.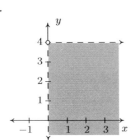

20.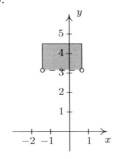

21. $A = \{(-4,-1),(-2,1),(0,3),(1,4)\}$

22. $B = \{(x,3) \mid x \geq -3\}$

23. $C = \{(2,y) \mid y > -3\}$

24. $D = \{(-2,y) \mid -4 \leq y < 3\}$

25. $E = \{(x,2) \mid -4 \leq x < 3\}$

26. $F = \{(x,y) \mid y \geq 0\}$

27. $G = \{(x,y) \mid x > -2\}$

28. $H = \{(x,y) \mid -3 < x \leq 2\}$

29. $I = \{(x,y) \mid x \geq 0, y \geq 0\}$

30. $J = \{(x,y) \mid -4 < x < 5,\ -3 < y < 2\}$

31.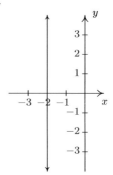

The line $x = -2$

32.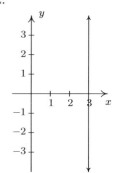

The line $x = 3$

33.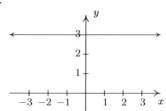

The line $y = 3$

34.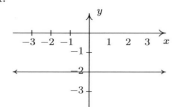

The line $y = -2$

35.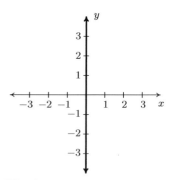

The line $x = 0$ is the y-axis

36.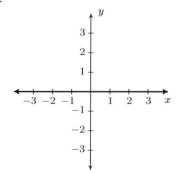

The line $y = 0$ is the x-axis

1.2 Relations

41. $y = x^2 + 1$

    The graph has no x-intercepts

    y-intercept: $(0, 1)$

x	y	(x, y)
-2	5	$(-2, 5)$
-1	2	$(-1, 2)$
0	1	$(0, 1)$
1	2	$(1, 2)$
2	5	$(2, 5)$

    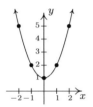

    The graph is not symmetric about the x-axis (e.g. $(2, 5)$ is on the graph but $(2, -5)$ is not)

    The graph is symmetric about the y-axis

    The graph is not symmetric about the origin (e.g. $(2, 5)$ is on the graph but $(-2, -5)$ is not)

42. $y = x^2 - 2x - 8$

    x-intercepts: $(4, 0), (-2, 0)$

    y-intercept: $(0, -8)$

x	y	(x, y)
-3	7	$(-3, 7)$
-2	0	$(-2, 0)$
-1	-5	$(-1, -5)$
0	-8	$(0, -8)$
1	-9	$(1, -9)$
2	-8	$(2, -8)$
3	-5	$(3, -5)$
4	0	$(4, 0)$
5	7	$(5, 7)$

    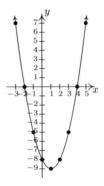

    The graph is not symmetric about the x-axis (e.g. $(-3, 7)$ is on the graph but $(-3, -7)$ is not)

    The graph is not symmetric about the y-axis (e.g. $(-3, 7)$ is on the graph but $(3, 7)$ is not)

    The graph is not symmetric about the origin (e.g. $(-3, 7)$ is on the graph but $(3, -7)$ is not)

43. $y = x^3 - x$

x-intercepts: $(-1,0), (0,0), (1,0)$

y-intercept: $(0,0)$

x	y	(x,y)
-2	-6	$(-2,-6)$
-1	0	$(-1,0)$
0	0	$(0,0)$
1	0	$(1,0)$
2	6	$(2,6)$

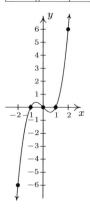

The graph is not symmetric about the x-axis. (e.g. $(2,6)$ is on the graph but $(2,-6)$ is not)

The graph is not symmetric about the y-axis. (e.g. $(2,6)$ is on the graph but $(-2,6)$ is not)

The graph is symmetric about the origin.

44. $y = \frac{x^3}{4} - 3x$

x-intercepts: $\left(\pm 2\sqrt{3}, 0\right), (0,0)$

y-intercept: $(0,0)$

x	y	(x,y)
-4	-4	$(-4,-4)$
-3	$\frac{9}{4}$	$\left(-3, \frac{9}{4}\right)$
-2	4	$(-2,4)$
-1	$\frac{11}{4}$	$\left(-1, \frac{11}{4}\right)$
0	0	$(0,0)$
1	$-\frac{11}{4}$	$\left(1, -\frac{11}{4}\right)$
2	-4	$(2,-4)$
3	$-\frac{9}{4}$	$\left(3, -\frac{9}{4}\right)$
4	4	$(4,4)$

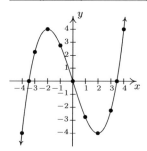

The graph is not symmetric about the x-axis (e.g. $(-4,-4)$ is on the graph but $(-4,4)$ is not)

The graph is not symmetric about the y-axis (e.g. $(-4,-4)$ is on the graph but $(4,-4)$ is not)

The graph is symmetric about the origin

1.2 Relations

45. $y = \sqrt{x-2}$

    x-intercept: $(2, 0)$

    The graph has no y-intercepts

x	y	(x, y)
2	0	$(2, 0)$
3	1	$(3, 1)$
6	2	$(6, 2)$
11	3	$(11, 3)$

    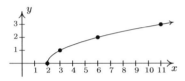

    The graph is not symmetric about the x-axis (e.g. $(3, 1)$ is on the graph but $(3, -1)$ is not)

    The graph is not symmetric about the y-axis (e.g. $(3, 1)$ is on the graph but $(-3, 1)$ is not)

    The graph is not symmetric about the origin (e.g. $(3, 1)$ is on the graph but $(-3, -1)$ is not)

46. $y = 2\sqrt{x+4} - 2$

    x-intercept: $(-3, 0)$

    y-intercept: $(0, 2)$

x	y	(x, y)
-4	-2	$(-4, -2)$
-3	0	$(-3, 0)$
-2	$2\sqrt{2} - 2$	$(-2, \sqrt{2} - 2)$
-1	$2\sqrt{3} - 2$	$(-2, \sqrt{3} - 2)$
0	2	$(0, 2)$
1	$2\sqrt{5} - 2$	$(-2, \sqrt{5} - 2)$

    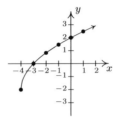

    The graph is not symmetric about the x-axis (e.g. $(-4, -2)$ is on the graph but $(-4, 2)$ is not)

    The graph is not symmetric about the y-axis (e.g. $(-4, -2)$ is on the graph but $(4, -2)$ is not)

    The graph is not symmetric about the origin (e.g. $(-4, -2)$ is on the graph but $(4, 2)$ is not)

47. $3x - y = 7$

Re-write as: $y = 3x - 7$.

x-intercept: $\left(\frac{7}{3}, 0\right)$

y-intercept: $(0, -7)$

x	y	(x, y)
-2	-13	$(-2, -13)$
-1	-10	$(-1, -10)$
0	-7	$(0, -7)$
1	-4	$(1, -4)$
2	-1	$(2, -1)$
3	2	$(3, 2)$

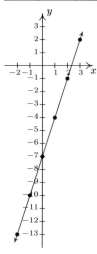

The graph is not symmetric about the x-axis (e.g. $(3, 2)$ is on the graph but $(3, -2)$ is not)

The graph is not symmetric about the y-axis (e.g. $(3, 2)$ is on the graph but $(-3, 2)$ is not)

The graph is not symmetric about the origin (e.g. $(3, 2)$ is on the graph but $(-3, -2)$ is not)

48. $3x - 2y = 10$

Re-write as: $y = \frac{3x - 10}{2}$.

x-intercepts: $\left(\frac{10}{3}, 0\right)$

y-intercept: $(0, -5)$

x	y	(x, y)
-2	-8	$(-2, -8)$
-1	$-\frac{13}{2}$	$\left(-1, -\frac{13}{2}\right)$
0	-5	$(0, -5)$
1	$-\frac{7}{2}$	$\left(1, -\frac{7}{2}\right)$
2	-2	$(2, -2)$

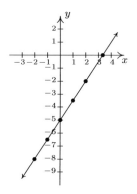

The graph is not symmetric about the x-axis (e.g. $(2, -2)$ is on the graph but $(2, 2)$ is not)

The graph is not symmetric about the y-axis (e.g. $(2, -2)$ is on the graph but $(-2, -2)$ is not)

The graph is not symmetric about the origin (e.g. $(2, -2)$ is on the graph but $(-2, 2)$ is not)

1.2 Relations

49. $(x+2)^2 + y^2 = 16$
Re-write as $y = \pm\sqrt{16-(x+2)^2}$.

x-intercepts: $(-6, 0), (2, 0)$

y-intercepts: $(0, \pm 2\sqrt{3})$

x	y	(x, y)
-6	0	$(-6, 0)$
-4	$\pm 2\sqrt{3}$	$(-4, \pm 2\sqrt{3})$
-2	± 4	$(-2, \pm 4)$
0	$\pm 2\sqrt{3}$	$(0, \pm 2\sqrt{3})$
2	0	$(2, 0)$

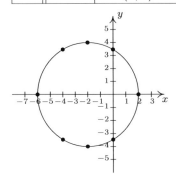

The graph is symmetric about the x-axis

The graph is not symmetric about the y-axis (e.g. $(-6, 0)$ is on the graph but $(6, 0)$ is not)

The graph is not symmetric about the origin (e.g. $(-6, 0)$ is on the graph but $(6, 0)$ is not)

50. $x^2 - y^2 = 1$
Re-write as: $y = \pm\sqrt{x^2 - 1}$.

x-intercepts: $(-1, 0), (1, 0)$

The graph has no y-intercepts

x	y	(x, y)
-3	$\pm\sqrt{8}$	$(-3, \pm\sqrt{8})$
-2	$\pm\sqrt{3}$	$(-2, \pm\sqrt{3})$
-1	0	$(-1, 0)$
1	0	$(1, 0)$
2	$\pm\sqrt{3}$	$(2, \pm\sqrt{3})$
3	$\pm\sqrt{8}$	$(3, \pm\sqrt{8})$

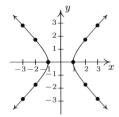

The graph is symmetric about the x-axis

The graph is symmetric about the y-axis

The graph is symmetric about the origin

51. $4y^2 - 9x^2 = 36$

Re-write as: $y = \pm \frac{\sqrt{9x^2+36}}{2}$.

The graph has no x-intercepts

y-intercepts: $(0, \pm 3)$

x	y	(x, y)
-4	$\pm 3\sqrt{5}$	$\left(-4, \pm 3\sqrt{5}\right)$
-2	$\pm 3\sqrt{2}$	$\left(-2, \pm 3\sqrt{2}\right)$
0	± 3	$(0, \pm 3)$
2	$\pm 3\sqrt{2}$	$\left(2, \pm 3\sqrt{2}\right)$
4	$\pm 3\sqrt{5}$	$\left(4, \pm 3\sqrt{5}\right)$

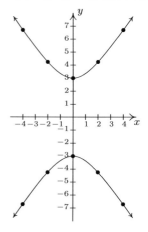

The graph is symmetric about the x-axis

The graph is symmetric about the y-axis

The graph is symmetric about the origin

52. $x^3 y = -4$

Re-write as: $y = -\dfrac{4}{x^3}$.

The graph has no x-intercepts

The graph has no y-intercepts

x	y	(x, y)
-2	$\frac{1}{2}$	$\left(-2, \frac{1}{2}\right)$
-1	4	$(-1, 4)$
$-\frac{1}{2}$	32	$\left(-\frac{1}{2}, 32\right)$
$\frac{1}{2}$	-32	$\left(\frac{1}{2}, -32\right)$
1	-4	$(1, -4)$
2	$-\frac{1}{2}$	$\left(2, -\frac{1}{2}\right)$

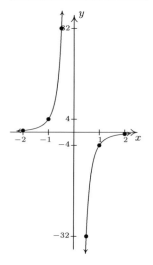

The graph is not symmetric about the x-axis (e.g. $(1, -4)$ is on the graph but $(1, 4)$ is not)

The graph is not symmetric about the y-axis (e.g. $(1, -4)$ is on the graph but $(-1, -4)$ is not)

The graph is symmetric about the origin

1.3 Introduction to Functions

One of the core concepts in College Algebra is the **function**. There are many ways to describe a function and we begin by defining a function as a special kind of relation.

> **Definition 1.6.** A relation in which each x-coordinate is matched with only one y-coordinate is said to describe y as a **function** of x.

Example 1.3.1. Which of the following relations describe y as a function of x?

1. $R_1 = \{(-2,1), (1,3), (1,4), (3,-1)\}$
2. $R_2 = \{(-2,1), (1,3), (2,3), (3,-1)\}$

Solution. A quick scan of the points in R_1 reveals that the x-coordinate 1 is matched with two *different* y-coordinates: namely 3 and 4. Hence in R_1, y is not a function of x. On the other hand, every x-coordinate in R_2 occurs only once which means each x-coordinate has only one corresponding y-coordinate. So, R_2 does represent y as a function of x. □

Note that in the previous example, the relation R_2 contained two different points with the same y-coordinates, namely $(1,3)$ and $(2,3)$. Remember, in order to say y is a function of x, we just need to ensure the same x-coordinate isn't used in more than one point.[1]

To see what the function concept means geometrically, we graph R_1 and R_2 in the plane.

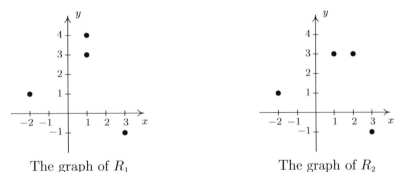

The graph of R_1 The graph of R_2

The fact that the x-coordinate 1 is matched with two different y-coordinates in R_1 presents itself graphically as the points $(1,3)$ and $(1,4)$ lying on the same vertical line, $x = 1$. If we turn our attention to the graph of R_2, we see that no two points of the relation lie on the same vertical line. We can generalize this idea as follows

> **Theorem 1.1. The Vertical Line Test:** A set of points in the plane represents y as a function of x if and only if no two points lie on the same vertical line.

---

[1] We will have occasion later in the text to concern ourselves with the concept of x being a function of y. In this case, R_1 represents x as a function of y; R_2 does not.

It is worth taking some time to meditate on the Vertical Line Test; it will check to see how well you understand the concept of 'function' as well as the concept of 'graph'.

Example 1.3.2. Use the Vertical Line Test to determine which of the following relations describes y as a function of x.

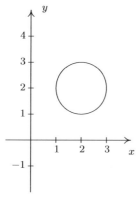

The graph of R

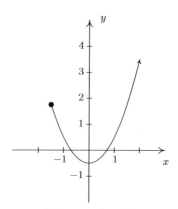

The graph of S

Solution. Looking at the graph of R, we can easily imagine a vertical line crossing the graph more than once. Hence, R does not represent y as a function of x. However, in the graph of S, every vertical line crosses the graph at most once, so S does represent y as a function of x. □

In the previous test, we say that the graph of the relation R **fails** the Vertical Line Test, whereas the graph of S **passes** the Vertical Line Test. Note that in the graph of R there are infinitely many vertical lines which cross the graph more than once. However, to fail the Vertical Line Test, all you need is one vertical line that fits the bill, as the next example illustrates.

Example 1.3.3. Use the Vertical Line Test to determine which of the following relations describes y as a function of x.

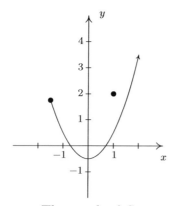

The graph of S_1

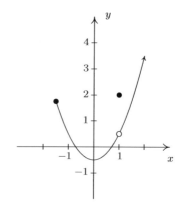

The graph of S_2

1.3 Introduction to Functions

Solution. Both S_1 and S_2 are slight modifications to the relation S in the previous example whose graph we determined passed the Vertical Line Test. In both S_1 and S_2, it is the addition of the point $(1, 2)$ which threatens to cause trouble. In S_1, there is a point on the curve with x-coordinate 1 just below $(1, 2)$, which means that both $(1, 2)$ and this point on the curve lie on the vertical line $x = 1$. (See the picture below and the left.) Hence, the graph of S_1 fails the Vertical Line Test, so y is not a function of x here. However, in S_2 notice that the point with x-coordinate 1 on the curve has been omitted, leaving an 'open circle' there. Hence, the vertical line $x = 1$ crosses the graph of S_2 only at the point $(1, 2)$. Indeed, any vertical line will cross the graph at most once, so we have that the graph of S_2 passes the Vertical Line Test. Thus it describes y as a function of x. □

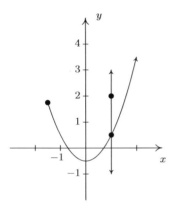

S_1 and the line $x = 1$

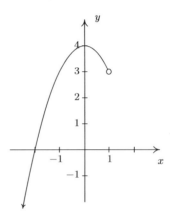

The graph of G for Ex. 1.3.4

Suppose a relation F describes y as a function of x. The sets of x- and y-coordinates are given special names which we define below.

> **Definition 1.7.** Suppose F is a relation which describes y as a function of x.
> - The set of the x-coordinates of the points in F is called the **domain** of F.
> - The set of the y-coordinates of the points in F is called the **range** of F.

We demonstrate finding the domain and range of functions given to us either graphically or via the roster method in the following example.

Example 1.3.4. Find the domain and range of the function $F = \{(-3, 2), (0, 1), (4, 2), (5, 2)\}$ and of the function G whose graph is given above on the right.

Solution. The domain of F is the set of the x-coordinates of the points in F, namely $\{-3, 0, 4, 5\}$ and the range of F is the set of the y-coordinates, namely $\{1, 2\}$.

To determine the domain and range of G, we need to determine which x and y values occur as coordinates of points on the given graph. To find the domain, it may be helpful to imagine collapsing the curve to the x-axis and determining the portion of the x-axis that gets covered. This is called **projecting** the curve to the x-axis. Before we start projecting, we need to pay attention to two subtle notations on the graph: the arrowhead on the lower left corner of the graph indicates that the graph continues to curve downwards to the left forever more; and the open circle at $(1,3)$ indicates that the point $(1,3)$ isn't on the graph, but all points on the curve leading up to that point are.

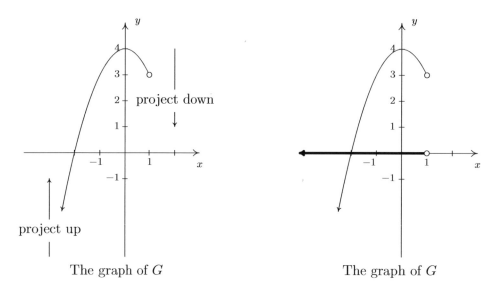

We see from the figure that if we project the graph of G to the x-axis, we get all real numbers less than 1. Using interval notation, we write the domain of G as $(-\infty, 1)$. To determine the range of G, we project the curve to the y-axis as follows:

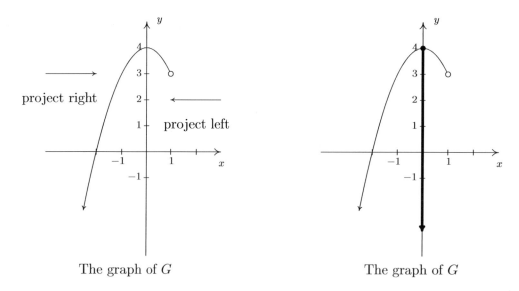

Note that even though there is an open circle at $(1,3)$, we still include the y value of 3 in our range, since the point $(-1, 3)$ is on the graph of G. We see that the range of G is all real numbers less than or equal to 4, or, in interval notation, $(-\infty, 4]$. □

1.3 Introduction to Functions

All functions are relations, but not all relations are functions. Thus the equations which described the relations in Section 1.2 may or may not describe y as a function of x. The algebraic representation of functions is possibly the most important way to view them so we need a process for determining whether or not an equation of a relation represents a function. (We delay the discussion of finding the domain of a function given algebraically until Section 1.4.)

Example 1.3.5. Determine which equations represent y as a function of x.

1. $x^3 + y^2 = 1$
2. $x^2 + y^3 = 1$
3. $x^2 y = 1 - 3y$

Solution. For each of these equations, we solve for y and determine whether each choice of x will determine only one corresponding value of y.

1.
$$\begin{aligned} x^3 + y^2 &= 1 \\ y^2 &= 1 - x^3 \\ \sqrt{y^2} &= \sqrt{1 - x^3} \quad \text{extract square roots} \\ y &= \pm\sqrt{1 - x^3} \end{aligned}$$

If we substitute $x = 0$ into our equation for y, we get $y = \pm\sqrt{1 - 0^3} = \pm 1$, so that $(0, 1)$ and $(0, -1)$ are on the graph of this equation. Hence, this equation does not represent y as a function of x.

2.
$$\begin{aligned} x^2 + y^3 &= 1 \\ y^3 &= 1 - x^2 \\ \sqrt[3]{y^3} &= \sqrt[3]{1 - x^2} \\ y &= \sqrt[3]{1 - x^2} \end{aligned}$$

For every choice of x, the equation $y = \sqrt[3]{1 - x^2}$ returns only **one** value of y. Hence, this equation describes y as a function of x.

3.
$$\begin{aligned} x^2 y &= 1 - 3y \\ x^2 y + 3y &= 1 \\ y\left(x^2 + 3\right) &= 1 \quad \text{factor} \\ y &= \frac{1}{x^2 + 3} \end{aligned}$$

For each choice of x, there is only one value for y, so this equation describes y as a function of x. □

We could try to use our graphing calculator to verify our responses to the previous example, but we immediately run into trouble. The calculator's "Y=" menu requires that the equation be of the form '$y = $ some expression of x'. If we wanted to verify that the first equation in Example 1.3.5

does not represent y as a function of x, we would need to enter two separate expressions into the calculator: one for the positive square root and one for the negative square root we found when solving the equation for y. As predicted, the resulting graph shown below clearly fails the Vertical Line Test, so the equation does not represent y as a function of x.

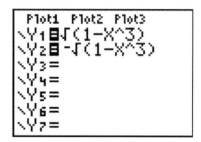 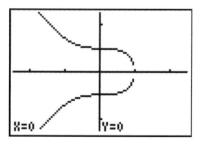

Thus in order to use the calculator to show that $x^3 + y^2 = 1$ does not represent y as a function of x we needed to know *analytically* that y was not a function of x so that we could use the calculator properly. There are more advanced graphing utilities out there which can do implicit function plots, but you need to know even more Algebra to make them work properly. Do you get the point we're trying to make here? We believe it is in your best interest to learn the analytic way of doing things so that you are always smarter than your calculator.

1.3 Introduction to Functions

1.3.1 Exercises

In Exercises 1 - 12, determine whether or not the relation represents y as a function of x. Find the domain and range of those relations which are functions.

1. $\{(-3,9), (-2,4), (-1,1), (0,0), (1,1), (2,4), (3,9)\}$

2. $\{(-3,0), (1,6), (2,-3), (4,2), (-5,6), (4,-9), (6,2)\}$

3. $\{(-3,0), (-7,6), (5,5), (6,4), (4,9), (3,0)\}$

4. $\{(1,2), (4,4), (9,6), (16,8), (25,10), (36,12), \ldots\}$

5. $\{(x,y) \,|\, x \text{ is an odd integer, and } y \text{ is an even integer}\}$

6. $\{(x,1) \,|\, x \text{ is an irrational number}\}$

7. $\{(1,0),\ (2,1),\ (4,2),\ (8,3),\ (16,4),\ (32,5),\ \ldots\}$

8. $\{\ldots, (-3,9), (-2,4), (-1,1), (0,0), (1,1), (2,4), (3,9), \ldots\}$

9. $\{(-2,y) \,|\, -3 < y < 4\}$

10. $\{(x,3) \,|\, -2 \leq x < 4\}$

11. $\{(x, x^2) \,|\, x \text{ is a real number}\}$

12. $\{(x^2, x) \,|\, x \text{ is a real number}\}$

In Exercises 13 - 32, determine whether or not the relation represents y as a function of x. Find the domain and range of those relations which are functions.

13.
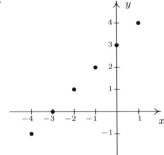

14.
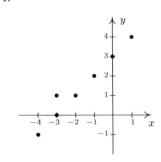

15.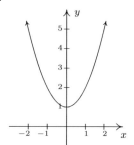

16.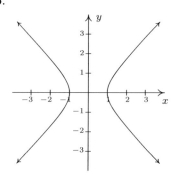

17.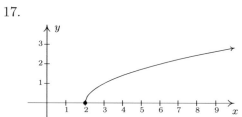

18.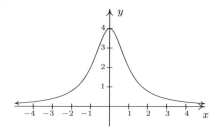

19.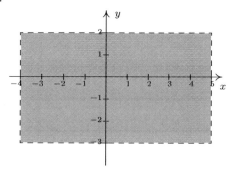

20.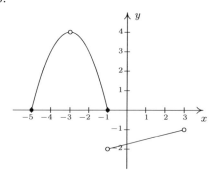

21.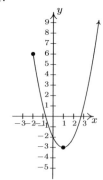

22.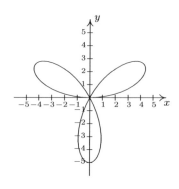

1.3 Introduction to Functions

23.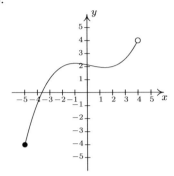

24.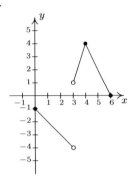

25.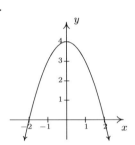

26.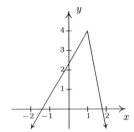

27.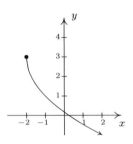

28.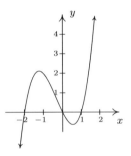

29.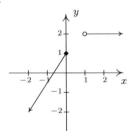

30.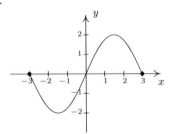

31.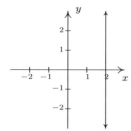

32.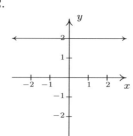

In Exercises 33 - 47, determine whether or not the equation represents y as a function of x.

33. $y = x^3 - x$

34. $y = \sqrt{x - 2}$

35. $x^3 y = -4$

36. $x^2 - y^2 = 1$

37. $y = \dfrac{x}{x^2 - 9}$

38. $x = -6$

39. $x = y^2 + 4$

40. $y = x^2 + 4$

41. $x^2 + y^2 = 4$

42. $y = \sqrt{4 - x^2}$

43. $x^2 - y^2 = 4$

44. $x^3 + y^3 = 4$

45. $2x + 3y = 4$

46. $2xy = 4$

47. $x^2 = y^2$

48. Explain why the population P of Sasquatch in a given area is a function of time t. What would be the range of this function?

49. Explain why the relation between your classmates and their email addresses may not be a function. What about phone numbers and Social Security Numbers?

The process given in Example 1.3.5 for determining whether an equation of a relation represents y as a function of x breaks down if we cannot solve the equation for y in terms of x. However, that does not prevent us from proving that an equation fails to represent y as a function of x. What we really need is two points with the same x-coordinate and different y-coordinates which both satisfy the equation so that the graph of the relation would fail the Vertical Line Test 1.1. Discuss with your classmates how you might find such points for the relations given in Exercises 50 - 53.

50. $x^3 + y^3 - 3xy = 0$

51. $x^4 = x^2 + y^2$

52. $y^2 = x^3 + 3x^2$

53. $(x^2 + y^2)^2 = x^3 + y^3$

1.3 Introduction to Functions

1.3.2 Answers

1. Function
   domain = $\{-3, -2, -1, 0, 1, 2, 3\}$
   range = $\{0, 1, 4, 9\}$

2. Not a function

3. Function
   domain = $\{-7, -3, 3, 4, 5, 6\}$
   range = $\{0, 4, 5, 6, 9\}$

4. Function
   domain = $\{1, 4, 9, 16, 25, 36, \ldots\}$
   = $\{x \,|\, x \text{ is a perfect square}\}$
   range = $\{2, 4, 6, 8, 10, 12, \ldots\}$
   = $\{y \,|\, y \text{ is a positive even integer}\}$

5. Not a function

6. Function
   domain = $\{x \,|\, x \text{ is irrational}\}$
   range = $\{1\}$

7. Function
   domain = $\{x \,|\, x = 2^n \text{ for some whole number } n\}$
   range = $\{y \,|\, y \text{ is any whole number}\}$

8. Function
   domain = $\{x \,|\, x \text{ is any integer}\}$
   range = $\{y \,|\, y = n^2 \text{ for some integer } n\}$

9. Not a function

10. Function
    domain = $[-2, 4)$, range = $\{3\}$

11. Function
    domain = $(-\infty, \infty)$
    range = $[0, \infty)$

12. Not a function

13. Function
    domain = $\{-4, -3, -2, -1, 0, 1\}$
    range = $\{-1, 0, 1, 2, 3, 4\}$

14. Not a function

15. Function
    domain = $(-\infty, \infty)$
    range = $[1, \infty)$

16. Not a function

17. Function
    domain = $[2, \infty)$
    range = $[0, \infty)$

18. Function
    domain = $(-\infty, \infty)$
    range = $(0, 4]$

19. Not a function

20. Function
    domain = $[-5, -3) \cup (-3, 3)$
    range = $(-2, -1) \cup [0, 4)$

21. Function
    domain = $[-2, \infty)$
    range = $[-3, \infty)$

22. Not a function

23. Function
    domain = $[-5, 4)$
    range = $[-4, 4)$

24. Function
    domain = $[0, 3) \cup (3, 6]$
    range = $(-4, -1] \cup [0, 4]$

25. Function
    domain = $(-\infty, \infty)$
    range = $(-\infty, 4]$

26. Function
    domain = $(-\infty, \infty)$
    range = $(-\infty, 4]$

27. Function
    domain = $[-2, \infty)$
    range = $(-\infty, 3]$

28. Function
    domain = $(-\infty, \infty)$
    range = $(-\infty, \infty)$

29. Function
    domain = $(-\infty, 0] \cup (1, \infty)$
    range = $(-\infty, 1] \cup \{2\}$

30. Function
    domain = $[-3, 3]$
    range = $[-2, 2]$

31. Not a function

32. Function
    domain = $(-\infty, \infty)$
    range = $\{2\}$

33. Function

34. Function

35. Function

36. Not a function

37. Function

38. Not a function

39. Not a function

40. Function

41. Not a function

42. Function

43. Not a function

44. Function

45. Function

46. Function

47. Not a function

1.4 Function Notation

In Definition 1.6, we described a function as a special kind of relation — one in which each x-coordinate is matched with only one y-coordinate. In this section, we focus more on the **process** by which the x is matched with the y. If we think of the domain of a function as a set of **inputs** and the range as a set of **outputs**, we can think of a function f as a process by which each input x is matched with only one output y. Since the output is completely determined by the input x and the process f, we symbolize the output with **function notation**: '$f(x)$', read 'f **of** x.' In other words, $f(x)$ is the output which results by applying the process f to the input x. In this case, the parentheses here do not indicate multiplication, as they do elsewhere in Algebra. This can cause confusion if the context is not clear, so you must read carefully. This relationship is typically visualized using a diagram similar to the one below.

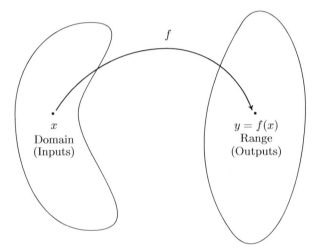

The value of y is completely dependent on the choice of x. For this reason, x is often called the **independent variable**, or **argument** of f, whereas y is often called the **dependent variable**.

As we shall see, the process of a function f is usually described using an algebraic formula. For example, suppose a function f takes a real number and performs the following two steps, in sequence

1. multiply by 3
2. add 4

If we choose 5 as our input, in step 1 we multiply by 3 to get $(5)(3) = 15$. In step 2, we add 4 to our result from step 1 which yields $15 + 4 = 19$. Using function notation, we would write $f(5) = 19$ to indicate that the result of applying the process f to the input 5 gives the output 19. In general, if we use x for the input, applying step 1 produces $3x$. Following with step 2 produces $3x + 4$ as our final output. Hence for an input x, we get the output $f(x) = 3x + 4$. Notice that to check our formula for the case $x = 5$, we replace the occurrence of x in the formula for $f(x)$ with 5 to get $f(5) = 3(5) + 4 = 15 + 4 = 19$, as required.

Example 1.4.1. Suppose a function g is described by applying the following steps, in sequence

1. add 4

2. multiply by 3

Determine $g(5)$ and find an expression for $g(x)$.

Solution. Starting with 5, step 1 gives $5 + 4 = 9$. Continuing with step 2, we get $(3)(9) = 27$. To find a formula for $g(x)$, we start with our input x. Step 1 produces $x + 4$. We now wish to multiply this entire quantity by 3, so we use a parentheses: $3(x + 4) = 3x + 12$. Hence, $g(x) = 3x + 12$. We can check our formula by replacing x with 5 to get $g(5) = 3(5) + 12 = 15 + 12 = 27 \checkmark$. □

Most of the functions we will encounter in College Algebra will be described using formulas like the ones we developed for $f(x)$ and $g(x)$ above. Evaluating formulas using this function notation is a key skill for success in this and many other Math courses.

Example 1.4.2. Let $f(x) = -x^2 + 3x + 4$

1. Find and simplify the following.

    (a) $f(-1)$, $f(0)$, $f(2)$
    (b) $f(2x)$, $2f(x)$
    (c) $f(x+2)$, $f(x) + 2$, $f(x) + f(2)$

2. Solve $f(x) = 4$.

Solution.

1. (a) To find $f(-1)$, we replace every occurrence of x in the expression $f(x)$ with -1

$$\begin{aligned} f(-1) &= -(-1)^2 + 3(-1) + 4 \\ &= -(1) + (-3) + 4 \\ &= 0 \end{aligned}$$

Similarly, $f(0) = -(0)^2 + 3(0) + 4 = 4$, and $f(2) = -(2)^2 + 3(2) + 4 = -4 + 6 + 4 = 6$.

(b) To find $f(2x)$, we replace every occurrence of x with the quantity $2x$

$$\begin{aligned} f(2x) &= -(2x)^2 + 3(2x) + 4 \\ &= -(4x^2) + (6x) + 4 \\ &= -4x^2 + 6x + 4 \end{aligned}$$

The expression $2f(x)$ means we multiply the expression $f(x)$ by 2

$$\begin{aligned} 2f(x) &= 2\left(-x^2 + 3x + 4\right) \\ &= -2x^2 + 6x + 8 \end{aligned}$$

(c) To find $f(x+2)$, we replace every occurrence of x with the quantity $x+2$

$$\begin{aligned} f(x+2) &= -(x+2)^2 + 3(x+2) + 4 \\ &= -\left(x^2 + 4x + 4\right) + (3x + 6) + 4 \\ &= -x^2 - 4x - 4 + 3x + 6 + 4 \\ &= -x^2 - x + 6 \end{aligned}$$

To find $f(x) + 2$, we add 2 to the expression for $f(x)$

$$\begin{aligned} f(x) + 2 &= \left(-x^2 + 3x + 4\right) + 2 \\ &= -x^2 + 3x + 6 \end{aligned}$$

From our work above, we see $f(2) = 6$ so that

$$\begin{aligned} f(x) + f(2) &= \left(-x^2 + 3x + 4\right) + 6 \\ &= -x^2 + 3x + 10 \end{aligned}$$

2. Since $f(x) = -x^2 + 3x + 4$, the equation $f(x) = 4$ is equivalent to $-x^2 + 3x + 4 = 4$. Solving we get $-x^2 + 3x = 0$, or $x(-x + 3) = 0$. We get $x = 0$ or $x = 3$, and we can verify these answers by checking that $f(0) = 4$ and $f(3) = 4$. □

A few notes about Example 1.4.2 are in order. First note the difference between the answers for $f(2x)$ and $2f(x)$. For $f(2x)$, we are multiplying the *input* by 2; for $2f(x)$, we are multiplying the *output* by 2. As we see, we get entirely different results. Along these lines, note that $f(x+2)$, $f(x)+2$ and $f(x)+f(2)$ are three *different* expressions as well. Even though function notation uses parentheses, as does multiplication, there is *no* general 'distributive property' of function notation. Finally, note the practice of using parentheses when substituting one algebraic expression into another; we highly recommend this practice as it will reduce careless errors.

Suppose now we wish to find $r(3)$ for $r(x) = \frac{2x}{x^2-9}$. Substitution gives

$$r(3) = \frac{2(3)}{(3)^2 - 9} = \frac{6}{0},$$

which is undefined. (Why is this, again?) The number 3 is not an allowable input to the function r; in other words, 3 is not in the domain of r. Which other real numbers are forbidden in this formula? We think back to arithmetic. The reason $r(3)$ is undefined is because substitution results in a division by 0. To determine which other numbers result in such a transgression, we set the denominator equal to 0 and solve

$$\begin{aligned} x^2 - 9 &= 0 \\ x^2 &= 9 \\ \sqrt{x^2} &= \sqrt{9} \quad \text{extract square roots} \\ x &= \pm 3 \end{aligned}$$

As long as we substitute numbers other than 3 and -3, the expression $r(x)$ is a real number. Hence, we write our domain in interval notation[1] as $(-\infty, -3) \cup (-3, 3) \cup (3, \infty)$. When a formula for a function is given, we assume that the function is valid for all real numbers which make arithmetic sense when substituted into the formula. This set of numbers is often called the **implied domain**[2] of the function. At this stage, there are only two mathematical sins we need to avoid: division by 0 and extracting even roots of negative numbers. The following example illustrates these concepts.

Example 1.4.3. Find the domain[3] of the following functions.

1. $g(x) = \sqrt{4 - 3x}$

2. $h(x) = \sqrt[5]{4 - 3x}$

3. $f(x) = \dfrac{2}{1 - \dfrac{4x}{x-3}}$

4. $F(x) = \dfrac{\sqrt[4]{2x+1}}{x^2 - 1}$

5. $r(t) = \dfrac{4}{6 - \sqrt{t+3}}$

6. $I(x) = \dfrac{3x^2}{x}$

Solution.

1. The potential disaster for g is if the radicand[4] is negative. To avoid this, we set $4 - 3x \geq 0$. From this, we get $3x \leq 4$ or $x \leq \frac{4}{3}$. What this shows is that as long as $x \leq \frac{4}{3}$, the expression $4 - 3x \geq 0$, and the formula $g(x)$ returns a real number. Our domain is $\left(-\infty, \frac{4}{3}\right]$.

2. The formula for $h(x)$ is hauntingly close to that of $g(x)$ with one key difference – whereas the expression for $g(x)$ includes an even indexed root (namely a square root), the formula for $h(x)$ involves an odd indexed root (the fifth root). Since odd roots of real numbers (even negative real numbers) are real numbers, there is no restriction on the inputs to h. Hence, the domain is $(-\infty, \infty)$.

3. In the expression for f, there are two denominators. We need to make sure neither of them is 0. To that end, we set each denominator equal to 0 and solve. For the 'small' denominator, we get $x - 3 = 0$ or $x = 3$. For the 'large' denominator

---

[1] See the Exercises for Section 1.1.
[2] or, 'implicit domain'
[3] The word 'implied' is, well, implied.
[4] The 'radicand' is the expression 'inside' the radical.

1.4 FUNCTION NOTATION

$$1 - \frac{4x}{x-3} = 0$$

$$1 = \frac{4x}{x-3}$$

$$(1)(x-3) = \left(\frac{4x}{x-3}\right)(x-3) \quad \text{clear denominators}$$

$$x - 3 = 4x$$

$$-3 = 3x$$

$$-1 = x$$

So we get two real numbers which make denominators 0, namely $x = -1$ and $x = 3$. Our domain is all real numbers except -1 and 3: $(-\infty, -1) \cup (-1, 3) \cup (3, \infty)$.

4. In finding the domain of F, we notice that we have two potentially hazardous issues: not only do we have a denominator, we have a fourth (even-indexed) root. Our strategy is to determine the restrictions imposed by each part and select the real numbers which satisfy both conditions. To satisfy the fourth root, we require $2x + 1 \geq 0$. From this we get $2x \geq -1$ or $x \geq -\frac{1}{2}$. Next, we round up the values of x which could cause trouble in the denominator by setting the denominator equal to 0. We get $x^2 - 1 = 0$, or $x = \pm 1$. Hence, in order for a real number x to be in the domain of F, $x \geq -\frac{1}{2}$ but $x \neq \pm 1$. In interval notation, this set is $\left[-\frac{1}{2}, 1\right) \cup (1, \infty)$.

5. Don't be put off by the 't' here. It is an independent variable representing a real number, just like x does, and is subject to the same restrictions. As in the previous problem, we have double danger here: we have a square root and a denominator. To satisfy the square root, we need a non-negative radicand so we set $t + 3 \geq 0$ to get $t \geq -3$. Setting the denominator equal to zero gives $6 - \sqrt{t+3} = 0$, or $\sqrt{t+3} = 6$. Squaring both sides gives $t + 3 = 36$, or $t = 33$. Since we squared both sides in the course of solving this equation, we need to check our answer.[5] Sure enough, when $t = 33$, $6 - \sqrt{t+3} = 6 - \sqrt{36} = 0$, so $t = 33$ will cause problems in the denominator. At last we can find the domain of r: we need $t \geq -3$, but $t \neq 33$. Our final answer is $[-3, 33) \cup (33, \infty)$.

6. It's tempting to simplify $I(x) = \frac{3x^2}{x} = 3x$, and, since there are no longer any denominators, claim that there are no longer any restrictions. However, in simplifying $I(x)$, we are assuming $x \neq 0$, since $\frac{0}{0}$ is undefined.[6] Proceeding as before, we find the domain of I to be all real numbers except 0: $(-\infty, 0) \cup (0, \infty)$. $\square$

It is worth reiterating the importance of finding the domain of a function *before* simplifying, as evidenced by the function I in the previous example. Even though the formula $I(x)$ simplifies to

---

[5] Do you remember why? Consider squaring both sides to 'solve' $\sqrt{t+1} = -2$.
[6] More precisely, the fraction $\frac{0}{0}$ is an 'indeterminant form'. Calculus is required tame such beasts.

$3x$, it would be inaccurate to write $I(x) = 3x$ without adding the stipulation that $x \neq 0$. It would be analogous to not reporting taxable income or some other sin of omission.

1.4.1 Modeling with Functions

The importance of Mathematics to our society lies in its value to approximate, or **model** real-world phenomenon. Whether it be used to predict the high temperature on a given day, determine the hours of daylight on a given day, or predict population trends of various and sundry real and mythical beasts,[7] Mathematics is second only to literacy in the importance humanity's development.[8]

It is important to keep in mind that anytime Mathematics is used to approximate reality, there are always limitations to the model. For example, suppose grapes are on sale at the local market for $1.50 per pound. Then one pound of grapes costs $1.50, two pounds of grapes cost $3.00, and so forth. Suppose we want to develop a formula which relates the cost of buying grapes to the amount of grapes being purchased. Since these two quantities vary from situation to situation, we assign them variables. Let c denote the cost of the grapes and let g denote the amount of grapes purchased. To find the cost c of the grapes, we multiply the amount of grapes g by the price $1.50 dollars per pound to get

$$c = 1.5g$$

In order for the units to be correct in the formula, g must be measured in *pounds* of grapes in which case the computed value of c is measured in *dollars*. Since we're interested in finding the cost c given an amount g, we think of g as the independent variable and c as the dependent variable. Using the language of function notation, we write

$$c(g) = 1.5g$$

where g is the amount of grapes purchased (in pounds) and $c(g)$ is the cost (in dollars). For example, $c(5)$ represents the cost, in dollars, to purchase 5 pounds of grapes. In this case, $c(5) = 1.5(5) = 7.5$, so it would cost $7.50. If, on the other hand, we wanted to find the *amount* of grapes we can purchase for $5, we would need to set $c(g) = 5$ and solve for g. In this case, $c(g) = 1.5g$, so solving $c(g) = 5$ is equivalent to solving $1.5g = 5$ Doing so gives $g = \frac{5}{1.5} = 3.\overline{3}$. This means we can purchase exactly $3.\overline{3}$ pounds of grapes for $5. Of course, you would be hard-pressed to buy exactly $3.\overline{3}$ pounds of grapes,[9] and this leads us to our next topic of discussion, the **applied domain**[10] of a function.

Even though, mathematically, $c(g) = 1.5g$ has no domain restrictions (there are no denominators and no even-indexed radicals), there are certain values of g that don't make any physical sense. For example, $g = -1$ corresponds to 'purchasing' -1 pounds of grapes.[11] Also, unless the 'local market' mentioned is the State of California (or some other exporter of grapes), it also doesn't make much sense for $g = 500{,}000{,}000$, either. So the reality of the situation limits what g can be, and

---

[7] See Sections 2.5, 11.1, and 6.5, respectively.
[8] In Carl's humble opinion, of course ...
[9] You could get close... within a certain specified margin of error, perhaps.
[10] or, 'explicit domain'
[11] Maybe this means *returning* a pound of grapes?

1.4 FUNCTION NOTATION   61

these limits determine the applied domain of g. Typically, an applied domain is stated explicitly. In this case, it would be common to see something like $c(g) = 1.5g$, $0 \leq g \leq 100$, meaning the number of pounds of grapes purchased is limited from 0 up to 100. The upper bound here, 100 may represent the inventory of the market, or some other limit as set by local policy or law. Even with this restriction, our model has its limitations. As we saw above, it is virtually impossible to buy exactly $3.\overline{3}$ pounds of grapes so that our cost is exactly $5. In this case, being sensible shoppers, we would most likely 'round down' and purchase 3 pounds of grapes or however close the market scale can read to $3.\overline{3}$ without being over. It is time for a more sophisticated example.

Example 1.4.4. The height h in feet of a model rocket above the ground t seconds after lift-off is given by

$$h(t) = \begin{cases} -5t^2 + 100t, & \text{if } 0 \leq t \leq 20 \\ 0, & \text{if } t > 20 \end{cases}$$

1. Find and interpret $h(10)$ and $h(60)$.

2. Solve $h(t) = 375$ and interpret your answers.

Solution.

1. We first note that the independent variable here is t, chosen because it represents time. Secondly, the function is broken up into two rules: one formula for values of t between 0 and 20 inclusive, and another for values of t greater than 20. Since $t = 10$ satisfies the inequality $0 \leq t \leq 20$, we use the first formula listed, $h(t) = -5t^2 + 100t$, to find $h(10)$. We get $h(10) = -5(10)^2 + 100(10) = 500$. Since t represents the number of seconds since lift-off and $h(t)$ is the height above the ground in feet, the equation $h(10) = 500$ means that 10 seconds after lift-off, the model rocket is 500 feet above the ground. To find $h(60)$, we note that $t = 60$ satisfies $t > 20$, so we use the rule $h(t) = 0$. This function returns a value of 0 regardless of what value is substituted in for t, so $h(60) = 0$. This means that 60 seconds after lift-off, the rocket is 0 feet above the ground; in other words, a minute after lift-off, the rocket has already returned to Earth.

2. Since the function h is defined in pieces, we need to solve $h(t) = 375$ in pieces. For $0 \leq t \leq 20$, $h(t) = -5t^2 + 100t$, so for these values of t, we solve $-5t^2 + 100t = 375$. Rearranging terms, we get $5t^2 - 100t + 375 = 0$, and factoring gives $5(t - 5)(t - 15) = 0$. Our answers are $t = 5$ and $t = 15$, and since both of these values of t lie between 0 and 20, we keep both solutions. For $t > 20$, $h(t) = 0$, and in this case, there are no solutions to $0 = 375$. In terms of the model rocket, solving $h(t) = 375$ corresponds to finding when, if ever, the rocket reaches 375 feet above the ground. Our two answers, $t = 5$ and $t = 15$ correspond to the rocket reaching this altitude *twice* – once 5 seconds after launch, and again 15 seconds after launch.[12]   □

---

[12]What goes up ...

The type of function in the previous example is called a **piecewise-defined** function, or 'piecewise' function for short. Many real-world phenomena, income tax formulas[13] for example, are modeled by such functions.

By the way, if we wanted to avoid using a piecewise function in Example 1.4.4, we could have used $h(t) = -5t^2 + 100t$ on the explicit domain $0 \leq t \leq 20$ because after 20 seconds, the rocket is on the ground and stops moving. In many cases, though, piecewise functions are your only choice, so it's best to understand them well.

Mathematical modeling is not a one-section topic. It's not even a one-*course* topic as is evidenced by undergraduate and graduate courses in mathematical modeling being offered at many universities. Thus our goal in this section cannot possibly be to tell you the whole story. What we can do is get you started. As we study new classes of functions, we will see what phenomena they can be used to model. In that respect, mathematical modeling cannot be a topic in a book, but rather, must be a theme of the book. For now, we have you explore some very basic models in the Exercises because you need to crawl to walk to run. As we learn more about functions, we'll help you build your own models and get you on your way to applying Mathematics to your world.

---

[13] See the Internal Revenue Service's website

1.4 Function Notation

1.4.2 Exercises

In Exercises 1 - 10, find an expression for $f(x)$ and state its domain.

1. f is a function that takes a real number x and performs the following three steps in the order given: (1) multiply by 2; (2) add 3; (3) divide by 4.

2. f is a function that takes a real number x and performs the following three steps in the order given: (1) add 3; (2) multiply by 2; (3) divide by 4.

3. f is a function that takes a real number x and performs the following three steps in the order given: (1) divide by 4; (2) add 3; (3) multiply by 2.

4. f is a function that takes a real number x and performs the following three steps in the order given: (1) multiply by 2; (2) add 3; (3) take the square root.

5. f is a function that takes a real number x and performs the following three steps in the order given: (1) add 3; (2) multiply by 2; (3) take the square root.

6. f is a function that takes a real number x and performs the following three steps in the order given: (1) add 3; (2) take the square root; (3) multiply by 2.

7. f is a function that takes a real number x and performs the following three steps in the order given: (1) take the square root; (2) subtract 13; (3) make the quantity the denominator of a fraction with numerator 4.

8. f is a function that takes a real number x and performs the following three steps in the order given: (1) subtract 13; (2) take the square root; (3) make the quantity the denominator of a fraction with numerator 4.

9. f is a function that takes a real number x and performs the following three steps in the order given: (1) take the square root; (2) make the quantity the denominator of a fraction with numerator 4; (3) subtract 13.

10. f is a function that takes a real number x and performs the following three steps in the order given: (1) make the quantity the denominator of a fraction with numerator 4; (2) take the square root; (3) subtract 13.

In Exercises 11 - 18, use the given function f to find and simplify the following:

- $f(3)$
- $f(-1)$
- $f\left(\frac{3}{2}\right)$
- $f(4x)$
- $4f(x)$
- $f(-x)$
- $f(x-4)$
- $f(x)-4$
- $f\left(x^2\right)$

11. $f(x) = 2x + 1$

12. $f(x) = 3 - 4x$

13. $f(x) = 2 - x^2$

14. $f(x) = x^2 - 3x + 2$

15. $f(x) = \dfrac{x}{x-1}$

16. $f(x) = \dfrac{2}{x^3}$

17. $f(x) = 6$

18. $f(x) = 0$

In Exercises 19 - 26, use the given function f to find and simplify the following:

- $f(2)$
- $f(-2)$
- $f(2a)$
- $2f(a)$
- $f(a+2)$
- $f(a) + f(2)$
- $f\left(\dfrac{2}{a}\right)$
- $\dfrac{f(a)}{2}$
- $f(a+h)$

19. $f(x) = 2x - 5$

20. $f(x) = 5 - 2x$

21. $f(x) = 2x^2 - 1$

22. $f(x) = 3x^2 + 3x - 2$

23. $f(x) = \sqrt{2x+1}$

24. $f(x) = 117$

25. $f(x) = \dfrac{x}{2}$

26. $f(x) = \dfrac{2}{x}$

In Exercises 27 - 34, use the given function f to find $f(0)$ and solve $f(x) = 0$

27. $f(x) = 2x - 1$

28. $f(x) = 3 - \tfrac{2}{5}x$

29. $f(x) = 2x^2 - 6$

30. $f(x) = x^2 - x - 12$

31. $f(x) = \sqrt{x+4}$

32. $f(x) = \sqrt{1-2x}$

33. $f(x) = \dfrac{3}{4-x}$

34. $f(x) = \dfrac{3x^2 - 12x}{4 - x^2}$

35. Let $f(x) = \begin{cases} x + 5 & \text{if } x \leq -3 \\ \sqrt{9 - x^2} & \text{if } -3 < x \leq 3 \\ -x + 5 & \text{if } x > 3 \end{cases}$ Compute the following function values.

   (a) $f(-4)$
   (b) $f(-3)$
   (c) $f(3)$
   (d) $f(3.001)$
   (e) $f(-3.001)$
   (f) $f(2)$

1.4 FUNCTION NOTATION

36. Let $f(x) = \begin{cases} x^2 & \text{if } x \leq -1 \\ \sqrt{1-x^2} & \text{if } -1 < x \leq 1 \\ x & \text{if } x > 1 \end{cases}$ Compute the following function values.

(a) $f(4)$ (b) $f(-3)$ (c) $f(1)$

(d) $f(0)$ (e) $f(-1)$ (f) $f(-0.999)$

In Exercises 37 - 62, find the (implied) domain of the function.

37. $f(x) = x^4 - 13x^3 + 56x^2 - 19$

38. $f(x) = x^2 + 4$

39. $f(x) = \dfrac{x-2}{x+1}$

40. $f(x) = \dfrac{3x}{x^2+x-2}$

41. $f(x) = \dfrac{2x}{x^2+3}$

42. $f(x) = \dfrac{2x}{x^2-3}$

43. $f(x) = \dfrac{x+4}{x^2-36}$

44. $f(x) = \dfrac{x-2}{x-2}$

45. $f(x) = \sqrt{3-x}$

46. $f(x) = \sqrt{2x+5}$

47. $f(x) = 9x\sqrt{x+3}$

48. $f(x) = \dfrac{\sqrt{7-x}}{x^2+1}$

49. $f(x) = \sqrt{6x-2}$

50. $f(x) = \dfrac{6}{\sqrt{6x-2}}$

51. $f(x) = \sqrt[3]{6x-2}$

52. $f(x) = \dfrac{6}{4-\sqrt{6x-2}}$

53. $f(x) = \dfrac{\sqrt{6x-2}}{x^2-36}$

54. $f(x) = \dfrac{\sqrt[3]{6x-2}}{x^2+36}$

55. $s(t) = \dfrac{t}{t-8}$

56. $Q(r) = \dfrac{\sqrt{r}}{r-8}$

57. $b(\theta) = \dfrac{\theta}{\sqrt{\theta-8}}$

58. $A(x) = \sqrt{x-7} + \sqrt{9-x}$

59. $\alpha(y) = \sqrt[3]{\dfrac{y}{y-8}}$

60. $g(v) = \dfrac{1}{4 - \dfrac{1}{v^2}}$

61. $T(t) = \dfrac{\sqrt{t}-8}{5-t}$

62. $u(w) = \dfrac{w-8}{5-\sqrt{w}}$

63. The area A enclosed by a square, in square inches, is a function of the length of one of its sides x, when measured in inches. This relation is expressed by the formula $A(x) = x^2$ for $x > 0$. Find $A(3)$ and solve $A(x) = 36$. Interpret your answers to each. Why is x restricted to $x > 0$?

64. The area A enclosed by a circle, in square meters, is a function of its radius r, when measured in meters. This relation is expressed by the formula $A(r) = \pi r^2$ for $r > 0$. Find $A(2)$ and solve $A(r) = 16\pi$. Interpret your answers to each. Why is r restricted to $r > 0$?

65. The volume V enclosed by a cube, in cubic centimeters, is a function of the length of one of its sides x, when measured in centimeters. This relation is expressed by the formula $V(x) = x^3$ for $x > 0$. Find $V(5)$ and solve $V(x) = 27$. Interpret your answers to each. Why is x restricted to $x > 0$?

66. The volume V enclosed by a sphere, in cubic feet, is a function of the radius of the sphere r, when measured in feet. This relation is expressed by the formula $V(r) = \frac{4\pi}{3} r^3$ for $r > 0$. Find $V(3)$ and solve $V(r) = \frac{32\pi}{3}$. Interpret your answers to each. Why is r restricted to $r > 0$?

67. The height of an object dropped from the roof of an eight story building is modeled by: $h(t) = -16t^2 + 64$, $0 \leq t \leq 2$. Here, h is the height of the object off the ground, in feet, t seconds after the object is dropped. Find $h(0)$ and solve $h(t) = 0$. Interpret your answers to each. Why is t restricted to $0 \leq t \leq 2$?

68. The temperature T in degrees Fahrenheit t hours after 6 AM is given by $T(t) = -\frac{1}{2}t^2 + 8t + 3$ for $0 \leq t \leq 12$. Find and interpret $T(0)$, $T(6)$ and $T(12)$.

69. The function $C(x) = x^2 - 10x + 27$ models the cost, in *hundreds* of dollars, to produce x *thousand* pens. Find and interpret $C(0)$, $C(2)$ and $C(5)$.

70. Using data from the Bureau of Transportation Statistics, the average fuel economy F in miles per gallon for passenger cars in the US can be modeled by $F(t) = -0.0076t^2 + 0.45t + 16$, $0 \leq t \leq 28$, where t is the number of years since 1980. Use your calculator to find $F(0)$, $F(14)$ and $F(28)$. Round your answers to two decimal places and interpret your answers to each.

71. The population of Sasquatch in Portage County can be modeled by the function $P(t) = \frac{150t}{t+15}$, where t represents the number of years since 1803. Find and interpret $P(0)$ and $P(205)$. Discuss with your classmates what the applied domain and range of P should be.

72. For n copies of the book *Me and my Sasquatch*, a print on-demand company charges $C(n)$ dollars, where $C(n)$ is determined by the formula

$$C(n) = \begin{cases} 15n & \text{if } 1 \leq n \leq 25 \\ 13.50n & \text{if } 25 < n \leq 50 \\ 12n & \text{if } n > 50 \end{cases}$$

(a) Find and interpret $C(20)$.

1.4 FUNCTION NOTATION

(b) How much does it cost to order 50 copies of the book? What about 51 copies?

(c) Your answer to 72b should get you thinking. Suppose a bookstore estimates it will sell 50 copies of the book. How many books can, in fact, be ordered for the same price as those 50 copies? (Round your answer to a whole number of books.)

73. An on-line comic book retailer charges shipping costs according to the following formula

$$S(n) = \begin{cases} 1.5n + 2.5 & \text{if } 1 \leq n \leq 14 \\ 0 & \text{if } n \geq 15 \end{cases}$$

where n is the number of comic books purchased and $S(n)$ is the shipping cost in dollars.

(a) What is the cost to ship 10 comic books?

(b) What is the significance of the formula $S(n) = 0$ for $n \geq 15$?

74. The cost C (in dollars) to talk m minutes a month on a mobile phone plan is modeled by

$$C(m) = \begin{cases} 25 & \text{if } 0 \leq m \leq 1000 \\ 25 + 0.1(m - 1000) & \text{if } m > 1000 \end{cases}$$

(a) How much does it cost to talk 750 minutes per month with this plan?

(b) How much does it cost to talk 20 hours a month with this plan?

(c) Explain the terms of the plan verbally.

75. In Section 1.1.1 we defined the set of **integers** as $\mathbb{Z} = \{\ldots, -3, -2, -1, 0, 1, 2, 3, \ldots\}$.[14] The **greatest integer of x**, denoted by $\lfloor x \rfloor$, is defined to be the largest integer k with $k \leq x$.

(a) Find $\lfloor 0.785 \rfloor$, $\lfloor 117 \rfloor$, $\lfloor -2.001 \rfloor$, and $\lfloor \pi + 6 \rfloor$

(b) Discuss with your classmates how $\lfloor x \rfloor$ may be described as a piecewise defined function.
HINT: There are infinitely many pieces!

(c) Is $\lfloor a + b \rfloor = \lfloor a \rfloor + \lfloor b \rfloor$ always true? What if a or b is an integer? Test some values, make a conjecture, and explain your result.

76. We have through our examples tried to convince you that, in general, $f(a + b) \neq f(a) + f(b)$. It has been our experience that students refuse to believe us so we'll try again with a different approach. With the help of your classmates, find a function f for which the following properties are always true.

(a) $f(0) = f(-1 + 1) = f(-1) + f(1)$

---
[14] The use of the letter $\mathbb{Z}$ for the integers is ostensibly because the German word *zahlen* means 'to count.'

(b) $f(5) = f(2+3) = f(2) + f(3)$

(c) $f(-6) = f(0-6) = f(0) - f(6)$

(d) $f(a+b) = f(a) + f(b)$ regardless of what two numbers we give you for a and b.

How many functions did you find that failed to satisfy the conditions above? Did $f(x) = x^2$ work? What about $f(x) = \sqrt{x}$ or $f(x) = 3x + 7$ or $f(x) = \dfrac{1}{x}$? Did you find an attribute common to those functions that did succeed? You should have, because there is only one extremely special family of functions that actually works here. Thus we return to our previous statement, **in general**, $f(a+b) \neq f(a) + f(b)$.

1.4 FUNCTION NOTATION

1.4.3 ANSWERS

1. $f(x) = \frac{2x+3}{4}$
   Domain: $(-\infty, \infty)$

2. $f(x) = \frac{2(x+3)}{4} = \frac{x+3}{2}$
   Domain: $(-\infty, \infty)$

3. $f(x) = 2\left(\frac{x}{4} + 3\right) = \frac{1}{2}x + 6$
   Domain: $(-\infty, \infty)$

4. $f(x) = \sqrt{2x+3}$
   Domain: $\left[-\frac{3}{2}, \infty\right)$

5. $f(x) = \sqrt{2(x+3)} = \sqrt{2x+6}$
   Domain: $[-3, \infty)$

6. $f(x) = 2\sqrt{x+3}$
   Domain: $[-3, \infty)$

7. $f(x) = \frac{4}{\sqrt{x}-13}$
   Domain: $[0, 169) \cup (169, \infty)$

8. $f(x) = \frac{4}{\sqrt{x-13}}$
   Domain: $(13, \infty)$

9. $f(x) = \frac{4}{\sqrt{x}} - 13$
   Domain: $(0, \infty)$

10. $f(x) = \sqrt{\frac{4}{x}} - 13 = \frac{2}{\sqrt{x}} - 13$
    Domain: $(0, \infty)$

11. For $f(x) = 2x + 1$

    - $f(3) = 7$
    - $f(-1) = -1$
    - $f\left(\frac{3}{2}\right) = 4$

    - $f(4x) = 8x + 1$
    - $4f(x) = 8x + 4$
    - $f(-x) = -2x + 1$

    - $f(x-4) = 2x - 7$
    - $f(x) - 4 = 2x - 3$
    - $f\left(x^2\right) = 2x^2 + 1$

12. For $f(x) = 3 - 4x$

    - $f(3) = -9$
    - $f(-1) = 7$
    - $f\left(\frac{3}{2}\right) = -3$

    - $f(4x) = 3 - 16x$
    - $4f(x) = 12 - 16x$
    - $f(-x) = 4x + 3$

    - $f(x-4) = 19 - 4x$
    - $f(x) - 4 = -4x - 1$
    - $f\left(x^2\right) = 3 - 4x^2$

13. For $f(x) = 2 - x^2$

    - $f(3) = -7$
    - $f(-1) = 1$
    - $f\left(\frac{3}{2}\right) = -\frac{1}{4}$
    - $f(4x) = 2 - 16x^2$
    - $4f(x) = 8 - 4x^2$
    - $f(-x) = 2 - x^2$
    - $f(x-4) = -x^2 + 8x - 14$
    - $f(x) - 4 = -x^2 - 2$
    - $f(x^2) = 2 - x^4$

14. For $f(x) = x^2 - 3x + 2$

    - $f(3) = 2$
    - $f(-1) = 6$
    - $f\left(\frac{3}{2}\right) = -\frac{1}{4}$
    - $f(4x) = 16x^2 - 12x + 2$
    - $4f(x) = 4x^2 - 12x + 8$
    - $f(-x) = x^2 + 3x + 2$
    - $f(x-4) = x^2 - 11x + 30$
    - $f(x) - 4 = x^2 - 3x - 2$
    - $f(x^2) = x^4 - 3x^2 + 2$

15. For $f(x) = \frac{x}{x-1}$

    - $f(3) = \frac{3}{2}$
    - $f(-1) = \frac{1}{2}$
    - $f\left(\frac{3}{2}\right) = 3$
    - $f(4x) = \frac{4x}{4x-1}$
    - $4f(x) = \frac{4x}{x-1}$
    - $f(-x) = \frac{x}{x+1}$
    - $f(x-4) = \frac{x-4}{x-5}$
    - $f(x) - 4 = \frac{x}{x-1} - 4 = \frac{4-3x}{x-1}$
    - $f(x^2) = \frac{x^2}{x^2-1}$

16. For $f(x) = \frac{2}{x^3}$

    - $f(3) = \frac{2}{27}$
    - $f(-1) = -2$
    - $f\left(\frac{3}{2}\right) = \frac{16}{27}$
    - $f(4x) = \frac{1}{32x^3}$
    - $4f(x) = \frac{8}{x^3}$
    - $f(-x) = -\frac{2}{x^3}$
    - $f(x-4) = \frac{2}{(x-4)^3} = \frac{2}{x^3 - 12x^2 + 48x - 64}$
    - $f(x) - 4 = \frac{2}{x^3} - 4 = \frac{2 - 4x^3}{x^3}$
    - $f(x^2) = \frac{2}{x^6}$

17. For $f(x) = 6$

    - $f(3) = 6$
    - $f(-1) = 6$
    - $f\left(\frac{3}{2}\right) = 6$
    - $f(4x) = 6$
    - $4f(x) = 24$
    - $f(-x) = 6$
    - $f(x-4) = 6$
    - $f(x) - 4 = 2$
    - $f(x^2) = 6$

1.4 Function Notation

18. For $f(x) = 0$

    - $f(3) = 0$
    - $f(-1) = 0$
    - $f\left(\frac{3}{2}\right) = 0$

    - $f(4x) = 0$
    - $4f(x) = 0$
    - $f(-x) = 0$

    - $f(x-4) = 0$
    - $f(x) - 4 = -4$
    - $f\left(x^2\right) = 0$

19. For $f(x) = 2x - 5$

    - $f(2) = -1$
    - $f(-2) = -9$
    - $f(2a) = 4a - 5$

    - $2f(a) = 4a - 10$
    - $f(a+2) = 2a - 1$
    - $f(a) + f(2) = 2a - 6$

    - $f\left(\frac{2}{a}\right) = \frac{4}{a} - 5$
      $= \frac{4-5a}{a}$
    - $\frac{f(a)}{2} = \frac{2a-5}{2}$
    - $f(a+h) = 2a + 2h - 5$

20. For $f(x) = 5 - 2x$

    - $f(2) = 1$
    - $f(-2) = 9$
    - $f(2a) = 5 - 4a$

    - $2f(a) = 10 - 4a$
    - $f(a+2) = 1 - 2a$
    - $f(a) + f(2) = 6 - 2a$

    - $f\left(\frac{2}{a}\right) = 5 - \frac{4}{a}$
      $= \frac{5a-4}{a}$
    - $\frac{f(a)}{2} = \frac{5-2a}{2}$
    - $f(a+h) = 5 - 2a - 2h$

21. For $f(x) = 2x^2 - 1$

    - $f(2) = 7$
    - $f(-2) = 7$
    - $f(2a) = 8a^2 - 1$

    - $2f(a) = 4a^2 - 2$
    - $f(a+2) = 2a^2 + 8a + 7$
    - $f(a) + f(2) = 2a^2 + 6$

    - $f\left(\frac{2}{a}\right) = \frac{8}{a^2} - 1$
      $= \frac{8-a^2}{a^2}$
    - $\frac{f(a)}{2} = \frac{2a^2-1}{2}$
    - $f(a+h) = 2a^2 + 4ah + 2h^2 - 1$

22. For $f(x) = 3x^2 + 3x - 2$

   - $f(2) = 16$
   - $f(-2) = 4$
   - $f(2a) = 12a^2 + 6a - 2$
   - $2f(a) = 6a^2 + 6a - 4$
   - $f(a+2) = 3a^2 + 15a + 16$
   - $f(a) + f(2) = 3a^2 + 3a + 14$
   - $f\left(\frac{2}{a}\right) = \frac{12}{a^2} + \frac{6}{a} - 2$
     $= \frac{12 + 6a - 2a^2}{a^2}$
   - $\frac{f(a)}{2} = \frac{3a^2 + 3a - 2}{2}$
   - $f(a+h) = 3a^2 + 6ah + 3h^2 + 3a + 3h - 2$

23. For $f(x) = \sqrt{2x+1}$

   - $f(2) = \sqrt{5}$
   - $f(-2)$ is not real
   - $f(2a) = \sqrt{4a+1}$
   - $2f(a) = 2\sqrt{2a+1}$
   - $f(a+2) = \sqrt{2a+5}$
   - $f(a) + f(2) = \sqrt{2a+1} + \sqrt{5}$
   - $f\left(\frac{2}{a}\right) = \sqrt{\frac{4}{a}+1}$
     $= \sqrt{\frac{a+4}{a}}$
   - $\frac{f(a)}{2} = \frac{\sqrt{2a+1}}{2}$
   - $f(a+h) = \sqrt{2a+2h+1}$

24. For $f(x) = 117$

   - $f(2) = 117$
   - $f(-2) = 117$
   - $f(2a) = 117$
   - $2f(a) = 234$
   - $f(a+2) = 117$
   - $f(a) + f(2) = 234$
   - $f\left(\frac{2}{a}\right) = 117$
   - $\frac{f(a)}{2} = \frac{117}{2}$
   - $f(a+h) = 117$

25. For $f(x) = \frac{x}{2}$

   - $f(2) = 1$
   - $f(-2) = -1$
   - $f(2a) = a$
   - $2f(a) = a$
   - $f(a+2) = \frac{a+2}{2}$
   - $f(a) + f(2) = \frac{a}{2} + 1$
     $= \frac{a+2}{2}$
   - $f\left(\frac{2}{a}\right) = \frac{1}{a}$
   - $\frac{f(a)}{2} = \frac{a}{4}$
   - $f(a+h) = \frac{a+h}{2}$

1.4 Function Notation

26. For $f(x) = \frac{2}{x}$

   - $f(2) = 1$
   - $f(-2) = -1$
   - $f(2a) = \frac{1}{a}$
   - $2f(a) = \frac{4}{a}$
   - $f(a+2) = \frac{2}{a+2}$
   - $f(a) + f(2) = \frac{2}{a} + 1 = \frac{a+2}{2}$
   - $f\left(\frac{2}{a}\right) = a$
   - $\frac{f(a)}{2} = \frac{1}{a}$
   - $f(a+h) = \frac{2}{a+h}$

27. For $f(x) = 2x - 1$, $f(0) = -1$ and $f(x) = 0$ when $x = \frac{1}{2}$

28. For $f(x) = 3 - \frac{2}{5}x$, $f(0) = 3$ and $f(x) = 0$ when $x = \frac{15}{2}$

29. For $f(x) = 2x^2 - 6$, $f(0) = -6$ and $f(x) = 0$ when $x = \pm\sqrt{3}$

30. For $f(x) = x^2 - x - 12$, $f(0) = -12$ and $f(x) = 0$ when $x = -3$ or $x = 4$

31. For $f(x) = \sqrt{x+4}$, $f(0) = 2$ and $f(x) = 0$ when $x = -4$

32. For $f(x) = \sqrt{1-2x}$, $f(0) = 1$ and $f(x) = 0$ when $x = \frac{1}{2}$

33. For $f(x) = \frac{3}{4-x}$, $f(0) = \frac{3}{4}$ and $f(x)$ is never equal to 0

34. For $f(x) = \frac{3x^2 - 12x}{4-x^2}$, $f(0) = 0$ and $f(x) = 0$ when $x = 0$ or $x = 4$

35. (a) $f(-4) = 1$ (b) $f(-3) = 2$ (c) $f(3) = 0$

    (d) $f(3.001) = 1.999$ (e) $f(-3.001) = 1.999$ (f) $f(2) = \sqrt{5}$

36. (a) $f(4) = 4$ (b) $f(-3) = 9$ (c) $f(1) = 0$

    (d) $f(0) = 1$ (e) $f(-1) = 1$ (f) $f(-0.999) \approx 0.0447$

37. $(-\infty, \infty)$

38. $(-\infty, \infty)$

39. $(-\infty, -1) \cup (-1, \infty)$

40. $(-\infty, -2) \cup (-2, 1) \cup (1, \infty)$

41. $(-\infty, \infty)$

42. $(-\infty, -\sqrt{3}) \cup (-\sqrt{3}, \sqrt{3}) \cup (\sqrt{3}, \infty)$

43. $(-\infty, -6) \cup (-6, 6) \cup (6, \infty)$

44. $(-\infty, 2) \cup (2, \infty)$

45. $(-\infty, 3]$

46. $\left[-\frac{5}{2}, \infty\right)$

47. $[-3, \infty)$ 48. $(-\infty, 7]$

49. $\left[\frac{1}{3}, \infty\right)$ 50. $\left(\frac{1}{3}, \infty\right)$

51. $(-\infty, \infty)$ 52. $\left[\frac{1}{3}, 3\right) \cup (3, \infty)$

53. $\left[\frac{1}{3}, 6\right) \cup (6, \infty)$ 54. $(-\infty, \infty)$

55. $(-\infty, 8) \cup (8, \infty)$ 56. $[0, 8) \cup (8, \infty)$

57. $(8, \infty)$ 58. $[7, 9]$

59. $(-\infty, 8) \cup (8, \infty)$ 60. $\left(-\infty, -\frac{1}{2}\right) \cup \left(-\frac{1}{2}, 0\right) \cup \left(0, \frac{1}{2}\right) \cup \left(\frac{1}{2}, \infty\right)$

61. $[0, 5) \cup (5, \infty)$ 62. $[0, 25) \cup (25, \infty)$

63. $A(3) = 9$, so the area enclosed by a square with a side of length 3 inches is 9 square inches. The solutions to $A(x) = 36$ are $x = \pm 6$. Since x is restricted to $x > 0$, we only keep $x = 6$. This means for the area enclosed by the square to be 36 square inches, the length of the side needs to be 6 inches. Since x represents a length, $x > 0$.

64. $A(2) = 4\pi$, so the area enclosed by a circle with radius 2 meters is 4π square meters. The solutions to $A(r) = 16\pi$ are $r = \pm 4$. Since r is restricted to $r > 0$, we only keep $r = 4$. This means for the area enclosed by the circle to be 16π square meters, the radius needs to be 4 meters. Since r represents a radius (length), $r > 0$.

65. $V(5) = 125$, so the volume enclosed by a cube with a side of length 5 centimeters is 125 cubic centimeters. The solution to $V(x) = 27$ is $x = 3$. This means for the volume enclosed by the cube to be 27 cubic centimeters, the length of the side needs to 3 centimeters. Since x represents a length, $x > 0$.

66. $V(3) = 36\pi$, so the volume enclosed by a sphere with radius 3 feet is 36π cubic feet. The solution to $V(r) = \frac{32\pi}{3}$ is $r = 2$. This means for the volume enclosed by the sphere to be $\frac{32\pi}{3}$ cubic feet, the radius needs to 2 feet. Since r represents a radius (length), $r > 0$.

67. $h(0) = 64$, so at the moment the object is dropped off the building, the object is 64 feet off of the ground. The solutions to $h(t) = 0$ are $t = \pm 2$. Since we restrict $0 \leq t \leq 2$, we only keep $t = 2$. This means 2 seconds after the object is dropped off the building, it is 0 feet off the ground. Said differently, the object hits the ground after 2 seconds. The restriction $0 \leq t \leq 2$ restricts the time to be between the moment the object is released and the moment it hits the ground.

68. $T(0) = 3$, so at 6 AM (0 hours after 6 AM), it is $3°$ Fahrenheit. $T(6) = 33$, so at noon (6 hours after 6 AM), the temperature is $33°$ Fahrenheit. $T(12) = 27$, so at 6 PM (12 hours after 6 AM), it is $27°$ Fahrenheit.

1.4 FUNCTION NOTATION

69. $C(0) = 27$, so to make 0 pens, it costs[15] $2700. $C(2) = 11$, so to make 2000 pens, it costs $1100. $C(5) = 2$, so to make 5000 pens, it costs $2000.

70. $F(0) = 16.00$, so in 1980 (0 years after 1980), the average fuel economy of passenger cars in the US was 16.00 miles per gallon. $F(14) = 20.81$, so in 1994 (14 years after 1980), the average fuel economy of passenger cars in the US was 20.81 miles per gallon. $F(28) = 22.64$, so in 2008 (28 years after 1980), the average fuel economy of passenger cars in the US was 22.64 miles per gallon.

71. $P(0) = 0$ which means in 1803 (0 years after 1803), there are no Sasquatch in Portage County. $P(205) = \frac{3075}{22} \approx 139.77$, so in 2008 (205 years after 1803), there were between 139 and 140 Sasquatch in Portage County.

72. (a) $C(20) = 300$. It costs $300 for 20 copies of the book.
    (b) $C(50) = 675$, so it costs $675 for 50 copies of the book. $C(51) = 612$, so it costs $612 for 51 copies of the book.
    (c) 56 books.

73. (a) $S(10) = 17.5$, so it costs $17.50 to ship 10 comic books.
    (b) There is free shipping on orders of 15 or more comic books.

74. (a) $C(750) = 25$, so it costs $25 to talk 750 minutes per month with this plan.
    (b) Since 20 hours = 1200 minutes, we substitute $m = 1200$ and get $C(1200) = 45$. It costs $45 to talk 20 hours per month with this plan.
    (c) It costs $25 for up to 1000 minutes and 10 cents per minute for each minute over 1000 minutes.

75. (a) $\lfloor 0.785 \rfloor = 0$, $\lfloor 117 \rfloor = 117$, $\lfloor -2.001 \rfloor = -3$, and $\lfloor \pi + 6 \rfloor = 9$

---

[15]This is called the 'fixed' or 'start-up' cost. We'll revisit this concept on page 82.

1.5 Function Arithmetic

In the previous section we used the newly defined function notation to make sense of expressions such as '$f(x) + 2$' and '$2f(x)$' for a given function f. It would seem natural, then, that functions should have their own arithmetic which is consistent with the arithmetic of real numbers. The following definitions allow us to add, subtract, multiply and divide functions using the arithmetic we already know for real numbers.

Function Arithmetic

Suppose f and g are functions and x is in both the domain of f and the domain of g.[a]

- The **sum** of f and g, denoted $f + g$, is the function defined by the formula

$$(f + g)(x) = f(x) + g(x)$$

- The **difference** of f and g, denoted $f - g$, is the function defined by the formula

$$(f - g)(x) = f(x) - g(x)$$

- The **product** of f and g, denoted fg, is the function defined by the formula

$$(fg)(x) = f(x)g(x)$$

- The **quotient** of f and g, denoted $\dfrac{f}{g}$, is the function defined by the formula

$$\left(\frac{f}{g}\right)(x) = \frac{f(x)}{g(x)},$$

provided $g(x) \neq 0$.

[a]Thus x is an element of the intersection of the two domains.

In other words, to add two functions, we add their outputs; to subtract two functions, we subtract their outputs, and so on. Note that while the formula $(f + g)(x) = f(x) + g(x)$ looks suspiciously like some kind of distributive property, it is nothing of the sort; the addition on the left hand side of the equation is *function* addition, and we are using this equation to *define* the output of the new function $f + g$ as the sum of the real number outputs from f and g.

Example 1.5.1. Let $f(x) = 6x^2 - 2x$ and $g(x) = 3 - \dfrac{1}{x}$.

1. Find $(f + g)(-1)$
2. Find $(fg)(2)$

3. Find the domain of $g - f$ then find and simplify a formula for $(g - f)(x)$.

1.5 FUNCTION ARITHMETIC   77

4. Find the domain of $\left(\frac{g}{f}\right)$ then find and simplify a formula for $\left(\frac{g}{f}\right)(x)$.

Solution.

1. To find $(f+g)(-1)$ we first find $f(-1) = 8$ and $g(-1) = 4$. By definition, we have that $(f+g)(-1) = f(-1) + g(-1) = 8 + 4 = 12$.

2. To find $(fg)(2)$, we first need $f(2)$ and $g(2)$. Since $f(2) = 20$ and $g(2) = \frac{5}{2}$, our formula yields $(fg)(2) = f(2)g(2) = (20)\left(\frac{5}{2}\right) = 50$.

3. One method to find the domain of $g - f$ is to find the domain of g and of f separately, then find the intersection of these two sets. Owing to the denominator in the expression $g(x) = 3 - \frac{1}{x}$, we get that the domain of g is $(-\infty, 0) \cup (0, \infty)$. Since $f(x) = 6x^2 - 2x$ is valid for all real numbers, we have no further restrictions. Thus the domain of $g - f$ matches the domain of g, namely, $(-\infty, 0) \cup (0, \infty)$.

   A second method is to analyze the formula for $(g - f)(x)$ *before simplifying* and look for the usual domain issues. In this case,

   $$(g - f)(x) = g(x) - f(x) = \left(3 - \frac{1}{x}\right) - \left(6x^2 - 2x\right),$$

   so we find, as before, the domain is $(-\infty, 0) \cup (0, \infty)$.

   Moving along, we need to simplify a formula for $(g - f)(x)$. In this case, we get common denominators and attempt to reduce the resulting fraction. Doing so, we get

   $$\begin{aligned}(g - f)(x) &= g(x) - f(x) \\ &= \left(3 - \frac{1}{x}\right) - \left(6x^2 - 2x\right) \\ &= 3 - \frac{1}{x} - 6x^2 + 2x \\ &= \frac{3x}{x} - \frac{1}{x} - \frac{6x^3}{x} + \frac{2x^2}{x} \qquad \text{get common denominators} \\ &= \frac{3x - 1 - 6x^3 - 2x^2}{x} \\ &= \frac{-6x^3 - 2x^2 + 3x - 1}{x}\end{aligned}$$

4. As in the previous example, we have two ways to approach finding the domain of $\frac{g}{f}$. First, we can find the domain of g and f separately, and find the intersection of these two sets. In addition, since $\left(\frac{g}{f}\right)(x) = \frac{g(x)}{f(x)}$, we are introducing a new denominator, namely $f(x)$, so we need to guard against this being 0 as well. Our previous work tells us that the domain of g is $(-\infty, 0) \cup (0, \infty)$ and the domain of f is $(-\infty, \infty)$. Setting $f(x) = 0$ gives $6x^2 - 2x = 0$

or $x = 0, \frac{1}{3}$. As a result, the domain of $\frac{g}{f}$ is all real numbers except $x = 0$ and $x = \frac{1}{3}$, or $(-\infty, 0) \cup \left(0, \frac{1}{3}\right) \cup \left(\frac{1}{3}, \infty\right)$.

Alternatively, we may proceed as above and analyze the expression $\left(\frac{g}{f}\right)(x) = \frac{g(x)}{f(x)}$ *before* simplifying. In this case,

$$\left(\frac{g}{f}\right)(x) = \frac{g(x)}{f(x)} = \frac{3 - \frac{1}{x}}{6x^2 - 2x}$$

We see immediately from the 'little' denominator that $x \neq 0$. To keep the 'big' denominator away from 0, we solve $6x^2 - 2x = 0$ and get $x = 0$ or $x = \frac{1}{3}$. Hence, as before, we find the domain of $\frac{g}{f}$ to be $(-\infty, 0) \cup \left(0, \frac{1}{3}\right) \cup \left(\frac{1}{3}, \infty\right)$.

Next, we find and simplify a formula for $\left(\frac{g}{f}\right)(x)$.

$$\begin{aligned}
\left(\frac{g}{f}\right)(x) &= \frac{g(x)}{f(x)} \\
&= \frac{3 - \frac{1}{x}}{6x^2 - 2x} \\
&= \frac{3 - \frac{1}{x}}{6x^2 - 2x} \cdot \frac{x}{x} \quad && \text{simplify compound fractions} \\
&= \frac{\left(3 - \frac{1}{x}\right)x}{(6x^2 - 2x)x} \\
&= \frac{3x - 1}{(6x^2 - 2x)x} \\
&= \frac{3x - 1}{2x^2(3x - 1)} \quad && \text{factor} \\
&= \frac{\cancel{(3x-1)}^1}{2x^2\cancel{(3x-1)}} \quad && \text{cancel} \\
&= \frac{1}{2x^2}
\end{aligned}$$

□

Please note the importance of finding the domain of a function *before* simplifying its expression. In number 4 in Example 1.5.1 above, had we waited to find the domain of $\frac{g}{f}$ until after simplifying, we'd just have the formula $\frac{1}{2x^2}$ to go by, and we would (incorrectly!) state the domain as $(-\infty, 0) \cup (0, \infty)$, since the other troublesome number, $x = \frac{1}{3}$, was canceled away.[1]

---

[1] We'll see what this means geometrically in Chapter 4.

1.5 Function Arithmetic

Next, we turn our attention to the **difference quotient** of a function.

> **Definition 1.8.** Given a function f, the **difference quotient** of f is the expression
> $$\frac{f(x+h) - f(x)}{h}$$

We will revisit this concept in Section 2.1, but for now, we use it as a way to practice function notation and function arithmetic. For reasons which will become clear in Calculus, 'simplifying' a difference quotient means rewriting it in a form where the 'h' in the definition of the difference quotient cancels from the denominator. Once that happens, we consider our work to be done.

Example 1.5.2. Find and simplify the difference quotients for the following functions

1. $f(x) = x^2 - x - 2$
2. $g(x) = \dfrac{3}{2x+1}$
3. $r(x) = \sqrt{x}$

Solution.

1. To find $f(x+h)$, we replace every occurrence of x in the formula $f(x) = x^2 - x - 2$ with the quantity $(x+h)$ to get

$$\begin{aligned} f(x+h) &= (x+h)^2 - (x+h) - 2 \\ &= x^2 + 2xh + h^2 - x - h - 2. \end{aligned}$$

So the difference quotient is

$$\begin{aligned} \frac{f(x+h) - f(x)}{h} &= \frac{\left(x^2 + 2xh + h^2 - x - h - 2\right) - \left(x^2 - x - 2\right)}{h} \\ &= \frac{x^2 + 2xh + h^2 - x - h - 2 - x^2 + x + 2}{h} \\ &= \frac{2xh + h^2 - h}{h} \\ &= \frac{h(2x + h - 1)}{h} \qquad \text{factor} \\ &= \frac{\cancel{h}(2x + h - 1)}{\cancel{h}} \qquad \text{cancel} \\ &= 2x + h - 1. \end{aligned}$$

2. To find $g(x+h)$, we replace every occurrence of x in the formula $g(x) = \frac{3}{2x+1}$ with the quantity $(x+h)$ to get

$$\begin{aligned} g(x+h) &= \frac{3}{2(x+h)+1} \\ &= \frac{3}{2x+2h+1}, \end{aligned}$$

which yields

$$\begin{aligned} \frac{g(x+h)-g(x)}{h} &= \frac{\dfrac{3}{2x+2h+1} - \dfrac{3}{2x+1}}{h} \\ &= \frac{\dfrac{3}{2x+2h+1} - \dfrac{3}{2x+1}}{h} \cdot \frac{(2x+2h+1)(2x+1)}{(2x+2h+1)(2x+1)} \\ &= \frac{3(2x+1) - 3(2x+2h+1)}{h(2x+2h+1)(2x+1)} \\ &= \frac{6x+3-6x-6h-3}{h(2x+2h+1)(2x+1)} \\ &= \frac{-6h}{h(2x+2h+1)(2x+1)} \\ &= \frac{-6\cancel{h}}{\cancel{h}(2x+2h+1)(2x+1)} \\ &= \frac{-6}{(2x+2h+1)(2x+1)}. \end{aligned}$$

Since we have managed to cancel the original 'h' from the denominator, we are done.

3. For $r(x) = \sqrt{x}$, we get $r(x+h) = \sqrt{x+h}$ so the difference quotient is

$$\frac{r(x+h)-r(x)}{h} = \frac{\sqrt{x+h}-\sqrt{x}}{h}$$

In order to cancel the 'h' from the denominator, we rationalize the *numerator* by multiplying by its conjugate.[2]

---

[2]Rationalizing the *numerator*!? How's that for a twist!

1.5 Function Arithmetic

$$\begin{aligned}
\frac{r(x+h) - r(x)}{h} &= \frac{\sqrt{x+h} - \sqrt{x}}{h} \\
&= \frac{\left(\sqrt{x+h} - \sqrt{x}\right)}{h} \cdot \frac{\left(\sqrt{x+h} + \sqrt{x}\right)}{\left(\sqrt{x+h} + \sqrt{x}\right)} \quad \text{Multiply by the conjugate.} \\
&= \frac{\left(\sqrt{x+h}\right)^2 - \left(\sqrt{x}\right)^2}{h\left(\sqrt{x+h} + \sqrt{x}\right)} \quad \text{Difference of Squares.} \\
&= \frac{(x+h) - x}{h\left(\sqrt{x+h} + \sqrt{x}\right)} \\
&= \frac{h}{h\left(\sqrt{x+h} + \sqrt{x}\right)} \\
&= \frac{\cancel{h}^1}{\cancel{h}\left(\sqrt{x+h} + \sqrt{x}\right)} \\
&= \frac{1}{\sqrt{x+h} + \sqrt{x}}
\end{aligned}$$

Since we have removed the original 'h' from the denominator, we are done. $\square$

As mentioned before, we will revisit difference quotients in Section 2.1 where we will explain them geometrically. For now, we want to move on to some classic applications of function arithmetic from Economics and for that, we need to think like an entrepreneur.[3]

Suppose you are a manufacturer making a certain product.[4] Let x be the **production level**, that is, the number of items produced in a given time period. It is customary to let $C(x)$ denote the function which calculates the total **cost** of producing the x items. The quantity $C(0)$, which represents the cost of producing no items, is called the **fixed** cost, and represents the amount of money required to begin production. Associated with the total cost $C(x)$ is cost per item, or **average cost**, denoted $\overline{C}(x)$ and read 'C-bar' of x. To compute $\overline{C}(x)$, we take the total cost $C(x)$ and divide by the number of items produced x to get

$$\overline{C}(x) = \frac{C(x)}{x}$$

On the retail end, we have the **price** p charged per item. To simplify the dialog and computations in this text, we assume that *the number of items sold equals the number of items produced*. From a

---
[3]Not really, but "entrepreneur" is the buzzword of the day and we're trying to be trendy.
[4]Poorly designed resin Sasquatch statues, for example. Feel free to choose your own entrepreneurial fantasy.

retail perspective, it seems natural to think of the number of items sold, x, as a function of the price charged, p. After all, the retailer can easily adjust the price to sell more product. In the language of functions, x would be the *dependent* variable and p would be the *independent* variable or, using function notation, we have a function $x(p)$. While we will adopt this convention later in the text,[5] we will hold with tradition at this point and consider the price p as a function of the number of items sold, x. That is, we regard x as the independent variable and p as the dependent variable and speak of the **price-demand** function, $p(x)$. Hence, $p(x)$ returns the price charged per item when x items are produced and sold. Our next function to consider is the **revenue** function, $R(x)$. The function $R(x)$ computes the amount of money collected as a result of selling x items. Since $p(x)$ is the price charged per item, we have $R(x) = xp(x)$. Finally, the **profit** function, $P(x)$ calculates how much money is earned after the costs are paid. That is, $P(x) = (R - C)(x) = R(x) - C(x)$. We summarize all of these functions below.

Summary of Common Economic Functions

Suppose x represents the quantity of items produced and sold.

- The price-demand function $p(x)$ calculates the price per item.

- The revenue function $R(x)$ calculates the total money collected by selling x items at a price $p(x)$, $R(x) = x\,p(x)$.

- The cost function $C(x)$ calculates the cost to produce x items. The value $C(0)$ is called the fixed cost or start-up cost.

- The average cost function $\overline{C}(x) = \frac{C(x)}{x}$ calculates the cost per item when making x items. Here, we necessarily assume $x > 0$.

- The profit function $P(x)$ calculates the money earned after costs are paid when x items are produced and sold, $P(x) = (R - C)(x) = R(x) - C(x)$.

It is high time for an example.

Example 1.5.3. Let x represent the number of dOpi media players ('dOpis'[6]) produced and sold in a typical week. Suppose the cost, in dollars, to produce x dOpis is given by $C(x) = 100x + 2000$, for $x \geq 0$, and the price, in dollars per dOpi, is given by $p(x) = 450 - 15x$ for $0 \leq x \leq 30$.

1. Find and interpret $C(0)$.

2. Find and interpret $\overline{C}(10)$.

3. Find and interpret $p(0)$ and $p(20)$.

4. Solve $p(x) = 0$ and interpret the result.

5. Find and simplify expressions for the revenue function $R(x)$ and the profit function $P(x)$.

6. Find and interpret $R(0)$ and $P(0)$.

7. Solve $P(x) = 0$ and interpret the result.

---

[5] See Example 5.2.4 in Section 5.2.
[6] Pronounced 'dopeys' ...

1.5 FUNCTION ARITHMETIC

Solution.

1. We substitute $x = 0$ into the formula for $C(x)$ and get $C(0) = 100(0) + 2000 = 2000$. This means to produce 0 dOpis, it costs $2000. In other words, the fixed (or start-up) costs are $2000. The reader is encouraged to contemplate what sorts of expenses these might be.

2. Since $\overline{C}(x) = \frac{C(x)}{x}$, $\overline{C}(10) = \frac{C(10)}{10} = \frac{3000}{10} = 300$. This means when 10 dOpis are produced, the cost to manufacture them amounts to $300 per dOpi.

3. Plugging $x = 0$ into the expression for $p(x)$ gives $p(0) = 450 - 15(0) = 450$. This means no dOpis are sold if the price is $450 per dOpi. On the other hand, $p(20) = 450 - 15(20) = 150$ which means to sell 20 dOpis in a typical week, the price should be set at $150 per dOpi.

4. Setting $p(x) = 0$ gives $450 - 15x = 0$. Solving gives $x = 30$. This means in order to sell 30 dOpis in a typical week, the price needs to be set to $0. What's more, this means that even if dOpis were given away for free, the retailer would only be able to move 30 of them.[7]

5. To find the revenue, we compute $R(x) = xp(x) = x(450 - 15x) = 450x - 15x^2$. Since the formula for $p(x)$ is valid only for $0 \leq x \leq 30$, our formula $R(x)$ is also restricted to $0 \leq x \leq 30$. For the profit, $P(x) = (R - C)(x) = R(x) - C(x)$. Using the given formula for $C(x)$ and the derived formula for $R(x)$, we get $P(x) = (450x - 15x^2) - (100x + 2000) = -15x^2 + 350x - 2000$. As before, the validity of this formula is for $0 \leq x \leq 30$ only.

6. We find $R(0) = 0$ which means if no dOpis are sold, we have no revenue, which makes sense. Turning to profit, $P(0) = -2000$ since $P(x) = R(x) - C(x)$ and $P(0) = R(0) - C(0) = -2000$. This means that if no dOpis are sold, more money ($2000 to be exact!) was put into producing the dOpis than was recouped in sales. In number 1, we found the fixed costs to be $2000, so it makes sense that if we sell no dOpis, we are out those start-up costs.

7. Setting $P(x) = 0$ gives $-15x^2 + 350x - 2000 = 0$. Factoring gives $-5(x - 10)(3x - 40) = 0$ so $x = 10$ or $x = \frac{40}{3}$. What do these values mean in the context of the problem? Since $P(x) = R(x) - C(x)$, solving $P(x) = 0$ is the same as solving $R(x) = C(x)$. This means that the solutions to $P(x) = 0$ are the production (and sales) figures for which the sales revenue exactly balances the total production costs. These are the so-called '**break even**' points. The solution $x = 10$ means 10 dOpis should be produced (and sold) during the week to recoup the cost of production. For $x = \frac{40}{3} = 13.\overline{3}$, things are a bit more complicated. Even though $x = 13.\overline{3}$ satisfies $0 \leq x \leq 30$, and hence is in the domain of P, it doesn't make sense in the context of this problem to produce a fractional part of a dOpi.[8] Evaluating $P(13) = 15$ and $P(14) = -40$, we see that producing and selling 13 dOpis per week makes a (slight) profit, whereas producing just one more puts us back into the red. While breaking even is nice, we ultimately would like to find what production level (and price) will result in the largest profit, and we'll do just that ... in Section 2.3. □

---

[7]Imagine that! Giving something away for free and hardly anyone taking advantage of it ...
[8]We've seen this sort of thing before in Section 1.4.1.

1.5.1 Exercises

In Exercises 1 - 10, use the pair of functions f and g to find the following values if they exist.

- $(f+g)(2)$
- $(f-g)(-1)$
- $(g-f)(1)$
- $(fg)\left(\frac{1}{2}\right)$
- $\left(\frac{f}{g}\right)(0)$
- $\left(\frac{g}{f}\right)(-2)$

1. $f(x) = 3x + 1$ and $g(x) = 4 - x$
2. $f(x) = x^2$ and $g(x) = -2x + 1$
3. $f(x) = x^2 - x$ and $g(x) = 12 - x^2$
4. $f(x) = 2x^3$ and $g(x) = -x^2 - 2x - 3$
5. $f(x) = \sqrt{x+3}$ and $g(x) = 2x - 1$
6. $f(x) = \sqrt{4-x}$ and $g(x) = \sqrt{x+2}$
7. $f(x) = 2x$ and $g(x) = \dfrac{1}{2x+1}$
8. $f(x) = x^2$ and $g(x) = \dfrac{3}{2x-3}$
9. $f(x) = x^2$ and $g(x) = \dfrac{1}{x^2}$
10. $f(x) = x^2 + 1$ and $g(x) = \dfrac{1}{x^2+1}$

In Exercises 11 - 20, use the pair of functions f and g to find the domain of the indicated function then find and simplify an expression for it.

- $(f+g)(x)$
- $(f-g)(x)$
- $(fg)(x)$
- $\left(\frac{f}{g}\right)(x)$

11. $f(x) = 2x + 1$ and $g(x) = x - 2$
12. $f(x) = 1 - 4x$ and $g(x) = 2x - 1$
13. $f(x) = x^2$ and $g(x) = 3x - 1$
14. $f(x) = x^2 - x$ and $g(x) = 7x$
15. $f(x) = x^2 - 4$ and $g(x) = 3x + 6$
16. $f(x) = -x^2 + x + 6$ and $g(x) = x^2 - 9$
17. $f(x) = \dfrac{x}{2}$ and $g(x) = \dfrac{2}{x}$
18. $f(x) = x - 1$ and $g(x) = \dfrac{1}{x-1}$
19. $f(x) = x$ and $g(x) = \sqrt{x+1}$
20. $f(x) = \sqrt{x-5}$ and $g(x) = f(x) = \sqrt{x-5}$

In Exercises 21 - 45, find and simplify the difference quotient $\dfrac{f(x+h) - f(x)}{h}$ for the given function.

21. $f(x) = 2x - 5$
22. $f(x) = -3x + 5$
23. $f(x) = 6$
24. $f(x) = 3x^2 - x$
25. $f(x) = -x^2 + 2x - 1$
26. $f(x) = 4x^2$

1.5 Function Arithmetic

27. $f(x) = x - x^2$

28. $f(x) = x^3 + 1$

29. $f(x) = mx + b$ where $m \neq 0$

30. $f(x) = ax^2 + bx + c$ where $a \neq 0$

31. $f(x) = \dfrac{2}{x}$

32. $f(x) = \dfrac{3}{1-x}$

33. $f(x) = \dfrac{1}{x^2}$

34. $f(x) = \dfrac{2}{x+5}$

35. $f(x) = \dfrac{1}{4x-3}$

36. $f(x) = \dfrac{3x}{x+1}$

37. $f(x) = \dfrac{x}{x-9}$

38. $f(x) = \dfrac{x^2}{2x+1}$

39. $f(x) = \sqrt{x-9}$

40. $f(x) = \sqrt{2x+1}$

41. $f(x) = \sqrt{-4x+5}$

42. $f(x) = \sqrt{4-x}$

43. $f(x) = \sqrt{ax+b}$, where $a \neq 0$.

44. $f(x) = x\sqrt{x}$

45. $f(x) = \sqrt[3]{x}$. **HINT:** $(a-b)\left(a^2 + ab + b^2\right) = a^3 - b^3$

In Exercises 46 - 50, $C(x)$ denotes the cost to produce x items and $p(x)$ denotes the price-demand function in the given economic scenario. In each Exercise, do the following:

- Find and interpret $C(0)$.
- Find and interpret $\overline{C}(10)$.
- Find and interpret $p(5)$
- Find and simplify $R(x)$.
- Find and simplify $P(x)$.
- Solve $P(x) = 0$ and interpret.

46. The cost, in dollars, to produce x "I'd rather be a Sasquatch" T-Shirts is $C(x) = 2x + 26$, $x \geq 0$ and the price-demand function, in dollars per shirt, is $p(x) = 30 - 2x$, $0 \leq x \leq 15$.

47. The cost, in dollars, to produce x bottles of 100% All-Natural Certified Free-Trade Organic Sasquatch Tonic is $C(x) = 10x + 100$, $x \geq 0$ and the price-demand function, in dollars per bottle, is $p(x) = 35 - x$, $0 \leq x \leq 35$.

48. The cost, in cents, to produce x cups of Mountain Thunder Lemonade at Junior's Lemonade Stand is $C(x) = 18x + 240$, $x \geq 0$ and the price-demand function, in cents per cup, is $p(x) = 90 - 3x$, $0 \leq x \leq 30$.

49. The daily cost, in dollars, to produce x Sasquatch Berry Pies $C(x) = 3x + 36$, $x \geq 0$ and the price-demand function, in dollars per pie, is $p(x) = 12 - 0.5x$, $0 \leq x \leq 24$.

50. The monthly cost, in hundreds of dollars, to produce x custom built electric scooters is $C(x) = 20x + 1000$, $x \geq 0$ and the price-demand function, in hundreds of dollars per scooter, is $p(x) = 140 - 2x$, $0 \leq x \leq 70$.

In Exercises 51 - 62, let f be the function defined by

$$f = \{(-3, 4), (-2, 2), (-1, 0), (0, 1), (1, 3), (2, 4), (3, -1)\}$$

and let g be the function defined

$$g = \{(-3, -2), (-2, 0), (-1, -4), (0, 0), (1, -3), (2, 1), (3, 2)\}$$

. Compute the indicated value if it exists.

51. $(f + g)(-3)$

52. $(f - g)(2)$

53. $(fg)(-1)$

54. $(g + f)(1)$

55. $(g - f)(3)$

56. $(gf)(-3)$

57. $\left(\frac{f}{g}\right)(-2)$

58. $\left(\frac{f}{g}\right)(-1)$

59. $\left(\frac{f}{g}\right)(2)$

60. $\left(\frac{g}{f}\right)(-1)$

61. $\left(\frac{g}{f}\right)(3)$

62. $\left(\frac{g}{f}\right)(-3)$

1.5 Function Arithmetic

1.5.2 Answers

1. For $f(x) = 3x + 1$ and $g(x) = 4 - x$

   - $(f+g)(2) = 9$
   - $(f-g)(-1) = -7$
   - $(g-f)(1) = -1$
   - $(fg)\left(\frac{1}{2}\right) = \frac{35}{4}$
   - $\left(\frac{f}{g}\right)(0) = \frac{1}{4}$
   - $\left(\frac{g}{f}\right)(-2) = -\frac{6}{5}$

2. For $f(x) = x^2$ and $g(x) = -2x + 1$

   - $(f+g)(2) = 1$
   - $(f-g)(-1) = -2$
   - $(g-f)(1) = -2$
   - $(fg)\left(\frac{1}{2}\right) = 0$
   - $\left(\frac{f}{g}\right)(0) = 0$
   - $\left(\frac{g}{f}\right)(-2) = \frac{5}{4}$

3. For $f(x) = x^2 - x$ and $g(x) = 12 - x^2$

   - $(f+g)(2) = 10$
   - $(f-g)(-1) = -9$
   - $(g-f)(1) = 11$
   - $(fg)\left(\frac{1}{2}\right) = -\frac{47}{16}$
   - $\left(\frac{f}{g}\right)(0) = 0$
   - $\left(\frac{g}{f}\right)(-2) = \frac{4}{3}$

4. For $f(x) = 2x^3$ and $g(x) = -x^2 - 2x - 3$

   - $(f+g)(2) = 5$
   - $(f-g)(-1) = 0$
   - $(g-f)(1) = -8$
   - $(fg)\left(\frac{1}{2}\right) = -\frac{17}{16}$
   - $\left(\frac{f}{g}\right)(0) = 0$
   - $\left(\frac{g}{f}\right)(-2) = \frac{3}{16}$

5. For $f(x) = \sqrt{x+3}$ and $g(x) = 2x - 1$

   - $(f+g)(2) = 3 + \sqrt{5}$
   - $(f-g)(-1) = 3 + \sqrt{2}$
   - $(g-f)(1) = -1$
   - $(fg)\left(\frac{1}{2}\right) = 0$
   - $\left(\frac{f}{g}\right)(0) = -\sqrt{3}$
   - $\left(\frac{g}{f}\right)(-2) = -5$

6. For $f(x) = \sqrt{4-x}$ and $g(x) = \sqrt{x+2}$

   - $(f+g)(2) = 2 + \sqrt{2}$
   - $(f-g)(-1) = -1 + \sqrt{5}$
   - $(g-f)(1) = 0$
   - $(fg)\left(\frac{1}{2}\right) = \frac{\sqrt{35}}{2}$
   - $\left(\frac{f}{g}\right)(0) = \sqrt{2}$
   - $\left(\frac{g}{f}\right)(-2) = 0$

7. For $f(x) = 2x$ and $g(x) = \frac{1}{2x+1}$

    - $(f+g)(2) = \frac{21}{5}$
    - $(f-g)(-1) = -1$
    - $(g-f)(1) = -\frac{5}{3}$
    - $(fg)\left(\frac{1}{2}\right) = \frac{1}{2}$
    - $\left(\frac{f}{g}\right)(0) = 0$
    - $\left(\frac{g}{f}\right)(-2) = \frac{1}{12}$

8. For $f(x) = x^2$ and $g(x) = \frac{3}{2x-3}$

    - $(f+g)(2) = 7$
    - $(f-g)(-1) = \frac{8}{5}$
    - $(g-f)(1) = -4$
    - $(fg)\left(\frac{1}{2}\right) = -\frac{3}{8}$
    - $\left(\frac{f}{g}\right)(0) = 0$
    - $\left(\frac{g}{f}\right)(-2) = -\frac{3}{28}$

9. For $f(x) = x^2$ and $g(x) = \frac{1}{x^2}$

    - $(f+g)(2) = \frac{17}{4}$
    - $(f-g)(-1) = 0$
    - $(g-f)(1) = 0$
    - $(fg)\left(\frac{1}{2}\right) = 1$
    - $\left(\frac{f}{g}\right)(0)$ is undefined.
    - $\left(\frac{g}{f}\right)(-2) = \frac{1}{16}$

10. For $f(x) = x^2 + 1$ and $g(x) = \frac{1}{x^2+1}$

    - $(f+g)(2) = \frac{26}{5}$
    - $(f-g)(-1) = \frac{3}{2}$
    - $(g-f)(1) = -\frac{3}{2}$
    - $(fg)\left(\frac{1}{2}\right) = 1$
    - $\left(\frac{f}{g}\right)(0) = 1$
    - $\left(\frac{g}{f}\right)(-2) = \frac{1}{25}$

11. For $f(x) = 2x+1$ and $g(x) = x-2$

    - $(f+g)(x) = 3x - 1$
      Domain: $(-\infty, \infty)$
    - $(f-g)(x) = x + 3$
      Domain: $(-\infty, \infty)$
    - $(fg)(x) = 2x^2 - 3x - 2$
      Domain: $(-\infty, \infty)$
    - $\left(\frac{f}{g}\right)(x) = \frac{2x+1}{x-2}$
      Domain: $(-\infty, 2) \cup (2, \infty)$

12. For $f(x) = 1 - 4x$ and $g(x) = 2x - 1$

    - $(f+g)(x) = -2x$
      Domain: $(-\infty, \infty)$
    - $(f-g)(x) = 2 - 6x$
      Domain: $(-\infty, \infty)$
    - $(fg)(x) = -8x^2 + 6x - 1$
      Domain: $(-\infty, \infty)$
    - $\left(\frac{f}{g}\right)(x) = \frac{1-4x}{2x-1}$
      Domain: $\left(-\infty, \frac{1}{2}\right) \cup \left(\frac{1}{2}, \infty\right)$

1.5 Function Arithmetic

13. For $f(x) = x^2$ and $g(x) = 3x - 1$

    - $(f+g)(x) = x^2 + 3x - 1$
      Domain: $(-\infty, \infty)$

    - $(fg)(x) = 3x^3 - x^2$
      Domain: $(-\infty, \infty)$

    - $(f-g)(x) = x^2 - 3x + 1$
      Domain: $(-\infty, \infty)$

    - $\left(\dfrac{f}{g}\right)(x) = \dfrac{x^2}{3x-1}$
      Domain: $\left(-\infty, \dfrac{1}{3}\right) \cup \left(\dfrac{1}{3}, \infty\right)$

14. For $f(x) = x^2 - x$ and $g(x) = 7x$

    - $(f+g)(x) = x^2 + 6x$
      Domain: $(-\infty, \infty)$

    - $(fg)(x) = 7x^3 - 7x^2$
      Domain: $(-\infty, \infty)$

    - $(f-g)(x) = x^2 - 8x$
      Domain: $(-\infty, \infty)$

    - $\left(\dfrac{f}{g}\right)(x) = \dfrac{x-1}{7}$
      Domain: $(-\infty, 0) \cup (0, \infty)$

15. For $f(x) = x^2 - 4$ and $g(x) = 3x + 6$

    - $(f+g)(x) = x^2 + 3x + 2$
      Domain: $(-\infty, \infty)$

    - $(fg)(x) = 3x^3 + 6x^2 - 12x - 24$
      Domain: $(-\infty, \infty)$

    - $(f-g)(x) = x^2 - 3x - 10$
      Domain: $(-\infty, \infty)$

    - $\left(\dfrac{f}{g}\right)(x) = \dfrac{x-2}{3}$
      Domain: $(-\infty, -2) \cup (-2, \infty)$

16. For $f(x) = -x^2 + x + 6$ and $g(x) = x^2 - 9$

    - $(f+g)(x) = x - 3$
      Domain: $(-\infty, \infty)$

    - $(fg)(x) = -x^4 + x^3 + 15x^2 - 9x - 54$
      Domain: $(-\infty, \infty)$

    - $(f-g)(x) = -2x^2 + x + 15$
      Domain: $(-\infty, \infty)$

    - $\left(\dfrac{f}{g}\right)(x) = -\dfrac{x+2}{x+3}$
      Domain: $(-\infty, -3) \cup (-3, 3) \cup (3, \infty)$

17. For $f(x) = \dfrac{x}{2}$ and $g(x) = \dfrac{2}{x}$

    - $(f+g)(x) = \dfrac{x^2+4}{2x}$
      Domain: $(-\infty, 0) \cup (0, \infty)$

    - $(fg)(x) = 1$
      Domain: $(-\infty, 0) \cup (0, \infty)$

    - $(f-g)(x) = \dfrac{x^2-4}{2x}$
      Domain: $(-\infty, 0) \cup (0, \infty)$

    - $\left(\dfrac{f}{g}\right)(x) = \dfrac{x^2}{4}$
      Domain: $(-\infty, 0) \cup (0, \infty)$

18. For $f(x) = x - 1$ and $g(x) = \frac{1}{x-1}$

   - $(f + g)(x) = \frac{x^2 - 2x + 2}{x-1}$
     Domain: $(-\infty, 1) \cup (1, \infty)$

   - $(f - g)(x) = \frac{x^2 - 2x}{x-1}$
     Domain: $(-\infty, 1) \cup (1, \infty)$

   - $(fg)(x) = 1$
     Domain: $(-\infty, 1) \cup (1, \infty)$

   - $\left(\frac{f}{g}\right)(x) = x^2 - 2x + 1$
     Domain: $(-\infty, 1) \cup (1, \infty)$

19. For $f(x) = x$ and $g(x) = \sqrt{x+1}$

   - $(f + g)(x) = x + \sqrt{x+1}$
     Domain: $[-1, \infty)$

   - $(f - g)(x) = x - \sqrt{x+1}$
     Domain: $[-1, \infty)$

   - $(fg)(x) = x\sqrt{x+1}$
     Domain: $[-1, \infty)$

   - $\left(\frac{f}{g}\right)(x) = \frac{x}{\sqrt{x+1}}$
     Domain: $(-1, \infty)$

20. For $f(x) = \sqrt{x-5}$ and $g(x) = f(x) = \sqrt{x-5}$

   - $(f + g)(x) = 2\sqrt{x-5}$
     Domain: $[5, \infty)$

   - $(f - g)(x) = 0$
     Domain: $[5, \infty)$

   - $(fg)(x) = x - 5$
     Domain: $[5, \infty)$

   - $\left(\frac{f}{g}\right)(x) = 1$
     Domain: $(5, \infty)$

21. 2

22. -3

23. 0

24. $6x + 3h - 1$

25. $-2x - h + 2$

26. $8x + 4h$

27. $-2x - h + 1$

28. $3x^2 + 3xh + h^2$

29. m

30. $2ax + ah + b$

31. $\dfrac{-2}{x(x+h)}$

32. $\dfrac{3}{(1-x-h)(1-x)}$

33. $\dfrac{-(2x+h)}{x^2(x+h)^2}$

34. $\dfrac{-2}{(x+5)(x+h+5)}$

35. $\dfrac{-4}{(4x-3)(4x+4h-3)}$

36. $\dfrac{3}{(x+1)(x+h+1)}$

1.5 Function Arithmetic

37. $\dfrac{-9}{(x-9)(x+h-9)}$

38. $\dfrac{2x^2 + 2xh + 2x + h}{(2x+1)(2x+2h+1)}$

39. $\dfrac{1}{\sqrt{x+h-9} + \sqrt{x-9}}$

40. $\dfrac{2}{\sqrt{2x+2h+1} + \sqrt{2x+1}}$

41. $\dfrac{-4}{\sqrt{-4x-4h+5} + \sqrt{-4x+5}}$

42. $\dfrac{-1}{\sqrt{4-x-h} + \sqrt{4-x}}$

43. $\dfrac{a}{\sqrt{ax+ah+b} + \sqrt{ax+b}}$

44. $\dfrac{3x^2 + 3xh + h^2}{(x+h)^{3/2} + x^{3/2}}$

45. $\dfrac{1}{(x+h)^{2/3} + (x+h)^{1/3}x^{1/3} + x^{2/3}}$

46.
    - $C(0) = 26$, so the fixed costs are $26.
    - $\overline{C}(10) = 4.6$, so when 10 shirts are produced, the cost per shirt is $4.60.
    - $p(5) = 20$, so to sell 5 shirts, set the price at $20 per shirt.
    - $R(x) = -2x^2 + 30x$, $0 \leq x \leq 15$
    - $P(x) = -2x^2 + 28x - 26$, $0 \leq x \leq 15$
    - $P(x) = 0$ when $x = 1$ and $x = 13$. These are the 'break even' points, so selling 1 shirt or 13 shirts will guarantee the revenue earned exactly recoups the cost of production.

47.
    - $C(0) = 100$, so the fixed costs are $100.
    - $\overline{C}(10) = 20$, so when 10 bottles of tonic are produced, the cost per bottle is $20.
    - $p(5) = 30$, so to sell 5 bottles of tonic, set the price at $30 per bottle.
    - $R(x) = -x^2 + 35x$, $0 \leq x \leq 35$
    - $P(x) = -x^2 + 25x - 100$, $0 \leq x \leq 35$
    - $P(x) = 0$ when $x = 5$ and $x = 20$. These are the 'break even' points, so selling 5 bottles of tonic or 20 bottles of tonic will guarantee the revenue earned exactly recoups the cost of production.

48.
    - $C(0) = 240$, so the fixed costs are 240¢ or $2.40.
    - $\overline{C}(10) = 42$, so when 10 cups of lemonade are made, the cost per cup is 42¢.
    - $p(5) = 75$, so to sell 5 cups of lemonade, set the price at 75¢ per cup.
    - $R(x) = -3x^2 + 90x$, $0 \leq x \leq 30$
    - $P(x) = -3x^2 + 72x - 240$, $0 \leq x \leq 30$
    - $P(x) = 0$ when $x = 4$ and $x = 20$. These are the 'break even' points, so selling 4 cups of lemonade or 20 cups of lemonade will guarantee the revenue earned exactly recoups the cost of production.

49.
- $C(0) = 36$, so the daily fixed costs are \$36.
- $\overline{C}(10) = 6.6$, so when 10 pies are made, the cost per pie is \$6.60.
- $p(5) = 9.5$, so to sell 5 pies a day, set the price at \$9.50 per pie.
- $R(x) = -0.5x^2 + 12x$, $0 \leq x \leq 24$
- $P(x) = -0.5x^2 + 9x - 36$, $0 \leq x \leq 24$
- $P(x) = 0$ when $x = 6$ and $x = 12$. These are the 'break even' points, so selling 6 pies or 12 pies a day will guarantee the revenue earned exactly recoups the cost of production.

50.
- $C(0) = 1000$, so the monthly fixed costs are 1000 *hundred* dollars, or \$100,000.
- $\overline{C}(10) = 120$, so when 10 scooters are made, the cost per scooter is 120 hundred dollars, or \$12,000.
- $p(5) = 130$, so to sell 5 scooters a month, set the price at 130 hundred dollars, or \$13,000 per scooter.
- $R(x) = -2x^2 + 140x$, $0 \leq x \leq 70$
- $P(x) = -2x^2 + 120x - 1000$, $0 \leq x \leq 70$
- $P(x) = 0$ when $x = 10$ and $x = 50$. These are the 'break even' points, so selling 10 scooters or 50 scooters a month will guarantee the revenue earned exactly recoups the cost of production.

51. $(f+g)(-3) = 2$

52. $(f-g)(2) = 3$

53. $(fg)(-1) = 0$

54. $(g+f)(1) = 0$

55. $(g-f)(3) = 3$

56. $(gf)(-3) = -8$

57. $\left(\frac{f}{g}\right)(-2)$ does not exist

58. $\left(\frac{f}{g}\right)(-1) = 0$

59. $\left(\frac{f}{g}\right)(2) = 4$

60. $\left(\frac{g}{f}\right)(-1)$ does not exist

61. $\left(\frac{g}{f}\right)(3) = -2$

62. $\left(\frac{g}{f}\right)(-3) = -\frac{1}{2}$

1.6 Graphs of Functions

In Section 1.3 we defined a function as a special type of relation; one in which each x-coordinate was matched with only one y-coordinate. We spent most of our time in that section looking at functions graphically because they were, after all, just sets of points in the plane. Then in Section 1.4 we described a function as a process and defined the notation necessary to work with functions algebraically. So now it's time to look at functions graphically again, only this time we'll do so with the notation defined in Section 1.4. We start with what should not be a surprising connection.

> **The Fundamental Graphing Principle for Functions**
>
> The graph of a function f is the set of points which satisfy the equation $y = f(x)$. That is, the point (x, y) is on the graph of f if and only if $y = f(x)$.

Example 1.6.1. Graph $f(x) = x^2 - x - 6$.

Solution. To graph f, we graph the equation $y = f(x)$. To this end, we use the techniques outlined in Section 1.2.1. Specifically, we check for intercepts, test for symmetry, and plot additional points as needed. To find the x-intercepts, we set $y = 0$. Since $y = f(x)$, this means $f(x) = 0$.

$$\begin{aligned} f(x) &= x^2 - x - 6 \\ 0 &= x^2 - x - 6 \\ 0 &= (x-3)(x+2) \quad \text{factor} \\ x - 3 = 0 \quad &\text{or} \quad x + 2 = 0 \\ x &= -2, 3 \end{aligned}$$

So we get $(-2, 0)$ and $(3, 0)$ as x-intercepts. To find the y-intercept, we set $x = 0$. Using function notation, this is the same as finding $f(0)$ and $f(0) = 0^2 - 0 - 6 = -6$. Thus the y-intercept is $(0, -6)$. As far as symmetry is concerned, we can tell from the intercepts that the graph possesses none of the three symmetries discussed thus far. (You should verify this.) We can make a table analogous to the ones we made in Section 1.2.1, plot the points and connect the dots in a somewhat pleasing fashion to get the graph below on the right.

x	$f(x)$	$(x, f(x))$
-3	6	$(-3, 6)$
-2	0	$(-2, 0)$
-1	-4	$(-1, -4)$
0	-6	$(0, -6)$
1	-6	$(1, -6)$
2	-4	$(2, -4)$
3	0	$(3, 0)$
4	6	$(4, 6)$

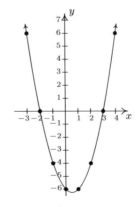

□

Graphing piecewise-defined functions is a bit more of a challenge.

Example 1.6.2. Graph: $f(x) = \begin{cases} 4 - x^2 & \text{if } x < 1 \\ x - 3, & \text{if } x \geq 1 \end{cases}$

Solution. We proceed as before – finding intercepts, testing for symmetry and then plotting additional points as needed. To find the x-intercepts, as before, we set $f(x) = 0$. The twist is that we have two formulas for $f(x)$. For $x < 1$, we use the formula $f(x) = 4 - x^2$. Setting $f(x) = 0$ gives $0 = 4 - x^2$, so that $x = \pm 2$. However, of these two answers, only $x = -2$ fits in the domain $x < 1$ for this piece. This means the only x-intercept for the $x < 1$ region of the x-axis is $(-2, 0)$. For $x \geq 1$, $f(x) = x - 3$. Setting $f(x) = 0$ gives $0 = x - 3$, or $x = 3$. Since $x = 3$ satisfies the inequality $x \geq 1$, we get $(3, 0)$ as another x-intercept. Next, we seek the y-intercept. Notice that $x = 0$ falls in the domain $x < 1$. Thus $f(0) = 4 - 0^2 = 4$ yields the y-intercept $(0, 4)$. As far as symmetry is concerned, you can check that the equation $y = 4 - x^2$ is symmetric about the y-axis; unfortunately, this equation (and its symmetry) is valid only for $x < 1$. You can also verify $y = x - 3$ possesses none of the symmetries discussed in the Section 1.2.1. When plotting additional points, it is important to keep in mind the restrictions on x for each piece of the function. The sticking point for this function is $x = 1$, since this is where the equations change. When $x = 1$, we use the formula $f(x) = x - 3$, so the point on the graph $(1, f(1))$ is $(1, -2)$. However, for all values less than 1, we use the formula $f(x) = 4 - x^2$. As we have discussed earlier in Section 1.2, there is no real number which immediately precedes $x = 1$ on the number line. Thus for the values $x = 0.9$, $x = 0.99$, $x = 0.999$, and so on, we find the corresponding y values using the formula $f(x) = 4 - x^2$. Making a table as before, we see that as the x values sneak up to $x = 1$ in this fashion, the $f(x)$ values inch closer and closer[1] to $4 - 1^2 = 3$. To indicate this graphically, we use an open circle at the point $(1, 3)$. Putting all of this information together and plotting additional points, we get

x	$f(x)$	$(x, f(x))$
0.9	3.19	$(0.9, 3.19)$
0.99	≈ 3.02	$(0.99, 3.02)$
0.999	≈ 3.002	$(0.999, 3.002)$

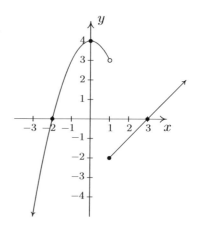

□

---

[1] We've just stepped into Calculus here!

1.6 GRAPHS OF FUNCTIONS

In the previous two examples, the x-coordinates of the x-intercepts of the graph of $y = f(x)$ were found by solving $f(x) = 0$. For this reason, they are called the **zeros** of f.

> **Definition 1.9.** The **zeros** of a function f are the solutions to the equation $f(x) = 0$. In other words, x is a zero of f if and only if $(x, 0)$ is an x-intercept of the graph of $y = f(x)$.

Of the three symmetries discussed in Section 1.2.1, only two are of significance to functions: symmetry about the y-axis and symmetry about the origin.[2] Recall that we can test whether the graph of an equation is symmetric about the y-axis by replacing x with $-x$ and checking to see if an equivalent equation results. If we are graphing the equation $y = f(x)$, substituting $-x$ for x results in the equation $y = f(-x)$. In order for this equation to be equivalent to the original equation $y = f(x)$ we need $f(-x) = f(x)$. In a similar fashion, we recall that to test an equation's graph for symmetry about the origin, we replace x and y with $-x$ and $-y$, respectively. Doing this substitution in the equation $y = f(x)$ results in $-y = f(-x)$. Solving the latter equation for y gives $y = -f(-x)$. In order for this equation to be equivalent to the original equation $y = f(x)$ we need $-f(-x) = f(x)$, or, equivalently, $f(-x) = -f(x)$. These results are summarized below.

> **Testing the Graph of a Function for Symmetry**
>
> The graph of a function f is symmetric
>
> - about the y-axis if and only if $f(-x) = f(x)$ for all x in the domain of f.
> - about the origin if and only if $f(-x) = -f(x)$ for all x in the domain of f.

For reasons which won't become clear until we study polynomials, we call a function **even** if its graph is symmetric about the y-axis or **odd** if its graph is symmetric about the origin. Apart from a very specialized family of functions which are both even and odd,[3] functions fall into one of three distinct categories: even, odd, or neither even nor odd.

Example 1.6.3. Determine analytically if the following functions are even, odd, or neither even nor odd. Verify your result with a graphing calculator.

1. $f(x) = \dfrac{5}{2 - x^2}$

2. $g(x) = \dfrac{5x}{2 - x^2}$

3. $h(x) = \dfrac{5x}{2 - x^3}$

4. $i(x) = \dfrac{5x}{2x - x^3}$

5. $j(x) = x^2 - \dfrac{x}{100} - 1$

6. $p(x) = \begin{cases} x + 3 & \text{if } x < 0 \\ -x + 3, & \text{if } x \geq 0 \end{cases}$

Solution. The first step in all of these problems is to replace x with $-x$ and simplify.

---
[2] Why are we so dismissive about symmetry about the x-axis for graphs of functions?
[3] Any ideas?

1.
$$f(x) = \frac{5}{2-x^2}$$
$$f(-x) = \frac{5}{2-(-x)^2}$$
$$f(-x) = \frac{5}{2-x^2}$$
$$f(-x) = f(x)$$

Hence, f is **even**. The graphing calculator furnishes the following.

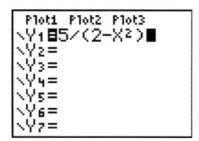 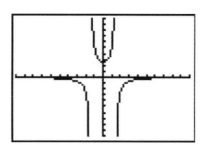

This suggests[4] that the graph of f is symmetric about the y-axis, as expected.

2.
$$g(x) = \frac{5x}{2-x^2}$$
$$g(-x) = \frac{5(-x)}{2-(-x)^2}$$
$$g(-x) = \frac{-5x}{2-x^2}$$

It doesn't appear that $g(-x)$ is equivalent to $g(x)$. To prove this, we check with an x value. After some trial and error, we see that $g(1) = 5$ whereas $g(-1) = -5$. This proves that g is not even, but it doesn't rule out the possibility that g is odd. (Why not?) To check if g is odd, we compare $g(-x)$ with $-g(x)$

$$-g(x) = -\frac{5x}{2-x^2}$$
$$= \frac{-5x}{2-x^2}$$
$$-g(x) = g(-x)$$

Hence, g is odd. Graphically,

---

[4]'Suggests' is about the extent of what it can do.

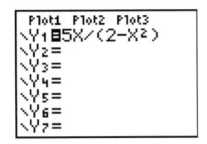 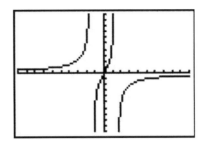

The calculator indicates the graph of g is symmetric about the origin, as expected.

3.
$$h(x) = \frac{5x}{2 - x^3}$$
$$h(-x) = \frac{5(-x)}{2 - (-x)^3}$$
$$h(-x) = \frac{-5x}{2 + x^3}$$

Once again, $h(-x)$ doesn't appear to be equivalent to $h(x)$. We check with an x value, for example, $h(1) = 5$ but $h(-1) = -\frac{5}{3}$. This proves that h is not even and it also shows h is not odd. (Why?) Graphically,

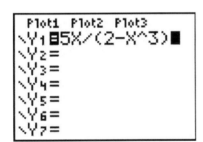 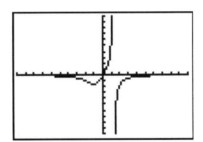

The graph of h appears to be neither symmetric about the y-axis nor the origin.

4.
$$i(x) = \frac{5x}{2x - x^3}$$
$$i(-x) = \frac{5(-x)}{2(-x) - (-x)^3}$$
$$i(-x) = \frac{-5x}{-2x + x^3}$$

The expression $i(-x)$ doesn't appear to be equivalent to $i(x)$. However, after checking some x values, for example $x = 1$ yields $i(1) = 5$ and $i(-1) = 5$, it appears that $i(-x)$ does, in fact, equal $i(x)$. However, while this suggests i is even, it doesn't prove it. (It does, however, prove

i is not odd.) To prove $i(-x) = i(x)$, we need to manipulate our expressions for $i(x)$ and $i(-x)$ and show that they are equivalent. A clue as to how to proceed is in the numerators: in the formula for $i(x)$, the numerator is $5x$ and in $i(-x)$ the numerator is $-5x$. To re-write $i(x)$ with a numerator of $-5x$, we need to multiply its numerator by -1. To keep the value of the fraction the same, we need to multiply the denominator by -1 as well. Thus

$$\begin{aligned} i(x) &= \frac{5x}{2x - x^3} \\ &= \frac{(-1)5x}{(-1)(2x - x^3)} \\ &= \frac{-5x}{-2x + x^3} \end{aligned}$$

Hence, $i(x) = i(-x)$, so i is even. The calculator supports our conclusion.

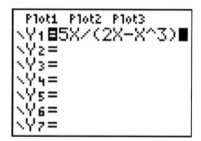 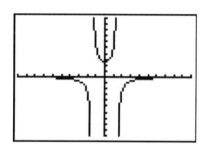

5.
$$\begin{aligned} j(x) &= x^2 - \frac{x}{100} - 1 \\ j(-x) &= (-x)^2 - \frac{-x}{100} - 1 \\ j(-x) &= x^2 + \frac{x}{100} - 1 \end{aligned}$$

The expression for $j(-x)$ doesn't seem to be equivalent to $j(x)$, so we check using $x = 1$ to get $j(1) = -\frac{1}{100}$ and $j(-1) = \frac{1}{100}$. This rules out j being even. However, it doesn't rule out j being odd. Examining $-j(x)$ gives

$$\begin{aligned} j(x) &= x^2 - \frac{x}{100} - 1 \\ -j(x) &= -\left(x^2 - \frac{x}{100} - 1\right) \\ -j(x) &= -x^2 + \frac{x}{100} + 1 \end{aligned}$$

The expression $-j(x)$ doesn't seem to match $j(-x)$ either. Testing $x = 2$ gives $j(2) = \frac{149}{50}$ and $j(-2) = \frac{151}{50}$, so j is not odd, either. The calculator gives:

1.6 GRAPHS OF FUNCTIONS

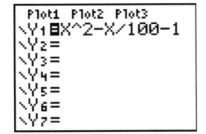
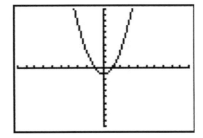

The calculator suggests that the graph of j is symmetric about the y-axis which would imply that j is even. However, we have proven that is not the case. □

6. Testing the graph of $y = p(x)$ for symmetry is complicated by the fact $p(x)$ is a piecewise-defined function. As always, we handle this by checking the condition for symmetry by checking it on each piece of the domain. We first consider the case when $x < 0$ and set about finding the correct expression for $p(-x)$. Even though $p(x) = x+3$ for $x < 0$, $p(-x) \neq -x+3$ here. The reason for this is that since $x < 0$, $-x > 0$ which means to find $p(-x)$, we need to use the *other* formula for $p(x)$, namely $p(x) = -x+3$. Hence, for $x < 0$, $p(-x) = -(-x)+3 = x + 3 = p(x)$. For $x \geq 0$, $p(x) = -x + 3$ and we have two cases. If $x > 0$, then $-x < 0$ so $p(-x) = (-x) + 3 = -x + 3 = p(x)$. If $x = 0$, then $p(0) = 3 = p(-0)$. Hence, in all cases, $p(-x) = p(x)$, so p is even. Since $p(0) = 3$ but $p(-0) = p(0) = 3 \neq -3$, we also have p is not odd. While graphing $y = p(x)$ is not onerous to do by hand, it is instructive to see how to enter this into our calculator. By using some of the logical commands,[5] we have:

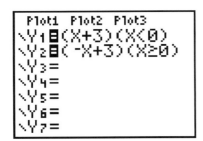
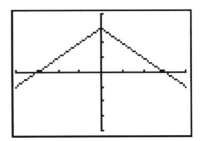

The calculator bears shows that the graph appears to be symmetric about the y-axis. □

There are two lessons to be learned from the last example. The first is that sampling function values at particular x values is not enough to prove that a function is even or odd – despite the fact that $j(-1) = -j(1)$, j turned out not to be odd. Secondly, while the calculator may *suggest* mathematical truths, it is the Algebra which *proves* mathematical truths.[6]

---

[5] Consult your owner's manual, instructor, or favorite video site!
[6] Or, in other words, don't rely too heavily on the machine!

1.6.1 General Function Behavior

The last topic we wish to address in this section is general function behavior. As you shall see in the next several chapters, each family of functions has its own unique attributes and we will study them all in great detail. The purpose of this section's discussion, then, is to lay the foundation for that further study by investigating aspects of function behavior which apply to all functions. To start, we will examine the concepts of **increasing**, **decreasing** and **constant**. Before defining the concepts algebraically, it is instructive to first look at them graphically. Consider the graph of the function f below.

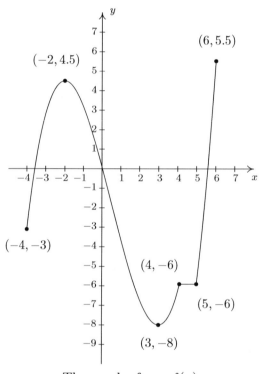

The graph of $y = f(x)$

Reading from left to right, the graph 'starts' at the point $(-4, -3)$ and 'ends' at the point $(6, 5.5)$. If we imagine walking from left to right on the graph, between $(-4, -3)$ and $(-2, 4.5)$, we are walking 'uphill'; then between $(-2, 4.5)$ and $(3, -8)$, we are walking 'downhill'; and between $(3, -8)$ and $(4, -6)$, we are walking 'uphill' once more. From $(4, -6)$ to $(5, -6)$, we 'level off', and then resume walking 'uphill' from $(5, -6)$ to $(6, 5.5)$. In other words, for the x values between -4 and -2 (inclusive), the y-coordinates on the graph are getting larger, or **increasing**, as we move from left to right. Since $y = f(x)$, the y values on the graph are the function values, and we say that the function f is **increasing** on the interval $[-4, -2]$. Analogously, we say that f is **decreasing** on the interval $[-2, 3]$ increasing once more on the interval $[3, 4]$, **constant** on $[4, 5]$, and finally increasing once again on $[5, 6]$. It is extremely important to notice that the behavior (increasing, decreasing or constant) occurs on an interval on the x-axis. When we say that the function f is increasing

1.6 GRAPHS OF FUNCTIONS

on $[-4, -2]$ we do not mention the actual y values that f attains along the way. Thus, we report *where* the behavior occurs, not to what extent the behavior occurs.[7] Also notice that we do not say that a function is increasing, decreasing or constant at a single x value. In fact, we would run into serious trouble in our previous example if we tried to do so because $x = -2$ is contained in an interval on which f was increasing and one on which it is decreasing. (There's more on this issue – and many others – in the Exercises.)

We're now ready for the more formal algebraic definitions of what it means for a function to be increasing, decreasing or constant.

> **Definition 1.10.** Suppose f is a function defined on an interval I. We say f is:
>
> - **increasing** on I if and only if $f(a) < f(b)$ for all real numbers a, b in I with $a < b$.
> - **decreasing** on I if and only if $f(a) > f(b)$ for all real numbers a, b in I with $a < b$.
> - **constant** on I if and only if $f(a) = f(b)$ for all real numbers a, b in I.

It is worth taking some time to see that the algebraic descriptions of increasing, decreasing and constant as stated in Definition 1.10 agree with our graphical descriptions given earlier. You should look back through the examples and exercise sets in previous sections where graphs were given to see if you can determine the intervals on which the functions are increasing, decreasing or constant. Can you find an example of a function for which none of the concepts in Definition 1.10 apply?

Now let's turn our attention to a few of the points on the graph. Clearly the point $(-2, 4.5)$ does not have the largest y value of all of the points on the graph of f – indeed that honor goes to $(6, 5.5)$ – but $(-2, 4.5)$ should get some sort of consolation prize for being 'the top of the hill' between $x = -4$ and $x = 3$. We say that the function f has a **local maximum**[8] at the point $(-2, 4.5)$, because the y-coordinate 4.5 is the largest y-value (hence, function value) on the curve 'near'[9] $x = -2$. Similarly, we say that the function f has a **local minimum**[10] at the point $(3, -8)$, since the y-coordinate -8 is the smallest function value near $x = 3$. Although it is tempting to say that local extrema[11] occur when the function changes from increasing to decreasing or vice versa, it is not a precise enough way to define the concepts for the needs of Calculus. At the risk of being pedantic, we will present the traditional definitions and thoroughly vet the pathologies they induce in the Exercises. We have one last observation to make before we proceed to the algebraic definitions and look at a fairly tame, yet helpful, example.

If we look at the entire graph, we see that the largest y value (the largest function value) is 5.5 at $x = 6$. In this case, we say the **maximum**[12] of f is 5.5; similarly, the **minimum**[13] of f is -8.

---

[7] The notions of how quickly or how slowly a function increases or decreases are explored in Calculus.
[8] Also called 'relative maximum'.
[9] We will make this more precise in a moment.
[10] Also called a 'relative minimum'.
[11] 'Maxima' is the plural of 'maximum' and 'mimima' is the plural of 'minimum'. 'Extrema' is the plural of 'extremum' which combines maximum and minimum.
[12] Sometimes called the 'absolute' or 'global' maximum.
[13] Again, 'absolute' or 'global' minimum can be used.

We formalize these concepts in the following definitions.

> **Definition 1.11.** Suppose f is a function with $f(a) = b$.
>
> - We say f has a **local maximum** at the point (a, b) if and only if there is an open interval I containing a for which $f(a) \geq f(x)$ for all x in I. The value $f(a) = b$ is called 'a local maximum value of f' in this case.
>
> - We say f has a **local minimum** at the point (a, b) if and only if there is an open interval I containing a for which $f(a) \leq f(x)$ for all x in I. The value $f(a) = b$ is called 'a local minimum value of f' in this case.
>
> - The value b is called the **maximum** of f if $b \geq f(x)$ for all x in the domain of f.
>
> - The value b is called the **minimum** of f if $b \leq f(x)$ for all x in the domain of f.

It's important to note that not every function will have all of these features. Indeed, it is possible to have a function with no local or absolute extrema at all! (Any ideas of what such a function's graph would have to look like?) We shall see examples of functions in the Exercises which have one or two, but not all, of these features, some that have instances of each type of extremum and some functions that seem to defy common sense. In all cases, though, we shall adhere to the algebraic definitions above as we explore the wonderful diversity of graphs that functions provide us.

Here is the 'tame' example which was promised earlier. It summarizes all of the concepts presented in this section as well as some from previous sections so you should spend some time thinking deeply about it before proceeding to the Exercises.

Example 1.6.4. Given the graph of $y = f(x)$ below, answer all of the following questions.

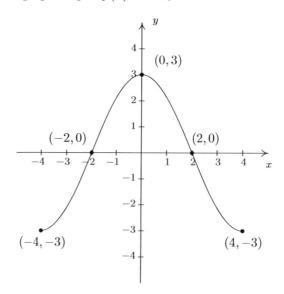

1.6 GRAPHS OF FUNCTIONS

1. Find the domain of f.

2. Find the range of f.

3. List the x-intercepts, if any exist.

4. List the y-intercepts, if any exist.

5. Find the zeros of f.

6. Solve $f(x) < 0$.

7. Determine $f(2)$.

8. Solve $f(x) = -3$.

9. Find the number of solutions to $f(x) = 1$.

10. Does f appear to be even, odd, or neither?

11. List the intervals on which f is increasing.

12. List the intervals on which f is decreasing.

13. List the local maximums, if any exist.

14. List the local minimums, if any exist.

15. Find the maximum, if it exists.

16. Find the minimum, if it exists.

Solution.

1. To find the domain of f, we proceed as in Section 1.3. By projecting the graph to the x-axis, we see that the portion of the x-axis which corresponds to a point on the graph is everything from -4 to 4, inclusive. Hence, the domain is $[-4, 4]$.

2. To find the range, we project the graph to the y-axis. We see that the y values from -3 to 3, inclusive, constitute the range of f. Hence, our answer is $[-3, 3]$.

3. The x-intercepts are the points on the graph with y-coordinate 0, namely $(-2, 0)$ and $(2, 0)$.

4. The y-intercept is the point on the graph with x-coordinate 0, namely $(0, 3)$.

5. The zeros of f are the x-coordinates of the x-intercepts of the graph of $y = f(x)$ which are $x = -2, 2$.

6. To solve $f(x) < 0$, we look for the x values of the points on the graph where the y-coordinate is less than 0. Graphically, we are looking for where the graph is below the x-axis. This happens for the x values from -4 to -2 and again from 2 to 4. So our answer is $[-4, -2) \cup (2, 4]$.

7. Since the graph of f is the graph of the equation $y = f(x)$, $f(2)$ is the y-coordinate of the point which corresponds to $x = 2$. Since the point $(2, 0)$ is on the graph, we have $f(2) = 0$.

8. To solve $f(x) = -3$, we look where $y = f(x) = -3$. We find two points with a y-coordinate of -3, namely $(-4, -3)$ and $(4, -3)$. Hence, the solutions to $f(x) = -3$ are $x = \pm 4$.

9. As in the previous problem, to solve $f(x) = 1$, we look for points on the graph where the y-coordinate is 1. Even though these points aren't specified, we see that the curve has two points with a y value of 1, as seen in the graph below. That means there are two solutions to $f(x) = 1$.

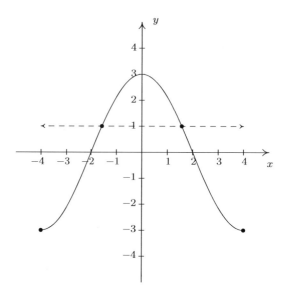

10. The graph appears to be symmetric about the y-axis. This suggests[14] that f is even.

11. As we move from left to right, the graph rises from $(-4, -3)$ to $(0, 3)$. This means f is increasing on the interval $[-4, 0]$. (Remember, the answer here is an interval on the x-axis.)

12. As we move from left to right, the graph falls from $(0, 3)$ to $(4, -3)$. This means f is decreasing on the interval $[0, 4]$. (Remember, the answer here is an interval on the x-axis.)

13. The function has its only local maximum at $(0, 3)$ so $f(0) = 3$ is the local minimum value.

14. There are no local minimums. Why don't $(-4, -3)$ and $(4, -3)$ count? Let's consider the point $(-4, -3)$ for a moment. Recall that, in the definition of local minimum, there needs to be an open interval I which contains $x = -4$ such that $f(-4) < f(x)$ for all x in I different from -4. But if we put an open interval around $x = -4$ a portion of that interval will lie outside of the domain of f. Because we are unable to fulfill the requirements of the definition for a local minimum, we cannot claim that f has one at $(-4, -3)$. The point $(4, -3)$ fails for the same reason − no open interval around $x = 4$ stays within the domain of f.

15. The maximum value of f is the largest y-coordinate which is 3.

16. The minimum value of f is the smallest y-coordinate which is -3. □

With few exceptions, we will not develop techniques in College Algebra which allow us to determine the intervals on which a function is increasing, decreasing or constant or to find the local maximums and local minimums analytically; this is the business of Calculus.[15] When we have need to find such beasts, we will resort to the calculator. Most graphing calculators have 'Minimum' and 'Maximum' features which can be used to approximate these values, as we now demonstrate.

---

[14] but does not prove

[15] Although, truth be told, there is only one step of Calculus involved, followed by several pages of algebra.

1.6 Graphs of Functions

Example 1.6.5. Let $f(x) = \dfrac{15x}{x^2+3}$. Use a graphing calculator to approximate the intervals on which f is increasing and those on which it is decreasing. Approximate all extrema.

Solution. Entering this function into the calculator gives

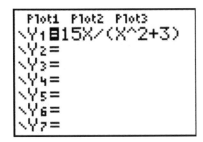
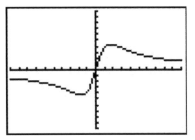

Using the Minimum and Maximum features, we get

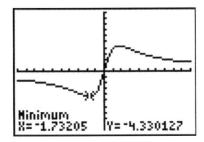
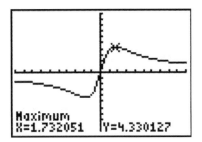

To two decimal places, f appears to have its only local minimum at $(-1.73, -4.33)$ and its only local maximum at $(1.73, 4.33)$. Given the symmetry about the origin suggested by the graph, the relation between these points shouldn't be too surprising. The function appears to be increasing on $[-1.73, 1.73]$ and decreasing on $(-\infty, -1.73] \cup [1.73, \infty)$. This makes -4.33 the (absolute) minimum and 4.33 the (absolute) maximum. $\square$

Example 1.6.6. Find the points on the graph of $y = (x-3)^2$ which are closest to the origin. Round your answers to two decimal places.

Solution. Suppose a point (x, y) is on the graph of $y = (x-3)^2$. Its distance to the origin $(0,0)$ is given by

$$\begin{aligned} d &= \sqrt{(x-0)^2 + (y-0)^2} \\ &= \sqrt{x^2 + y^2} \\ &= \sqrt{x^2 + [(x-3)^2]^2} \qquad \text{Since } y = (x-3)^2 \\ &= \sqrt{x^2 + (x-3)^4} \end{aligned}$$

Given a value for x, the formula $d = \sqrt{x^2 + (x-3)^4}$ is the distance from $(0,0)$ to the point (x, y) on the curve $y = (x-3)^2$. What we have defined, then, is a function $d(x)$ which we wish to minimize over all values of x. To accomplish this task analytically would require Calculus so as we've mentioned before, we can use a graphing calculator to find an approximate solution. Using the calculator, we enter the function $d(x)$ as shown below and graph.

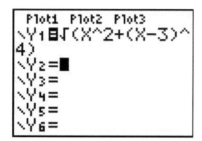 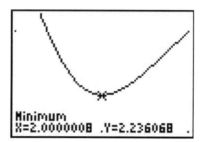

Using the Minimum feature, we see above on the right that the (absolute) minimum occurs near $x = 2$. Rounding to two decimal places, we get that the minimum distance occurs when $x = 2.00$. To find the y value on the parabola associated with $x = 2.00$, we substitute 2.00 into the equation to get $y = (x-3)^2 = (2.00-3)^2 = 1.00$. So, our final answer is $(2.00, 1.00)$.[16] (What does the y value listed on the calculator screen mean in this problem?) □

---

[16]It seems silly to list a final answer as $(2.00, 1.00)$. Indeed, Calculus confirms that the *exact* answer to this problem is, in fact, $(2, 1)$. As you are well aware by now, the authors are overly pedantic, and as such, use the decimal places to remind the reader that *any* result garnered from a calculator in this fashion is an approximation, and should be treated as such.

1.6.2 Exercises

In Exercises 1 - 12, sketch the graph of the given function. State the domain of the function, identify any intercepts and test for symmetry.

1. $f(x) = 2 - x$
2. $f(x) = \dfrac{x-2}{3}$
3. $f(x) = x^2 + 1$
4. $f(x) = 4 - x^2$
5. $f(x) = 2$
6. $f(x) = x^3$
7. $f(x) = x(x-1)(x+2)$
8. $f(x) = \sqrt{x-2}$
9. $f(x) = \sqrt{5-x}$
10. $f(x) = 3 - 2\sqrt{x+2}$
11. $f(x) = \sqrt[3]{x}$
12. $f(x) = \dfrac{1}{x^2+1}$

In Exercises 13 - 20, sketch the graph of the given piecewise-defined function.

13. $f(x) = \begin{cases} 4-x & \text{if } x \leq 3 \\ 2 & \text{if } x > 3 \end{cases}$

14. $f(x) = \begin{cases} x^2 & \text{if } x \leq 0 \\ 2x & \text{if } x > 0 \end{cases}$

15. $f(x) = \begin{cases} -3 & \text{if } x < 0 \\ 2x - 3 & \text{if } 0 \leq x \leq 3 \\ 3 & \text{if } x > 3 \end{cases}$

16. $f(x) = \begin{cases} x^2 - 4 & \text{if } x \leq -2 \\ 4 - x^2 & \text{if } -2 < x < 2 \\ x^2 - 4 & \text{if } x \geq 2 \end{cases}$

17. $f(x) = \begin{cases} -2x - 4 & \text{if } x < 0 \\ 3x & \text{if } x \geq 0 \end{cases}$

18. $f(x) = \begin{cases} \sqrt{x+4} & \text{if } -4 \leq x < 5 \\ \sqrt{x-1} & \text{if } x \geq 5 \end{cases}$

19. $f(x) = \begin{cases} x^2 & \text{if } x \leq -2 \\ 3 - x & \text{if } -2 < x < 2 \\ 4 & \text{if } x \geq 2 \end{cases}$

20. $f(x) = \begin{cases} \dfrac{1}{x} & \text{if } -6 < x < -1 \\ x & \text{if } -1 < x < 1 \\ \sqrt{x} & \text{if } 1 < x < 9 \end{cases}$

In Exercises 21 - 41, determine analytically if the following functions are even, odd or neither.

21. $f(x) = 7x$
22. $f(x) = 7x + 2$
23. $f(x) = 7$
24. $f(x) = 3x^2 - 4$
25. $f(x) = 4 - x^2$
26. $f(x) = x^2 - x - 6$
27. $f(x) = 2x^3 - x$
28. $f(x) = -x^5 + 2x^3 - x$
29. $f(x) = x^6 - x^4 + x^2 + 9$
30. $f(x) = x^3 + x^2 + x + 1$
31. $f(x) = \sqrt{1-x}$
32. $f(x) = \sqrt{1-x^2}$
33. $f(x) = 0$
34. $f(x) = \sqrt[3]{x}$
35. $f(x) = \sqrt[3]{x^2}$

36. $f(x) = \dfrac{3}{x^2}$

37. $f(x) = \dfrac{2x-1}{x+1}$

38. $f(x) = \dfrac{3x}{x^2+1}$

39. $f(x) = \dfrac{x^2-3}{x-4x^3}$

40. $f(x) = \dfrac{9}{\sqrt{4-x^2}}$

41. $f(x) = \dfrac{\sqrt[3]{x^3+x}}{5x}$

In Exercises 42 - 57, use the graph of $y = f(x)$ given below to answer the question.

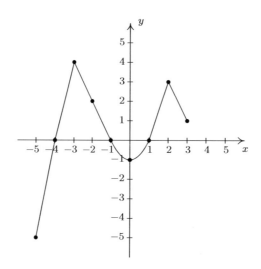

42. Find the domain of f.

43. Find the range of f.

44. Determine $f(-2)$.

45. Solve $f(x) = 4$.

46. List the x-intercepts, if any exist.

47. List the y-intercepts, if any exist.

48. Find the zeros of f.

49. Solve $f(x) \geq 0$.

50. Find the number of solutions to $f(x) = 1$.

51. Does f appear to be even, odd, or neither?

52. List the intervals where f is increasing.

53. List the intervals where f is decreasing.

54. List the local maximums, if any exist.

55. List the local minimums, if any exist.

56. Find the maximum, if it exists.

57. Find the minimum, if it exists.

1.6 Graphs of Functions

In Exercises 58 - 73, use the graph of $y = f(x)$ given below to answer the question.

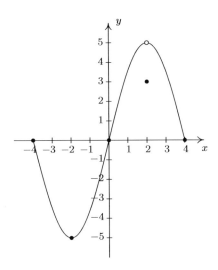

58. Find the domain of f.

59. Find the range of f.

60. Determine $f(2)$.

61. Solve $f(x) = -5$.

62. List the x-intercepts, if any exist.

63. List the y-intercepts, if any exist.

64. Find the zeros of f.

65. Solve $f(x) \leq 0$.

66. Find the number of solutions to $f(x) = 3$.

67. Does f appear to be even, odd, or neither?

68. List the intervals where f is increasing.

69. List the intervals where f is decreasing.

70. List the local maximums, if any exist.

71. List the local minimums, if any exist.

72. Find the maximum, if it exists.

73. Find the minimum, if it exists.

In Exercises 74 - 77, use your graphing calculator to approximate the local and absolute extrema of the given function. Approximate the intervals on which the function is increasing and those on which it is decreasing. Round your answers to two decimal places.

74. $f(x) = x^4 - 3x^3 - 24x^2 + 28x + 48$

75. $f(x) = x^{2/3}(x - 4)$

76. $f(x) = \sqrt{9 - x^2}$

77. $f(x) = x\sqrt{9 - x^2}$

In Exercises 78 - 85, use the graphs of $y = f(x)$ and $y = g(x)$ below to find the function value.

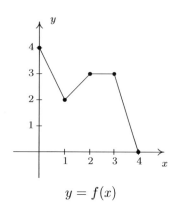

$y = f(x)$

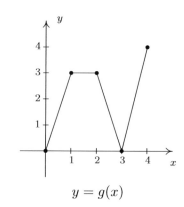

$y = g(x)$

78. $(f + g)(0)$

79. $(f + g)(1)$

80. $(f - g)(1)$

81. $(g - f)(2)$

82. $(fg)(2)$

83. $(fg)(1)$

84. $\left(\frac{f}{g}\right)(4)$

85. $\left(\frac{g}{f}\right)(2)$

The graph below represents the height h of a Sasquatch (in feet) as a function of its age N in years. Use it to answer the questions in Exercises 86 - 90.

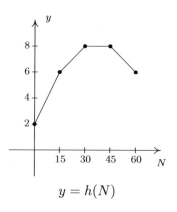

$y = h(N)$

86. Find and interpret $h(0)$.

87. How tall is the Sasquatch when she is 15 years old?

88. Solve $h(N) = 6$ and interpret.

89. List the interval over which h is constant and interpret your answer.

90. List the interval over which h is decreasing and interpret your answer.

1.6 GRAPHS OF FUNCTIONS

For Exercises 91 - 93, let $f(x) = \lfloor x \rfloor$ be the greatest integer function as defined in Exercise 75 in Section 1.4.

91. Graph $y = f(x)$. Be careful to correctly describe the behavior of the graph near the integers.

92. Is f even, odd, or neither? Explain.

93. Discuss with your classmates which points on the graph are local minimums, local maximums or both. Is f ever increasing? Decreasing? Constant?

In Exercises 94 - 95, use your graphing calculator to show that the given function does not have any extrema, neither local nor absolute.

94. $f(x) = x^3 + x - 12$
95. $f(x) = -5x + 2$

96. In Exercise 71 in Section 1.4, we saw that the population of Sasquatch in Portage County could be modeled by the function $P(t) = \dfrac{150t}{t+15}$, where $t = 0$ represents the year 1803. Use your graphing calculator to analyze the general function behavior of P. Will there ever be a time when 200 Sasquatch roam Portage County?

97. Suppose f and g are both even functions. What can be said about the functions $f+g$, $f-g$, fg and $\frac{f}{g}$? What if f and g are both odd? What if f is even but g is odd?

98. One of the most important aspects of the Cartesian Coordinate Plane is its ability to put Algebra into geometric terms and Geometry into algebraic terms. We've spent most of this chapter looking at this very phenomenon and now you should spend some time with your classmates reviewing what we've done. What major results do we have that tie Algebra and Geometry together? What concepts from Geometry have we not yet described algebraically? What topics from Intermediate Algebra have we not yet discussed geometrically?

It's now time to "thoroughly vet the pathologies induced" by the precise definitions of local maximum and local minimum. We'll do this by providing you and your classmates a series of Exercises to discuss. You will need to refer back to Definition 1.10 (Increasing, Decreasing and Constant) and Definition 1.11 (Maximum and Minimum) during the discussion.

99. Consider the graph of the function f given below.

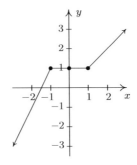

(a) Show that f has a local maximum but not a local minimum at the point $(-1, 1)$.

(b) Show that f has a local minimum but not a local maximum at the point $(1, 1)$.

(c) Show that f has a local maximum AND a local minimum at the point $(0, 1)$.

(d) Show that f is constant on the interval $[-1, 1]$ and thus has both a local maximum AND a local minimum at every point $(x, f(x))$ where $-1 < x < 1$.

100. Using Example 1.6.4 as a guide, show that the function g whose graph is given below does <u>not</u> have a local maximum at $(-3, 5)$ nor does it have a local minimum at $(3, -3)$. Find its extrema, both local and absolute. What's unique about the point $(0, -4)$ on this graph? Also find the intervals on which g is increasing and those on which g is decreasing.

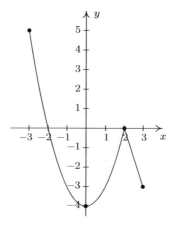

101. We said earlier in the section that it is not good enough to say local extrema exist where a function changes from increasing to decreasing or vice versa. As a previous exercise showed, we could have local extrema when a function is constant so now we need to examine some functions whose graphs do indeed change direction. Consider the functions graphed below. Notice that all four of them change direction at an open circle on the graph. Examine each for local extrema. What is the effect of placing the "dot" on the y-axis above or below the open circle? What could you say if no function value were assigned to $x = 0$?

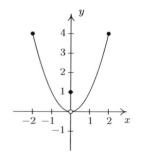

(a) Function I

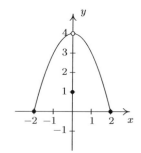

(b) Function II

1.6 GRAPHS OF FUNCTIONS

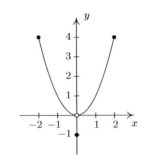

(c) Function III

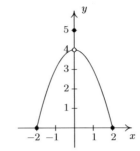

(d) Function IV

1.6.3 Answers

1. $f(x) = 2 - x$
   Domain: $(-\infty, \infty)$
   x-intercept: $(2, 0)$
   y-intercept: $(0, 2)$
   No symmetry

   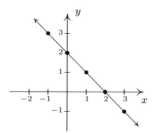

2. $f(x) = \dfrac{x-2}{3}$
   Domain: $(-\infty, \infty)$
   x-intercept: $(2, 0)$
   y-intercept: $\left(0, -\dfrac{2}{3}\right)$
   No symmetry

   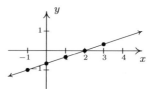

3. $f(x) = x^2 + 1$
   Domain: $(-\infty, \infty)$
   x-intercept: None
   y-intercept: $(0, 1)$
   Even

   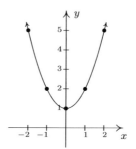

4. $f(x) = 4 - x^2$
   Domain: $(-\infty, \infty)$
   x-intercepts: $(-2, 0)$, $(2, 0)$
   y-intercept: $(0, 4)$
   Even

   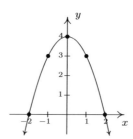

5. $f(x) = 2$
   Domain: $(-\infty, \infty)$
   x-intercept: None
   y-intercept: $(0, 2)$
   Even

   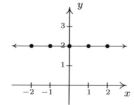

6. $f(x) = x^3$

   Domain: $(-\infty, \infty)$

   x-intercept: $(0, 0)$

   y-intercept: $(0, 0)$

   Odd

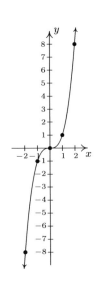

7. $f(x) = x(x-1)(x+2)$

   Domain: $(-\infty, \infty)$

   x-intercepts: $(-2, 0)$, $(0, 0)$, $(1, 0)$

   y-intercept: $(0, 0)$

   No symmetry

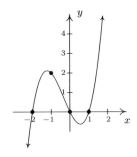

8. $f(x) = \sqrt{x-2}$

   Domain: $[2, \infty)$

   x-intercept: $(2, 0)$

   y-intercept: None

   No symmetry

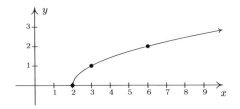

9. $f(x) = \sqrt{5-x}$

   Domain: $(-\infty, 5]$

   x-intercept: $(5, 0)$

   y-intercept: $(0, \sqrt{5})$

   No symmetry

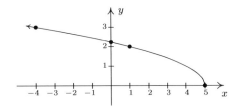

10. $f(x) = 3 - 2\sqrt{x+2}$
    Domain: $[-2, \infty)$
    x-intercept: $\left(\frac{1}{4}, 0\right)$
    y-intercept: $(0, 3 - 2\sqrt{2})$
    No symmetry

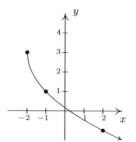

11. $f(x) = \sqrt[3]{x}$
    Domain: $(-\infty, \infty)$
    x-intercept: $(0, 0)$
    y-intercept: $(0, 0)$
    Odd

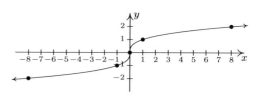

12. $f(x) = \dfrac{1}{x^2 + 1}$
    Domain: $(-\infty, \infty)$
    x-intercept: None
    y-intercept: $(0, 1)$
    Even

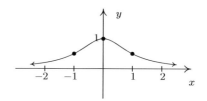

13.
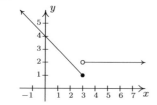

14.
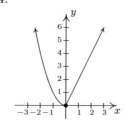

15.
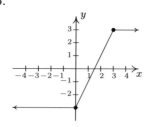

16.
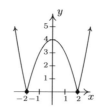

1.6 Graphs of Functions

17.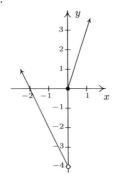

18.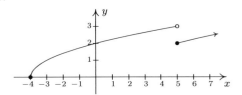

19.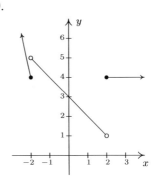

20.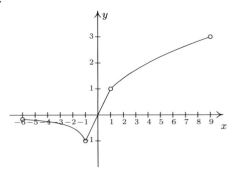

21. odd

22. neither

23. even

24. even

25. even

26. neither

27. odd

28. odd

29. even

30. neither

31. neither

32. even

33. even **and** odd

34. odd

35. even

36. even

37. neither

38. odd

39. odd

40. even

41. even

42. $[-5, 3]$

43. $[-5, 4]$

44. $f(-2) = 2$

45. $x = -3$

46. $(-4, 0), (-1, 0), (1, 0)$

47. $(0, -1)$

48. $-4, -1, 1$

49. $[-4, -1] \cup [1, 3]$

50. 4

51. neither   52. $[-5, -3]$, $[0, 2]$ 53. $[-3, 0]$, $[2, 3]$

54. $f(-3) = 4$, $f(2) = 3$ 55. $f(0) = -1$

56. $f(-3) = 4$ 57. $f(-5) = -5$

58. $[-4, 4]$ 59. $[-5, 5)$ 60. $f(2) = 3$

61. $x = -2$ 62. $(-4, 0), (0, 0), (4, 0)$ 63. $(0, 0)$

64. $-4, 0, 4$ 65. $[-4, 0] \cup \{4\}$ 66. 3

67. neither   68. $(-2, 2)$ 69. $[-4, -2]$, $(2, 4]$

70. none   71. $f(-2) = -5$, $f(2) = 3$

72. none   73. $f(-2) = -5$

74. No absolute maximum
    Absolute minimum $f(4.55) \approx -175.46$
    Local minimum at $(-2.84, -91.32)$
    Local maximum at $(0.54, 55.73)$
    Local minimum at $(4.55, -175.46)$
    Increasing on $[-2.84, 0.54], [4.55, \infty)$
    Decreasing on $(-\infty, -2.84], [0.54, 4.55]$

75. No absolute maximum
    No absolute minimum
    Local maximum at $(0, 0)$
    Local minimum at $(1.60, -3.28)$
    Increasing on $(-\infty, 0], [1.60, \infty)$
    Decreasing on $[0, 1.60]$

76. Absolute maximum $f(0) = 3$
    Absolute minimum $f(\pm 3) = 0$
    Local maximum at $(0, 3)$
    No local minimum
    Increasing on $[-3, 0]$
    Decreasing on $[0, 3]$

77. Absolute maximum $f(2.12) \approx 4.50$
    Absolute minimum $f(-2.12) \approx -4.50$
    Local maximum $(2.12, 4.50)$
    Local minimum $(-2.12, -4.50)$
    Increasing on $[-2.12, 2.12]$
    Decreasing on $[-3, -2.12], [2.12, 3]$

78. $(f + g)(0) = 4$ 79. $(f + g)(1) = 5$ 80. $(f - g)(1) = -1$ 81. $(g - f)(2) = 0$

82. $(fg)(2) = 9$ 83. $(fg)(1) = 6$ 84. $\left(\frac{f}{g}\right)(4) = 0$ 85. $\left(\frac{g}{f}\right)(2) = 1$

86. $h(0) = 2$, so the Sasquatch is 2 feet tall at birth.

87. $h(15) = 6$, so the Saquatch is 6 feet tall when she is 15 years old.

88. $h(N) = 6$ when $N = 15$ and $N = 60$. This means the Sasquatch is 6 feet tall when she is 15 and 60 years old.

89. h is constant on $[30, 45]$. This means the Sasquatch's height is constant (at 8 feet) for these years.

1.6 Graphs of Functions

90. h is decreasing on $[45, 60]$. This means the Sasquatch is getting shorter from the age of 45 to the age of 60. (Sasquatchteoporosis, perhaps?)

91.

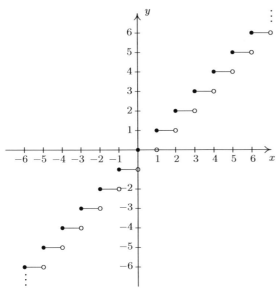

The graph of $f(x) = \lfloor x \rfloor$.

92. Note that $f(1.1) = 1$, but $f(-1.1) = -2$, so f is neither even nor odd.

1.7 Transformations

In this section, we study how the graphs of functions change, or **transform**, when certain specialized modifications are made to their formulas. The transformations we will study fall into three broad categories: shifts, reflections and scalings, and we will present them in that order. Suppose the graph below is the complete graph of a function f.

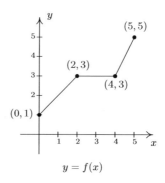

$y = f(x)$

The Fundamental Graphing Principle for Functions says that for a point (a, b) to be on the graph, $f(a) = b$. In particular, we know $f(0) = 1$, $f(2) = 3$, $f(4) = 3$ and $f(5) = 5$. Suppose we wanted to graph the function defined by the formula $g(x) = f(x) + 2$. Let's take a minute to remind ourselves of what g is doing. We start with an input x to the function f and we obtain the output $f(x)$. The function g takes the output $f(x)$ and adds 2 to it. In order to graph g, we need to graph the points $(x, g(x))$. How are we to find the values for $g(x)$ without a formula for $f(x)$? The answer is that we don't need a *formula* for $f(x)$, we just need the *values* of $f(x)$. The values of $f(x)$ are the y values on the graph of $y = f(x)$. For example, using the points indicated on the graph of f, we can make the following table.

x	$(x, f(x))$	$f(x)$	$g(x) = f(x) + 2$	$(x, g(x))$
0	$(0, 1)$	1	3	$(0, 3)$
2	$(2, 3)$	3	5	$(2, 5)$
4	$(4, 3)$	3	5	$(4, 5)$
5	$(5, 5)$	5	7	$(5, 7)$

In general, if (a, b) is on the graph of $y = f(x)$, then $f(a) = b$, so $g(a) = f(a) + 2 = b + 2$. Hence, $(a, b+2)$ is on the graph of g. In other words, to obtain the graph of g, we add 2 to the y-coordinate of each point on the graph of f. Geometrically, adding 2 to the y-coordinate of a point moves the point 2 units above its previous location. Adding 2 to every y-coordinate on a graph *en masse* is usually described as 'shifting the graph up 2 units'. Notice that the graph retains the same basic shape as before, it is just 2 units above its original location. In other words, we connect the four points we moved in the same manner in which they were connected before. We have the results side-by-side at the top of the next page.

1.7 Transformations 121

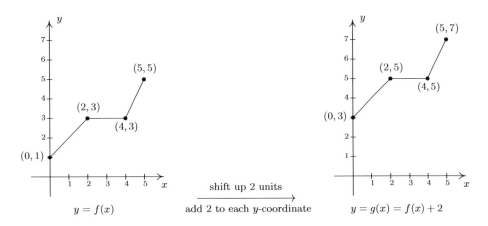

You'll note that the domain of f and the domain of g are the same, namely $[0,5]$, but that the range of f is $[1,5]$ while the range of g is $[3,7]$. In general, shifting a function vertically like this will leave the domain unchanged, but could very well affect the range. You can easily imagine what would happen if we wanted to graph the function $j(x) = f(x) - 2$. Instead of adding 2 to each of the y-coordinates on the graph of f, we'd be subtracting 2. Geometrically, we would be moving the graph down 2 units. We leave it to the reader to verify that the domain of j is the same as f, but the range of j is $[-1,3]$. What we have discussed is generalized in the following theorem.

> **Theorem 1.2. Vertical Shifts.** Suppose f is a function and k is a positive number.
>
> - To graph $y = f(x) + k$, shift the graph of $y = f(x)$ up k units by adding k to the y-coordinates of the points on the graph of f.
>
> - To graph $y = f(x) - k$, shift the graph of $y = f(x)$ down k units by subtracting k from the y-coordinates of the points on the graph of f.

The key to understanding Theorem 1.2 and, indeed, all of the theorems in this section comes from an understanding of the Fundamental Graphing Principle for Functions. If (a,b) is on the graph of f, then $f(a) = b$. Substituting $x = a$ into the equation $y = f(x) + k$ gives $y = f(a) + k = b + k$. Hence, $(a, b+k)$ is on the graph of $y = f(x) + k$, and we have the result. In the language of 'inputs' and 'outputs', Theorem 1.2 can be paraphrased as "Adding to, or subtracting from, the *output* of a function causes the graph to shift up or down, respectively." So what happens if we add to or subtract from the *input* of the function?

Keeping with the graph of $y = f(x)$ above, suppose we wanted to graph $g(x) = f(x+2)$. In other words, we are looking to see what happens when we add 2 to the input of the function.[1] Let's try to generate a table of values of g based on those we know for f. We quickly find that we run into some difficulties.

---
[1] We have spent a lot of time in this text showing you that $f(x+2)$ and $f(x)+2$ are, in general, wildly different algebraic animals. We will see momentarily that their geometry is also dramatically different.

x	$(x, f(x))$	$f(x)$	$g(x) = f(x+2)$	$(x, g(x))$
0	(0, 1)	1	$f(0+2) = f(2) = 3$	(0, 3)
2	(2, 3)	3	$f(2+2) = f(4) = 3$	(2, 3)
4	(4, 3)	3	$f(4+2) = f(6) = ?$	
5	(5, 5)	5	$f(5+2) = f(7) = ?$	

When we substitute $x = 4$ into the formula $g(x) = f(x+2)$, we are asked to find $f(4+2) = f(6)$ which doesn't exist because the domain of f is only $[0, 5]$. The same thing happens when we attempt to find $g(5)$. What we need here is a new strategy. We know, for instance, $f(0) = 1$. To determine the corresponding point on the graph of g, we need to figure out what value of x we must substitute into $g(x) = f(x+2)$ so that the quantity $x + 2$, works out to be 0. Solving $x + 2 = 0$ gives $x = -2$, and $g(-2) = f((-2) + 2) = f(0) = 1$ so $(-2, 1)$ is on the graph of g. To use the fact $f(2) = 3$, we set $x + 2 = 2$ to get $x = 0$. Substituting gives $g(0) = f(0+2) = f(2) = 3$. Continuing in this fashion, we get

x	$x+2$	$g(x) = f(x+2)$	$(x, g(x))$
-2	0	$g(-2) = f(0) = 1$	$(-2, 1)$
0	2	$g(0) = f(2) = 3$	(0, 3)
2	4	$g(2) = f(4) = 3$	(2, 3)
3	5	$g(3) = f(5) = 5$	(3, 5)

In summary, the points $(0, 1)$, $(2, 3)$, $(4, 3)$ and $(5, 5)$ on the graph of $y = f(x)$ give rise to the points $(-2, 1)$, $(0, 3)$, $(2, 3)$ and $(3, 5)$ on the graph of $y = g(x)$, respectively. In general, if (a, b) is on the graph of $y = f(x)$, then $f(a) = b$. Solving $x + 2 = a$ gives $x = a - 2$ so that $g(a - 2) = f((a - 2) + 2) = f(a) = b$. As such, $(a - 2, b)$ is on the graph of $y = g(x)$. The point $(a - 2, b)$ is exactly 2 units to the *left* of the point (a, b) so the graph of $y = g(x)$ is obtained by shifting the graph $y = f(x)$ to the left 2 units, as pictured below.

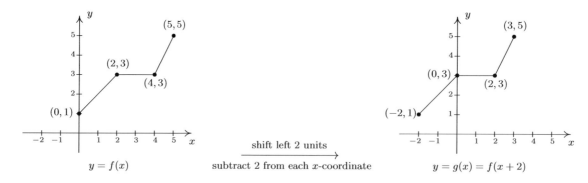

Note that while the ranges of f and g are the same, the domain of g is $[-2, 3]$ whereas the domain of f is $[0, 5]$. In general, when we shift the graph horizontally, the range will remain the same, but the domain could change. If we set out to graph $j(x) = f(x - 2)$, we would find ourselves *adding*

1.7 Transformations

2 to all of the x values of the points on the graph of $y = f(x)$ to effect a shift to the *right* 2 units. Generalizing these notions produces the following result.

Theorem 1.3. Horizontal Shifts. Suppose f is a function and h is a positive number.

- To graph $y = f(x+h)$, shift the graph of $y = f(x)$ left h units by subtracting h from the x-coordinates of the points on the graph of f.

- To graph $y = f(x-h)$, shift the graph of $y = f(x)$ right h units by adding h to the x-coordinates of the points on the graph of f.

In other words, Theorem 1.3 says that adding to or subtracting from the *input* to a function amounts to shifting the graph left or right, respectively. Theorems 1.2 and 1.3 present a theme which will run common throughout the section: changes to the outputs from a function affect the y-coordinates of the graph, resulting in some kind of vertical change; changes to the inputs to a function affect the x-coordinates of the graph, resulting in some kind of horizontal change.

Example 1.7.1.

1. Graph $f(x) = \sqrt{x}$. Plot at least three points.

2. Use your graph in 1 to graph $g(x) = \sqrt{x} - 1$.

3. Use your graph in 1 to graph $j(x) = \sqrt{x-1}$.

4. Use your graph in 1 to graph $m(x) = \sqrt{x+3} - 2$.

Solution.

1. Owing to the square root, the domain of f is $x \geq 0$, or $[0, \infty)$. We choose perfect squares to build our table and graph below. From the graph we verify the domain of f is $[0, \infty)$ and the range of f is also $[0, \infty)$.

x	$f(x)$	$(x, f(x))$
0	0	$(0,0)$
1	1	$(1,1)$
4	2	$(4,2)$

   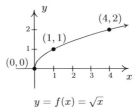

   $y = f(x) = \sqrt{x}$

2. The domain of g is the same as the domain of f, since the only condition on both functions is that $x \geq 0$. If we compare the formula for $g(x)$ with $f(x)$, we see that $g(x) = f(x) - 1$. In other words, we have subtracted 1 from the output of the function f. By Theorem 1.2, we know that in order to graph g, we shift the graph of f down one unit by subtracting 1 from each of the y-coordinates of the points on the graph of f. Applying this to the three points we have specified on the graph, we move $(0, 0)$ to $(0, -1)$, $(1, 1)$ to $(1, 0)$, and $(4, 2)$ to $(4, 1)$.

The rest of the points follow suit, and we connect them with the same basic shape as before. We confirm the domain of g is $[0, \infty)$ and find the range of g to be $[-1, \infty)$.

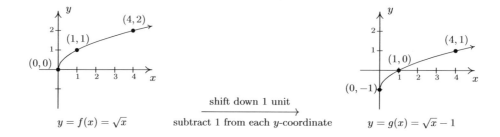

3. Solving $x - 1 \geq 0$ gives $x \geq 1$, so the domain of j is $[1, \infty)$. To graph j, we note that $j(x) = f(x - 1)$. In other words, we are subtracting 1 from the *input* of f. According to Theorem 1.3, this induces a shift to the right of the graph of f. We add 1 to the x-coordinates of the points on the graph of f and get the result below. The graph reaffirms that the domain of j is $[1, \infty)$ and tells us that the range of j is $[0, \infty)$.

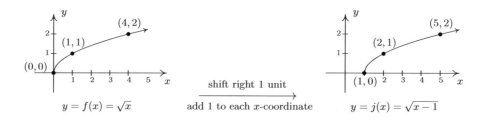

4. To find the domain of m, we solve $x + 3 \geq 0$ and get $[-3, \infty)$. Comparing the formulas of $f(x)$ and $m(x)$, we have $m(x) = f(x + 3) - 2$. We have 3 being added to an input, indicating a horizontal shift, and 2 being subtracted from an output, indicating a vertical shift. We leave it to the reader to verify that, in this particular case, the order in which we perform these transformations is immaterial; we will arrive at the same graph regardless as to which transformation we apply first.[2] We follow the convention 'inputs first',[3] and to that end we first tackle the horizontal shift. Letting $m_1(x) = f(x + 3)$ denote this intermediate step, Theorem 1.3 tells us that the graph of $y = m_1(x)$ is the graph of f shifted to the left 3 units. Hence, we subtract 3 from each of the x-coordinates of the points on the graph of f.

---

[2]We shall see in the next example that order is generally important when applying more than one transformation to a graph.
[3]We could equally have chosen the convention 'outputs first'.

1.7 Transformations

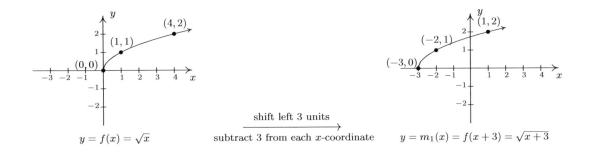

Since $m(x) = f(x+3) - 2$ and $f(x+3) = m_1(x)$, we have $m(x) = m_1(x) - 2$. We can apply Theorem 1.2 and obtain the graph of m by subtracting 2 from the y-coordinates of each of the points on the graph of $m_1(x)$. The graph verifies that the domain of m is $[-3, \infty)$ and we find the range of m to be $[-2, \infty)$.

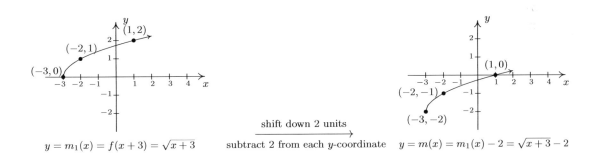

Keep in mind that we can check our answer to any of these kinds of problems by showing that any of the points we've moved lie on the graph of our final answer. For example, we can check that $(-3, -2)$ is on the graph of m by computing $m(-3) = \sqrt{(-3)+3} - 2 = \sqrt{0} - 2 = -2$ ✓ □

We now turn our attention to reflections. We know from Section 1.1 that to reflect a point (x, y) across the x-axis, we replace y with $-y$. If (x, y) is on the graph of f, then $y = f(x)$, so replacing y with $-y$ is the same as replacing $f(x)$ with $-f(x)$. Hence, the graph of $y = -f(x)$ is the graph of f reflected across the x-axis. Similarly, the graph of $y = f(-x)$ is the graph of f reflected across the y-axis. Returning to the language of inputs and outputs, multiplying the output from a function by -1 reflects its graph across the x-axis, while multiplying the input to a function by -1 reflects the graph across the y-axis.[4]

---

[4]The expressions $-f(x)$ and $f(-x)$ should look familiar - they are the quantities we used in Section 1.6 to test if a function was even, odd or neither. The interested reader is invited to explore the role of reflections and symmetry of functions. What happens if you reflect an even function across the y-axis? What happens if you reflect an odd function across the y-axis? What about the x-axis?

> **Theorem 1.4. Reflections.** Suppose f is a function.
>
> - To graph $y = -f(x)$, reflect the graph of $y = f(x)$ across the x-axis by multiplying the y-coordinates of the points on the graph of f by -1.
>
> - To graph $y = f(-x)$, reflect the graph of $y = f(x)$ across the y-axis by multiplying the x-coordinates of the points on the graph of f by -1.

Applying Theorem 1.4 to the graph of $y = f(x)$ given at the beginning of the section, we can graph $y = -f(x)$ by reflecting the graph of f about the x-axis

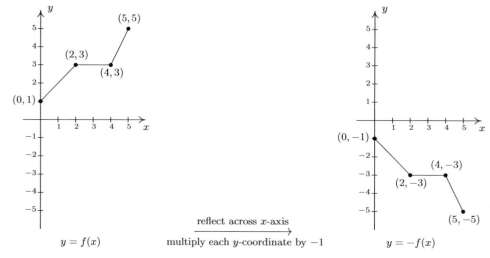

By reflecting the graph of f across the y-axis, we obtain the graph of $y = f(-x)$.

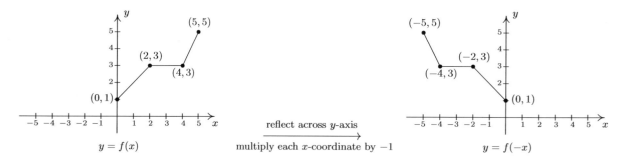

With the addition of reflections, it is now more important than ever to consider the order of transformations, as the next example illustrates.

Example 1.7.2. Let $f(x) = \sqrt{x}$. Use the graph of f from Example 1.7.1 to graph the following functions. Also, state their domains and ranges.

1. $g(x) = \sqrt{-x}$
2. $j(x) = \sqrt{3-x}$
3. $m(x) = 3 - \sqrt{x}$

1.7 Transformations

Solution.

1. The mere sight of $\sqrt{-x}$ usually causes alarm, if not panic. When we discussed domains in Section 1.4, we clearly banished negatives from the radicands of even roots. However, we must remember that x is a variable, and as such, the quantity $-x$ isn't always negative. For example, if $x = -4$, $-x = 4$, thus $\sqrt{-x} = \sqrt{-(-4)} = 2$ is perfectly well-defined. To find the domain analytically, we set $-x \geq 0$ which gives $x \leq 0$, so that the domain of g is $(-\infty, 0]$. Since $g(x) = f(-x)$, Theorem 1.4 tells us that the graph of g is the reflection of the graph of f across the y-axis. We accomplish this by multiplying each x-coordinate on the graph of f by -1, so that the points $(0, 0)$, $(1, 1)$, and $(4, 2)$ move to $(0, 0)$, $(-1, 1)$, and $(-4, 2)$, respectively. Graphically, we see that the domain of g is $(-\infty, 0]$ and the range of g is the same as the range of f, namely $[0, \infty)$.

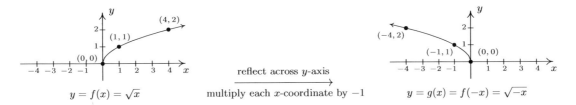

2. To determine the domain of $j(x) = \sqrt{3 - x}$, we solve $3 - x \geq 0$ and get $x \leq 3$, or $(-\infty, 3]$. To determine which transformations we need to apply to the graph of f to obtain the graph of j, we rewrite $j(x) = \sqrt{-x + 3} = f(-x + 3)$. Comparing this formula with $f(x) = \sqrt{x}$, we see that not only are we multiplying the input x by -1, which results in a reflection across the y-axis, but also we are adding 3, which indicates a horizontal shift to the left. Does it matter in which order we do the transformations? If so, which order is the correct order? Let's consider the point $(4, 2)$ on the graph of f. We refer to the discussion leading up to Theorem 1.3. We know $f(4) = 2$ and wish to find the point on $y = j(x) = f(-x + 3)$ which corresponds to $(4, 2)$. We set $-x + 3 = 4$ and solve. Our first step is to subtract 3 from both sides to get $-x = 1$. Subtracting 3 from the x-coordinate 4 is shifting the point $(4, 2)$ to the left. From $-x = 1$, we then multiply[5] both sides by -1 to get $x = -1$. Multiplying the x-coordinate by -1 corresponds to reflecting the point about the y-axis. Hence, we perform the horizontal shift first, then follow it with the reflection about the y-axis. Starting with $f(x) = \sqrt{x}$, we let $j_1(x)$ be the intermediate function which shifts the graph of f 3 units to the left, $j_1(x) = f(x + 3)$.

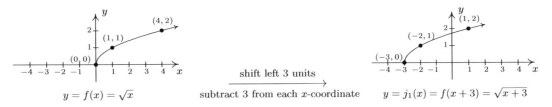

---
[5] Or divide - it amounts to the same thing.

To obtain the function j, we reflect the graph of j_1 about y-axis. Theorem 1.4 tells us we have $j(x) = j_1(-x)$. Putting it all together, we have $j(x) = j_1(-x) = f(-x+3) = \sqrt{-x+3}$, which is what we want.[6] From the graph, we confirm the domain of j is $(-\infty, 3]$ and we get that the range is $[0, \infty)$.

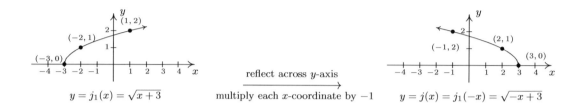

3. The domain of m works out to be the domain of f, $[0, \infty)$. Rewriting $m(x) = -\sqrt{x} + 3$, we see $m(x) = -f(x) + 3$. Since we are multiplying the output of f by -1 and then adding 3, we once again have two transformations to deal with: a reflection across the x-axis and a vertical shift. To determine the correct order in which to apply the transformations, we imagine trying to determine the point on the graph of m which corresponds to $(4, 2)$ on the graph of f. Since in the formula for $m(x)$, the input to f is just x, we substitute to find $m(4) = -f(4) + 3 = -2 + 3 = 1$. Hence, $(4, 1)$ is the corresponding point on the graph of m. If we closely examine the arithmetic, we see that we first multiply $f(4)$ by -1, which corresponds to the reflection across the x-axis, and then we add 3, which corresponds to the vertical shift. If we define an intermediate function $m_1(x) = -f(x)$ to take care of the reflection, we get

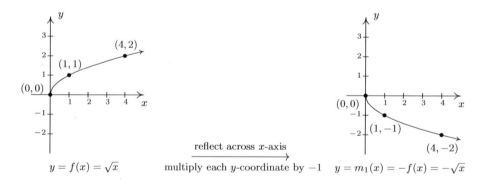

To shift the graph of m_1 up 3 units, we set $m(x) = m_1(x) + 3$. Since $m_1(x) = -f(x)$, when we put it all together, we get $m(x) = m_1(x) + 3 = -f(x) + 3 = -\sqrt{x} + 3$. We see from the graph that the range of m is $(-\infty, 3]$.

---

[6]If we had done the reflection first, then $j_1(x) = f(-x)$. Following this by a shift left would give us $j(x) = j_1(x+3) = f(-(x+3)) = f(-x-3) = \sqrt{-x-3}$ which isn't what we want. However, if we did the reflection first and followed it by a shift to the right 3 units, we would have arrived at the function $j(x)$. We leave it to the reader to verify the details.

1.7 Transformations

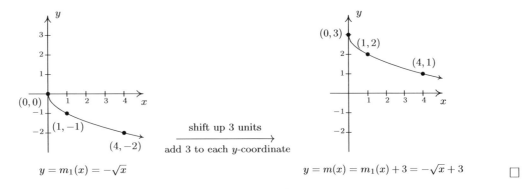

We now turn our attention to our last class of transformations known as **scalings**. A thorough discussion of scalings can get complicated because they are not as straight-forward as the previous transformations. A quick review of what we've covered so far, namely vertical shifts, horizontal shifts and reflections, will show you why those transformations are known as **rigid transformations**. Simply put, they do not change the *shape* of the graph, only its position and orientation in the plane. If, however, we wanted to make a new graph twice as tall as a given graph, or one-third as wide, we would be changing the shape of the graph. This type of transformation is called **non-rigid** for obvious reasons. Not only will it be important for us to differentiate between modifying inputs versus outputs, we must also pay close attention to the magnitude of the changes we make. As you will see shortly, the Mathematics turns out to be easier than the associated grammar.

Suppose we wish to graph the function $g(x) = 2f(x)$ where $f(x)$ is the function whose graph is given at the beginning of the section. From its graph, we can build a table of values for g as before.

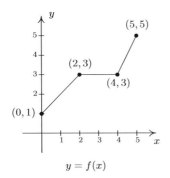

x	$(x, f(x))$	$f(x)$	$g(x) = 2f(x)$	$(x, g(x))$
0	$(0, 1)$	1	2	$(0, 2)$
2	$(2, 3)$	3	6	$(2, 6)$
4	$(4, 3)$	3	6	$(4, 6)$
5	$(5, 5)$	5	10	$(5, 10)$

In general, if (a, b) is on the graph of f, then $f(a) = b$ so that $g(a) = 2f(a) = 2b$ puts $(a, 2b)$ on the graph of g. In other words, to obtain the graph of g, we multiply all of the y-coordinates of the points on the graph of f by 2. Multiplying all of the y-coordinates of all of the points on the graph of f by 2 causes what is known as a 'vertical scaling[7] by a factor of 2', and the results are given on the next page.

---

[7] Also called a 'vertical stretching', 'vertical expansion' or 'vertical dilation' by a factor of 2.

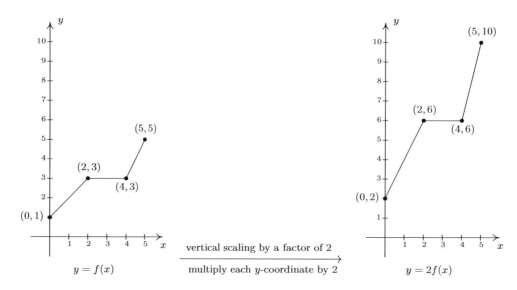

If we wish to graph $y = \frac{1}{2}f(x)$, we multiply the all of the y-coordinates of the points on the graph of f by $\frac{1}{2}$. This creates a 'vertical scaling[8] by a factor of $\frac{1}{2}$' as seen below.

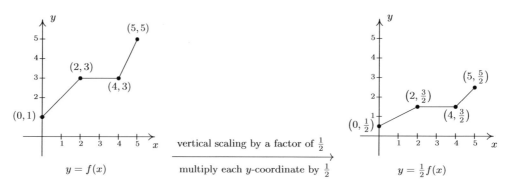

These results are generalized in the following theorem.

Theorem 1.5. Vertical Scalings. Suppose f is a function and $a > 0$. To graph $y = af(x)$, multiply all of the y-coordinates of the points on the graph of f by a. We say the graph of f has been vertically scaled by a factor of a.

- If $a > 1$, we say the graph of f has undergone a vertical stretching (expansion, dilation) by a factor of a.

- If $0 < a < 1$, we say the graph of f has undergone a vertical shrinking (compression, contraction) by a factor of $\frac{1}{a}$.

---

[8] Also called 'vertical shrinking', 'vertical compression' or 'vertical contraction' by a factor of 2.

1.7 Transformations

A few remarks about Theorem 1.5 are in order. First, a note about the verbiage. To the authors, the words 'stretching', 'expansion', and 'dilation' all indicate something getting bigger. Hence, 'stretched by a factor of 2' makes sense if we are scaling something by multiplying it by 2. Similarly, we believe words like 'shrinking', 'compression' and 'contraction' all indicate something getting smaller, so if we scale something by a factor of $\frac{1}{2}$, we would say it 'shrinks by a factor of 2' - not 'shrinks by a factor of $\frac{1}{2}$'. This is why we have written the descriptions 'stretching by a factor of a' and 'shrinking by a factor of $\frac{1}{a}$' in the statement of the theorem. Second, in terms of inputs and outputs, Theorem 1.5 says multiplying the *outputs* from a function by positive number a causes the graph to be vertically scaled by a factor of a. It is natural to ask what would happen if we multiply the *inputs* of a function by a positive number. This leads us to our last transformation of the section.

Referring to the graph of f given at the beginning of this section, suppose we want to graph $g(x) = f(2x)$. In other words, we are looking to see what effect multiplying the inputs to f by 2 has on its graph. If we attempt to build a table directly, we quickly run into the same problem we had in our discussion leading up to Theorem 1.3, as seen in the table on the left below. We solve this problem in the same way we solved this problem before. For example, if we want to determine the point on g which corresponds to the point $(2,3)$ on the graph of f, we set $2x = 2$ so that $x = 1$. Substituting $x = 1$ into $g(x)$, we obtain $g(1) = f(2 \cdot 1) = f(2) = 3$, so that $(1,3)$ is on the graph of g. Continuing in this fashion, we obtain the table on the lower right.

x	$(x, f(x))$	$f(x)$	$g(x) = f(2x)$	$(x, g(x))$	x	$2x$	$g(x) = f(2x)$	$(x, g(x))$
0	$(0,1)$	1	$f(2 \cdot 0) = f(0) = 1$	$(0,1)$	0	0	$g(0) = f(0) = 1$	$(0,0)$
2	$(2,3)$	3	$f(2 \cdot 2) = f(4) = 3$	$(2,3)$	1	2	$g(1) = f(2) = 3$	$(1,3)$
4	$(4,3)$	3	$f(2 \cdot 4) = f(8) = ?$		2	4	$g(2) = f(4) = 3$	$(2,3)$
5	$(5,5)$	5	$f(2 \cdot 5) = f(10) = ?$		$\frac{5}{2}$	5	$g\left(\frac{5}{2}\right) = f(5) = 5$	$\left(\frac{5}{2}, 5\right)$

In general, if (a, b) is on the graph of f, then $f(a) = b$. Hence $g\left(\frac{a}{2}\right) = f\left(2 \cdot \frac{a}{2}\right) = f(a) = b$ so that $\left(\frac{a}{2}, b\right)$ is on the graph of g. In other words, to graph g we divide the x-coordinates of the points on the graph of f by 2. This results in a horizontal scaling[9] by a factor of $\frac{1}{2}$.

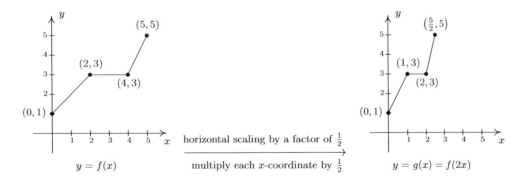

---

[9] Also called 'horizontal shrinking', 'horizontal compression' or 'horizontal contraction' by a factor of 2.

If, on the other hand, we wish to graph $y = f\left(\frac{1}{2}x\right)$, we end up multiplying the x-coordinates of the points on the graph of f by 2 which results in a horizontal scaling[10] by a factor of 2, as demonstrated below.

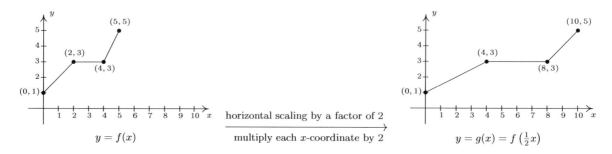

We have the following theorem.

Theorem 1.6. Horizontal Scalings. Suppose f is a function and $b > 0$. To graph $y = f(bx)$, divide all of the x-coordinates of the points on the graph of f by b. We say the graph of f has been horizontally scaled by a factor of $\frac{1}{b}$.

- If $0 < b < 1$, we say the graph of f has undergone a horizontal stretching (expansion, dilation) by a factor of $\frac{1}{b}$.

- If $b > 1$, we say the graph of f has undergone a horizontal shrinking (compression, contraction) by a factor of b.

Theorem 1.6 tells us that if we multiply the input to a function by b, the resulting graph is scaled horizontally by a factor of $\frac{1}{b}$ since the x-values are divided by b to produce corresponding points on the graph of $y = f(bx)$. The next example explores how vertical and horizontal scalings sometimes interact with each other and with the other transformations introduced in this section.

Example 1.7.3. Let $f(x) = \sqrt{x}$. Use the graph of f from Example 1.7.1 to graph the following functions. Also, state their domains and ranges.

1. $g(x) = 3\sqrt{x}$
2. $j(x) = \sqrt{9x}$
3. $m(x) = 1 - \sqrt{\frac{x+3}{2}}$

Solution.

1. First we note that the domain of g is $[0, \infty)$ for the usual reason. Next, we have $g(x) = 3f(x)$ so by Theorem 1.5, we obtain the graph of g by multiplying all of the y-coordinates of the points on the graph of f by 3. The result is a vertical scaling of the graph of f by a factor of 3. We find the range of g is also $[0, \infty)$.

---
[10]Also called 'horizontal stretching', 'horizontal expansion' or 'horizontal dilation' by a factor of 2.

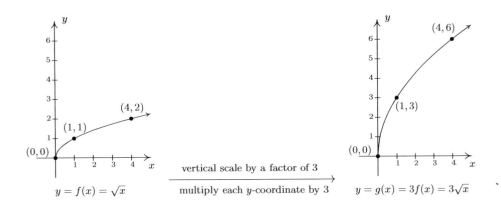

2. To determine the domain of j, we solve $9x \geq 0$ to find $x \geq 0$. Our domain is once again $[0,\infty)$. We recognize $j(x) = f(9x)$ and by Theorem 1.6, we obtain the graph of j by dividing the x-coordinates of the points on the graph of f by 9. From the graph, we see the range of j is also $[0,\infty)$.

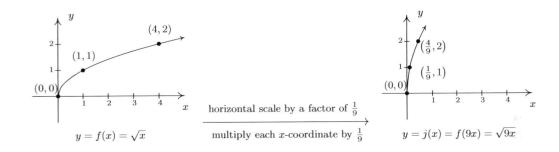

3. Solving $\frac{x+3}{2} \geq 0$ gives $x \geq -3$, so the domain of m is $[-3,\infty)$. To take advantage of what we know of transformations, we rewrite $m(x) = -\sqrt{\frac{1}{2}x + \frac{3}{2}} + 1$, or $m(x) = -f\left(\frac{1}{2}x + \frac{3}{2}\right) + 1$. Focusing on the inputs first, we note that the input to f in the formula for $m(x)$ is $\frac{1}{2}x + \frac{3}{2}$. Multiplying the x by $\frac{1}{2}$ corresponds to a horizontal stretching by a factor of 2, and adding the $\frac{3}{2}$ corresponds to a shift to the left by $\frac{3}{2}$. As before, we resolve which to perform first by thinking about how we would find the point on m corresponding to a point on f, in this case, $(4,2)$. To use $f(4) = 2$, we solve $\frac{1}{2}x + \frac{3}{2} = 4$. Our first step is to subtract the $\frac{3}{2}$ (the horizontal shift) to obtain $\frac{1}{2}x = \frac{5}{2}$. Next, we multiply by 2 (the horizontal stretching) and obtain $x = 5$. We define two intermediate functions to handle first the shift, then the stretching. In accordance with Theorem 1.3, $m_1(x) = f\left(x + \frac{3}{2}\right) = \sqrt{x + \frac{3}{2}}$ will shift the graph of f to the left $\frac{3}{2}$ units.

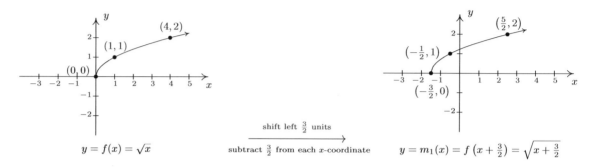

Next, $m_2(x) = m_1\left(\frac{1}{2}x\right) = \sqrt{\frac{1}{2}x + \frac{3}{2}}$ will, according to Theorem 1.6, horizontally stretch the graph of m_1 by a factor of 2.

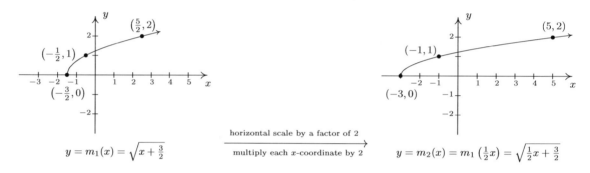

We now examine what's happening to the outputs. From $m(x) = -f\left(\frac{1}{2}x + \frac{3}{2}\right) + 1$, we see that the output from f is being multiplied by -1 (a reflection about the x-axis) and then a 1 is added (a vertical shift up 1). As before, we can determine the correct order by looking at how the point $(4, 2)$ is moved. We already know that to make use of the equation $f(4) = 2$, we need to substitute $x = 5$. We get $m(5) = -f\left(\frac{1}{2}(5) + \frac{3}{2}\right) + 1 = -f(4) + 1 = -2 + 1 = -1$. We see that $f(4)$ (the output from f) is first multiplied by -1 then the 1 is added meaning we first reflect the graph about the x-axis then shift up 1. Theorem 1.4 tells us $m_3(x) = -m_2(x)$ will handle the reflection.

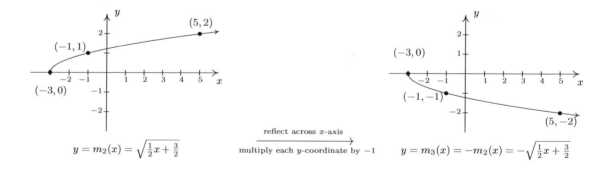

1.7 Transformations

Finally, to handle the vertical shift, Theorem 1.2 gives $m(x) = m_3(x) + 1$, and we see that the range of m is $(-\infty, 1]$.

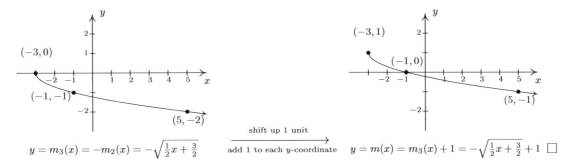

Some comments about Example 1.7.3 are in order. First, recalling the properties of radicals from Intermediate Algebra, we know that the functions g and j are the same, since j and g have the same domains and $j(x) = \sqrt{9x} = \sqrt{9}\sqrt{x} = 3\sqrt{x} = g(x)$. (We invite the reader to verify that all of the points we plotted on the graph of g lie on the graph of j and vice-versa.) Hence, for $f(x) = \sqrt{x}$, a vertical stretch by a factor of 3 and a horizontal shrinking by a factor of 9 result in the same transformation. While this kind of phenomenon is not universal, it happens commonly enough with some of the families of functions studied in College Algebra that it is worthy of note. Secondly, to graph the function m, we applied a series of four transformations. While it would have been easier on the authors to simply inform the reader of which steps to take, we have strived to explain why the order in which the transformations were applied made sense. We generalize the procedure in the theorem below.

Theorem 1.7. Transformations. Suppose f is a function. If $A \neq 0$ and $B \neq 0$, then to graph

$$g(x) = Af(Bx + H) + K$$

1. Subtract H from each of the x-coordinates of the points on the graph of f. This results in a horizontal shift to the left if $H > 0$ or right if $H < 0$.

2. Divide the x-coordinates of the points on the graph obtained in Step 1 by B. This results in a horizontal scaling, but may also include a reflection about the y-axis if $B < 0$.

3. Multiply the y-coordinates of the points on the graph obtained in Step 2 by A. This results in a vertical scaling, but may also include a reflection about the x-axis if $A < 0$.

4. Add K to each of the y-coordinates of the points on the graph obtained in Step 3. This results in a vertical shift up if $K > 0$ or down if $K < 0$.

Theorem 1.7 can be established by generalizing the techniques developed in this section. Suppose (a, b) is on the graph of f. Then $f(a) = b$, and to make good use of this fact, we set $Bx + H = a$ and solve. We first subtract the H (causing the horizontal shift) and then divide by B. If B

is a positive number, this induces only a horizontal scaling by a factor of $\frac{1}{B}$. If $B < 0$, then we have a factor of -1 in play, and dividing by it induces a reflection about the y-axis. So we have $x = \frac{a-H}{B}$ as the input to g which corresponds to the input $x = a$ to f. We now evaluate $g\left(\frac{a-H}{B}\right) = Af\left(B \cdot \frac{a-H}{B} + H\right) + K = Af(a) + K = Ab + K$. We notice that the output from f is first multiplied by A. As with the constant B, if $A > 0$, this induces only a vertical scaling. If $A < 0$, then the -1 induces a reflection across the x-axis. Finally, we add K to the result, which is our vertical shift. A less precise, but more intuitive way to paraphrase Theorem 1.7 is to think of the quantity $Bx + H$ is the 'inside' of the function f. What's happening inside f affects the inputs or x-coordinates of the points on the graph of f. To find the x-coordinates of the corresponding points on g, we undo what has been done to x in the same way we would solve an equation. What's happening to the output can be thought of as things happening 'outside' the function, f. Things happening outside affect the outputs or y-coordinates of the points on the graph of f. Here, we follow the usual order of operations agreement: we first multiply by A then add K to find the corresponding y-coordinates on the graph of g.

Example 1.7.4. Below is the complete graph of $y = f(x)$. Use it to graph $g(x) = \frac{4-3f(1-2x)}{2}$.

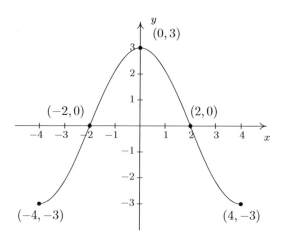

Solution. We use Theorem 1.7 to track the five 'key points' $(-4, -3)$, $(-2, 0)$, $(0, 3)$, $(2, 0)$ and $(4, -3)$ indicated on the graph of f to their new locations. We first rewrite $g(x)$ in the form presented in Theorem 1.7, $g(x) = -\frac{3}{2}f(-2x + 1) + 2$. We set $-2x + 1$ equal to the x-coordinates of the key points and solve. For example, solving $-2x + 1 = -4$, we first subtract 1 to get $-2x = -5$ then divide by -2 to get $x = \frac{5}{2}$. Subtracting the 1 is a horizontal shift to the left 1 unit. Dividing by -2 can be thought of as a two step process: dividing by 2 which compresses the graph horizontally by a factor of 2 followed by dividing (multiplying) by -1 which causes a reflection across the y-axis. We summarize the results in the table on the next page.

1.7 Transformations

$(a, f(a))$	a	$-2x+1=a$	x
$(-4,-3)$	-4	$-2x+1=-4$	$x=\frac{5}{2}$
$(-2,0)$	-2	$-2x+1=-2$	$x=\frac{3}{2}$
$(0,3)$	0	$-2x+1=0$	$x=\frac{1}{2}$
$(2,0)$	2	$-2x+1=2$	$x=-\frac{1}{2}$
$(4,-3)$	4	$-2x+1=4$	$x=-\frac{3}{2}$

Next, we take each of the x values and substitute them into $g(x) = -\frac{3}{2}f(-2x+1) + 2$ to get the corresponding y-values. Substituting $x = \frac{5}{2}$, and using the fact that $f(-4) = -3$, we get

$$g\left(\frac{5}{2}\right) = -\frac{3}{2}f\left(-2\left(\frac{5}{2}\right)+1\right)+2 = -\frac{3}{2}f(-4)+2 = -\frac{3}{2}(-3)+2 = \frac{9}{2}+2 = \frac{13}{2}$$

We see that the output from f is first multiplied by $-\frac{3}{2}$. Thinking of this as a two step process, multiplying by $\frac{3}{2}$ then by -1, we have a vertical stretching by a factor of $\frac{3}{2}$ followed by a reflection across the x-axis. Adding 2 results in a vertical shift up 2 units. Continuing in this manner, we get the table below.

x	$g(x)$	$(x, g(x))$
$\frac{5}{2}$	$\frac{13}{2}$	$\left(\frac{5}{2}, \frac{13}{2}\right)$
$\frac{3}{2}$	2	$\left(\frac{3}{2}, 2\right)$
$\frac{1}{2}$	$-\frac{5}{2}$	$\left(\frac{1}{2}, -\frac{5}{2}\right)$
$-\frac{1}{2}$	2	$\left(-\frac{1}{2}, 2\right)$
$-\frac{3}{2}$	$\frac{13}{2}$	$\left(-\frac{3}{2}, \frac{13}{2}\right)$

To graph g, we plot each of the points in the table above and connect them in the same order and fashion as the points to which they correspond. Plotting f and g side-by-side gives

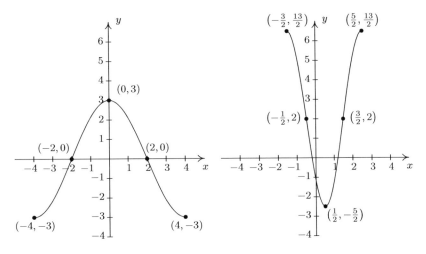

138   RELATIONS AND FUNCTIONS

The reader is strongly encouraged[11] to graph the series of functions which shows the gradual transformation of the graph of f into the graph of g. We have outlined the sequence of transformations in the above exposition; all that remains is to plot the five intermediate stages. □

Our last example turns the tables and asks for the formula of a function given a desired sequence of transformations. If nothing else, it is a good review of function notation.

Example 1.7.5. Let $f(x) = x^2$. Find and simplify the formula of the function $g(x)$ whose graph is the result of f undergoing the following sequence of transformations. Check your answer using a graphing calculator.

1. Vertical shift up 2 units

2. Reflection across the x-axis

3. Horizontal shift right 1 unit

4. Horizontal stretching by a factor of 2

Solution. We build up to a formula for $g(x)$ using intermediate functions as we've seen in previous examples. We let g_1 take care of our first step. Theorem 1.2 tells us $g_1(x) = f(x)+2 = x^2+2$. Next, we reflect the graph of g_1 about the x-axis using Theorem 1.4: $g_2(x) = -g_1(x) = -\left(x^2 + 2\right) = -x^2 - 2$. We shift the graph to the right 1 unit, according to Theorem 1.3, by setting $g_3(x) = g_2(x-1) = -(x-1)^2 - 2 = -x^2 + 2x - 3$. Finally, we induce a horizontal stretch by a factor of 2 using Theorem 1.6 to get $g(x) = g_3\left(\frac{1}{2}x\right) = -\left(\frac{1}{2}x\right)^2 + 2\left(\frac{1}{2}x\right) - 3$ which yields $g(x) = -\frac{1}{4}x^2 + x - 3$. We use the calculator to graph the stages below to confirm our result.

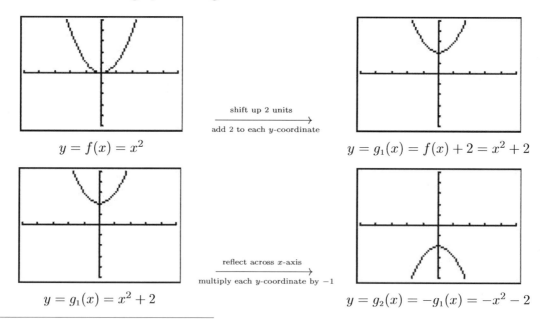

---

[11]You really should do this once in your life.

1.7 Transformations

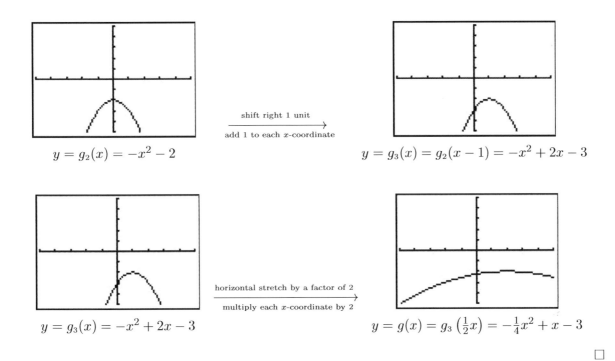

We have kept the viewing window the same in all of the graphs above. This had the undesirable consequence of making the last graph look 'incomplete' in that we cannot see the original shape of $f(x) = x^2$. Altering the viewing window results in a more complete graph of the transformed function as seen below.

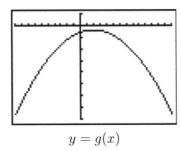

$y = g(x)$

This example brings our first chapter to a close. In the chapters which lie ahead, be on the lookout for the concepts developed here to resurface as we study different families of functions.

1.7.1 Exercises

Suppose $(2, -3)$ is on the graph of $y = f(x)$. In Exercises 1 - 18, use Theorem 1.7 to find a point on the graph of the given transformed function.

1. $y = f(x) + 3$
2. $y = f(x + 3)$
3. $y = f(x) - 1$

4. $y = f(x - 1)$
5. $y = 3f(x)$
6. $y = f(3x)$

7. $y = -f(x)$
8. $y = f(-x)$
9. $y = f(x - 3) + 1$

10. $y = 2f(x + 1)$
11. $y = 10 - f(x)$
12. $y = 3f(2x) - 1$

13. $y = \frac{1}{2}f(4 - x)$
14. $y = 5f(2x + 1) + 3$
15. $y = 2f(1 - x) - 1$

16. $y = f\left(\dfrac{7 - 2x}{4}\right)$
17. $y = \dfrac{f(3x) - 1}{2}$
18. $y = \dfrac{4 - f(3x - 1)}{7}$

The complete graph of $y = f(x)$ is given below. In Exercises 19 - 27, use it and Theorem 1.7 to graph the given transformed function.

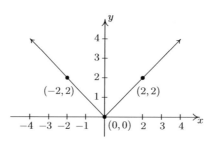

The graph for Ex. 19 - 27

19. $y = f(x) + 1$
20. $y = f(x) - 2$
21. $y = f(x + 1)$

22. $y = f(x - 2)$
23. $y = 2f(x)$
24. $y = f(2x)$

25. $y = 2 - f(x)$
26. $y = f(2 - x)$
27. $y = 2 - f(2 - x)$

28. Some of the answers to Exercises 19 - 27 above should be the same. Which ones match up? What properties of the graph of $y = f(x)$ contribute to the duplication?

1.7 Transformations

The complete graph of $y = f(x)$ is given below. In Exercises 29 - 37, use it and Theorem 1.7 to graph the given transformed function.

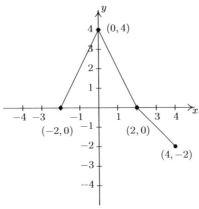

The graph for Ex. 29 - 37

29. $y = f(x) - 1$

30. $y = f(x+1)$

31. $y = \frac{1}{2}f(x)$

32. $y = f(2x)$

33. $y = -f(x)$

34. $y = f(-x)$

35. $y = f(x+1) - 1$

36. $y = 1 - f(x)$

37. $y = \frac{1}{2}f(x+1) - 1$

The complete graph of $y = f(x)$ is given below. In Exercises 38 - 49, use it and Theorem 1.7 to graph the given transformed function.

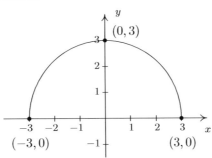

The graph for Ex. 38 - 49

38. $g(x) = f(x) + 3$

39. $h(x) = f(x) - \frac{1}{2}$

40. $j(x) = f\left(x - \frac{2}{3}\right)$

41. $a(x) = f(x+4)$

42. $b(x) = f(x+1) - 1$

43. $c(x) = \frac{3}{5}f(x)$

44. $d(x) = -2f(x)$

45. $k(x) = f\left(\frac{2}{3}x\right)$

46. $m(x) = -\frac{1}{4}f(3x)$

47. $n(x) = 4f(x-3) - 6$

48. $p(x) = 4 + f(1 - 2x)$

49. $q(x) = -\frac{1}{2}f\left(\frac{x+4}{2}\right) - 3$

The complete graph of $y = S(x)$ is given below.

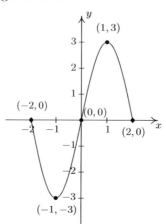

The graph of $y = S(x)$

The purpose of Exercises 50 - 53 is to graph $y = \frac{1}{2}S(-x+1)+1$ by graphing each transformation, one step at a time.

50. $y = S_1(x) = S(x+1)$

51. $y = S_2(x) = S_1(-x) = S(-x+1)$

52. $y = S_3(x) = \frac{1}{2}S_2(x) = \frac{1}{2}S(-x+1)$

53. $y = S_4(x) = S_3(x) + 1 = \frac{1}{2}S(-x+1) + 1$

Let $f(x) = \sqrt{x}$. Find a formula for a function g whose graph is obtained from f from the given sequence of transformations.

54. (1) shift right 2 units; (2) shift down 3 units

55. (1) shift down 3 units; (2) shift right 2 units

56. (1) reflect across the x-axis; (2) shift up 1 unit

57. (1) shift up 1 unit; (2) reflect across the x-axis

58. (1) shift left 1 unit; (2) reflect across the y-axis; (3) shift up 2 units

59. (1) reflect across the y-axis; (2) shift left 1 unit; (3) shift up 2 units

60. (1) shift left 3 units; (2) vertical stretch by a factor of 2; (3) shift down 4 units

61. (1) shift left 3 units; (2) shift down 4 units; (3) vertical stretch by a factor of 2

62. (1) shift right 3 units; (2) horizontal shrink by a factor of 2; (3) shift up 1 unit

63. (1) horizontal shrink by a factor of 2; (2) shift right 3 units; (3) shift up 1 unit

1.7 Transformations

64. The graph of $y = f(x) = \sqrt[3]{x}$ is given below on the left and the graph of $y = g(x)$ is given on the right. Find a formula for g based on transformations of the graph of f. Check your answer by confirming that the points shown on the graph of g satisfy the equation $y = g(x)$.

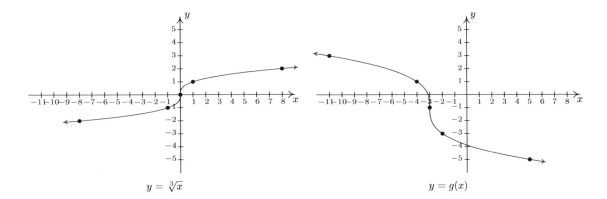

65. For many common functions, the properties of Algebra make a horizontal scaling the same as a vertical scaling by (possibly) a different factor. For example, we stated earlier that $\sqrt{9x} = 3\sqrt{x}$. With the help of your classmates, find the equivalent vertical scaling produced by the horizontal scalings $y = (2x)^3$, $y = |5x|$, $y = \sqrt[3]{27x}$ and $y = \left(\frac{1}{2}x\right)^2$. What about $y = (-2x)^3$, $y = |-5x|$, $y = \sqrt[3]{-27x}$ and $y = \left(-\frac{1}{2}x\right)^2$?

66. We mentioned earlier in the section that, in general, the order in which transformations are applied matters, yet in our first example with two transformations the order did not matter. (You could perform the shift to the left followed by the shift down or you could shift down and then left to achieve the same result.) With the help of your classmates, determine the situations in which order does matter and those in which it does not.

67. What happens if you reflect an even function across the y-axis?

68. What happens if you reflect an odd function across the y-axis?

69. What happens if you reflect an even function across the x-axis?

70. What happens if you reflect an odd function across the x-axis?

71. How would you describe symmetry about the origin in terms of reflections?

72. As we saw in Example 1.7.5, the viewing window on the graphing calculator affects how we see the transformations done to a graph. Using two different calculators, find viewing windows so that $f(x) = x^2$ on the one calculator looks like $g(x) = 3x^2$ on the other.

144   Relations and Functions

1.7.2 Answers

1. $(2, 0)$ 2. $(-1, -3)$ 3. $(2, -4)$

4. $(3, -3)$ 5. $(2, -9)$ 6. $\left(\frac{2}{3}, -3\right)$

7. $(2, 3)$ 8. $(-2, -3)$ 9. $(5, -2)$

10. $(1, -6)$ 11. $(2, 13)$ 12. $y = (1, -10)$

13. $\left(2, -\frac{3}{2}\right)$ 14. $\left(\frac{1}{2}, -12\right)$ 15. $(-1, -7)$

16. $\left(-\frac{1}{2}, -3\right)$ 17. $\left(\frac{2}{3}, -2\right)$ 18. $(1, 1)$

19. $y = f(x) + 1$

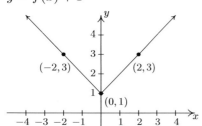

20. $y = f(x) - 2$

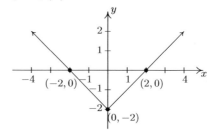

21. $y = f(x + 1)$

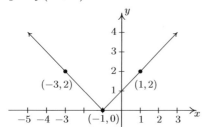

22. $y = f(x - 2)$

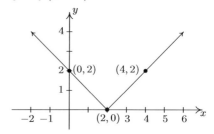

23. $y = 2f(x)$

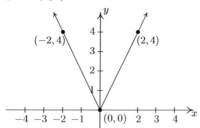

24. $y = f(2x)$

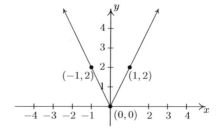

1.7 Transformations

25. $y = 2 - f(x)$

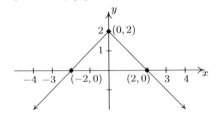

26. $y = f(2 - x)$

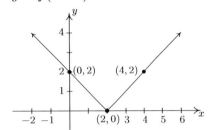

27. $y = 2 - f(2 - x)$

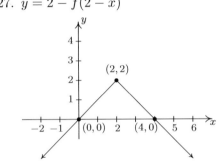

29. $y = f(x) - 1$

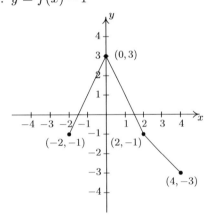

30. $y = f(x + 1)$

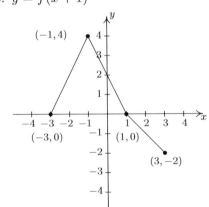

31. $y = \frac{1}{2}f(x)$

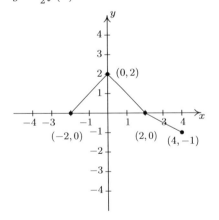

32. $y = f(2x)$

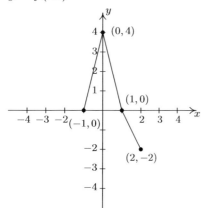

33. $y = -f(x)$

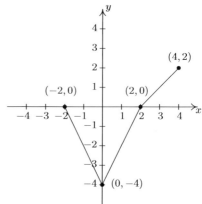

34. $y = f(-x)$

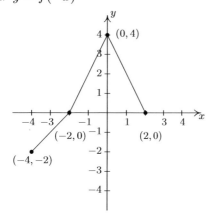

35. $y = f(x+1) - 1$

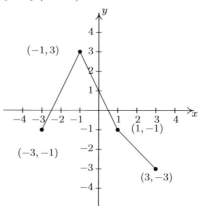

36. $y = 1 - f(x)$

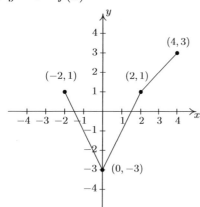

37. $y = \frac{1}{2}f(x+1) - 1$

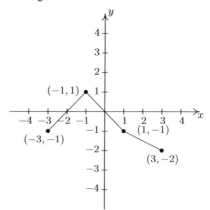

1.7 Transformations

38. $g(x) = f(x) + 3$

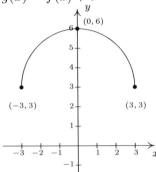

39. $h(x) = f(x) - \frac{1}{2}$

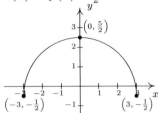

40. $j(x) = f\left(x - \frac{2}{3}\right)$

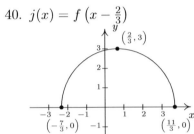

41. $a(x) = f(x+4)$

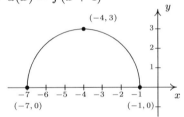

42. $b(x) = f(x+1) - 1$

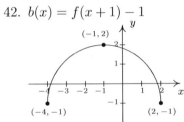

43. $c(x) = \frac{3}{5}f(x)$

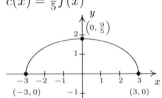

44. $d(x) = -2f(x)$

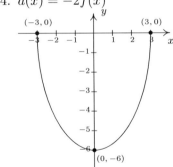

45. $k(x) = f\left(\frac{2}{3}x\right)$

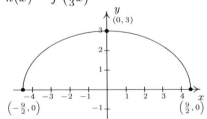

46. $m(x) = -\frac{1}{4}f(3x)$

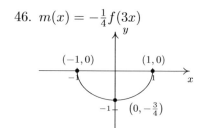

47. $n(x) = 4f(x-3) - 6$

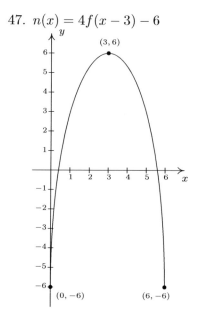

48. $p(x) = 4 + f(1-2x) = f(-2x+1) + 4$

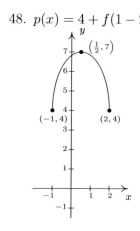

49. $q(x) = -\frac{1}{2}f\left(\frac{x+4}{2}\right) - 3 = -\frac{1}{2}f\left(\frac{1}{2}x+2\right) - 3$

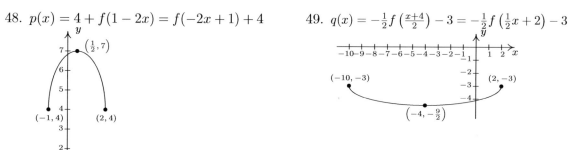

1.7 Transformations

50. $y = S_1(x) = S(x+1)$

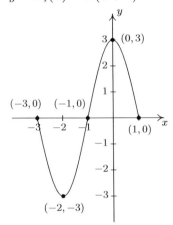

51. $y = S_2(x) = S_1(-x) = S(-x+1)$

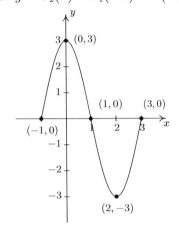

52. $y = S_3(x) = \frac{1}{2}S_2(x) = \frac{1}{2}S(-x+1)$

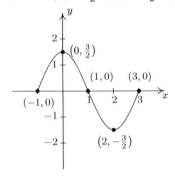

53. $y = S_4(x) = S_3(x) + 1 = \frac{1}{2}S(-x+1) + 1$

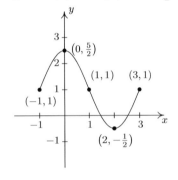

54. $g(x) = \sqrt{x-2} - 3$

55. $g(x) = \sqrt{x-2} - 3$

56. $g(x) = -\sqrt{x} + 1$

57. $g(x) = -(\sqrt{x} + 1) = -\sqrt{x} - 1$

58. $g(x) = \sqrt{-x+1} + 2$

59. $g(x) = \sqrt{-(x+1)} + 2 = \sqrt{-x-1} + 2$

60. $g(x) = 2\sqrt{x+3} - 4$

61. $g(x) = 2\left(\sqrt{x+3} - 4\right) = 2\sqrt{x+3} - 8$

62. $g(x) = \sqrt{2x-3} + 1$

63. $g(x) = \sqrt{2(x-3)} + 1 = \sqrt{2x-6} + 1$

64. $g(x) = -2\sqrt[3]{x+3} - 1$ or $g(x) = 2\sqrt[3]{-x-3} - 1$

Chapter 2

Linear and Quadratic Functions

2.1 Linear Functions

We now begin the study of families of functions. Our first family, linear functions, are old friends as we shall soon see. Recall from Geometry that two distinct points in the plane determine a unique line containing those points, as indicated below.

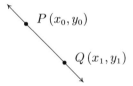

To give a sense of the 'steepness' of the line, we recall that we can compute the **slope** of the line using the formula below.

Equation 2.1. The **slope** m of the line containing the points $P(x_0, y_0)$ and $Q(x_1, y_1)$ is:

$$m = \frac{y_1 - y_0}{x_1 - x_0},$$

provided $x_1 \neq x_0$.

A couple of notes about Equation 2.1 are in order. First, don't ask why we use the letter 'm' to represent slope. There are many explanations out there, but apparently no one really knows for sure.[1] Secondly, the stipulation $x_1 \neq x_0$ ensures that we aren't trying to divide by zero. The reader is invited to pause to think about what is happening geometrically; the anxious reader can skip along to the next example.

Example 2.1.1. Find the slope of the line containing the following pairs of points, if it exists. Plot each pair of points and the line containing them.

---

[1] See www.mathforum.org or www.mathworld.wolfram.com for discussions on this topic.

1. $P(0,0)$, $Q(2,4)$
2. $P(-1,2)$, $Q(3,4)$
3. $P(-2,3)$, $Q(2,-3)$
4. $P(-3,2)$, $Q(4,2)$
5. $P(2,3)$, $Q(2,-1)$
6. $P(2,3)$, $Q(2.1,-1)$

Solution. In each of these examples, we apply the slope formula, Equation 2.1.

1. $m = \dfrac{4-0}{2-0} = \dfrac{4}{2} = 2$

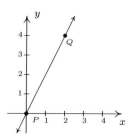

2. $m = \dfrac{4-2}{3-(-1)} = \dfrac{2}{4} = \dfrac{1}{2}$

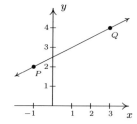

3. $m = \dfrac{-3-3}{2-(-2)} = \dfrac{-6}{4} = -\dfrac{3}{2}$

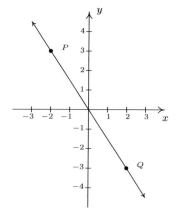

4. $m = \dfrac{2-2}{4-(-3)} = \dfrac{0}{7} = 0$

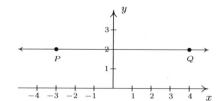

2.1 Linear Functions

5. $m = \dfrac{-1-3}{2-2} = \dfrac{-4}{0}$, which is undefined

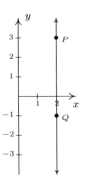

6. $m = \dfrac{-1-3}{2.1-2} = \dfrac{-4}{0.1} = -40$

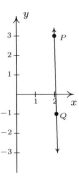

□

A few comments about Example 2.1.1 are in order. First, for reasons which will be made clear soon, if the slope is positive then the resulting line is said to be increasing. If it is negative, we say the line is decreasing. A slope of 0 results in a horizontal line which we say is constant, and an undefined slope results in a vertical line.[2] Second, the larger the slope is in absolute value, the steeper the line. You may recall from Intermediate Algebra that slope can be described as the ratio '$\frac{\text{rise}}{\text{run}}$'. For example, in the second part of Example 2.1.1, we found the slope to be $\frac{1}{2}$. We can interpret this as a rise of 1 unit upward for every 2 units to the right we travel along the line, as shown below.

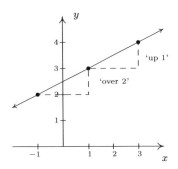

---

[2] Some authors use the unfortunate moniker 'no slope' when a slope is undefined. It's easy to confuse the notions of 'no slope' with 'slope of 0'. For this reason, we will describe slopes of vertical lines as 'undefined'.

Using more formal notation, given points (x_0, y_0) and (x_1, y_1), we use the Greek letter delta 'Δ' to write $\Delta y = y_1 - y_0$ and $\Delta x = x_1 - x_0$. In most scientific circles, the symbol Δ means 'change in'. Hence, we may write

$$m = \frac{\Delta y}{\Delta x},$$

which describes the slope as the **rate of change** of y with respect to x. Rates of change abound in the 'real world', as the next example illustrates.

Example 2.1.2. Suppose that two separate temperature readings were taken at the ranger station on the top of Mt. Sasquatch: at 6 AM the temperature was 24°F and at 10 AM it was 32°F.

1. Find the slope of the line containing the points $(6, 24)$ and $(10, 32)$.

2. Interpret your answer to the first part in terms of temperature and time.

3. Predict the temperature at noon.

Solution.

1. For the slope, we have $m = \frac{32-24}{10-6} = \frac{8}{4} = 2$.

2. Since the values in the numerator correspond to the temperatures in °F, and the values in the denominator correspond to time in hours, we can interpret the slope as $2 = \frac{2}{1} = \frac{2°\text{F}}{1\text{ hour}}$, or 2°F per hour. Since the slope is positive, we know this corresponds to an increasing line. Hence, the temperature is increasing at a rate of 2°F per hour.

3. Noon is two hours after 10 AM. Assuming a temperature increase of 2°F per hour, in two hours the temperature should rise 4°F. Since the temperature at 10 AM is 32°F, we would expect the temperature at noon to be $32 + 4 = 36$°F. $\square$

Now it may well happen that in the previous scenario, at noon the temperature is only 33°F. This doesn't mean our calculations are incorrect, rather, it means that the temperature change throughout the day isn't a constant 2°F per hour. As discussed in Section 1.4.1, mathematical models are just that: models. The predictions we get out of the models may be mathematically accurate, but may not resemble what happens in the real world.

In Section 1.2, we discussed the equations of vertical and horizontal lines. Using the concept of slope, we can develop equations for the other varieties of lines. Suppose a line has a slope of m and contains the point (x_0, y_0). Suppose (x, y) is another point on the line, as indicated below.

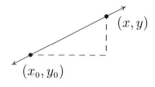

2.1 LINEAR FUNCTIONS

Equation 2.1 yields

$$\begin{aligned} m &= \frac{y - y_0}{x - x_0} \\ m(x - x_0) &= y - y_0 \\ y - y_0 &= m(x - x_0) \end{aligned}$$

We have just derived the **point-slope form** of a line.[3]

Equation 2.2. The **point-slope form** of the line with slope m containing the point (x_0, y_0) is the equation $y - y_0 = m(x - x_0)$.

Example 2.1.3. Write the equation of the line containing the points $(-1, 3)$ and $(2, 1)$.

Solution. In order to use Equation 2.2 we need to find the slope of the line in question so we use Equation 2.1 to get $m = \frac{\Delta y}{\Delta x} = \frac{1-3}{2-(-1)} = -\frac{2}{3}$. We are spoiled for choice for a point (x_0, y_0). We'll use $(-1, 3)$ and leave it to the reader to check that using $(2, 1)$ results in the same equation. Substituting into the point-slope form of the line, we get

$$\begin{aligned} y - y_0 &= m(x - x_0) \\ y - 3 &= -\frac{2}{3}(x - (-1)) \\ y - 3 &= -\frac{2}{3}(x + 1) \\ y - 3 &= -\frac{2}{3}x - \frac{2}{3} \\ y &= -\frac{2}{3}x + \frac{7}{3}. \end{aligned}$$

We can check our answer by showing that both $(-1, 3)$ and $(2, 1)$ are on the graph of $y = -\frac{2}{3}x + \frac{7}{3}$ algebraically, as we did in Section 1.2.1. □

In simplifying the equation of the line in the previous example, we produced another form of a line, the **slope-intercept form**. This is the familiar $y = mx + b$ form you have probably seen in Intermediate Algebra. The 'intercept' in 'slope-intercept' comes from the fact that if we set $x = 0$, we get $y = b$. In other words, the y-intercept of the line $y = mx + b$ is $(0, b)$.

Equation 2.3. The **slope-intercept form** of the line with slope m and y-intercept $(0, b)$ is the equation $y = mx + b$.

Note that if we have slope $m = 0$, we get the equation $y = b$ which matches our formula for a horizontal line given in Section 1.2. The formula given in Equation 2.3 can be used to describe all lines except vertical lines. All lines except vertical lines are functions (Why is this?) so we have finally reached a good point to introduce **linear functions**.

---
[3]We can also understand this equation in terms of applying transformations to the function $I(x) = x$. See the Exercises.

Definition 2.1. A **linear function** is a function of the form

$$f(x) = mx + b,$$

where m and b are real numbers with $m \neq 0$. The domain of a linear function is $(-\infty, \infty)$.

For the case $m = 0$, we get $f(x) = b$. These are given their own classification.

Definition 2.2. A **constant function** is a function of the form

$$f(x) = b,$$

where b is real number. The domain of a constant function is $(-\infty, \infty)$.

Recall that to graph a function, f, we graph the equation $y = f(x)$. Hence, the graph of a linear function is a line with slope m and y-intercept $(0, b)$; the graph of a constant function is a horizontal line (a line with slope $m = 0$) and a y-intercept of $(0, b)$. Now think back to Section 1.6.1, specifically Definition 1.10 concerning increasing, decreasing and constant functions. A line with positive slope was called an increasing line because a linear function with $m > 0$ is an increasing function. Similarly, a line with a negative slope was called a decreasing line because a linear function with $m < 0$ is a decreasing function. And horizontal lines were called constant because, well, we hope you've already made the connection.

Example 2.1.4. Graph the following functions. Identify the slope and y-intercept.

1. $f(x) = 3$

2. $f(x) = 3x - 1$

3. $f(x) = \dfrac{3 - 2x}{4}$

4. $f(x) = \dfrac{x^2 - 4}{x - 2}$

Solution.

1. To graph $f(x) = 3$, we graph $y = 3$. This is a horizontal line ($m = 0$) through $(0, 3)$.

2. The graph of $f(x) = 3x - 1$ is the graph of the line $y = 3x - 1$. Comparison of this equation with Equation 2.3 yields $m = 3$ and $b = -1$. Hence, our slope is 3 and our y-intercept is $(0, -1)$. To get another point on the line, we can plot $(1, f(1)) = (1, 2)$.

2.1 Linear Functions

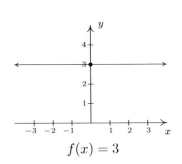

$f(x) = 3$

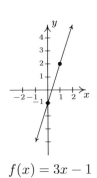

$f(x) = 3x - 1$

3. At first glance, the function $f(x) = \frac{3-2x}{4}$ does not fit the form in Definition 2.1 but after some rearranging we get $f(x) = \frac{3-2x}{4} = \frac{3}{4} - \frac{2x}{4} = -\frac{1}{2}x + \frac{3}{4}$. We identify $m = -\frac{1}{2}$ and $b = \frac{3}{4}$. Hence, our graph is a line with a slope of $-\frac{1}{2}$ and a y-intercept of $\left(0, \frac{3}{4}\right)$. Plotting an additional point, we can choose $(1, f(1))$ to get $\left(1, \frac{1}{4}\right)$.

4. If we simplify the expression for f, we get

$$f(x) = \frac{x^2 - 4}{x - 2} = \frac{(x-2)(x+2)}{(x-2)} = x + 2.$$

If we were to state $f(x) = x + 2$, we would be committing a sin of omission. Remember, to find the domain of a function, we do so **before** we simplify! In this case, f has big problems when $x = 2$, and as such, the domain of f is $(-\infty, 2) \cup (2, \infty)$. To indicate this, we write $f(x) = x + 2$, $x \neq 2$. So, except at $x = 2$, we graph the line $y = x + 2$. The slope $m = 1$ and the y-intercept is $(0, 2)$. A second point on the graph is $(1, f(1)) = (1, 3)$. Since our function f is not defined at $x = 2$, we put an open circle at the point that would be on the line $y = x + 2$ when $x = 2$, namely $(2, 4)$.

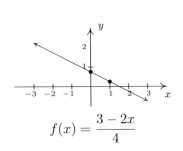

$f(x) = \dfrac{3 - 2x}{4}$

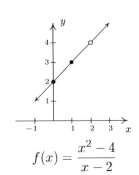

$f(x) = \dfrac{x^2 - 4}{x - 2}$

The last two functions in the previous example showcase some of the difficulty in defining a linear function using the phrase 'of the form' as in Definition 2.1, since some algebraic manipulations may be needed to rewrite a given function to match 'the form'. Keep in mind that the domains of linear and constant functions are all real numbers $(-\infty, \infty)$, so while $f(x) = \frac{x^2-4}{x-2}$ simplified to a formula $f(x) = x + 2$, f is not considered a linear function since its domain excludes $x = 2$. However, we would consider

$$f(x) = \frac{2x^2 + 2}{x^2 + 1}$$

to be a constant function since its domain is all real numbers (Can you tell us why?) and

$$f(x) = \frac{2x^2 + 2}{x^2 + 1} = \frac{2\cancel{(x^2+1)}}{\cancel{(x^2+1)}} = 2$$

The following example uses linear functions to model some basic economic relationships.

Example 2.1.5. The cost C, in dollars, to produce x PortaBoy[4] game systems for a local retailer is given by $C(x) = 80x + 150$ for $x \geq 0$.

1. Find and interpret $C(10)$.

2. How many PortaBoys can be produced for $15,000?

3. Explain the significance of the restriction on the domain, $x \geq 0$.

4. Find and interpret $C(0)$.

5. Find and interpret the slope of the graph of $y = C(x)$.

Solution.

1. To find $C(10)$, we replace every occurrence of x with 10 in the formula for $C(x)$ to get $C(10) = 80(10) + 150 = 950$. Since x represents the number of PortaBoys produced, and $C(x)$ represents the cost, in dollars, $C(10) = 950$ means it costs $950 to produce 10 PortaBoys for the local retailer.

2. To find how many PortaBoys can be produced for $15,000, we solve $C(x) = 15000$, or $80x + 150 = 15000$. Solving, we get $x = \frac{14850}{80} = 185.625$. Since we can only produce a whole number amount of PortaBoys, we can produce 185 PortaBoys for $15,000.

3. The restriction $x \geq 0$ is the applied domain, as discussed in Section 1.4.1. In this context, x represents the number of PortaBoys produced. It makes no sense to produce a negative quantity of game systems.[5]

---

[4]The similarity of this name to PortaJohn is deliberate.

[5]Actually, it makes no sense to produce a fractional part of a game system, either, as we saw in the previous part of this example. This absurdity, however, seems quite forgivable in some textbooks but not to us.

2.1 LINEAR FUNCTIONS

4. We find $C(0) = 80(0) + 150 = 150$. This means it costs \$150 to produce 0 PortaBoys. As mentioned on page 82, this is the fixed, or start-up cost of this venture.

5. If we were to graph $y = C(x)$, we would be graphing the portion of the line $y = 80x + 150$ for $x \geq 0$. We recognize the slope, $m = 80$. Like any slope, we can interpret this as a rate of change. Here, $C(x)$ is the cost in dollars, while x measures the number of PortaBoys so

$$m = \frac{\Delta y}{\Delta x} = \frac{\Delta C}{\Delta x} = 80 = \frac{80}{1} = \frac{\$80}{1\,\text{PortaBoy}}.$$

In other words, the cost is increasing at a rate of \$80 per PortaBoy produced. This is often called the **variable cost** for this venture. □

The next example asks us to find a linear function to model a related economic problem.

Example 2.1.6. The local retailer in Example 2.1.5 has determined that the number x of PortaBoy game systems sold in a week is related to the price p in dollars of each system. When the price was \$220, 20 game systems were sold in a week. When the systems went on sale the following week, 40 systems were sold at \$190 a piece.

1. Find a linear function which fits this data. Use the weekly sales x as the independent variable and the price p as the dependent variable.

2. Find a suitable applied domain.

3. Interpret the slope.

4. If the retailer wants to sell 150 PortaBoys next week, what should the price be?

5. What would the weekly sales be if the price were set at \$150 per system?

Solution.

1. We recall from Section 1.4 the meaning of 'independent' and 'dependent' variable. Since x is to be the independent variable, and p the dependent variable, we treat x as the input variable and p as the output variable. Hence, we are looking for a function of the form $p(x) = mx + b$. To determine m and b, we use the fact that 20 PortaBoys were sold during the week when the price was 220 dollars and 40 units were sold when the price was 190 dollars. Using function notation, these two facts can be translated as $p(20) = 220$ and $p(40) = 190$. Since m represents the rate of change of p with respect to x, we have

$$m = \frac{\Delta p}{\Delta x} = \frac{190 - 220}{40 - 20} = \frac{-30}{20} = -1.5.$$

We now have determined $p(x) = -1.5x + b$. To determine b, we can use our given data again. Using $p(20) = 220$, we substitute $x = 20$ into $p(x) = 1.5x + b$ and set the result equal to 220: $-1.5(20) + b = 220$. Solving, we get $b = 250$. Hence, we get $p(x) = -1.5x + 250$. We can check our formula by computing $p(20)$ and $p(40)$ to see if we get 220 and 190, respectively. You may recall from page 82 that the function $p(x)$ is called the price-demand (or simply demand) function for this venture.

2. To determine the applied domain, we look at the physical constraints of the problem. Certainly, we can't sell a negative number of PortaBoys, so $x \geq 0$. However, we also note that the slope of this linear function is negative, and as such, the price is decreasing as more units are sold. Thus another constraint on the price is $p(x) \geq 0$. Solving $-1.5x + 250 \geq 0$ results in $-1.5x \geq -250$ or $x \leq \frac{500}{3} = 166.\overline{6}$. Since x represents the number of PortaBoys sold in a week, we round down to 166. As a result, a reasonable applied domain for p is $[0, 166]$.

3. The slope $m = -1.5$, once again, represents the rate of change of the price of a system with respect to weekly sales of PortaBoys. Since the slope is negative, we have that the price is decreasing at a rate of $1.50 per PortaBoy sold. (Said differently, you can sell one more PortaBoy for every $1.50 drop in price.)

4. To determine the price which will move 150 PortaBoys, we find $p(150) = -1.5(150)+250 = 25$. That is, the price would have to be $25.

5. If the price of a PortaBoy were set at $150, we have $p(x) = 150$, or, $-1.5x+250 = 150$. Solving, we get $-1.5x = -100$ or $x = 66.\overline{6}$. This means you would be able to sell 66 PortaBoys a week if the price were $150 per system. □

Not all real-world phenomena can be modeled using linear functions. Nevertheless, it is possible to use the concept of slope to help analyze non-linear functions using the following.

Definition 2.3. Let f be a function defined on the interval $[a, b]$. The **average rate of change** of f over $[a, b]$ is defined as:
$$\frac{\Delta f}{\Delta x} = \frac{f(b) - f(a)}{b - a}$$

Geometrically, if we have the graph of $y = f(x)$, the average rate of change over $[a, b]$ is the slope of the line which connects $(a, f(a))$ and $(b, f(b))$. This is called the **secant line** through these points. For that reason, some textbooks use the notation m_{sec} for the average rate of change of a function. Note that for a linear function $m = m_{\text{sec}}$, or in other words, its rate of change over an interval is the same as its average rate of change.

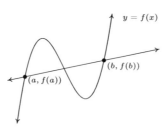

The graph of $y = f(x)$ and its secant line through $(a, f(a))$ and $(b, f(b))$

The interested reader may question the adjective 'average' in the phrase 'average rate of change'. In the figure above, we can see that the function changes wildly on $[a, b]$, yet the slope of the secant line only captures a snapshot of the action at a and b. This situation is entirely analogous to the

2.1 LINEAR FUNCTIONS

average speed on a trip. Suppose it takes you 2 hours to travel 100 miles. Your average speed is $\frac{100 \text{ miles}}{2 \text{ hours}} = 50$ miles per hour. However, it is entirely possible that at the start of your journey, you traveled 25 miles per hour, then sped up to 65 miles per hour, and so forth. The average rate of change is akin to your average speed on the trip. Your speedometer measures your speed at any one instant along the trip, your **instantaneous rate of change**, and this is one of the central themes of Calculus.[6]

When interpreting rates of change, we interpret them the same way we did slopes. In the context of functions, it may be helpful to think of the average rate of change as:

$$\frac{\text{change in outputs}}{\text{change in inputs}}$$

Example 2.1.7. Recall from page 82, the revenue from selling x units at a price p per unit is given by the formula $R = xp$. Suppose we are in the scenario of Examples 2.1.5 and 2.1.6.

1. Find and simplify an expression for the weekly revenue $R(x)$ as a function of weekly sales x.

2. Find and interpret the average rate of change of $R(x)$ over the interval $[0, 50]$.

3. Find and interpret the average rate of change of $R(x)$ as x changes from 50 to 100 and compare that to your result in part 2.

4. Find and interpret the average rate of change of weekly revenue as weekly sales increase from 100 PortaBoys to 150 PortaBoys.

Solution.

1. Since $R = xp$, we substitute $p(x) = -1.5x + 250$ from Example 2.1.6 to get $R(x) = x(-1.5x + 250) = -1.5x^2 + 250x$. Since we determined the price-demand function $p(x)$ is restricted to $0 \leq x \leq 166$, $R(x)$ is restricted to these values of x as well.

2. Using Definition 2.3, we get that the average rate of change is

$$\frac{\Delta R}{\Delta x} = \frac{R(50) - R(0)}{50 - 0} = \frac{8750 - 0}{50 - 0} = 175.$$

Interpreting this slope as we have in similar situations, we conclude that for every additional PortaBoy sold during a given week, the weekly revenue increases $175.

3. The wording of this part is slightly different than that in Definition 2.3, but its meaning is to find the average rate of change of R over the interval $[50, 100]$. To find this rate of change, we compute

$$\frac{\Delta R}{\Delta x} = \frac{R(100) - R(50)}{100 - 50} = \frac{10000 - 8750}{50} = 25.$$

---

[6]Here we go again...

In other words, for each additional PortaBoy sold, the revenue increases by $25. Note that while the revenue is still increasing by selling more game systems, we aren't getting as much of an increase as we did in part 2 of this example. (Can you think of why this would happen?)

4. Translating the English to the mathematics, we are being asked to find the average rate of change of R over the interval $[100, 150]$. We find

$$\frac{\Delta R}{\Delta x} = \frac{R(150) - R(100)}{150 - 100} = \frac{3750 - 10000}{50} = -125.$$

This means that we are losing $125 dollars of weekly revenue for each additional PortaBoy sold. (Can you think why this is possible?) □

We close this section with a new look at difference quotients which were first introduced in Section 1.4. If we wish to compute the average rate of change of a function f over the interval $[x, x+h]$, then we would have

$$\frac{\Delta f}{\Delta x} = \frac{f(x+h) - f(x)}{(x+h) - x} = \frac{f(x+h) - f(x)}{h}$$

As we have indicated, the rate of change of a function (average or otherwise) is of great importance in Calculus.[7] Also, we have the geometric interpretation of difference quotients which was promised to you back on page 81 – a difference quotient yields the slope of a secant line.

---

[7] So we are not torturing you with these for nothing.

2.1 LINEAR FUNCTIONS

2.1.1 EXERCISES

In Exercises 1 - 10, find both the point-slope form and the slope-intercept form of the line with the given slope which passes through the given point.

1. $m = 3$, $P(3, -1)$
2. $m = -2$, $P(-5, 8)$
3. $m = -1$, $P(-7, -1)$
4. $m = \frac{2}{3}$, $P(-2, 1)$
5. $m = -\frac{1}{5}$, $P(10, 4)$
6. $m = \frac{1}{7}$, $P(-1, 4)$
7. $m = 0$, $P(3, 117)$
8. $m = -\sqrt{2}$, $P(0, -3)$
9. $m = -5$, $P(\sqrt{3}, 2\sqrt{3})$
10. $m = 678$, $P(-1, -12)$

In Exercises 11 - 20, find the slope-intercept form of the line which passes through the given points.

11. $P(0, 0)$, $Q(-3, 5)$
12. $P(-1, -2)$, $Q(3, -2)$
13. $P(5, 0)$, $Q(0, -8)$
14. $P(3, -5)$, $Q(7, 4)$
15. $P(-1, 5)$, $Q(7, 5)$
16. $P(4, -8)$, $Q(5, -8)$
17. $P\left(\frac{1}{2}, \frac{3}{4}\right)$, $Q\left(\frac{5}{2}, -\frac{7}{4}\right)$
18. $P\left(\frac{2}{3}, \frac{7}{2}\right)$, $Q\left(-\frac{1}{3}, \frac{3}{2}\right)$
19. $P\left(\sqrt{2}, -\sqrt{2}\right)$, $Q\left(-\sqrt{2}, \sqrt{2}\right)$
20. $P\left(-\sqrt{3}, -1\right)$, $Q\left(\sqrt{3}, 1\right)$

In Exercises 21 - 26, graph the function. Find the slope, y-intercept and x-intercept, if any exist.

21. $f(x) = 2x - 1$
22. $f(x) = 3 - x$
23. $f(x) = 3$
24. $f(x) = 0$
25. $f(x) = \frac{2}{3}x + \frac{1}{3}$
26. $f(x) = \dfrac{1 - x}{2}$

27. Find all of the points on the line $y = 2x + 1$ which are 4 units from the point $(-1, 3)$.

28. Jeff can walk comfortably at 3 miles per hour. Find a linear function d that represents the total distance Jeff can walk in t hours, assuming he doesn't take any breaks.

29. Carl can stuff 6 envelopes per *minute*. Find a linear function E that represents the total number of envelopes Carl can stuff after t *hours*, assuming he doesn't take any breaks.

30. A landscaping company charges $45 per cubic yard of mulch plus a delivery charge of $20. Find a linear function which computes the total cost C (in dollars) to deliver x cubic yards of mulch.

31. A plumber charges $50 for a service call plus $80 per hour. If she spends no longer than 8 hours a day at any one site, find a linear function that represents her total daily charges C (in dollars) as a function of time t (in hours) spent at any one given location.

32. A salesperson is paid $200 per week plus 5% commission on her weekly sales of x dollars. Find a linear function that represents her total weekly pay, W (in dollars) in terms of x. What must her weekly sales be in order for her to earn $475.00 for the week?

33. An on-demand publisher charges $22.50 to print a 600 page book and $15.50 to print a 400 page book. Find a linear function which models the cost of a book C as a function of the number of pages p. Interpret the slope of the linear function and find and interpret $C(0)$.

34. The Topology Taxi Company charges $2.50 for the first fifth of a mile and $0.45 for each additional fifth of a mile. Find a linear function which models the taxi fare F as a function of the number of miles driven, m. Interpret the slope of the linear function and find and interpret $F(0)$.

35. Water freezes at $0°$ Celsius and $32°$ Fahrenheit and it boils at $100°C$ and $212°F$.

    (a) Find a linear function F that expresses temperature in the Fahrenheit scale in terms of degrees Celsius. Use this function to convert $20°C$ into Fahrenheit.

    (b) Find a linear function C that expresses temperature in the Celsius scale in terms of degrees Fahrenheit. Use this function to convert $110°F$ into Celsius.

    (c) Is there a temperature n such that $F(n) = C(n)$?

36. Legend has it that a bull Sasquatch in rut will howl approximately 9 times per hour when it is $40°F$ outside and only 5 times per hour if it's $70°F$. Assuming that the number of howls per hour, N, can be represented by a linear function of temperature Fahrenheit, find the number of howls per hour he'll make when it's only $20°F$ outside. What is the applied domain of this function? Why?

37. Economic forces beyond anyone's control have changed the cost function for PortaBoys to $C(x) = 105x + 175$. Rework Example 2.1.5 with this new cost function.

38. In response to the economic forces in Exercise 37 above, the local retailer sets the selling price of a PortaBoy at $250. Remarkably, 30 units were sold each week. When the systems went on sale for $220, 40 units per week were sold. Rework Examples 2.1.6 and 2.1.7 with this new data. What difficulties do you encounter?

39. A local pizza store offers medium two-topping pizzas delivered for $6.00 per pizza plus a $1.50 delivery charge per order. On weekends, the store runs a 'game day' special: if six or more medium two-topping pizzas are ordered, they are $5.50 each with no delivery charge. Write a piecewise-defined linear function which calculates the cost C (in dollars) of p medium two-topping pizzas delivered during a weekend.

2.1 Linear Functions

40. A restaurant offers a buffet which costs $15 per person. For parties of 10 or more people, a group discount applies, and the cost is $12.50 per person. Write a piecewise-defined linear function which calculates the total bill T of a party of n people who all choose the buffet.

41. A mobile plan charges a base monthly rate of $10 for the first 500 minutes of air time plus a charge of 15¢ for each additional minute. Write a piecewise-defined linear function which calculates the monthly cost C (in dollars) for using m minutes of air time.

    HINT: You may want to revisit Exercise 74 in Section 1.4

42. The local pet shop charges 12¢ per cricket up to 100 crickets, and 10¢ per cricket thereafter. Write a piecewise-defined linear function which calculates the price P, in dollars, of purchasing c crickets.

43. The cross-section of a swimming pool is below. Write a piecewise-defined linear function which describes the depth of the pool, D (in feet) as a function of:

    (a) the distance (in feet) from the edge of the shallow end of the pool, d.

    (b) the distance (in feet) from the edge of the deep end of the pool, s.

    (c) Graph each of the functions in (a) and (b). Discuss with your classmates how to transform one into the other and how they relate to the diagram of the pool.

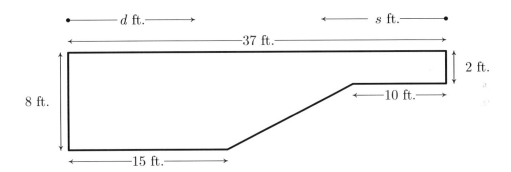

In Exercises 44 - 49, compute the average rate of change of the function over the specified interval.

44. $f(x) = x^3$, $[-1, 2]$

45. $f(x) = \dfrac{1}{x}$, $[1, 5]$

46. $f(x) = \sqrt{x}$, $[0, 16]$

47. $f(x) = x^2$, $[-3, 3]$

48. $f(x) = \dfrac{x+4}{x-3}$, $[5, 7]$

49. $f(x) = 3x^2 + 2x - 7$, $[-4, 2]$

In Exercises 50 - 53, compute the average rate of change of the given function over the interval $[x, x+h]$. Here we assume $[x, x+h]$ is in the domain of the function.

50. $f(x) = x^3$

51. $f(x) = \dfrac{1}{x}$

52. $f(x) = \dfrac{x+4}{x-3}$

53. $f(x) = 3x^2 + 2x - 7$

54. The height of an object dropped from the roof of an eight story building is modeled by: $h(t) = -16t^2 + 64$, $0 \leq t \leq 2$. Here, h is the height of the object off the ground in feet, t seconds after the object is dropped. Find and interpret the average rate of change of h over the interval $[0, 2]$.

55. Using data from Bureau of Transportation Statistics, the average fuel economy F in miles per gallon for passenger cars in the US can be modeled by $F(t) = -0.0076t^2 + 0.45t + 16$, $0 \leq t \leq 28$, where t is the number of years since 1980. Find and interpret the average rate of change of F over the interval $[0, 28]$.

56. The temperature T in degrees Fahrenheit t hours after 6 AM is given by:

$$T(t) = -\frac{1}{2}t^2 + 8t + 32, \quad 0 \leq t \leq 12$$

(a) Find and interpret $T(4)$, $T(8)$ and $T(12)$.

(b) Find and interpret the average rate of change of T over the interval $[4, 8]$.

(c) Find and interpret the average rate of change of T from $t = 8$ to $t = 12$.

(d) Find and interpret the average rate of temperature change between 10 AM and 6 PM.

57. Suppose $C(x) = x^2 - 10x + 27$ represents the costs, in *hundreds*, to produce x *thousand* pens. Find and interpret the average rate of change as production is increased from making 3000 to 5000 pens.

58. With the help of your classmates find several other "real-world" examples of rates of change that are used to describe non-linear phenomena.

(Parallel Lines) Recall from Intermediate Algebra that parallel lines have the same slope. (Please note that two vertical lines are also parallel to one another even though they have an undefined slope.) In Exercises 59 - 64, you are given a line and a point which is not on that line. Find the line parallel to the given line which passes through the given point.

59. $y = 3x + 2$, $P(0, 0)$

60. $y = -6x + 5$, $P(3, 2)$

2.1 Linear Functions

61. $y = \frac{2}{3}x - 7$, $P(6, 0)$

62. $y = \dfrac{4-x}{3}$, $P(1, -1)$

63. $y = 6$, $P(3, -2)$

64. $x = 1$, $P(-5, 0)$

(Perpendicular Lines) Recall from Intermediate Algebra that two non-vertical lines are perpendicular if and only if they have negative reciprocal slopes. That is to say, if one line has slope m_1 and the other has slope m_2 then $m_1 \cdot m_2 = -1$. (You will be guided through a proof of this result in Exercise 71.) Please note that a horizontal line is perpendicular to a vertical line and vice versa, so we assume $m_1 \neq 0$ and $m_2 \neq 0$. In Exercises 65 - 70, you are given a line and a point which is not on that line. Find the line perpendicular to the given line which passes through the given point.

65. $y = \frac{1}{3}x + 2$, $P(0, 0)$

66. $y = -6x + 5$, $P(3, 2)$

67. $y = \frac{2}{3}x - 7$, $P(6, 0)$

68. $y = \dfrac{4-x}{3}$, $P(1, -1)$

69. $y = 6$, $P(3, -2)$

70. $x = 1$, $P(-5, 0)$

71. We shall now prove that $y = m_1 x + b_1$ is perpendicular to $y = m_2 x + b_2$ if and only if $m_1 \cdot m_2 = -1$. To make our lives easier we shall assume that $m_1 > 0$ and $m_2 < 0$. We can also "move" the lines so that their point of intersection is the origin without messing things up, so we'll assume $b_1 = b_2 = 0$. (Take a moment with your classmates to discuss why this is okay.) Graphing the lines and plotting the points $O(0,0)$, $P(1, m_1)$ and $Q(1, m_2)$ gives us the following set up.

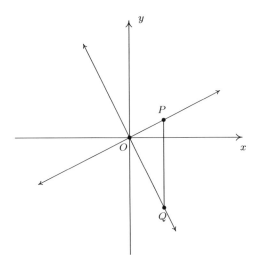

The line $y = m_1 x$ will be perpendicular to the line $y = m_2 x$ if and only if $\triangle OPQ$ is a right triangle. Let d_1 be the distance from O to P, let d_2 be the distance from O to Q and let d_3 be the distance from P to Q. Use the Pythagorean Theorem to show that $\triangle OPQ$ is a right triangle if and only if $m_1 \cdot m_2 = -1$ by showing $d_1^2 + d_2^2 = d_3^2$ if and only if $m_1 \cdot m_2 = -1$.

72. Show that if $a \neq b$, the line containing the points (a, b) and (b, a) is perpendicular to the line $y = x$. (Coupled with the result from Example 1.1.7 on page 13, we have now shown that the line $y = x$ is a *perpendicular* bisector of the line segment connecting (a, b) and (b, a). This means the points (a, b) and (b, a) are symmetric about the line $y = x$. We will revisit this symmetry in section 5.2.)

73. The function defined by $I(x) = x$ is called the Identity Function.

    (a) Discuss with your classmates why this name makes sense.
    (b) Show that the point-slope form of a line (Equation 2.2) can be obtained from I using a sequence of the transformations defined in Section 1.7.

2.1.2 Answers

1. $y + 1 = 3(x - 3)$
   $y = 3x - 10$

2. $y - 8 = -2(x + 5)$
   $y = -2x - 2$

3. $y + 1 = -(x + 7)$
   $y = -x - 8$

4. $y - 1 = \frac{2}{3}(x + 2)$
   $y = \frac{2}{3}x + \frac{7}{3}$

5. $y - 4 = -\frac{1}{5}(x - 10)$
   $y = -\frac{1}{5}x + 6$

6. $y - 4 = \frac{1}{7}(x + 1)$
   $y = \frac{1}{7}x + \frac{29}{7}$

7. $y - 117 = 0$
   $y = 117$

8. $y + 3 = -\sqrt{2}(x - 0)$
   $y = -\sqrt{2}x - 3$

9. $y - 2\sqrt{3} = -5(x - \sqrt{3})$
   $y = -5x + 7\sqrt{3}$

10. $y + 12 = 678(x + 1)$
    $y = 678x + 666$

11. $y = -\frac{5}{3}x$

12. $y = -2$

13. $y = \frac{8}{5}x - 8$

14. $y = \frac{9}{4}x - \frac{47}{4}$

15. $y = 5$

16. $y = -8$

17. $y = -\frac{5}{4}x + \frac{11}{8}$

18. $y = 2x + \frac{13}{6}$

19. $y = -x$

20. $y = \frac{\sqrt{3}}{3}x$

21. $f(x) = 2x - 1$
    slope: $m = 2$
    y-intercept: $(0, -1)$
    x-intercept: $\left(\frac{1}{2}, 0\right)$

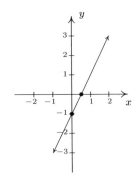

22. $f(x) = 3 - x$
    slope: $m = -1$
    y-intercept: $(0, 3)$
    x-intercept: $(3, 0)$

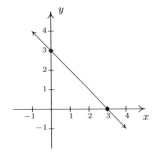

23. $f(x) = 3$

slope: $m = 0$

y-intercept: $(0, 3)$

x-intercept: none

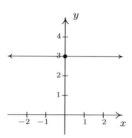

24. $f(x) = 0$

slope: $m = 0$

y-intercept: $(0, 0)$

x-intercept: $\{(x, 0) \,|\, x \text{ is a real number}\}$

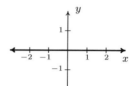

25. $f(x) = \frac{2}{3}x + \frac{1}{3}$

slope: $m = \frac{2}{3}$

y-intercept: $\left(0, \frac{1}{3}\right)$

x-intercept: $\left(-\frac{1}{2}, 0\right)$

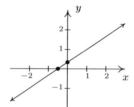

26. $f(x) = \dfrac{1-x}{2}$

slope: $m = -\frac{1}{2}$

y-intercept: $\left(0, \frac{1}{2}\right)$

x-intercept: $(1, 0)$

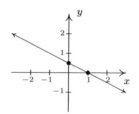

27. $(-1, -1)$ and $\left(\frac{11}{5}, \frac{27}{5}\right)$

28. $d(t) = 3t$, $t \geq 0$.

29. $E(t) = 360t$, $t \geq 0$.

30. $C(x) = 45x + 20$, $x \geq 0$.

31. $C(t) = 80t + 50$, $0 \leq t \leq 8$.

32. $W(x) = 200 + .05x$, $x \geq 0$ She must make $5500 in weekly sales.

33. $C(p) = 0.035p + 1.5$ The slope 0.035 means it costs 3.5¢ per page. $C(0) = 1.5$ means there is a fixed, or start-up, cost of $1.50 to make each book.

34. $F(m) = 2.25m + 2.05$ The slope 2.25 means it costs an additional $2.25 for each mile beyond the first 0.2 miles. $F(0) = 2.05$, so according to the model, it would cost $2.05 for a trip of 0 miles. Would this ever really happen? Depends on the driver and the passenger, we suppose.

2.1 Linear Functions

35. (a) $F(C) = \frac{9}{5}C + 32$ (b) $C(F) = \frac{5}{9}(F - 32) = \frac{5}{9}F - \frac{160}{9}$

    (c) $F(-40) = -40 = C(-40)$.

36. $N(T) = -\frac{2}{15}T + \frac{43}{3}$ and $N(20) = \frac{35}{3} \approx 12$ howls per hour.

    Having a negative number of howls makes no sense and since $N(107.5) = 0$ we can put an upper bound of $107.5°F$ on the domain. The lower bound is trickier because there's nothing other than common sense to go on. As it gets colder, he howls more often. At some point it will either be so cold that he freezes to death or he's howling non-stop. So we're going to say that he can withstand temperatures no lower than $-60°F$ so that the applied domain is $[-60, 107.5]$.

39. $C(p) = \begin{cases} 6p + 1.5 & \text{if } 1 \leq p \leq 5 \\ 5.5p & \text{if } p \geq 6 \end{cases}$

40. $T(n) = \begin{cases} 15n & \text{if } 1 \leq n \leq 9 \\ 12.5n & \text{if } n \geq 10 \end{cases}$

41. $C(m) = \begin{cases} 10 & \text{if } 0 \leq m \leq 500 \\ 10 + 0.15(m - 500) & \text{if } m > 500 \end{cases}$

42. $P(c) = \begin{cases} 0.12c & \text{if } 1 \leq c \leq 100 \\ 12 + 0.1(c - 100) & \text{if } c > 100 \end{cases}$

43. (a)
$$D(d) = \begin{cases} 8 & \text{if } 0 \leq d \leq 15 \\ -\frac{1}{2}d + \frac{31}{2} & \text{if } 15 \leq d \leq 27 \\ 2 & \text{if } 27 \leq d \leq 37 \end{cases}$$

    (b)
$$D(s) = \begin{cases} 2 & \text{if } 0 \leq s \leq 10 \\ \frac{1}{2}s - 3 & \text{if } 10 \leq s \leq 22 \\ 8 & \text{if } 22 \leq s \leq 37 \end{cases}$$

    (c)

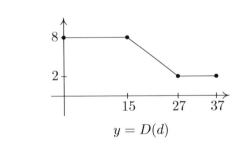

$y = D(d)$

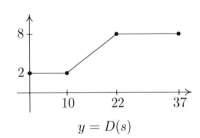

$y = D(s)$

172   LINEAR AND QUADRATIC FUNCTIONS

44. $\dfrac{2^3 - (-1)^3}{2 - (-1)} = 3$

45. $\dfrac{\frac{1}{5} - \frac{1}{1}}{5 - 1} = -\dfrac{1}{5}$

46. $\dfrac{\sqrt{16} - \sqrt{0}}{16 - 0} = \dfrac{1}{4}$

47. $\dfrac{3^2 - (-3)^2}{3 - (-3)} = 0$

48. $\dfrac{\frac{7+4}{7-3} - \frac{5+4}{5-3}}{7 - 5} = -\dfrac{7}{8}$

49. $\dfrac{(3(2)^2 + 2(2) - 7) - (3(-4)^2 + 2(-4) - 7)}{2 - (-4)} = -4$

50. $3x^2 + 3xh + h^2$

51. $\dfrac{-1}{x(x+h)}$

52. $\dfrac{-7}{(x-3)(x+h-3)}$

53. $6x + 3h + 2$

54. The average rate of change is $\dfrac{h(2)-h(0)}{2-0} = -32$. During the first two seconds after it is dropped, the object has fallen at an average rate of 32 feet per second. (This is called the *average velocity* of the object.)

55. The average rate of change is $\dfrac{F(28)-F(0)}{28-0} = 0.2372$. During the years from 1980 to 2008, the average fuel economy of passenger cars in the US increased, on average, at a rate of 0.2372 miles per gallon per year.

56. (a) $T(4) = 56$, so at 10 AM (4 hours after 6 AM), it is 56°F. $T(8) = 64$, so at 2 PM (8 hours after 6 AM), it is 64°F. $T(12) = 56$, so at 6 PM (12 hours after 6 AM), it is 56°F.

   (b) The average rate of change is $\dfrac{T(8)-T(4)}{8-4} = 2$. Between 10 AM and 2 PM, the temperature increases, on average, at a rate of 2°F per hour.

   (c) The average rate of change is $\dfrac{T(12)-T(8)}{12-8} = -2$. Between 2 PM and 6 PM, the temperature decreases, on average, at a rate of 2°F per hour.

   (d) The average rate of change is $\dfrac{T(12)-T(4)}{12-4} = 0$. Between 10 AM and 6 PM, the temperature, on average, remains constant.

57. The average rate of change is $\dfrac{C(5)-C(3)}{5-3} = -2$. As production is increased from 3000 to 5000 pens, the cost decreases at an average rate of $200 per 1000 pens produced (20¢ per pen.)

59. $y = 3x$

60. $y = -6x + 20$

61. $y = \frac{2}{3}x - 4$

62. $y = -\frac{1}{3}x - \frac{2}{3}$

63. $y = -2$

64. $x = -5$

65. $y = -3x$

66. $y = \frac{1}{6}x + \frac{3}{2}$

67. $y = -\frac{3}{2}x + 9$

68. $y = 3x - 4$

69. $x = 3$

70. $y = 0$

2.2 Absolute Value Functions

There are a few ways to describe what is meant by the absolute value $|x|$ of a real number x. You may have been taught that $|x|$ is the distance from the real number x to 0 on the number line. So, for example, $|5| = 5$ and $|-5| = 5$, since each is 5 units from 0 on the number line.

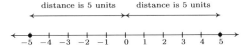

Another way to define absolute value is by the equation $|x| = \sqrt{x^2}$. Using this definition, we have $|5| = \sqrt{(5)^2} = \sqrt{25} = 5$ and $|-5| = \sqrt{(-5)^2} = \sqrt{25} = 5$. The long and short of both of these procedures is that $|x|$ takes negative real numbers and assigns them to their positive counterparts while it leaves positive numbers alone. This last description is the one we shall adopt, and is summarized in the following definition.

Definition 2.4. The **absolute value** of a real number x, denoted $|x|$, is given by

$$|x| = \begin{cases} -x, & \text{if } x < 0 \\ x, & \text{if } x \geq 0 \end{cases}$$

In Definition 2.4, we define $|x|$ using a piecewise-defined function. (See page 62 in Section 1.4.) To check that this definition agrees with what we previously understood as absolute value, note that since $5 \geq 0$, to find $|5|$ we use the rule $|x| = x$, so $|5| = 5$. Similarly, since $-5 < 0$, we use the rule $|x| = -x$, so that $|-5| = -(-5) = 5$. This is one of the times when it's best to interpret the expression '$-x$' as 'the opposite of x' as opposed to 'negative x'. Before we begin studying absolute value functions, we remind ourselves of the properties of absolute value.

Theorem 2.1. Properties of Absolute Value: Let a, b and x be real numbers and let n be an integer.[a] Then

- **Product Rule:** $|ab| = |a||b|$

- **Power Rule:** $|a^n| = |a|^n$ whenever a^n is defined

- **Quotient Rule:** $\left|\dfrac{a}{b}\right| = \dfrac{|a|}{|b|}$, provided $b \neq 0$

Equality Properties:

- $|x| = 0$ if and only if $x = 0$.

- For $c > 0$, $|x| = c$ if and only if $x = c$ or $-x = c$.

- For $c < 0$, $|x| = c$ has no solution.

---
[a] See page 2 if you don't remember what an integer is.

The proofs of the Product and Quotient Rules in Theorem 2.1 boil down to checking four cases: when both a and b are positive; when they are both negative; when one is positive and the other is negative; and when one or both are zero.

For example, suppose we wish to show that $|ab| = |a||b|$. We need to show that this equation is true for all real numbers a and b. If a and b are both positive, then so is ab. Hence, $|a| = a$, $|b| = b$ and $|ab| = ab$. Hence, the equation $|ab| = |a||b|$ is the same as $ab = ab$ which is true. If both a and b are negative, then ab is positive. Hence, $|a| = -a$, $|b| = -b$ and $|ab| = ab$. The equation $|ab| = |a||b|$ becomes $ab = (-a)(-b)$, which is true. Suppose a is positive and b is negative. Then ab is negative, and we have $|ab| = -ab$, $|a| = a$ and $|b| = -b$. The equation $|ab| = |a||b|$ reduces to $-ab = a(-b)$ which is true. A symmetric argument shows the equation $|ab| = |a||b|$ holds when a is negative and b is positive. Finally, if either a or b (or both) are zero, then both sides of $|ab| = |a||b|$ are zero, so the equation holds in this case, too. All of this rhetoric has shown that the equation $|ab| = |a||b|$ holds true in all cases.

The proof of the Quotient Rule is very similar, with the exception that $b \neq 0$. The Power Rule can be shown by repeated application of the Product Rule. The 'Equality Properties' can be proved using Definition 2.4 and by looking at the cases when $x \geq 0$, in which case $|x| = x$, or when $x < 0$, in which case $|x| = -x$. For example, if $c > 0$, and $|x| = c$, then if $x \geq 0$, we have $x = |x| = c$. If, on the other hand, $x < 0$, then $-x = |x| = c$, so $x = -c$. The remaining properties are proved similarly and are left for the Exercises. Our first example reviews how to solve basic equations involving absolute value using the properties listed in Theorem 2.1.

Example 2.2.1. Solve each of the following equations.

1. $|3x - 1| = 6$
2. $3 - |x + 5| = 1$
3. $3|2x + 1| - 5 = 0$
4. $4 - |5x + 3| = 5$
5. $|x| = x^2 - 6$
6. $|x - 2| + 1 = x$

Solution.

1. The equation $|3x - 1| = 6$ is of the form $|x| = c$ for $c > 0$, so by the Equality Properties, $|3x - 1| = 6$ is equivalent to $3x - 1 = 6$ or $3x - 1 = -6$. Solving the former, we arrive at $x = \frac{7}{3}$, and solving the latter, we get $x = -\frac{5}{3}$. We may check both of these solutions by substituting them into the original equation and showing that the arithmetic works out.

2. To use the Equality Properties to solve $3 - |x + 5| = 1$, we first isolate the absolute value.

$$\begin{aligned} 3 - |x + 5| &= 1 \\ -|x + 5| &= -2 \quad \text{subtract 3} \\ |x + 5| &= 2 \quad \text{divide by } -1 \end{aligned}$$

From the Equality Properties, we have $x + 5 = 2$ or $x + 5 = -2$, and get our solutions to be $x = -3$ or $x = -7$. We leave it to the reader to check both answers in the original equation.

2.2 Absolute Value Functions

3. As in the previous example, we first isolate the absolute value in the equation $3|2x+1|-5=0$ and get $|2x+1|=\frac{5}{3}$. Using the Equality Properties, we have $2x+1=\frac{5}{3}$ or $2x+1=-\frac{5}{3}$. Solving the former gives $x=\frac{1}{3}$ and solving the latter gives $x=-\frac{4}{3}$. As usual, we may substitute both answers in the original equation to check.

4. Upon isolating the absolute value in the equation $4-|5x+3|=5$, we get $|5x+3|=-1$. At this point, we know there cannot be any real solution, since, by definition, the absolute value of *anything* is never negative. We are done.

5. The equation $|x|=x^2-6$ presents us with some difficulty, since x appears both inside and outside of the absolute value. Moreover, there are values of x for which x^2-6 is positive, negative and zero, so we cannot use the Equality Properties without the risk of introducing extraneous solutions, or worse, losing solutions. For this reason, we break equations like this into cases by rewriting the term in absolute values, $|x|$, using Definition 2.4. For $x<0$, $|x|=-x$, so for $x<0$, the equation $|x|=x^2-6$ is equivalent to $-x=x^2-6$. Rearranging this gives us $x^2+x-6=0$, or $(x+3)(x-2)=0$. We get $x=-3$ or $x=2$. Since only $x=-3$ satisfies $x<0$, this is the answer we keep. For $x\geq 0$, $|x|=x$, so the equation $|x|=x^2-6$ becomes $x=x^2-6$. From this, we get $x^2-x-6=0$ or $(x-3)(x+2)=0$. Our solutions are $x=3$ or $x=-2$, and since only $x=3$ satisfies $x\geq 0$, this is the one we keep. Hence, our two solutions to $|x|=x^2-6$ are $x=-3$ and $x=3$.

6. To solve $|x-2|+1=x$, we first isolate the absolute value and get $|x-2|=x-1$. Since we see x both inside and outside of the absolute value, we break the equation into cases. The term with absolute values here is $|x-2|$, so we replace 'x' with the quantity '$(x-2)$' in Definition 2.4 to get

$$|x-2| = \begin{cases} -(x-2), & \text{if } (x-2)<0 \\ (x-2), & \text{if } (x-2)\geq 0 \end{cases}$$

Simplifying yields

$$|x-2| = \begin{cases} -x+2, & \text{if } x<2 \\ x-2, & \text{if } x\geq 2 \end{cases}$$

So, for $x<2$, $|x-2|=-x+2$ and our equation $|x-2|=x-1$ becomes $-x+2=x-1$, which gives $x=\frac{3}{2}$. Since this solution satisfies $x<2$, we keep it. Next, for $x\geq 2$, $|x-2|=x-2$, so the equation $|x-2|=x-1$ becomes $x-2=x-1$. Here, the equation reduces to $-2=-1$, which signifies we have no solutions here. Hence, our only solution is $x=\frac{3}{2}$. $\square$

Next, we turn our attention to graphing absolute value functions. Our strategy in the next example is to make liberal use of Definition 2.4 along with what we know about graphing linear functions (from Section 2.1) and piecewise-defined functions (from Section 1.4).

Example 2.2.2. Graph each of the following functions.

1. $f(x)=|x|$
2. $g(x)=|x-3|$
3. $h(x)=|x|-3$
4. $i(x)=4-2|3x+1|$

Find the zeros of each function and the x- and y-intercepts of each graph, if any exist. From the graph, determine the domain and range of each function, list the intervals on which the function is increasing, decreasing or constant, and find the relative and absolute extrema, if they exist.

Solution.

1. To find the zeros of f, we set $f(x) = 0$. We get $|x| = 0$, which, by Theorem 2.1 gives us $x = 0$. Since the zeros of f are the x-coordinates of the x-intercepts of the graph of $y = f(x)$, we get $(0, 0)$ as our only x-intercept. To find the y-intercept, we set $x = 0$, and find $y = f(0) = 0$, so that $(0, 0)$ is our y-intercept as well.[1] Using Definition 2.4, we get

$$f(x) = |x| = \begin{cases} -x, & \text{if } x < 0 \\ x, & \text{if } x \geq 0 \end{cases}$$

Hence, for $x < 0$, we are graphing the line $y = -x$; for $x \geq 0$, we have the line $y = x$. Proceeding as we did in Section 1.6, we get

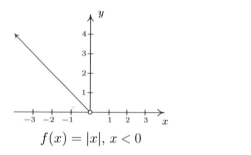
$f(x) = |x|, x < 0$

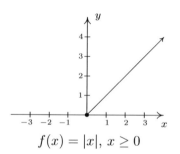
$f(x) = |x|, x \geq 0$

Notice that we have an 'open circle' at $(0, 0)$ in the graph when $x < 0$. As we have seen before, this is due to the fact that the points on $y = -x$ approach $(0, 0)$ as the x-values approach 0. Since x is required to be strictly less than zero on this stretch, the open circle is drawn at the origin. However, notice that when $x \geq 0$, we get to fill in the point at $(0, 0)$, which effectively 'plugs' the hole indicated by the open circle. Thus we get,

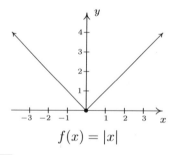
$f(x) = |x|$

---

[1] Actually, since functions can have at most one y-intercept (Do you know why?), as soon as we found $(0, 0)$ as the x-intercept, we knew this was also the y-intercept.

2.2 ABSOLUTE VALUE FUNCTIONS

By projecting the graph to the x-axis, we see that the domain is $(-\infty, \infty)$. Projecting to the y-axis gives us the range $[0, \infty)$. The function is increasing on $[0, \infty)$ and decreasing on $(-\infty, 0]$. The relative minimum value of f is the same as the absolute minimum, namely 0 which occurs at $(0, 0)$. There is no relative maximum value of f. There is also no absolute maximum value of f, since the y values on the graph extend infinitely upwards.

2. To find the zeros of g, we set $g(x) = |x - 3| = 0$. By Theorem 2.1, we get $x - 3 = 0$ so that $x = 3$. Hence, the x-intercept is $(3, 0)$. To find our y-intercept, we set $x = 0$ so that $y = g(0) = |0 - 3| = 3$, which yields $(0, 3)$ as our y-intercept. To graph $g(x) = |x - 3|$, we use Definition 2.4 to rewrite g as

$$g(x) = |x - 3| = \begin{cases} -(x - 3), & \text{if } (x - 3) < 0 \\ (x - 3), & \text{if } (x - 3) \geq 0 \end{cases}$$

Simplifying, we get

$$g(x) = \begin{cases} -x + 3, & \text{if } x < 3 \\ x - 3, & \text{if } x \geq 3 \end{cases}$$

As before, the open circle we introduce at $(3, 0)$ from the graph of $y = -x + 3$ is filled by the point $(3, 0)$ from the line $y = x - 3$. We determine the domain as $(-\infty, \infty)$ and the range as $[0, \infty)$. The function g is increasing on $[3, \infty)$ and decreasing on $(-\infty, 3]$. The relative and absolute minimum value of g is 0 which occurs at $(3, 0)$. As before, there is no relative or absolute maximum value of g.

3. Setting $h(x) = 0$ to look for zeros gives $|x| - 3 = 0$. As in Example 2.2.1, we isolate the absolute value to get $|x| = 3$ so that $x = 3$ or $x = -3$. As a result, we have a pair of x-intercepts: $(-3, 0)$ and $(3, 0)$. Setting $x = 0$ gives $y = h(0) = |0| - 3 = -3$, so our y-intercept is $(0, -3)$. As before, we rewrite the absolute value in h to get

$$h(x) = \begin{cases} -x - 3, & \text{if } x < 0 \\ x - 3, & \text{if } x \geq 0 \end{cases}$$

Once again, the open circle at $(0, -3)$ from one piece of the graph of h is filled by the point $(0, -3)$ from the other piece of h. From the graph, we determine the domain of h is $(-\infty, \infty)$ and the range is $[-3, \infty)$. On $[0, \infty)$, h is increasing; on $(-\infty, 0]$ it is decreasing. The relative minimum occurs at the point $(0, -3)$ on the graph, and we see -3 is both the relative and absolute minimum value of h. Also, h has no relative or absolute maximum value.

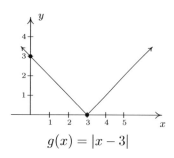
$g(x) = |x-3|$

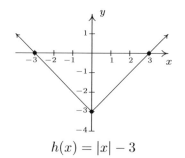
$h(x) = |x| - 3$

4. As before, we set $i(x) = 0$ to find the zeros of i and get $4 - 2|3x+1| = 0$. Isolating the absolute value term gives $|3x+1| = 2$, so either $3x+1 = 2$ or $3x+1 = -2$. We get $x = \frac{1}{3}$ or $x = -1$, so our x-intercepts are $\left(\frac{1}{3}, 0\right)$ and $(-1, 0)$. Substituting $x = 0$ gives $y = i(0) = 4 - 2|3(0) + 1| = 2$, for a y-intercept of $(0, 2)$. Rewriting the formula for $i(x)$ without absolute values gives

$$i(x) = \begin{cases} 4 - 2(-(3x+1)), & \text{if } (3x+1) < 0 \\ 4 - 2(3x+1), & \text{if } (3x+1) \geq 0 \end{cases} = \begin{cases} 6x + 6, & \text{if } x < -\frac{1}{3} \\ -6x + 2, & \text{if } x \geq -\frac{1}{3} \end{cases}$$

The usual analysis near the trouble spot $x = -\frac{1}{3}$ gives the 'corner' of this graph is $\left(-\frac{1}{3}, 4\right)$, and we get the distinctive 'V' shape:

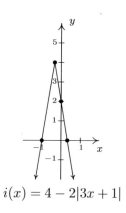
$i(x) = 4 - 2|3x+1|$

The domain of i is $(-\infty, \infty)$ while the range is $(-\infty, 4]$. The function i is increasing on $\left(-\infty, -\frac{1}{3}\right]$ and decreasing on $\left[-\frac{1}{3}, \infty\right)$. The relative maximum occurs at the point $\left(-\frac{1}{3}, 4\right)$ and the relative and absolute maximum value of i is 4. Since the graph of i extends downwards forever more, there is no absolute minimum value. As we can see from the graph, there is no relative minimum, either. □

Note that all of the functions in the previous example bear the characteristic 'V' shape of the graph of $y = |x|$. We could have graphed the functions g, h and i in Example 2.2.2 starting with the graph of $f(x) = |x|$ and applying transformations as in Section 1.7 as our next example illustrates.

2.2 Absolute Value Functions

Example 2.2.3. Graph the following functions starting with the graph of $f(x) = |x|$ and using transformations.

1. $g(x) = |x - 3|$
2. $h(x) = |x| - 3$
3. $i(x) = 4 - 2|3x + 1|$

Solution. We begin by graphing $f(x) = |x|$ and labeling three points, $(-1, 1)$, $(0, 0)$ and $(1, 1)$.

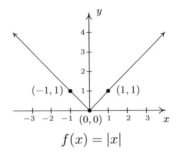

$f(x) = |x|$

1. Since $g(x) = |x - 3| = f(x - 3)$, Theorem 1.7 tells us to *add* 3 to each of the x-values of the points on the graph of $y = f(x)$ to obtain the graph of $y = g(x)$. This shifts the graph of $y = f(x)$ to the *right* 3 units and moves the point $(-1, 1)$ to $(2, 1)$, $(0, 0)$ to $(3, 0)$ and $(1, 1)$ to $(4, 1)$. Connecting these points in the classic 'V' fashion produces the graph of $y = g(x)$.

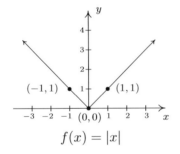 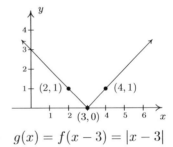

$f(x) = |x|$ shift right 3 units $g(x) = f(x - 3) = |x - 3|$
           add 3 to each x-coordinate

2. For $h(x) = |x| - 3 = f(x) - 3$, Theorem 1.7 tells us to *subtract* 3 from each of the y-values of the points on the graph of $y = f(x)$ to obtain the graph of $y = h(x)$. This shifts the graph of $y = f(x)$ *down* 3 units and moves $(-1, 1)$ to $(-1, -2)$, $(0, 0)$ to $(0, -3)$ and $(1, 1)$ to $(1, -2)$. Connecting these points with the 'V' shape produces our graph of $y = h(x)$.

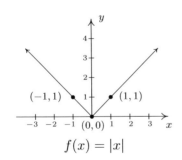 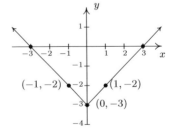

$f(x) = |x|$ shift down 3 units $h(x) = f(x) - 3 = |x| - 3$
         subtract 3 from each y-coordinate

3. We re-write $i(x) = 4 - 2|3x + 1| = 4 - 2f(3x + 1) = -2f(3x + 1) + 4$ and apply Theorem 1.7. First, we take care of the changes on the 'inside' of the absolute value. Instead of $|x|$, we have $|3x + 1|$, so, in accordance with Theorem 1.7, we first *subtract* 1 from each of the x-values of points on the graph of $y = f(x)$, then *divide* each of those new values by 3. This effects a horizontal shift *left* 1 unit followed by a horizontal *shrink* by a factor of 3. These transformations move $(-1, 1)$ to $\left(-\frac{2}{3}, 1\right)$, $(0, 0)$ to $\left(-\frac{1}{3}, 0\right)$ and $(1, 1)$ to $(0, 1)$. Next, we take care of what's happening 'outside of' the absolute value. Theorem 1.7 instructs us to first *multiply* each y-value of these new points by -2 then *add* 4. Geometrically, this corresponds to a vertical *stretch* by a factor of 2, a reflection across the x-axis and finally, a vertical shift *up* 4 units. These transformations move $\left(-\frac{2}{3}, 1\right)$ to $\left(-\frac{2}{3}, 2\right)$, $\left(-\frac{1}{3}, 0\right)$ to $\left(-\frac{1}{3}, 4\right)$, and $(0, 1)$ to $(0, 2)$. Connecting these points with the usual 'V' shape produces our graph of $y = i(x)$.

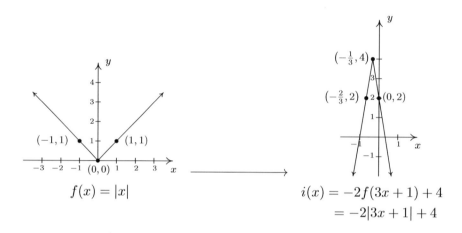

$$f(x) = |x|$$

$$i(x) = -2f(3x + 1) + 4$$
$$= -2|3x + 1| + 4$$

$\square$

While the methods in Section 1.7 can be used to graph an entire family of absolute value functions, not all functions involving absolute values posses the characteristic 'V' shape. As the next example illustrates, often there is no substitute for appealing directly to the definition.

Example 2.2.4. Graph each of the following functions. Find the zeros of each function and the x- and y-intercepts of each graph, if any exist. From the graph, determine the domain and range of each function, list the intervals on which the function is increasing, decreasing or constant, and find the relative and absolute extrema, if they exist.

1. $f(x) = \dfrac{|x|}{x}$

2. $g(x) = |x + 2| - |x - 3| + 1$

Solution.

1. We first note that, due to the fraction in the formula of $f(x)$, $x \neq 0$. Thus the domain is $(-\infty, 0) \cup (0, \infty)$. To find the zeros of f, we set $f(x) = \frac{|x|}{x} = 0$. This last equation implies $|x| = 0$, which, from Theorem 2.1, implies $x = 0$. However, $x = 0$ is not in the domain of f,

which means we have, in fact, no x-intercepts. We have no y-intercepts either, since $f(0)$ is undefined. Re-writing the absolute value in the function gives

$$f(x) = \begin{cases} \dfrac{-x}{x}, & \text{if } x < 0 \\ \dfrac{x}{x}, & \text{if } x > 0 \end{cases} = \begin{cases} -1, & \text{if } x < 0 \\ 1, & \text{if } x > 0 \end{cases}$$

To graph this function, we graph two horizontal lines: $y = -1$ for $x < 0$ and $y = 1$ for $x > 0$. We have open circles at $(0, -1)$ and $(0, 1)$ (Can you explain why?) so we get

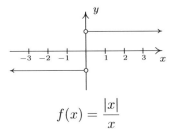

$$f(x) = \dfrac{|x|}{x}$$

As we found earlier, the domain is $(-\infty, 0) \cup (0, \infty)$. The range consists of just two y-values: $\{-1, 1\}$. The function f is constant on $(-\infty, 0)$ and $(0, \infty)$. The local minimum value of f is the absolute minimum value of f, namely -1; the local maximum and absolute maximum values for f also coincide — they both are 1. Every point on the graph of f is simultaneously a relative maximum and a relative minimum. (Can you remember why in light of Definition 1.11? This was explored in the Exercises in Section 1.6.2.)

2. To find the zeros of g, we set $g(x) = 0$. The result is $|x + 2| - |x - 3| + 1 = 0$. Attempting to isolate the absolute value term is complicated by the fact that there are **two** terms with absolute values. In this case, it easier to proceed using cases by re-writing the function g with two separate applications of Definition 2.4 to remove each instance of the absolute values, one at a time. In the first round we get

$$g(x) = \begin{cases} -(x + 2) - |x - 3| + 1, & \text{if } (x + 2) < 0 \\ (x + 2) - |x - 3| + 1, & \text{if } (x + 2) \geq 0 \end{cases} = \begin{cases} -x - 1 - |x - 3|, & \text{if } x < -2 \\ x + 3 - |x - 3|, & \text{if } x \geq -2 \end{cases}$$

Given that

$$|x - 3| = \begin{cases} -(x - 3), & \text{if } (x - 3) < 0 \\ x - 3, & \text{if } (x - 3) \geq 0 \end{cases} = \begin{cases} -x + 3, & \text{if } x < 3 \\ x - 3, & \text{if } x \geq 3 \end{cases},$$

we need to break up the domain again at $x = 3$. Note that if $x < -2$, then $x < 3$, so we replace $|x - 3|$ with $-x + 3$ for that part of the domain, too. Our completed revision of the form of g yields

$$g(x) = \begin{cases} -x - 1 - (-x + 3), & \text{if } x < -2 \\ x + 3 - (-x + 3), & \text{if } x \geq -2 \text{ and } x < 3 \\ x + 3 - (x - 3), & \text{if } x \geq 3 \end{cases} = \begin{cases} -4, & \text{if } x < -2 \\ 2x, & \text{if } -2 \leq x < 3 \\ 6, & \text{if } x \geq 3 \end{cases}$$

To solve $g(x) = 0$, we see that the only piece which contains a variable is $g(x) = 2x$ for $-2 \leq x < 3$. Solving $2x = 0$ gives $x = 0$. Since $x = 0$ is in the interval $[-2, 3)$, we keep this solution and have $(0,0)$ as our only x-intercept. Accordingly, the y-intercept is also $(0,0)$. To graph g, we start with $x < -2$ and graph the horizontal line $y = -4$ with an open circle at $(-2, -4)$. For $-2 \leq x < 3$, we graph the line $y = 2x$ and the point $(-2, -4)$ patches the hole left by the previous piece. An open circle at $(3, 6)$ completes the graph of this part. Finally, we graph the horizontal line $y = 6$ for $x \geq 3$, and the point $(3, 6)$ fills in the open circle left by the previous part of the graph. The finished graph is

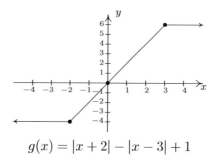

$g(x) = |x + 2| - |x - 3| + 1$

The domain of g is all real numbers, $(-\infty, \infty)$, and the range of g is all real numbers between -4 and 6 inclusive, $[-4, 6]$. The function is increasing on $[-2, 3]$ and constant on $(-\infty, -2]$ and $[3, \infty)$. The relative minimum value of f is -4 which matches the absolute minimum. The relative and absolute maximum values also coincide at 6. Every point on the graph of $y = g(x)$ for $x < -2$ and $x > 3$ yields both a relative minimum and relative maximum. The point $(-2, -4)$, however, gives only a relative minimum and the point $(3, 6)$ yields only a relative maximum. (Recall the Exercises in Section 1.6.2 which dealt with constant functions.) □

Many of the applications that the authors are aware of involving absolute values also involve absolute value inequalities. For that reason, we save our discussion of applications for Section 2.4.

2.2 Absolute Value Functions

2.2.1 Exercises

In Exercises 1 - 15, solve the equation.

1. $|x| = 6$
2. $|3x - 1| = 10$
3. $|4 - x| = 7$
4. $4 - |x| = 3$
5. $2|5x + 1| - 3 = 0$
6. $|7x - 1| + 2 = 0$
7. $\dfrac{5 - |x|}{2} = 1$
8. $\frac{2}{3}|5 - 2x| - \frac{1}{2} = 5$
9. $|x| = x + 3$
10. $|2x - 1| = x + 1$
11. $4 - |x| = 2x + 1$
12. $|x - 4| = x - 5$
13. $|x| = x^2$
14. $|x| = 12 - x^2$
15. $|x^2 - 1| = 3$

Prove that if $|f(x)| = |g(x)|$ then either $f(x) = g(x)$ or $f(x) = -g(x)$. Use that result to solve the equations in Exercises 16 - 21.

16. $|3x - 2| = |2x + 7|$
17. $|3x + 1| = |4x|$
18. $|1 - 2x| = |x + 1|$
19. $|4 - x| - |x + 2| = 0$
20. $|2 - 5x| = 5|x + 1|$
21. $3|x - 1| = 2|x + 1|$

In Exercises 22 - 33, graph the function. Find the zeros of each function and the x- and y-intercepts of each graph, if any exist. From the graph, determine the domain and range of each function, list the intervals on which the function is increasing, decreasing or constant, and find the relative and absolute extrema, if they exist.

22. $f(x) = |x + 4|$
23. $f(x) = |x| + 4$
24. $f(x) = |4x|$
25. $f(x) = -3|x|$
26. $f(x) = 3|x + 4| - 4$
27. $f(x) = \dfrac{1}{3}|2x - 1|$
28. $f(x) = \dfrac{|x + 4|}{x + 4}$
29. $f(x) = \dfrac{|2 - x|}{2 - x}$
30. $f(x) = x + |x| - 3$
31. $f(x) = |x + 2| - x$
32. $f(x) = |x + 2| - |x|$
33. $f(x) = |x + 4| + |x - 2|$

34. With the help of your classmates, find an absolute value function whose graph is given below.

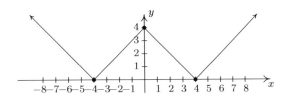

35. With help from your classmates, prove the second, third and fifth parts of Theorem 2.1.

36. Prove **The Triangle Inequality:** For all real numbers a and b, $|a + b| \leq |a| + |b|$.

2.2.2 Answers

1. $x = -6$ or $x = 6$
2. $x = -3$ or $x = \frac{11}{3}$
3. $x = -3$ or $x = 11$
4. $x = -1$ or $x = 1$
5. $x = -\frac{1}{2}$ or $x = \frac{1}{10}$
6. no solution
7. $x = -3$ or $x = 3$
8. $x = -\frac{13}{8}$ or $x = \frac{53}{8}$
9. $x = -\frac{3}{2}$
10. $x = 0$ or $x = 2$
11. $x = 1$
12. no solution
13. $x = -1$, $x = 0$ or $x = 1$
14. $x = -3$ or $x = 3$
15. $x = -2$ or $x = 2$
16. $x = -1$ or $x = 9$
17. $x = -\frac{1}{7}$ or $x = 1$
18. $x = 0$ or $x = 2$
19. $x = 1$
20. $x = -\frac{3}{10}$
21. $x = \frac{1}{5}$ or $x = 5$

22. $f(x) = |x + 4|$
    $f(-4) = 0$
    x-intercept $(-4, 0)$
    y-intercept $(0, 4)$
    Domain $(-\infty, \infty)$
    Range $[0, \infty)$
    Decreasing on $(-\infty, -4]$
    Increasing on $[-4, \infty)$
    Relative and absolute min. at $(-4, 0)$
    No relative or absolute maximum

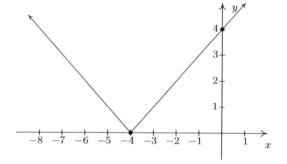

23. $f(x) = |x| + 4$
    No zeros
    No x-intercepts
    y-intercept $(0, 4)$
    Domain $(-\infty, \infty)$
    Range $[4, \infty)$
    Decreasing on $(-\infty, 0]$
    Increasing on $[0, \infty)$
    Relative and absolute minimum at $(0, 4)$
    No relative or absolute maximum

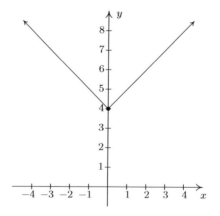

2.2 Absolute Value Functions

24. $f(x) = |4x|$
    $f(0) = 0$
    x-intercept $(0, 0)$
    y-intercept $(0, 0)$
    Domain $(-\infty, \infty)$
    Range $[0, \infty)$
    Decreasing on $(-\infty, 0]$
    Increasing on $[0, \infty)$
    Relative and absolute minimum at $(0, 0)$
    No relative or absolute maximum

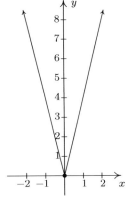

25. $f(x) = -3|x|$
    $f(0) = 0$
    x-intercept $(0, 0)$
    y-intercept $(0, 0)$
    Domain $(-\infty, \infty)$
    Range $(-\infty, 0]$
    Increasing on $(-\infty, 0]$
    Decreasing on $[0, \infty)$
    Relative and absolute maximum at $(0, 0)$
    No relative or absolute minimum

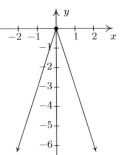

26. $f(x) = 3|x + 4| - 4$
    $f\left(-\frac{16}{3}\right) = 0$, $f\left(-\frac{8}{3}\right) = 0$
    x-intercepts $\left(-\frac{16}{3}, 0\right)$, $\left(-\frac{8}{3}, 0\right)$
    y-intercept $(0, 8)$
    Domain $(-\infty, \infty)$
    Range $[-4, \infty)$
    Decreasing on $(-\infty, -4]$
    Increasing on $[-4, \infty)$
    Relative and absolute min. at $(-4, -4)$
    No relative or absolute maximum

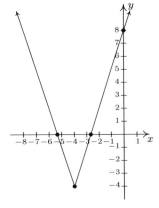

27. $f(x) = \frac{1}{3}|2x - 1|$
    $f\left(\frac{1}{2}\right) = 0$
    x-intercepts $\left(\frac{1}{2}, 0\right)$
    y-intercept $\left(0, \frac{1}{3}\right)$
    Domain $(-\infty, \infty)$
    Range $[0, \infty)$
    Decreasing on $\left(-\infty, \frac{1}{2}\right]$
    Increasing on $\left[\frac{1}{2}, \infty\right)$

    Relative and absolute min. at $\left(\frac{1}{2}, 0\right)$
    No relative or absolute maximum

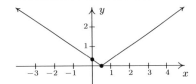

28. $f(x) = \dfrac{|x+4|}{x+4}$
No zeros
No x-intercept
y-intercept $(0, 1)$
Domain $(-\infty, -4) \cup (-4, \infty)$
Range $\{-1, 1\}$
Constant on $(-\infty, -4)$
Constant on $(-4, \infty)$
Absolute minimum at every point $(x, -1)$ where $x < -4$
Absolute maximum at every point $(x, 1)$ where $x > -4$
Relative maximum AND minimum at every point on the graph

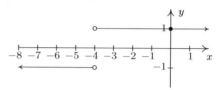

29. $f(x) = \dfrac{|2-x|}{2-x}$
No zeros
No x-intercept
y-intercept $(0, 1)$
Domain $(-\infty, 2) \cup (2, \infty)$
Range $\{-1, 1\}$
Constant on $(-\infty, 2)$
Constant on $(2, \infty)$
Absolute minimum at every point $(x, -1)$ where $x > 2$
Absolute maximum at every point $(x, 1)$ where $x < 2$
Relative maximum AND minimum at every point on the graph

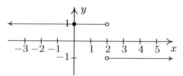

30. Re-write $f(x) = x + |x| - 3$ as
$$f(x) = \begin{cases} -3 & \text{if } x < 0 \\ 2x - 3 & \text{if } x \geq 0 \end{cases}$$
$f\left(\tfrac{3}{2}\right) = 0$
x-intercept $\left(\tfrac{3}{2}, 0\right)$
y-intercept $(0, -3)$
Domain $(-\infty, \infty)$
Range $[-3, \infty)$
Increasing on $[0, \infty)$
Constant on $(-\infty, 0]$
Absolute minimum at every point $(x, -3)$ where $x \leq 0$
No absolute maximum
Relative minimum at every point $(x, -3)$ where $x \leq 0$
Relative maximum at every point $(x, -3)$ where $x < 0$

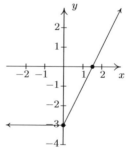

2.2 Absolute Value Functions

31. Re-write $f(x) = |x+2| - x$ as
$$f(x) = \begin{cases} -2x - 2 & \text{if } x < -2 \\ 2 & \text{if } x \geq -2 \end{cases}$$
No zeros
No x-intercepts
y-intercept $(0, 2)$
Domain $(-\infty, \infty)$
Range $[2, \infty)$
Decreasing on $(-\infty, -2]$
Constant on $[-2, \infty)$
Absolute minimum at every point $(x, 2)$ where $x \geq -2$

No absolute maximum
Relative minimum at every point $(x, 2)$ where $x \geq -2$
Relative maximum at every point $(x, 2)$ where $x > -2$

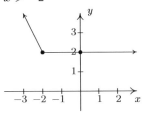

32. Re-write $f(x) = |x+2| - |x|$ as
$$f(x) = \begin{cases} -2 & \text{if } x < -2 \\ 2x + 2 & \text{if } -2 \leq x < 0 \\ 2 & \text{if } x \geq 0 \end{cases}$$
$f(-1) = 0$
x-intercept $(-1, 0)$
y-intercept $(0, 2)$
Domain $(-\infty, \infty)$
Range $[-2, 2]$
Increasing on $[-2, 0]$
Constant on $(-\infty, -2]$
Constant on $[0, \infty)$
Absolute minimum at every point $(x, -2)$ where $x \leq -2$

Absolute maximum at every point $(x, 2)$ where $x \geq 0$
Relative minimum at every point $(x, -2)$ where $x \leq -2$ and at every point $(x, 2)$ where $x > 0$
Relative maximum at every point $(x, -2)$ where $x < -2$ and at every point $(x, 2)$ where $x \geq 0$

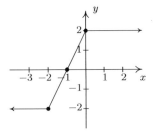

33. Re-write $f(x) = |x+4| + |x-2|$ as
$$f(x) = \begin{cases} -2x - 2 & \text{if } x < -4 \\ 6 & \text{if } -4 \leq x < 2 \\ 2x + 2 & \text{if } x \geq 2 \end{cases}$$
No zeros
No x-intercept
y-intercept $(0, 6)$
Domain $(-\infty, \infty)$
Range $[6, \infty)$
Decreasing on $(-\infty, -4]$
Constant on $[-4, 2]$
Increasing on $[2, \infty)$
Absolute minimum at every point $(x, 6)$ where $-4 \leq x \leq 2$
No absolute maximum
Relative minimum at every point $(x, 6)$ where $-4 \leq x \leq 2$
Relative maximum at every point $(x, 6)$ where $-4 < x < 2$

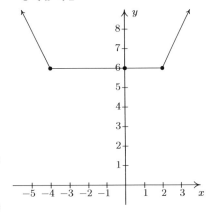

35. $f(x) = ||x| - 4|$

2.3 Quadratic Functions

You may recall studying quadratic equations in Intermediate Algebra. In this section, we review those equations in the context of our next family of functions: the quadratic functions.

> **Definition 2.5.** A **quadratic function** is a function of the form
>
> $$f(x) = ax^2 + bx + c,$$
>
> where a, b and c are real numbers with $a \neq 0$. The domain of a quadratic function is $(-\infty, \infty)$.

The most basic quadratic function is $f(x) = x^2$, whose graph appears below. Its shape should look familiar from Intermediate Algebra – it is called a **parabola**. The point $(0,0)$ is called the **vertex** of the parabola. In this case, the vertex is a relative minimum and is also the where the absolute minimum value of f can be found.

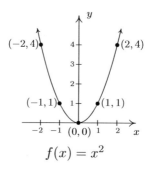

$f(x) = x^2$

Much like many of the absolute value functions in Section 2.2, knowing the graph of $f(x) = x^2$ enables us to graph an entire family of quadratic functions using transformations.

Example 2.3.1. Graph the following functions starting with the graph of $f(x) = x^2$ and using transformations. Find the vertex, state the range and find the x- and y-intercepts, if any exist.

1. $g(x) = (x+2)^2 - 3$
2. $h(x) = -2(x-3)^2 + 1$

Solution.

1. Since $g(x) = (x+2)^2 - 3 = f(x+2) - 3$, Theorem 1.7 instructs us to first *subtract* 2 from each of the x-values of the points on $y = f(x)$. This shifts the graph of $y = f(x)$ to the *left* 2 units and moves $(-2, 4)$ to $(-4, 4)$, $(-1, 1)$ to $(-3, 1)$, $(0, 0)$ to $(-2, 0)$, $(1, 1)$ to $(-1, 1)$ and $(2, 4)$ to $(0, 4)$. Next, we *subtract* 3 from each of the y-values of these new points. This moves the graph *down* 3 units and moves $(-4, 4)$ to $(-4, 1)$, $(-3, 1)$ to $(-3, -2)$, $(-2, 0)$ to $(-2, 3)$, $(-1, 1)$ to $(-1, -2)$ and $(0, 4)$ to $(0, 1)$. We connect the dots in parabolic fashion to get

2.3 QUADRATIC FUNCTIONS

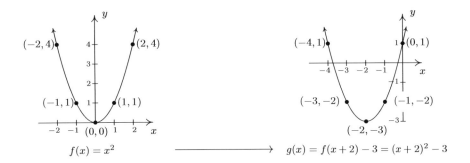

$$f(x) = x^2 \quad\longrightarrow\quad g(x) = f(x+2) - 3 = (x+2)^2 - 3$$

From the graph, we see that the vertex has moved from $(0,0)$ on the graph of $y = f(x)$ to $(-2,-3)$ on the graph of $y = g(x)$. This sets $[-3, \infty)$ as the range of g. We see that the graph of $y = g(x)$ crosses the x-axis twice, so we expect two x-intercepts. To find these, we set $y = g(x) = 0$ and solve. Doing so yields the equation $(x+2)^2 - 3 = 0$, or $(x+2)^2 = 3$. Extracting square roots gives $x + 2 = \pm\sqrt{3}$, or $x = -2 \pm \sqrt{3}$. Our x-intercepts are $(-2 - \sqrt{3}, 0) \approx (-3.73, 0)$ and $(-2 + \sqrt{3}, 0) \approx (-0.27, 0)$. The y-intercept of the graph, $(0,1)$ was one of the points we originally plotted, so we are done.

2. Following Theorem 1.7 once more, to graph $h(x) = -2(x-3)^2 + 1 = -2f(x-3) + 1$, we first start by *adding* 3 to each of the x-values of the points on the graph of $y = f(x)$. This effects a horizontal shift *right* 3 units and moves $(-2, 4)$ to $(1, 4)$, $(-1, 1)$ to $(2, 1)$, $(0, 0)$ to $(3, 0)$, $(1, 1)$ to $(4, 1)$ and $(2, 4)$ to $(5, 4)$. Next, we *multiply* each of our y-values first by -2 and then *add* 1 to that result. Geometrically, this is a vertical *stretch* by a factor of 2, followed by a reflection about the x-axis, followed by a vertical shift *up* 1 unit. This moves $(1, 4)$ to $(1, -7)$, $(2, 1)$ to $(2, -1)$, $(3, 0)$ to $(3, 1)$, $(4, 1)$ to $(4, -1)$ and $(5, 4)$ to $(5, -7)$.

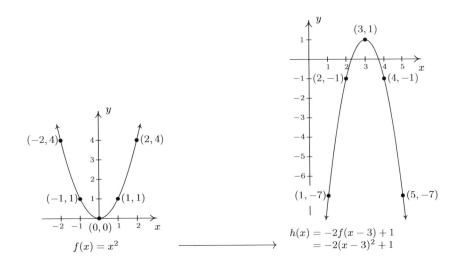

The vertex is $(3, 1)$ which makes the range of h $(-\infty, 1]$. From our graph, we know that there are two x-intercepts, so we set $y = h(x) = 0$ and solve. We get $-2(x-3)^2 + 1 = 0$

which gives $(x-3)^2 = \frac{1}{2}$. Extracting square roots[1] gives $x - 3 = \pm\frac{\sqrt{2}}{2}$, so that when we add 3 to each side,[2] we get $x = \frac{6\pm\sqrt{2}}{2}$. Hence, our x-intercepts are $\left(\frac{6-\sqrt{2}}{2}, 0\right) \approx (2.29, 0)$ and $\left(\frac{6+\sqrt{2}}{2}, 0\right) \approx (3.71, 0)$. Although our graph doesn't show it, there is a y-intercept which can be found by setting $x = 0$. With $h(0) = -2(0-3)^2 + 1 = -17$, we have that our y-intercept is $(0, -17)$. $\square$

A few remarks about Example 2.3.1 are in order. First note that neither the formula given for $g(x)$ nor the one given for $h(x)$ match the form given in Definition 2.5. We could, of course, convert both $g(x)$ and $h(x)$ into that form by expanding and collecting like terms. Doing so, we find $g(x) = (x+2)^2 - 3 = x^2 + 4x + 1$ and $h(x) = -2(x-3)^2 + 1 = -2x^2 + 12x - 17$. While these 'simplified' formulas for $g(x)$ and $h(x)$ satisfy Definition 2.5, they do not lend themselves to graphing easily. For that reason, the form of g and h presented in Example 2.3.2 is given a special name, which we list below, along with the form presented in Definition 2.5.

Definition 2.6. Standard and General Form of Quadratic Functions: Suppose f is a quadratic function.

- The **general form** of the quadratic function f is $f(x) = ax^2 + bx + c$, where a, b and c are real numbers with $a \neq 0$.

- The **standard form** of the quadratic function f is $f(x) = a(x-h)^2 + k$, where a, h and k are real numbers with $a \neq 0$.

It is important to note at this stage that we have no guarantees that *every* quadratic function can be written in standard form. This is actually true, and we prove this later in the exposition, but for now we celebrate the advantages of the standard form, starting with the following theorem.

Theorem 2.2. Vertex Formula for Quadratics in Standard Form: For the quadratic function $f(x) = a(x-h)^2 + k$, where a, h and k are real numbers with $a \neq 0$, the vertex of the graph of $y = f(x)$ is (h, k).

We can readily verify the formula given Theorem 2.2 with the two functions given in Example 2.3.1. After a (slight) rewrite, $g(x) = (x+2)^2 - 3 = (x - (-2))^2 + (-3)$, and we identify $h = -2$ and $k = -3$. Sure enough, we found the vertex of the graph of $y = g(x)$ to be $(-2, -3)$. For $h(x) = -2(x-3)^2 + 1$, no rewrite is needed. We can directly identify $h = 3$ and $k = 1$ and, sure enough, we found the vertex of the graph of $y = h(x)$ to be $(3, 1)$.

To see why the formula in Theorem 2.2 produces the vertex, consider the graph of the equation $y = a(x-h)^2 + k$. When we substitute $x = h$, we get $y = k$, so (h, k) is on the graph. If $x \neq h$, then $x - h \neq 0$ so $(x-h)^2$ is a positive number. If $a > 0$, then $a(x-h)^2$ is positive, thus $y = a(x-h)^2 + k$ is always a number larger than k. This means that when $a > 0$, (h, k) is the lowest point on the graph and thus the parabola must open upwards, making (h, k) the vertex. A similar argument

---

[1] and rationalizing denominators!
[2] and get common denominators!

2.3 Quadratic Functions

shows that if $a < 0$, (h, k) is the highest point on the graph, so the parabola opens downwards, and (h, k) is also the vertex in this case.

Alternatively, we can apply the machinery in Section 1.7. Since the vertex of $y = x^2$ is $(0, 0)$, we can determine the vertex of $y = a(x-h)^2 + k$ by determining the final destination of $(0, 0)$ as it is moved through each transformation. To obtain the formula $f(x) = a(x-h)^2 + k$, we start with $g(x) = x^2$ and first define $g_1(x) = ag(x) = ax^2$. This is results in a vertical scaling and/or reflection.[3] Since we multiply the output by a, we multiply the y-coordinates on the graph of g by a, so the point $(0, 0)$ remains $(0, 0)$ and remains the vertex. Next, we define $g_2(x) = g_1(x-h) = a(x-h)^2$. This induces a horizontal shift right or left h units[4] moves the vertex, in either case, to $(h, 0)$. Finally, $f(x) = g_2(x) + k = a(x-h)^2 + k$ which effects a vertical shift up or down k units[5] resulting in the vertex moving from $(h, 0)$ to (h, k).

In addition to verifying Theorem 2.2, the arguments in the two preceding paragraphs have also shown us the role of the number a in the graphs of quadratic functions. The graph of $y = a(x-h)^2 + k$ is a parabola 'opening upwards' if $a > 0$, and 'opening downwards' if $a < 0$. Moreover, the symmetry enjoyed by the graph of $y = x^2$ about the y-axis is translated to a symmetry about the vertical line $x = h$ which is the vertical line through the vertex.[6] This line is called the **axis of symmetry** of the parabola and is dashed in the figures below.

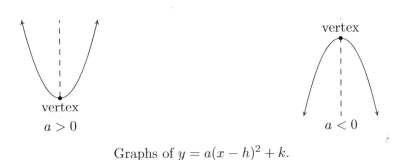

Graphs of $y = a(x-h)^2 + k$.

Without a doubt, the standard form of a quadratic function, coupled with the machinery in Section 1.7, allows us to list the attributes of the graphs of such functions quickly and elegantly. What remains to be shown, however, is the fact that every quadratic function *can be written* in standard form. To convert a quadratic function given in general form into standard form, we employ the ancient rite of 'Completing the Square'. We remind the reader how this is done in our next example.

Example 2.3.2. Convert the functions below from general form to standard form. Find the vertex, axis of symmetry and any x- or y-intercepts. Graph each function and determine its range.

1. $f(x) = x^2 - 4x + 3$.

2. $g(x) = 6 - x - x^2$

---

[3] Just a scaling if $a > 0$. If $a < 0$, there is a reflection involved.
[4] Right if $h > 0$, left if $h < 0$.
[5] Up if $k > 0$, down if $k < 0$
[6] You should use transformations to verify this!

Solution.

1. To convert from general form to standard form, we complete the square.[7] First, we verify that the coefficient of x^2 is 1. Next, we find the coefficient of x, in this case -4, and take half of it to get $\frac{1}{2}(-4) = -2$. This tells us that our target perfect square quantity is $(x-2)^2$. To get an expression equivalent to $(x-2)^2$, we need to add $(-2)^2 = 4$ to the $x^2 - 4x$ to create a perfect square trinomial, but to keep the balance, we must also subtract it. We collect the terms which create the perfect square and gather the remaining constant terms. Putting it all together, we get

$$\begin{aligned} f(x) &= x^2 - 4x + 3 & \text{(Compute } \tfrac{1}{2}(-4) = -2.\text{)} \\ &= \left(x^2 - 4x + \underline{4} - \underline{4}\right) + 3 & \text{(Add and subtract } (-2)^2 = 4 \text{ to } (x^2 + 4x).\text{)} \\ &= \left(x^2 - 4x + 4\right) - 4 + 3 & \text{(Group the perfect square trinomial.)} \\ &= (x-2)^2 - 1 & \text{(Factor the perfect square trinomial.)} \end{aligned}$$

Of course, we can always check our answer by multiplying out $f(x) = (x-2)^2 - 1$ to see that it simplifies to $f(x) = x^2 - 4x - 1$. In the form $f(x) = (x-2)^2 - 1$, we readily find the vertex to be $(2, -1)$ which makes the axis of symmetry $x = 2$. To find the x-intercepts, we set $y = f(x) = 0$. We are spoiled for choice, since we have *two* formulas for $f(x)$. Since we recognize $f(x) = x^2 - 4x + 3$ to be easily factorable,[8] we proceed to solve $x^2 - 4x + 3 = 0$. Factoring gives $(x-3)(x-1) = 0$ so that $x = 3$ or $x = 1$. The x-intercepts are then $(1, 0)$ and $(3, 0)$. To find the y-intercept, we set $x = 0$. Once again, the general form $f(x) = x^2 - 4x + 3$ is easiest to work with here, and we find $y = f(0) = 3$. Hence, the y-intercept is $(0, 3)$. With the vertex, axis of symmetry and the intercepts, we get a pretty good graph without the need to plot additional points. We see that the range of f is $[-1, \infty)$ and we are done.

2. To get started, we rewrite $g(x) = 6 - x - x^2 = -x^2 - x + 6$ and note that the coefficient of x^2 is -1, not 1. This means our first step is to factor out the (-1) from both the x^2 and x terms. We then follow the completing the square recipe as above.

$$\begin{aligned} g(x) &= -x^2 - x + 6 \\ &= (-1)\left(x^2 + x\right) + 6 & \text{(Factor the coefficient of } x^2 \text{ from } x^2 \text{ and } x.\text{)} \\ &= (-1)\left(x^2 + x + \tfrac{1}{4} - \tfrac{1}{4}\right) + 6 \\ &= (-1)\left(x^2 + x + \tfrac{1}{4}\right) + (-1)\left(-\tfrac{1}{4}\right) + 6 & \text{(Group the perfect square trinomial.)} \\ &= -\left(x + \tfrac{1}{2}\right)^2 + \tfrac{25}{4} \end{aligned}$$

---

[7] If you forget why we do what we do to complete the square, start with $a(x-h)^2 + k$, multiply it out, step by step, and then reverse the process.
[8] Experience pays off, here!

From $g(x) = -\left(x + \frac{1}{2}\right)^2 + \frac{25}{4}$, we get the vertex to be $\left(-\frac{1}{2}, \frac{25}{4}\right)$ and the axis of symmetry to be $x = -\frac{1}{2}$. To get the x-intercepts, we opt to set the given formula $g(x) = 6 - x - x^2 = 0$. Solving, we get $x = -3$ and $x = 2$, so the x-intercepts are $(-3, 0)$ and $(2, 0)$. Setting $x = 0$, we find $g(0) = 6$, so the y-intercept is $(0, 6)$. Plotting these points gives us the graph below. We see that the range of g is $\left(-\infty, \frac{25}{4}\right]$.

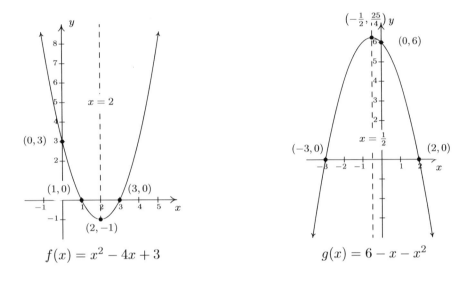

□

With Example 2.3.2 fresh in our minds, we are now in a position to show that every quadratic function can be written in standard form. We begin with $f(x) = ax^2 + bx + c$, assume $a \neq 0$, and complete the square in *complete* generality.

$$
\begin{aligned}
f(x) &= ax^2 + bx + c \\
&= a\left(x^2 + \frac{b}{a}x\right) + c && \text{(Factor out coefficient of } x^2 \text{ from } x^2 \text{ and } x.\text{)} \\
&= a\left(x^2 + \frac{b}{a}x + \frac{b^2}{4a^2} - \frac{b^2}{4a^2}\right) + c \\
&= a\left(x^2 + \frac{b}{a}x + \frac{b^2}{4a^2}\right) - a\left(\frac{b^2}{4a^2}\right) + c && \text{(Group the perfect square trinomial.)} \\
&= a\left(x + \frac{b}{2a}\right)^2 + \frac{4ac - b^2}{4a} && \text{(Factor and get a common denominator.)}
\end{aligned}
$$

Comparing this last expression with the standard form, we identify $(x - h)$ with $\left(x + \frac{b}{2a}\right)$ so that $h = -\frac{b}{2a}$. Instead of memorizing the value $k = \frac{4ac - b^2}{4a}$, we see that $f\left(-\frac{b}{2a}\right) = \frac{4ac - b^2}{4a}$. As such, we have derived a vertex formula for the general form. We summarize both vertex formulas in the box at the top of the next page.

> **Equation 2.4. Vertex Formulas for Quadratic Functions:** Suppose a, b, c, h and k are real numbers with $a \neq 0$.
>
> - If $f(x) = a(x-h)^2 + k$, the vertex of the graph of $y = f(x)$ is the point (h, k).
> - If $f(x) = ax^2 + bx + c$, the vertex of the graph of $y = f(x)$ is the point $\left(-\dfrac{b}{2a}, f\left(-\dfrac{b}{2a}\right)\right)$.

There are two more results which can be gleaned from the completed-square form of the general form of a quadratic function,

$$f(x) = ax^2 + bx + c = a\left(x + \frac{b}{2a}\right)^2 + \frac{4ac - b^2}{4a}$$

We have seen that the number a in the standard form of a quadratic function determines whether the parabola opens upwards (if $a > 0$) or downwards (if $a < 0$). We see here that this number a is none other than the coefficient of x^2 in the general form of the quadratic function. In other words, it is the coefficient of x^2 alone which determines this behavior – a result that is generalized in Section 3.1. The second treasure is a re-discovery of the **quadratic formula**.

> **Equation 2.5. The Quadratic Formula:** If a, b and c are real numbers with $a \neq 0$, then the solutions to $ax^2 + bx + c = 0$ are
>
> $$x = \frac{-b \pm \sqrt{b^2 - 4ac}}{2a}.$$

Assuming the conditions of Equation 2.5, the solutions to $ax^2 + bx + c = 0$ are precisely the zeros of $f(x) = ax^2 + bx + c$. Since

$$f(x) = ax^2 + bx + c = a\left(x + \frac{b}{2a}\right)^2 + \frac{4ac - b^2}{4a}$$

the equation $ax^2 + bx + c = 0$ is equivalent to

$$a\left(x + \frac{b}{2a}\right)^2 + \frac{4ac - b^2}{4a} = 0.$$

Solving gives

2.3 Quadratic Functions

$$a\left(x + \frac{b}{2a}\right)^2 + \frac{4ac - b^2}{4a} = 0$$

$$a\left(x + \frac{b}{2a}\right)^2 = -\frac{4ac - b^2}{4a}$$

$$\frac{1}{a}\left[a\left(x + \frac{b}{2a}\right)^2\right] = \frac{1}{a}\left(\frac{b^2 - 4ac}{4a}\right)$$

$$\left(x + \frac{b}{2a}\right)^2 = \frac{b^2 - 4ac}{4a^2}$$

$$x + \frac{b}{2a} = \pm\sqrt{\frac{b^2 - 4ac}{4a^2}} \qquad \text{extract square roots}$$

$$x + \frac{b}{2a} = \pm\frac{\sqrt{b^2 - 4ac}}{2a}$$

$$x = -\frac{b}{2a} \pm \frac{\sqrt{b^2 - 4ac}}{2a}$$

$$x = \frac{-b \pm \sqrt{b^2 - 4ac}}{2a}$$

In our discussions of domain, we were warned against having negative numbers underneath the square root. Given that $\sqrt{b^2 - 4ac}$ is part of the Quadratic Formula, we will need to pay special attention to the radicand $b^2 - 4ac$. It turns out that the quantity $b^2 - 4ac$ plays a critical role in determining the nature of the solutions to a quadratic equation. It is given a special name.

Definition 2.7. If a, b and c are real numbers with $a \neq 0$, then the **discriminant** of the quadratic equation $ax^2 + bx + c = 0$ is the quantity $b^2 - 4ac$.

The discriminant 'discriminates' between the kinds of solutions we get from a quadratic equation. These cases, and their relation to the discriminant, are summarized below.

Theorem 2.3. Discriminant Trichotomy: Let a, b and c be real numbers with $a \neq 0$.

- If $b^2 - 4ac < 0$, the equation $ax^2 + bx + c = 0$ has no real solutions.
- If $b^2 - 4ac = 0$, the equation $ax^2 + bx + c = 0$ has exactly one real solution.
- If $b^2 - 4ac > 0$, the equation $ax^2 + bx + c = 0$ has exactly two real solutions.

The proof of Theorem 2.3 stems from the position of the discriminant in the quadratic equation, and is left as a good mental exercise for the reader. The next example exploits the fruits of all of our labor in this section thus far.

Example 2.3.3. Recall that the profit (defined on page 82) for a product is defined by the equation Profit = Revenue − Cost, or $P(x) = R(x) - C(x)$. In Example 2.1.7 the weekly revenue, in dollars, made by selling x PortaBoy Game Systems was found to be $R(x) = -1.5x^2 + 250x$ with the restriction (carried over from the price-demand function) that $0 \leq x \leq 166$. The cost, in dollars, to produce x PortaBoy Game Systems is given in Example 2.1.5 as $C(x) = 80x + 150$ for $x \geq 0$.

1. Determine the weekly profit function $P(x)$.

2. Graph $y = P(x)$. Include the x- and y-intercepts as well as the vertex and axis of symmetry.

3. Interpret the zeros of P.

4. Interpret the vertex of the graph of $y = P(x)$.

5. Recall that the weekly price-demand equation for PortaBoys is $p(x) = -1.5x + 250$, where $p(x)$ is the price per PortaBoy, in dollars, and x is the weekly sales. What should the price per system be in order to maximize profit?

Solution.

1. To find the profit function $P(x)$, we subtract

$$P(x) = R(x) - C(x) = \left(-1.5x^2 + 250x\right) - (80x + 150) = -1.5x^2 + 170x - 150.$$

Since the revenue function is valid when $0 \leq x \leq 166$, P is also restricted to these values.

2. To find the x-intercepts, we set $P(x) = 0$ and solve $-1.5x^2 + 170x - 150 = 0$. The mere thought of trying to factor the left hand side of this equation could do serious psychological damage, so we resort to the quadratic formula, Equation 2.5. Identifying $a = -1.5$, $b = 170$, and $c = -150$, we obtain

$$\begin{aligned} x &= \frac{-b \pm \sqrt{b^2 - 4ac}}{2a} \\ &= \frac{-170 \pm \sqrt{170^2 - 4(-1.5)(-150)}}{2(-1.5)} \\ &= \frac{-170 \pm \sqrt{28000}}{-3} \\ &= \frac{170 \pm 20\sqrt{70}}{3} \end{aligned}$$

We get two x-intercepts: $\left(\frac{170-20\sqrt{70}}{3}, 0\right)$ and $\left(\frac{170+20\sqrt{70}}{3}, 0\right)$. To find the y-intercept, we set $x = 0$ and find $y = P(0) = -150$ for a y-intercept of $(0, -150)$. To find the vertex, we use the fact that $P(x) = -1.5x^2 + 170x - 150$ is in the general form of a quadratic function and appeal to Equation 2.4. Substituting $a = -1.5$ and $b = 170$, we get $x = -\frac{170}{2(-1.5)} = \frac{170}{3}$.

To find the y-coordinate of the vertex, we compute $P\left(\frac{170}{3}\right) = \frac{14000}{3}$ and find that our vertex is $\left(\frac{170}{3}, \frac{14000}{3}\right)$. The axis of symmetry is the vertical line passing through the vertex so it is the line $x = \frac{170}{3}$. To sketch a reasonable graph, we approximate the x-intercepts, $(0.89, 0)$ and $(112.44, 0)$, and the vertex, $(56.67, 4666.67)$. (Note that in order to get the x-intercepts and the vertex to show up in the same picture, we had to scale the x-axis differently than the y-axis. This results in the left-hand x-intercept and the y-intercept being uncomfortably close to each other and to the origin in the picture.)

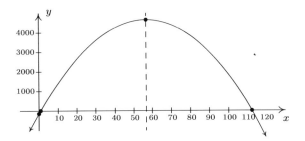

3. The zeros of P are the solutions to $P(x) = 0$, which we have found to be approximately 0.89 and 112.44. As we saw in Example 1.5.3, these are the 'break-even' points of the profit function, where enough product is sold to recover the cost spent to make the product. More importantly, we see from the graph that as long as x is between 0.89 and 112.44, the graph $y = P(x)$ is above the x-axis, meaning $y = P(x) > 0$ there. This means that for these values of x, a profit is being made. Since x represents the weekly sales of PortaBoy Game Systems, we round the zeros to positive integers and have that as long as 1, but no more than 112 game systems are sold weekly, the retailer will make a profit.

4. From the graph, we see that the maximum value of P occurs at the vertex, which is approximately $(56.67, 4666.67)$. As above, x represents the weekly sales of PortaBoy systems, so we can't sell 56.67 game systems. Comparing $P(56) = 4666$ and $P(57) = 4666.5$, we conclude that we will make a maximum profit of $4666.50 if we sell 57 game systems.

5. In the previous part, we found that we need to sell 57 PortaBoys per week to maximize profit. To find the price per PortaBoy, we substitute $x = 57$ into the price-demand function to get $p(57) = -1.5(57) + 250 = 164.5$. The price should be set at $164.50. □

Our next example is another classic application of quadratic functions.

Example 2.3.4. Much to Donnie's surprise and delight, he inherits a large parcel of land in Ashtabula County from one of his (e)strange(d) relatives. The time is finally right for him to pursue his dream of farming alpaca. He wishes to build a rectangular pasture, and estimates that he has enough money for 200 linear feet of fencing material. If he makes the pasture adjacent to a stream (so no fencing is required on that side), what are the dimensions of the pasture which maximize the area? What is the maximum area? If an average alpaca needs 25 square feet of grazing area, how many alpaca can Donnie keep in his pasture?

Solution. It is always helpful to sketch the problem situation, so we do so below.

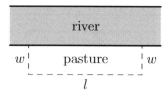

We are tasked to find the dimensions of the pasture which would give a maximum area. We let w denote the width of the pasture and we let l denote the length of the pasture. Since the units given to us in the statement of the problem are feet, we assume w and l are measured in feet. The area of the pasture, which we'll call A, is related to w and l by the equation $A = wl$. Since w and l are both measured in feet, A has units of feet^2, or square feet. We are given the total amount of fencing available is 200 feet, which means $w + l + w = 200$, or, $l + 2w = 200$. We now have two equations, $A = wl$ and $l + 2w = 200$. In order to use the tools given to us in this section to *maximize* A, we need to use the information given to write A as a function of just *one* variable, either w or l. This is where we use the equation $l + 2w = 200$. Solving for l, we find $l = 200 - 2w$, and we substitute this into our equation for A. We get $A = wl = w(200 - 2w) = 200w - 2w^2$. We now have A as a function of w, $A(w) = 200w - 2w^2 = -2w^2 + 200w$.

Before we go any further, we need to find the applied domain of A so that we know what values of w make sense in this problem situation.[9] Since w represents the width of the pasture, $w > 0$. Likewise, l represents the length of the pasture, so $l = 200 - 2w > 0$. Solving this latter inequality, we find $w < 100$. Hence, the function we wish to maximize is $A(w) = -2w^2 + 200w$ for $0 < w < 100$. Since A is a quadratic function (of w), we know that the graph of $y = A(w)$ is a parabola. Since the coefficient of w^2 is -2, we know that this parabola opens downwards. This means that there is a maximum value to be found, and we know it occurs at the vertex. Using the vertex formula, we find $w = -\frac{200}{2(-2)} = 50$, and $A(50) = -2(50)^2 + 200(50) = 5000$. Since $w = 50$ lies in the applied domain, $0 < w < 100$, we have that the area of the pasture is maximized when the width is 50 feet. To find the length, we use $l = 200 - 2w$ and find $l = 200 - 2(50) = 100$, so the length of the pasture is 100 feet. The maximum area is $A(50) = 5000$, or 5000 square feet. If an average alpaca requires 25 square feet of pasture, Donnie can raise $\frac{5000}{25} = 200$ average alpaca. □

We conclude this section with the graph of a more complicated absolute value function.

Example 2.3.5. Graph $f(x) = |x^2 - x - 6|$.

Solution. Using the definition of absolute value, Definition 2.4, we have

$$f(x) = \begin{cases} -(x^2 - x - 6), & \text{if } x^2 - x - 6 < 0 \\ x^2 - x - 6, & \text{if } x^2 - x - 6 \geq 0 \end{cases}$$

The trouble is that we have yet to develop any analytic techniques to solve nonlinear inequalities such as $x^2 - x - 6 < 0$. You won't have to wait long; this is one of the main topics of Section 2.4.

---

[9]Donnie would be very upset if, for example, we told him the width of the pasture needs to be -50 feet.

2.3 QUADRATIC FUNCTIONS

Nevertheless, we can attack this problem graphically. To that end, we graph $y = g(x) = x^2 - x - 6$ using the intercepts and the vertex. To find the x-intercepts, we solve $x^2 - x - 6 = 0$. Factoring gives $(x - 3)(x + 2) = 0$ so $x = -2$ or $x = 3$. Hence, $(-2, 0)$ and $(3, 0)$ are x-intercepts. The y-intercept $(0, -6)$ is found by setting $x = 0$. To plot the vertex, we find $x = -\frac{b}{2a} = -\frac{-1}{2(1)} = \frac{1}{2}$, and $y = \left(\frac{1}{2}\right)^2 - \left(\frac{1}{2}\right) - 6 = -\frac{25}{4} = -6.25$. Plotting, we get the parabola seen below on the left. To obtain points on the graph of $y = f(x) = |x^2 - x - 6|$, we can take points on the graph of $g(x) = x^2 - x - 6$ and apply the absolute value to each of the y values on the parabola. We see from the graph of g that for $x \leq -2$ or $x \geq 3$, the y values on the parabola are greater than or equal to zero (since the graph is on or above the x-axis), so the absolute value leaves these portions of the graph alone. For x between -2 and 3, however, the y values on the parabola are negative. For example, the point $(0, -6)$ on $y = x^2 - x - 6$ would result in the point $(0, |-6|) = (0, -(-6)) = (0, 6)$ on the graph of $f(x) = |x^2 - x - 6|$. Proceeding in this manner for all points with x-coordinates between -2 and 3 results in the graph seen below on the right.

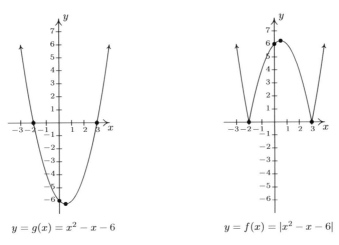

$y = g(x) = x^2 - x - 6$ \qquad $y = f(x) = |x^2 - x - 6|$

□

If we take a step back and look at the graphs of g and f in the last example, we notice that to obtain the graph of f from the graph of g, we reflect a *portion* of the graph of g about the x-axis. We can see this analytically by substituting $g(x) = x^2 - x - 6$ into the formula for $f(x)$ and calling to mind Theorem 1.4 from Section 1.7.

$$f(x) = \begin{cases} -g(x), & \text{if } g(x) < 0 \\ g(x), & \text{if } g(x) \geq 0 \end{cases}$$

The function f is defined so that when $g(x)$ is negative (i.e., when its graph is below the x-axis), the graph of f is its refection across the x-axis. This is a general template to graph functions of the form $f(x) = |g(x)|$. From this perspective, the graph of $f(x) = |x|$ can be obtained by reflecting the portion of the line $g(x) = x$ which is below the x-axis back above the x-axis creating the characteristic 'V' shape.

2.3.1 Exercises

In Exercises 1 - 9, graph the quadratic function. Find the x- and y-intercepts of each graph, if any exist. If it is given in general form, convert it into standard form; if it is given in standard form, convert it into general form. Find the domain and range of the function and list the intervals on which the function is increasing or decreasing. Identify the vertex and the axis of symmetry and determine whether the vertex yields a relative and absolute maximum or minimum.

1. $f(x) = x^2 + 2$
2. $f(x) = -(x+2)^2$
3. $f(x) = x^2 - 2x - 8$
4. $f(x) = -2(x+1)^2 + 4$
5. $f(x) = 2x^2 - 4x - 1$
6. $f(x) = -3x^2 + 4x - 7$
7. $f(x) = x^2 + x + 1$
8. $f(x) = -3x^2 + 5x + 4$
9.[10] $f(x) = x^2 - \frac{1}{100}x - 1$

In Exercises 10 - 14, the cost and price-demand functions are given for different scenarios. For each scenario,

- Find the profit function $P(x)$.
- Find the number of items which need to be sold in order to maximize profit.
- Find the maximum profit.
- Find the price to charge per item in order to maximize profit.
- Find and interpret break-even points.

10. The cost, in dollars, to produce x "I'd rather be a Sasquatch" T-Shirts is $C(x) = 2x + 26$, $x \geq 0$ and the price-demand function, in dollars per shirt, is $p(x) = 30 - 2x$, $0 \leq x \leq 15$.

11. The cost, in dollars, to produce x bottles of 100% All-Natural Certified Free-Trade Organic Sasquatch Tonic is $C(x) = 10x + 100$, $x \geq 0$ and the price-demand function, in dollars per bottle, is $p(x) = 35 - x$, $0 \leq x \leq 35$.

12. The cost, in cents, to produce x cups of Mountain Thunder Lemonade at Junior's Lemonade Stand is $C(x) = 18x + 240$, $x \geq 0$ and the price-demand function, in cents per cup, is $p(x) = 90 - 3x$, $0 \leq x \leq 30$.

13. The daily cost, in dollars, to produce x Sasquatch Berry Pies is $C(x) = 3x + 36$, $x \geq 0$ and the price-demand function, in dollars per pie, is $p(x) = 12 - 0.5x$, $0 \leq x \leq 24$.

14. The monthly cost, in *hundreds* of dollars, to produce x custom built electric scooters is $C(x) = 20x + 1000$, $x \geq 0$ and the price-demand function, in *hundreds* of dollars per scooter, is $p(x) = 140 - 2x$, $0 \leq x \leq 70$.

---

[10]We have already seen the graph of this function. It was used as an example in Section 1.6 to show how the graphing calculator can be misleading.

15. The International Silver Strings Submarine Band holds a bake sale each year to fund their trip to the National Sasquatch Convention. It has been determined that the cost in dollars of baking x cookies is $C(x) = 0.1x + 25$ and that the demand function for their cookies is $p = 10 - .01x$. How many cookies should they bake in order to maximize their profit?

16. Using data from Bureau of Transportation Statistics, the average fuel economy F in miles per gallon for passenger cars in the US can be modeled by $F(t) = -0.0076t^2 + 0.45t + 16$, $0 \leq t \leq 28$, where t is the number of years since 1980. Find and interpret the coordinates of the vertex of the graph of $y = F(t)$.

17. The temperature T, in degrees Fahrenheit, t hours after 6 AM is given by:

$$T(t) = -\frac{1}{2}t^2 + 8t + 32, \quad 0 \leq t \leq 12$$

What is the warmest temperature of the day? When does this happen?

18. Suppose $C(x) = x^2 - 10x + 27$ represents the costs, in *hundreds*, to produce x *thousand* pens. How many pens should be produced to minimize the cost? What is this minimum cost?

19. Skippy wishes to plant a vegetable garden along one side of his house. In his garage, he found 32 linear feet of fencing. Since one side of the garden will border the house, Skippy doesn't need fencing along that side. What are the dimensions of the garden which will maximize the area of the garden? What is the maximum area of the garden?

20. In the situation of Example 2.3.4, Donnie has a nightmare that one of his alpaca herd fell into the river and drowned. To avoid this, he wants to move his rectangular pasture *away* from the river. This means that all four sides of the pasture require fencing. If the total amount of fencing available is still 200 linear feet, what dimensions maximize the area of the pasture now? What is the maximum area? Assuming an average alpaca requires 25 square feet of pasture, how many alpaca can he raise now?

21. What is the largest rectangular area one can enclose with 14 inches of string?

22. The height of an object dropped from the roof of an eight story building is modeled by $h(t) = -16t^2 + 64$, $0 \leq t \leq 2$. Here, h is the height of the object off the ground, in feet, t seconds after the object is dropped. How long before the object hits the ground?

23. The height h in feet of a model rocket above the ground t seconds after lift-off is given by $h(t) = -5t^2 + 100t$, for $0 \leq t \leq 20$. When does the rocket reach its maximum height above the ground? What is its maximum height?

24. Carl's friend Jason participates in the Highland Games. In one event, the hammer throw, the height h in feet of the hammer above the ground t seconds after Jason lets it go is modeled by $h(t) = -16t^2 + 22.08t + 6$. What is the hammer's maximum height? What is the hammer's total time in the air? Round your answers to two decimal places.

25. Assuming no air resistance or forces other than the Earth's gravity, the height above the ground at time t of a falling object is given by $s(t) = -4.9t^2 + v_0 t + s_0$ where s is in meters, t is in seconds, v_0 is the object's initial velocity in meters per second and s_0 is its initial position in meters.

    (a) What is the applied domain of this function?

    (b) Discuss with your classmates what each of $v_0 > 0$, $v_0 = 0$ and $v_0 < 0$ would mean.

    (c) Come up with a scenario in which $s_0 < 0$.

    (d) Let's say a slingshot is used to shoot a marble straight up from the ground ($s_0 = 0$) with an initial velocity of 15 meters per second. What is the marble's maximum height above the ground? At what time will it hit the ground?

    (e) Now shoot the marble from the top of a tower which is 25 meters tall. When does it hit the ground?

    (f) What would the height function be if instead of shooting the marble up off of the tower, you were to shoot it straight DOWN from the top of the tower?

26. The two towers of a suspension bridge are 400 feet apart. The parabolic cable[11] attached to the tops of the towers is 10 feet above the point on the bridge deck that is midway between the towers. If the towers are 100 feet tall, find the height of the cable directly above a point of the bridge deck that is 50 feet to the right of the left-hand tower.

27. Graph $f(x) = |1 - x^2|$

28. Find all of the points on the line $y = 1 - x$ which are 2 units from $(1, -1)$.

29. Let L be the line $y = 2x + 1$. Find a function $D(x)$ which measures the distance *squared* from a point on L to $(0, 0)$. Use this to find the point on L closest to $(0, 0)$.

30. With the help of your classmates, show that if a quadratic function $f(x) = ax^2 + bx + c$ has two real zeros then the x-coordinate of the vertex is the midpoint of the zeros.

In Exercises 31 - 36, solve the quadratic equation for the indicated variable.

31. $x^2 - 10y^2 = 0$ for x

32. $y^2 - 4y = x^2 - 4$ for x

33. $x^2 - mx = 1$ for x

34. $y^2 - 3y = 4x$ for y

35. $y^2 - 4y = x^2 - 4$ for y

36. $-gt^2 + v_0 t + s_0 = 0$ for t (Assume $g \neq 0$.)

---

[11] The weight of the bridge deck forces the bridge cable into a parabola and a free hanging cable such as a power line does not form a parabola. We shall see in Exercise 35 in Section 6.5 what shape a free hanging cable makes.

2.3.2 Answers

1. $f(x) = x^2 + 2$ (this is both forms!)
   No x-intercepts
   y-intercept $(0, 2)$
   Domain: $(-\infty, \infty)$
   Range: $[2, \infty)$
   Decreasing on $(-\infty, 0]$
   Increasing on $[0, \infty)$
   Vertex $(0, 2)$ is a minimum
   Axis of symmetry $x = 0$

   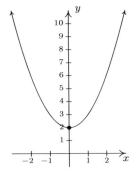

2. $f(x) = -(x+2)^2 = -x^2 - 4x - 4$
   x-intercept $(-2, 0)$
   y-intercept $(0, -4)$
   Domain: $(-\infty, \infty)$
   Range: $(-\infty, 0]$
   Increasing on $(-\infty, -2]$
   Decreasing on $[-2, \infty)$
   Vertex $(-2, 0)$ is a maximum
   Axis of symmetry $x = -2$

   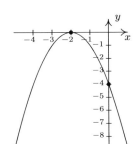

3. $f(x) = x^2 - 2x - 8 = (x-1)^2 - 9$
   x-intercepts $(-2, 0)$ and $(4, 0)$
   y-intercept $(0, -8)$
   Domain: $(-\infty, \infty)$
   Range: $[-9, \infty)$
   Decreasing on $(-\infty, 1]$
   Increasing on $[1, \infty)$
   Vertex $(1, -9)$ is a minimum
   Axis of symmetry $x = 1$

   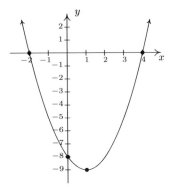

4. $f(x) = -2(x+1)^2 + 4 = -2x^2 - 4x + 2$
   x-intercepts $(-1 - \sqrt{2}, 0)$ and $(-1 + \sqrt{2}, 0)$
   y-intercept $(0, 2)$
   Domain: $(-\infty, \infty)$
   Range: $(-\infty, 4]$
   Increasing on $(-\infty, -1]$
   Decreasing on $[-1, \infty)$
   Vertex $(-1, 4)$ is a maximum
   Axis of symmetry $x = -1$

   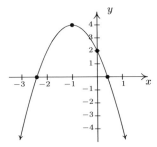

5. $f(x) = 2x^2 - 4x - 1 = 2(x-1)^2 - 3$
   x-intercepts $\left(\frac{2-\sqrt{6}}{2}, 0\right)$ and $\left(\frac{2+\sqrt{6}}{2}, 0\right)$
   y-intercept $(0, -1)$
   Domain: $(-\infty, \infty)$
   Range: $[-3, \infty)$
   Increasing on $[1, \infty)$
   Decreasing on $(-\infty, 1]$
   Vertex $(1, -3)$ is a minimum
   Axis of symmetry $x = 1$

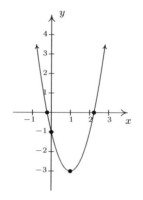

6. $f(x) = -3x^2 + 4x - 7 = -3\left(x - \frac{2}{3}\right)^2 - \frac{17}{3}$
   No x-intercepts
   y-intercept $(0, -7)$
   Domain: $(-\infty, \infty)$
   Range: $\left(-\infty, -\frac{17}{3}\right]$
   Increasing on $\left(-\infty, \frac{2}{3}\right]$
   Decreasing on $\left[\frac{2}{3}, \infty\right)$
   Vertex $\left(\frac{2}{3}, -\frac{17}{3}\right)$ is a maximum
   Axis of symmetry $x = \frac{2}{3}$

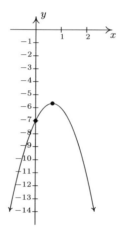

7. $f(x) = x^2 + x + 1 = \left(x + \frac{1}{2}\right)^2 + \frac{3}{4}$
   No x-intercepts
   y-intercept $(0, 1)$
   Domain: $(-\infty, \infty)$
   Range: $\left[\frac{3}{4}, \infty\right)$
   Increasing on $\left[-\frac{1}{2}, \infty\right)$
   Decreasing on $\left(-\infty, -\frac{1}{2}\right]$
   Vertex $\left(-\frac{1}{2}, \frac{3}{4}\right)$ is a minimum
   Axis of symmetry $x = -\frac{1}{2}$

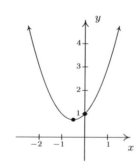

8. $f(x) = -3x^2 + 5x + 4 = -3\left(x - \frac{5}{6}\right)^2 + \frac{73}{12}$
   x-intercepts $\left(\frac{5-\sqrt{73}}{6}, 0\right)$ and $\left(\frac{5+\sqrt{73}}{6}, 0\right)$
   y-intercept $(0, 4)$
   Domain: $(-\infty, \infty)$
   Range: $\left(-\infty, \frac{73}{12}\right]$
   Increasing on $\left(-\infty, \frac{5}{6}\right]$
   Decreasing on $\left[\frac{5}{6}, \infty\right)$
   Vertex $\left(\frac{5}{6}, \frac{73}{12}\right)$ is a maximum
   Axis of symmetry $x = \frac{5}{6}$

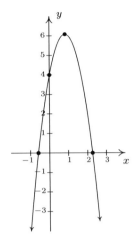

9. $f(x) = x^2 - \frac{1}{100}x - 1 = \left(x - \frac{1}{200}\right)^2 - \frac{40001}{40000}$
   x-intercepts $\left(\frac{1+\sqrt{40001}}{200}\right)$ and $\left(\frac{1-\sqrt{40001}}{200}\right)$
   y-intercept $(0, -1)$
   Domain: $(-\infty, \infty)$
   Range: $\left[-\frac{40001}{40000}, \infty\right)$
   Decreasing on $\left(-\infty, \frac{1}{200}\right]$
   Increasing on $\left[\frac{1}{200}, \infty\right)$
   Vertex $\left(\frac{1}{200}, -\frac{40001}{40000}\right)$ is a minimum[12]
   Axis of symmetry $x = \frac{1}{200}$

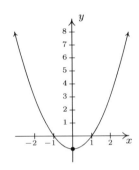

10.
    - $P(x) = -2x^2 + 28x - 26$, for $0 \le x \le 15$.
    - 7 T-shirts should be made and sold to maximize profit.
    - The maximum profit is $72.
    - The price per T-shirt should be set at $16 to maximize profit.
    - The break even points are $x = 1$ and $x = 13$, so to make a profit, between 1 and 13 T-shirts need to be made and sold.

11.
    - $P(x) = -x^2 + 25x - 100$, for $0 \le x \le 35$
    - Since the vertex occurs at $x = 12.5$, and it is impossible to make or sell 12.5 bottles of tonic, maximum profit occurs when either 12 or 13 bottles of tonic are made and sold.
    - The maximum profit is $56.
    - The price per bottle can be either $23 (to sell 12 bottles) or $22 (to sell 13 bottles.) Both will result in the maximum profit.
    - The break even points are $x = 5$ and $x = 20$, so to make a profit, between 5 and 20 bottles of tonic need to be made and sold.

---

[12] You'll need to use your calculator to zoom in far enough to see that the vertex is not the y-intercept.

12.
- $P(x) = -3x^2 + 72x - 240$, for $0 \leq x \leq 30$
- 12 cups of lemonade need to be made and sold to maximize profit.
- The maximum profit is 192¢ or $1.92.
- The price per cup should be set at 54¢ per cup to maximize profit.
- The break even points are $x = 4$ and $x = 20$, so to make a profit, between 4 and 20 cups of lemonade need to be made and sold.

13.
- $P(x) = -0.5x^2 + 9x - 36$, for $0 \leq x \leq 24$
- 9 pies should be made and sold to maximize the daily profit.
- The maximum daily profit is $4.50.
- The price per pie should be set at $7.50 to maximize profit.
- The break even points are $x = 6$ and $x = 12$, so to make a profit, between 6 and 12 pies need to be made and sold daily.

14.
- $P(x) = -2x^2 + 120x - 1000$, for $0 \leq x \leq 70$
- 30 scooters need to be made and sold to maximize profit.
- The maximum monthly profit is 800 hundred dollars, or $80,000.
- The price per scooter should be set at 80 hundred dollars, or $8000 per scooter.
- The break even points are $x = 10$ and $x = 50$, so to make a profit, between 10 and 50 scooters need to be made and sold monthly.

15. 495 cookies

16. The vertex is (approximately) $(29.60, 22.66)$, which corresponds to a maximum fuel economy of 22.66 miles per gallon, reached sometime between 2009 and 2010 (29 – 30 years after 1980.) Unfortunately, the model is only valid up until 2008 (28 years after 1908.) So, at this point, we are using the model to *predict* the maximum fuel economy.

17. 64° at 2 PM (8 hours after 6 AM.)

18. 5000 pens should be produced for a cost of $200.

19. 8 feet by 16 feet; maximum area is 128 square feet.

20. 50 feet by 50 feet; maximum area is 2500 feet; he can raise 100 average alpacas.

21. The largest rectangle has area 12.25 square inches.

22. 2 seconds.

23. The rocket reaches its maximum height of 500 feet 10 seconds after lift-off.

24. The hammer reaches a maximum height of approximately 13.62 feet. The hammer is in the air approximately 1.61 seconds.

2.3 Quadratic Functions

25. (a) The applied domain is $[0, \infty)$.

    (d) The height function is this case is $s(t) = -4.9t^2 + 15t$. The vertex of this parabola is approximately $(1.53, 11.48)$ so the maximum height reached by the marble is 11.48 meters. It hits the ground again when $t \approx 3.06$ seconds.

    (e) The revised height function is $s(t) = -4.9t^2 + 15t + 25$ which has zeros at $t \approx -1.20$ and $t \approx 4.26$. We ignore the negative value and claim that the marble will hit the ground after 4.26 seconds.

    (f) Shooting down means the initial velocity is negative so the height functions becomes $s(t) = -4.9t^2 - 15t + 25$.

26. Make the vertex of the parabola $(0, 10)$ so that the point on the top of the left-hand tower where the cable connects is $(-200, 100)$ and the point on the top of the right-hand tower is $(200, 100)$. Then the parabola is given by $p(x) = \frac{9}{4000}x^2 + 10$. Standing 50 feet to the right of the left-hand tower means you're standing at $x = -150$ and $p(-150) = 60.625$. So the cable is 60.625 feet above the bridge deck there.

27. $y = |1 - x^2|$

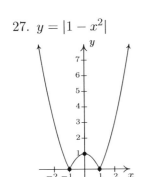

28. $\left(\dfrac{3 - \sqrt{7}}{2}, \dfrac{-1 + \sqrt{7}}{2}\right)$, $\left(\dfrac{3 + \sqrt{7}}{2}, \dfrac{-1 - \sqrt{7}}{2}\right)$

29. $D(x) = x^2 + (2x+1)^2 = 5x^2 + 4x + 1$, D is minimized when $x = -\frac{2}{5}$, so the point on $y = 2x+1$ closest to $(0,0)$ is $\left(-\frac{2}{5}, \frac{1}{5}\right)$

31. $x = \pm y\sqrt{10}$

32. $x = \pm(y - 2)$

33. $x = \dfrac{m \pm \sqrt{m^2 + 4}}{2}$

34. $y = \dfrac{3 \pm \sqrt{16x + 9}}{2}$

35. $y = 2 \pm x$

36. $t = \dfrac{v_0 \pm \sqrt{v_0^2 + 4gs_0}}{2g}$

208

2.4 Inequalities with Absolute Value and Quadratic Functions

In this section, not only do we develop techniques for solving various classes of inequalities analytically, we also look at them graphically. The first example motivates the core ideas.

Example 2.4.1. Let $f(x) = 2x - 1$ and $g(x) = 5$.

1. Solve $f(x) = g(x)$.

2. Solve $f(x) < g(x)$.

3. Solve $f(x) > g(x)$.

4. Graph $y = f(x)$ and $y = g(x)$ on the same set of axes and interpret your solutions to parts 1 through 3 above.

Solution.

1. To solve $f(x) = g(x)$, we replace $f(x)$ with $2x - 1$ and $g(x)$ with 5 to get $2x - 1 = 5$. Solving for x, we get $x = 3$.

2. The inequality $f(x) < g(x)$ is equivalent to $2x - 1 < 5$. Solving gives $x < 3$ or $(-\infty, 3)$.

3. To find where $f(x) > g(x)$, we solve $2x - 1 > 5$. We get $x > 3$, or $(3, \infty)$.

4. To graph $y = f(x)$, we graph $y = 2x - 1$, which is a line with a y-intercept of $(0, -1)$ and a slope of 2. The graph of $y = g(x)$ is $y = 5$ which is a horizontal line through $(0, 5)$.

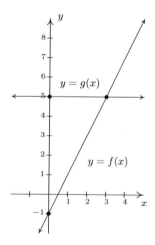

To see the connection between the graph and the Algebra, we recall the Fundamental Graphing Principle for Functions in Section 1.6: the point (a, b) is on the graph of f if and only if $f(a) = b$. In other words, a generic point on the graph of $y = f(x)$ is $(x, f(x))$, and a generic point on the graph of $y = g(x)$ is $(x, g(x))$. When we seek solutions to $f(x) = g(x)$, we are looking for x values whose y values on the graphs of f and g are the same. In part 1, we found $x = 3$ is the solution to $f(x) = g(x)$. Sure enough, $f(3) = 5$ and $g(3) = 5$ so that the point $(3, 5)$ is on both graphs. In other words, the graphs of f and g *intersect* at $(3, 5)$. In part 2, we set $f(x) < g(x)$ and solved to find $x < 3$. For $x < 3$, the point $(x, f(x))$ is *below* $(x, g(x))$ since the y values on the graph of f are less than the y values on the graph of g there. Analogously, in part 3, we solved $f(x) > g(x)$ and found $x > 3$. For $x > 3$, note that the graph of f is *above* the graph of g, since the y values on the graph of f are greater than the y values on the graph of g for those values of x.

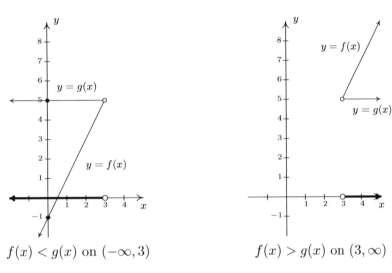

$f(x) < g(x)$ on $(-\infty, 3)$ $f(x) > g(x)$ on $(3, \infty)$

□

The preceding example demonstrates the following, which is a consequence of the Fundamental Graphing Principle for Functions.

Graphical Interpretation of Equations and Inequalities

Suppose f and g are functions.

- The solutions to $f(x) = g(x)$ are the x values where the graphs of $y = f(x)$ and $y = g(x)$ intersect.

- The solution to $f(x) < g(x)$ is the set of x values where the graph of $y = f(x)$ is *below* the graph of $y = g(x)$.

- The solution to $f(x) > g(x)$ is the set of x values where the graph of $y = f(x)$ *above* the graph of $y = g(x)$.

The next example turns the tables and furnishes the graphs of two functions and asks for solutions to equations and inequalities.

Example 2.4.2. The graphs of f and g are below. (The graph of $y = g(x)$ is bolded.) Use these graphs to answer the following questions.

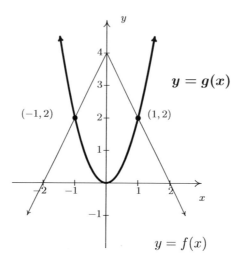

1. Solve $f(x) = g(x)$.
2. Solve $f(x) < g(x)$.
3. Solve $f(x) \geq g(x)$.

Solution.

1. To solve $f(x) = g(x)$, we look for where the graphs of f and g intersect. These appear to be at the points $(-1, 2)$ and $(1, 2)$, so our solutions to $f(x) = g(x)$ are $x = -1$ and $x = 1$.

2. To solve $f(x) < g(x)$, we look for where the graph of f is below the graph of g. This appears to happen for the x values less than -1 and greater than 1. Our solution is $(-\infty, -1) \cup (1, \infty)$.

3. To solve $f(x) \geq g(x)$, we look for solutions to $f(x) = g(x)$ as well as $f(x) > g(x)$. We solved the former equation and found $x = \pm 1$. To solve $f(x) > g(x)$, we look for where the graph of f is above the graph of g. This appears to happen between $x = -1$ and $x = 1$, on the interval $(-1, 1)$. Hence, our solution to $f(x) \geq g(x)$ is $[-1, 1]$.

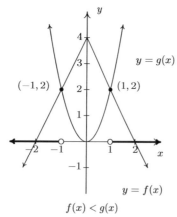

$f(x) < g(x)$

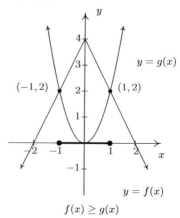

$f(x) \geq g(x)$

□

2.4 Inequalities with Absolute Value and Quadratic Functions

We now turn our attention to solving inequalities involving the absolute value. We have the following theorem from Intermediate Algebra to help us.

> **Theorem 2.4. Inequalities Involving the Absolute Value:** Let c be a real number.
>
> - For $c > 0$, $|x| < c$ is equivalent to $-c < x < c$.
> - For $c > 0$, $|x| \leq c$ is equivalent to $-c \leq x \leq c$.
> - For $c \leq 0$, $|x| < c$ has no solution, and for $c < 0$, $|x| \leq c$ has no solution.
> - For $c \geq 0$, $|x| > c$ is equivalent to $x < -c$ or $x > c$.
> - For $c \geq 0$, $|x| \geq c$ is equivalent to $x \leq -c$ or $x \geq c$.
> - For $c < 0$, $|x| > c$ and $|x| \geq c$ are true for all real numbers.

As with Theorem 2.1 in Section 2.2, we could argue Theorem 2.4 using cases. However, in light of what we have developed in this section, we can understand these statements graphically. For instance, if $c > 0$, the graph of $y = c$ is a horizontal line which lies above the x-axis through $(0, c)$. To solve $|x| < c$, we are looking for the x values where the graph of $y = |x|$ is below the graph of $y = c$. We know that the graphs intersect when $|x| = c$, which, from Section 2.2, we know happens when $x = c$ or $x = -c$. Graphing, we get

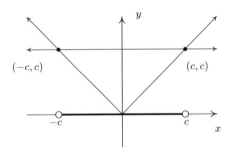

We see that the graph of $y = |x|$ is below $y = c$ for x between $-c$ and c, and hence we get $|x| < c$ is equivalent to $-c < x < c$. The other properties in Theorem 2.4 can be shown similarly.

Example 2.4.3. Solve the following inequalities analytically; check your answers graphically.

1. $|x - 1| \geq 3$
2. $4 - 3|2x + 1| > -2$
3. $2 < |x - 1| \leq 5$
4. $|x + 1| \geq \dfrac{x + 4}{2}$

Solution.

1. From Theorem 2.4, $|x - 1| \geq 3$ is equivalent to $x - 1 \leq -3$ or $x - 1 \geq 3$. Solving, we get $x \leq -2$ or $x \geq 4$, which, in interval notation is $(-\infty, -2] \cup [4, \infty)$. Graphically, we have

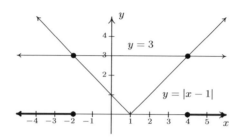

We see that the graph of $y = |x-1|$ is above the horizontal line $y = 3$ for $x < -2$ and $x > 4$ hence this is where $|x-1| > 3$. The two graphs intersect when $x = -2$ and $x = 4$, so we have graphical confirmation of our analytic solution.

2. To solve $4 - 3|2x+1| > -2$ analytically, we first isolate the absolute value before applying Theorem 2.4. To that end, we get $-3|2x+1| > -6$ or $|2x+1| < 2$. Rewriting, we now have $-2 < 2x+1 < 2$ so that $-\frac{3}{2} < x < \frac{1}{2}$. In interval notation, we write $\left(-\frac{3}{2}, \frac{1}{2}\right)$. Graphically we see that the graph of $y = 4 - 3|2x+1|$ is above $y = -2$ for x values between $-\frac{3}{2}$ and $\frac{1}{2}$.

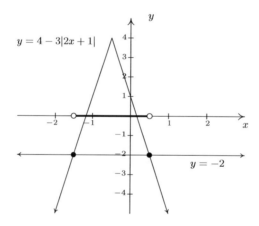

3. Rewriting the compound inequality $2 < |x-1| \leq 5$ as '$2 < |x-1|$ and $|x-1| \leq 5$' allows us to solve each piece using Theorem 2.4. The first inequality, $2 < |x-1|$ can be re-written as $|x-1| > 2$ so $x-1 < -2$ or $x-1 > 2$. We get $x < -1$ or $x > 3$. Our solution to the first inequality is then $(-\infty, -1) \cup (3, \infty)$. For $|x-1| \leq 5$, we combine results in Theorems 2.1 and 2.4 to get $-5 \leq x - 1 \leq 5$ so that $-4 \leq x \leq 6$, or $[-4, 6]$. Our solution to $2 < |x-1| \leq 5$ is comprised of values of x which satisfy both parts of the inequality, so we take the intersection[1] of $(-\infty, -1) \cup (3, \infty)$ and $[-4, 6]$ to get $[-4, -1) \cup (3, 6]$. Graphically, we see that the graph of $y = |x-1|$ is 'between' the horizontal lines $y = 2$ and $y = 5$ for x values between -4 and -1 as well as those between 3 and 6. Including the x values where $y = |x-1|$ and $y = 5$ intersect, we get

---

[1] See Definition 1.2 in Section 1.1.1.

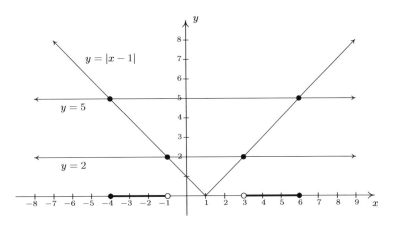

4. We need to exercise some special caution when solving $|x+1| \geq \frac{x+4}{2}$. As we saw in Example 2.2.1 in Section 2.2, when variables are both inside and outside of the absolute value, it's usually best to refer to the definition of absolute value, Definition 2.4, to remove the absolute values and proceed from there. To that end, we have $|x+1| = -(x+1)$ if $x < -1$ and $|x+1| = x+1$ if $x \geq -1$. We break the inequality into cases, the first case being when $x < -1$. For these values of x, our inequality becomes $-(x+1) \geq \frac{x+4}{2}$. Solving, we get $-2x - 2 \geq x + 4$, so that $-3x \geq 6$, which means $x \leq -2$. Since all of these solutions fall into the category $x < -1$, we keep them all. For the second case, we assume $x \geq -1$. Our inequality becomes $x + 1 \geq \frac{x+4}{2}$, which gives $2x + 2 \geq x + 4$ or $x \geq 2$. Since all of these values of x are greater than or equal to -1, we accept all of these solutions as well. Our final answer is $(-\infty, -2] \cup [2, \infty)$.

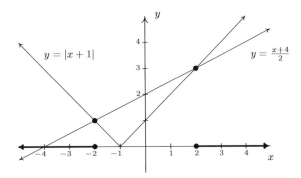

□

We now turn our attention to quadratic inequalities. In the last example of Section 2.3, we needed to determine the solution to $x^2 - x - 6 < 0$. We will now re-visit this problem using some of the techniques developed in this section not only to reinforce our solution in Section 2.3, but to also help formulate a general analytic procedure for solving all quadratic inequalities. If we consider $f(x) = x^2 - x - 6$ and $g(x) = 0$, then solving $x^2 - x - 6 < 0$ corresponds graphically to finding the values of x for which the graph of $y = f(x) = x^2 - x - 6$ (the parabola) is below the graph of $y = g(x) = 0$ (the x-axis). We've provided the graph again for reference.

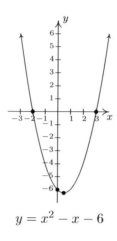

$$y = x^2 - x - 6$$

We can see that the graph of f does dip below the x-axis between its two x-intercepts. The zeros of f are $x = -2$ and $x = 3$ in this case and they divide the domain (the x-axis) into three intervals: $(-\infty, -2)$, $(-2, 3)$ and $(3, \infty)$. For every number in $(-\infty, -2)$, the graph of f is above the x-axis; in other words, $f(x) > 0$ for all x in $(-\infty, -2)$. Similarly, $f(x) < 0$ for all x in $(-2, 3)$, and $f(x) > 0$ for all x in $(3, \infty)$. We can schematically represent this with the **sign diagram** below.

$$(+) \; 0 \; (-) \; 0 \; (+)$$
$$\leftarrow\!\!\!\!\dashv\!\!\!\!\dashv\!\!\!\!\rightarrow$$
$$\quad\;\; -2 \quad\;\; 3$$

Here, the $(+)$ above a portion of the number line indicates $f(x) > 0$ for those values of x; the $(-)$ indicates $f(x) < 0$ there. The numbers labeled on the number line are the zeros of f, so we place 0 above them. We see at once that the solution to $f(x) < 0$ is $(-2, 3)$.

Our next goal is to establish a procedure by which we can generate the sign diagram without graphing the function. An important property[2] of quadratic functions is that if the function is positive at one point and negative at another, the function must have at least one zero in between. Graphically, this means that a parabola can't be above the x-axis at one point and below the x-axis at another point without crossing the x-axis. This allows us to determine the sign of *all* of the function values on a given interval by testing the function at just *one* value in the interval. This gives us the following.

---

[2] We will give this property a name in Chapter 3 and revisit this concept then.

2.4 INEQUALITIES WITH ABSOLUTE VALUE AND QUADRATIC FUNCTIONS   215

Steps for Solving a Quadratic Inequality

1. Rewrite the inequality, if necessary, as a quadratic function $f(x)$ on one side of the inequality and 0 on the other.

2. Find the zeros of f and place them on the number line with the number 0 above them.

3. Choose a real number, called a **test value**, in each of the intervals determined in step 2.

4. Determine the sign of $f(x)$ for each test value in step 3, and write that sign above the corresponding interval.

5. Choose the intervals which correspond to the correct sign to solve the inequality.

Example 2.4.4. Solve the following inequalities analytically using sign diagrams. Verify your answer graphically.

1. $2x^2 \leq 3 - x$

2. $x^2 - 2x > 1$

3. $x^2 + 1 \leq 2x$

4. $2x - x^2 \geq |x - 1| - 1$

Solution.

1. To solve $2x^2 \leq 3 - x$, we first get 0 on one side of the inequality which yields $2x^2 + x - 3 \leq 0$. We find the zeros of $f(x) = 2x^2 + x - 3$ by solving $2x^2 + x - 3 = 0$ for x. Factoring gives $(2x + 3)(x - 1) = 0$, so $x = -\frac{3}{2}$ or $x = 1$. We place these values on the number line with 0 above them and choose test values in the intervals $\left(-\infty, -\frac{3}{2}\right)$, $\left(-\frac{3}{2}, 1\right)$ and $(1, \infty)$. For the interval $\left(-\infty, -\frac{3}{2}\right)$, we choose[3] $x = -2$; for $\left(-\frac{3}{2}, 1\right)$, we pick $x = 0$; and for $(1, \infty)$, $x = 2$. Evaluating the function at the three test values gives us $f(-2) = 3 > 0$, so we place $(+)$ above $\left(-\infty, -\frac{3}{2}\right)$; $f(0) = -3 < 0$, so $(-)$ goes above the interval $\left(-\frac{3}{2}, 1\right)$; and, $f(2) = 7$, which means $(+)$ is placed above $(1, \infty)$. Since we are solving $2x^2 + x - 3 \leq 0$, we look for solutions to $2x^2 + x - 3 < 0$ as well as solutions for $2x^2 + x - 3 = 0$. For $2x^2 + x - 3 < 0$, we need the intervals which we have a $(-)$. Checking the sign diagram, we see this is $\left(-\frac{3}{2}, 1\right)$. We know $2x^2 + x - 3 = 0$ when $x = -\frac{3}{2}$ and $x = 1$, so our final answer is $\left[-\frac{3}{2}, 1\right]$.

To verify our solution graphically, we refer to the original inequality, $2x^2 \leq 3 - x$. We let $g(x) = 2x^2$ and $h(x) = 3 - x$. We are looking for the x values where the graph of g is below that of h (the solution to $g(x) < h(x)$) as well as the points of intersection (the solutions to $g(x) = h(x)$). The graphs of g and h are given on the right with the sign chart on the left.

---

[3]We have to choose something in each interval. If you don't like our choices, please feel free to choose different numbers. You'll get the same sign chart.

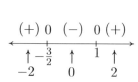 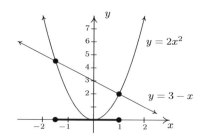

2. Once again, we re-write $x^2 - 2x > 1$ as $x^2 - 2x - 1 > 0$ and we identify $f(x) = x^2 - 2x - 1$. When we go to find the zeros of f, we find, to our chagrin, that the quadratic $x^2 - 2x - 1$ doesn't factor nicely. Hence, we resort to the quadratic formula to solve $x^2 - 2x - 1 = 0$, and arrive at $x = 1 \pm \sqrt{2}$. As before, these zeros divide the number line into three pieces. To help us decide on test values, we approximate $1 - \sqrt{2} \approx -0.4$ and $1 + \sqrt{2} \approx 2.4$. We choose $x = -1$, $x = 0$ and $x = 3$ as our test values and find $f(-1) = 2$, which is $(+)$; $f(0) = -1$ which is $(-)$; and $f(3) = 2$ which is $(+)$ again. Our solution to $x^2 - 2x - 1 > 0$ is where we have $(+)$, so, in interval notation $(-\infty, 1 - \sqrt{2}) \cup (1 + \sqrt{2}, \infty)$. To check the inequality $x^2 - 2x > 1$ graphically, we set $g(x) = x^2 - 2x$ and $h(x) = 1$. We are looking for the x values where the graph of g is above the graph of h. As before we present the graphs on the right and the sign chart on the left.

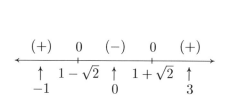 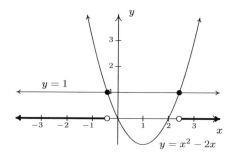

3. To solve $x^2 + 1 \leq 2x$, as before, we solve $x^2 - 2x + 1 \leq 0$. Setting $f(x) = x^2 - 2x + 1 = 0$, we find the only one zero of f, $x = 1$. This one x value divides the number line into two intervals, from which we choose $x = 0$ and $x = 2$ as test values. We find $f(0) = 1 > 0$ and $f(2) = 1 > 0$. Since we are looking for solutions to $x^2 - 2x + 1 \leq 0$, we are looking for x values where $x^2 - 2x + 1 < 0$ as well as where $x^2 - 2x + 1 = 0$. Looking at our sign diagram, there are no places where $x^2 - 2x + 1 < 0$ (there are no $(-)$), so our solution is only $x = 1$ (where $x^2 - 2x + 1 = 0$). We write this as $\{1\}$. Graphically, we solve $x^2 + 1 \leq 2x$ by graphing $g(x) = x^2 + 1$ and $h(x) = 2x$. We are looking for the x values where the graph of g is below the graph of h (for $x^2 + 1 < 2x$) and where the two graphs intersect ($x^2 + 1 = 2x$). Notice that the line and the parabola touch at $(1, 2)$, but the parabola is always above the line otherwise.[4]

---

[4]In this case, we say the line $y = 2x$ is **tangent** to $y = x^2 + 1$ at $(1, 2)$. Finding tangent lines to arbitrary functions is a fundamental problem solved, in general, with Calculus.

2.4 INEQUALITIES WITH ABSOLUTE VALUE AND QUADRATIC FUNCTIONS 217

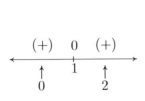

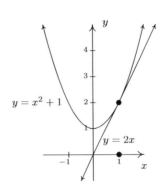

4. To solve our last inequality, $2x - x^2 \geq |x-1| - 1$, we re-write the absolute value using cases. For $x < 1$, $|x-1| = -(x-1) = 1-x$, so we get $2x - x^2 \geq 1 - x - 1$, or $x^2 - 3x \leq 0$. Finding the zeros of $f(x) = x^2 - 3x$, we get $x = 0$ and $x = 3$. However, we are only concerned with the portion of the number line where $x < 1$, so the only zero that we concern ourselves with is $x = 0$. This divides the interval $x < 1$ into two intervals: $(-\infty, 0)$ and $(0, 1)$. We choose $x = -1$ and $x = \frac{1}{2}$ as our test values. We find $f(-1) = 4$ and $f\left(\frac{1}{2}\right) = -\frac{5}{4}$. Hence, our solution to $x^2 - 3x \leq 0$ for $x < 1$ is $[0, 1)$. Next, we turn our attention to the case $x \geq 1$. Here, $|x-1| = x-1$, so our original inequality becomes $2x - x^2 \geq x - 1 - 1$, or $x^2 - x - 2 \leq 0$. Setting $g(x) = x^2 - x - 2$, we find the zeros of g to be $x = -1$ and $x = 2$. Of these, only $x = 2$ lies in the region $x \geq 1$, so we ignore $x = -1$. Our test intervals are now $[1, 2)$ and $(2, \infty)$. We choose $x = 1$ and $x = 3$ as our test values and find $g(1) = -2$ and $g(3) = 4$. Hence, our solution to $g(x) = x^2 - x - 2 \leq 0$, in this region is $[1, 2]$.

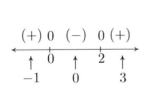

Combining these into one sign diagram, we have that our solution is $[0, 2]$. Graphically, to check $2x - x^2 \geq |x - 1| - 1$, we set $h(x) = 2x - x^2$ and $i(x) = |x - 1| - 1$ and look for the x values where the graph of h is above the the graph of i (the solution of $h(x) > i(x)$) as well as the x-coordinates of the intersection points of both graphs (where $h(x) = i(x)$). The combined sign chart is given on the left and the graphs are on the right.

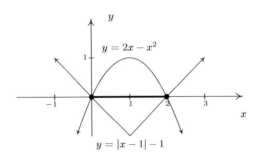

One of the classic applications of inequalities is the notion of tolerances.[5] Recall that for real numbers x and c, the quantity $|x - c|$ may be interpreted as the distance from x to c. Solving inequalities of the form $|x - c| \leq d$ for $d \geq 0$ can then be interpreted as finding all numbers x which lie within d units of c. We can think of the number d as a 'tolerance' and our solutions x as being within an accepted tolerance of c. We use this principle in the next example.

Example 2.4.5. The area A (in square inches) of a square piece of particle board which measures x inches on each side is $A(x) = x^2$. Suppose a manufacturer needs to produce a 24 inch by 24 inch square piece of particle board as part of a home office desk kit. How close does the side of the piece of particle board need to be cut to 24 inches to guarantee that the area of the piece is within a tolerance of 0.25 square inches of the target area of 576 square inches?

Solution. Mathematically, we express the desire for the area $A(x)$ to be within 0.25 square inches of 576 as $|A - 576| \leq 0.25$. Since $A(x) = x^2$, we get $|x^2 - 576| \leq 0.25$, which is equivalent to $-0.25 \leq x^2 - 576 \leq 0.25$. One way to proceed at this point is to solve the two inequalities $-0.25 \leq x^2 - 576$ and $x^2 - 576 \leq 0.25$ individually using sign diagrams and then taking the intersection of the solution sets. While this way will (eventually) lead to the correct answer, we take this opportunity to showcase the increasing property of the square root: if $0 \leq a \leq b$, then $\sqrt{a} \leq \sqrt{b}$. To use this property, we proceed as follows

$$
\begin{array}{rcll}
-0.25 \leq & x^2 - 576 & \leq 0.25 & \\
575.75 \leq & x^2 & \leq 576.25 & \text{(add 576 across the inequalities.)} \\
\sqrt{575.75} \leq & \sqrt{x^2} & \leq \sqrt{576.25} & \text{(take square roots.)} \\
\sqrt{575.75} \leq & |x| & \leq \sqrt{576.25} & (\sqrt{x^2} = |x|)
\end{array}
$$

By Theorem 2.4, we find the solution to $\sqrt{575.75} \leq |x|$ to be $\left(-\infty, -\sqrt{575.75}\right] \cup \left[\sqrt{575.75}, \infty\right)$ and the solution to $|x| \leq \sqrt{576.25}$ to be $\left[-\sqrt{576.25}, \sqrt{576.25}\right]$. To solve $\sqrt{575.75} \leq |x| \leq \sqrt{576.25}$, we intersect these two sets to get $\left[-\sqrt{576.25}, -\sqrt{575.75}\right] \cup \left[\sqrt{575.75}, \sqrt{576.25}\right]$. Since x represents a length, we discard the negative answers and get $\left[\sqrt{575.75}, \sqrt{576.25}\right]$. This means that the side of the piece of particle board must be cut between $\sqrt{575.75} \approx 23.995$ and $\sqrt{576.25} \approx 24.005$ inches, a tolerance of (approximately) 0.005 inches of the target length of 24 inches. □

Our last example in the section demonstrates how inequalities can be used to describe regions in the plane, as we saw earlier in Section 1.2.

Example 2.4.6. Sketch the following relations.

1. $R = \{(x, y) : y > |x|\}$

2. $S = \{(x, y) : y \leq 2 - x^2\}$

3. $T = \{(x, y) : |x| < y \leq 2 - x^2\}$

---

[5] The underlying concept of Calculus can be phrased in terms of tolerances, so this is well worth your attention.

Solution.

1. The relation R consists of all points (x, y) whose y-coordinate is greater than $|x|$. If we graph $y = |x|$, then we want all of the points in the plane *above* the points on the graph. Dotting the graph of $y = |x|$ as we have done before to indicate that the points on the graph itself are not in the relation, we get the shaded region below on the left.

2. For a point to be in S, its y-coordinate must be less than or equal to the y-coordinate on the parabola $y = 2 - x^2$. This is the set of all points *below* or *on* the parabola $y = 2 - x^2$.

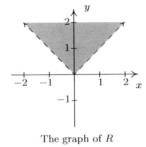

The graph of R

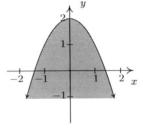

The graph of S

3. Finally, the relation T takes the points whose y-coordinates satisfy both the conditions given in R and those of S. Thus we shade the region between $y = |x|$ and $y = 2 - x^2$, keeping those points on the parabola, but not the points on $y = |x|$. To get an accurate graph, we need to find where these two graphs intersect, so we set $|x| = 2 - x^2$. Proceeding as before, breaking this equation into cases, we get $x = -1, 1$. Graphing yields

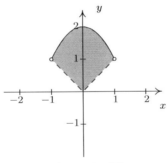

The graph of T

2.4.1 Exercises

In Exercises 1 - 32, solve the inequality. Write your answer using interval notation.

1. $|3x - 5| \leq 4$
2. $|7x + 2| > 10$
3. $|2x + 1| - 5 < 0$
4. $|2 - x| - 4 \geq -3$
5. $|3x + 5| + 2 < 1$
6. $2|7 - x| + 4 > 1$
7. $2 \leq |4 - x| < 7$
8. $1 < |2x - 9| \leq 3$
9. $|x + 3| \geq |6x + 9|$
10. $|x - 3| - |2x + 1| < 0$
11. $|1 - 2x| \geq x + 5$
12. $x + 5 < |x + 5|$
13. $x \geq |x + 1|$
14. $|2x + 1| \leq 6 - x$
15. $x + |2x - 3| < 2$
16. $|3 - x| \geq x - 5$
17. $x^2 + 2x - 3 \geq 0$
18. $16x^2 + 8x + 1 > 0$
19. $x^2 + 9 < 6x$
20. $9x^2 + 16 \geq 24x$
21. $x^2 + 4 \leq 4x$
22. $x^2 + 1 < 0$
23. $3x^2 \leq 11x + 4$
24. $x > x^2$
25. $2x^2 - 4x - 1 > 0$
26. $5x + 4 \leq 3x^2$
27. $2 \leq |x^2 - 9| < 9$
28. $x^2 \leq |4x - 3|$
29. $x^2 + x + 1 \geq 0$
30. $x^2 \geq |x|$
31. $x|x + 5| \geq -6$
32. $x|x - 3| < 2$

33. The profit, in dollars, made by selling x bottles of 100% All-Natural Certified Free-Trade Organic Sasquatch Tonic is given by $P(x) = -x^2 + 25x - 100$, for $0 \leq x \leq 35$. How many bottles of tonic must be sold to make at least $50 in profit?

34. Suppose $C(x) = x^2 - 10x + 27$, $x \geq 0$ represents the costs, in *hundreds* of dollars, to produce x *thousand* pens. Find the number of pens which can be produced for no more than $1100.

35. The temperature T, in degrees Fahrenheit, t hours after 6 AM is given by $T(t) = -\frac{1}{2}t^2 + 8t + 32$, for $0 \leq t \leq 12$. When is it warmer than $42°$ Fahrenheit?

2.4 Inequalities with Absolute Value and Quadratic Functions

36. The height h in feet of a model rocket above the ground t seconds after lift-off is given by $h(t) = -5t^2 + 100t$, for $0 \leq t \leq 20$. When is the rocket at least 250 feet off the ground? Round your answer to two decimal places.

37. If a slingshot is used to shoot a marble straight up into the air from 2 meters above the ground with an initial velocity of 30 meters per second, for what values of time t will the marble be over 35 meters above the ground? (Refer to Exercise 25 in Section 2.3 for assistance if needed.) Round your answers to two decimal places.

38. What temperature values in degrees Celsius are equivalent to the temperature range $50°F$ to $95°F$? (Refer to Exercise 35 in Section 2.1 for assistance if needed.)

In Exercises 39 - 42, write and solve an inequality involving absolute values for the given statement.

39. Find all real numbers x so that x is within 4 units of 2.

40. Find all real numbers x so that $3x$ is within 2 units of -1.

41. Find all real numbers x so that x^2 is within 1 unit of 3.

42. Find all real numbers x so that x^2 is at least 7 units away from 4.

43. The surface area S of a cube with edge length x is given by $S(x) = 6x^2$ for $x > 0$. Suppose the cubes your company manufactures are supposed to have a surface area of exactly 42 square centimeters, but the machines you own are old and cannot always make a cube with the precise surface area desired. Write an inequality using absolute value that says the surface area of a given cube is no more than 3 square centimeters away (high or low) from the target of 42 square centimeters. Solve the inequality and write your answer using interval notation.

44. Suppose f is a function, L is a real number and ε is a positive number. Discuss with your classmates what the inequality $|f(x) - L| < \varepsilon$ means algebraically and graphically.[6]

In Exercises 45 - 50, sketch the graph of the relation.

45. $R = \{(x, y) : y \leq x - 1\}$

46. $R = \{(x, y) : y > x^2 + 1\}$

47. $R = \{(x, y) : -1 < y \leq 2x + 1\}$

48. $R = \{(x, y) : x^2 \leq y < x + 2\}$

49. $R = \{(x, y) : |x| - 4 < y < 2 - x\}$

50. $R = \{(x, y) : x^2 < y \leq |4x - 3|\}$

51. Prove the second, third and fourth parts of Theorem 2.4.

---

[6]Understanding this type of inequality is really important in Calculus.

2.4.2 Answers

1. $\left[\frac{1}{3}, 3\right]$

2. $\left(-\infty, -\frac{12}{7}\right) \cup \left(\frac{8}{7}, \infty\right)$

3. $(-3, 2)$

4. $(-\infty, 1] \cup [3, \infty)$

5. No solution

6. $(-\infty, \infty)$

7. $(-3, 2] \cup [6, 11)$

8. $[3, 4) \cup (5, 6]$

9. $\left[-\frac{12}{7}, -\frac{6}{5}\right]$

10. $(-\infty, -4) \cup \left(\frac{2}{3}, \infty\right)$

11. $\left(-\infty, -\frac{4}{3}\right] \cup [6, \infty)$

12. $(-\infty, -5)$

13. No Solution.

14. $\left[-7, \frac{5}{3}\right]$

15. $\left(1, \frac{5}{3}\right)$

16. $(-\infty, \infty)$

17. $(-\infty, -3] \cup [1, \infty)$

18. $\left(-\infty, -\frac{1}{4}\right) \cup \left(-\frac{1}{4}, \infty\right)$

19. No solution

20. $(-\infty, \infty)$

21. $\{2\}$

22. No solution

23. $\left[-\frac{1}{3}, 4\right]$

24. $(0, 1)$

25. $\left(-\infty, 1 - \frac{\sqrt{6}}{2}\right) \cup \left(1 + \frac{\sqrt{6}}{2}, \infty\right)$

26. $\left(-\infty, \frac{5-\sqrt{73}}{6}\right] \cup \left[\frac{5+\sqrt{73}}{6}, \infty\right)$

27. $\left(-3\sqrt{2}, -\sqrt{11}\right] \cup \left[-\sqrt{7}, 0\right) \cup \left(0, \sqrt{7}\right] \cup \left[\sqrt{11}, 3\sqrt{2}\right)$

28. $\left[-2 - \sqrt{7}, -2 + \sqrt{7}\right] \cup [1, 3]$

29. $(-\infty, \infty)$

30. $(-\infty, -1] \cup \{0\} \cup [1, \infty)$

31. $[-6, -3] \cup [-2, \infty)$

32. $(-\infty, 1) \cup \left(2, \frac{3+\sqrt{17}}{2}\right)$

33. $P(x) \geq 50$ on $[10, 15]$. This means anywhere between 10 and 15 bottles of tonic need to be sold to earn at least \$50 in profit.

34. $C(x) \leq 11$ on $[2, 8]$. This means anywhere between 2000 and 8000 pens can be produced and the cost will not exceed \$1100.

35. $T(t) > 42$ on $(8 - 2\sqrt{11}, 8 + 2\sqrt{11}) \approx (1.37, 14.63)$, which corresponds to between 7:22 AM (1.37 hours after 6 AM) to 8:38 PM (14.63 hours after 6 AM.) However, since the model is valid only for t, $0 \leq t \leq 12$, we restrict our answer and find it is warmer than $42°$ Fahrenheit from 7:22 AM to 6 PM.

2.4 Inequalities with Absolute Value and Quadratic Functions

36. $h(t) \geq 250$ on $[10 - 5\sqrt{2}, 10 + 5\sqrt{2}] \approx [2.93, 17.07]$. This means the rocket is at least 250 feet off the ground between 2.93 and 17.07 seconds after lift off.

37. $s(t) = -4.9t^2 + 30t + 2$. $s(t) > 35$ on (approximately) $(1.44, 4.68)$. This means between 1.44 and 4.68 seconds after it is launched into the air, the marble is more than 35 feet off the ground.

38. From our previous work $C(F) = \frac{5}{9}(F - 32)$ so $50 \leq F \leq 95$ becomes $10 \leq C \leq 35$.

39. $|x - 2| \leq 4$, $[-2, 6]$

40. $|3x + 1| \leq 2$, $\left[-1, \frac{1}{3}\right]$

41. $|x^2 - 3| \leq 1$, $[-2, -\sqrt{2}] \cup [\sqrt{2}, 2]$

42. $|x^2 - 4| \geq 7$, $(-\infty, -\sqrt{11}] \cup [\sqrt{11}, \infty)$

43. Solving $|S(x) - 42| \leq 3$, and disregarding the negative solutions yields $\left[\sqrt{\frac{13}{2}}, \sqrt{\frac{15}{2}}\right] \approx [2.550, 2.739]$. The edge length must be within 2.550 and 2.739 centimeters.

45.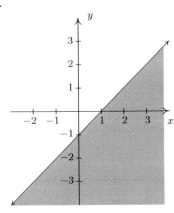

46.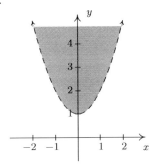

47.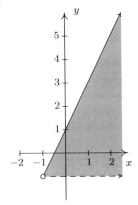

48.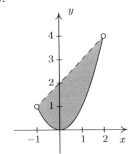

49.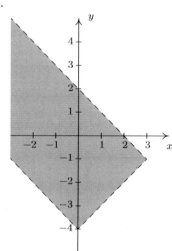

50.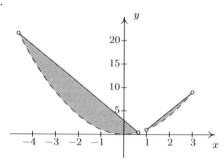

2.5 Regression

We have seen examples already in the text where linear and quadratic functions are used to model a wide variety of real world phenomena ranging from production costs to the height of a projectile above the ground. In this section, we use some basic tools from statistical analysis to quantify linear and quadratic trends that we may see in real world data in order to generate linear and quadratic models. Our goal is to give the reader an understanding of the basic processes involved, but we are quick to refer the reader to a more advanced course[1] for a complete exposition of this material. Suppose we collected three data points: $\{(1, 2), (3, 1), (4, 3)\}$. By plotting these points, we can clearly see that they do not lie along the same line. If we pick any two of the points, we can find a line containing both which completely misses the third, but our aim is to find a line which is in some sense 'close' to all the points, even though it may go through none of them. The way we measure 'closeness' in this case is to find the **total squared error** between the data points and the line. Consider our three data points and the line $y = \frac{1}{2}x + \frac{1}{2}$. For each of our data points, we find the vertical distance between the point and the line. To accomplish this, we need to find a point on the line directly above or below each data point - in other words, a point on the line with the same x-coordinate as our data point. For example, to find the point on the line directly below $(1, 2)$, we plug $x = 1$ into $y = \frac{1}{2}x + \frac{1}{2}$ and we get the point $(1, 1)$. Similarly, we get $(3, 1)$ to correspond to $(3, 2)$ and $\left(4, \frac{5}{2}\right)$ for $(4, 3)$.

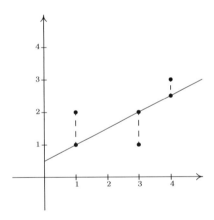

We find the total squared error E by taking the sum of the squares of the differences of the y-coordinates of each data point and its corresponding point on the line. For the data and line above $E = (2-1)^2 + (1-2)^2 + \left(3 - \frac{5}{2}\right)^2 = \frac{9}{4}$. Using advanced mathematical machinery,[2] it is possible to find the line which results in the lowest value of E. This line is called the **least squares regression line**, or sometimes the 'line of best fit'. The formula for the line of best fit requires notation we won't present until Chapter 9.1, so we will revisit it then. The graphing calculator can come to our assistance here, since it has a built-in feature to compute the regression line. We enter the data and perform the Linear Regression feature and we get

---

[1] and authors with more expertise in this area,
[2] Like Calculus and Linear Algebra

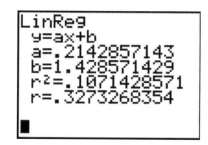

The calculator tells us that the line of best fit is $y = ax + b$ where the slope is $a \approx 0.214$ and the y-coordinate of the y-intercept is $b \approx 1.428$. (We will stick to using three decimal places for our approximations.) Using this line, we compute the total squared error for our data to be $E \approx 1.786$. The value r is the **correlation coefficient** and is a measure of how close the data is to being on the same line. The closer $|r|$ is to 1, the better the linear fit. Since $r \approx 0.327$, this tells us that the line of best fit doesn't fit all that well - in other words, our data points aren't close to being linear. The value r^2 is called the **coefficient of determination** and is also a measure of the goodness of fit.[3] Plotting the data with its regression line results in the picture below.

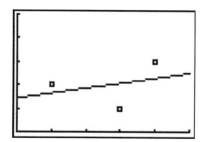

Our first example looks at energy consumption in the US over the past 50 years.[4]

Year	Energy Usage, in Quads[5]
1950	34.6
1960	45.1
1970	67.8
1980	78.3
1990	84.6
2000	98.9

Example 2.5.1. Using the energy consumption data given above,

1. Plot the data using a graphing calculator.

---

[3] We refer the interested reader to a course in Statistics to explore the significance of r and r^2.
[4] See this Department of Energy activity
[5] The unit 1 Quad is 1 Quadrillion = 10^{15} BTUs, which is enough heat to raise Lake Erie roughly 1°F

2. Find the least squares regression line and comment on the goodness of fit.

3. Interpret the slope of the line of best fit.

4. Use the regression line to predict the annual US energy consumption in the year 2013.

5. Use the regression line to predict when the annual consumption will reach 120 Quads.

Solution.

1. Entering the data into the calculator gives

   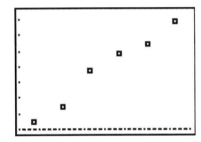

   The data certainly appears to be linear in nature.

2. Performing a linear regression produces

   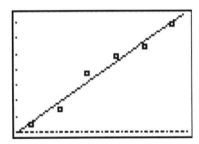

   We can tell both from the correlation coefficient as well as the graph that the regression line is a good fit to the data.

3. The slope of the regression line is $a \approx 1.287$. To interpret this, recall that the slope is the rate of change of the y-coordinates with respect to the x-coordinates. Since the y-coordinates represent the energy usage in Quads, and the x-coordinates represent years, a slope of positive 1.287 indicates an increase in annual energy usage at the rate of 1.287 Quads per year.

4. To predict the energy needs in 2013, we substitute $x = 2013$ into the equation of the line of best fit to get $y = 1.287(2013) - 2473.890 \approx 116.841$. The predicted annual energy usage of the US in 2013 is approximately 116.841 Quads.

5. To predict when the annual US energy usage will reach 120 Quads, we substitute $y = 120$ into the equation of the line of best fit to get $120 = 1.287x - 2473.908$. Solving for x yields $x \approx 2015.454$. Since the regression line is increasing, we interpret this result as saying the annual usage in 2015 won't yet be 120 Quads, but that in 2016, the demand will be more than 120 Quads. □

Our next example gives us an opportunity to find a nonlinear model to fit the data. According to the National Weather Service, the predicted hourly temperatures for Painesville on March 3, 2009 were given as summarized below.

Time	Temperature, °F
10AM	17
11AM	19
12PM	21
1PM	23
2PM	24
3PM	24
4PM	23

To enter this data into the calculator, we need to adjust the x values, since just entering the numbers could cause confusion. (Do you see why?) We have a few options available to us. Perhaps the easiest is to convert the times into the 24 hour clock time so that 1 PM is 13, 2 PM is 14, etc.. If we enter these data into the graphing calculator and plot the points we get

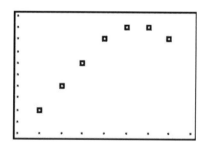

While the beginning of the data looks linear, the temperature begins to fall in the afternoon hours. This sort of behavior reminds us of parabolas, and, sure enough, it is possible to find a parabola of best fit in the same way we found a line of best fit. The process is called **quadratic regression** and its goal is to minimize the least square error of the data with their corresponding points on the parabola. The calculator has a built in feature for this as well which yields

2.5 Regression

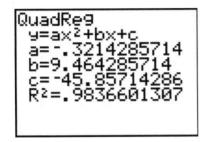 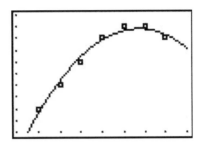

The coefficient of determination R^2 seems reasonably close to 1, and the graph visually seems to be a decent fit. We use this model in our next example.

Example 2.5.2. Using the quadratic model for the temperature data above, predict the warmest temperature of the day. When will this occur?

Solution. The maximum temperature will occur at the vertex of the parabola. Recalling the Vertex Formula, Equation 2.4, $x = -\frac{b}{2a} \approx -\frac{9.464}{2(-0.321)} \approx 14.741$. This corresponds to roughly 2:45 PM. To find the temperature, we substitute $x = 14.741$ into $y = -0.321x^2 + 9.464x - 45.857$ to get $y \approx 23.899$, or $23.899°F$. $\square$

The results of the last example should remind you that regression models are just that, models. Our predicted warmest temperature was found to be $23.899°F$, but our data says it will warm to $24°F$. It's all well and good to observe trends and guess at a model, but a more thorough investigation into *why* certain data should be linear or quadratic in nature is usually in order - and that, most often, is the business of scientists.

2.5.1 Exercises

1. According to this website[6], the census data for Lake County, Ohio is:

Year	1970	1980	1990	2000
Population	197200	212801	215499	227511

   (a) Find the least squares regression line for these data and comment on the goodness of fit.[7] Interpret the slope of the line of best fit.

   (b) Use the regression line to predict the population of Lake County in 2010. (The recorded figure from the 2010 census is 230,041)

   (c) Use the regression line to predict when the population of Lake County will reach 250,000.

2. According to this website[8], the census data for Lorain County, Ohio is:

Year	1970	1980	1990	2000
Population	256843	274909	271126	284664

   (a) Find the least squares regression line for these data and comment on the goodness of fit. Interpret the slope of the line of best fit.

   (b) Use the regression line to predict the population of Lorain County in 2010. (The recorded figure from the 2010 census is 301,356)

   (c) Use the regression line to predict when the population of Lake County will reach 325,000.

3. Using the energy production data given below

Year	1950	1960	1970	1980	1990	2000
Production (in Quads)	35.6	42.8	63.5	67.2	70.7	71.2

   (a) Plot the data using a graphing calculator and explain why it does not appear to be linear.

   (b) Discuss with your classmates why ignoring the first two data points may be justified from a historical perspective.

   (c) Find the least squares regression line for the last four data points and comment on the goodness of fit. Interpret the slope of the line of best fit.

   (d) Use the regression line to predict the annual US energy production in the year 2010.

   (e) Use the regression line to predict when the annual US energy production will reach 100 Quads.

---

[6] http://www.ohiobiz.com/census/Lake.pdf

[7] We'll develop more sophisticated models for the growth of populations in Chapter 6. For the moment, we use a theorem from Calculus to approximate those functions with lines.

[8] http://www.ohiobiz.com/census/Lorain.pdf

2.5 REGRESSION

4. The chart below contains a portion of the fuel consumption information for a 2002 Toyota Echo that I (Jeff) used to own. The first row is the cumulative number of gallons of gasoline that I had used and the second row is the odometer reading when I refilled the gas tank. So, for example, the fourth entry is the point (28.25, 1051) which says that I had used a total of 28.25 gallons of gasoline when the odometer read 1051 miles.

Gasoline Used (Gallons)	0	9.26	19.03	28.25	36.45	44.64	53.57	62.62	71.93	81.69	90.43
Odometer (Miles)	41	356	731	1051	1347	1631	1966	2310	2670	3030	3371

Find the least squares line for this data. Is it a good fit? What does the slope of the line represent? Do you and your classmates believe this model would have held for ten years had I not crashed the car on the Turnpike a few years ago? (I'm keeping a fuel log for my 2006 Scion xA for future College Algebra books so I hope not to crash it, too.)

5. On New Year's Day, I (Jeff, again) started weighing myself every morning in order to have an interesting data set for this section of the book. (Discuss with your classmates if that makes me a nerd or a geek. Also, the professionals in the field of weight management strongly discourage weighing yourself every day. When you focus on the number and not your overall health, you tend to lose sight of your objectives. I was making a noble sacrifice for science, but you should not try this at home.) The whole chart would be too big to put into the book neatly, so I've decided to give only a small portion of the data to you. This then becomes a Civics lesson in honesty, as you shall soon see. There are two charts given below. One has my weight for the first eight Thursdays of the year (January 1, 2009 was a Thursday and we'll count it as Day 1.) and the other has my weight for the first 10 Saturdays of the year.

Day # (Thursday)	1	8	15	22	29	36	43	50
My weight in pounds	238.2	237.0	235.6	234.4	233.0	233.8	232.8	232.0

Day # (Saturday)	3	10	17	24	31	38	45	52	59	66
My weight in pounds	238.4	235.8	235.0	234.2	236.2	236.2	235.2	233.2	236.8	238.2

(a) Find the least squares line for the Thursday data and comment on its goodness of fit.

(b) Find the least squares line for the Saturday data and comment on its goodness of fit.

(c) Use Quadratic Regression to find a parabola which models the Saturday data and comment on its goodness of fit.

(d) Compare and contrast the predictions the three models make for my weight on January 1, 2010 (Day #366). Can any of these models be used to make a prediction of my weight 20 years from now? Explain your answer.

(e) Why is this a Civics lesson in honesty? Well, compare the two linear models you obtained above. One was a good fit and the other was not, yet both came from careful selections of real data. In presenting the tables to you, I have not lied about my weight, nor have you used any bad math to falsify the predictions. The word we're looking for here is 'disingenuous'. Look it up and then discuss the implications this type of data manipulation could have in a larger, more complex, politically motivated setting. (Even Obi-Wan presented the truth to Luke only "from a certain point of view.")

6. (Data that is neither linear nor quadratic.) We'll close this exercise set with two data sets that, for reasons presented later in the book, cannot be modeled correctly by lines or parabolas. It is a good exercise, though, to see what happens when you attempt to use a linear or quadratic model when it's not appropriate.

   (a) This first data set came from a Summer 2003 publication of the Portage County Animal Protective League called "Tattle Tails". They make the following statement and then have a chart of data that supports it. "It doesn't take long for two cats to turn into 80 million. If two cats and their surviving offspring reproduced for ten years, you'd end up with 80,399,780 cats." We assume $N(0) = 2$.

Year x	1	2	3	4	5	6	7	8	9	10
Number of Cats $N(x)$	12	66	382	2201	12680	73041	420715	2423316	13968290	80399780

   Use Quadratic Regression to find a parabola which models this data and comment on its goodness of fit. (Spoiler Alert: Does anyone know what type of function we need here?)

   (b) This next data set comes from the U.S. Naval Observatory. That site has loads of awesome stuff on it, but for this exercise I used the sunrise/sunset times in Fairbanks, Alaska for 2009 to give you a chart of the number of hours of daylight they get on the 21st of each month. We'll let $x = 1$ represent January 21, 2009, $x = 2$ represent February 21, 2009, and so on.

Month Number	1	2	3	4	5	6	7	8	9	10	11	12
Hours of Daylight	5.8	9.3	12.4	15.9	19.4	21.8	19.4	15.6	12.4	9.1	5.6	3.3

   Use Quadratic Regression to find a parabola which models this data and comment on its goodness of fit. (Spoiler Alert: Does anyone know what type of function we need here?)

2.5 REGRESSION

2.5.2 ANSWERS

1. (a) $y = 936.31x - 1645322.6$ with $r = 0.9696$ which indicates a good fit. The slope 936.31 indicates Lake County's population is increasing at a rate of (approximately) 936 people per year.

   (b) According to the model, the population in 2010 will be 236,660.

   (c) According to the model, the population of Lake County will reach 250,000 sometime between 2024 and 2025.

2. (a) $y = 796.8x - 1309762.5$ with $r = 0.8916$ which indicates a reasonable fit. The slope 796.8 indicates Lorain County's population is increasing at a rate of (approximately) 797 people per year.

   (b) According to the model, the population in 2010 will be 291,805.

   (c) According to the model, the population of Lake County will reach 325,000 sometime between 2051 and 2052.

3. (c) $y = 0.266x - 459.86$ with $r = 0.9607$ which indicates a good fit. The slope 0.266 indicates the country's energy production is increasing at a rate of 0.266 Quad per year.

   (d) According to the model, the production in 2010 will be 74.8 Quad.

   (e) According to the model, the production will reach 100 Quad in the year 2105.

4. The line is $y = 36.8x + 16.39$. We have $r = .99987$ and $r^2 = .9997$ so this is an excellent fit to the data. The slope 36.8 represents miles per gallon.

5. (a) The line for the Thursday data is $y = -.12x + 237.69$. We have $r = -.9568$ and $r^2 = .9155$ so this is a really good fit.

   (b) The line for the Saturday data is $y = -0.000693x + 235.94$. We have $r = -0.008986$ and $r^2 = 0.0000807$ which is horrible. This data is not even close to linear.

   (c) The parabola for the Saturday data is $y = 0.003x^2 - 0.21x + 238.30$. We have $R^2 = .47497$ which isn't good. Thus the data isn't modeled well by a quadratic function, either.

   (d) The Thursday linear model had my weight on January 1, 2010 at 193.77 pounds. The Saturday models give 235.69 and 563.31 pounds, respectively. The Thursday line has my weight going below 0 pounds in about five and a half years, so that's no good. The quadratic has a positive leading coefficient which would mean unbounded weight gain for the rest of my life. The Saturday line, which mathematically does not fit the data at all, yields a plausible weight prediction in the end. I think this is why grown-ups talk about "Lies, Damned Lies and Statistics."

6. (a) The quadratic model for the cats in Portage county is $y = 1917803.54x^2 - 16036408.29x + 24094857.7$. Although $R^2 = .70888$ this is not a good model because it's so far off for small values of x. Case in point, the model gives us 24,094,858 cats when $x = 0$ but we know $N(0) = 2$.

(b) The quadratic model for the hours of daylight in Fairbanks, Alaska is $y = .51x^2 + 6.23x - .36$. Even with $R^2 = .92295$ we should be wary of making predictions beyond the data. Case in point, the model gives -4.84 hours of daylight when $x = 13$. So January 21, 2010 will be "extra dark"? Obviously a parabola pointing down isn't telling us the whole story.

CHAPTER 3

POLYNOMIAL FUNCTIONS

3.1 GRAPHS OF POLYNOMIALS

Three of the families of functions studied thus far – constant, linear and quadratic – belong to a much larger group of functions called **polynomials**. We begin our formal study of general polynomials with a definition and some examples.

> **Definition 3.1.** A **polynomial function** is a function of the form
> $$f(x) = a_n x^n + a_{n-1} x^{n-1} + \ldots + a_2 x^2 + a_1 x + a_0,$$
> where a_0, a_1, ..., a_n are real numbers and $n \geq 1$ is a natural number. The domain of a polynomial function is $(-\infty, \infty)$.

There are several things about Definition 3.1 that may be off-putting or downright frightening. The best thing to do is look at an example. Consider $f(x) = 4x^5 - 3x^2 + 2x - 5$. Is this a polynomial function? We can re-write the formula for f as $f(x) = 4x^5 + 0x^4 + 0x^3 + (-3)x^2 + 2x + (-5)$. Comparing this with Definition 3.1, we identify $n = 5$, $a_5 = 4$, $a_4 = 0$, $a_3 = 0$, $a_2 = -3$, $a_1 = 2$ and $a_0 = -5$. In other words, a_5 is the coefficient of x^5, a_4 is the coefficient of x^4, and so forth; the subscript on the a's merely indicates to which power of x the coefficient belongs. The business of restricting n to be a natural number lets us focus on well-behaved algebraic animals.[1]

Example 3.1.1. Determine if the following functions are polynomials. Explain your reasoning.

1. $g(x) = \dfrac{4 + x^3}{x}$

2. $p(x) = \dfrac{4x + x^3}{x}$

3. $q(x) = \dfrac{4x + x^3}{x^2 + 4}$

4. $f(x) = \sqrt[3]{x}$

5. $h(x) = |x|$

6. $z(x) = 0$

---

[1] Enjoy this while it lasts. Before we're through with the book, you'll have been exposed to the most terrible of algebraic beasts. We will tame them all, in time.

Solution.

1. We note directly that the domain of $g(x) = \frac{x^3+4}{x}$ is $x \neq 0$. By definition, a polynomial has all real numbers as its domain. Hence, g can't be a polynomial.

2. Even though $p(x) = \frac{x^3+4x}{x}$ simplifies to $p(x) = x^2 + 4$, which certainly looks like the form given in Definition 3.1, the domain of p, which, as you may recall, we determine *before* we simplify, excludes 0. Alas, p is not a polynomial function for the same reason g isn't.

3. After what happened with p in the previous part, you may be a little shy about simplifying $q(x) = \frac{x^3+4x}{x^2+4}$ to $q(x) = x$, which certainly fits Definition 3.1. If we look at the domain of q before we simplified, we see that it is, indeed, all real numbers. A function which can be written in the form of Definition 3.1 whose domain is all real numbers is, in fact, a polynomial.

4. We can rewrite $f(x) = \sqrt[3]{x}$ as $f(x) = x^{\frac{1}{3}}$. Since $\frac{1}{3}$ is not a natural number, f is not a polynomial.

5. The function $h(x) = |x|$ isn't a polynomial, since it can't be written as a combination of powers of x even though it can be written as a piecewise function involving polynomials. As we shall see in this section, graphs of polynomials possess a quality[2] that the graph of h does not.

6. There's nothing in Definition 3.1 which prevents all the coefficients a_n, etc., from being 0. Hence, $z(x) = 0$, is an honest-to-goodness polynomial.

Definition 3.2. Suppose f is a polynomial function.

- Given $f(x) = a_n x^n + a_{n-1} x^{n-1} + \ldots + a_2 x^2 + a_1 x + a_0$ with $a_n \neq 0$, we say
    - The natural number n is called the **degree** of the polynomial f.
    - The term $a_n x^n$ is called the **leading term** of the polynomial f.
    - The real number a_n is called the **leading coefficient** of the polynomial f.
    - The real number a_0 is called the **constant term** of the polynomial f.
- If $f(x) = a_0$, and $a_0 \neq 0$, we say f has degree 0.
- If $f(x) = 0$, we say f has no degree.[a]

[a]Some authors say $f(x) = 0$ has degree $-\infty$ for reasons not even we will go into.

The reader may well wonder why we have chosen to separate off constant functions from the other polynomials in Definition 3.2. Why not just lump them all together and, instead of forcing n to be a natural number, $n = 1, 2, \ldots$, allow n to be a whole number, $n = 0, 1, 2, \ldots$. We could unify all

---
[2]One which really relies on Calculus to verify.

3.1 Graphs of Polynomials

of the cases, since, after all, isn't $a_0 x^0 = a_0$? The answer is 'yes, as long as $x \neq 0$.' The function $f(x) = 3$ and $g(x) = 3x^0$ are different, because their domains are different. The number $f(0) = 3$ is defined, whereas $g(0) = 3(0)^0$ is not.[3] Indeed, much of the theory we will develop in this chapter doesn't include the constant functions, so we might as well treat them as outsiders from the start. One good thing that comes from Definition 3.2 is that we can now think of linear functions as degree 1 (or 'first degree') polynomial functions and quadratic functions as degree 2 (or 'second degree') polynomial functions.

Example 3.1.2. Find the degree, leading term, leading coefficient and constant term of the following polynomial functions.

1. $f(x) = 4x^5 - 3x^2 + 2x - 5$
2. $g(x) = 12x + x^3$
3. $h(x) = \dfrac{4-x}{5}$
4. $p(x) = (2x-1)^3(x-2)(3x+2)$

Solution.

1. There are no surprises with $f(x) = 4x^5 - 3x^2 + 2x - 5$. It is written in the form of Definition 3.2, and we see that the degree is 5, the leading term is $4x^5$, the leading coefficient is 4 and the constant term is -5.

2. The form given in Definition 3.2 has the highest power of x first. To that end, we re-write $g(x) = 12x + x^3 = x^3 + 12x$, and see that the degree of g is 3, the leading term is x^3, the leading coefficient is 1 and the constant term is 0.

3. We need to rewrite the formula for h so that it resembles the form given in Definition 3.2: $h(x) = \frac{4-x}{5} = \frac{4}{5} - \frac{x}{5} = -\frac{1}{5}x + \frac{4}{5}$. The degree of h is 1, the leading term is $-\frac{1}{5}x$, the leading coefficient is $-\frac{1}{5}$ and the constant term is $\frac{4}{5}$.

4. It may seem that we have some work ahead of us to get p in the form of Definition 3.2. However, it is possible to glean the information requested about p without multiplying out the entire expression $(2x-1)^3(x-2)(3x+2)$. The leading term of p will be the term which has the highest power of x. The way to get this term is to multiply the terms with the highest power of x from each factor together - in other words, the leading term of $p(x)$ is the product of the leading terms of the factors of $p(x)$. Hence, the leading term of p is $(2x)^3(x)(3x) = 24x^5$. This means that the degree of p is 5 and the leading coefficient is 24. As for the constant term, we can perform a similar trick. The constant term is obtained by multiplying the constant terms from each of the factors $(-1)^3(-2)(2) = 4$. □

Our next example shows how polynomials of higher degree arise 'naturally'[4] in even the most basic geometric applications.

---

[3] Technically, 0^0 is an indeterminant form, which is a special case of being undefined. The authors realize this is beyond pedantry, but we wouldn't mention it if we didn't feel it was neccessary.

[4] this is a dangerous word...

Example 3.1.3. A box with no top is to be fashioned from a 10 inch × 12 inch piece of cardboard by cutting out congruent squares from each corner of the cardboard and then folding the resulting tabs. Let x denote the length of the side of the square which is removed from each corner.

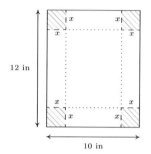
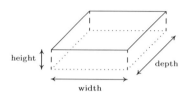

1. Find the volume V of the box as a function of x. Include an appropriate applied domain.

2. Use a graphing calculator to graph $y = V(x)$ on the domain you found in part 1 and approximate the dimensions of the box with maximum volume to two decimal places. What is the maximum volume?

Solution.

1. From Geometry, we know that Volume = width × height × depth. The key is to find each of these quantities in terms of x. From the figure, we see that the height of the box is x itself. The cardboard piece is initially 10 inches wide. Removing squares with a side length of x inches from each corner leaves $10 - 2x$ inches for the width.[5] As for the depth, the cardboard is initially 12 inches long, so after cutting out x inches from each side, we would have $12 - 2x$ inches remaining. As a function[6] of x, the volume is

$$V(x) = x(10 - 2x)(12 - 2x) = 4x^3 - 44x^2 + 120x$$

To find a suitable applied domain, we note that to make a box at all we need $x > 0$. Also the shorter of the two dimensions of the cardboard is 10 inches, and since we are removing $2x$ inches from this dimension, we also require $10 - 2x > 0$ or $x < 5$. Hence, our applied domain is $0 < x < 5$.

2. Using a graphing calculator, we see that the graph of $y = V(x)$ has a relative maximum. For $0 < x < 5$, this is also the absolute maximum. Using the 'Maximum' feature of the calculator, we get $x \approx 1.81$, $y \approx 96.77$. This yields a height of $x \approx 1.81$ inches, a width of $10 - 2x \approx 6.38$ inches, and a depth of $12 - 2x \approx 8.38$ inches. The y-coordinate is the maximum volume, which is approximately 96.77 cubic inches (also written in^3).

---

[5]There's no harm in taking an extra step here and making sure this makes sense. If we chopped out a 1 inch square from each side, then the width would be 8 inches, so chopping out x inches would leave $10 - 2x$ inches.
[6]When we write $V(x)$, it is in the context of function notation, not the volume V times the quantity x.

3.1 GRAPHS OF POLYNOMIALS

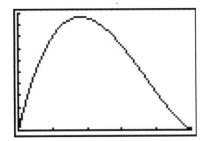 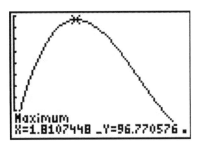

In order to solve Example 3.1.3, we made good use of the graph of the polynomial $y = V(x)$, so we ought to turn our attention to graphs of polynomials in general. Below are the graphs of $y = x^2$, $y = x^4$ and $y = x^6$, side-by-side. We have omitted the axes to allow you to see that as the exponent increases, the 'bottom' becomes 'flatter' and the 'sides' become 'steeper.' If you take the the time to graph these functions by hand,[7] you will see why.

$y = x^2$ $y = x^4$ $y = x^6$

All of these functions are even, (Do you remember how to show this?) and it is exactly because the exponent is even.[8] This symmetry is important, but we want to explore a different yet equally important feature of these functions which we can be seen graphically – their **end behavior**.

The end behavior of a function is a way to describe what is happening to the function values (the y-values) as the x-values approach the 'ends' of the x-axis.[9] That is, what happens to y as x becomes small without bound[10] (written $x \to -\infty$) and, on the flip side, as x becomes large without bound[11] (written $x \to \infty$).

For example, given $f(x) = x^2$, as $x \to -\infty$, we imagine substituting $x = -100$, $x = -1000$, etc., into f to get $f(-100) = 10000$, $f(-1000) = 1000000$, and so on. Thus the function values are becoming larger and larger positive numbers (without bound). To describe this behavior, we write: as $x \to -\infty$, $f(x) \to \infty$. If we study the behavior of f as $x \to \infty$, we see that in this case, too, $f(x) \to \infty$. (We told you that the symmetry was important!) The same can be said for any function of the form $f(x) = x^n$ where n is an even natural number. If we generalize just a bit to include vertical scalings and reflections across the x-axis,[12] we have

---

[7]Make sure you choose some x-values between -1 and 1.
[8]Herein lies one of the possible origins of the term 'even' when applied to functions.
[9]Of course, there are no ends to the x-axis.
[10]We think of x as becoming a very large (in the sense of its absolute value) *negative* number far to the left of zero.
[11]We think of x as moving far to the right of zero and becoming a very large *positive* number.
[12]See Theorems 1.4 and 1.5 in Section 1.7.

End Behavior of functions $f(x) = ax^n$, n even.

Suppose $f(x) = ax^n$ where $a \neq 0$ is a real number and n is an even natural number. The end behavior of the graph of $y = f(x)$ matches one of the following:

- for $a > 0$, as $x \to -\infty$, $f(x) \to \infty$ and as $x \to \infty$, $f(x) \to \infty$
- for $a < 0$, as $x \to -\infty$, $f(x) \to -\infty$ and as $x \to \infty$, $f(x) \to -\infty$

Graphically:

$a > 0 \qquad a < 0$

We now turn our attention to functions of the form $f(x) = x^n$ where $n \geq 3$ is an odd natural number. (We ignore the case when $n = 1$, since the graph of $f(x) = x$ is a line and doesn't fit the general pattern of higher-degree odd polynomials.) Below we have graphed $y = x^3$, $y = x^5$, and $y = x^7$. The 'flattening' and 'steepening' that we saw with the even powers presents itself here as well, and, it should come as no surprise that all of these functions are odd.[13] The end behavior of these functions is all the same, with $f(x) \to -\infty$ as $x \to -\infty$ and $f(x) \to \infty$ as $x \to \infty$.

$y = x^3 \qquad\qquad y = x^5 \qquad\qquad y = x^7$

As with the even degreed functions we studied earlier, we can generalize their end behavior.

End Behavior of functions $f(x) = ax^n$, n odd.

Suppose $f(x) = ax^n$ where $a \neq 0$ is a real number and $n \geq 3$ is an odd natural number. The end behavior of the graph of $y = f(x)$ matches one of the following:

- for $a > 0$, as $x \to -\infty$, $f(x) \to -\infty$ and as $x \to \infty$, $f(x) \to \infty$
- for $a < 0$, as $x \to -\infty$, $f(x) \to \infty$ and as $x \to \infty$, $f(x) \to -\infty$

Graphically:

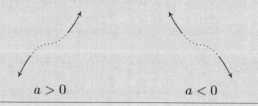

$a > 0 \qquad a < 0$

---

[13] And are, perhaps, the inspiration for the moniker 'odd function'.

3.1 Graphs of Polynomials

Despite having different end behavior, all functions of the form $f(x) = ax^n$ for natural numbers n share two properties which help distinguish them from other animals in the algebra zoo: they are **continuous** and **smooth**. While these concepts are formally defined using Calculus,[14] informally, graphs of continuous functions have no 'breaks' or 'holes' in them, and the graphs of smooth functions have no 'sharp turns'. It turns out that these traits are preserved when functions are added together, so general polynomial functions inherit these qualities. Below we find the graph of a function which is neither smooth nor continuous, and to its right we have a graph of a polynomial, for comparison. The function whose graph appears on the left fails to be continuous where it has a 'break' or 'hole' in the graph; everywhere else, the function is continuous. The function is continuous at the 'corner' and the 'cusp', but we consider these 'sharp turns', so these are places where the function fails to be smooth. Apart from these four places, the function is smooth and continuous. Polynomial functions are smooth and continuous everywhere, as exhibited in the graph on the right.

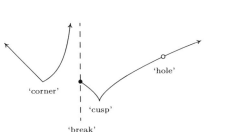
Pathologies not found on graphs of polynomials

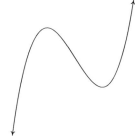
The graph of a polynomial

The notion of smoothness is what tells us graphically that, for example, $f(x) = |x|$, whose graph is the characteristic 'V' shape, cannot be a polynomial. The notion of continuity is what allowed us to construct the sign diagram for quadratic inequalities as we did in Section 2.4. This last result is formalized in the following theorem.

> **Theorem 3.1. The Intermediate Value Theorem (Zero Version):** Suppose f is a continuous function on an interval containing $x = a$ and $x = b$ with $a < b$. If $f(a)$ and $f(b)$ have different signs, then f has at least one zero between $x = a$ and $x = b$; that is, for at least one real number c such that $a < c < b$, we have $f(c) = 0$.

The Intermediate Value Theorem is extremely profound; it gets to the heart of what it means to be a real number, and is one of the most often used and under appreciated theorems in Mathematics. With that being said, most students see the result as common sense since it says, geometrically, that the graph of a polynomial function cannot be above the x-axis at one point and below the x-axis at another point without crossing the x-axis somewhere in between. The following example uses the Intermediate Value Theorem to establish a fact that that most students take for granted. Many students, and sadly some instructors, will find it silly.

---

[14] In fact, if you take Calculus, you'll find that smooth functions are automatically continuous, so that saying 'polynomials are continuous and smooth' is redundant.

Example 3.1.4. Use the Intermediate Value Theorem to establish that $\sqrt{2}$ is a real number.

Solution. Consider the polynomial function $f(x) = x^2 - 2$. Then $f(1) = -1$ and $f(3) = 7$. Since $f(1)$ and $f(3)$ have different signs, the Intermediate Value Theorem guarantees us a real number c between 1 and 3 with $f(c) = 0$. If $c^2 - 2 = 0$ then $c = \pm\sqrt{2}$. Since c is between 1 and 3, c is positive, so $c = \sqrt{2}$. $\square$

Our primary use of the Intermediate Value Theorem is in the construction of sign diagrams, as in Section 2.4, since it guarantees us that polynomial functions are always positive (+) or always negative (−) on intervals which do not contain any of its zeros. The general algorithm for polynomials is given below.

Steps for Constructing a Sign Diagram for a Polynomial Function

Suppose f is a polynomial function.

1. Find the zeros of f and place them on the number line with the number 0 above them.

2. Choose a real number, called a **test value**, in each of the intervals determined in step 1.

3. Determine the sign of $f(x)$ for each test value in step 2, and write that sign above the corresponding interval.

Example 3.1.5. Construct a sign diagram for $f(x) = x^3(x-3)^2(x+2)\left(x^2+1\right)$. Use it to give a rough sketch of the graph of $y = f(x)$.

Solution. First, we find the zeros of f by solving $x^3(x-3)^2(x+2)\left(x^2+1\right) = 0$. We get $x = 0$, $x = 3$ and $x = -2$. (The equation $x^2+1 = 0$ produces no real solutions.) These three points divide the real number line into four intervals: $(-\infty, -2)$, $(-2, 0)$, $(0, 3)$ and $(3, \infty)$. We select the test values $x = -3$, $x = -1$, $x = 1$ and $x = 4$. We find $f(-3)$ is (+), $f(-1)$ is (−) and $f(1)$ is (+) as is $f(4)$. Wherever f is (+), its graph is above the x-axis; wherever f is (−), its graph is below the x-axis. The x-intercepts of the graph of f are $(-2, 0)$, $(0, 0)$ and $(3, 0)$. Knowing f is smooth and continuous allows us to sketch its graph.

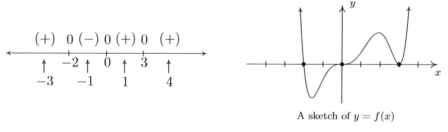

A sketch of $y = f(x)$ $\square$

A couple of notes about the Example 3.1.5 are in order. First, note that we purposefully did not label the y-axis in the sketch of the graph of $y = f(x)$. This is because the sign diagram gives us the zeros and the relative position of the graph - it doesn't give us any information as to how high or low the graph strays from the x-axis. Furthermore, as we have mentioned earlier in the text, without Calculus, the values of the relative maximum and minimum can only be found approximately using a calculator. If we took the time to find the leading term of f, we would find it to be x^8. Looking

3.1 Graphs of Polynomials

at the end behavior of f, we notice that it matches the end behavior of $y = x^8$. This is no accident, as we find out in the next theorem.

Theorem 3.2. End Behavior for Polynomial Functions: The end behavior of a polynomial $f(x) = a_n x^n + a_{n-1} x^{n-1} + \ldots + a_2 x^2 + a_1 x + a_0$ with $a_n \neq 0$ matches the end behavior of $y = a_n x^n$.

To see why Theorem 3.2 is true, let's first look at a specific example. Consider $f(x) = 4x^3 - x + 5$. If we wish to examine end behavior, we look to see the behavior of f as $x \to \pm\infty$. Since we're concerned with x's far down the x-axis, we are far away from $x = 0$ so can rewrite $f(x)$ for these values of x as

$$f(x) = 4x^3 \left(1 - \frac{1}{4x^2} + \frac{5}{4x^3}\right)$$

As x becomes unbounded (in either direction), the terms $\frac{1}{4x^2}$ and $\frac{5}{4x^3}$ become closer and closer to 0, as the table below indicates.

x	$\frac{1}{4x^2}$	$\frac{5}{4x^3}$
-1000	0.00000025	-0.00000000125
-100	0.000025	-0.00000125
-10	0.0025	-0.00125
10	0.0025	0.00125
100	0.000025	0.00000125
1000	0.00000025	0.00000000125

In other words, as $x \to \pm\infty$, $f(x) \approx 4x^3 (1 - 0 + 0) = 4x^3$, which is the leading term of f. The formal proof of Theorem 3.2 works in much the same way. Factoring out the leading term leaves

$$f(x) = a_n x^n \left(1 + \frac{a_{n-1}}{a_n x} + \ldots + \frac{a_2}{a_n x^{n-2}} + \frac{a_1}{a_n x^{n-1}} + \frac{a_0}{a_n x^n}\right)$$

As $x \to \pm\infty$, any term with an x in the denominator becomes closer and closer to 0, and we have $f(x) \approx a_n x^n$. Geometrically, Theorem 3.2 says that if we graph $y = f(x)$ using a graphing calculator, and continue to 'zoom out', the graph of it and its leading term become indistinguishable. Below are the graphs of $y = 4x^3 - x + 5$ (the thicker line) and $y = 4x^3$ (the thinner line) in two different windows.

A view 'close' to the origin.

A 'zoomed out' view.

Let's return to the function in Example 3.1.5, $f(x) = x^3(x-3)^2(x+2)\left(x^2+1\right)$, whose sign diagram and graph are reproduced below for reference. Theorem 3.2 tells us that the end behavior is the same as that of its leading term x^8. This tells us that the graph of $y = f(x)$ starts and ends above the x-axis. In other words, $f(x)$ is $(+)$ as $x \to \pm\infty$, and as a result, we no longer need to evaluate f at the test values $x = -3$ and $x = 4$. Is there a way to eliminate the need to evaluate f at the other test values? What we would really need to know is how the function behaves near its zeros - does it cross through the x-axis at these points, as it does at $x = -2$ and $x = 0$, or does it simply touch and rebound like it does at $x = 3$. From the sign diagram, the graph of f will cross the x-axis whenever the signs on either side of the zero switch (like they do at $x = -2$ and $x = 0$); it will touch when the signs are the same on either side of the zero (as is the case with $x = 3$). What we need to determine is the reason behind whether or not the sign change occurs.

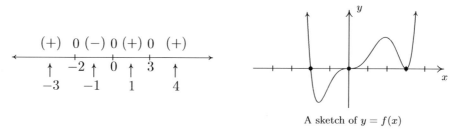

A sketch of $y = f(x)$

Fortunately, f was given to us in factored form: $f(x) = x^3(x-3)^2(x+2)$. When we attempt to determine the sign of $f(-4)$, we are attempting to find the sign of the number $(-4)^3(-7)^2(-2)$, which works out to be $(-)(+)(-)$ which is $(+)$. If we move to the other side of $x = -2$, and find the sign of $f(-1)$, we are determining the sign of $(-1)^3(-4)^2(+1)$, which is $(-)(+)(+)$ which gives us the $(-)$. Notice that signs of the first two factors in both expressions are the same in $f(-4)$ and $f(-1)$. The only factor which switches sign is the third factor, $(x+2)$, precisely the factor which gave us the zero $x = -2$. If we move to the other side of 0 and look closely at $f(1)$, we get the sign pattern $(+1)^3(-2)^2(+3)$ or $(+)(+)(+)$ and we note that, once again, going from $f(-1)$ to $f(1)$, the only factor which changed sign was the first factor, x^3, which corresponds to the zero $x = 0$. Finally, to find $f(4)$, we substitute to get $(+4)^3(+2)^2(+5)$ which is $(+)(+)(+)$ or $(+)$. The sign didn't change for the middle factor $(x-3)^2$. Even though this is the factor which corresponds to the zero $x = 3$, the fact that the quantity is *squared* kept the sign of the middle factor the same on either side of 3. If we look back at the exponents on the factors $(x+2)$ and x^3, we see that they are both odd, so as we substitute values to the left and right of the corresponding zeros, the signs of the corresponding factors change which results in the sign of the function value changing. This is the key to the behavior of the function near the zeros. We need a definition and then a theorem.

Definition 3.3. Suppose f is a polynomial function and m is a natural number. If $(x-c)^m$ is a factor of $f(x)$ but $(x-c)^{m+1}$ is not, then we say $x = c$ is a zero of **multiplicity** m.

Hence, rewriting $f(x) = x^3(x-3)^2(x+2)$ as $f(x) = (x-0)^3(x-3)^2(x-(-2))^1$, we see that $x = 0$ is a zero of multiplicity 3, $x = 3$ is a zero of multiplicity 2 and $x = -2$ is a zero of multiplicity 1.

3.1 Graphs of Polynomials

Theorem 3.3. The Role of Multiplicity: Suppose f is a polynomial function and $x = c$ is a zero of multiplicity m.

- If m is even, the graph of $y = f(x)$ touches and rebounds from the x-axis at $(c, 0)$.
- If m is odd, the graph of $y = f(x)$ crosses through the x-axis at $(c, 0)$.

Our last example shows how end behavior and multiplicity allow us to sketch a decent graph without appealing to a sign diagram.

Example 3.1.6. Sketch the graph of $f(x) = -3(2x - 1)(x + 1)^2$ using end behavior and the multiplicity of its zeros.

Solution. The end behavior of the graph of f will match that of its leading term. To find the leading term, we multiply by the leading terms of each factor to get $(-3)(2x)(x)^2 = -6x^3$. This tells us that the graph will start above the x-axis, in Quadrant II, and finish below the x-axis, in Quadrant IV. Next, we find the zeros of f. Fortunately for us, f is factored.[15] Setting each factor equal to zero gives is $x = \frac{1}{2}$ and $x = -1$ as zeros. To find the multiplicity of $x = \frac{1}{2}$ we note that it corresponds to the factor $(2x - 1)$. This isn't strictly in the form required in Definition 3.3. If we factor out the 2, however, we get $(2x - 1) = 2\left(x - \frac{1}{2}\right)$, and we see that the multiplicity of $x = \frac{1}{2}$ is 1. Since 1 is an odd number, we know from Theorem 3.3 that the graph of f will cross through the x-axis at $\left(\frac{1}{2}, 0\right)$. Since the zero $x = -1$ corresponds to the factor $(x + 1)^2 = (x - (-1))^2$, we find its multiplicity to be 2 which is an even number. As such, the graph of f will touch and rebound from the x-axis at $(-1, 0)$. Though we're not asked to, we can find the y-intercept by finding $f(0) = -3(2(0) - 1)(0 + 1)^2 = 3$. Thus $(0, 3)$ is an additional point on the graph. Putting this together gives us the graph below.

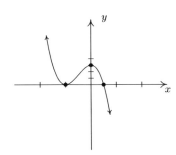

---

[15] Obtaining the factored form of a polynomial is the main focus of the next few sections.

3.1.1 Exercises

In Exercises 1 - 10, find the degree, the leading term, the leading coefficient, the constant term and the end behavior of the given polynomial.

1. $f(x) = 4 - x - 3x^2$

2. $g(x) = 3x^5 - 2x^2 + x + 1$

3. $q(r) = 1 - 16r^4$

4. $Z(b) = 42b - b^3$

5. $f(x) = \sqrt{3}x^{17} + 22.5x^{10} - \pi x^7 + \frac{1}{3}$

6. $s(t) = -4.9t^2 + v_0 t + s_0$

7. $P(x) = (x-1)(x-2)(x-3)(x-4)$

8. $p(t) = -t^2(3-5t)(t^2+t+4)$

9. $f(x) = -2x^3(x+1)(x+2)^2$

10. $G(t) = 4(t-2)^2 \left(t + \frac{1}{2}\right)$

In Exercises 11 - 20, find the real zeros of the given polynomial and their corresponding multiplicities. Use this information along with a sign chart to provide a rough sketch of the graph of the polynomial. Compare your answer with the result from a graphing utility.

11. $a(x) = x(x+2)^2$

12. $g(x) = x(x+2)^3$

13. $f(x) = -2(x-2)^2(x+1)$

14. $g(x) = (2x+1)^2(x-3)$

15. $F(x) = x^3(x+2)^2$

16. $P(x) = (x-1)(x-2)(x-3)(x-4)$

17. $Q(x) = (x+5)^2(x-3)^4$

18. $h(x) = x^2(x-2)^2(x+2)^2$

19. $H(t) = (3-t)(t^2+1)$

20. $Z(b) = b(42 - b^2)$

In Exercises 21 - 26, given the pair of functions f and g, sketch the graph of $y = g(x)$ by starting with the graph of $y = f(x)$ and using transformations. Track at least three points of your choice through the transformations. State the domain and range of g.

21. $f(x) = x^3$, $g(x) = (x+2)^3 + 1$

22. $f(x) = x^4$, $g(x) = (x+2)^4 + 1$

23. $f(x) = x^4$, $g(x) = 2 - 3(x-1)^4$

24. $f(x) = x^5$, $g(x) = -x^5 - 3$

25. $f(x) = x^5$, $g(x) = (x+1)^5 + 10$

26. $f(x) = x^6$, $g(x) = 8 - x^6$

27. Use the Intermediate Value Theorem to prove that $f(x) = x^3 - 9x + 5$ has a real zero in each of the following intervals: $[-4, -3], [0, 1]$ and $[2, 3]$.

28. Rework Example 3.1.3 assuming the box is to be made from an 8.5 inch by 11 inch sheet of paper. Using scissors and tape, construct the box. Are you surprised?[16]

---

[16] Consider decorating the box and presenting it to your instructor. If done well enough, maybe your instructor will issue you some bonus points. Or maybe not.

3.1 GRAPHS OF POLYNOMIALS

In Exercises 29 - 31, suppose the revenue R, in *thousands* of dollars, from producing and selling x *hundred* LCD TVs is given by $R(x) = -5x^3 + 35x^2 + 155x$ for $0 \leq x \leq 10.07$.

29. Use a graphing utility to graph $y = R(x)$ and determine the number of TVs which should be sold to maximize revenue. What is the maximum revenue?

30. Assume that the cost, in *thousands* of dollars, to produce x *hundred* LCD TVs is given by $C(x) = 200x + 25$ for $x \geq 0$. Find and simplify an expression for the profit function $P(x)$. (Remember: Profit = Revenue - Cost.)

31. Use a graphing utility to graph $y = P(x)$ and determine the number of TVs which should be sold to maximize profit. What is the maximum profit?

32. While developing their newest game, Sasquatch Attack!, the makers of the PortaBoy (from Example 2.1.5) revised their cost function and now use $C(x) = .03x^3 - 4.5x^2 + 225x + 250$, for $x \geq 0$. As before, $C(x)$ is the cost to make x PortaBoy Game Systems. Market research indicates that the demand function $p(x) = -1.5x + 250$ remains unchanged. Use a graphing utility to find the production level x that maximizes the *profit* made by producing and selling x PortaBoy game systems.

33. According to US Postal regulations, a rectangular shipping box must satisfy the inequality "Length + Girth $\leq$ 130 inches" for Parcel Post and "Length + Girth $\leq$ 108 inches" for other services. Let's assume we have a closed rectangular box with a square face of side length x as drawn below. The length is the longest side and is clearly labeled. The girth is the distance around the box in the other two dimensions so in our case it is the sum of the four sides of the square, $4x$.

   (a) Assuming that we'll be mailing a box via Parcel Post where Length + Girth = 130 inches, express the length of the box in terms of x and then express the volume V of the box in terms of x.

   (b) Find the dimensions of the box of maximum volume that can be shipped via Parcel Post.

   (c) Repeat parts 33a and 33b if the box is shipped using "other services".

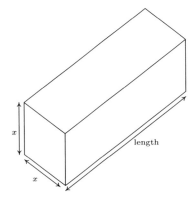

34. We now revisit the data set from Exercise 6b in Section 2.5. In that exercise, you were given a chart of the number of hours of daylight they get on the 21st of each month in Fairbanks, Alaska based on the 2009 sunrise and sunset data found on the U.S. Naval Observatory website. We let $x = 1$ represent January 21, 2009, $x = 2$ represent February 21, 2009, and so on. The chart is given again for reference.

Month Number	1	2	3	4	5	6	7	8	9	10	11	12
Hours of Daylight	5.8	9.3	12.4	15.9	19.4	21.8	19.4	15.6	12.4	9.1	5.6	3.3

Find cubic (third degree) and quartic (fourth degree) polynomials which model this data and comment on the goodness of fit for each. What can we say about using either model to make predictions about the year 2020? (Hint: Think about the end behavior of polynomials.) Use the models to see how many hours of daylight they got on your birthday and then check the website to see how accurate the models are. Knowing that Sasquatch are largely nocturnal, what days of the year according to your models are going to allow for at least 14 hours of darkness for field research on the elusive creatures?

35. An electric circuit is built with a variable resistor installed. For each of the following resistance values (measured in kilo-ohms, $k\Omega$), the corresponding power to the load (measured in milliwatts, mW) is given in the table below. [17]

Resistance: ($k\Omega$)	1.012	2.199	3.275	4.676	6.805	9.975
Power: (mW)	1.063	1.496	1.610	1.613	1.505	1.314

(a) Make a scatter diagram of the data using the Resistance as the independent variable and Power as the dependent variable.

(b) Use your calculator to find quadratic (2nd degree), cubic (3rd degree) and quartic (4th degree) regression models for the data and judge the reasonableness of each.

(c) For each of the models found above, find the predicted maximum power that can be delivered to the load. What is the corresponding resistance value?

(d) Discuss with your classmates the limitations of these models - in particular, discuss the end behavior of each.

36. Show that the end behavior of a linear function $f(x) = mx + b$ is as it should be according to the results we've established in the section for polynomials of odd degree.[18] (That is, show that the graph of a linear function is "up on one side and down on the other" just like the graph of $y = a_n x^n$ for odd numbers n.)

---

[17] The authors wish to thank Don Anthan and Ken White of Lakeland Community College for devising this problem and generating the accompanying data set.

[18] Remember, to be a linear function, $m \neq 0$.

3.1 GRAPHS OF POLYNOMIALS

37. There is one subtlety about the role of multiplicity that we need to discuss further; specifically we need to see 'how' the graph crosses the x-axis at a zero of odd multiplicity. In the section, we deliberately excluded the function $f(x) = x$ from the discussion of the end behavior of $f(x) = x^n$ for odd numbers n and we said at the time that it was due to the fact that $f(x) = x$ didn't fit the pattern we were trying to establish. You just showed in the previous exercise that the end behavior of a linear function behaves like every other polynomial of odd degree, so what doesn't $f(x) = x$ do that $g(x) = x^3$ does? It's the 'flattening' for values of x near zero. It is this local behavior that will distinguish between a zero of multiplicity 1 and one of higher odd multiplicity. Look again closely at the graphs of $a(x) = x(x+2)^2$ and $F(x) = x^3(x+2)^2$ from Exercise 3.1.1. Discuss with your classmates how the graphs are fundamentally different at the origin. It might help to use a graphing calculator to zoom in on the origin to see the different crossing behavior. Also compare the behavior of $a(x) = x(x+2)^2$ to that of $g(x) = x(x+2)^3$ near the point $(-2, 0)$. What do you predict will happen at the zeros of $f(x) = (x-1)(x-2)^2(x-3)^3(x-4)^4(x-5)^5$?

38. Here are a few other questions for you to discuss with your classmates.

    (a) How many local extrema could a polynomial of degree n have? How few local extrema can it have?

    (b) Could a polynomial have two local maxima but no local minima?

    (c) If a polynomial has two local maxima and two local minima, can it be of odd degree? Can it be of even degree?

    (d) Can a polynomial have local extrema without having any real zeros?

    (e) Why must every polynomial of odd degree have at least one real zero?

    (f) Can a polynomial have two distinct real zeros and no local extrema?

    (g) Can an x-intercept yield a local extrema? Can it yield an absolute extrema?

    (h) If the y-intercept yields an absolute minimum, what can we say about the degree of the polynomial and the sign of the leading coefficient?

3.1.2 Answers

1. $f(x) = 4 - x - 3x^2$
   Degree 2
   Leading term $-3x^2$
   Leading coefficient -3
   Constant term 4
   As $x \to -\infty$, $f(x) \to -\infty$
   As $x \to \infty$, $f(x) \to -\infty$

2. $g(x) = 3x^5 - 2x^2 + x + 1$
   Degree 5
   Leading term $3x^5$
   Leading coefficient 3
   Constant term 1
   As $x \to -\infty$, $g(x) \to -\infty$
   As $x \to \infty$, $g(x) \to \infty$

3. $q(r) = 1 - 16r^4$
   Degree 4
   Leading term $-16r^4$
   Leading coefficient -16
   Constant term 1
   As $r \to -\infty$, $q(r) \to -\infty$
   As $r \to \infty$, $q(r) \to -\infty$

4. $Z(b) = 42b - b^3$
   Degree 3
   Leading term $-b^3$
   Leading coefficient -1
   Constant term 0
   As $b \to -\infty$, $Z(b) \to \infty$
   As $b \to \infty$, $Z(b) \to -\infty$

5. $f(x) = \sqrt{3}x^{17} + 22.5x^{10} - \pi x^7 + \frac{1}{3}$
   Degree 17
   Leading term $\sqrt{3}x^{17}$
   Leading coefficient $\sqrt{3}$
   Constant term $\frac{1}{3}$
   As $x \to -\infty$, $f(x) \to -\infty$
   As $x \to \infty$, $f(x) \to \infty$

6. $s(t) = -4.9t^2 + v_0 t + s_0$
   Degree 2
   Leading term $-4.9t^2$
   Leading coefficient -4.9
   Constant term s_0
   As $t \to -\infty$, $s(t) \to -\infty$
   As $t \to \infty$, $s(t) \to -\infty$

7. $P(x) = (x-1)(x-2)(x-3)(x-4)$
   Degree 4
   Leading term x^4
   Leading coefficient 1
   Constant term 24
   As $x \to -\infty$, $P(x) \to \infty$
   As $x \to \infty$, $P(x) \to \infty$

8. $p(t) = -t^2(3 - 5t)(t^2 + t + 4)$
   Degree 5
   Leading term $5t^5$
   Leading coefficient 5
   Constant term 0
   As $t \to -\infty$, $p(t) \to -\infty$
   As $t \to \infty$, $p(t) \to \infty$

3.1 Graphs of Polynomials

9. $f(x) = -2x^3(x+1)(x+2)^2$
   Degree 6
   Leading term $-2x^6$
   Leading coefficient -2
   Constant term 0
   As $x \to -\infty$, $f(x) \to -\infty$
   As $x \to \infty$, $f(x) \to -\infty$

10. $G(t) = 4(t-2)^2\left(t+\frac{1}{2}\right)$
    Degree 3
    Leading term $4t^3$
    Leading coefficient 4
    Constant term 8
    As $t \to -\infty$, $G(t) \to -\infty$
    As $t \to \infty$, $G(t) \to \infty$

11. $a(x) = x(x+2)^2$
    $x = 0$ multiplicity 1
    $x = -2$ multiplicity 2

12. $g(x) = x(x+2)^3$
    $x = 0$ multiplicity 1
    $x = -2$ multiplicity 3

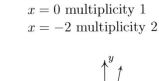
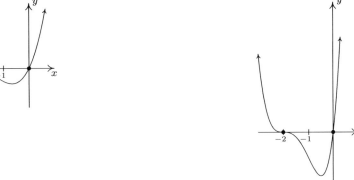

13. $f(x) = -2(x-2)^2(x+1)$
    $x = 2$ multiplicity 2
    $x = -1$ multiplicity 1

14. $g(x) = (2x+1)^2(x-3)$
    $x = -\frac{1}{2}$ multiplicity 2
    $x = 3$ multiplicity 1

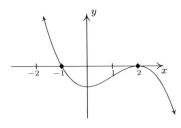
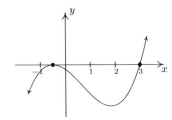

15. $F(x) = x^3(x+2)^2$
    $x = 0$ multiplicity 3
    $x = -2$ multiplicity 2

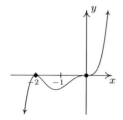

16. $P(x) = (x-1)(x-2)(x-3)(x-4)$
    $x = 1$ multiplicity 1
    $x = 2$ multiplicity 1
    $x = 3$ multiplicity 1
    $x = 4$ multiplicity 1

17. $Q(x) = (x+5)^2(x-3)^4$
    $x = -5$ multiplicity 2
    $x = 3$ multiplicity 4

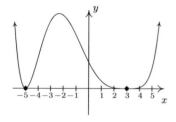

18. $f(x) = x^2(x-2)^2(x+2)^2$
    $x = -2$ multiplicity 2
    $x = 0$ multiplicity 2
    $x = 2$ multiplicity 2

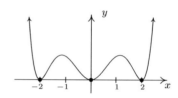

19. $H(t) = (3-t)\left(t^2+1\right)$
    $x = 3$ multiplicity 1

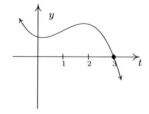

20. $Z(b) = b(42 - b^2)$
    $b = -\sqrt{42}$ multiplicity 1
    $b = 0$ multiplicity 1
    $b = \sqrt{42}$ multiplicity 1

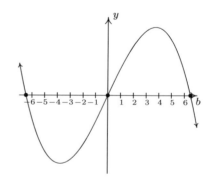

3.1 GRAPHS OF POLYNOMIALS

21. $g(x) = (x+2)^3 + 1$
    domain: $(-\infty, \infty)$
    range: $(-\infty, \infty)$

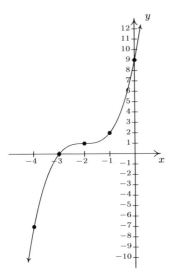

22. $g(x) = (x+2)^4 + 1$
    domain: $(-\infty, \infty)$
    range: $[1, \infty)$

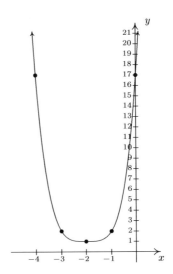

23. $g(x) = 2 - 3(x-1)^4$
    domain: $(-\infty, \infty)$
    range: $(-\infty, 2]$

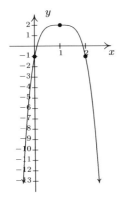

24. $g(x) = -x^5 - 3$
    domain: $(-\infty, \infty)$
    range: $(-\infty, \infty)$

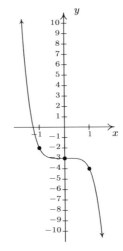

25. $g(x) = (x+1)^5 + 10$
    domain: $(-\infty, \infty)$
    range: $(-\infty, \infty)$

26. $g(x) = 8 - x^6$
    domain: $(-\infty, \infty)$
    range: $(-\infty, 8]$

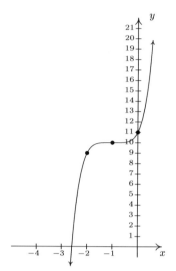

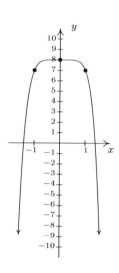

27. We have $f(-4) = -23$, $f(-3) = 5$, $f(0) = 5$, $f(1) = -3$, $f(2) = -5$ and $f(3) = 5$ so the Intermediate Value Theorem tells us that $f(x) = x^3 - 9x + 5$ has real zeros in the intervals $[-4, -3], [0, 1]$ and $[2, 3]$.

28. $V(x) = x(8.5 - 2x)(11 - 2x) = 4x^3 - 39x^2 + 93.5x$, $0 < x < 4.25$. Volume is maximized when $x \approx 1.58$, so the dimensions of the box with maximum volume are: height ≈ 1.58 inches, width ≈ 5.34 inches, and depth ≈ 7.84 inches. The maximum volume is ≈ 66.15 cubic inches.

29. The calculator gives the location of the absolute maximum (rounded to three decimal places) as $x \approx 6.305$ and $y \approx 1115.417$. Since x represents the number of TVs sold in hundreds, $x = 6.305$ corresponds to 630.5 TVs. Since we can't sell half of a TV, we compare $R(6.30) \approx 1115.415$ and $R(6.31) \approx 1115.416$, so selling 631 TVs results in a (slightly) higher revenue. Since y represents the revenue in *thousands* of dollars, the maximum revenue is \$1,115,416.

30. $P(x) = R(x) - C(x) = -5x^3 + 35x^2 - 45x - 25$, $0 \leq x \leq 10.07$.

31. The calculator gives the location of the absolute maximum (rounded to three decimal places) as $x \approx 3.897$ and $y \approx 35.255$. Since x represents the number of TVs sold in hundreds, $x = 3.897$ corresponds to 389.7 TVs. Since we can't sell 0.7 of a TV, we compare $P(3.89) \approx 35.254$ and $P(3.90) \approx 35.255$, so selling 390 TVs results in a (slightly) higher revenue. Since y represents the revenue in *thousands* of dollars, the maximum revenue is \$35,255.

32. Making and selling 71 PortaBoys yields a maximized profit of \$5910.67.

33. (a) Our ultimate goal is to maximize the volume, so we'll start with the maximum Length + Girth of 130. This means the length is $130 - 4x$. The volume of a rectangular box is always length × width × height so we get $V(x) = x^2(130 - 4x) = -4x^3 + 130x^2$.

   (b) Graphing $y = V(x)$ on $[0, 33] \times [0, 21000]$ shows a maximum at $(21.67, 20342.59)$ so the dimensions of the box with maximum volume are 21.67in. × 21.67in. × 43.32in. for a volume of 20342.59in.^{3}.

   (c) If we start with Length + Girth = 108 then the length is $108 - 4x$ and the volume is $V(x) = -4x^3 + 108x^2$. Graphing $y = V(x)$ on $[0, 27] \times [0, 11700]$ shows a maximum at $(18.00, 11664.00)$ so the dimensions of the box with maximum volume are 18.00in. × 18.00in. × 36in. for a volume of 11664.00in.^{3}. (Calculus will confirm that the measurements which maximize the volume are <u>exactly</u> 18in. by 18in. by 36in., however, as I'm sure you are aware by now, we treat all calculator results as approximations and list them as such.)

34. The cubic regression model is $p_3(x) = 0.0226x^3 - 0.9508x^2 + 8.615x - 3.446$. It has $R^2 = 0.93765$ which isn't bad. The graph of $y = p_3(x)$ in the viewing window $[-1, 13] \times [0, 24]$ along with the scatter plot is shown below on the left. Notice that p_3 hits the x-axis at about $x = 12.45$ making this a bad model for future predictions. To use the model to approximate the number of hours of sunlight on your birthday, you'll have to figure out what decimal value of x is close enough to your birthday and then plug it into the model. My (Jeff's) birthday is July 31 which is 10 days after July 21 ($x = 7$). Assuming 30 days in a month, I think $x = 7.33$ should work for my birthday and $p_3(7.33) \approx 17.5$. The website says there will be about 18.25 hours of daylight that day. To have 14 hours of darkness we need 10 hours of daylight. We see that $p_3(1.96) \approx 10$ and $p_3(10.05) \approx 10$ so it seems reasonable to say that we'll have at least 14 hours of darkness from December 21, 2008 ($x = 0$) to February 21, 2009 ($x = 2$) and then again from October 21, 2009 ($x = 10$) to December 21, 2009 ($x = 12$).

The quartic regression model is $p_4(x) = 0.0144x^4 - 0.3507x^3 + 2.259x^2 - 1.571x + 5.513$. It has $R^2 = 0.98594$ which is good. The graph of $y = p_4(x)$ in the viewing window $[-1, 15] \times [0, 35]$ along with the scatter plot is shown below on the right. Notice that $p_4(15)$ is above 24 making this a bad model as well for future predictions. However, $p_4(7.33) \approx 18.71$ making it much better at predicting the hours of daylight on July 31 (my birthday). This model says we'll have at least 14 hours of darkness from December 21, 2008 ($x = 0$) to about March 1, 2009 ($x = 2.30$) and then again from October 10, 2009 ($x = 9.667$) to December 21, 2009 ($x = 12$).

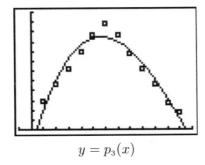
$y = p_3(x)$

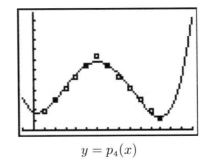
$y = p_4(x)$

35. (a) The scatter plot is shown below with each of the three regression models.

(b) The quadratic model is $P_2(x) = -0.02x^2 + 0.241x + 0.956$ with $R^2 = 0.77708$.
The cubic model is $P_3(x) = 0.005x^3 - 0.103x^2 + 0.602x + 0.573$ with $R^2 = 0.98153$.
The quartic model is $P_4(x) = -0.000969x^4 + 0.0253x^3 - 0.240x^2 + 0.944x + 0.330$ with $R^2 = 0.99929$.

(c) The maximums predicted by the three models are $P_2(5.737) \approx 1.648$, $P_3(4.232) \approx 1.657$ and $P_4(3.784) \approx 1.630$, respectively.

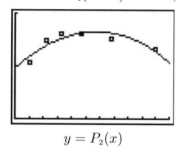

$y = P_2(x)$

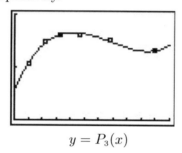

$y = P_3(x)$

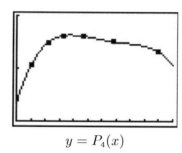

$y = P_4(x)$

3.2 The Factor Theorem and the Remainder Theorem

Suppose we wish to find the zeros of $f(x) = x^3 + 4x^2 - 5x - 14$. Setting $f(x) = 0$ results in the polynomial equation $x^3 + 4x^2 - 5x - 14 = 0$. Despite all of the factoring techniques we learned[1] in Intermediate Algebra, this equation foils[2] us at every turn. If we graph f using the graphing calculator, we get

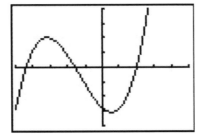

The graph suggests that the function has three zeros, one of which is $x = 2$. It's easy to show that $f(2) = 0$, but the other two zeros seem to be less friendly. Even though we could use the 'Zero' command to find decimal approximations for these, we seek a method to find the remaining zeros *exactly*. Based on our experience, if $x = 2$ is a zero, it seems that there should be a factor of $(x - 2)$ lurking around in the factorization of $f(x)$. In other words, we should expect that $x^3 + 4x^2 - 5x - 14 = (x - 2)\, q(x)$, where $q(x)$ is some other polynomial. How could we find such a $q(x)$, if it even exists? The answer comes from our old friend, polynomial division. Dividing $x^3 + 4x^2 - 5x - 14$ by $x - 2$ gives

$$
\begin{array}{r}
x^2 + 6x + 7 \\
x-2 \overline{\smash{\big)}\, x^3 + 4x^2 - 5x - 14} \\
-\underline{(x^3 - 2x^2)} \\
6x^2 - 5x \\
-\underline{(6x^2 - 12x)} \\
7x - 14 \\
-\underline{(7x - 14)} \\
0
\end{array}
$$

As you may recall, this means $x^3 + 4x^2 - 5x - 14 = (x - 2)\left(x^2 + 6x + 7\right)$, so to find the zeros of f, we now solve $(x - 2)\left(x^2 + 6x + 7\right) = 0$. We get $x - 2 = 0$ (which gives us our known zero, $x = 2$) as well as $x^2 + 6x + 7 = 0$. The latter doesn't factor nicely, so we apply the Quadratic Formula to get $x = -3 \pm \sqrt{2}$. The point of this section is to generalize the technique applied here. First up is a friendly reminder of what we can expect when we divide polynomials.

---

[1] and probably forgot
[2] pun intended

> **Theorem 3.4. Polynomial Division:** Suppose $d(x)$ and $p(x)$ are nonzero polynomials where the degree of p is greater than or equal to the degree of d. There exist two unique polynomials, $q(x)$ and $r(x)$, such that $p(x) = d(x)\, q(x) + r(x)$, where either $r(x) = 0$ or the degree of r is strictly less than the degree of d.

As you may recall, all of the polynomials in Theorem 3.4 have special names. The polynomial p is called the **dividend**; d is the **divisor**; q is the **quotient**; r is the **remainder**. If $r(x) = 0$ then d is called a **factor** of p. The proof of Theorem 3.4 is usually relegated to a course in Abstract Algebra,[3] but we can still use the result to establish two important facts which are the basis of the rest of the chapter.

> **Theorem 3.5. The Remainder Theorem:** Suppose p is a polynomial of degree at least 1 and c is a real number. When $p(x)$ is divided by $x - c$ the remainder is $p(c)$.

The proof of Theorem 3.5 is a direct consequence of Theorem 3.4. When a polynomial is divided by $x - c$, the remainder is either 0 or has degree less than the degree of $x - c$. Since $x - c$ is degree 1, the degree of the remainder must be 0, which means the remainder is a constant. Hence, in either case, $p(x) = (x - c)\, q(x) + r$, where r, the remainder, is a real number, possibly 0. It follows that $p(c) = (c - c)\, q(c) + r = 0 \cdot q(c) + r = r$, so we get $r = p(c)$ as required. There is one last 'low hanging fruit'[4] to collect which we present below.

> **Theorem 3.6. The Factor Theorem:** Suppose p is a nonzero polynomial. The real number c is a zero of p if and only if $(x - c)$ is a factor of $p(x)$.

The proof of The Factor Theorem is a consequence of what we already know. If $(x - c)$ is a factor of $p(x)$, this means $p(x) = (x - c)\, q(x)$ for some polynomial q. Hence, $p(c) = (c - c)\, q(c) = 0$, so c is a zero of p. Conversely, if c is a zero of p, then $p(c) = 0$. In this case, The Remainder Theorem tells us the remainder when $p(x)$ is divided by $(x - c)$, namely $p(c)$, is 0, which means $(x - c)$ is a factor of p. What we have established is the fundamental connection between zeros of polynomials and factors of polynomials.

Of the things The Factor Theorem tells us, the most pragmatic is that we had better find a more efficient way to divide polynomials by quantities of the form $x - c$. Fortunately, people like Ruffini and Horner have already blazed this trail. Let's take a closer look at the long division we performed at the beginning of the section and try to streamline it. First off, let's change all of the subtractions into additions by distributing through the -1s.

---

[3] Yes, Virginia, there are Algebra courses more abstract than this one.
[4] Jeff hates this expression and Carl included it just to annoy him.

3.2 THE FACTOR THEOREM AND THE REMAINDER THEOREM

$$
\begin{array}{r}
x^2 + 6x + 7 \\
x-2 \overline{\smash{\big)}\, x^3 + 4x^2 - 5x - 14}\\
-x^3 + 2x^2 \\
\hline
6x^2 - 5x \\
-6x^2 + 12x \\
\hline
7x - 14\\
-7x + 14\\
\hline
0
\end{array}
$$

Next, observe that the terms $-x^3$, $-6x^2$ and $-7x$ are the exact opposite of the terms above them. The algorithm we use ensures this is always the case, so we can omit them without losing any information. Also note that the terms we 'bring down' (namely the $-5x$ and -14) aren't really necessary to recopy, so we omit them, too.

$$
\begin{array}{r}
x^2 + 6x + 7 \\
x-2 \overline{\smash{\big)}\, x^3 + 4x^2 - 5x - 14}\\
2x^2 \\
\hline
6x^2 \\
12x \\
\hline
7x \\
14\\
\hline
0
\end{array}
$$

Now, let's move things up a bit and, for reasons which will become clear in a moment, copy the x^3 into the last row.

$$
\begin{array}{r}
x^2 + 6x + 7 \\
x-2 \overline{\smash{\big)}\, x^3 + 4x^2 - 5x - 14}\\
2x^2 \quad 12x \quad 14\\
\hline
x^3 \quad 6x^2 \quad 7x \quad 0
\end{array}
$$

Note that by arranging things in this manner, each term in the last row is obtained by adding the two terms above it. Notice also that the quotient polynomial can be obtained by dividing each of the first three terms in the last row by x and adding the results. If you take the time to work back through the original division problem, you will find that this is exactly the way we determined the quotient polynomial. This means that we no longer need to write the quotient polynomial down, nor the x in the divisor, to determine our answer.

$$
\begin{array}{r}
-2 \,\big|\, x^3 + 4x^2 - 5x - 14\\
2x^2 \quad 12x \quad 14\\
\hline
x^3 \quad 6x^2 \quad 7x \quad 0
\end{array}
$$

We've streamlined things quite a bit so far, but we can still do more. Let's take a moment to remind ourselves where the $2x^2$, $12x$ and 14 came from in the second row. Each of these terms was obtained by multiplying the terms in the quotient, x^2, $6x$ and 7, respectively, by the -2 in $x-2$, then by -1 when we changed the subtraction to addition. Multiplying by -2 then by -1 is the same as multiplying by 2, so we replace the -2 in the divisor by 2. Furthermore, the coefficients of the quotient polynomial match the coefficients of the first three terms in the last row, so we now take the plunge and write only the coefficients of the terms to get

$$\begin{array}{r|rrrr} 2 & 1 & 4 & -5 & -14 \\ & & 2 & 12 & 14 \\ \hline & 1 & 6 & 7 & 0 \end{array}$$

We have constructed a **synthetic division tableau** for this polynomial division problem. Let's rework our division problem using this tableau to see how it greatly streamlines the division process. To divide $x^3 + 4x^2 - 5x - 14$ by $x - 2$, we write 2 in the place of the divisor and the coefficients of $x^3 + 4x^2 - 5x - 14$ in for the dividend. Then 'bring down' the first coefficient of the dividend.

$$\begin{array}{r|rrrr} 2 & 1 & 4 & -5 & -14 \\ & & & & \\ \hline & & & & \end{array} \qquad \begin{array}{r|rrrr} 2 & 1 & 4 & -5 & -14 \\ & \downarrow & & & \\ \hline & 1 & & & \end{array}$$

Next, take the 2 from the divisor and multiply by the 1 that was 'brought down' to get 2. Write this underneath the 4, then add to get 6.

$$\begin{array}{r|rrrr} 2 & 1 & 4 & -5 & -14 \\ & \downarrow & 2 & & \\ \hline & 1 & & & \end{array} \qquad \begin{array}{r|rrrr} 2 & 1 & 4 & -5 & -14 \\ & \downarrow & 2 & & \\ \hline & 1 & 6 & & \end{array}$$

Now take the 2 from the divisor times the 6 to get 12, and add it to the -5 to get 7.

$$\begin{array}{r|rrrr} 2 & 1 & 4 & -5 & -14 \\ & \downarrow & 2 & 12 & \\ \hline & 1 & 6 & & \end{array} \qquad \begin{array}{r|rrrr} 2 & 1 & 4 & -5 & -14 \\ & \downarrow & 2 & 12 & \\ \hline & 1 & 6 & 7 & \end{array}$$

Finally, take the 2 in the divisor times the 7 to get 14, and add it to the -14 to get 0.

$$\begin{array}{r|rrrr} 2 & 1 & 4 & -5 & -14 \\ & \downarrow & 2 & 12 & 14 \\ \hline & 1 & 6 & 7 & \end{array} \qquad \begin{array}{r|rrrr} 2 & 1 & 4 & -5 & -14 \\ & \downarrow & 2 & 12 & 14 \\ \hline & 1 & 6 & 7 & \boxed{0} \end{array}$$

3.2 The Factor Theorem and the Remainder Theorem

The first three numbers in the last row of our tableau are the coefficients of the quotient polynomial. Remember, we started with a third degree polynomial and divided by a first degree polynomial, so the quotient is a second degree polynomial. Hence the quotient is $x^2 + 6x + 7$. The number in the box is the remainder. Synthetic division is our tool of choice for dividing polynomials by divisors of the form $x - c$. It is important to note that it works *only* for these kinds of divisors.[5] Also take note that when a polynomial (of degree at least 1) is divided by $x - c$, the result will be a polynomial of exactly one less degree. Finally, it is worth the time to trace each step in synthetic division back to its corresponding step in long division. While the authors have done their best to indicate where the algorithm comes from, there is no substitute for working through it yourself.

Example 3.2.1. Use synthetic division to perform the following polynomial divisions. Find the quotient and the remainder polynomials, then write the dividend, quotient and remainder in the form given in Theorem 3.4.

1. $\left(5x^3 - 2x^2 + 1\right) \div (x - 3)$ 2. $\left(x^3 + 8\right) \div (x + 2)$ 3. $\dfrac{4 - 8x - 12x^2}{2x - 3}$

Solution.

1. When setting up the synthetic division tableau, we need to enter 0 for the coefficient of x in the dividend. Doing so gives

$$\begin{array}{r|rrrr} 3 & 5 & -2 & 0 & 1 \\ & \downarrow & 15 & 39 & 117 \\ \hline & 5 & 13 & 39 & \boxed{118} \end{array}$$

   Since the dividend was a third degree polynomial, the quotient is a quadratic polynomial with coefficients 5, 13 and 39. Our quotient is $q(x) = 5x^2 + 13x + 39$ and the remainder is $r(x) = 118$. According to Theorem 3.4, we have $5x^3 - 2x^2 + 1 = (x-3)\left(5x^2 + 13x + 39\right) + 118$.

2. For this division, we rewrite $x + 2$ as $x - (-2)$ and proceed as before

$$\begin{array}{r|rrrr} -2 & 1 & 0 & 0 & 8 \\ & \downarrow & -2 & 4 & -8 \\ \hline & 1 & -2 & 4 & \boxed{0} \end{array}$$

   We get the quotient $q(x) = x^2 - 2x + 4$ and the remainder $r(x) = 0$. Relating the dividend, quotient and remainder gives $x^3 + 8 = (x+2)\left(x^2 - 2x + 4\right)$.

3. To divide $4 - 8x - 12x^2$ by $2x - 3$, two things must be done. First, we write the dividend in descending powers of x as $-12x^2 - 8x + 4$. Second, since synthetic division works only for factors of the form $x - c$, we factor $2x - 3$ as $2\left(x - \frac{3}{2}\right)$. Our strategy is to first divide $-12x^2 - 8x + 4$ by 2, to get $-6x^2 - 4x + 2$. Next, we divide by $\left(x - \frac{3}{2}\right)$. The tableau becomes

---
[5]You'll need to use good old-fashioned polynomial long division for divisors of degree larger than 1.

$$\begin{array}{r|rrr} \frac{3}{2} & -6 & -4 & 2 \\ & \downarrow & -9 & -\frac{39}{2} \\ \hline & -6 & -13 & \boxed{-\frac{35}{2}} \end{array}$$

From this, we get $-6x^2 - 4x + 2 = \left(x - \frac{3}{2}\right)(-6x - 13) - \frac{35}{2}$. Multiplying both sides by 2 and distributing gives $-12x^2 - 8x + 4 = (2x - 3)(-6x - 13) - 35$. At this stage, we have written $-12x^2 - 8x + 4$ in the **form** $(2x-3)q(x) + r(x)$, but how can we be sure the quotient polynomial is $-6x - 13$ and the remainder is -35? The answer is the word 'unique' in Theorem 3.4. The theorem states that there is only one way to decompose $-12x^2 - 8x + 4$ into a multiple of $(2x - 3)$ plus a constant term. Since we have found such a way, we can be sure it is the only way. □

The next example pulls together all of the concepts discussed in this section.

Example 3.2.2. Let $p(x) = 2x^3 - 5x + 3$.

1. Find $p(-2)$ using The Remainder Theorem. Check your answer by substitution.

2. Use the fact that $x = 1$ is a zero of p to factor $p(x)$ and then find all of the real zeros of p.

Solution.

1. The Remainder Theorem states $p(-2)$ is the remainder when $p(x)$ is divided by $x - (-2)$. We set up our synthetic division tableau below. We are careful to record the coefficient of x^2 as 0, and proceed as above.

$$\begin{array}{r|rrrr} -2 & 2 & 0 & -5 & 3 \\ & \downarrow & -4 & 8 & -6 \\ \hline & 2 & -4 & 3 & \boxed{-3} \end{array}$$

According to the Remainder Theorem, $p(-2) = -3$. We can check this by direct substitution into the formula for $p(x)$: $p(-2) = 2(-2)^3 - 5(-2) + 3 = -16 + 10 + 3 = -3$.

2. The Factor Theorem tells us that since $x = 1$ is a zero of p, $x - 1$ is a factor of $p(x)$. To factor $p(x)$, we divide

$$\begin{array}{r|rrrr} 1 & 2 & 0 & -5 & 3 \\ & \downarrow & 2 & 2 & -3 \\ \hline & 2 & 2 & -3 & \boxed{0} \end{array}$$

We get a remainder of 0 which verifies that, indeed, $p(1) = 0$. Our quotient polynomial is a second degree polynomial with coefficients 2, 2, and -3. So $q(x) = 2x^2 + 2x - 3$. Theorem 3.4 tells us $p(x) = (x-1)\left(2x^2 + 2x - 3\right)$. To find the remaining real zeros of p, we need to solve $2x^2 + 2x - 3 = 0$ for x. Since this doesn't factor nicely, we use the quadratic formula to find that the remaining zeros are $x = \frac{-1 \pm \sqrt{7}}{2}$. □

3.2 The Factor Theorem and the Remainder Theorem

In Section 3.1, we discussed the notion of the multiplicity of a zero. Roughly speaking, a zero with multiplicity 2 can be divided twice into a polynomial; multiplicity 3, three times and so on. This is illustrated in the next example.

Example 3.2.3. Let $p(x) = 4x^4 - 4x^3 - 11x^2 + 12x - 3$. Given that $x = \frac{1}{2}$ is a zero of multiplicity 2, find all of the real zeros of p.

Solution. We set up for synthetic division. Since we are told the multiplicity of $\frac{1}{2}$ is two, we continue our tableau and divide $\frac{1}{2}$ into the quotient polynomial

$$
\begin{array}{c|ccccc}
\frac{1}{2} & 4 & -4 & -11 & 12 & -3 \\
 & \downarrow & 2 & -1 & -6 & 3 \\
\hline
\frac{1}{2} & 4 & -2 & -12 & 6 & \boxed{0} \\
 & \downarrow & 2 & 0 & -6 & \\
\hline
 & 4 & 0 & -12 & \boxed{0} &
\end{array}
$$

From the first division, we get $4x^4 - 4x^3 - 11x^2 + 12x - 3 = \left(x - \frac{1}{2}\right)\left(4x^3 - 2x^2 - 12x + 6\right)$. The second division tells us $4x^3 - 2x^2 - 12x + 6 = \left(x - \frac{1}{2}\right)\left(4x^2 - 12\right)$. Combining these results, we have $4x^4 - 4x^3 - 11x^2 + 12x - 3 = \left(x - \frac{1}{2}\right)^2 \left(4x^2 - 12\right)$. To find the remaining zeros of p, we set $4x^2 - 12 = 0$ and get $x = \pm\sqrt{3}$. □

A couple of things about the last example are worth mentioning. First, the extension of the synthetic division tableau for repeated divisions will be a common site in the sections to come. Typically, we will start with a higher order polynomial and peel off one zero at a time until we are left with a quadratic, whose roots can always be found using the Quadratic Formula. Secondly, we found $x = \pm\sqrt{3}$ are zeros of p. The Factor Theorem guarantees $\left(x - \sqrt{3}\right)$ and $\left(x - \left(-\sqrt{3}\right)\right)$ are both factors of p. We can certainly put the Factor Theorem to the test and continue the synthetic division tableau from above to see what happens.

$$
\begin{array}{c|ccccc}
\frac{1}{2} & 4 & -4 & -11 & 12 & -3 \\
 & \downarrow & 2 & -1 & -6 & 3 \\
\hline
\frac{1}{2} & 4 & -2 & -12 & 6 & \boxed{0} \\
 & \downarrow & 2 & 0 & -6 & \\
\hline
\sqrt{3} & 4 & 0 & -12 & \boxed{0} & \\
 & \downarrow & 4\sqrt{3} & 12 & & \\
\hline
-\sqrt{3} & 4 & 4\sqrt{3} & \boxed{0} & & \\
 & \downarrow & -4\sqrt{3} & & & \\
\hline
 & 4 & \boxed{0} & & &
\end{array}
$$

This gives us $4x^4 - 4x^3 - 11x^2 + 12x - 3 = \left(x - \frac{1}{2}\right)^2 \left(x - \sqrt{3}\right)\left(x - \left(-\sqrt{3}\right)\right)(4)$, or, when written with the constant in front

$$p(x) = 4\left(x - \frac{1}{2}\right)^2 \left(x - \sqrt{3}\right)\left(x - \left(-\sqrt{3}\right)\right)$$

We have shown that p is a product of its leading coefficient times linear factors of the form $(x - c)$ where c are zeros of p. It may surprise and delight the reader that, in theory, all polynomials can be reduced to this kind of factorization. We leave that discussion to Section 3.4, because the zeros may not be real numbers. Our final theorem in the section gives us an upper bound on the number of real zeros.

> **Theorem 3.7.** Suppose f is a polynomial of degree $n \geq 1$. Then f has at most n real zeros, counting multiplicities.

Theorem 3.7 is a consequence of the Factor Theorem and polynomial multiplication. Every zero c of f gives us a factor of the form $(x - c)$ for $f(x)$. Since f has degree n, there can be at most n of these factors. The next section provides us some tools which not only help us determine where the real zeros are to be found, but which real numbers they may be.

We close this section with a summary of several concepts previously presented. You should take the time to look back through the text to see where each concept was first introduced and where each connection to the other concepts was made.

> **Connections Between Zeros, Factors and Graphs of Polynomial Functions**
>
> Suppose p is a polynomial function of degree $n \geq 1$. The following statements are equivalent:
>
> - The real number c is a zero of p
> - $p(c) = 0$
> - $x = c$ is a solution to the polynomial equation $p(x) = 0$
> - $(x - c)$ is a factor of $p(x)$
> - The point $(c, 0)$ is an x-intercept of the graph of $y = p(x)$

3.2 The Factor Theorem and the Remainder Theorem

3.2.1 Exercises

In Exercises 1 - 6, use polynomial long division to perform the indicated division. Write the polynomial in the form $p(x) = d(x)q(x) + r(x)$.

1. $\left(4x^2 + 3x - 1\right) \div (x - 3)$
2. $\left(2x^3 - x + 1\right) \div \left(x^2 + x + 1\right)$
3. $\left(5x^4 - 3x^3 + 2x^2 - 1\right) \div \left(x^2 + 4\right)$
4. $\left(-x^5 + 7x^3 - x\right) \div \left(x^3 - x^2 + 1\right)$
5. $\left(9x^3 + 5\right) \div (2x - 3)$
6. $\left(4x^2 - x - 23\right) \div \left(x^2 - 1\right)$

In Exercises 7 - 20 use synthetic division to perform the indicated division. Write the polynomial in the form $p(x) = d(x)q(x) + r(x)$.

7. $\left(3x^2 - 2x + 1\right) \div (x - 1)$
8. $\left(x^2 - 5\right) \div (x - 5)$
9. $\left(3 - 4x - 2x^2\right) \div (x + 1)$
10. $\left(4x^2 - 5x + 3\right) \div (x + 3)$
11. $\left(x^3 + 8\right) \div (x + 2)$
12. $\left(4x^3 + 2x - 3\right) \div (x - 3)$
13. $\left(18x^2 - 15x - 25\right) \div \left(x - \frac{5}{3}\right)$
14. $\left(4x^2 - 1\right) \div \left(x - \frac{1}{2}\right)$
15. $\left(2x^3 + x^2 + 2x + 1\right) \div \left(x + \frac{1}{2}\right)$
16. $\left(3x^3 - x + 4\right) \div \left(x - \frac{2}{3}\right)$
17. $\left(2x^3 - 3x + 1\right) \div \left(x - \frac{1}{2}\right)$
18. $\left(4x^4 - 12x^3 + 13x^2 - 12x + 9\right) \div \left(x - \frac{3}{2}\right)$
19. $\left(x^4 - 6x^2 + 9\right) \div \left(x - \sqrt{3}\right)$
20. $\left(x^6 - 6x^4 + 12x^2 - 8\right) \div \left(x + \sqrt{2}\right)$

In Exercises 21 - 30, determine $p(c)$ using the Remainder Theorem for the given polynomial functions and value of c. If $p(c) = 0$, factor $p(x) = (x - c)q(x)$.

21. $p(x) = 2x^2 - x + 1$, $c = 4$
22. $p(x) = 4x^2 - 33x - 180$, $c = 12$
23. $p(x) = 2x^3 - x + 6$, $c = -3$
24. $p(x) = x^3 + 2x^2 + 3x + 4$, $c = -1$
25. $p(x) = 3x^3 - 6x^2 + 4x - 8$, $c = 2$
26. $p(x) = 8x^3 + 12x^2 + 6x + 1$, $c = -\frac{1}{2}$
27. $p(x) = x^4 - 2x^2 + 4$, $c = \frac{3}{2}$
28. $p(x) = 6x^4 - x^2 + 2$, $c = -\frac{2}{3}$
29. $p(x) = x^4 + x^3 - 6x^2 - 7x - 7$, $c = -\sqrt{7}$
30. $p(x) = x^2 - 4x + 1$, $c = 2 - \sqrt{3}$

In Exercises 31 - 40, you are given a polynomial and one of its zeros. Use the techniques in this section to find the rest of the real zeros and factor the polynomial.

31. $x^3 - 6x^2 + 11x - 6$, $c = 1$

32. $x^3 - 24x^2 + 192x - 512$, $c = 8$

33. $3x^3 + 4x^2 - x - 2$, $c = \frac{2}{3}$

34. $2x^3 - 3x^2 - 11x + 6$, $c = \frac{1}{2}$

35. $x^3 + 2x^2 - 3x - 6$, $c = -2$

36. $2x^3 - x^2 - 10x + 5$, $c = \frac{1}{2}$

37. $4x^4 - 28x^3 + 61x^2 - 42x + 9$, $c = \frac{1}{2}$ is a zero of multiplicity 2

38. $x^5 + 2x^4 - 12x^3 - 38x^2 - 37x - 12$, $c = -1$ is a zero of multiplicity 3

39. $125x^5 - 275x^4 - 2265x^3 - 3213x^2 - 1728x - 324$, $c = -\frac{3}{5}$ is a zero of multiplicity 3

40. $x^2 - 2x - 2$, $c = 1 - \sqrt{3}$

In Exercises 41 - 45, create a polynomial p which has the desired characteristics. You may leave the polynomial in factored form.

41.
- The zeros of p are $c = \pm 2$ and $c = \pm 1$
- The leading term of $p(x)$ is $117x^4$.

42.
- The zeros of p are $c = 1$ and $c = 3$
- $c = 3$ is a zero of multiplicity 2.
- The leading term of $p(x)$ is $-5x^3$

43.
- The solutions to $p(x) = 0$ are $x = \pm 3$ and $x = 6$
- The leading term of $p(x)$ is $7x^4$
- The point $(-3, 0)$ is a local minimum on the graph of $y = p(x)$.

44.
- The solutions to $p(x) = 0$ are $x = \pm 3$, $x = -2$, and $x = 4$.
- The leading term of $p(x)$ is $-x^5$.
- The point $(-2, 0)$ is a local maximum on the graph of $y = p(x)$.

45.
- p is degree 4.
- as $x \to \infty$, $p(x) \to -\infty$
- p has exactly three x-intercepts: $(-6, 0)$, $(1, 0)$ and $(117, 0)$
- The graph of $y = p(x)$ crosses through the x-axis at $(1, 0)$.

46. Find a quadratic polynomial with integer coefficients which has $x = \dfrac{3}{5} \pm \dfrac{\sqrt{29}}{5}$ as its real zeros.

3.2 The Factor Theorem and the Remainder Theorem

3.2.2 Answers

1. $4x^2 + 3x - 1 = (x - 3)(4x + 15) + 44$

2. $2x^3 - x + 1 = \left(x^2 + x + 1\right)(2x - 2) + (-x + 3)$

3. $5x^4 - 3x^3 + 2x^2 - 1 = \left(x^2 + 4\right)\left(5x^2 - 3x - 18\right) + (12x + 71)$

4. $-x^5 + 7x^3 - x = \left(x^3 - x^2 + 1\right)\left(-x^2 - x + 6\right) + \left(7x^2 - 6\right)$

5. $9x^3 + 5 = (2x - 3)\left(\frac{9}{2}x^2 + \frac{27}{4}x + \frac{81}{8}\right) + \frac{283}{8}$

6. $4x^2 - x - 23 = \left(x^2 - 1\right)(4) + (-x - 19)$

7. $\left(3x^2 - 2x + 1\right) = (x - 1)(3x + 1) + 2$

8. $\left(x^2 - 5\right) = (x - 5)(x + 5) + 20$

9. $\left(3 - 4x - 2x^2\right) = (x + 1)(-2x - 2) + 5$

10. $\left(4x^2 - 5x + 3\right) = (x + 3)(4x - 17) + 54$

11. $\left(x^3 + 8\right) = (x + 2)\left(x^2 - 2x + 4\right) + 0$

12. $\left(4x^3 + 2x - 3\right) = (x - 3)\left(4x^2 + 12x + 38\right) + 111$

13. $\left(18x^2 - 15x - 25\right) = \left(x - \frac{5}{3}\right)(18x + 15) + 0$

14. $\left(4x^2 - 1\right) = \left(x - \frac{1}{2}\right)(4x + 2) + 0$

15. $\left(2x^3 + x^2 + 2x + 1\right) = \left(x + \frac{1}{2}\right)\left(2x^2 + 2\right) + 0$

16. $\left(3x^3 - x + 4\right) = \left(x - \frac{2}{3}\right)\left(3x^2 + 2x + \frac{1}{3}\right) + \frac{38}{9}$

17. $\left(2x^3 - 3x + 1\right) = \left(x - \frac{1}{2}\right)\left(2x^2 + x - \frac{5}{2}\right) - \frac{1}{4}$

18. $\left(4x^4 - 12x^3 + 13x^2 - 12x + 9\right) = \left(x - \frac{3}{2}\right)\left(4x^3 - 6x^2 + 4x - 6\right) + 0$

19. $\left(x^4 - 6x^2 + 9\right) = \left(x - \sqrt{3}\right)\left(x^3 + \sqrt{3}\,x^2 - 3x - 3\sqrt{3}\right) + 0$

20. $\left(x^6 - 6x^4 + 12x^2 - 8\right) = \left(x + \sqrt{2}\right)\left(x^5 - \sqrt{2}\,x^4 - 4x^3 + 4\sqrt{2}\,x^2 + 4x - 4\sqrt{2}\right) + 0$

21. $p(4) = 29$

22. $p(12) = 0$, $p(x) = (x - 12)(4x + 15)$

23. $p(-3) = -45$

24. $p(-1) = 2$

25. $p(2) = 0$, $p(x) = (x - 2)\left(3x^2 + 4\right)$

26. $p\left(-\frac{1}{2}\right) = 0$, $p(x) = \left(x + \frac{1}{2}\right)\left(8x^2 + 8x + 2\right)$

27. $p\left(\frac{3}{2}\right) = \frac{73}{16}$
28. $p\left(-\frac{2}{3}\right) = \frac{74}{27}$

29. $p(-\sqrt{7}) = 0$, $p(x) = (x+\sqrt{7})\left(x^3 + (1-\sqrt{7})x^2 + (1-\sqrt{7})x - \sqrt{7}\right)$

30. $p(2-\sqrt{3}) = 0$, $p(x) = (x-(2-\sqrt{3}))(x-(2+\sqrt{3}))$

31. $x^3 - 6x^2 + 11x - 6 = (x-1)(x-2)(x-3)$

32. $x^3 - 24x^2 + 192x - 512 = (x-8)^3$

33. $3x^3 + 4x^2 - x - 2 = 3\left(x - \frac{2}{3}\right)(x+1)^2$

34. $2x^3 - 3x^2 - 11x + 6 = 2\left(x - \frac{1}{2}\right)(x+2)(x-3)$

35. $x^3 + 2x^2 - 3x - 6 = (x+2)(x+\sqrt{3})(x-\sqrt{3})$

36. $2x^3 - x^2 - 10x + 5 = 2\left(x - \frac{1}{2}\right)(x+\sqrt{5})(x-\sqrt{5})$

37. $4x^4 - 28x^3 + 61x^2 - 42x + 9 = 4\left(x - \frac{1}{2}\right)^2(x-3)^2$

38. $x^5 + 2x^4 - 12x^3 - 38x^2 - 37x - 12 = (x+1)^3(x+3)(x-4)$

39. $125x^5 - 275x^4 - 2265x^3 - 3213x^2 - 1728x - 324 = 125\left(x + \frac{3}{5}\right)^3(x+2)(x-6)$

40. $x^2 - 2x - 2 = (x-(1-\sqrt{3}))(x-(1+\sqrt{3}))$

41. $p(x) = 117(x+2)(x-2)(x+1)(x-1)$

42. $p(x) = -5(x-1)(x-3)^2$

43. $p(x) = 7(x+3)^2(x-3)(x-6)$

44. $p(x) = -(x+2)^2(x-3)(x+3)(x-4)$

45. $p(x) = a(x+6)^2(x-1)(x-117)$ or $p(x) = a(x+6)(x-1)(x-117)^2$ where a can be any negative real number

46. $p(x) = 5x^2 - 6x - 4$

3.3 Real Zeros of Polynomials

In Section 3.2, we found that we can use synthetic division to determine if a given real number is a zero of a polynomial function. This section presents results which will help us determine good candidates to test using synthetic division. There are two approaches to the topic of finding the real zeros of a polynomial. The first approach (which is gaining popularity) is to use a little bit of Mathematics followed by a good use of technology like graphing calculators. The second approach (for purists) makes good use of mathematical machinery (theorems) only. For completeness, we include the two approaches but in separate subsections.[1] Both approaches benefit from the following two theorems, the first of which is due to the famous mathematician Augustin Cauchy. It gives us an interval on which all of the real zeros of a polynomial can be found.

> **Theorem 3.8. Cauchy's Bound:** Suppose $f(x) = a_n x^n + a_{n-1} x^{n-1} + \ldots + a_1 x + a_0$ is a polynomial of degree n with $n \geq 1$. Let M be the largest of the numbers: $\frac{|a_0|}{|a_n|}, \frac{|a_1|}{|a_n|}, \ldots, \frac{|a_{n-1}|}{|a_n|}$. Then all the real zeros of f lie in in the interval $[-(M+1), M+1]$.

The proof of this fact is not easily explained within the confines of this text. This paper contains the result and gives references to its proof. Like many of the results in this section, Cauchy's Bound is best understood with an example.

Example 3.3.1. Let $f(x) = 2x^4 + 4x^3 - x^2 - 6x - 3$. Determine an interval which contains all of the real zeros of f.

Solution. To find the M stated in Cauchy's Bound, we take the absolute value of the leading coefficient, in this case $|2| = 2$ and divide it into the largest (in absolute value) of the remaining coefficients, in this case $|-6| = 6$. This yields $M = 3$ so it is guaranteed that all of the real zeros of f lie in the interval $[-4, 4]$. □

Whereas the previous result tells us *where* we can find the real zeros of a polynomial, the next theorem gives us a list of *possible* real zeros.

> **Theorem 3.9. Rational Zeros Theorem:** Suppose $f(x) = a_n x^n + a_{n-1} x^{n-1} + \ldots + a_1 x + a_0$ is a polynomial of degree n with $n \geq 1$, and $a_0, a_1, \ldots a_n$ are integers. If r is a rational zero of f, then r is of the form $\pm \frac{p}{q}$, where p is a factor of the constant term a_0, and q is a factor of the leading coefficient a_n.

The Rational Zeros Theorem gives us a list of numbers to try in our synthetic division and that is a lot nicer than simply guessing. If none of the numbers in the list are zeros, then either the polynomial has no real zeros at all, or all of the real zeros are irrational numbers. To see why the Rational Zeros Theorem works, suppose c is a zero of f and $c = \frac{p}{q}$ in lowest terms. This means p and q have no common factors. Since $f(c) = 0$, we have

$$a_n \left(\frac{p}{q}\right)^n + a_{n-1} \left(\frac{p}{q}\right)^{n-1} + \ldots + a_1 \left(\frac{p}{q}\right) + a_0 = 0.$$

---

[1] Carl is the purist and is responsible for all of the theorems in this section. Jeff, on the other hand, has spent too much time in school politics and has been polluted with notions of 'compromise.' You can blame the slow decline of civilization on him and those like him who mingle Mathematics with technology.

Multiplying both sides of this equation by q^n, we clear the denominators to get

$$a_n p^n + a_{n-1} p^{n-1} q + \ldots + a_1 p q^{n-1} + a_0 q^n = 0$$

Rearranging this equation, we get

$$a_n p^n = -a_{n-1} p^{n-1} q - \ldots - a_1 p q^{n-1} - a_0 q^n$$

Now, the left hand side is an integer multiple of p, and the right hand side is an integer multiple of q. (Can you see why?) This means $a_n p^n$ is both a multiple of p and a multiple of q. Since p and q have no common factors, a_n must be a multiple of q. If we rearrange the equation

$$a_n p^n + a_{n-1} p^{n-1} q + \ldots + a_1 p q^{n-1} + a_0 q^n = 0$$

as

$$a_0 q^n = -a_n p^n - a_{n-1} p^{n-1} q - \ldots - a_1 p q^{n-1}$$

we can play the same game and conclude a_0 is a multiple of p, and we have the result.

Example 3.3.2. Let $f(x) = 2x^4 + 4x^3 - x^2 - 6x - 3$. Use the Rational Zeros Theorem to list all of the possible rational zeros of f.

Solution. To generate a complete list of rational zeros, we need to take each of the factors of constant term, $a_0 = -3$, and divide them by each of the factors of the leading coefficient $a_4 = 2$. The factors of -3 are ± 1 and ± 3. Since the Rational Zeros Theorem tacks on a $\pm$ anyway, for the moment, we consider only the positive factors 1 and 3. The factors of 2 are 1 and 2, so the Rational Zeros Theorem gives the list $\{\pm \frac{1}{1}, \pm \frac{1}{2}, \pm \frac{3}{1}, \pm \frac{3}{2}\}$ or $\{\pm \frac{1}{2}, \pm 1, \pm \frac{3}{2}, \pm 3\}$. □

Our discussion now diverges between those who wish to use technology and those who do not.

3.3.1 For Those Wishing to use a Graphing Calculator

At this stage, we know not only the interval in which all of the zeros of $f(x) = 2x^4 + 4x^3 - x^2 - 6x - 3$ are located, but we also know some potential candidates. We can now use our calculator to help us determine all of the real zeros of f, as illustrated in the next example.

Example 3.3.3. Let $f(x) = 2x^4 + 4x^3 - x^2 - 6x - 3$.

1. Graph $y = f(x)$ on the calculator using the interval obtained in Example 3.3.1 as a guide.

2. Use the graph to shorten the list of possible rational zeros obtained in Example 3.3.2.

3. Use synthetic division to find the real zeros of f, and state their multiplicities.

Solution.

1. In Example 3.3.1, we determined all of the real zeros of f lie in the interval $[-4, 4]$. We set our window accordingly and get

3.3 REAL ZEROS OF POLYNOMIALS

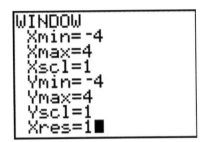 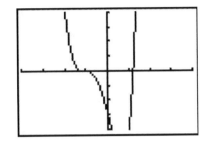

2. In Example 3.3.2, we learned that any rational zero of f must be in the list $\{\pm \frac{1}{2}, \pm 1, \pm \frac{3}{2}, \pm 3\}$. From the graph, it looks as if we can rule out any of the positive rational zeros, since the graph seems to cross the x-axis at a value just a little greater than 1. On the negative side, -1 looks good, so we try that for our synthetic division.

$$
\begin{array}{r|rrrrr}
-1 & 2 & 4 & -1 & -6 & -3 \\
  & \downarrow & -2 & -2 & 3 & 3 \\ \hline
  & 2 & 2 & -3 & -3 & \boxed{0}
\end{array}
$$

We have a winner! Remembering that f was a fourth degree polynomial, we know that our quotient is a third degree polynomial. If we can do one more successful division, we will have knocked the quotient down to a quadratic, and, if all else fails, we can use the quadratic formula to find the last two zeros. Since there seems to be no other rational zeros to try, we continue with -1. Also, the shape of the crossing at $x = -1$ leads us to wonder if the zero $x = -1$ has multiplicity 3.

$$
\begin{array}{r|rrrrr}
-1 & 2 & 4 & -1 & -6 & -3 \\
  & \downarrow & -2 & -2 & 3 & 3 \\ \hline
-1 & 2 & 2 & -3 & -3 & \boxed{0} \\
  & \downarrow & -2 & 0 & 3 & \\ \hline
  & 2 & 0 & -3 & \boxed{0} &
\end{array}
$$

Success! Our quotient polynomial is now $2x^2 - 3$. Setting this to zero gives $2x^2 - 3 = 0$, or $x^2 = \frac{3}{2}$, which gives us $x = \pm \frac{\sqrt{6}}{2}$. Concerning multiplicities, based on our division, we have that -1 has a multiplicity of at least 2. The Factor Theorem tells us our remaining zeros, $\pm \frac{\sqrt{6}}{2}$, each have multiplicity at least 1. However, Theorem 3.7 tells us f can have at most 4 real zeros, counting multiplicity, and so we conclude that -1 is of multiplicity exactly 2 and $\pm \frac{\sqrt{6}}{2}$ each has multiplicity 1. (Thus, we were wrong to think that -1 had multiplicity 3.) □

It is interesting to note that we could greatly improve on the graph of $y = f(x)$ in the previous example given to us by the calculator. For instance, from our determination of the zeros of f and their multiplicities, we know the graph crosses at $x = -\frac{\sqrt{6}}{2} \approx -1.22$ then turns back upwards to touch the x-axis at $x = -1$. This tells us that, despite what the calculator showed us the first time, there is a relative maximum occurring at $x = -1$ and not a 'flattened crossing' as we originally believed. After resizing the window, we see not only the relative maximum but also a relative minimum[2] just to the left of $x = -1$ which shows us, once again, that Mathematics enhances the technology, instead of vice-versa.

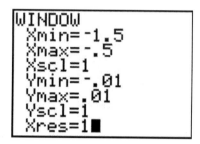 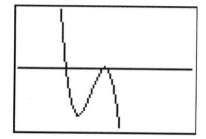

Our next example shows how even a mild-mannered polynomial can cause problems.

Example 3.3.4. Let $f(x) = x^4 + x^2 - 12$.

1. Use Cauchy's Bound to determine an interval in which all of the real zeros of f lie.

2. Use the Rational Zeros Theorem to determine a list of possible rational zeros of f.

3. Graph $y = f(x)$ using your graphing calculator.

4. Find all of the real zeros of f and their multiplicities.

Solution.

1. Applying Cauchy's Bound, we find $M = 12$, so all of the real zeros lie in the interval $[-13, 13]$.

2. Applying the Rational Zeros Theorem with constant term $a_0 = -12$ and leading coefficient $a_4 = 1$, we get the list $\{\pm 1, \pm 2, \pm 3, \pm 4, \pm 6, \pm 12\}$.

3. Graphing $y = f(x)$ on the interval $[-13, 13]$ produces the graph below on the left. Zooming in a bit gives the graph below on the right. Based on the graph, none of our rational zeros will work. (Do you see why not?)

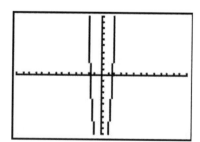 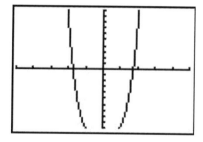

---

[2]This is an example of what is called 'hidden behavior.'

4. From the graph, we know f has two real zeros, one positive, and one negative. Our only hope at this point is to try and find the zeros of f by setting $f(x) = x^4 + x^2 - 12 = 0$ and solving. If we stare at this equation long enough, we may recognize it as a 'quadratic in disguise' or 'quadratic in form'. In other words, we have three terms: x^4, x^2 and 12, and the exponent on the first term, x^4, is exactly twice that of the second term, x^2. We may rewrite this as $\left(x^2\right)^2 + \left(x^2\right) - 12 = 0$. To better see the forest for the trees, we momentarily replace x^2 with the variable u. In terms of u, our equation becomes $u^2 + u - 12 = 0$, which we can readily factor as $(u+4)(u-3) = 0$. In terms of x, this means $x^4 + x^2 - 12 = \left(x^2 - 3\right)\left(x^2 + 4\right) = 0$. We get $x^2 = 3$, which gives us $x = \pm\sqrt{3}$, or $x^2 = -4$, which admits no real solutions. Since $\sqrt{3} \approx 1.73$, the two zeros match what we expected from the graph. In terms of multiplicity, the Factor Theorem guarantees $\left(x - \sqrt{3}\right)$ and $\left(x + \sqrt{3}\right)$ are factors of $f(x)$. Since $f(x)$ can be factored as $f(x) = \left(x^2 - 3\right)\left(x^2 + 4\right)$, and $x^2 + 4$ has no real zeros, the quantities $\left(x - \sqrt{3}\right)$ and $\left(x + \sqrt{3}\right)$ must both be factors of $x^2 - 3$. According to Theorem 3.7, $x^2 - 3$ can have at most 2 zeros, counting multiplicity, hence each of $\pm\sqrt{3}$ is a zero of f of multiplicity 1. □

The technique used to factor $f(x)$ in Example 3.3.4 is called **u-substitution**. We shall see more of this technique in Section 5.3. In general, substitution can help us identify a 'quadratic in disguise' provided that there are exactly three terms and the exponent of the first term is exactly twice that of the second. It is entirely possible that a polynomial has no real roots at all, or worse, it has real roots but none of the techniques discussed in this section can help us find them exactly. In the latter case, we are forced to approximate, which in this subsection means we use the 'Zero' command on the graphing calculator.

3.3.2 For Those Wishing NOT to use a Graphing Calculator

Suppose we wish to find the zeros of $f(x) = 2x^4 + 4x^3 - x^2 - 6x - 3$ *without* using the calculator. In this subsection, we present some more advanced mathematical tools (theorems) to help us. Our first result is due to René Descartes.

> **Theorem 3.10. Descartes' Rule of Signs:** Suppose $f(x)$ is the formula for a polynomial function written with descending powers of x.
>
> - If P denotes the number of variations of sign in the formula for $f(x)$, then the number of positive real zeros (counting multiplicity) is one of the numbers $\{P, P-2, P-4, \ldots\}$.
>
> - If N denotes the number of variations of sign in the formula for $f(-x)$, then the number of negative real zeros (counting multiplicity) is one of the numbers $\{N, N-2, N-4, \ldots\}$.

A few remarks are in order. First, to use Descartes' Rule of Signs, we need to understand what is meant by a '**variation in sign**' of a polynomial function. Consider $f(x) = 2x^4 + 4x^3 - x^2 - 6x - 3$. If we focus on only the *signs* of the coefficients, we start with a $(+)$, followed by another $(+)$, then switch to $(-)$, and stay $(-)$ for the remaining two coefficients. Since the signs of the coefficients switched *once* as we read from left to right, we say that $f(x)$ has *one* variation in sign. When we speak of the variations in sign of a polynomial function f we assume the formula for $f(x)$ is written with descending powers of x, as in Definition 3.1, and concern ourselves only with the nonzero coefficients. Second, unlike the Rational Zeros Theorem, Descartes' Rule of Signs gives us an estimate to the *number* of positive and negative real zeros, not the actual *value* of the zeros. Lastly, Descartes' Rule of Signs counts multiplicities. This means that, for example, if one of the zeros has multiplicity 2, Descsartes' Rule of Signs would count this as *two* zeros. Lastly, note that the number of positive or negative real zeros always starts with the number of sign changes and decreases by an even number. For example, if $f(x)$ has 7 sign changes, then, counting multplicities, f has either 7, 5, 3 or 1 positive real zero. This implies that the graph of $y = f(x)$ crosses the positive x-axis at least once. If $f(-x)$ results in 4 sign changes, then, counting multiplicities, f has 4, 2 or 0 negative real zeros; hence, the graph of $y = f(x)$ may not cross the negative x-axis at all. The proof of Descartes' Rule of Signs is a bit technical, and can be found here.

Example 3.3.5. Let $f(x) = 2x^4 + 4x^3 - x^2 - 6x - 3$. Use Descartes' Rule of Signs to determine the possible number and location of the real zeros of f.

Solution. As noted above, the variations of sign of $f(x)$ is 1. This means, counting multiplicities, f has exactly 1 positive real zero. Since $f(-x) = 2(-x)^4 + 4(-x)^3 - (-x)^2 - 6(-x) - 3 = 2x^4 - 4x^3 - x^2 + 6x - 3$ has 3 variations in sign, f has either 3 negative real zeros or 1 negative real zero, counting multiplicities. □

Cauchy's Bound gives us a general bound on the zeros of a polynomial function. Our next result helps us determine bounds on the real zeros of a polynomial as we synthetically divide which are often sharper[3] bounds than Cauchy's Bound.

> **Theorem 3.11. Upper and Lower Bounds:** Suppose f is a polynomial of degree $n \geq 1$.
>
> - If $c > 0$ is synthetically divided into f and all of the numbers in the final line of the division tableau have the same signs, then c is an upper bound for the real zeros of f. That is, there are no real zeros greater than c.
>
> - If $c < 0$ is synthetically divided into f and the numbers in the final line of the division tableau alternate signs, then c is a lower bound for the real zeros of f. That is, there are no real zeros less than c.
>
>   **NOTE:** If the number 0 occurs in the final line of the division tableau in either of the above cases, it can be treated as $(+)$ or $(-)$ as needed.

The Upper and Lower Bounds Theorem works because of Theorem 3.4. For the upper bound part of the theorem, suppose $c > 0$ is divided into f and the resulting line in the division tableau contains, for example, all nonnegative numbers. This means $f(x) = (x - c)q(x) + r$, where the coefficients of the quotient polynomial and the remainder are nonnegative. (Note that the leading coefficient of q is the same as f so $q(x)$ is not the zero polynomial.) If $b > c$, then $f(b) = (b - c)q(b) + r$, where $(b - c)$ and $q(b)$ are both positive and $r \geq 0$. Hence $f(b) > 0$ which shows b cannot be a zero of f. Thus no real number $b > c$ can be a zero of f, as required. A similar argument proves

---
[3] That is, better, or more accurate.

3.3 REAL ZEROS OF POLYNOMIALS   275

$f(b) < 0$ if all of the numbers in the final line of the synthetic division tableau are non-positive. To prove the lower bound part of the theorem, we note that a lower bound for the negative real zeros of $f(x)$ is an upper bound for the positive real zeros of $f(-x)$. Applying the upper bound portion to $f(-x)$ gives the result. (Do you see where the alternating signs come in?) With the additional mathematical machinery of Descartes' Rule of Signs and the Upper and Lower Bounds Theorem, we can find the real zeros of $f(x) = 2x^4 + 4x^3 - x^2 - 6x - 3$ without the use of a graphing calculator.

Example 3.3.6. Let $f(x) = 2x^4 + 4x^3 - x^2 - 6x - 3$.

1. Find all of the real zeros of f and their multiplicities.

2. Sketch the graph of $y = f(x)$.

Solution.

1. We know from Cauchy's Bound that all of the real zeros lie in the interval $[-4, 4]$ and that our possible rational zeros are $\pm \frac{1}{2}$, ± 1, $\pm \frac{3}{2}$ and ± 3. Descartes' Rule of Signs guarantees us at least one negative real zero and exactly one positive real zero, counting multiplicity. We try our positive rational zeros, starting with the smallest, $\frac{1}{2}$. Since the remainder isn't zero, we know $\frac{1}{2}$ isn't a zero. Sadly, the final line in the division tableau has both positive and negative numbers, so $\frac{1}{2}$ is not an upper bound. The only information we get from this division is courtesy of the Remainder Theorem which tells us $f\left(\frac{1}{2}\right) = -\frac{45}{8}$ so the point $\left(\frac{1}{2}, -\frac{45}{8}\right)$ is on the graph of f. We continue to our next possible zero, 1. As before, the only information we can glean from this is that $(1, -4)$ is on the graph of f. When we try our next possible zero, $\frac{3}{2}$, we get that it is not a zero, and we also see that it is an upper bound on the zeros of f, since all of the numbers in the final line of the division tableau are positive. This means there is no point trying our last possible rational zero, 3. Descartes' Rule of Signs guaranteed us a positive real zero, and at this point we have shown this zero is irrational. Furthermore, the Intermediate Value Theorem, Theorem 3.1, tells us the zero lies between 1 and $\frac{3}{2}$, since $f(1) < 0$ and $f\left(\frac{3}{2}\right) > 0$.

$$
\begin{array}{r|rrrrr}
\frac{1}{2} & 2 & 4 & -1 & -6 & -3 \\
 & \downarrow & 1 & \frac{5}{2} & \frac{3}{4} & -\frac{21}{8} \\
\hline
 & 2 & 5 & \frac{3}{2} & -\frac{21}{4} & \boxed{-\frac{45}{8}}
\end{array}
\qquad
\begin{array}{r|rrrrr}
1 & 2 & 4 & -1 & -6 & -3 \\
 & \downarrow & 2 & 6 & 5 & -1 \\
\hline
 & 2 & 6 & 5 & -1 & \boxed{-4}
\end{array}
\qquad
\begin{array}{r|rrrrr}
\frac{3}{2} & 2 & 4 & -1 & -6 & -3 \\
 & \downarrow & 3 & \frac{21}{2} & \frac{57}{4} & \frac{99}{8} \\
\hline
 & 2 & 7 & \frac{19}{2} & \frac{33}{4} & \boxed{\frac{75}{8}}
\end{array}
$$

We now turn our attention to negative real zeros. We try the largest possible zero, $-\frac{1}{2}$. Synthetic division shows us it is not a zero, nor is it a lower bound (since the numbers in the final line of the division tableau do not alternate), so we proceed to -1. This division shows -1 is a zero. Descartes' Rule of Signs told us that we may have up to three negative real zeros, counting multiplicity, so we try -1 again, and it works once more. At this point, we have taken f, a fourth degree polynomial, and performed two successful divisions. Our quotient polynomial is quadratic, so we look at it to find the remaining zeros.

$$\begin{array}{r|rrrrr} -\frac{1}{2} & 2 & 4 & -1 & -6 & -3 \\ & \downarrow & -1 & -\frac{3}{2} & \frac{5}{4} & \frac{19}{8} \\ \hline & 2 & 3 & -\frac{5}{2} & -\frac{19}{4} & \boxed{-\frac{5}{8}} \end{array}$$

$$\begin{array}{r|rrrrr} -1 & 2 & 4 & -1 & -6 & -3 \\ & \downarrow & -2 & -2 & 3 & 3 \\ \hline -1 & 2 & 2 & -3 & -3 & \boxed{0} \\ & \downarrow & -2 & 0 & 3 & \\ \hline & 2 & 0 & -3 & \boxed{0} & \end{array}$$

Setting the quotient polynomial equal to zero yields $2x^2 - 3 = 0$, so that $x^2 = \frac{3}{2}$, or $x = \pm \frac{\sqrt{6}}{2}$. Descartes' Rule of Signs tells us that the positive real zero we found, $\frac{\sqrt{6}}{2}$, has multiplicity 1. Descartes also tells us the total multiplicity of negative real zeros is 3, which forces -1 to be a zero of multiplicity 2 and $-\frac{\sqrt{6}}{2}$ to have multiplicity 1.

2. We know the end behavior of $y = f(x)$ resembles that of its leading term $y = 2x^4$. This means that the graph enters the scene in Quadrant II and exits in Quadrant I. Since $\pm \frac{\sqrt{6}}{2}$ are zeros of odd multiplicity, we have that the graph crosses through the x-axis at the points $\left(-\frac{\sqrt{6}}{2}, 0\right)$ and $\left(\frac{\sqrt{6}}{2}, 0\right)$. Since -1 is a zero of multiplicity 2, the graph of $y = f(x)$ touches and rebounds off the x-axis at $(-1, 0)$. Putting this together, we get

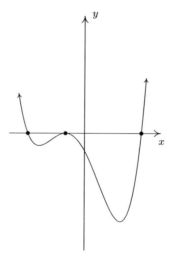

□

You can see why the 'no calculator' approach is not very popular these days. It requires more computation and more theorems than the alternative.[4] In general, no matter how many theorems you throw at a polynomial, it may well be impossible[5] to find their zeros exactly. The polynomial $f(x) = x^5 - x - 1$ is one such beast.[6] According to Descartes' Rule of Signs, f has exactly one positive real zero, and it could have two negative real zeros, or none at all. The Rational Zeros

---

[4] This is apparently a bad thing.

[5] We don't use this word lightly; it can be proven that the zeros of some polynomials cannot be expressed using the usual algebraic symbols.

[6] See this page.

Test gives us ± 1 as rational zeros to try but neither of these work since $f(1) = f(-1) = -1$. If we try the substitution technique we used in Example 3.3.4, we find $f(x)$ has three terms, but the exponent on the x^5 isn't exactly twice the exponent on x. How could we go about approximating the positive zero without resorting to the 'Zero' command of a graphing calculator? We use the **Bisection Method**. The first step in the Bisection Method is to find an interval on which f changes sign. We know $f(1) = -1$ and we find $f(2) = 29$. By the Intermediate Value Theorem, we know that the zero of f lies in the interval $[1, 2]$. Next, we 'bisect' this interval and find the midpoint is 1.5. We have that $f(1.5) \approx 5.09$. This means that our zero is between 1 and 1.5, since f changes sign on this interval. Now, we 'bisect' the interval $[1, 1.5]$ and find $f(1.25) \approx 0.80$, so now we have the zero between 1 and 1.25. Bisecting $[1, 1.25]$, we find $f(1.125) \approx -0.32$, which means the zero of f is between 1.125 and 1.25. We continue in this fashion until we have 'sandwiched' the zero between two numbers which differ by no more than a desired accuracy. You can think of the Bisection Method as reversing the sign diagram process: instead of finding the zeros and checking the sign of f using test values, we are using test values to determine where the signs switch to find the zeros. It is a slow and tedious, yet fool-proof, method for approximating a real zero.

Our next example reminds us of the role finding zeros plays in solving equations and inequalities.

Example 3.3.7.

1. Find all of the real solutions to the equation $2x^5 + 6x^3 + 3 = 3x^4 + 8x^2$.

2. Solve the inequality $2x^5 + 6x^3 + 3 \leq 3x^4 + 8x^2$.

3. Interpret your answer to part 2 graphically, and verify using a graphing calculator.

Solution.

1. Finding the real solutions to $2x^5 + 6x^3 + 3 = 3x^4 + 8x^2$ is the same as finding the real solutions to $2x^5 - 3x^4 + 6x^3 - 8x^2 + 3 = 0$. In other words, we are looking for the real zeros of $p(x) = 2x^5 - 3x^4 + 6x^3 - 8x^2 + 3$. Using the techniques developed in this section, we get

    $$\begin{array}{r|rrrrrr} 1 & 2 & -3 & 6 & -8 & 0 & 3 \\ & \downarrow & 2 & -1 & 5 & -3 & -3 \\ \hline 1 & 2 & -1 & 5 & -3 & -3 & \boxed{0} \\ & \downarrow & 2 & 1 & 6 & 3 & \\ \hline -\frac{1}{2} & 2 & 1 & 6 & 3 & \boxed{0} & \\ & \downarrow & -1 & 0 & -3 & & \\ \hline & 2 & 0 & 6 & \boxed{0} & & \end{array}$$

    The quotient polynomial is $2x^2 + 6$ which has no real zeros so we get $x = -\frac{1}{2}$ and $x = 1$.

2. To solve this nonlinear inequality, we follow the same guidelines set forth in Section 2.4: we get 0 on one side of the inequality and construct a sign diagram. Our original inequality can be rewritten as $2x^5 - 3x^4 + 6x^3 - 8x^2 + 3 \leq 0$. We found the zeros of $p(x) = 2x^5 - 3x^4 + 6x^3 - 8x^2 + 3$ in part 1 to be $x = -\frac{1}{2}$ and $x = 1$. We construct our sign diagram as before.

```
         (−) 0  (+)  0 (+)
    ←——————+————————+——————→
       ↑ −½  ↑    1   ↑
      −1    0        2
```

The solution to $p(x) < 0$ is $\left(-\infty, -\frac{1}{2}\right)$, and we know $p(x) = 0$ at $x = -\frac{1}{2}$ and $x = 1$. Hence, the solution to $p(x) \leq 0$ is $\left(-\infty, -\frac{1}{2}\right] \cup \{1\}$.

3. To interpret this solution graphically, we set $f(x) = 2x^5 + 6x^3 + 3$ and $g(x) = 3x^4 + 8x^2$. We recall that the solution to $f(x) \leq g(x)$ is the set of x values for which the graph of f is below the graph of g (where $f(x) < g(x)$) along with the x values where the two graphs intersect ($f(x) = g(x)$). Graphing f and g on the calculator produces the picture on the lower left. (The end behavior should tell you which is which.) We see that the graph of f is below the graph of g on $\left(-\infty, -\frac{1}{2}\right)$. However, it is difficult to see what is happening near $x = 1$. Zooming in (and making the graph of g thicker), we see that the graphs of f and g do intersect at $x = 1$, but the graph of g remains below the graph of f on either side of $x = 1$.

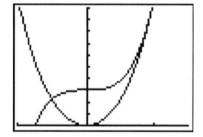 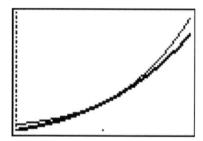

□

Our last example revisits an application from page 247 in the Exercises of Section 3.1.

Example 3.3.8. Suppose the profit P, in *thousands* of dollars, from producing and selling x *hundred* LCD TVs is given by $P(x) = -5x^3 + 35x^2 - 45x - 25$, $0 \leq x \leq 10.07$. How many TVs should be produced to make a profit? Check your answer using a graphing utility.

Solution. To 'make a profit' means to solve $P(x) = -5x^3 + 35x^2 - 45x - 25 > 0$, which we do analytically using a sign diagram. To simplify things, we first factor out the -5 common to all the coefficients to get $-5\left(x^3 - 7x^2 + 9x - 5\right) > 0$, so we can just focus on finding the zeros of $f(x) = x^3 - 7x^2 + 9x + 5$. The possible rational zeros of f are ± 1 and ± 5, and going through the usual computations, we find $x = 5$ is the only rational zero. Using this, we factor $f(x) = x^3 - 7x^2 + 9x + 5 = (x - 5)\left(x^2 - 2x - 1\right)$, and we find the remaining zeros by applying the Quadratic Formula to $x^2 - 2x - 1 = 0$. We find three real zeros, $x = 1 - \sqrt{2} = -0.414\ldots$, $x = 1 + \sqrt{2} = 2.414\ldots$, and $x = 5$, of which only the last two fall in the applied domain of $[0, 10.07]$. We choose $x = 0$, $x = 3$ and $x = 10.07$ as our test values and plug them into the function $P(x) = -5x^3 + 35x^2 - 45x - 25$ (not $f(x) = x^3 - 7x^2 + 9x - 5$) to get the sign diagram below.

3.3 Real Zeros of Polynomials

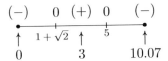

We see immediately that $P(x) > 0$ on $(1+\sqrt{2}, 5)$. Since x measures the number of TVs in *hundreds*, $x = 1+\sqrt{2}$ corresponds to $241.4\ldots$ TVs. Since we can't produce a fractional part of a TV, we need to choose between producing 241 and 242 TVs. From the sign diagram, we see that $P(2.41) < 0$ but $P(2.42) > 0$ so, in this case we take the next *larger* integer value and set the minimum production to 242 TVs. At the other end of the interval, we have $x = 5$ which corresponds to 500 TVs. Here, we take the next *smaller* integer value, 499 TVs to ensure that we make a profit. Hence, in order to make a profit, at least 242, but no more than 499 TVs need to be produced. To check our answer using a calculator, we graph $y = P(x)$ and make use of the 'Zero' command. We see that the calculator approximations bear out our analysis.[7]

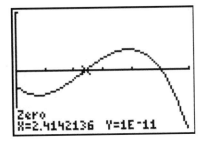
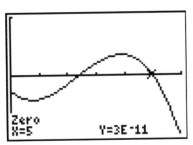

□

---

[7] Note that the y-coordinates of the points here aren't registered as 0. They are expressed in Scientific Notation. For instance, $1E - 11$ corresponds to 0.00000000001, which is pretty close in the calculator's eyes[8] to 0.

[8] but not a Mathematician's

3.3.3 Exercises

In Exercises 1 - 10, for the given polynomial:

- Use Cauchy's Bound to find an interval containing all of the real zeros.

- Use the Rational Zeros Theorem to make a list of possible rational zeros.

- Use Descartes' Rule of Signs to list the possible number of positive and negative real zeros, counting multiplicities.

1. $f(x) = x^3 - 2x^2 - 5x + 6$

2. $f(x) = x^4 + 2x^3 - 12x^2 - 40x - 32$

3. $f(x) = x^4 - 9x^2 - 4x + 12$

4. $f(x) = x^3 + 4x^2 - 11x + 6$

5. $f(x) = x^3 - 7x^2 + x - 7$

6. $f(x) = -2x^3 + 19x^2 - 49x + 20$

7. $f(x) = -17x^3 + 5x^2 + 34x - 10$

8. $f(x) = 36x^4 - 12x^3 - 11x^2 + 2x + 1$

9. $f(x) = 3x^3 + 3x^2 - 11x - 10$

10. $f(x) = 2x^4 + x^3 - 7x^2 - 3x + 3$

In Exercises 11 - 30, find the real zeros of the polynomial using the techniques specified by your instructor. State the multiplicity of each real zero.

11. $f(x) = x^3 - 2x^2 - 5x + 6$

12. $f(x) = x^4 + 2x^3 - 12x^2 - 40x - 32$

13. $f(x) = x^4 - 9x^2 - 4x + 12$

14. $f(x) = x^3 + 4x^2 - 11x + 6$

15. $f(x) = x^3 - 7x^2 + x - 7$

16. $f(x) = -2x^3 + 19x^2 - 49x + 20$

17. $f(x) = -17x^3 + 5x^2 + 34x - 10$

18. $f(x) = 36x^4 - 12x^3 - 11x^2 + 2x + 1$

19. $f(x) = 3x^3 + 3x^2 - 11x - 10$

20. $f(x) = 2x^4 + x^3 - 7x^2 - 3x + 3$

21. $f(x) = 9x^3 - 5x^2 - x$

22. $f(x) = 6x^4 - 5x^3 - 9x^2$

23. $f(x) = x^4 + 2x^2 - 15$

24. $f(x) = x^4 - 9x^2 + 14$

25. $f(x) = 3x^4 - 14x^2 - 5$

26. $f(x) = 2x^4 - 7x^2 + 6$

27. $f(x) = x^6 - 3x^3 - 10$

28. $f(x) = 2x^6 - 9x^3 + 10$

29. $f(x) = x^5 - 2x^4 - 4x + 8$

30. $f(x) = 2x^5 + 3x^4 - 18x - 27$

3.3 REAL ZEROS OF POLYNOMIALS

In Exercises 31 - 33, use your calculator,[9] to help you find the real zeros of the polynomial. State the multiplicity of each real zero.

31. $f(x) = x^5 - 60x^3 - 80x^2 + 960x + 2304$

32. $f(x) = 25x^5 - 105x^4 + 174x^3 - 142x^2 + 57x - 9$

33. $f(x) = 90x^4 - 399x^3 + 622x^2 - 399x + 90$

34. Find the real zeros of $f(x) = x^3 - \frac{1}{12}x^2 - \frac{7}{72}x + \frac{1}{72}$ by first finding a polynomial $q(x)$ with integer coefficients such that $q(x) = N \cdot f(x)$ for some integer N. (Recall that the Rational Zeros Theorem required the polynomial in question to have integer coefficients.) Show that f and q have the same real zeros.

In Exercises 35 - 44, find the real solutions of the polynomial equation. (See Example 3.3.7.)

35. $9x^3 = 5x^2 + x$

36. $9x^2 + 5x^3 = 6x^4$

37. $x^3 + 6 = 2x^2 + 5x$

38. $x^4 + 2x^3 = 12x^2 + 40x + 32$

39. $x^3 - 7x^2 = 7 - x$

40. $2x^3 = 19x^2 - 49x + 20$

41. $x^3 + x^2 = \dfrac{11x + 10}{3}$

42. $x^4 + 2x^2 = 15$

43. $14x^2 + 5 = 3x^4$

44. $2x^5 + 3x^4 = 18x + 27$

In Exercises 45 - 54, solve the polynomial inequality and state your answer using interval notation.

45. $-2x^3 + 19x^2 - 49x + 20 > 0$

46. $x^4 - 9x^2 \leq 4x - 12$

47. $(x-1)^2 \geq 4$

48. $4x^3 \geq 3x + 1$

49. $x^4 \leq 16 + 4x - x^3$

50. $3x^2 + 2x < x^4$

51. $\dfrac{x^3 + 2x^2}{2} < x + 2$

52. $\dfrac{x^3 + 20x}{8} \geq x^2 + 2$

53. $2x^4 > 5x^2 + 3$

54. $x^6 + x^3 \geq 6$

55. In Example 3.1.3 in Section 3.1, a box with no top is constructed from a 10 inch × 12 inch piece of cardboard by cutting out congruent squares from each corner of the cardboard and then folding the resulting tabs. We determined the volume of that box (in cubic inches) is given by $V(x) = 4x^3 - 44x^2 + 120x$, where x denotes the length of the side of the square which is removed from each corner (in inches), $0 < x < 5$. Solve the inequality $V(x) \geq 80$ analytically and interpret your answer in the context of that example.

---

[9] You *can* do these without your calculator, but it may test your mettle!

56. From Exercise 32 in Section 3.1, $C(x) = .03x^3 - 4.5x^2 + 225x + 250$, for $x \geq 0$ models the cost, in dollars, to produce x PortaBoy game systems. If the production budget is $5000, find the number of game systems which can be produced and still remain under budget.

57. Let $f(x) = 5x^7 - 33x^6 + 3x^5 - 71x^4 - 597x^3 + 2097x^2 - 1971x + 567$. With the help of your classmates, find the x- and y- intercepts of the graph of f. Find the intervals on which the function is increasing, the intervals on which it is decreasing and the local extrema. Sketch the graph of f, using more than one picture if necessary to show all of the important features of the graph.

58. With the help of your classmates, create a list of five polynomials with different degrees whose real zeros cannot be found using any of the techniques in this section.

3.3 REAL ZEROS OF POLYNOMIALS

3.3.4 Answers

1. For $f(x) = x^3 - 2x^2 - 5x + 6$

   - All of the real zeros lie in the interval $[-7, 7]$
   - Possible rational zeros are ± 1, ± 2, ± 3, ± 6
   - There are 2 or 0 positive real zeros; there is 1 negative real zero

2. For $f(x) = x^4 + 2x^3 - 12x^2 - 40x - 32$

   - All of the real zeros lie in the interval $[-41, 41]$
   - Possible rational zeros are ± 1, ± 2, ± 4, ± 8, ± 16, ± 32
   - There is 1 positive real zero; there are 3 or 1 negative real zeros

3. For $f(x) = x^4 - 9x^2 - 4x + 12$

   - All of the real zeros lie in the interval $[-13, 13]$
   - Possible rational zeros are ± 1, ± 2, ± 3, ± 4, ± 6, ± 12
   - There are 2 or 0 positive real zeros; there are 2 or 0 negative real zeros

4. For $f(x) = x^3 + 4x^2 - 11x + 6$

   - All of the real zeros lie in the interval $[-12, 12]$
   - Possible rational zeros are ± 1, ± 2, ± 3, ± 6
   - There are 2 or 0 positive real zeros; there is 1 negative real zero

5. For $f(x) = x^3 - 7x^2 + x - 7$

   - All of the real zeros lie in the interval $[-8, 8]$
   - Possible rational zeros are ± 1, ± 7
   - There are 3 or 1 positive real zeros; there are no negative real zeros

6. For $f(x) = -2x^3 + 19x^2 - 49x + 20$

   - All of the real zeros lie in the interval $\left[-\frac{51}{2}, \frac{51}{2}\right]$
   - Possible rational zeros are $\pm \frac{1}{2}$, ± 1, ± 2, $\pm \frac{5}{2}$, ± 4, ± 5, ± 10, ± 20
   - There are 3 or 1 positive real zeros; there are no negative real zeros

7. For $f(x) = -17x^3 + 5x^2 + 34x - 10$

   - All of the real zeros lie in the interval $[-3, 3]$
   - Possible rational zeros are $\pm \frac{1}{17}$, $\pm \frac{2}{17}$, $\pm \frac{5}{17}$, $\pm \frac{10}{17}$, ± 1, ± 2, ± 5, ± 10
   - There are 2 or 0 positive real zeros; there is 1 negative real zero

8. For $f(x) = 36x^4 - 12x^3 - 11x^2 + 2x + 1$

    - All of the real zeros lie in the interval $\left[-\frac{4}{3}, \frac{4}{3}\right]$
    - Possible rational zeros are $\pm\frac{1}{36}, \pm\frac{1}{18}, \pm\frac{1}{12}, \pm\frac{1}{9}, \pm\frac{1}{6}, \pm\frac{1}{4}, \pm\frac{1}{3}, \pm\frac{1}{2}, \pm 1$
    - There are 2 or 0 positive real zeros; there are 2 or 0 negative real zeros

9. For $f(x) = 3x^3 + 3x^2 - 11x - 10$

    - All of the real zeros lie in the interval $\left[-\frac{14}{3}, \frac{14}{3}\right]$
    - Possible rational zeros are $\pm\frac{1}{3}, \pm\frac{2}{3}, \pm\frac{5}{3}, \pm\frac{10}{3}, \pm 1, \pm 2, \pm 5, \pm 10$
    - There is 1 positive real zero; there are 2 or 0 negative real zeros

10. For $f(x) = 2x^4 + x^3 - 7x^2 - 3x + 3$

    - All of the real zeros lie in the interval $\left[-\frac{9}{2}, \frac{9}{2}\right]$
    - Possible rational zeros are $\pm\frac{1}{2}, \pm 1, \pm\frac{3}{2}, \pm 3$
    - There are 2 or 0 positive real zeros; there are 2 or 0 negative real zeros

11. $f(x) = x^3 - 2x^2 - 5x + 6$
    $x = -2$, $x = 1$, $x = 3$ (each has mult. 1)

12. $f(x) = x^4 + 2x^3 - 12x^2 - 40x - 32$
    $x = -2$ (mult. 3), $x = 4$ (mult. 1)

13. $f(x) = x^4 - 9x^2 - 4x + 12$
    $x = -2$ (mult. 2), $x = 1$ (mult. 1), $x = 3$ (mult. 1)

14. $f(x) = x^3 + 4x^2 - 11x + 6$
    $x = -6$ (mult. 1), $x = 1$ (mult. 2)

15. $f(x) = x^3 - 7x^2 + x - 7$
    $x = 7$ (mult. 1)

16. $f(x) = -2x^3 + 19x^2 - 49x + 20$
    $x = \frac{1}{2}$, $x = 4$, $x = 5$ (each has mult. 1)

17. $f(x) = -17x^3 + 5x^2 + 34x - 10$
    $x = \frac{5}{17}$, $x = \pm\sqrt{2}$ (each has mult. 1)

18. $f(x) = 36x^4 - 12x^3 - 11x^2 + 2x + 1$
    $x = \frac{1}{2}$ (mult. 2), $x = -\frac{1}{3}$ (mult. 2)

19. $f(x) = 3x^3 + 3x^2 - 11x - 10$
    $x = -2$, $x = \frac{3 \pm \sqrt{69}}{6}$ (each has mult. 1)

3.3 Real Zeros of Polynomials

20. $f(x) = 2x^4 + x^3 - 7x^2 - 3x + 3$
    $x = -1$, $x = \frac{1}{2}$, $x = \pm\sqrt{3}$ (each mult. 1)

21. $f(x) = 9x^3 - 5x^2 - x$
    $x = 0$, $x = \frac{5\pm\sqrt{61}}{18}$ (each has mult. 1)

22. $f(x) = 6x^4 - 5x^3 - 9x^2$
    $x = 0$ (mult. 2), $x = \frac{5\pm\sqrt{241}}{12}$ (each has mult. 1)

23. $f(x) = x^4 + 2x^2 - 15$
    $x = \pm\sqrt{3}$ (each has mult. 1)

24. $f(x) = x^4 - 9x^2 + 14$
    $x = \pm\sqrt{2}$, $x = \pm\sqrt{7}$ (each has mult. 1)

25. $f(x) = 3x^4 - 14x^2 - 5$
    $x = \pm\sqrt{5}$ (each has mult. 1)

26. $f(x) = 2x^4 - 7x^2 + 6$
    $x = \pm\frac{\sqrt{6}}{2}$, $x = \pm\sqrt{2}$ (each has mult. 1)

27. $f(x) = x^6 - 3x^3 - 10$
    $x = \sqrt[3]{-2} = -\sqrt[3]{2}$, $x = \sqrt[3]{5}$ (each has mult. 1)

28. $f(x) = 2x^6 - 9x^3 + 10$
    $x = \frac{\sqrt[3]{20}}{2}$, $x = \sqrt[3]{2}$ (each has mult. 1)

29. $f(x) = x^5 - 2x^4 - 4x + 8$
    $x = 2$, $x = \pm\sqrt{2}$ (each has mult. 1)

30. $f(x) = 2x^5 + 3x^4 - 18x - 27$
    $x = -\frac{3}{2}$, $x = \pm\sqrt{3}$ (each has mult. 1)

31. $f(x) = x^5 - 60x^3 - 80x^2 + 960x + 2304$
    $x = -4$ (mult. 3), $x = 6$ (mult. 2)

32. $f(x) = 25x^5 - 105x^4 + 174x^3 - 142x^2 + 57x - 9$
    $x = \frac{3}{5}$ (mult. 2), $x = 1$ (mult. 3)

33. $f(x) = 90x^4 - 399x^3 + 622x^2 - 399x + 90$
    $x = \frac{2}{3}$, $x = \frac{3}{2}$, $x = \frac{5}{3}$, $x = \frac{3}{5}$ (each has mult. 1)

34. We choose $q(x) = 72x^3 - 6x^2 - 7x + 1 = 72 \cdot f(x)$. Clearly $f(x) = 0$ if and only if $q(x) = 0$ so they have the same real zeros. In this case, $x = -\frac{1}{3}$, $x = \frac{1}{6}$ and $x = \frac{1}{4}$ are the real zeros of both f and q.

35. $x = 0, \frac{5 \pm \sqrt{61}}{18}$

36. $x = 0, \frac{5 \pm \sqrt{241}}{12}$

37. $x = -2, 1, 3$

38. $x = -2, 4$

39. $x = 7$

40. $x = \frac{1}{2}, 4, 5$

41. $x = -2, \frac{3 \pm \sqrt{69}}{6}$

42. $x = \pm\sqrt{3}$

43. $x = \pm\sqrt{5}$

44. $x = -\frac{3}{2}, \pm\sqrt{3}$

45. $(-\infty, \frac{1}{2}) \cup (4, 5)$

46. $\{-2\} \cup [1, 3]$

47. $(-\infty, -1] \cup [3, \infty)$

48. $\left\{-\frac{1}{2}\right\} \cup [1, \infty)$

49. $[-2, 2]$

50. $(-\infty, -1) \cup (-1, 0) \cup (2, \infty)$

51. $(-\infty, -2) \cup (-\sqrt{2}, \sqrt{2})$

52. $\{2\} \cup [4, \infty)$

53. $(-\infty, -\sqrt{3}) \cup (\sqrt{3}, \infty)$

54. $(-\infty, -\sqrt[3]{3}) \cup (\sqrt[3]{2}, \infty)$

55. $V(x) \geq 80$ on $[1, 5-\sqrt{5}] \cup [5+\sqrt{5}, \infty)$. Only the portion $[1, 5-\sqrt{5}]$ lies in the applied domain, however. In the context of the problem, this says for the volume of the box to be at least 80 cubic inches, the square removed from each corner needs to have a side length of at least 1 inch, but no more than $5 - \sqrt{5} \approx 2.76$ inches.

56. $C(x) \leq 5000$ on (approximately) $(-\infty, 82.18]$. The portion of this which lies in the applied domain is $(0, 82.18]$. Since x represents the number of game systems, we check $C(82) = 4983.04$ and $C(83) = 5078.11$, so to remain within the production budget, anywhere between 1 and 82 game systems can be produced.

3.4 Complex Zeros and the Fundamental Theorem of Algebra

In Section 3.3, we were focused on finding the real zeros of a polynomial function. In this section, we expand our horizons and look for the non-real zeros as well. Consider the polynomial $p(x) = x^2 + 1$. The zeros of p are the solutions to $x^2 + 1 = 0$, or $x^2 = -1$. This equation has no real solutions, but you may recall from Intermediate Algebra that we can formally extract the square roots of both sides to get $x = \pm\sqrt{-1}$. The quantity $\sqrt{-1}$ is usually re-labeled i, the so-called **imaginary unit**.[1] The number i, while not a real number, plays along well with real numbers, and acts very much like any other radical expression. For instance, $3(2i) = 6i$, $7i - 3i = 4i$, $(2 - 7i) + (3 + 4i) = 5 - 3i$, and so forth. The key properties which distinguish i from the real numbers are listed below.

> **Definition 3.4.** The imaginary unit i satisfies the two following properties
>
> 1. $i^2 = -1$
>
> 2. If c is a real number with $c \geq 0$ then $\sqrt{-c} = i\sqrt{c}$

Property 1 in Definition 3.4 establishes that i does act as a square root[2] of -1, and property 2 establishes what we mean by the 'principal square root' of a negative real number. In property 2, it is important to remember the restriction on c. For example, it is perfectly acceptable to say $\sqrt{-4} = i\sqrt{4} = i(2) = 2i$. However, $\sqrt{-(-4)} \neq i\sqrt{-4}$, otherwise, we'd get

$$2 = \sqrt{4} = \sqrt{-(-4)} = i\sqrt{-4} = i(2i) = 2i^2 = 2(-1) = -2,$$

which is unacceptable.[3] We are now in the position to define the **complex numbers**.

> **Definition 3.5.** A **complex number** is a number of the form $a + bi$, where a and b are real numbers and i is the imaginary unit.

Complex numbers include things you'd normally expect, like $3 + 2i$ and $\frac{2}{5} - i\sqrt{3}$. However, don't forget that a or b could be zero, which means numbers like $3i$ and 6 are also complex numbers. In other words, don't forget that the complex numbers *include* the real numbers, so 0 and $\pi - \sqrt{21}$ are both considered complex numbers.[4] The arithmetic of complex numbers is as you would expect. The only things you need to remember are the two properties in Definition 3.4. The next example should help recall how these animals behave.

---

[1] Some Technical Mathematics textbooks label it 'j'.

[2] Note the use of the indefinite article 'a'. Whatever beast is chosen to be i, $-i$ is the other square root of -1.

[3] We want to enlarge the number system so we can solve things like $x^2 = -1$, but not at the cost of the established rules already set in place. For that reason, the general properties of radicals simply do not apply for even roots of negative quantities.

[4] See the remarks in Section 1.1.1.

Example 3.4.1. Perform the indicated operations. Write your answer in the form[5] $a + bi$.

1. $(1 - 2i) - (3 + 4i)$
2. $(1 - 2i)(3 + 4i)$
3. $\dfrac{1 - 2i}{3 - 4i}$
4. $\sqrt{-3}\sqrt{-12}$
5. $\sqrt{(-3)(-12)}$
6. $(x - [1 + 2i])(x - [1 - 2i])$

Solution.

1. As mentioned earlier, we treat expressions involving i as we would any other radical. We combine like terms to get $(1 - 2i) - (3 + 4i) = 1 - 2i - 3 - 4i = -2 - 6i$.

2. Using the distributive property, we get $(1 - 2i)(3 + 4i) = (1)(3) + (1)(4i) - (2i)(3) - (2i)(4i) = 3 + 4i - 6i - 8i^2$. Since $i^2 = -1$, we get $3 + 4i - 6i - 8i^2 = 3 - 2i - (-8) = 11 - 2i$.

3. How in the world are we supposed to simplify $\frac{1-2i}{3-4i}$? Well, we deal with the denominator $3 - 4i$ as we would any other denominator containing a radical, and multiply both numerator and denominator by $3 + 4i$ (the conjugate of $3 - 4i$).[6] Doing so produces

$$\frac{1 - 2i}{3 - 4i} \cdot \frac{3 + 4i}{3 + 4i} = \frac{(1 - 2i)(3 + 4i)}{(3 - 4i)(3 + 4i)} = \frac{11 - 2i}{25} = \frac{11}{25} - \frac{2}{25}i$$

4. We use property 2 of Definition 3.4 first, then apply the rules of radicals applicable to real radicals to get $\sqrt{-3}\sqrt{-12} = (i\sqrt{3})(i\sqrt{12}) = i^2\sqrt{3 \cdot 12} = -\sqrt{36} = -6$.

5. We adhere to the order of operations here and perform the multiplication before the radical to get $\sqrt{(-3)(-12)} = \sqrt{36} = 6$.

6. We can brute force multiply using the distributive property and see that

$$\begin{aligned}(x - [1 + 2i])(x - [1 - 2i]) &= x^2 - x[1 - 2i] - x[1 + 2i] + [1 - 2i][1 + 2i] \\ &= x^2 - x + 2ix - x - 2ix + 1 - 2i + 2i - 4i^2 \\ &= x^2 - 2x + 5\end{aligned}$$

□

A couple of remarks about the last example are in order. First, the **conjugate** of a complex number $a + bi$ is the number $a - bi$. The notation commonly used for conjugation is a 'bar': $\overline{a + bi} = a - bi$. For example, $\overline{3 + 2i} = 3 - 2i$, $\overline{3 - 2i} = 3 + 2i$, $\overline{6} = 6$, $\overline{4i} = -4i$, and $\overline{3 + \sqrt{5}} = 3 + \sqrt{5}$. The properties of the conjugate are summarized in the following theorem.

---

[5]OK, we'll accept things like $3 - 2i$ even though it can be written as $3 + (-2)i$.
[6]We will talk more about this in a moment.

3.4 COMPLEX ZEROS AND THE FUNDAMENTAL THEOREM OF ALGEBRA   289

> **Theorem 3.12. Properties of the Complex Conjugate:** Let z and w be complex numbers.
>
> - $\overline{\overline{z}} = z$
> - $\overline{z + w} = \overline{z} + \overline{w}$
> - $\overline{zw} = \overline{z}\,\overline{w}$
> - $(\overline{z})^n = \overline{z^n}$, for any natural number n
> - z is a real number if and only if $\overline{z} = z$.

Essentially, Theorem 3.12 says that complex conjugation works well with addition, multiplication and powers. The proof of these properties can best be achieved by writing out $z = a + bi$ and $w = c + di$ for real numbers a, b, c and d. Next, we compute the left and right hand sides of each equation and check to see that they are the same. The proof of the first property is a very quick exercise.[7] To prove the second property, we compare $\overline{z} + \overline{w}$ and $\overline{z+w}$. We have $\overline{z} + \overline{w} = \overline{a+bi} + \overline{c+di} = a - bi + c - di$. To find $\overline{z+w}$, we first compute

$$z + w = (a+bi) + (c+di) = (a+c) + (b+d)i$$

so

$$\overline{z+w} = \overline{(a+c) + (b+d)i} = (a+c) - (b+d)i = a - bi + c - di$$

As such, we have established $\overline{z} + \overline{w} = \overline{z+w}$. The proof for multiplication works similarly. The proof that the conjugate works well with powers can be viewed as a repeated application of the product rule, and is best proved using a technique called Mathematical Induction.[8] The last property is a characterization of real numbers. If z is real, then $z = a + 0i$, so $\overline{z} = a - 0i = a = z$. On the other hand, if $z = \overline{z}$, then $a + bi = a - bi$ which means $b = -b$ so $b = 0$. Hence, $z = a + 0i = a$ and is real.

We now return to the business of zeros. Suppose we wish to find the zeros of $f(x) = x^2 - 2x + 5$. To solve the equation $x^2 - 2x + 5 = 0$, we note that the quadratic doesn't factor nicely, so we resort to the Quadratic Formula, Equation 2.5 and obtain

$$x = \frac{-(-2) \pm \sqrt{(-2)^2 - 4(1)(5)}}{2(1)} = \frac{2 \pm \sqrt{-16}}{2} = \frac{2 \pm 4i}{2} = 1 \pm 2i.$$

Two things are important to note. First, the zeros $1 + 2i$ and $1 - 2i$ are complex conjugates. If ever we obtain non-real zeros to a quadratic function with <u>real</u> coefficients, the zeros will be a complex conjugate pair. (Do you see why?) Next, we note that in Example 3.4.1, part 6, we found $(x - [1+2i])(x - [1-2i]) = x^2 - 2x + 5$. This demonstrates that the factor theorem holds even for non-real zeros, i.e, $x = 1 + 2i$ is a zero of f, and, sure enough, $(x - [1+2i])$ is a factor of $f(x)$. It turns out that polynomial division works the same way for all complex numbers, real and non-real alike, so the Factor and Remainder Theorems hold as well. But how do we know if a

---
[7]Trust us on this.
[8]See Section 9.3.

general polynomial has any complex zeros at all? We have many examples of polynomials with no real zeros. Can there be polynomials with no zeros whatsoever? The answer to that last question is "No." and the theorem which provides that answer is The Fundamental Theorem of Algebra.

> **Theorem 3.13. The Fundamental Theorem of Algebra:** Suppose f is a polynomial function with complex number coefficients of degree $n \geq 1$, then f has at least one complex zero.

The Fundamental Theorem of Algebra is an example of an 'existence' theorem in Mathematics. Like the Intermediate Value Theorem, Theorem 3.1, the Fundamental Theorem of Algebra guarantees the existence of at least one zero, but gives us no algorithm to use in finding it. In fact, as we mentioned in Section 3.3, there are polynomials whose real zeros, though they exist, cannot be expressed using the 'usual' combinations of arithmetic symbols, and must be approximated. The authors are fully aware that the full impact and profound nature of the Fundamental Theorem of Algebra is lost on most students studying College Algebra, and that's fine. It took mathematicians literally hundreds of years to prove the theorem in its full generality, and some of that history is recorded here. Note that the Fundamental Theorem of Algebra applies to not only polynomial functions with real coefficients, but to those with complex number coefficients as well.

Suppose f is a polynomial of degree $n \geq 1$. The Fundamental Theorem of Algebra guarantees us at least one complex zero, z_1, and as such, the Factor Theorem guarantees that $f(x)$ factors as $f(x) = (x - z_1) q_1(x)$ for a polynomial function q_1, of degree exactly $n - 1$. If $n - 1 \geq 1$, then the Fundamental Theorem of Algebra guarantees a complex zero of q_1 as well, say z_2, so then the Factor Theorem gives us $q_1(x) = (x - z_2) q_2(x)$, and hence $f(x) = (x - z_1)(x - z_2) q_2(x)$. We can continue this process exactly n times, at which point our quotient polynomial q_n has degree 0 so it's a constant. This argument gives us the following factorization theorem.

> **Theorem 3.14. Complex Factorization Theorem:** Suppose f is a polynomial function with complex number coefficients. If the degree of f is n and $n \geq 1$, then f has exactly n complex zeros, counting multiplicity. If $z_1, z_2, \ldots, z_k$ are the distinct zeros of f, with multiplicities $m_1, m_2, \ldots, m_k$, respectively, then $f(x) = a(x - z_1)^{m_1} (x - z_2)^{m_2} \cdots (x - z_k)^{m_k}$.

Note that the value a in Theorem 3.14 is the leading coefficient of $f(x)$ (Can you see why?) and as such, we see that a polynomial is completely determined by its zeros, their multiplicities, and its leading coefficient. We put this theorem to good use in the next example.

Example 3.4.2. Let $f(x) = 12x^5 - 20x^4 + 19x^3 - 6x^2 - 2x + 1$.

1. Find all of the complex zeros of f and state their multiplicities.

2. Factor $f(x)$ using Theorem 3.14

Solution.

1. Since f is a fifth degree polynomial, we know that we need to perform at least three successful divisions to get the quotient down to a quadratic function. At that point, we can find the remaining zeros using the Quadratic Formula, if necessary. Using the techniques developed in Section 3.3, we get

3.4 Complex Zeros and the Fundamental Theorem of Algebra

$$\begin{array}{r|rrrrrr}
\frac{1}{2} & 12 & -20 & 19 & -6 & -2 & 1 \\
 & \downarrow & 6 & -7 & 6 & 0 & -1 \\
\hline
\frac{1}{2} & 12 & -14 & 12 & 0 & -2 & \boxed{0} \\
 & \downarrow & 6 & -4 & 4 & 2 & \\
\hline
-\frac{1}{3} & 12 & -8 & 8 & 4 & \boxed{0} & \\
 & \downarrow & -4 & 4 & -4 & & \\
\hline
 & 12 & -12 & 12 & \boxed{0} & &
\end{array}$$

Our quotient is $12x^2 - 12x + 12$, whose zeros we find to be $\frac{1 \pm i\sqrt{3}}{2}$. From Theorem 3.14, we know f has exactly 5 zeros, counting multiplicities, and as such we have the zero $\frac{1}{2}$ with multiplicity 2, and the zeros $-\frac{1}{3}$, $\frac{1+i\sqrt{3}}{2}$ and $\frac{1-i\sqrt{3}}{2}$, each of multiplicity 1.

2. Applying Theorem 3.14, we are guaranteed that f factors as

$$f(x) = 12\left(x - \frac{1}{2}\right)^2 \left(x + \frac{1}{3}\right)\left(x - \left[\frac{1+i\sqrt{3}}{2}\right]\right)\left(x - \left[\frac{1-i\sqrt{3}}{2}\right]\right) \qquad \square$$

A true test of Theorem 3.14 (and a student's mettle!) would be to take the factored form of $f(x)$ in the previous example and multiply it out[9] to see that it really does reduce to the original formula $f(x) = 12x^5 - 20x^4 + 19x^3 - 6x^2 - 2x + 1$. When factoring a polynomial using Theorem 3.14, we say that it is **factored completely over the complex numbers**, meaning that it is impossible to factor the polynomial any further using complex numbers. If we wanted to completely factor $f(x)$ over the **real numbers** then we would have stopped short of finding the nonreal zeros of f and factored f using our work from the synthetic division to write $f(x) = \left(x - \frac{1}{2}\right)^2 \left(x + \frac{1}{3}\right)(12x^2 - 12x + 12)$, or $f(x) = 12\left(x - \frac{1}{2}\right)^2 \left(x + \frac{1}{3}\right)(x^2 - x + 1)$. Since the zeros of $x^2 - x + 1$ are nonreal, we call $x^2 - x + 1$ an **irreducible quadratic** meaning it is impossible to break it down any further using *real* numbers.

The last two results of the section show us that, at least in theory, if we have a polynomial function with real coefficients, we can always factor it down enough so that any nonreal zeros come from irreducible quadratics.

Theorem 3.15. Conjugate Pairs Theorem: If f is a polynomial function with real number coefficients and z is a zero of f, then so is $\bar{z}$.

To prove the theorem, suppose f is a polynomial with real number coefficients. Specifically, let $f(x) = a_n x^n + a_{n-1} x^{n-1} + \ldots + a_2 x^2 + a_1 x + a_0$. If z is a zero of f, then $f(z) = 0$, which means $a_n z^n + a_{n-1} z^{n-1} + \ldots + a_2 z^2 + a_1 z + a_0 = 0$. Next, we consider $f(\bar{z})$ and apply Theorem 3.12 below.

---
[9] You really should do this once in your life to convince yourself that all of the theory actually does work!

$$\begin{aligned}
f(\overline{z}) &= a_n(\overline{z})^n + a_{n-1}(\overline{z})^{n-1} + \ldots + a_2(\overline{z})^2 + a_1\overline{z} + a_0 \\
&= a_n\overline{z^n} + a_{n-1}\overline{z^{n-1}} + \ldots + a_2\overline{z^2} + a_1\overline{z} + a_0 && \text{since } (\overline{z})^n = \overline{z^n} \\
&= \overline{a_n}\,\overline{z^n} + \overline{a_{n-1}}\,\overline{z^{n-1}} + \ldots + \overline{a_2}\,\overline{z^2} + \overline{a_1}\,\overline{z} + \overline{a_0} && \text{since the coefficients are real} \\
&= \overline{a_n z^n} + \overline{a_{n-1} z^{n-1}} + \ldots + \overline{a_2 z^2} + \overline{a_1 z} + \overline{a_0} && \text{since } \overline{z}\,\overline{w} = \overline{zw} \\
&= \overline{a_n z^n + a_{n-1} z^{n-1} + \ldots + a_2 z^2 + a_1 z + a_0} && \text{since } \overline{z} + \overline{w} = \overline{z+w} \\
&= \overline{f(z)} \\
&= \overline{0} \\
&= 0
\end{aligned}$$

This shows that $\overline{z}$ is a zero of f. So, if f is a polynomial function with real number coefficients, Theorem 3.15 tells us that if $a+bi$ is a nonreal zero of f, then so is $a-bi$. In other words, nonreal zeros of f come in conjugate pairs. The Factor Theorem kicks in to give us both $(x - [a+bi])$ and $(x - [a-bi])$ as factors of $f(x)$ which means $(x - [a+bi])(x - [a-bi]) = x^2 + 2ax + (a^2 + b^2)$ is an irreducible quadratic factor of f. As a result, we have our last theorem of the section.

Theorem 3.16. Real Factorization Theorem: Suppose f is a polynomial function with real number coefficients. Then $f(x)$ can be factored into a product of linear factors corresponding to the real zeros of f and irreducible quadratic factors which give the nonreal zeros of f.

We now present an example which pulls together all of the major ideas of this section.

Example 3.4.3. Let $f(x) = x^4 + 64$.

1. Use synthetic division to show that $x = 2 + 2i$ is a zero of f.

2. Find the remaining complex zeros of f.

3. Completely factor $f(x)$ over the complex numbers.

4. Completely factor $f(x)$ over the real numbers.

Solution.

1. Remembering to insert the 0's in the synthetic division tableau we have

$$\begin{array}{r|rrrrr}
2+2i & 1 & 0 & 0 & 0 & 64 \\
& \downarrow & 2+2i & 8i & -16+16i & -64 \\
\hline
& 1 & 2+2i & 8i & -16+16i & \boxed{0}
\end{array}$$

2. Since f is a fourth degree polynomial, we need to make two successful divisions to get a quadratic quotient. Since $2 + 2i$ is a zero, we know from Theorem 3.15 that $2 - 2i$ is also a zero. We continue our synthetic division tableau.

3.4 COMPLEX ZEROS AND THE FUNDAMENTAL THEOREM OF ALGEBRA

$$
\begin{array}{r|rrrrr}
2+2i & 1 & 0 & 0 & 0 & 64 \\
 & \downarrow & 2+2i & 8i & -16+16i & -64 \\ \hline
2-2i & 1 & 2+2i & 8i & -16+16i & \boxed{0} \\
 & \downarrow & 2-2i & 8-8i & 16-16i & \\ \hline
 & 1 & 4 & 8 & \boxed{0} &
\end{array}
$$

Our quotient polynomial is x^2+4x+8. Using the quadratic formula, we obtain the remaining zeros $-2+2i$ and $-2-2i$.

3. Using Theorem 3.14, we get $f(x) = (x - [2-2i])(x - [2+2i])(x - [-2+2i])(x - [-2-2i])$.

4. We multiply the linear factors of $f(x)$ which correspond to complex conjugate pairs. We find $(x - [2-2i])(x - [2+2i]) = x^2 - 4x + 8$, and $(x - [-2+2i])(x - [-2-2i]) = x^2 + 4x + 8$. Our final answer is $f(x) = \left(x^2 - 4x + 8\right)\left(x^2 + 4x + 8\right)$. □

Our last example turns the tables and asks us to manufacture a polynomial with certain properties of its graph and zeros.

Example 3.4.4. Find a polynomial p of lowest degree that has integer coefficients and satisfies all of the following criteria:

- the graph of $y = p(x)$ touches (but doesn't cross) the x-axis at $\left(\frac{1}{3}, 0\right)$
- $x = 3i$ is a zero of p.
- as $x \to -\infty$, $p(x) \to -\infty$
- as $x \to \infty$, $p(x) \to -\infty$

Solution. To solve this problem, we will need a good understanding of the relationship between the x-intercepts of the graph of a function and the zeros of a function, the Factor Theorem, the role of multiplicity, complex conjugates, the Complex Factorization Theorem, and end behavior of polynomial functions. (In short, you'll need most of the major concepts of this chapter.) Since the graph of p touches the x-axis at $\left(\frac{1}{3}, 0\right)$, we know $x = \frac{1}{3}$ is a zero of even multiplicity. Since we are after a polynomial of lowest degree, we need $x = \frac{1}{3}$ to have multiplicity exactly 2. The Factor Theorem now tells us $\left(x - \frac{1}{3}\right)^2$ is a factor of $p(x)$. Since $x = 3i$ is a zero and our final answer is to have integer (real) coefficients, $x = -3i$ is also a zero. The Factor Theorem kicks in again to give us $(x-3i)$ and $(x+3i)$ as factors of $p(x)$. We are given no further information about zeros or intercepts so we conclude, by the Complex Factorization Theorem that $p(x) = a\left(x - \frac{1}{3}\right)^2 (x - 3i)(x + 3i)$ for some real number a. Expanding this, we get $p(x) = ax^4 - \frac{2a}{3}x^3 + \frac{82a}{9}x^2 - 6ax + a$. In order to obtain integer coefficients, we know a must be an integer multiple of 9. Our last concern is end behavior. Since the leading term of $p(x)$ is ax^4, we need $a < 0$ to get $p(x) \to -\infty$ as $x \to \pm\infty$. Hence, if we choose $x = -9$, we get $p(x) = -9x^4 + 6x^3 - 82x^2 + 54x - 9$. We can verify our handiwork using the techniques developed in this chapter. □

This example concludes our study of polynomial functions.[10] The last few sections have contained what is considered by many to be 'heavy' Mathematics. Like a heavy meal, heavy Mathematics takes time to digest. Don't be overly concerned if it doesn't seem to sink in all at once, and pace yourself in the Exercises or you're liable to get mental cramps. But before we get to the Exercises, we'd like to offer a bit of an epilogue.

Our main goal in presenting the material on the complex zeros of a polynomial was to give the chapter a sense of completeness. Given that it can be shown that some polynomials have real zeros which cannot be expressed using the usual algebraic operations, and still others have no real zeros at all, it was nice to discover that every polynomial of degree $n \geq 1$ has n complex zeros. So like we said, it gives us a sense of closure. But the observant reader will note that we did not give any examples of applications which involve complex numbers. Students often wonder when complex numbers will be used in 'real-world' applications. After all, didn't we call i the imaginary unit? How can imaginary things be used in reality? It turns out that complex numbers are very useful in many applied fields such as fluid dynamics, electromagnetism and quantum mechanics, but most of the applications require Mathematics well beyond College Algebra to fully understand them. That does not mean you'll never be be able to understand them; in fact, it is the authors' sincere hope that all of you will reach a point in your studies when the glory, awe and splendor of complex numbers are revealed to you. For now, however, the really good stuff is beyond the scope of this text. We invite you and your classmates to find a few examples of complex number applications and see what you can make of them. A simple Internet search with the phrase 'complex numbers in real life' should get you started. Basic electronics classes are another place to look, but remember, they might use the letter j where we have used i.

For the remainder of the text, with the exception of Section 11.7 and a few exploratory exercises scattered about, we will restrict our attention to real numbers. We do this primarily because the first Calculus sequence you will take, ostensibly the one that this text is preparing you for, studies only functions of real variables. Also, lots of really cool scientific things don't require any deep understanding of complex numbers to study them, but they do need more Mathematics like exponential, logarithmic and trigonometric functions. We believe it makes more sense pedagogically for you to learn about those functions now then take a course in Complex Function Theory in your junior or senior year once you've completed the Calculus sequence. It is in that course that the true power of the complex numbers is released. But for now, in order to fully prepare you for life immediately after College Algebra, we will say that functions like $f(x) = \frac{1}{x^2+1}$ have a domain of all real numbers, even though we know $x^2 + 1 = 0$ has two complex solutions, namely $x = \pm i$. Because $x^2 + 1 > 0$ for all *real* numbers x, the fraction $\frac{1}{x^2+1}$ is never undefined in the real variable setting.

---

[10] With the exception of the Exercises on the next page, of course.

3.4.1 Exercises

In Exercises 1 - 10, use the given complex numbers z and w to find and simplify the following. Write your answers in the form $a + bi$.

- $z + w$
- zw
- z^2
- $\dfrac{1}{z}$
- $\dfrac{z}{w}$
- $\dfrac{w}{z}$
- $\bar{z}$
- $z\bar{z}$
- $(\bar{z})^2$

1. $z = 2 + 3i$, $w = 4i$
2. $z = 1 + i$, $w = -i$
3. $z = i$, $w = -1 + 2i$
4. $z = 4i$, $w = 2 - 2i$
5. $z = 3 - 5i$, $w = 2 + 7i$
6. $z = -5 + i$, $w = 4 + 2i$
7. $z = \sqrt{2} - i\sqrt{2}$, $w = \sqrt{2} + i\sqrt{2}$
8. $z = 1 - i\sqrt{3}$, $w = -1 - i\sqrt{3}$
9. $z = \dfrac{1}{2} + \dfrac{\sqrt{3}}{2}i$, $w = -\dfrac{1}{2} + \dfrac{\sqrt{3}}{2}i$
10. $z = -\dfrac{\sqrt{2}}{2} + \dfrac{\sqrt{2}}{2}i$, $w = -\dfrac{\sqrt{2}}{2} - \dfrac{\sqrt{2}}{2}i$

In Exercises 11 - 18, simplify the quantity.

11. $\sqrt{-49}$
12. $\sqrt{-9}$
13. $\sqrt{-25}\sqrt{-4}$
14. $\sqrt{(-25)(-4)}$
15. $\sqrt{-9}\sqrt{-16}$
16. $\sqrt{(-9)(-16)}$
17. $\sqrt{-(-9)}$
18. $-\sqrt{(-9)}$

We know that $i^2 = -1$ which means $i^3 = i^2 \cdot i = (-1) \cdot i = -i$ and $i^4 = i^2 \cdot i^2 = (-1)(-1) = 1$. In Exercises 19 - 26, use this information to simplify the given power of i.

19. i^5
20. i^6
21. i^7
22. i^8
23. i^{15}
24. i^{26}
25. i^{117}
26. i^{304}

In Exercises 27 - 48, find all of the zeros of the polynomial then completely factor it over the real numbers and completely factor it over the complex numbers.

27. $f(x) = x^2 - 4x + 13$
28. $f(x) = x^2 - 2x + 5$
29. $f(x) = 3x^2 + 2x + 10$
30. $f(x) = x^3 - 2x^2 + 9x - 18$
31. $f(x) = x^3 + 6x^2 + 6x + 5$
32. $f(x) = 3x^3 - 13x^2 + 43x - 13$

33. $f(x) = x^3 + 3x^2 + 4x + 12$

34. $f(x) = 4x^3 - 6x^2 - 8x + 15$

35. $f(x) = x^3 + 7x^2 + 9x - 2$

36. $f(x) = 9x^3 + 2x + 1$

37. $f(x) = 4x^4 - 4x^3 + 13x^2 - 12x + 3$

38. $f(x) = 2x^4 - 7x^3 + 14x^2 - 15x + 6$

39. $f(x) = x^4 + x^3 + 7x^2 + 9x - 18$

40. $f(x) = 6x^4 + 17x^3 - 55x^2 + 16x + 12$

41. $f(x) = -3x^4 - 8x^3 - 12x^2 - 12x - 5$

42. $f(x) = 8x^4 + 50x^3 + 43x^2 + 2x - 4$

43. $f(x) = x^4 + 9x^2 + 20$

44. $f(x) = x^4 + 5x^2 - 24$

45. $f(x) = x^5 - x^4 + 7x^3 - 7x^2 + 12x - 12$

46. $f(x) = x^6 - 64$

47. $f(x) = x^4 - 2x^3 + 27x^2 - 2x + 26$ (Hint: $x = i$ is one of the zeros.)

48. $f(x) = 2x^4 + 5x^3 + 13x^2 + 7x + 5$ (Hint: $x = -1 + 2i$ is a zero.)

In Exercises 49 - 53, create a polynomial f with real number coefficients which has all of the desired characteristics. You may leave the polynomial in factored form.

49.
- The zeros of f are $c = \pm 1$ and $c = \pm i$
- The leading term of $f(x)$ is $42x^4$

50.
- $c = 2i$ is a zero.
- the point $(-1, 0)$ is a local minimum on the graph of $y = f(x)$
- the leading term of $f(x)$ is $117x^4$

51.
- The solutions to $f(x) = 0$ are $x = \pm 2$ and $x = \pm 7i$
- The leading term of $f(x)$ is $-3x^5$
- The point $(2, 0)$ is a local maximum on the graph of $y = f(x)$.

52.
- f is degree 5.
- $x = 6$, $x = i$ and $x = 1 - 3i$ are zeros of f
- as $x \to -\infty$, $f(x) \to \infty$

53.
- The leading term of $f(x)$ is $-2x^3$
- $c = 2i$ is a zero
- $f(0) = -16$

54. Let z and w be arbitrary complex numbers. Show that $\overline{z}\,\overline{w} = \overline{zw}$ and $\overline{\overline{z}} = z$.

3.4 Complex Zeros and the Fundamental Theorem of Algebra

3.4.2 Answers

1. For $z = 2 + 3i$ and $w = 4i$

   - $z + w = 2 + 7i$
   - $zw = -12 + 8i$
   - $z^2 = -5 + 12i$
   - $\frac{1}{z} = \frac{2}{13} - \frac{3}{13}i$
   - $\frac{z}{w} = \frac{3}{4} - \frac{1}{2}i$
   - $\frac{w}{z} = \frac{12}{13} + \frac{8}{13}i$
   - $\overline{z} = 2 - 3i$
   - $z\overline{z} = 13$
   - $(\overline{z})^2 = -5 - 12i$

2. For $z = 1 + i$ and $w = -i$

   - $z + w = 1$
   - $zw = 1 - i$
   - $z^2 = 2i$
   - $\frac{1}{z} = \frac{1}{2} - \frac{1}{2}i$
   - $\frac{z}{w} = -1 + i$
   - $\frac{w}{z} = -\frac{1}{2} - \frac{1}{2}i$
   - $\overline{z} = 1 - i$
   - $z\overline{z} = 2$
   - $(\overline{z})^2 = -2i$

3. For $z = i$ and $w = -1 + 2i$

   - $z + w = -1 + 3i$
   - $zw = -2 - i$
   - $z^2 = -1$
   - $\frac{1}{z} = -i$
   - $\frac{z}{w} = \frac{2}{5} - \frac{1}{5}i$
   - $\frac{w}{z} = 2 + i$
   - $\overline{z} = -i$
   - $z\overline{z} = 1$
   - $(\overline{z})^2 = -1$

4. For $z = 4i$ and $w = 2 - 2i$

   - $z + w = 2 + 2i$
   - $zw = 8 + 8i$
   - $z^2 = -16$
   - $\frac{1}{z} = -\frac{1}{4}i$
   - $\frac{z}{w} = -1 + i$
   - $\frac{w}{z} = -\frac{1}{2} - \frac{1}{2}i$
   - $\overline{z} = -4i$
   - $z\overline{z} = 16$
   - $(\overline{z})^2 = -16$

5. For $z = 3 - 5i$ and $w = 2 + 7i$

   - $z + w = 5 + 2i$
   - $zw = 41 + 11i$
   - $z^2 = -16 - 30i$
   - $\frac{1}{z} = \frac{3}{34} + \frac{5}{34}i$
   - $\frac{z}{w} = -\frac{29}{53} - \frac{31}{53}i$
   - $\frac{w}{z} = -\frac{29}{34} + \frac{31}{34}i$
   - $\overline{z} = 3 + 5i$
   - $z\overline{z} = 34$
   - $(\overline{z})^2 = -16 + 30i$

6. For $z = -5 + i$ and $w = 4 + 2i$

   - $z + w = -1 + 3i$
   - $\frac{1}{z} = -\frac{5}{26} - \frac{1}{26}i$
   - $\overline{z} = -5 - i$
   - $zw = -22 - 6i$
   - $\frac{z}{w} = -\frac{9}{10} + \frac{7}{10}i$
   - $z\overline{z} = 26$
   - $z^2 = 24 - 10i$
   - $\frac{w}{z} = -\frac{9}{13} - \frac{7}{13}i$
   - $(\overline{z})^2 = 24 + 10i$

7. For $z = \sqrt{2} - i\sqrt{2}$ and $w = \sqrt{2} + i\sqrt{2}$

   - $z + w = 2\sqrt{2}$
   - $\frac{1}{z} = \frac{\sqrt{2}}{4} + \frac{\sqrt{2}}{4}i$
   - $\overline{z} = \sqrt{2} + i\sqrt{2}$
   - $zw = 4$
   - $\frac{z}{w} = -i$
   - $z\overline{z} = 4$
   - $z^2 = -4i$
   - $\frac{w}{z} = i$
   - $(\overline{z})^2 = 4i$

8. For $z = 1 - i\sqrt{3}$ and $w = -1 - i\sqrt{3}$

   - $z + w = -2i\sqrt{3}$
   - $\frac{1}{z} = \frac{1}{4} + \frac{\sqrt{3}}{4}i$
   - $\overline{z} = 1 + i\sqrt{3}$
   - $zw = -4$
   - $\frac{z}{w} = \frac{1}{2} + \frac{\sqrt{3}}{2}i$
   - $z\overline{z} = 4$
   - $z^2 = -2 - 2i\sqrt{3}$
   - $\frac{w}{z} = \frac{1}{2} - \frac{\sqrt{3}}{2}i$
   - $(\overline{z})^2 = -2 + 2i\sqrt{3}$

9. For $z = \frac{1}{2} + \frac{\sqrt{3}}{2}i$ and $w = -\frac{1}{2} + \frac{\sqrt{3}}{2}i$

   - $z + w = i\sqrt{3}$
   - $\frac{1}{z} = \frac{1}{2} - \frac{\sqrt{3}}{2}i$
   - $\overline{z} = \frac{1}{2} - \frac{\sqrt{3}}{2}i$
   - $zw = -1$
   - $\frac{z}{w} = \frac{1}{2} - \frac{\sqrt{3}}{2}i$
   - $z\overline{z} = 1$
   - $z^2 = -\frac{1}{2} + \frac{\sqrt{3}}{2}i$
   - $\frac{w}{z} = \frac{1}{2} + \frac{\sqrt{3}}{2}i$
   - $(\overline{z})^2 = -\frac{1}{2} - \frac{\sqrt{3}}{2}i$

10. For $z = -\frac{\sqrt{2}}{2} + \frac{\sqrt{2}}{2}i$ and $w = -\frac{\sqrt{2}}{2} - \frac{\sqrt{2}}{2}i$

    - $-\sqrt{2}$
    - $\frac{1}{z} = -\frac{\sqrt{2}}{2} - \frac{\sqrt{2}}{2}i$
    - $\overline{z} = -\frac{\sqrt{2}}{2} - \frac{\sqrt{2}}{2}i$
    - $zw = 1$
    - $\frac{z}{w} = -i$
    - $z\overline{z} = 1$
    - $z^2 = -i$
    - $\frac{w}{z} = i$
    - $(\overline{z})^2 = i$

11. $7i$ 12. $3i$ 13. -10 14. 10

3.4 COMPLEX ZEROS AND THE FUNDAMENTAL THEOREM OF ALGEBRA

15. -12 16. 12 17. 3 18. $-3i$

19. $i^5 = i^4 \cdot i = 1 \cdot i = i$ 20. $i^6 = i^4 \cdot i^2 = 1 \cdot (-1) = -1$

21. $i^7 = i^4 \cdot i^3 = 1 \cdot (-i) = -i$ 22. $i^8 = i^4 \cdot i^4 = (i^4)^2 = (1)^2 = 1$

23. $i^{15} = (i^4)^3 \cdot i^3 = 1 \cdot (-i) = -i$ 24. $i^{26} = (i^4)^6 \cdot i^2 = 1 \cdot (-1) = -1$

25. $i^{117} = (i^4)^{29} \cdot i = 1 \cdot i = i$ 26. $i^{304} = (i^4)^{76} = 1^{76} = 1$

27. $f(x) = x^2 - 4x + 13 = (x - (2+3i))(x - (2-3i))$
Zeros: $x = 2 \pm 3i$

28. $f(x) = x^2 - 2x + 5 = (x - (1+2i))(x - (1-2i))$
Zeros: $x = 1 \pm 2i$

29. $f(x) = 3x^2 + 2x + 10 = 3\left(x - \left(-\frac{1}{3} + \frac{\sqrt{29}}{3}i\right)\right)\left(x - \left(-\frac{1}{3} - \frac{\sqrt{29}}{3}i\right)\right)$
Zeros: $x = -\frac{1}{3} \pm \frac{\sqrt{29}}{3}i$

30. $f(x) = x^3 - 2x^2 + 9x - 18 = (x-2)(x^2+9) = (x-2)(x-3i)(x+3i)$
Zeros: $x = 2, \pm 3i$

31. $f(x) = x^3 + 6x^2 + 6x + 5 = (x+5)(x^2+x+1) = (x+5)\left(x - \left(-\frac{1}{2} + \frac{\sqrt{3}}{2}i\right)\right)\left(x - \left(-\frac{1}{2} - \frac{\sqrt{3}}{2}i\right)\right)$
Zeros: $x = -5$, $x = -\frac{1}{2} \pm \frac{\sqrt{3}}{2}i$

32. $f(x) = 3x^3 - 13x^2 + 43x - 13 = (3x-1)(x^2-4x+13) = (3x-1)(x-(2+3i))(x-(2-3i))$
Zeros: $x = \frac{1}{3}$, $x = 2 \pm 3i$

33. $f(x) = x^3 + 3x^2 + 4x + 12 = (x+3)(x^2+4) = (x+3)(x+2i)(x-2i)$
Zeros: $x = -3, \pm 2i$

34. $f(x) = 4x^3 - 6x^2 - 8x + 15 = \left(x + \frac{3}{2}\right)\left(4x^2 - 12x + 10\right)$
$= 4\left(x + \frac{3}{2}\right)\left(x - \left(\frac{3}{2} + \frac{1}{2}i\right)\right)\left(x - \left(\frac{3}{2} - \frac{1}{2}i\right)\right)$
Zeros: $x = -\frac{3}{2}$, $x = \frac{3}{2} \pm \frac{1}{2}i$

35. $f(x) = x^3 + 7x^2 + 9x - 2 = (x+2)\left(x - \left(-\frac{5}{2} + \frac{\sqrt{29}}{2}\right)\right)\left(x - \left(-\frac{5}{2} - \frac{\sqrt{29}}{2}\right)\right)$
Zeros: $x = -2$, $x = -\frac{5}{2} \pm \frac{\sqrt{29}}{2}$

36. $f(x) = 9x^3 + 2x + 1 = \left(x + \frac{1}{3}\right)\left(9x^2 - 3x + 3\right)$
$= 9\left(x + \frac{1}{3}\right)\left(x - \left(\frac{1}{6} + \frac{\sqrt{11}}{6}i\right)\right)\left(x - \left(\frac{1}{6} - \frac{\sqrt{11}}{6}i\right)\right)$
Zeros: $x = -\frac{1}{3}$, $x = \frac{1}{6} \pm \frac{\sqrt{11}}{6}i$

37. $f(x) = 4x^4 - 4x^3 + 13x^2 - 12x + 3 = \left(x - \frac{1}{2}\right)^2 (4x^2 + 12) = 4\left(x - \frac{1}{2}\right)^2 (x + i\sqrt{3})(x - i\sqrt{3})$
Zeros: $x = \frac{1}{2}$, $x = \pm\sqrt{3}i$

38. $f(x) = 2x^4 - 7x^3 + 14x^2 - 15x + 6 = (x-1)^2 \left(2x^2 - 3x + 6\right)$
    $= 2(x-1)^2 \left(x - \left(\frac{3}{4} + \frac{\sqrt{39}}{4}i\right)\right)\left(x - \left(\frac{3}{4} - \frac{\sqrt{39}}{4}i\right)\right)$
    Zeros: $x = 1$, $x = \frac{3}{4} \pm \frac{\sqrt{39}}{4}i$

39. $f(x) = x^4 + x^3 + 7x^2 + 9x - 18 = (x+2)(x-1)\left(x^2 + 9\right) = (x+2)(x-1)(x+3i)(x-3i)$
    Zeros: $x = -2,\ 1,\ \pm 3i$

40. $f(x) = 6x^4 + 17x^3 - 55x^2 + 16x + 12 = 6\left(x + \frac{1}{3}\right)\left(x - \frac{3}{2}\right)\left(x - (-2 + 2\sqrt{2})\right)\left(x - (-2 - 2\sqrt{2})\right)$
    Zeros: $x = -\frac{1}{3}$, $x = \frac{3}{2}$, $x = -2 \pm 2\sqrt{2}$

41. $f(x) = -3x^4 - 8x^3 - 12x^2 - 12x - 5 = (x+1)^2\left(-3x^2 - 2x - 5\right)$
    $= -3(x+1)^2 \left(x - \left(-\frac{1}{3} + \frac{\sqrt{14}}{3}i\right)\right)\left(x - \left(-\frac{1}{3} - \frac{\sqrt{14}}{3}i\right)\right)$
    Zeros: $x = -1$, $x = -\frac{1}{3} \pm \frac{\sqrt{14}}{3}i$

42. $f(x) = 8x^4 + 50x^3 + 43x^2 + 2x - 4 = 8\left(x + \frac{1}{2}\right)\left(x - \frac{1}{4}\right)(x - (-3 + \sqrt{5}))(x - (-3 - \sqrt{5}))$
    Zeros: $x = -\frac{1}{2},\ \frac{1}{4}$, $x = -3 \pm \sqrt{5}$

43. $f(x) = x^4 + 9x^2 + 20 = \left(x^2 + 4\right)\left(x^2 + 5\right) = (x - 2i)(x + 2i)\left(x - i\sqrt{5}\right)\left(x + i\sqrt{5}\right)$
    Zeros: $x = \pm 2i, \pm i\sqrt{5}$

44. $f(x) = x^4 + 5x^2 - 24 = \left(x^2 - 3\right)\left(x^2 + 8\right) = (x - \sqrt{3})(x + \sqrt{3})\left(x - 2i\sqrt{2}\right)\left(x + 2i\sqrt{2}\right)$
    Zeros: $x = \pm\sqrt{3}, \pm 2i\sqrt{2}$

45. $f(x) = x^5 - x^4 + 7x^3 - 7x^2 + 12x - 12 = (x - 1)\left(x^2 + 3\right)\left(x^2 + 4\right)$
    $= (x-1)(x - i\sqrt{3})(x + i\sqrt{3})(x - 2i)(x + 2i)$
    Zeros: $x = 1,\ \pm\sqrt{3}i,\ \pm 2i$

46. $f(x) = x^6 - 64 = (x-2)(x+2)\left(x^2 + 2x + 4\right)\left(x^2 - 2x + 4\right)$
    $= (x-2)(x+2)\left(x - (-1 + i\sqrt{3})\right)\left(x - (-1 - i\sqrt{3})\right)\left(x - (1 + i\sqrt{3})\right)\left(x - (1 - i\sqrt{3})\right)$
    Zeros: $x = \pm 2$, $x = -1 \pm i\sqrt{3}$, $x = 1 \pm i\sqrt{3}$

47. $f(x) = x^4 - 2x^3 + 27x^2 - 2x + 26 = (x^2 - 2x + 26)(x^2 + 1) = (x - (1+5i))(x - (1-5i))(x+i)(x-i)$
    Zeros: $x = 1 \pm 5i$, $x = \pm i$

48. $f(x) = 2x^4 + 5x^3 + 13x^2 + 7x + 5 = \left(x^2 + 2x + 5\right)\left(2x^2 + x + 1\right)$
    $= 2(x - (-1 + 2i))(x - (-1 - 2i))\left(x - \left(-\frac{1}{4} + i\frac{\sqrt{7}}{4}\right)\right)\left(x - \left(-\frac{1}{4} - i\frac{\sqrt{7}}{4}\right)\right)$
    Zeros: $x = -1 \pm 2i, -\frac{1}{4} \pm i\frac{\sqrt{7}}{4}$

49. $f(x) = 42(x-1)(x+1)(x-i)(x+i)$

50. $f(x) = 117(x+1)^2(x-2i)(x+2i)$

51. $f(x) = -3(x-2)^2(x+2)(x-7i)(x+7i)$

52. $f(x) = a(x-6)(x-i)(x+i)(x - (1 - 3i))(x - (1 + 3i))$ where a is any real number, $a < 0$

53. $f(x) = -2(x - 2i)(x + 2i)(x + 2)$

Chapter 4

Rational Functions

4.1 Introduction to Rational Functions

If we add, subtract or multiply polynomial functions according to the function arithmetic rules defined in Section 1.5, we will produce another polynomial function. If, on the other hand, we divide two polynomial functions, the result may not be a polynomial. In this chapter we study **rational functions** - functions which are ratios of polynomials.

> **Definition 4.1.** A **rational function** is a function which is the ratio of polynomial functions. Said differently, r is a rational function if it is of the form
> $$r(x) = \frac{p(x)}{q(x)},$$
> where p and q are polynomial functions.[a]
>
> ---
> [a] According to this definition, all polynomial functions are also rational functions. (Take $q(x) = 1$).

As we recall from Section 1.4, we have domain issues anytime the denominator of a fraction is zero. In the example below, we review this concept as well as some of the arithmetic of rational expressions.

Example 4.1.1. Find the domain of the following rational functions. Write them in the form $\frac{p(x)}{q(x)}$ for polynomial functions p and q and simplify.

1. $f(x) = \dfrac{2x - 1}{x + 1}$

2. $g(x) = 2 - \dfrac{3}{x + 1}$

3. $h(x) = \dfrac{2x^2 - 1}{x^2 - 1} - \dfrac{3x - 2}{x^2 - 1}$

4. $r(x) = \dfrac{2x^2 - 1}{x^2 - 1} \div \dfrac{3x - 2}{x^2 - 1}$

Solution.

1. To find the domain of f, we proceed as we did in Section 1.4: we find the zeros of the denominator and exclude them from the domain. Setting $x + 1 = 0$ results in $x = -1$. Hence,

our domain is $(-\infty, -1) \cup (-1, \infty)$. The expression $f(x)$ is already in the form requested and when we check for common factors among the numerator and denominator we find none, so we are done.

2. Proceeding as before, we determine the domain of g by solving $x + 1 = 0$. As before, we find the domain of g is $(-\infty, -1) \cup (-1, \infty)$. To write $g(x)$ in the form requested, we need to get a common denominator

$$\begin{aligned} g(x) &= 2 - \frac{3}{x+1} = \frac{2}{1} - \frac{3}{x+1} = \frac{(2)(x+1)}{(1)(x+1)} - \frac{3}{x+1} \\ &= \frac{(2x+2) - 3}{x+1} = \frac{2x-1}{x+1} \end{aligned}$$

This formula is now completely simplified.

3. The denominators in the formula for $h(x)$ are both $x^2 - 1$ whose zeros are $x = \pm 1$. As a result, the domain of h is $(-\infty, -1) \cup (-1, 1) \cup (1, \infty)$. We now proceed to simplify $h(x)$. Since we have the same denominator in both terms, we subtract the numerators. We then factor the resulting numerator and denominator, and cancel out the common factor.

$$\begin{aligned} h(x) &= \frac{2x^2 - 1}{x^2 - 1} - \frac{3x - 2}{x^2 - 1} = \frac{(2x^2 - 1) - (3x - 2)}{x^2 - 1} \\ &= \frac{2x^2 - 1 - 3x + 2}{x^2 - 1} = \frac{2x^2 - 3x + 1}{x^2 - 1} \\ &= \frac{(2x - 1)(x - 1)}{(x + 1)(x - 1)} = \frac{(2x - 1)\cancel{(x-1)}}{(x+1)\cancel{(x-1)}} \\ &= \frac{2x - 1}{x + 1} \end{aligned}$$

4. To find the domain of r, it may help to temporarily rewrite $r(x)$ as

$$r(x) = \frac{\dfrac{2x^2 - 1}{x^2 - 1}}{\dfrac{3x - 2}{x^2 - 1}}$$

We need to set all of the denominators equal to zero which means we need to solve not only $x^2 - 1 = 0$, but also $\frac{3x-2}{x^2-1} = 0$. We find $x = \pm 1$ for the former and $x = \frac{2}{3}$ for the latter. Our domain is $(-\infty, -1) \cup \left(-1, \frac{2}{3}\right) \cup \left(\frac{2}{3}, 1\right) \cup (1, \infty)$. We simplify $r(x)$ by rewriting the division as multiplication by the reciprocal and then by canceling the common factor

4.1 Introduction to Rational Functions

$$r(x) = \frac{2x^2-1}{x^2-1} \div \frac{3x-2}{x^2-1} = \frac{2x^2-1}{x^2-1} \cdot \frac{x^2-1}{3x-2} = \frac{(2x^2-1)(x^2-1)}{(x^2-1)(3x-2)}$$

$$= \frac{(2x^2-1)\cancel{(x^2-1)}}{\cancel{(x^2-1)}(3x-2)} = \frac{2x^2-1}{3x-2} \qquad \square$$

A few remarks about Example 4.1.1 are in order. Note that the expressions for $f(x)$, $g(x)$ and $h(x)$ work out to be the same. However, only two of these functions are actually equal. Recall that functions are ultimately sets of ordered pairs,[1] so for two functions to be equal, they need, among other things, to have the same domain. Since $f(x) = g(x)$ and f and g have the same domain, they are equal functions. Even though the formula $h(x)$ is the same as $f(x)$, the domain of h is different than the domain of f, and thus they are different functions.

We now turn our attention to the graphs of rational functions. Consider the function $f(x) = \frac{2x-1}{x+1}$ from Example 4.1.1. Using a graphing calculator, we obtain

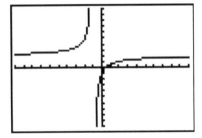

Two behaviors of the graph are worthy of further discussion. First, note that the graph appears to 'break' at $x = -1$. We know from our last example that $x = -1$ is not in the domain of f which means $f(-1)$ is undefined. When we make a table of values to study the behavior of f near $x = -1$ we see that we can get 'near' $x = -1$ from two directions. We can choose values a little less than -1, for example $x = -1.1$, $x = -1.01$, $x = -1.001$, and so on. These values are said to 'approach -1 from the *left*.' Similarly, the values $x = -0.9$, $x = -0.99$, $x = -0.999$, etc., are said to 'approach -1 from the *right*.' If we make two tables, we find that the numerical results confirm what we see graphically.

x	$f(x)$	$(x, f(x))$
-1.1	32	$(-1.1, 32)$
-1.01	302	$(-1.01, 302)$
-1.001	3002	$(-1.001, 3002)$
-1.0001	30002	$(-1.001, 30002)$

x	$f(x)$	$(x, f(x))$
-0.9	-28	$(-0.9, -28)$
-0.99	-298	$(-0.99, -298)$
-0.999	-2998	$(-0.999, -2998)$
-0.9999	-29998	$(-0.9999, -29998)$

As the x values approach -1 from the left, the function values become larger and larger positive numbers.[2] We express this symbolically by stating as $x \to -1^-$, $f(x) \to \infty$. Similarly, using analogous notation, we conclude from the table that as $x \to -1^+$, $f(x) \to -\infty$. For this type of

---

[1] You should review Sections 1.2 and 1.3 if this statement caught you off guard.
[2] We would need Calculus to confirm this analytically.

unbounded behavior, we say the graph of $y = f(x)$ has a **vertical asymptote** of $x = -1$. Roughly speaking, this means that near $x = -1$, the graph looks very much like the vertical line $x = -1$.

The other feature worthy of note about the graph of $y = f(x)$ is that it seems to 'level off' on the left and right hand sides of the screen. This is a statement about the end behavior of the function. As we discussed in Section 3.1, the end behavior of a function is its behavior as x attains larger[3] and larger negative values without bound, $x \to -\infty$, and as x becomes large without bound, $x \to \infty$. Making tables of values, we find

x	$f(x)$	$(x, f(x))$
-10	≈ 2.3333	$\approx (-10, 2.3333)$
-100	≈ 2.0303	$\approx (-100, 2.0303)$
-1000	≈ 2.0030	$\approx (-1000, 2.0030)$
-10000	≈ 2.0003	$\approx (-10000, 2.0003)$

x	$f(x)$	$(x, f(x))$
10	≈ 1.7273	$\approx (10, 1.7273)$
100	≈ 1.9703	$\approx (100, 1.9703)$
1000	≈ 1.9970	$\approx (1000, 1.9970)$
10000	≈ 1.9997	$\approx (10000, 1.9997)$

From the tables, we see that as $x \to -\infty$, $f(x) \to 2^+$ and as $x \to \infty$, $f(x) \to 2^-$. Here the '+' means 'from above' and the '−' means 'from below'. In this case, we say the graph of $y = f(x)$ has a **horizontal asymptote** of $y = 2$. This means that the end behavior of f resembles the horizontal line $y = 2$, which explains the 'leveling off' behavior we see in the calculator's graph. We formalize the concepts of vertical and horizontal asymptotes in the following definitions.

Definition 4.2. The line $x = c$ is called a **vertical asymptote** of the graph of a function $y = f(x)$ if as $x \to c^-$ or as $x \to c^+$, either $f(x) \to \infty$ or $f(x) \to -\infty$.

Definition 4.3. The line $y = c$ is called a **horizontal asymptote** of the graph of a function $y = f(x)$ if as $x \to -\infty$ or as $x \to \infty$, $f(x) \to c$.

Note that in Definition 4.3, we write $f(x) \to c$ (not $f(x) \to c^+$ or $f(x) \to c^-$) because we are unconcerned from which direction the values $f(x)$ approach the value c, just as long as they do so.[4] In our discussion following Example 4.1.1, we determined that, despite the fact that the formula for $h(x)$ reduced to the same formula as $f(x)$, the functions f and h are different, since $x = 1$ is in the domain of f, but $x = 1$ is not in the domain of h. If we graph $h(x) = \frac{2x^2-1}{x^2-1} - \frac{3x-2}{x^2-1}$ using a graphing calculator, we are surprised to find that the graph looks identical to the graph of $y = f(x)$. There is a vertical asymptote at $x = -1$, but near $x = 1$, everything seem fine. Tables of values provide numerical evidence which supports the graphical observation.

---

[3]Here, the word 'larger' means larger in absolute value.

[4]As we shall see in the next section, the graphs of rational functions may, in fact, *cross* their horizontal asymptotes. If this happens, however, it does so only a *finite* number of times, and so for each choice of $x \to -\infty$ and $x \to \infty$, $f(x)$ will approach c from either below (in the case $f(x) \to c^-$) or above (in the case $f(x) \to c^+$.) We leave $f(x) \to c$ generic in our definition, however, to allow this concept to apply to less tame specimens in the Precalculus zoo, such as Exercise 50 in Section 10.5.

4.1 Introduction to Rational Functions

x	$h(x)$	$(x, h(x))$
0.9	≈ 0.4210	$\approx (0.9, 0.4210)$
0.99	≈ 0.4925	$\approx (0.99, 0.4925)$
0.999	≈ 0.4992	$\approx (0.999, 0.4992)$
0.9999	≈ 0.4999	$\approx (0.9999, 0.4999)$

x	$h(x)$	$(x, h(x))$
1.1	≈ 0.5714	$\approx (1.1, 0.5714)$
1.01	≈ 0.5075	$\approx (1.01, 0.5075)$
1.001	≈ 0.5007	$\approx (1.001, 0.5007)$
1.0001	≈ 0.5001	$\approx (1.0001, 0.5001)$

We see that as $x \to 1^-$, $h(x) \to 0.5^-$ and as $x \to 1^+$, $h(x) \to 0.5^+$. In other words, the points on the graph of $y = h(x)$ are approaching $(1, 0.5)$, but since $x = 1$ is not in the domain of h, it would be inaccurate to fill in a point at $(1, 0.5)$. As we've done in past sections when something like this occurs,[5] we put an open circle (also called a **hole** in this case[6]) at $(1, 0.5)$. Below is a detailed graph of $y = h(x)$, with the vertical and horizontal asymptotes as dashed lines.

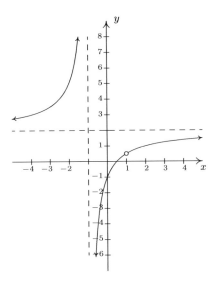

Neither $x = -1$ nor $x = 1$ are in the domain of h, yet the behavior of the graph of $y = h(x)$ is drastically different near these x-values. The reason for this lies in the second to last step when we simplified the formula for $h(x)$ in Example 4.1.1, where we had $h(x) = \frac{(2x-1)(x-1)}{(x+1)(x-1)}$. The reason $x = -1$ is not in the domain of h is because the factor $(x + 1)$ appears in the denominator of $h(x)$; similarly, $x = 1$ is not in the domain of h because of the factor $(x - 1)$ in the denominator of $h(x)$. The major difference between these two factors is that $(x - 1)$ cancels with a factor in the numerator whereas $(x + 1)$ does not. Loosely speaking, the trouble caused by $(x - 1)$ in the denominator is canceled away while the factor $(x + 1)$ remains to cause mischief. This is why the graph of $y = h(x)$ has a vertical asymptote at $x = -1$ but only a hole at $x = 1$. These observations are generalized and summarized in the theorem below, whose proof is found in Calculus.

---

[5] For instance, graphing piecewise defined functions in Section 1.6.
[6] In Calculus, we will see how these 'holes' can be 'plugged' when embarking on a more advanced study of continuity.

Theorem 4.1. Location of Vertical Asymptotes and Holes:[a] Suppose r is a rational function which can be written as $r(x) = \frac{p(x)}{q(x)}$ where p and q have no common zeros.[b] Let c be a real number which is not in the domain of r.

- If $q(c) \neq 0$, then the graph of $y = r(x)$ has a hole at $\left(c, \frac{p(c)}{q(c)}\right)$.

- If $q(c) = 0$, then the line $x = c$ is a vertical asymptote of the graph of $y = r(x)$.

[a]Or, 'How to tell your asymptote from a hole in the graph.'
[b]In other words, $r(x)$ is in lowest terms.

In English, Theorem 4.1 says that if $x = c$ is not in the domain of r but, when we simplify $r(x)$, it no longer makes the denominator 0, then we have a hole at $x = c$. Otherwise, the line $x = c$ is a vertical asymptote of the graph of $y = r(x)$.

Example 4.1.2. Find the vertical asymptotes of, and/or holes in, the graphs of the following rational functions. Verify your answers using a graphing calculator, and describe the behavior of the graph near them using proper notation.

1. $f(x) = \dfrac{2x}{x^2 - 3}$

2. $g(x) = \dfrac{x^2 - x - 6}{x^2 - 9}$

3. $h(x) = \dfrac{x^2 - x - 6}{x^2 + 9}$

4. $r(x) = \dfrac{x^2 - x - 6}{x^2 + 4x + 4}$

Solution.

1. To use Theorem 4.1, we first find all of the real numbers which aren't in the domain of f. To do so, we solve $x^2 - 3 = 0$ and get $x = \pm\sqrt{3}$. Since the expression $f(x)$ is in lowest terms, there is no cancellation possible, and we conclude that the lines $x = -\sqrt{3}$ and $x = \sqrt{3}$ are vertical asymptotes to the graph of $y = f(x)$. The calculator verifies this claim, and from the graph, we see that as $x \to -\sqrt{3}^-$, $f(x) \to -\infty$, as $x \to -\sqrt{3}^+$, $f(x) \to \infty$, as $x \to \sqrt{3}^-$, $f(x) \to -\infty$, and finally as $x \to \sqrt{3}^+$, $f(x) \to \infty$.

2. Solving $x^2 - 9 = 0$ gives $x = \pm 3$. In lowest terms $g(x) = \frac{x^2-x-6}{x^2-9} = \frac{(x-3)(x+2)}{(x-3)(x+3)} = \frac{x+2}{x+3}$. Since $x = -3$ continues to make trouble in the denominator, we know the line $x = -3$ is a vertical asymptote of the graph of $y = g(x)$. Since $x = 3$ no longer produces a 0 in the denominator, we have a hole at $x = 3$. To find the y-coordinate of the hole, we substitute $x = 3$ into $\frac{x+2}{x+3}$ and find the hole is at $\left(3, \frac{5}{6}\right)$. When we graph $y = g(x)$ using a calculator, we clearly see the vertical asymptote at $x = -3$, but everything seems calm near $x = 3$. Hence, as $x \to -3^-$, $g(x) \to \infty$, as $x \to -3^+$, $g(x) \to -\infty$, as $x \to 3^-$, $g(x) \to \frac{5}{6}^-$, and as $x \to 3^+$, $g(x) \to \frac{5}{6}^+$.

4.1 Introduction to Rational Functions

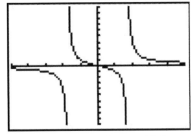

The graph of $y = f(x)$

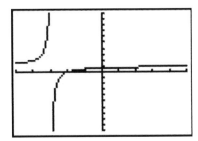

The graph of $y = g(x)$

3. The domain of h is all real numbers, since $x^2 + 9 = 0$ has no real solutions. Accordingly, the graph of $y = h(x)$ is devoid of both vertical asymptotes and holes.

4. Setting $x^2 + 4x + 4 = 0$ gives us $x = -2$ as the only real number of concern. Simplifying, we see $r(x) = \frac{x^2-x-6}{x^2+4x+4} = \frac{(x-3)(x+2)}{(x+2)^2} = \frac{x-3}{x+2}$. Since $x = -2$ continues to produce a 0 in the denominator of the reduced function, we know $x = -2$ is a vertical asymptote to the graph. The calculator bears this out, and, moreover, we see that as $x \to -2^-$, $r(x) \to \infty$ and as $x \to -2^+$, $r(x) \to -\infty$.

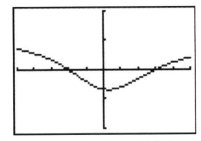

The graph of $y = h(x)$

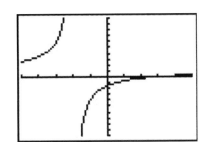

The graph of $y = r(x)$

□

Our next example gives us a physical interpretation of a vertical asymptote. This type of model arises from a family of equations cheerily named 'doomsday' equations.[7]

Example 4.1.3. A mathematical model for the population P, in thousands, of a certain species of bacteria, t days after it is introduced to an environment is given by $P(t) = \frac{100}{(5-t)^2}$, $0 \leq t < 5$.

1. Find and interpret $P(0)$.

2. When will the population reach 100,000?

3. Determine the behavior of P as $t \to 5^-$. Interpret this result graphically and within the context of the problem.

---
[7]These functions arise in Differential Equations. The unfortunate name will make sense shortly.

Solution.

1. Substituting $t = 0$ gives $P(0) = \frac{100}{(5-0)^2} = 4$, which means 4000 bacteria are initially introduced into the environment.

2. To find when the population reaches 100,000, we first need to remember that $P(t)$ is measured in *thousands*. In other words, 100,000 bacteria corresponds to $P(t) = 100$. Substituting for $P(t)$ gives the equation $\frac{100}{(5-t)^2} = 100$. Clearing denominators and dividing by 100 gives $(5-t)^2 = 1$, which, after extracting square roots, produces $t = 4$ or $t = 6$. Of these two solutions, only $t = 4$ in our domain, so this is the solution we keep. Hence, it takes 4 days for the population of bacteria to reach 100,000.

3. To determine the behavior of P as $t \to 5^-$, we can make a table

t	$P(t)$
4.9	10000
4.99	1000000
4.999	100000000
4.9999	10000000000

   In other words, as $t \to 5^-$, $P(t) \to \infty$. Graphically, the line $t = 5$ is a vertical asymptote of the graph of $y = P(t)$. Physically, this means that the population of bacteria is increasing without bound as we near 5 days, which cannot actually happen. For this reason, $t = 5$ is called the 'doomsday' for this population. There is no way any environment can support infinitely many bacteria, so shortly before $t = 5$ the environment would collapse. □

Now that we have thoroughly investigated vertical asymptotes, we can turn our attention to horizontal asymptotes. The next theorem tells us when to expect horizontal asymptotes.

Theorem 4.2. Location of Horizontal Asymptotes: Suppose r is a rational function and $r(x) = \frac{p(x)}{q(x)}$, where p and q are polynomial functions with leading coefficients a and b, respectively.

- If the degree of $p(x)$ is the same as the degree of $q(x)$, then $y = \frac{a}{b}$ is the[a] horizontal asymptote of the graph of $y = r(x)$.

- If the degree of $p(x)$ is less than the degree of $q(x)$, then $y = 0$ is the horizontal asymptote of the graph of $y = r(x)$.

- If the degree of $p(x)$ is greater than the degree of $q(x)$, then the graph of $y = r(x)$ has no horizontal asymptotes.

---
[a]The use of the definite article will be justified momentarily.

Like Theorem 4.1, Theorem 4.2 is proved using Calculus. Nevertheless, we can understand the idea behind it using our example $f(x) = \frac{2x-1}{x+1}$. If we interpret $f(x)$ as a division problem, $(2x-1) \div (x+1)$,

4.1 INTRODUCTION TO RATIONAL FUNCTIONS

we find that the quotient is 2 with a remainder of -3. Using what we know about polynomial division, specifically Theorem 3.4, we get $2x - 1 = 2(x + 1) - 3$. Dividing both sides by $(x + 1)$ gives $\frac{2x-1}{x+1} = 2 - \frac{3}{x+1}$. (You may remember this as the formula for $g(x)$ in Example 4.1.1.) As x becomes unbounded in either direction, the quantity $\frac{3}{x+1}$ gets closer and closer to 0 so that the values of $f(x)$ become closer and closer[8] to 2. In symbols, as $x \to \pm\infty$, $f(x) \to 2$, and we have the result.[9] Notice that the graph gets close to the same y value as $x \to -\infty$ or $x \to \infty$. This means that the graph can have only <u>one</u> horizontal asymptote if it is going to have one at all. Thus we were justified in using 'the' in the previous theorem.

Alternatively, we can use what we know about end behavior of polynomials to help us understand this theorem. From Theorem 3.2, we know the end behavior of a polynomial is determined by its leading term. Applying this to the numerator and denominator of $f(x)$, we get that as $x \to \pm\infty$, $f(x) = \frac{2x-1}{x+1} \approx \frac{2x}{x} = 2$. This last approach is useful in Calculus, and, indeed, is made rigorous there. (Keep this in mind for the remainder of this paragraph.) Applying this reasoning to the general case, suppose $r(x) = \frac{p(x)}{q(x)}$ where a is the leading coefficient of $p(x)$ and b is the leading coefficient of $q(x)$. As $x \to \pm\infty$, $r(x) \approx \frac{ax^n}{bx^m}$, where n and m are the degrees of $p(x)$ and $q(x)$, respectively. If the degree of $p(x)$ and the degree of $q(x)$ are the same, then $n = m$ so that $r(x) \approx \frac{a}{b}$, which means $y = \frac{a}{b}$ is the horizontal asymptote in this case. If the degree of $p(x)$ is less than the degree of $q(x)$, then $n < m$, so $m - n$ is a positive number, and hence, $r(x) \approx \frac{a}{bx^{m-n}} \to 0$ as $x \to \pm\infty$. If the degree of $p(x)$ is greater than the degree of $q(x)$, then $n > m$, and hence $n - m$ is a positive number and $r(x) \approx \frac{ax^{n-m}}{b}$, which becomes unbounded as $x \to \pm\infty$. As we said before, if a rational function has a horizontal asymptote, then it will have only one. (This is not true for other types of functions we shall see in later chapters.)

Example 4.1.4. List the horizontal asymptotes, if any, of the graphs of the following functions. Verify your answers using a graphing calculator, and describe the behavior of the graph near them using proper notation.

1. $f(x) = \dfrac{5x}{x^2 + 1}$
2. $g(x) = \dfrac{x^2 - 4}{x + 1}$
3. $h(x) = \dfrac{6x^3 - 3x + 1}{5 - 2x^3}$

Solution.

1. The numerator of $f(x)$ is $5x$, which has degree 1. The denominator of $f(x)$ is $x^2 + 1$, which has degree 2. Applying Theorem 4.2, $y = 0$ is the horizontal asymptote. Sure enough, we see from the graph that as $x \to -\infty$, $f(x) \to 0^-$ and as $x \to \infty$, $f(x) \to 0^+$.

2. The numerator of $g(x)$, $x^2 - 4$, has degree 2, but the degree of the denominator, $x + 1$, has degree 1. By Theorem 4.2, there is no horizontal asymptote. From the graph, we see that the graph of $y = g(x)$ doesn't appear to level off to a constant value, so there is no horizontal asymptote.[10]

---

[8] As seen in the tables immediately preceding Definition 4.2.
[9] More specifically, as $x \to -\infty$, $f(x) \to 2^+$, and as $x \to \infty$, $f(x) \to 2^-$.
[10] Sit tight! We'll revisit this function and its end behavior shortly.

3. The degrees of the numerator and denominator of $h(x)$ are both three, so Theorem 4.2 tells us $y = \frac{6}{-2} = -3$ is the horizontal asymptote. We see from the calculator's graph that as $x \to -\infty$, $h(x) \to -3^+$, and as $x \to \infty$, $h(x) \to -3^-$.

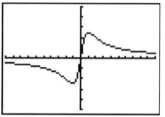
The graph of $y = f(x)$

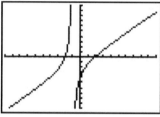
The graph of $y = g(x)$

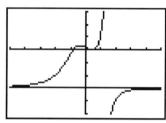
The graph of $y = h(x)$

Our next example of the section gives us a real-world application of a horizontal asymptote.[11]

Example 4.1.5. The number of students N at local college who have had the flu t months after the semester begins can be modeled by the formula $N(t) = 500 - \frac{450}{1+3t}$ for $t \geq 0$.

1. Find and interpret $N(0)$.

2. How long will it take until 300 students will have had the flu?

3. Determine the behavior of N as $t \to \infty$. Interpret this result graphically and within the context of the problem.

Solution.

1. $N(0) = 500 - \frac{450}{1+3(0)} = 50$. This means that at the beginning of the semester, 50 students have had the flu.

2. We set $N(t) = 300$ to get $500 - \frac{450}{1+3t} = 300$ and solve. Isolating the fraction gives $\frac{450}{1+3t} = 200$. Clearing denominators gives $450 = 200(1 + 3t)$. Finally, we get $t = \frac{5}{12}$. This means it will take $\frac{5}{12}$ months, or about 13 days, for 300 students to have had the flu.

3. To determine the behavior of N as $t \to \infty$, we can use a table.

t	$N(t)$
10	≈ 485.48
100	≈ 498.50
1000	≈ 499.85
10000	≈ 499.98

The table suggests that as $t \to \infty$, $N(t) \to 500$. (More specifically, 500^-.) This means as time goes by, only a total of 500 students will have ever had the flu.

---

[11] Though the population below is more accurately modeled with the functions in Chapter 6, we approximate it (using Calculus, of course!) using a rational function.

We close this section with a discussion of the *third* (and final!) kind of asymptote which can be associated with the graphs of rational functions. Let us return to the function $g(x) = \frac{x^2-4}{x+1}$ in Example 4.1.4. Performing long division,[12] we get $g(x) = \frac{x^2-4}{x+1} = x - 1 - \frac{3}{x+1}$. Since the term $\frac{3}{x+1} \to 0$ as $x \to \pm\infty$, it stands to reason that as x becomes unbounded, the function values $g(x) = x - 1 - \frac{3}{x+1} \approx x - 1$. Geometrically, this means that the graph of $y = g(x)$ should resemble the line $y = x - 1$ as $x \to \pm\infty$. We see this play out both numerically and graphically below.

x	$g(x)$	$x - 1$
-10	≈ -10.6667	-11
-100	≈ -100.9697	-101
-1000	≈ -1000.9970	-1001
-10000	≈ -10000.9997	-10001

x	$g(x)$	$x - 1$
10	≈ 8.7273	9
100	≈ 98.9703	99
1000	≈ 998.9970	999
10000	≈ 9998.9997	9999

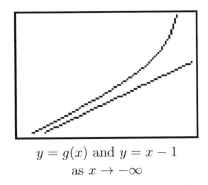

$y = g(x)$ and $y = x - 1$
as $x \to -\infty$

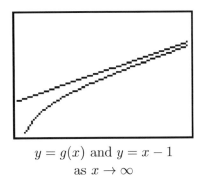

$y = g(x)$ and $y = x - 1$
as $x \to \infty$

The way we symbolize the relationship between the end behavior of $y = g(x)$ with that of the line $y = x - 1$ is to write 'as $x \to \pm\infty$, $g(x) \to x - 1$.' In this case, we say the line $y = x - 1$ is a **slant asymptote**[13] to the graph of $y = g(x)$. Informally, the graph of a rational function has a slant asymptote if, as $x \to \infty$ or as $x \to -\infty$, the graph resembles a non-horizontal, or 'slanted' line. Formally, we define a slant asymptote as follows.

Definition 4.4. The line $y = mx + b$ where $m \neq 0$ is called a **slant asymptote** of the graph of a function $y = f(x)$ if as $x \to -\infty$ or as $x \to \infty$, $f(x) \to mx + b$.

A few remarks are in order. First, note that the stipulation $m \neq 0$ in Definition 4.4 is what makes the 'slant' asymptote 'slanted' as opposed to the case when $m = 0$ in which case we'd have a horizontal asymptote. Secondly, while we have motivated what me mean intuitively by the notation '$f(x) \to mx + b$,' like so many ideas in this section, the formal definition requires Calculus. Another way to express this sentiment, however, is to rephrase '$f(x) \to mx + b$' as '$f(x) - (mx + b) \to 0$.' In other words, the graph of $y = f(x)$ has the *slant* asymptote $y = mx + b$ if and only if the graph of $y = f(x) - (mx + b)$ has a *horizontal* asymptote $y = 0$.

---
[12]See the remarks following Theorem 4.2.
[13]Also called an 'oblique' asymptote in some, ostensibly higher class (and more expensive), texts.

Our next task is to determine the conditions under which the graph of a rational function has a slant asymptote, and if it does, how to find it. In the case of $g(x) = \frac{x^2-4}{x+1}$, the degree of the numerator $x^2 - 4$ is 2, which is *exactly one more* than the degree if its denominator $x + 1$ which is 1. This results in a *linear* quotient polynomial, and it is this quotient polynomial which is the slant asymptote. Generalizing this situation gives us the following theorem.[14]

> **Theorem 4.3. Determination of Slant Asymptotes:** Suppose r is a rational function and $r(x) = \frac{p(x)}{q(x)}$, where the degree of p is *exactly* one more than the degree of q. Then the graph of $y = r(x)$ has the slant asymptote $y = L(x)$ where $L(x)$ is the quotient obtained by dividing $p(x)$ by $q(x)$.

In the same way that Theorem 4.2 gives us an easy way to see if the graph of a rational function $r(x) = \frac{p(x)}{q(x)}$ has a horizontal asymptote by comparing the degrees of the numerator and denominator, Theorem 4.3 gives us an easy way to check for slant asymptotes. Unlike Theorem 4.2, which gives us a quick way to *find* the horizontal asymptotes (if any exist), Theorem 4.3 gives us no such 'short-cut'. If a slant asymptote exists, we have no recourse but to use long division to find it.[15]

Example 4.1.6. Find the slant asymptotes of the graphs of the following functions if they exist. Verify your answers using a graphing calculator and describe the behavior of the graph near them using proper notation.

1. $f(x) = \dfrac{x^2 - 4x + 2}{1 - x}$
2. $g(x) = \dfrac{x^2 - 4}{x - 2}$
3. $h(x) = \dfrac{x^3 + 1}{x^2 - 4}$

Solution.

1. The degree of the numerator is 2 and the degree of the denominator is 1, so Theorem 4.3 guarantees us a slant asymptote. To find it, we divide $1 - x = -x + 1$ into $x^2 - 4x + 2$ and get a quotient of $-x + 3$, so our slant asymptote is $y = -x + 3$. We confirm this graphically, and we see that as $x \to -\infty$, the graph of $y = f(x)$ approaches the asymptote from below, and as $x \to \infty$, the graph of $y = f(x)$ approaches the asymptote from above.[16]

2. As with the previous example, the degree of the numerator $g(x) = \frac{x^2-4}{x-2}$ is 2 and the degree of the denominator is 1, so Theorem 4.3 applies. In this case,

$$g(x) = \frac{x^2 - 4}{x - 2} = \frac{(x+2)(x-2)}{(x-2)} = \frac{(x+2)\cancel{(x-2)}}{\cancel{(x-2)}} = x + 2, \quad x \neq 2$$

---

[14]Once again, this theorem is brought to you courtesy of Theorem 3.4 and Calculus.

[15]That's OK, though. In the next section, we'll use long division to analyze end behavior and it's worth the effort!

[16]Note that we are purposefully avoiding notation like 'as $x \to \infty$, $f(x) \to (-x+3)^+$. While it is possible to define these notions formally with Calculus, it is not standard to do so. Besides, with the introduction of the symbol '?' in the next section, the authors feel we are in enough trouble already.

4.1 Introduction to Rational Functions

so we have that the slant asymptote $y = x + 2$ is identical to the graph of $y = g(x)$ except at $x = 2$ (where the latter has a 'hole' at $(2, 4)$.) The calculator supports this claim.[17]

3. For $h(x) = \frac{x^3+1}{x^2-4}$, the degree of the numerator is 3 and the degree of the denominator is 2 so again, we are guaranteed the existence of a slant asymptote. The long division $(x^3 + 1) \div (x^2 - 4)$ gives a quotient of just x, so our slant asymptote is the line $y = x$. The calculator confirms this, and we find that as $x \to -\infty$, the graph of $y = h(x)$ approaches the asymptote from below, and as $x \to \infty$, the graph of $y = h(x)$ approaches the asymptote from above.

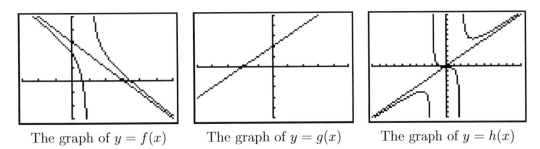

The graph of $y = f(x)$ The graph of $y = g(x)$ The graph of $y = h(x)$

□

The reader may be a bit disappointed with the authors at this point owing to the fact that in Examples 4.1.2, 4.1.4, and 4.1.6, we used the *calculator* to determine function behavior near asymptotes. We rectify that in the next section where we, in excruciating detail, demonstrate the usefulness of 'number sense' to reveal this behavior analytically.

---

[17] While the word 'asymptote' has the connotation of 'approaching but not equaling,' Definitions 4.3 and 4.4 invite the same kind of pathologies we saw with Definitions 1.11 in Section 1.6.

4.1.1 Exercises

In Exercises 1 - 18, for the given rational function f:

- Find the domain of f.
- Identify any vertical asymptotes of the graph of $y = f(x)$.
- Identify any holes in the graph.
- Find the horizontal asymptote, if it exists.
- Find the slant asymptote, if it exists.
- Graph the function using a graphing utility and describe the behavior near the asymptotes.

1. $f(x) = \dfrac{x}{3x - 6}$
2. $f(x) = \dfrac{3 + 7x}{5 - 2x}$
3. $f(x) = \dfrac{x}{x^2 + x - 12}$
4. $f(x) = \dfrac{x}{x^2 + 1}$
5. $f(x) = \dfrac{x + 7}{(x + 3)^2}$
6. $f(x) = \dfrac{x^3 + 1}{x^2 - 1}$
7. $f(x) = \dfrac{4x}{x^2 + 4}$
8. $f(x) = \dfrac{4x}{x^2 - 4}$
9. $f(x) = \dfrac{x^2 - x - 12}{x^2 + x - 6}$
10. $f(x) = \dfrac{3x^2 - 5x - 2}{x^2 - 9}$
11. $f(x) = \dfrac{x^3 + 2x^2 + x}{x^2 - x - 2}$
12. $f(x) = \dfrac{x^3 - 3x + 1}{x^2 + 1}$
13. $f(x) = \dfrac{2x^2 + 5x - 3}{3x + 2}$
14. $f(x) = \dfrac{-x^3 + 4x}{x^2 - 9}$
15. $f(x) = \dfrac{-5x^4 - 3x^3 + x^2 - 10}{x^3 - 3x^2 + 3x - 1}$
16. $f(x) = \dfrac{x^3}{1 - x}$
17. $f(x) = \dfrac{18 - 2x^2}{x^2 - 9}$
18. $f(x) = \dfrac{x^3 - 4x^2 - 4x - 5}{x^2 + x + 1}$

19. The cost C in dollars to remove $p\%$ of the invasive species of Ippizuti fish from Sasquatch Pond is given by
$$C(p) = \dfrac{1770p}{100 - p}, \quad 0 \leq p < 100$$

   (a) Find and interpret $C(25)$ and $C(95)$.
   (b) What does the vertical asymptote at $x = 100$ mean within the context of the problem?
   (c) What percentage of the Ippizuti fish can you remove for $40000?

20. In Exercise 71 in Section 1.4, the population of Sasquatch in Portage County was modeled by the function
$$P(t) = \dfrac{150t}{t + 15},$$
where $t = 0$ represents the year 1803. Find the horizontal asymptote of the graph of $y = P(t)$ and explain what it means.

21. Recall from Example 1.5.3 that the cost C (in dollars) to make x dOpi media players is $C(x) = 100x + 2000$, $x \geq 0$.

    (a) Find a formula for the average cost $\overline{C}(x)$. Recall: $\overline{C}(x) = \frac{C(x)}{x}$.
    (b) Find and interpret $\overline{C}(1)$ and $\overline{C}(100)$.
    (c) How many dOpis need to be produced so that the average cost per dOpi is $200?
    (d) Interpret the behavior of $\overline{C}(x)$ as $x \to 0^+$. (HINT: You may want to find the fixed cost $C(0)$ to help in your interpretation.)
    (e) Interpret the behavior of $\overline{C}(x)$ as $x \to \infty$. (HINT: You may want to find the variable cost (defined in Example 2.1.5 in Section 2.1) to help in your interpretation.)

22. In Exercise 35 in Section 3.1, we fit a few polynomial models to the following electric circuit data. (The circuit was built with a variable resistor. For each of the following resistance values (measured in kilo-ohms, $k\Omega$), the corresponding power to the load (measured in milliwatts, mW) is given in the table below.)[18]

Resistance: ($k\Omega$)	1.012	2.199	3.275	4.676	6.805	9.975
Power: (mW)	1.063	1.496	1.610	1.613	1.505	1.314

    Using some fundamental laws of circuit analysis mixed with a healthy dose of algebra, we can derive the actual formula relating power to resistance. For this circuit, it is $P(x) = \frac{25x}{(x+3.9)^2}$, where x is the resistance value, $x \geq 0$.

    (a) Graph the data along with the function $y = P(x)$ on your calculator.
    (b) Use your calculator to approximate the maximum power that can be delivered to the load. What is the corresponding resistance value?
    (c) Find and interpret the end behavior of $P(x)$ as $x \to \infty$.

23. In his now famous 1919 dissertation The Learning Curve Equation, Louis Leon Thurstone presents a rational function which models the number of words a person can type in four minutes as a function of the number of pages of practice one has completed. (This paper, which is now in the public domain and can be found here, is from a bygone era when students at business schools took typing classes on manual typewriters.) Using his original notation and original language, we have $Y = \frac{L(X+P)}{(X+P)+R}$ where L is the predicted practice limit in terms of speed units, X is pages written, Y is writing speed in terms of words in four minutes, P is equivalent previous practice in terms of pages and R is the rate of learning. In Figure 5 of the paper, he graphs a scatter plot and the curve $Y = \frac{216(X+19)}{X+148}$. Discuss this equation with your classmates. How would you update the notation? Explain what the horizontal asymptote of the graph means. You should take some time to look at the original paper. Skip over the computations you don't understand yet and try to get a sense of the time and place in which the study was conducted.

---

[18] The authors wish to thank Don Anthan and Ken White of Lakeland Community College for devising this problem and generating the accompanying data set.

4.1.2 Answers

1. $f(x) = \dfrac{x}{3x-6}$
   Domain: $(-\infty, 2) \cup (2, \infty)$
   Vertical asymptote: $x = 2$
   As $x \to 2^-, f(x) \to -\infty$
   As $x \to 2^+, f(x) \to \infty$
   No holes in the graph
   Horizontal asymptote: $y = \frac{1}{3}$
   As $x \to -\infty, f(x) \to \frac{1}{3}^-$
   As $x \to \infty, f(x) \to \frac{1}{3}^+$

2. $f(x) = \dfrac{3+7x}{5-2x}$
   Domain: $\left(-\infty, \frac{5}{2}\right) \cup \left(\frac{5}{2}, \infty\right)$
   Vertical asymptote: $x = \frac{5}{2}$
   As $x \to \frac{5}{2}^-, f(x) \to \infty$
   As $x \to \frac{5}{2}^+, f(x) \to -\infty$
   No holes in the graph
   Horizontal asymptote: $y = -\frac{7}{2}$
   As $x \to -\infty, f(x) \to -\frac{7}{2}^+$
   As $x \to \infty, f(x) \to -\frac{7}{2}^-$

3. $f(x) = \dfrac{x}{x^2+x-12} = \dfrac{x}{(x+4)(x-3)}$
   Domain: $(-\infty, -4) \cup (-4, 3) \cup (3, \infty)$
   Vertical asymptotes: $x = -4, x = 3$
   As $x \to -4^-, f(x) \to -\infty$
   As $x \to -4^+, f(x) \to \infty$
   As $x \to 3^-, f(x) \to -\infty$
   As $x \to 3^+, f(x) \to \infty$
   No holes in the graph
   Horizontal asymptote: $y = 0$
   As $x \to -\infty, f(x) \to 0^-$
   As $x \to \infty, f(x) \to 0^+$

4. $f(x) = \dfrac{x}{x^2+1}$
   Domain: $(-\infty, \infty)$
   No vertical asymptotes
   No holes in the graph
   Horizontal asymptote: $y = 0$
   As $x \to -\infty, f(x) \to 0^-$
   As $x \to \infty, f(x) \to 0^+$

5. $f(x) = \dfrac{x+7}{(x+3)^2}$
   Domain: $(-\infty, -3) \cup (-3, \infty)$
   Vertical asymptote: $x = -3$
   As $x \to -3^-, f(x) \to \infty$
   As $x \to -3^+, f(x) \to \infty$
   No holes in the graph
   Horizontal asymptote: $y = 0$
   [19]As $x \to -\infty, f(x) \to 0^-$
   As $x \to \infty, f(x) \to 0^+$

6. $f(x) = \dfrac{x^3+1}{x^2-1} = \dfrac{x^2-x+1}{x-1}$
   Domain: $(-\infty, -1) \cup (-1, 1) \cup (1, \infty)$
   Vertical asymptote: $x = 1$
   As $x \to 1^-, f(x) \to -\infty$
   As $x \to 1^+, f(x) \to \infty$
   Hole at $\left(-1, -\frac{3}{2}\right)$
   Slant asymptote: $y = x$
   As $x \to -\infty$, the graph is below $y = x$
   As $x \to \infty$, the graph is above $y = x$

---

[19]This is hard to see on the calculator, but trust me, the graph is below the x-axis to the left of $x = -7$.

4.1 Introduction to Rational Functions

7. $f(x) = \dfrac{4x}{x^2+4}$
   Domain: $(-\infty, \infty)$
   No vertical asymptotes
   No holes in the graph
   Horizontal asymptote: $y = 0$
   As $x \to -\infty, f(x) \to 0^-$
   As $x \to \infty, f(x) \to 0^+$

8. $f(x) = \dfrac{4x}{x^2-4} = \dfrac{4x}{(x+2)(x-2)}$
   Domain: $(-\infty, -2) \cup (-2, 2) \cup (2, \infty)$
   Vertical asymptotes: $x = -2, x = 2$
   As $x \to -2^-, f(x) \to -\infty$
   As $x \to -2^+, f(x) \to \infty$
   As $x \to 2^-, f(x) \to -\infty$
   As $x \to 2^+, f(x) \to \infty$
   No holes in the graph
   Horizontal asymptote: $y = 0$
   As $x \to -\infty, f(x) \to 0^-$
   As $x \to \infty, f(x) \to 0^+$

9. $f(x) = \dfrac{x^2 - x - 12}{x^2 + x - 6} = \dfrac{x-4}{x-2}$
   Domain: $(-\infty, -3) \cup (-3, 2) \cup (2, \infty)$
   Vertical asymptote: $x = 2$
   As $x \to 2^-, f(x) \to \infty$
   As $x \to 2^+, f(x) \to -\infty$
   Hole at $\left(-3, \tfrac{7}{5}\right)$
   Horizontal asymptote: $y = 1$
   As $x \to -\infty, f(x) \to 1^+$
   As $x \to \infty, f(x) \to 1^-$

10. $f(x) = \dfrac{3x^2 - 5x - 2}{x^2 - 9} = \dfrac{(3x+1)(x-2)}{(x+3)(x-3)}$
    Domain: $(-\infty, -3) \cup (-3, 3) \cup (3, \infty)$
    Vertical asymptotes: $x = -3, x = 3$
    As $x \to -3^-, f(x) \to \infty$
    As $x \to -3^+, f(x) \to -\infty$
    As $x \to 3^-, f(x) \to -\infty$
    As $x \to 3^+, f(x) \to \infty$
    No holes in the graph
    Horizontal asymptote: $y = 3$
    As $x \to -\infty, f(x) \to 3^+$
    As $x \to \infty, f(x) \to 3^-$

11. $f(x) = \dfrac{x^3 + 2x^2 + x}{x^2 - x - 2} = \dfrac{x(x+1)}{x-2}$
    Domain: $(-\infty, -1) \cup (-1, 2) \cup (2, \infty)$
    Vertical asymptote: $x = 2$
    As $x \to 2^-, f(x) \to -\infty$
    As $x \to 2^+, f(x) \to \infty$
    Hole at $(-1, 0)$
    Slant asymptote: $y = x + 3$
    As $x \to -\infty$, the graph is below $y = x + 3$
    As $x \to \infty$, the graph is above $y = x + 3$

12. $f(x) = \dfrac{x^3 - 3x + 1}{x^2 + 1}$
    Domain: $(-\infty, \infty)$
    No vertical asymptotes
    No holes in the graph
    Slant asymptote: $y = x$
    As $x \to -\infty$, the graph is above $y = x$
    As $x \to \infty$, the graph is below $y = x$

13. $f(x) = \dfrac{2x^2 + 5x - 3}{3x + 2}$
Domain: $\left(-\infty, -\tfrac{2}{3}\right) \cup \left(-\tfrac{2}{3}, \infty\right)$
Vertical asymptote: $x = -\tfrac{2}{3}$
As $x \to -\tfrac{2}{3}^-$, $f(x) \to \infty$
As $x \to -\tfrac{2}{3}^+$, $f(x) \to -\infty$
No holes in the graph
Slant asymptote: $y = \tfrac{2}{3}x + \tfrac{11}{9}$
As $x \to -\infty$, the graph is above $y = \tfrac{2}{3}x + \tfrac{11}{9}$
As $x \to \infty$, the graph is below $y = \tfrac{2}{3}x + \tfrac{11}{9}$

14. $f(x) = \dfrac{-x^3 + 4x}{x^2 - 9} = \dfrac{-x^3 + 4x}{(x-3)(x+3)}$
Domain: $(-\infty, -3) \cup (-3, 3) \cup (3, \infty)$
Vertical asymptotes: $x = -3$, $x = 3$
As $x \to -3^-$, $f(x) \to \infty$
As $x \to -3^+$, $f(x) \to -\infty$
As $x \to 3^-$, $f(x) \to \infty$
As $x \to 3^+$, $f(x) \to -\infty$
No holes in the graph
Slant asymptote: $y = -x$
As $x \to -\infty$, the graph is above $y = -x$
As $x \to \infty$, the graph is below $y = -x$

15. $f(x) = \dfrac{-5x^4 - 3x^3 + x^2 - 10}{x^3 - 3x^2 + 3x - 1}$
$= \dfrac{-5x^4 - 3x^3 + x^2 - 10}{(x-1)^3}$
Domain: $(-\infty, 1) \cup (1, \infty)$
Vertical asymptotes: $x = 1$
As $x \to 1^-$, $f(x) \to \infty$
As $x \to 1^+$, $f(x) \to -\infty$
No holes in the graph
Slant asymptote: $y = -5x - 18$
As $x \to -\infty$, the graph is above $y = -5x - 18$
As $x \to \infty$, the graph is below $y = -5x - 18$

16. $f(x) = \dfrac{x^3}{1 - x}$
Domain: $(-\infty, 1) \cup (1, \infty)$
Vertical asymptote: $x = 1$
As $x \to 1^-$, $f(x) \to \infty$
As $x \to 1^+$, $f(x) \to -\infty$
No holes in the graph
No horizontal or slant asymptote
As $x \to -\infty$, $f(x) \to -\infty$
As $x \to \infty$, $f(x) \to -\infty$

17. $f(x) = \dfrac{18 - 2x^2}{x^2 - 9} = -2$
Domain: $(-\infty, -3) \cup (-3, 3) \cup (3, \infty)$
No vertical asymptotes
Holes in the graph at $(-3, -2)$ and $(3, -2)$
Horizontal asymptote $y = -2$
As $x \to \pm\infty$, $f(x) = -2$

18. $f(x) = \dfrac{x^3 - 4x^2 - 4x - 5}{x^2 + x + 1} = x - 5$
Domain: $(-\infty, \infty)$
No vertical asymptotes
No holes in the graph
Slant asymptote: $y = x - 5$
$f(x) = x - 5$ everywhere.

19. (a) $C(25) = 590$ means it costs \$590 to remove 25% of the fish and and $C(95) = 33630$ means it would cost \$33630 to remove 95% of the fish from the pond.

(b) The vertical asymptote at $x = 100$ means that as we try to remove 100% of the fish from the pond, the cost increases without bound; i.e., it's impossible to remove all of the fish.

(c) For \$40000 you could remove about 95.76% of the fish.

20. The horizontal asymptote of the graph of $P(t) = \frac{150t}{t+15}$ is $y = 150$ and it means that the model predicts the population of Sasquatch in Portage County will never exceed 150.

21. (a) $\overline{C}(x) = \frac{100x + 2000}{x}$, $x > 0$.

    (b) $\overline{C}(1) = 2100$ and $\overline{C}(100) = 120$. When just 1 dOpi is produced, the cost per dOpi is $2100, but when 100 dOpis are produced, the cost per dOpi is $120.

    (c) $\overline{C}(x) = 200$ when $x = 20$. So to get the cost per dOpi to $200, 20 dOpis need to be produced.

    (d) As $x \to 0^+$, $\overline{C}(x) \to \infty$. This means that as fewer and fewer dOpis are produced, the cost per dOpi becomes unbounded. In this situation, there is a fixed cost of $2000 ($C(0) = 2000$), we are trying to spread that $2000 over fewer and fewer dOpis.

    (e) As $x \to \infty$, $\overline{C}(x) \to 100^+$. This means that as more and more dOpis are produced, the cost per dOpi approaches $100, but is always a little more than $100. Since $100 is the variable cost per dOpi ($C(x) = \underline{100x} + 2000$), it means that no matter how many dOpis are produced, the average cost per dOpi will always be a bit higher than the variable cost to produce a dOpi. As before, we can attribute this to the $2000 fixed cost, which factors into the average cost per dOpi no matter how many dOpis are produced.

22. (a)

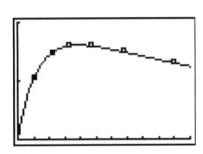

    (b) The maximum power is approximately $1.603\ mW$ which corresponds to $3.9\ k\Omega$.

    (c) As $x \to \infty$, $P(x) \to 0^+$ which means as the resistance increases without bound, the power diminishes to zero.

4.2 Graphs of Rational Functions

In this section, we take a closer look at graphing rational functions. In Section 4.1, we learned that the graphs of rational functions may have holes in them and could have vertical, horizontal and slant asymptotes. Theorems 4.1, 4.2 and 4.3 tell us exactly when and where these behaviors will occur, and if we combine these results with what we already know about graphing functions, we will quickly be able to generate reasonable graphs of rational functions.

One of the standard tools we will use is the sign diagram which was first introduced in Section 2.4, and then revisited in Section 3.1. In those sections, we operated under the belief that a function couldn't change its sign without its graph crossing through the x-axis. The major theorem we used to justify this belief was the Intermediate Value Theorem, Theorem 3.1. It turns out the Intermediate Value Theorem applies to all *continuous* functions,[1] not just polynomials. Although rational functions are continuous on their domains,[2] Theorem 4.1 tells us that vertical asymptotes and holes occur at the values excluded from their domains. In other words, rational functions aren't continuous at these excluded values which leaves open the possibility that the function could change sign *without* crossing through the x-axis. Consider the graph of $y = h(x)$ from Example 4.1.1, recorded below for convenience. We have added its x-intercept at $\left(\frac{1}{2}, 0\right)$ for the discussion that follows. Suppose we wish to construct a sign diagram for $h(x)$. Recall that the intervals where $h(x) > 0$, or $(+)$, correspond to the x-values where the graph of $y = h(x)$ is *above* the x-axis; the intervals on which $h(x) < 0$, or $(-)$ correspond to where the graph is *below* the x-axis.

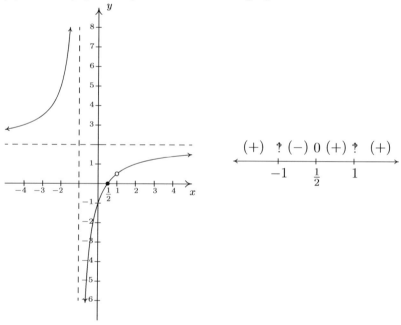

As we examine the graph of $y = h(x)$, reading from left to right, we note that from $(-\infty, -1)$, the graph is above the x-axis, so $h(x)$ is $(+)$ there. At $x = -1$, we have a vertical asymptote, at which point the graph 'jumps' across the x-axis. On the interval $\left(-1, \frac{1}{2}\right)$, the graph is below the

---

[1] Recall that, for our purposes, this means the graphs are devoid of any breaks, jumps or holes
[2] Another result from Calculus.

4.2 Graphs of Rational Functions

x-axis, so $h(x)$ is $(-)$ there. The graph crosses through the x-axis at $\left(\frac{1}{2}, 0\right)$ and remains above the x-axis until $x = 1$, where we have a 'hole' in the graph. Since $h(1)$ is undefined, there is no sign here. So we have $h(x)$ as $(+)$ on the interval $\left(\frac{1}{2}, 1\right)$. Continuing, we see that on $(1, \infty)$, the graph of $y = h(x)$ is above the x-axis, so we mark $(+)$ there. To construct a sign diagram from this information, we not only need to denote the zero of h, but also the places not in the domain of h. As is our custom, we write '0' above $\frac{1}{2}$ on the sign diagram to remind us that it is a zero of h. We need a different notation for -1 and 1, and we have chosen to use '?' - a nonstandard symbol called the interrobang. We use this symbol to convey a sense of surprise, caution and wonderment - an appropriate attitude to take when approaching these points. The moral of the story is that when constructing sign diagrams for rational functions, we include the zeros as well as the values excluded from the domain.

Steps for Constructing a Sign Diagram for a Rational Function

Suppose r is a rational function.

1. Place any values excluded from the domain of r on the number line with an '?' above them.

2. Find the zeros of r and place them on the number line with the number 0 above them.

3. Choose a test value in each of the intervals determined in steps 1 and 2.

4. Determine the sign of $r(x)$ for each test value in step 3, and write that sign above the corresponding interval.

We now present our procedure for graphing rational functions and apply it to a few exhaustive examples. Please note that we decrease the amount of detail given in the explanations as we move through the examples. The reader should be able to fill in any details in those steps which we have abbreviated.

Steps for Graphing Rational Functions

Suppose r is a rational function.

1. Find the domain of r.

2. Reduce $r(x)$ to lowest terms, if applicable.

3. Find the x- and y-intercepts of the graph of $y = r(x)$, if they exist.

4. Determine the location of any vertical asymptotes or holes in the graph, if they exist. Analyze the behavior of r on either side of the vertical asymptotes, if applicable.

5. Analyze the end behavior of r. Find the horizontal or slant asymptote, if one exists.

6. Use a sign diagram and plot additional points, as needed, to sketch the graph of $y = r(x)$.

Example 4.2.1. Sketch a detailed graph of $f(x) = \dfrac{3x}{x^2 - 4}$.

Solution. We follow the six step procedure outlined above.

1. As usual, we set the denominator equal to zero to get $x^2 - 4 = 0$. We find $x = \pm 2$, so our domain is $(-\infty, -2) \cup (-2, 2) \cup (2, \infty)$.

2. To reduce $f(x)$ to lowest terms, we factor the numerator and denominator which yields $f(x) = \dfrac{3x}{(x-2)(x+2)}$. There are no common factors which means $f(x)$ is already in lowest terms.

3. To find the x-intercepts of the graph of $y = f(x)$, we set $y = f(x) = 0$. Solving $\dfrac{3x}{(x-2)(x+2)} = 0$ results in $x = 0$. Since $x = 0$ is in our domain, $(0, 0)$ is the x-intercept. To find the y-intercept, we set $x = 0$ and find $y = f(0) = 0$, so that $(0, 0)$ is our y-intercept as well.[3]

4. The two numbers excluded from the domain of f are $x = -2$ and $x = 2$. Since $f(x)$ didn't reduce at all, both of these values of x still cause trouble in the denominator. Thus by Theorem 4.1, $x = -2$ and $x = 2$ are vertical asymptotes of the graph. We can actually go a step further at this point and determine exactly how the graph approaches the asymptote near each of these values. Though not absolutely necessary,[4] it is good practice for those heading off to Calculus. For the discussion that follows, it is best to use the factored form of $f(x) = \dfrac{3x}{(x-2)(x+2)}$.

   - The behavior of $y = f(x)$ as $x \to -2$: Suppose $x \to -2^-$. If we were to build a table of values, we'd use x-values a little less than -2, say -2.1, -2.01 and -2.001. While there is no harm in actually building a table like we did in Section 4.1, we want to develop a 'number sense' here. Let's think about each factor in the formula of $f(x)$ as we imagine substituting a number like $x = -2.000001$ into $f(x)$. The quantity $3x$ would be very close to -6, the quantity $(x - 2)$ would be very close to -4, and the factor $(x + 2)$ would be very close to 0. More specifically, $(x + 2)$ would be a little less than 0, in this case, -0.000001. We will call such a number a 'very small $(-)$', 'very small' meaning close to zero in absolute value. So, mentally, as $x \to -2^-$, we estimate

   $$f(x) = \frac{3x}{(x-2)(x+2)} \approx \frac{-6}{(-4)\,(\text{very small }(-))} = \frac{3}{2\,(\text{very small }(-))}$$

   Now, the closer x gets to -2, the smaller $(x + 2)$ will become, so even though we are multiplying our 'very small $(-)$' by 2, the denominator will continue to get smaller and smaller, and remain negative. The result is a fraction whose numerator is positive, but whose denominator is very small and negative. Mentally,

   $$f(x) \approx \frac{3}{2\,(\text{very small }(-))} \approx \frac{3}{\text{very small }(-)} \approx \text{very big }(-)$$

---

[3] As we mentioned at least once earlier, since functions can have at most one y-intercept, once we find that $(0, 0)$ is on the graph, we know it is the y-intercept.

[4] The sign diagram in step 6 will also determine the behavior near the vertical asymptotes.

The term 'very big $(-)$' means a number with a large absolute value which is negative.[5] What all of this means is that as $x \to -2^-$, $f(x) \to -\infty$. Now suppose we wanted to determine the behavior of $f(x)$ as $x \to -2^+$. If we imagine substituting something a little larger than -2 in for x, say -1.999999, we mentally estimate

$$f(x) \approx \frac{-6}{(-4)\,(\text{very small }(+))} = \frac{3}{2\,(\text{very small }(+))} \approx \frac{3}{\text{very small }(+)} \approx \text{very big }(+)$$

We conclude that as $x \to -2^+$, $f(x) \to \infty$.

- *The behavior of $y = f(x)$ as $x \to 2$:* Consider $x \to 2^-$. We imagine substituting $x = 1.999999$. Approximating $f(x)$ as we did above, we get

$$f(x) \approx \frac{6}{(\text{very small }(-))\,(4)} = \frac{3}{2\,(\text{very small }(-))} \approx \frac{3}{\text{very small }(-)} \approx \text{very big }(-)$$

We conclude that as $x \to 2^-$, $f(x) \to -\infty$. Similarly, as $x \to 2^+$, we imagine substituting $x = 2.000001$ to get $f(x) \approx \frac{3}{\text{very small }(+)} \approx$ very big $(+)$. So as $x \to 2^+$, $f(x) \to \infty$.

Graphically, we have that near $x = -2$ and $x = 2$ the graph of $y = f(x)$ looks like[6]

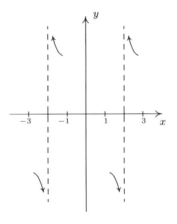

5. Next, we determine the end behavior of the graph of $y = f(x)$. Since the degree of the numerator is 1, and the degree of the denominator is 2, Theorem 4.2 tells us that $y = 0$ is the horizontal asymptote. As with the vertical asymptotes, we can glean more detailed information using 'number sense'. For the discussion below, we use the formula $f(x) = \frac{3x}{x^2-4}$.

   - *The behavior of $y = f(x)$ as $x \to -\infty$:* If we were to make a table of values to discuss the behavior of f as $x \to -\infty$, we would substitute very 'large' negative numbers in for x, say for example, $x = -1$ billion. The numerator $3x$ would then be -3 billion, whereas

---

[5] The actual retail value of $f(-2.000001)$ is approximately $-1,500,000$.
[6] We have deliberately left off the labels on the y-axis because we know only the behavior near $x = \pm 2$, not the actual function values.

the denominator $x^2 - 4$ would be $(-1 \text{ billion})^2 - 4$, which is pretty much the same as $1(\text{billion})^2$. Hence,

$$f(-1 \text{ billion}) \approx \frac{-3 \text{ billion}}{1(\text{billion})^2} \approx -\frac{3}{\text{billion}} \approx \text{very small } (-)$$

Notice that if we substituted in $x = -1$ trillion, essentially the same kind of cancellation would happen, and we would be left with an even 'smaller' negative number. This not only confirms the fact that as $x \to -\infty$, $f(x) \to 0$, it tells us that $f(x) \to 0^-$. In other words, the graph of $y = f(x)$ is a little bit *below* the x-axis as we move to the far left.

- *The behavior of $y = f(x)$ as $x \to \infty$:* On the flip side, we can imagine substituting very large positive numbers in for x and looking at the behavior of $f(x)$. For example, let $x = 1$ billion. Proceeding as before, we get

$$f(1 \text{ billion}) \approx \frac{3 \text{ billion}}{1(\text{billion})^2} \approx \frac{3}{\text{billion}} \approx \text{very small } (+)$$

The larger the number we put in, the smaller the positive number we would get out. In other words, as $x \to \infty$, $f(x) \to 0^+$, so the graph of $y = f(x)$ is a little bit *above* the x-axis as we look toward the far right.

Graphically, we have[7]

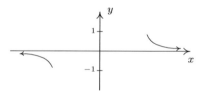

6. Lastly, we construct a sign diagram for $f(x)$. The x-values excluded from the domain of f are $x = \pm 2$, and the only zero of f is $x = 0$. Displaying these appropriately on the number line gives us four test intervals, and we choose the test values[8] $x = -3$, $x = -1$, $x = 1$ and $x = 3$. We find $f(-3)$ is $(-)$, $f(-1)$ is $(+)$, $f(1)$ is $(-)$ and $f(3)$ is $(+)$. Combining this with our previous work, we get the graph of $y = f(x)$ below.

---

[7]As with the vertical asymptotes in the previous step, we know only the behavior of the graph as $x \to \pm\infty$. For that reason, we provide no x-axis labels.

[8]In this particular case, we can eschew test values, since our analysis of the behavior of f near the vertical asymptotes and our end behavior analysis have given us the signs on each of the test intervals. In general, however, this won't always be the case, so for demonstration purposes, we continue with our usual construction.

4.2 GRAPHS OF RATIONAL FUNCTIONS

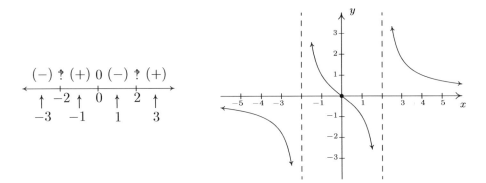

A couple of notes are in order. First, the graph of $y = f(x)$ certainly seems to possess symmetry with respect to the origin. In fact, we can check $f(-x) = -f(x)$ to see that f is an odd function. In some textbooks, checking for symmetry is part of the standard procedure for graphing rational functions; but since it happens comparatively rarely[9] we'll just point it out when we see it. Also note that while $y = 0$ is the horizontal asymptote, the graph of f actually crosses the x-axis at $(0, 0)$. The myth that graphs of rational functions can't cross their horizontal asymptotes is completely false,[10] as we shall see again in our next example.

Example 4.2.2. Sketch a detailed graph of $g(x) = \dfrac{2x^2 - 3x - 5}{x^2 - x - 6}$.

Solution.

1. Setting $x^2 - x - 6 = 0$ gives $x = -2$ and $x = 3$. Our domain is $(-\infty, -2) \cup (-2, 3) \cup (3, \infty)$.

2. Factoring $g(x)$ gives $g(x) = \dfrac{(2x-5)(x+1)}{(x-3)(x+2)}$. There is no cancellation, so $g(x)$ is in lowest terms.

3. To find the x-intercept we set $y = g(x) = 0$. Using the factored form of $g(x)$ above, we find the zeros to be the solutions of $(2x - 5)(x + 1) = 0$. We obtain $x = \frac{5}{2}$ and $x = -1$. Since both of these numbers are in the domain of g, we have two x-intercepts, $\left(\frac{5}{2}, 0\right)$ and $(-1, 0)$. To find the y-intercept, we set $x = 0$ and find $y = g(0) = \frac{5}{6}$, so our y-intercept is $\left(0, \frac{5}{6}\right)$.

4. Since $g(x)$ was given to us in lowest terms, we have, once again by Theorem 4.1 vertical asymptotes $x = -2$ and $x = 3$. Keeping in mind $g(x) = \dfrac{(2x-5)(x+1)}{(x-3)(x+2)}$, we proceed to our analysis near each of these values.

    - *The behavior of $y = g(x)$ as $x \to -2$:* As $x \to -2^-$, we imagine substituting a number a little bit less than -2. We have
    $$g(x) \approx \frac{(-9)(-1)}{(-5)(\text{very small } (-))} \approx \frac{9}{\text{very small } (+)} \approx \text{very big } (+)$$

---

[9] And Jeff doesn't think much of it to begin with...
[10] That's why we called it a MYTH!

so as $x \to -2^-$, $g(x) \to \infty$. On the flip side, as $x \to -2^+$, we get

$$g(x) \approx \frac{9}{\text{very small } (-)} \approx \text{very big } (-)$$

so $g(x) \to -\infty$.

- *The behavior of $y = g(x)$ as $x \to 3$:* As $x \to 3^-$, we imagine plugging in a number just shy of 3. We have

$$g(x) \approx \frac{(1)(4)}{(\text{ very small } (-))(5)} \approx \frac{4}{\text{very small } (-)} \approx \text{very big } (-)$$

Hence, as $x \to 3^-$, $g(x) \to -\infty$. As $x \to 3^+$, we get

$$g(x) \approx \frac{4}{\text{very small } (+)} \approx \text{very big } (+)$$

so $g(x) \to \infty$.

Graphically, we have (again, without labels on the y-axis)

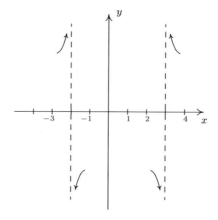

5. Since the degrees of the numerator and denominator of $g(x)$ are the same, we know from Theorem 4.2 that we can find the horizontal asymptote of the graph of g by taking the ratio of the leading terms coefficients, $y = \frac{2}{1} = 2$. However, if we take the time to do a more detailed analysis, we will be able to reveal some 'hidden' behavior which would be lost otherwise.[11] As in the discussion following Theorem 4.2, we use the result of the long division $(2x^2 - 3x - 5) \div (x^2 - x - 6)$ to rewrite $g(x) = \frac{2x^2-3x-5}{x^2-x-6}$ as $g(x) = 2 - \frac{x-7}{x^2-x-6}$. We focus our attention on the term $\frac{x-7}{x^2-x-6}$.

---

[11]That is, if you use a calculator to graph. Once again, Calculus is the ultimate graphing power tool.

4.2 Graphs of Rational Functions

- *The behavior of $y = g(x)$ as $x \to -\infty$:* If imagine substituting $x = -1$ billion into $\frac{x-7}{x^2-x-6}$, we estimate $\frac{x-7}{x^2-x-6} \approx \frac{-1 \text{ billion}}{1 \text{billion}^2} \approx$ very small $(-)$.[12] Hence,

$$g(x) = 2 - \frac{x-7}{x^2-x-6} \approx 2 - \text{very small } (-) = 2 + \text{very small } (+)$$

In other words, as $x \to -\infty$, the graph of $y = g(x)$ is a little bit *above* the line $y = 2$.

- *The behavior of $y = g(x)$ as $x \to \infty$.* To consider $\frac{x-7}{x^2-x-6}$ as $x \to \infty$, we imagine substituting $x = 1$ billion and, going through the usual mental routine, find

$$\frac{x-7}{x^2-x-6} \approx \text{very small } (+)$$

Hence, $g(x) \approx 2-$ very small $(+)$, in other words, the graph of $y = g(x)$ is just *below* the line $y = 2$ as $x \to \infty$.

On $y = g(x)$, we have (again, without labels on the x-axis)

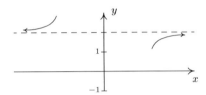

6. Finally we construct our sign diagram. We place an '?' above $x = -2$ and $x = 3$, and a '0' above $x = \frac{5}{2}$ and $x = -1$. Choosing test values in the test intervals gives us $f(x)$ is $(+)$ on the intervals $(-\infty, -2)$, $\left(-1, \frac{5}{2}\right)$ and $(3, \infty)$, and $(-)$ on the intervals $(-2, -1)$ and $\left(\frac{5}{2}, 3\right)$. As we piece together all of the information, we note that the graph must cross the horizontal asymptote at some point after $x = 3$ in order for it to approach $y = 2$ from underneath. This is the subtlety that we would have missed had we skipped the long division and subsequent end behavior analysis. We can, in fact, find exactly when the graph crosses $y = 2$. As a result of the long division, we have $g(x) = 2 - \frac{x-7}{x^2-x-6}$. For $g(x) = 2$, we would need $\frac{x-7}{x^2-x-6} = 0$. This gives $x - 7 = 0$, or $x = 7$. Note that $x - 7$ is the remainder when $2x^2 - 3x - 5$ is divided by $x^2 - x - 6$, so it makes sense that for $g(x)$ to equal the quotient 2, the remainder from the division must be 0. Sure enough, we find $g(7) = 2$. Moreover, it stands to reason that g must attain a relative minimum at some point past $x = 7$. Calculus verifies that at $x = 13$, we have such a minimum at exactly $(13, 1.96)$. The reader is challenged to find calculator windows which show the graph crossing its horizontal asymptote on one window, and the relative minimum in the other.

---

[12] In the denominator, we would have $(1 \text{billion})^2 - 1 \text{billion} - 6$. It's easy to see why the 6 is insignificant, but to ignore the 1 billion seems criminal. However, compared to $(1 \text{ billion})^2$, it's on the insignificant side; it's 10^{18} versus 10^9. We are once again using the fact that for polynomials, end behavior is determined by the leading term, so in the denominator, the x^2 term wins out over the x term.

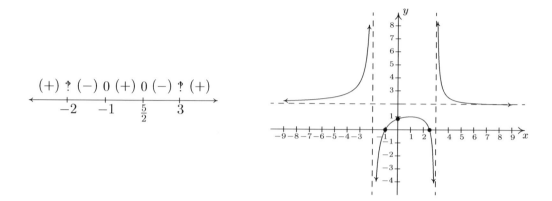

Our next example gives us an opportunity to more thoroughly analyze a slant asymptote.

Example 4.2.3. Sketch a detailed graph of $h(x) = \dfrac{2x^3 + 5x^2 + 4x + 1}{x^2 + 3x + 2}$.

Solution.

1. For domain, you know the drill. Solving $x^2 + 3x + 2 = 0$ gives $x = -2$ and $x = -1$. Our answer is $(-\infty, -2) \cup (-2, -1) \cup (-1, \infty)$.

2. To reduce $h(x)$, we need to factor the numerator and denominator. To factor the numerator, we use the techniques[13] set forth in Section 3.3 and we get

$$h(x) = \frac{2x^3 + 5x^2 + 4x + 1}{x^2 + 3x + 2} = \frac{(2x+1)(x+1)^2}{(x+2)(x+1)} = \frac{(2x+1)(x+1)^{\cancel{2}^1}}{(x+2)\cancel{(x+1)}} = \frac{(2x+1)(x+1)}{x+2}$$

We will use this reduced formula for $h(x)$ as long as we're not substituting $x = -1$. To make this exclusion specific, we write $h(x) = \frac{(2x+1)(x+1)}{x+2}$, $x \neq -1$.

3. To find the x-intercepts, as usual, we set $h(x) = 0$ and solve. Solving $\frac{(2x+1)(x+1)}{x+2} = 0$ yields $x = -\frac{1}{2}$ and $x = -1$. The latter isn't in the domain of h, so we exclude it. Our only x-intercept is $\left(-\frac{1}{2}, 0\right)$. To find the y-intercept, we set $x = 0$. Since $0 \neq -1$, we can use the reduced formula for $h(x)$ and we get $h(0) = \frac{1}{2}$ for a y-intercept of $\left(0, \frac{1}{2}\right)$.

4. From Theorem 4.1, we know that since $x = -2$ still poses a threat in the denominator of the reduced function, we have a vertical asymptote there. As for $x = -1$, the factor $(x+1)$ was canceled from the denominator when we reduced $h(x)$, so it no longer causes trouble there. This means that we get a hole when $x = -1$. To find the y-coordinate of the hole, we substitute $x = -1$ into $\frac{(2x+1)(x+1)}{x+2}$, per Theorem 4.1 and get 0. Hence, we have a hole on

---

[13] Bet you never thought you'd never see *that* stuff again before the Final Exam!

the x-axis at $(-1, 0)$. It should make you uncomfortable plugging $x = -1$ into the reduced formula for $h(x)$, especially since we've made such a big deal concerning the stipulation about not letting $x = -1$ for that formula. What we are really doing is taking a Calculus short-cut to the more detailed kind of analysis near $x = -1$ which we will show below. Speaking of which, for the discussion that follows, we will use the formula $h(x) = \frac{(2x+1)(x+1)}{x+2}$, $x \neq -1$.

- *The behavior of $y = h(x)$ as $x \to -2$:* As $x \to -2^-$, we imagine substituting a number a little bit less than -2. We have $h(x) \approx \frac{(-3)(-1)}{\text{(very small }(-))} \approx \frac{3}{\text{(very small }(-))} \approx$ very big $(-)$ thus as $x \to -2^-$, $h(x) \to -\infty$. On the other side of -2, as $x \to -2^+$, we find that $h(x) \approx \frac{3}{\text{very small }(+)} \approx$ very big $(+)$, so $h(x) \to \infty$.

- *The behavior of $y = h(x)$ as $x \to -1$.* As $x \to -1^-$, we imagine plugging in a number a bit less than $x = -1$. We have $h(x) \approx \frac{(-1)(\text{very small }(-))}{1}$ = very small $(+)$ Hence, as $x \to -1^-$, $h(x) \to 0^+$. This means that as $x \to -1^-$, the graph is a bit above the point $(-1, 0)$. As $x \to -1^+$, we get $h(x) \approx \frac{(-1)(\text{very small }(+))}{1}$ = very small $(-)$. This gives us that as $x \to -1^+$, $h(x) \to 0^-$, so the graph is a little bit lower than $(-1, 0)$ here.

Graphically, we have

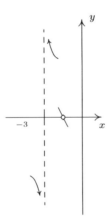

5. For end behavior, we note that the degree of the numerator of $h(x)$, $2x^3 + 5x^2 + 4x + 1$, is 3 and the degree of the denominator, $x^2 + 3x + 2$, is 2 so by Theorem 4.3, the graph of $y = h(x)$ has a slant asymptote. For $x \to \pm\infty$, we are far enough away from $x = -1$ to use the reduced formula, $h(x) = \frac{(2x+1)(x+1)}{x+2}$, $x \neq -1$. To perform long division, we multiply out the numerator and get $h(x) = \frac{2x^2+3x+1}{x+2}$, $x \neq -1$, and rewrite $h(x) = 2x - 1 + \frac{3}{x+2}$, $x \neq -1$. By Theorem 4.3, the slant asymptote is $y = 2x - 1$, and to better see *how* the graph approaches the asymptote, we focus our attention on the term generated from the remainder, $\frac{3}{x+2}$.

- *The behavior of $y = h(x)$ as $x \to -\infty$:* Substituting $x = -1$ billion into $\frac{3}{x+2}$, we get the estimate $\frac{3}{-1 \text{ billion}} \approx$ very small $(-)$. Hence, $h(x) = 2x - 1 + \frac{3}{x+2} \approx 2x - 1 +$ very small $(-)$. This means the graph of $y = h(x)$ is a little bit *below* the line $y = 2x - 1$ as $x \to -\infty$.

- *The behavior of $y = h(x)$ as $x \to \infty$:* If $x \to \infty$, then $\frac{3}{x+2} \approx$ very small $(+)$. This means $h(x) \approx 2x - 1 +$ very small $(+)$, or that the graph of $y = h(x)$ is a little bit *above* the line $y = 2x - 1$ as $x \to \infty$.

Graphically we have

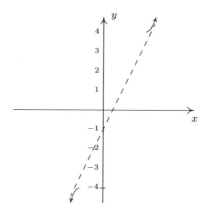

6. To make our sign diagram, we place an '?' above $x = -2$ and $x = -1$ and a '0' above $x = -\frac{1}{2}$. On our four test intervals, we find $h(x)$ is $(+)$ on $(-2, -1)$ and $\left(-\frac{1}{2}, \infty\right)$ and $h(x)$ is $(-)$ on $(-\infty, -2)$ and $\left(-1, -\frac{1}{2}\right)$. Putting all of our work together yields the graph below.

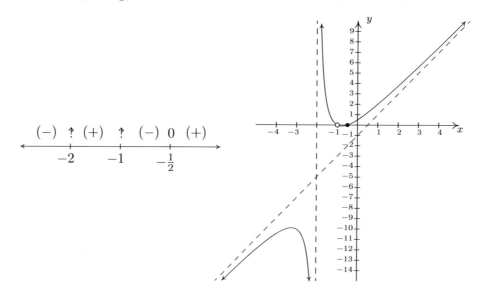

We could ask whether the graph of $y = h(x)$ crosses its slant asymptote. From the formula $h(x) = 2x - 1 + \frac{3}{x+2}$, $x \neq -1$, we see that if $h(x) = 2x - 1$, we would have $\frac{3}{x+2} = 0$. Since this will never happen, we conclude the graph never crosses its slant asymptote.[14] □

---
[14]But rest assured, some graphs do!

4.2 Graphs of Rational Functions

We end this section with an example that shows it's not all pathological weirdness when it comes to rational functions and technology still has a role to play in studying their graphs at this level.

Example 4.2.4. Sketch the graph of $r(x) = \dfrac{x^4 + 1}{x^2 + 1}$.

Solution.

1. The denominator $x^2 + 1$ is never zero so the domain is $(-\infty, \infty)$.

2. With no real zeros in the denominator, $x^2 + 1$ is an irreducible quadratic. Our only hope of reducing $r(x)$ is if $x^2 + 1$ is a factor of $x^4 + 1$. Performing long division gives us

$$\frac{x^4 + 1}{x^2 + 1} = x^2 - 1 + \frac{2}{x^2 + 1}$$

   The remainder is not zero so $r(x)$ is already reduced.

3. To find the x-intercept, we'd set $r(x) = 0$. Since there are no real solutions to $\frac{x^4+1}{x^2+1} = 0$, we have no x-intercepts. Since $r(0) = 1$, we get $(0, 1)$ as the y-intercept.

4. This step doesn't apply to r, since its domain is all real numbers.

5. For end behavior, we note that since the degree of the numerator is exactly *two* more than the degree of the denominator, neither Theorems 4.2 nor 4.3 apply.[15] We know from our attempt to reduce $r(x)$ that we can rewrite $r(x) = x^2 - 1 + \frac{2}{x^2+1}$, so we focus our attention on the term corresponding to the remainder, $\frac{2}{x^2+1}$ It should be clear that as $x \to \pm\infty$, $\frac{2}{x^2+1} \approx$ very small $(+)$, which means $r(x) \approx x^2 - 1 +$ very small $(+)$. So the graph $y = r(x)$ is a little bit *above* the graph of the parabola $y = x^2 - 1$ as $x \to \pm\infty$. Graphically,

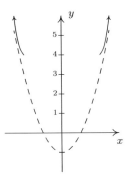

6. There isn't much work to do for a sign diagram for $r(x)$, since its domain is all real numbers and it has no zeros. Our sole test interval is $(-\infty, \infty)$, and since we know $r(0) = 1$, we conclude $r(x)$ is $(+)$ for all real numbers. At this point, we don't have much to go on for

---
[15]This won't stop us from giving it the old community college try, however!

a graph.[16] Below is a comparison of what we have determined analytically versus what the calculator shows us. We have no way to detect the relative extrema analytically[17] apart from brute force plotting of points, which is done more efficiently by the calculator.

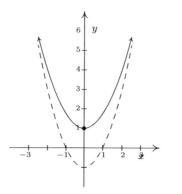
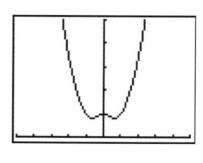

□

As usual, the authors offer no apologies for what may be construed as 'pedantry' in this section. We feel that the detail presented in this section is necessary to obtain a firm grasp of the concepts presented here and it also serves as an introduction to the methods employed in Calculus. As we have said many times in the past, your instructor will decide how much, if any, of the kinds of details presented here are 'mission critical' to your understanding of Precalculus. Without further delay, we present you with this section's Exercises.

---

[16] So even Jeff at this point may check for symmetry! We leave it to the reader to show $r(-x) = r(x)$ so r is even, and, hence, its graph is symmetric about the y-axis.

[17] Without appealing to Calculus, of course.

4.2 Graphs of Rational Functions

4.2.1 Exercises

In Exercises 1 - 16, use the six-step procedure to graph the rational function. Be sure to draw any asymptotes as dashed lines.

1. $f(x) = \dfrac{4}{x+2}$

2. $f(x) = \dfrac{5x}{6-2x}$

3. $f(x) = \dfrac{1}{x^2}$

4. $f(x) = \dfrac{1}{x^2+x-12}$

5. $f(x) = \dfrac{2x-1}{-2x^2-5x+3}$

6. $f(x) = \dfrac{x}{x^2+x-12}$

7. $f(x) = \dfrac{4x}{x^2+4}$

8. $f(x) = \dfrac{4x}{x^2-4}$

9. $f(x) = \dfrac{x^2-x-12}{x^2+x-6}$

10. $f(x) = \dfrac{3x^2-5x-2}{x^2-9}$

11. $f(x) = \dfrac{x^2-x-6}{x+1}$

12. $f(x) = \dfrac{x^2-x}{3-x}$

13. $f(x) = \dfrac{x^3+2x^2+x}{x^2-x-2}$

14. $f(x) = \dfrac{-x^3+4x}{x^2-9}$

15. $f(x) = \dfrac{x^3-2x^2+3x}{2x^2+2}$

16.[18] $f(x) = \dfrac{x^2-2x+1}{x^3+x^2-2x}$

In Exercises 17 - 20, graph the rational function by applying transformations to the graph of $y = \dfrac{1}{x}$.

17. $f(x) = \dfrac{1}{x-2}$

18. $g(x) = 1 - \dfrac{3}{x}$

19. $h(x) = \dfrac{-2x+1}{x}$ (Hint: Divide)

20. $j(x) = \dfrac{3x-7}{x-2}$ (Hint: Divide)

21. Discuss with your classmates how you would graph $f(x) = \dfrac{ax+b}{cx+d}$. What restrictions must be placed on a, b, c and d so that the graph is indeed a transformation of $y = \dfrac{1}{x}$?

22. In Example 3.1.1 in Section 3.1 we showed that $p(x) = \dfrac{4x+x^3}{x}$ is not a polynomial even though its formula reduced to $4 + x^2$ for $x \neq 0$. However, it is a rational function similar to those studied in the section. With the help of your classmates, graph $p(x)$.

---

[18]Once you've done the six-step procedure, use your calculator to graph this function on the viewing window $[0, 12] \times [0, 0.25]$. What do you see?

23. Let $g(x) = \dfrac{x^4 - 8x^3 + 24x^2 - 72x + 135}{x^3 - 9x^2 + 15x - 7}$. With the help of your classmates, find the x- and y- intercepts of the graph of g. Find the intervals on which the function is increasing, the intervals on which it is decreasing and the local extrema. Find all of the asymptotes of the graph of g and any holes in the graph, if they exist. Be sure to show all of your work including any polynomial or synthetic division. Sketch the graph of g, using more than one picture if necessary to show all of the important features of the graph.

Example 4.2.4 showed us that the six-step procedure cannot tell us everything of importance about the graph of a rational function. Without Calculus, we need to use our graphing calculators to reveal the hidden mysteries of rational function behavior. Working with your classmates, use a graphing calculator to examine the graphs of the rational functions given in Exercises 24 - 27. Compare and contrast their features. Which features can the six-step process reveal and which features cannot be detected by it?

24. $f(x) = \dfrac{1}{x^2 + 1}$
25. $f(x) = \dfrac{x}{x^2 + 1}$
26. $f(x) = \dfrac{x^2}{x^2 + 1}$
27. $f(x) = \dfrac{x^3}{x^2 + 1}$

4.2.2 Answers

1. $f(x) = \dfrac{4}{x+2}$
   Domain: $(-\infty, -2) \cup (-2, \infty)$
   No x-intercepts
   y-intercept: $(0, 2)$
   Vertical asymptote: $x = -2$
   As $x \to -2^-$, $f(x) \to -\infty$
   As $x \to -2^+$, $f(x) \to \infty$
   Horizontal asymptote: $y = 0$
   As $x \to -\infty$, $f(x) \to 0^-$
   As $x \to \infty$, $f(x) \to 0^+$

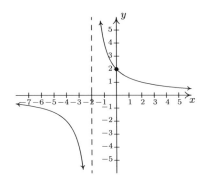

2. $f(x) = \dfrac{5x}{6 - 2x}$
   Domain: $(-\infty, 3) \cup (3, \infty)$
   x-intercept: $(0, 0)$
   y-intercept: $(0, 0)$
   Vertical asymptote: $x = 3$
   As $x \to 3^-$, $f(x) \to \infty$
   As $x \to 3^+$, $f(x) \to -\infty$
   Horizontal asymptote: $y = -\dfrac{5}{2}$
   As $x \to -\infty$, $f(x) \to -\dfrac{5}{2}^+$
   As $x \to \infty$, $f(x) \to -\dfrac{5}{2}^-$

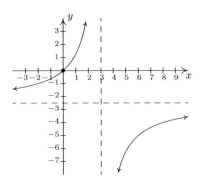

3. $f(x) = \dfrac{1}{x^2}$
   Domain: $(-\infty, 0) \cup (0, \infty)$
   No x-intercepts
   No y-intercepts
   Vertical asymptote: $x = 0$
   As $x \to 0^-$, $f(x) \to \infty$
   As $x \to 0^+$, $f(x) \to \infty$
   Horizontal asymptote: $y = 0$
   As $x \to -\infty$, $f(x) \to 0^+$
   As $x \to \infty$, $f(x) \to 0^+$

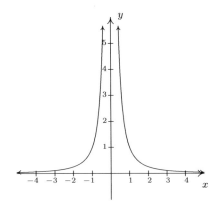

4. $f(x) = \dfrac{1}{x^2 + x - 12} = \dfrac{1}{(x-3)(x+4)}$
Domain: $(-\infty, -4) \cup (-4, 3) \cup (3, \infty)$
No x-intercepts
y-intercept: $\left(0, -\frac{1}{12}\right)$
Vertical asymptotes: $x = -4$ and $x = 3$
As $x \to -4^-$, $f(x) \to \infty$
As $x \to -4^+$, $f(x) \to -\infty$
As $x \to 3^-$, $f(x) \to -\infty$
As $x \to 3^+$, $f(x) \to \infty$
Horizontal asymptote: $y = 0$
As $x \to -\infty$, $f(x) \to 0^+$
As $x \to \infty$, $f(x) \to 0^+$

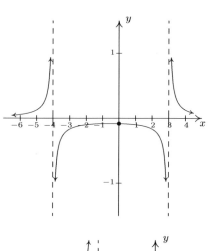

5. $f(x) = \dfrac{2x - 1}{-2x^2 - 5x + 3} = -\dfrac{2x - 1}{(2x - 1)(x + 3)}$
Domain: $(-\infty, -3) \cup \left(-3, \frac{1}{2}\right) \cup \left(\frac{1}{2}, \infty\right)$
No x-intercepts
y-intercept: $\left(0, -\frac{1}{3}\right)$
$f(x) = \dfrac{-1}{x + 3}$, $x \neq \frac{1}{2}$
Hole in the graph at $\left(\frac{1}{2}, -\frac{2}{7}\right)$
Vertical asymptote: $x = -3$
As $x \to -3^-$, $f(x) \to \infty$
As $x \to -3^+$, $f(x) \to -\infty$
Horizontal asymptote: $y = 0$
As $x \to -\infty$, $f(x) \to 0^+$
As $x \to \infty$, $f(x) \to 0^-$

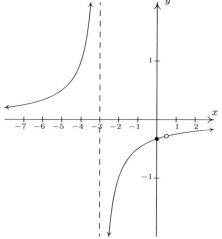

6. $f(x) = \dfrac{x}{x^2 + x - 12} = \dfrac{x}{(x-3)(x+4)}$
Domain: $(-\infty, -4) \cup (-4, 3) \cup (3, \infty)$
x-intercept: $(0, 0)$
y-intercept: $(0, 0)$
Vertical asymptotes: $x = -4$ and $x = 3$
As $x \to -4^-$, $f(x) \to -\infty$
As $x \to -4^+$, $f(x) \to \infty$
As $x \to 3^-$, $f(x) \to -\infty$
As $x \to 3^+$, $f(x) \to \infty$
Horizontal asymptote: $y = 0$
As $x \to -\infty$, $f(x) \to 0^-$
As $x \to \infty$, $f(x) \to 0^+$

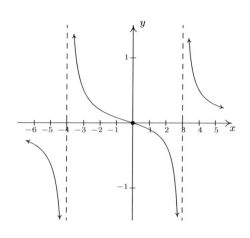

4.2 Graphs of Rational Functions

7. $f(x) = \dfrac{4x}{x^2 + 4}$
   Domain: $(-\infty, \infty)$
   x-intercept: $(0, 0)$
   y-intercept: $(0, 0)$
   No vertical asymptotes
   No holes in the graph
   Horizontal asymptote: $y = 0$
   As $x \to -\infty$, $f(x) \to 0^-$
   As $x \to \infty$, $f(x) \to 0^+$

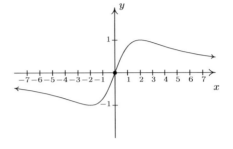

8. $f(x) = \dfrac{4x}{x^2 - 4} = \dfrac{4x}{(x+2)(x-2)}$
   Domain: $(-\infty, -2) \cup (-2, 2) \cup (2, \infty)$
   x-intercept: $(0, 0)$
   y-intercept: $(0, 0)$
   Vertical asymptotes: $x = -2$, $x = 2$
   As $x \to -2^-$, $f(x) \to -\infty$
   As $x \to -2^+$, $f(x) \to \infty$
   As $x \to 2^-$, $f(x) \to -\infty$
   As $x \to 2^+$, $f(x) \to \infty$
   No holes in the graph
   Horizontal asymptote: $y = 0$
   As $x \to -\infty$, $f(x) \to 0^-$
   As $x \to \infty$, $f(x) \to 0^+$

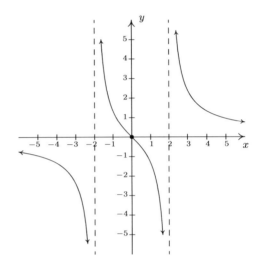

9. $f(x) = \dfrac{x^2 - x - 12}{x^2 + x - 6} = \dfrac{x - 4}{x - 2}$, $x \neq -3$
   Domain: $(-\infty, -3) \cup (-3, 2) \cup (2, \infty)$
   x-intercept: $(4, 0)$
   y-intercept: $(0, 2)$
   Vertical asymptote: $x = 2$
   As $x \to 2^-$, $f(x) \to \infty$
   As $x \to 2^+$, $f(x) \to -\infty$
   Hole at $\left(-3, \tfrac{7}{5}\right)$
   Horizontal asymptote: $y = 1$
   As $x \to -\infty$, $f(x) \to 1^+$
   As $x \to \infty$, $f(x) \to 1^-$

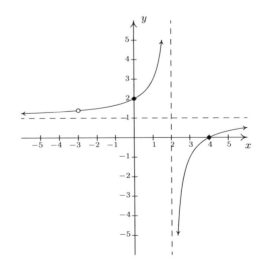

10. $f(x) = \dfrac{3x^2 - 5x - 2}{x^2 - 9} = \dfrac{(3x+1)(x-2)}{(x+3)(x-3)}$
    Domain: $(-\infty, -3) \cup (-3, 3) \cup (3, \infty)$
    x-intercepts: $\left(-\frac{1}{3}, 0\right)$, $(2, 0)$
    y-intercept: $\left(0, \frac{2}{9}\right)$
    Vertical asymptotes: $x = -3, x = 3$
    As $x \to -3^-, f(x) \to \infty$
    As $x \to -3^+, f(x) \to -\infty$
    As $x \to 3^-, f(x) \to -\infty$
    As $x \to 3^+, f(x) \to \infty$
    No holes in the graph
    Horizontal asymptote: $y = 3$
    As $x \to -\infty, f(x) \to 3^+$
    As $x \to \infty, f(x) \to 3^-$

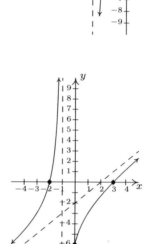

11. $f(x) = \dfrac{x^2 - x - 6}{x + 1} = \dfrac{(x-3)(x+2)}{x+1}$
    Domain: $(-\infty, -1) \cup (-1, \infty)$
    x-intercepts: $(-2, 0), (3, 0)$
    y-intercept: $(0, -6)$
    Vertical asymptote: $x = -1$
    As $x \to -1^-, f(x) \to \infty$
    As $x \to -1^+, f(x) \to -\infty$
    Slant asymptote: $y = x - 2$
    As $x \to -\infty$, the graph is above $y = x - 2$
    As $x \to \infty$, the graph is below $y = x - 2$

12. $f(x) = \dfrac{x^2 - x}{3 - x} = \dfrac{x(x-1)}{3-x}$
    Domain: $(-\infty, 3) \cup (3, \infty)$
    x-intercepts: $(0, 0), (1, 0)$
    y-intercept: $(0, 0)$
    Vertical asymptote: $x = 3$
    As $x \to 3^-, f(x) \to \infty$
    As $x \to 3^+, f(x) \to -\infty$
    Slant asymptote: $y = -x - 2$
    As $x \to -\infty$, the graph is above $y = -x - 2$
    As $x \to \infty$, the graph is below $y = -x - 2$

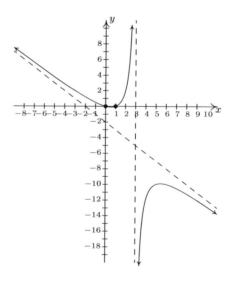

4.2 Graphs of Rational Functions

13. $f(x) = \dfrac{x^3 + 2x^2 + x}{x^2 - x - 2} = \dfrac{x(x+1)}{x-2}$ $x \neq -1$
    Domain: $(-\infty, -1) \cup (-1, 2) \cup (2, \infty)$
    x-intercept: $(0, 0)$
    y-intercept: $(0, 0)$
    Vertical asymptote: $x = 2$
    As $x \to 2^-, f(x) \to -\infty$
    As $x \to 2^+, f(x) \to \infty$
    Hole at $(-1, 0)$
    Slant asymptote: $y = x + 3$
    As $x \to -\infty$, the graph is below $y = x + 3$
    As $x \to \infty$, the graph is above $y = x + 3$

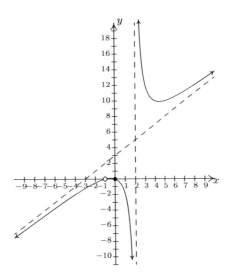

14. $f(x) = \dfrac{-x^3 + 4x}{x^2 - 9}$
    Domain: $(-\infty, -3) \cup (-3, 3) \cup (3, \infty)$
    x-intercepts: $(-2, 0), (0, 0), (2, 0)$
    y-intercept: $(0, 0)$
    Vertical asymptotes: $x = -3, x = 3$
    As $x \to -3^-$, $f(x) \to \infty$
    As $x \to -3^+$, $f(x) \to -\infty$
    As $x \to 3^-$, $f(x) \to \infty$
    As $x \to 3^+$, $f(x) \to -\infty$
    Slant asymptote: $y = -x$
    As $x \to -\infty$, the graph is above $y = -x$
    As $x \to \infty$, the graph is below $y = -x$

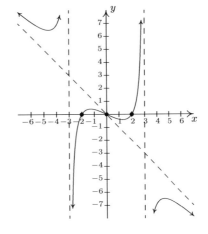

15. $f(x) = \dfrac{x^3 - 2x^2 + 3x}{2x^2 + 2}$
    Domain: $(-\infty, \infty)$
    x-intercept: $(0, 0)$
    y-intercept: $(0, 0)$
    Slant asymptote: $y = \frac{1}{2}x - 1$
    As $x \to -\infty$, the graph is below $y = \frac{1}{2}x - 1$
    As $x \to \infty$, the graph is above $y = \frac{1}{2}x - 1$

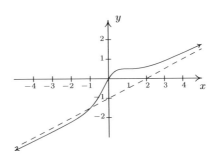

16. $f(x) = \dfrac{x^2 - 2x + 1}{x^3 + x^2 - 2x}$
Domain: $(-\infty, -2) \cup (-2, 0) \cup (0, 1) \cup (1, \infty)$
$f(x) = \dfrac{x - 1}{x(x + 2)}$, $x \neq 1$
No x-intercepts
No y-intercepts
Vertical asymptotes: $x = -2$ and $x = 0$
As $x \to -2^-$, $f(x) \to -\infty$
As $x \to -2^+$, $f(x) \to \infty$
As $x \to 0^-$, $f(x) \to \infty$
As $x \to 0^+$, $f(x) \to -\infty$
Hole in the graph at $(1, 0)$
Horizontal asymptote: $y = 0$
As $x \to -\infty$, $f(x) \to 0^-$
As $x \to \infty$, $f(x) \to 0^+$

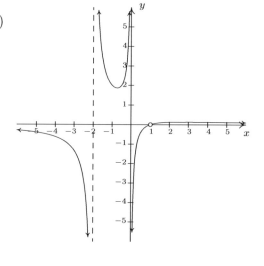

17. $f(x) = \dfrac{1}{x - 2}$

Shift the graph of $y = \dfrac{1}{x}$
to the right 2 units.

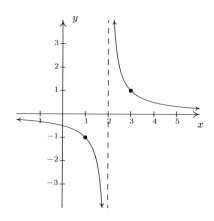

18. $g(x) = 1 - \dfrac{3}{x}$

Vertically stretch the graph of $y = \dfrac{1}{x}$
by a factor of 3.
Reflect the graph of $y = \dfrac{3}{x}$
about the x-axis.
Shift the graph of $y = -\dfrac{3}{x}$
up 1 unit.

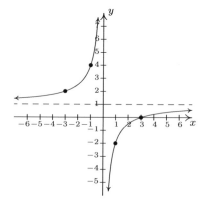

4.2 Graphs of Rational Functions

19. $h(x) = \dfrac{-2x+1}{x} = -2 + \dfrac{1}{x}$

    Shift the graph of $y = \dfrac{1}{x}$
    down 2 units.

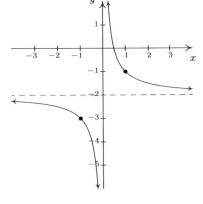

20. $j(x) = \dfrac{3x-7}{x-2} = 3 - \dfrac{1}{x-2}$

    Shift the graph of $y = \dfrac{1}{x}$
    to the right 2 units.
    Reflect the graph of $y = \dfrac{1}{x-2}$
    about the x-axis.
    Shift the graph of $y = -\dfrac{1}{x-2}$
    up 3 units.

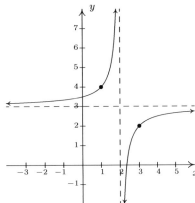

4.3 Rational Inequalities and Applications

In this section, we solve equations and inequalities involving rational functions and explore associated application problems. Our first example showcases the critical difference in procedure between solving a rational equation and a rational inequality.

Example 4.3.1.

1. Solve $\dfrac{x^3 - 2x + 1}{x - 1} = \dfrac{1}{2}x - 1.$

2. Solve $\dfrac{x^3 - 2x + 1}{x - 1} \geq \dfrac{1}{2}x - 1.$

3. Use your calculator to graphically check your answers to 1 and 2.

Solution.

1. To solve the equation, we clear denominators

$$\begin{aligned}
\dfrac{x^3 - 2x + 1}{x - 1} &= \dfrac{1}{2}x - 1 \\
\left(\dfrac{x^3 - 2x + 1}{x - 1}\right) \cdot 2(x - 1) &= \left(\dfrac{1}{2}x - 1\right) \cdot 2(x - 1) \\
2x^3 - 4x + 2 &= x^2 - 3x + 2 \qquad \text{expand} \\
2x^3 - x^2 - x &= 0 \\
x(2x + 1)(x - 1) &= 0 \qquad \text{factor} \\
x &= -\tfrac{1}{2},\, 0,\, 1
\end{aligned}$$

Since we cleared denominators, we need to check for extraneous solutions. Sure enough, we see that $x = 1$ does not satisfy the original equation and must be discarded. Our solutions are $x = -\tfrac{1}{2}$ and $x = 0$.

2. To solve the inequality, it may be tempting to begin as we did with the equation – namely by multiplying both sides by the quantity $(x - 1)$. The problem is that, depending on x, $(x - 1)$ may be positive (which doesn't affect the inequality) or $(x - 1)$ could be negative (which would reverse the inequality). Instead of working by cases, we collect all of the terms on one side of the inequality with 0 on the other and make a sign diagram using the technique given on page 321 in Section 4.2.

$$\begin{aligned}
\dfrac{x^3 - 2x + 1}{x - 1} &\geq \dfrac{1}{2}x - 1 \\
\dfrac{x^3 - 2x + 1}{x - 1} - \dfrac{1}{2}x + 1 &\geq 0 \\
\dfrac{2\left(x^3 - 2x + 1\right) - x(x - 1) + 1(2(x - 1))}{2(x - 1)} &\geq 0 \qquad \text{get a common denominator} \\
\dfrac{2x^3 - x^2 - x}{2x - 2} &\geq 0 \qquad \text{expand}
\end{aligned}$$

Viewing the left hand side as a rational function $r(x)$ we make a sign diagram. The only value excluded from the domain of r is $x = 1$ which is the solution to $2x - 2 = 0$. The zeros of r are the solutions to $2x^3 - x^2 - x = 0$, which we have already found to be $x = 0$, $x = -\frac{1}{2}$ and $x = 1$, the latter was discounted as a zero because it is not in the domain. Choosing test values in each test interval, we construct the sign diagram below.

$$(+) \; 0 \; (-) \; 0 \; (+) \; ? \; (+)$$
$$\xleftarrow{\quad\quad\quad\; \Big|_{-\frac{1}{2}} \quad\quad\; \Big|_{0} \quad\quad \Big|_{1} \quad\quad\;}\rightarrow$$

We are interested in where $r(x) \geq 0$. We find $r(x) > 0$, or $(+)$, on the intervals $\left(-\infty, -\frac{1}{2}\right)$, $(0,1)$ and $(1,\infty)$. We add to these intervals the zeros of r, $-\frac{1}{2}$ and 0, to get our final solution: $\left(-\infty, -\frac{1}{2}\right] \cup [0,1) \cup (1,\infty)$.

3. Geometrically, if we set $f(x) = \frac{x^3 - 2x + 1}{x-1}$ and $g(x) = \frac{1}{2}x - 1$, the solutions to $f(x) = g(x)$ are the x-coordinates of the points where the graphs of $y = f(x)$ and $y = g(x)$ intersect. The solution to $f(x) \geq g(x)$ represents not only where the graphs meet, but the intervals over which the graph of $y = f(x)$ is above $(>)$ the graph of $g(x)$. We obtain the graphs below.

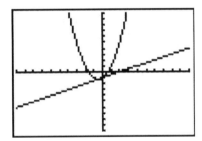
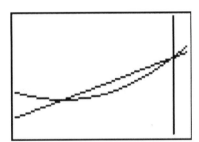

The 'Intersect' command confirms that the graphs cross when $x = -\frac{1}{2}$ and $x = 0$. It is clear from the calculator that the graph of $y = f(x)$ is above the graph of $y = g(x)$ on $\left(-\infty, -\frac{1}{2}\right)$ as well as on $(0, \infty)$. According to the calculator, our solution is then $\left(-\infty, -\frac{1}{2}\right] \cup [0, \infty)$ which *almost* matches the answer we found analytically. We have to remember that f is not defined at $x = 1$, and, even though it isn't shown on the calculator, there is a hole[1] in the graph of $y = f(x)$ when $x = 1$ which is why $x = 1$ is not part of our final answer. □

Next, we explore how rational equations can be used to solve some classic problems involving rates.

Example 4.3.2. Carl decides to explore the Meander River, the location of several recent Sasquatch sightings. From camp, he canoes downstream five miles to check out a purported Sasquatch nest. Finding nothing, he immediately turns around, retraces his route (this time traveling upstream),

---

[1]There is no asymptote at $x = 1$ since the graph is well behaved near $x = 1$. According to Theorem 4.1, there must be a hole there.

and returns to camp 3 hours after he left. If Carl canoes at a rate of 6 miles per hour in still water, how fast was the Meander River flowing on that day?

Solution. We are given information about distances, rates (speeds) and times. The basic principle relating these quantities is:

$$\text{distance} = \text{rate} \cdot \text{time}$$

The first observation to make, however, is that the distance, rate and time given to us aren't 'compatible': the distance given is the distance for only *part* of the trip, the rate given is the speed Carl can canoe in still water, not in a flowing river, and the time given is the duration of the *entire* trip. Ultimately, we are after the speed of the river, so let's call that R measured in miles per hour to be consistent with the other rate given to us. To get started, let's divide the trip into its two parts: the initial trip downstream and the return trip upstream. For the downstream trip, all we know is that the distance traveled is 5 miles.

$$\begin{aligned} \text{distance downstream} &= \text{rate traveling downstream} \cdot \text{time traveling downstream} \\ 5\,\text{miles} &= \text{rate traveling downstream} \cdot \text{time traveling downstream} \end{aligned}$$

Since the return trip upstream followed the same route as the trip downstream, we know that the distance traveled upstream is also 5 miles.

$$\begin{aligned} \text{distance upstream} &= \text{rate traveling upstream} \cdot \text{time traveling upstream} \\ 5\,\text{miles} &= \text{rate traveling upstream} \cdot \text{time traveling upstream} \end{aligned}$$

We are told Carl can canoe at a rate of 6 miles per hour in still water. How does this figure into the rates traveling upstream and downstream? The speed the canoe travels in the river is a combination of the speed at which Carl can propel the canoe in still water, 6 miles per hour, and the speed of the river, which we're calling R. When traveling downstream, the river is helping Carl along, so we *add* these two speeds:

$$\begin{aligned} \text{rate traveling downstream} &= \text{rate Carl propels the canoe} + \text{speed of the river} \\ &= 6\,\tfrac{\text{miles}}{\text{hour}} + R\,\tfrac{\text{miles}}{\text{hour}} \end{aligned}$$

So our downstream speed is $(6+R)\tfrac{\text{miles}}{\text{hour}}$. Substituting this into our 'distance-rate-time' equation for the downstream part of the trip, we get:

$$\begin{aligned} 5\,\text{miles} &= \text{rate traveling downstream} \cdot \text{time traveling downstream} \\ 5\,\text{miles} &= (6+R)\tfrac{\text{miles}}{\text{hour}} \cdot \text{time traveling downstream} \end{aligned}$$

When traveling upstream, Carl works against the current. Since the canoe manages to travel upstream, the speed Carl can canoe in still water is greater than the river's speed, so we *subtract* the river's speed *from* Carl's canoing speed to get:

$$\begin{aligned} \text{rate traveling upstream} &= \text{rate Carl propels the canoe} - \text{river speed} \\ &= 6\,\tfrac{\text{miles}}{\text{hour}} - R\,\tfrac{\text{miles}}{\text{hour}} \end{aligned}$$

Proceeding as before, we get

4.3 RATIONAL INEQUALITIES AND APPLICATIONS

$$5 \text{ miles} = \text{rate traveling upstream} \cdot \text{time traveling upstream}$$
$$5 \text{ miles} = (6 - R)\tfrac{\text{miles}}{\text{hour}} \cdot \text{time traveling upstream}$$

The last piece of information given to us is that the total trip lasted 3 hours. If we let t_{down} denote the time of the downstream trip and t_{up} the time of the upstream trip, we have: $t_{\text{down}} + t_{\text{up}} = 3$ hours. Substituting t_{down} and t_{up} into the 'distance-rate-time' equations, we get (suppressing the units) *three* equations in *three* unknowns:[2]

$$\begin{cases} E1 & (6+R)\, t_{\text{down}} = 5 \\ E2 & (6-R)\, t_{\text{up}} = 5 \\ E3 & t_{\text{down}} + t_{\text{up}} = 3 \end{cases}$$

Since we are ultimately after R, we need to use these three equations to get at least one equation involving only R. To that end, we solve $E1$ for t_{down} by dividing both sides[3] by the quantity $(6+R)$ to get $t_{\text{down}} = \frac{5}{6+R}$. Similarly, we solve $E2$ for t_{up} and get $t_{\text{up}} = \frac{5}{6-R}$. Substituting these into $E3$, we get:[4]

$$\frac{5}{6+R} + \frac{5}{6-R} = 3.$$

Clearing denominators, we get $5(6-R) + 5(6+R) = 3(6+R)(6-R)$ which reduces to $R^2 = 16$. We find $R = \pm 4$, and since R represents the speed of the river, we choose $R = 4$. On the day in question, the Meander River is flowing at a rate of 4 miles per hour. □

One of the important lessons to learn from Example 4.3.2 is that speeds, and more generally, rates, are additive. As we see in our next example, the concept of rate and its associated principles can be applied to a wide variety of problems - not just 'distance-rate-time' scenarios.

Example 4.3.3. Working alone, Taylor can weed the garden in 4 hours. If Carl helps, they can weed the garden in 3 hours. How long would it take for Carl to weed the garden on his own?

Solution. The key relationship between work and time which we use in this problem is:

$$\text{amount of work done} = \text{rate of work} \cdot \text{time spent working}$$

We are told that, working alone, Taylor can weed the garden in 4 hours. In Taylor's case then:

$$\text{amount of work Taylor does} = \text{rate of Taylor working} \cdot \text{time Taylor spent working}$$
$$1 \text{ garden} = (\text{rate of Taylor working}) \cdot (4 \text{ hours})$$

So we have that the rate Taylor works is $\frac{1 \text{ garden}}{4 \text{ hours}} = \frac{1}{4} \frac{\text{garden}}{\text{hour}}$. We are also told that when working together, Taylor and Carl can weed the garden in just 3 hours. We have:

---

[2] This is called a *system* of equations. No doubt, you've had experience with these things before, and we will study systems in greater detail in Chapter 8.

[3] While we usually discourage dividing both sides of an equation by a variable expression, we know $(6+R) \neq 0$ since otherwise we couldn't possibly multiply it by t_{down} and get 5.

[4] The reader is encouraged to verify that the units in this equation are the same on both sides. To get you started, the units on the '3' is 'hours.'

$$\text{amount of work done together} = \text{rate of working together} \cdot \text{time spent working together}$$
$$1\,\text{garden} = (\text{rate of working together}) \cdot (3\,\text{hours})$$

From this, we find that the rate of Taylor and Carl working together is $\frac{1\,\text{garden}}{3\,\text{hours}} = \frac{1}{3}\frac{\text{garden}}{\text{hour}}$. We are asked to find out how long it would take for Carl to weed the garden on his own. Let us call this unknown t, measured in hours to be consistent with the other times given to us in the problem. Then:

$$\text{amount of work Carl does} = \text{rate of Carl working} \cdot \text{time Carl spent working}$$
$$1\,\text{garden} = (\text{rate of Carl working}) \cdot (t\,\text{hours})$$

In order to find t, we need to find the rate of Carl working, so let's call this quantity R, with units $\frac{\text{garden}}{\text{hour}}$. Using the fact that rates are additive, we have:

$$\text{rate working together} = \text{rate of Taylor working} + \text{rate of Carl working}$$
$$\frac{1}{3}\frac{\text{garden}}{\text{hour}} = \frac{1}{4}\frac{\text{garden}}{\text{hour}} + R\frac{\text{garden}}{\text{hour}}$$

so that $R = \frac{1}{12}\frac{\text{garden}}{\text{hour}}$. Substituting this into our 'work-rate-time' equation for Carl, we get:

$$1\,\text{garden} = (\text{rate of Carl working}) \cdot (t\,\text{hours})$$
$$1\,\text{garden} = \left(\frac{1}{12}\frac{\text{garden}}{\text{hour}}\right) \cdot (t\,\text{hours})$$

Solving $1 = \frac{1}{12}t$, we get $t = 12$, so it takes Carl 12 hours to weed the garden on his own.[5] □

As is common with 'word problems' like Examples 4.3.2 and 4.3.3, there is no short-cut to the answer. We encourage the reader to carefully think through and apply the basic principles of rate to each (potentially different!) situation. It is time well spent. We also encourage the tracking of units, especially in the early stages of the problem. Not only does this promote uniformity in the units, it also serves as a quick means to check if an equation makes sense.[6]

Our next example deals with the average cost function, first introduced on page 82, as applied to PortaBoy Game systems from Example 2.1.5 in Section 2.1.

Example 4.3.4. Given a cost function $C(x)$, which returns the total cost of producing x items, recall that the average cost function, $\overline{C}(x) = \frac{C(x)}{x}$ computes the cost per item when x items are produced. Suppose the cost C, in dollars, to produce x PortaBoy game systems for a local retailer is $C(x) = 80x + 150$, $x \geq 0$.

1. Find an expression for the average cost function $\overline{C}(x)$.

2. Solve $\overline{C}(x) < 100$ and interpret.

---
[5] Carl would much rather spend his time writing open-source Mathematics texts than gardening anyway.
[6] In other words, make sure you don't try to add apples to oranges!

4.3 RATIONAL INEQUALITIES AND APPLICATIONS   347

3. Determine the behavior of $\overline{C}(x)$ as $x \to \infty$ and interpret.

Solution.

1. From $\overline{C}(x) = \frac{C(x)}{x}$, we obtain $\overline{C}(x) = \frac{80x+150}{x}$. The domain of C is $x \geq 0$, but since $x = 0$ causes problems for $\overline{C}(x)$, we get our domain to be $x > 0$, or $(0, \infty)$.

2. Solving $\overline{C}(x) < 100$ means we solve $\frac{80x+150}{x} < 100$. We proceed as in the previous example.

$$\frac{80x + 150}{x} < 100$$

$$\frac{80x + 150}{x} - 100 < 0$$

$$\frac{80x + 150 - 100x}{x} < 0 \quad \text{common denominator}$$

$$\frac{150 - 20x}{x} < 0$$

If we take the left hand side to be a rational function $r(x)$, we need to keep in mind that the applied domain of the problem is $x > 0$. This means we consider only the positive half of the number line for our sign diagram. On $(0, \infty)$, r is defined everywhere so we need only look for zeros of r. Setting $r(x) = 0$ gives $150 - 20x = 0$, so that $x = \frac{15}{2} = 7.5$. The test intervals on our domain are $(0, 7.5)$ and $(7.5, \infty)$. We find $r(x) < 0$ on $(7.5, \infty)$.

```
     ? (+) 0 (−)
    ├────┼────→
     0   7.5
```

In the context of the problem, x represents the number of PortaBoy games systems produced and $\overline{C}(x)$ is the average cost to produce each system. Solving $\overline{C}(x) < 100$ means we are trying to find how many systems we need to produce so that the average cost is less than \$100 per system. Our solution, $(7.5, \infty)$ tells us that we need to produce more than 7.5 systems to achieve this. Since it doesn't make sense to produce half a system, our final answer is $[8, \infty)$.

3. When we apply Theorem 4.2 to $\overline{C}(x)$ we find that $y = 80$ is a horizontal asymptote to the graph of $y = \overline{C}(x)$. To more precisely determine the behavior of $\overline{C}(x)$ as $x \to \infty$, we first use long division[7] and rewrite $\overline{C}(x) = 80 + \frac{150}{x}$. As $x \to \infty$, $\frac{150}{x} \to 0^+$, which means $\overline{C}(x) \approx 80 +$ very small $(+)$. Thus the average cost per system is getting closer to \$80 per system. If we set $\overline{C}(x) = 80$, we get $\frac{150}{x} = 0$, which is impossible, so we conclude that $\overline{C}(x) > 80$ for all $x > 0$. This means that the average cost per system is always greater than \$80 per system, but the average cost is approaching this amount as more and more systems are produced. Looking back at Example 2.1.5, we realize \$80 is the variable cost per system −

---

[7]In this case, long division amounts to term-by-term division.

the cost per system above and beyond the fixed initial cost of $150. Another way to interpret our answer is that 'infinitely' many systems would need to be produced to effectively 'zero out' the fixed cost. □

Our next example is another classic 'box with no top' problem.

Example 4.3.5. A box with a square base and no top is to be constructed so that it has a volume of 1000 cubic centimeters. Let x denote the width of the box, in centimeters as seen below.

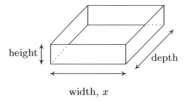

1. Express the height h in centimeters as a function of the width x and state the applied domain.

2. Solve $h(x) \geq x$ and interpret.

3. Find and interpret the behavior of $h(x)$ as $x \to 0^+$ and as $x \to \infty$.

4. Express the surface area S of the box as a function of x and state the applied domain.

5. Use a calculator to approximate (to two decimal places) the dimensions of the box which minimize the surface area.

Solution.

1. We are told that the volume of the box is 1000 cubic centimeters and that x represents the width, in centimeters. From geometry, we know Volume = width × height × depth. Since the base of the box is a square, the width and the depth are both x centimeters. Using h for the height, we have $1000 = x^2 h$, so that $h = \frac{1000}{x^2}$. Using function notation,[8] $h(x) = \frac{1000}{x^2}$ As for the applied domain, in order for there to be a box at all, $x > 0$, and since every such choice of x will return a positive number for the height h we have no other restrictions and conclude our domain is $(0, \infty)$.

2. To solve $h(x) \geq x$, we proceed as before and collect all nonzero terms on one side of the inequality in order to use a sign diagram.

---

[8]That is, $h(x)$ means 'h of x', not 'h times x' here.

4.3 RATIONAL INEQUALITIES AND APPLICATIONS

$$h(x) \geq x$$

$$\frac{1000}{x^2} \geq x$$

$$\frac{1000}{x^2} - x \geq 0$$

$$\frac{1000 - x^3}{x^2} \geq 0 \quad \text{common denominator}$$

We consider the left hand side of the inequality as our rational function $r(x)$. We see r is undefined at $x = 0$, but, as in the previous example, the applied domain of the problem is $x > 0$, so we are considering only the behavior of r on $(0, \infty)$. The sole zero of r comes when $1000 - x^3 = 0$, which is $x = 10$. Choosing test values in the intervals $(0, 10)$ and $(10, \infty)$ gives the following diagram.

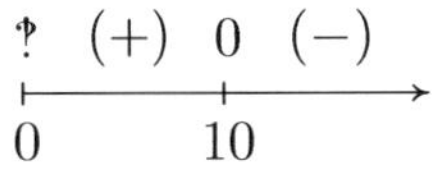

$$\begin{array}{c} ? \quad (+) \quad 0 \quad (-) \\ \vdash\!\!\!\!-\!\!\!\!-\!\!\!\!-\!\!\!\!-\!\!\!\!-\!\!\!\!-\!\!\!\!\!\!+\!\!\!\!-\!\!\!\!-\!\!\!\!-\!\!\!\!-\!\!\!\!\rightarrow \\ 0 \quad\quad 10 \end{array}$$

We see $r(x) > 0$ on $(0, 10)$, and since $r(x) = 0$ at $x = 10$, our solution is $(0, 10]$. In the context of the problem, h represents the height of the box while x represents the width (and depth) of the box. Solving $h(x) \geq x$ is tantamount to finding the values of x which result in a box where the height is at least as big as the width (and, in this case, depth.) Our answer tells us the width of the box can be at most 10 centimeters for this to happen.

3. As $x \to 0^+$, $h(x) = \frac{1000}{x^2} \to \infty$. This means that the smaller the width x (and, in this case, depth), the larger the height h has to be in order to maintain a volume of 1000 cubic centimeters. As $x \to \infty$, we find $h(x) \to 0^+$, which means that in order to maintain a volume of 1000 cubic centimeters, the width and depth must get bigger as the height becomes smaller.

4. Since the box has no top, the surface area can be found by adding the area of each of the sides to the area of the base. The base is a square of dimensions x by x, and each side has dimensions x by h. We get the surface area, $S = x^2 + 4xh$. To get S as a function of x, we substitute $h = \frac{1000}{x^2}$ to obtain $S = x^2 + 4x\left(\frac{1000}{x^2}\right)$. Hence, as a function of x, $S(x) = x^2 + \frac{4000}{x}$. The domain of S is the same as h, namely $(0, \infty)$, for the same reasons as above.

5. A first attempt at the graph of $y = S(x)$ on the calculator may lead to frustration. Chances are good that the first window chosen to view the graph will suggest $y = S(x)$ has the x-axis as a horizontal asymptote. From the formula $S(x) = x^2 + \frac{4000}{x}$, however, we get $S(x) \approx x^2$ as $x \to \infty$, so $S(x) \to \infty$. Readjusting the window, we find S does possess a relative minimum at $x \approx 12.60$. As far as we can tell,[9] this is the only relative extremum, so it is the absolute minimum as well. This means that the width and depth of the box should each measure

---

[9] without Calculus, that is...

approximately 12.60 centimeters. To determine the height, we find $h(12.60) \approx 6.30$, so the height of the box should be approximately 6.30 centimeters.

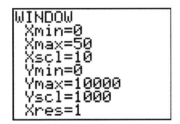 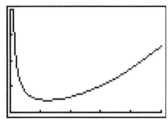 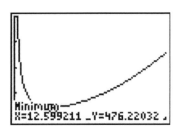

□

4.3.1 Variation

In many instances in the sciences, rational functions are encountered as a result of fundamental natural laws which are typically a result of assuming certain basic relationships between variables. These basic relationships are summarized in the definition below.

Definition 4.5. Suppose x, y and z are variable quantities. We say

- y **varies directly with** (or is **directly proportional to**) x if there is a constant k such that $y = kx$.

- y **varies inversely with** (or is **inversely proportional to**) x if there is a constant k such that $y = \frac{k}{x}$.

- z **varies jointly with** (or is **jointly proportional to**) x and y if there is a constant k such that $z = kxy$.

The constant k in the above definitions is called the **constant of proportionality**.

Example 4.3.6. Translate the following into mathematical equations using Definition 4.5.

1. Hooke's Law: The force F exerted on a spring is directly proportional the extension x of the spring.

2. Boyle's Law: At a constant temperature, the pressure P of an ideal gas is inversely proportional to its volume V.

3. The volume V of a right circular cone varies jointly with the height h of the cone and the square of the radius r of the base.

4. Ohm's Law: The current I through a conductor between two points is directly proportional to the voltage V between the two points and inversely proportional to the resistance R between the two points.

4.3 RATIONAL INEQUALITIES AND APPLICATIONS 351

5. Newton's Law of Universal Gravitation: Suppose two objects, one of mass m and one of mass M, are positioned so that the distance between their centers of mass is r. The gravitational force F exerted on the two objects varies directly with the product of the two masses and inversely with the square of the distance between their centers of mass.

Solution.

1. Applying the definition of direct variation, we get $F = kx$ for some constant k.

2. Since P and V are inversely proportional, we write $P = \frac{k}{V}$.

3. There is a bit of ambiguity here. It's clear that the volume and the height of the cone are represented by the quantities V and h, respectively, but does r represent the radius of the base or the square of the radius of the base? It is the former. Usually, if an algebraic operation is specified (like squaring), it is meant to be expressed in the formula. We apply Definition 4.5 to get $V = khr^2$.

4. Even though the problem doesn't use the phrase 'varies jointly', it is implied by the fact that the current I is related to two different quantities. Since I varies directly with V but inversely with R, we write $I = \frac{kV}{R}$.

5. We write the product of the masses mM and the square of the distance as r^2. We have that F varies directly with mM and inversely with r^2, so $F = \frac{kmM}{r^2}$. □

In many of the formulas in the previous example, more than two varying quantities are related. In practice, however, usually all but two quantities are held constant in an experiment and the data collected is used to relate just two of the variables. Comparing just two varying quantities allows us to view the relationship between them as functional, as the next example illustrates.

Example 4.3.7. According to this website the actual data relating the volume V of a gas and its pressure P used by Boyle and his assistant in 1662 to verify the gas law that bears his name is given below.

V	48	46	44	42	40	38	36	34	32	30	28	26	24
P	29.13	30.56	31.94	33.5	35.31	37	39.31	41.63	44.19	47.06	50.31	54.31	58.81

V	23	22	21	20	19	18	17	16	15	14	13	12
P	61.31	64.06	67.06	70.69	74.13	77.88	82.75	87.88	93.06	100.44	107.81	117.56

1. Use your calculator to generate a scatter diagram for these data using V as the independent variable and P as the dependent variable. Does it appear from the graph that P is inversely proportional to V? Explain.

2. Assuming that P and V do vary inversely, use the data to approximate the constant of proportionality.

3. Use your calculator to determine a 'Power Regression' for this data[10] and use it verify your results in 1 and 2.

Solution.

1. If P really does vary inversely with V, then $P = \frac{k}{V}$ for some constant k. From the data plot, the points do seem to lie along a curve like $y = \frac{k}{x}$.

2. To determine the constant of proportionality, we note that from $P = \frac{k}{V}$, we get $k = PV$. Multiplying each of the volume numbers times each of the pressure numbers,[11] we produce a number which is always approximately 1400. We suspect that $P = \frac{1400}{V}$. Graphing $y = \frac{1400}{x}$ along with the data gives us good reason to believe our hypotheses that P and V are, in fact, inversely related.

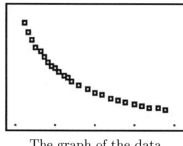

The graph of the data   The data with $y = \frac{1400}{x}$

3. After performing a 'Power Regression', the calculator fits the data to the curve $y = ax^b$ where $a \approx 1400$ and $b \approx -1$ with a correlation coefficient which is darned near perfect.[12] In other words, $y = 1400x^{-1}$ or $y = \frac{1400}{x}$, as we guessed.

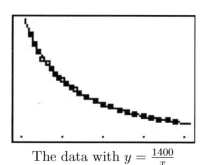

□

---

[10]We will talk more about this in the coming chapters.

[11]You can use tell the calculator to do this arithmetic on the lists and save yourself some time.

[12]We will revisit this example once we have developed logarithms in Chapter 6 to see how we can actually 'linearize' this data and do a linear regression to obtain the same result.

4.3 RATIONAL INEQUALITIES AND APPLICATIONS

4.3.2 Exercises

In Exercises 1 - 6, solve the rational equation. Be sure to check for extraneous solutions.

1. $\dfrac{x}{5x+4} = 3$

2. $\dfrac{3x-1}{x^2+1} = 1$

3. $\dfrac{1}{x+3} + \dfrac{1}{x-3} = \dfrac{x^2-3}{x^2-9}$

4. $\dfrac{2x+17}{x+1} = x+5$

5. $\dfrac{x^2-2x+1}{x^3+x^2-2x} = 1$

6. $\dfrac{-x^3+4x}{x^2-9} = 4x$

In Exercises 7 - 20, solve the rational inequality. Express your answer using interval notation.

7. $\dfrac{1}{x+2} \geq 0$

8. $\dfrac{x-3}{x+2} \leq 0$

9. $\dfrac{x}{x^2-1} > 0$

10. $\dfrac{4x}{x^2+4} \geq 0$

11. $\dfrac{x^2-x-12}{x^2+x-6} > 0$

12. $\dfrac{3x^2-5x-2}{x^2-9} < 0$

13. $\dfrac{x^3+2x^2+x}{x^2-x-2} \geq 0$

14. $\dfrac{x^2+5x+6}{x^2-1} > 0$

15. $\dfrac{3x-1}{x^2+1} \leq 1$

16. $\dfrac{2x+17}{x+1} > x+5$

17. $\dfrac{-x^3+4x}{x^2-9} \geq 4x$

18. $\dfrac{1}{x^2+1} < 0$

19. $\dfrac{x^4-4x^3+x^2-2x-15}{x^3-4x^2} \geq x$

20. $\dfrac{5x^3-12x^2+9x+10}{x^2-1} \geq 3x-1$

21. Carl and Mike start a 3 mile race at the same time. If Mike ran the race at 6 miles per hour and finishes the race 10 minutes before Carl, how fast does Carl run?

22. One day, Donnie observes that the wind is blowing at 6 miles per hour. A unladen swallow nesting near Donnie's house flies three quarters of a mile down the road (in the direction of the wind), turns around, and returns exactly 4 minutes later. What is the airspeed of the unladen swallow? (Here, 'airspeed' is the speed that the swallow can fly in still air.)

23. In order to remove water from a flooded basement, two pumps, each rated at 40 gallons per minute, are used. After half an hour, the one pump burns out, and the second pump finishes removing the water half an hour later. How many gallons of water were removed from the basement?

24. A faucet can fill a sink in 5 minutes while a drain will empty the same sink in 8 minutes. If the faucet is turned on and the drain is left open, how long will it take to fill the sink?

25. Working together, Daniel and Donnie can clean the llama pen in 45 minutes. On his own, Daniel can clean the pen in an hour. How long does it take Donnie to clean the llama pen on his own?

26. In Exercise 32, the function $C(x) = .03x^3 - 4.5x^2 + 225x + 250$, for $x \geq 0$ was used to model the cost (in dollars) to produce x PortaBoy game systems. Using this cost function, find the number of PortaBoys which should be produced to minimize the average cost $\overline{C}$. Round your answer to the nearest number of systems.

27. Suppose we are in the same situation as Example 4.3.5. If the volume of the box is to be 500 cubic centimeters, use your calculator to find the dimensions of the box which minimize the surface area. What is the minimum surface area? Round your answers to two decimal places.

28. The box for the new Sasquatch-themed cereal, 'Crypt-Os', is to have a volume of 140 cubic inches. For aesthetic reasons, the height of the box needs to be 1.62 times the width of the base of the box.[13] Find the dimensions of the box which will minimize the surface area of the box. What is the minimum surface area? Round your answers to two decimal places.

29. Sally is Skippy's neighbor from Exercise 19 in Section 2.3. Sally also wants to plant a vegetable garden along the side of her home. She doesn't have any fencing, but wants to keep the size of the garden to 100 square feet. What are the dimensions of the garden which will minimize the amount of fencing she needs to buy? What is the minimum amount of fencing she needs to buy? Round your answers to the nearest foot. (Note: Since one side of the garden will border the house, Sally doesn't need fencing along that side.)

30. Another Classic Problem: A can is made in the shape of a right circular cylinder and is to hold one pint. (For dry goods, one pint is equal to 33.6 cubic inches.)[14]

    (a) Find an expression for the volume V of the can in terms of the height h and the base radius r.

    (b) Find an expression for the surface area S of the can in terms of the height h and the base radius r. (Hint: The top and bottom of the can are circles of radius r and the side of the can is really just a rectangle that has been bent into a cylinder.)

    (c) Using the fact that $V = 33.6$, write S as a function of r and state its applied domain.

    (d) Use your graphing calculator to find the dimensions of the can which has minimal surface area.

31. A right cylindrical drum is to hold 7.35 cubic feet of liquid. Find the dimensions (radius of the base and height) of the drum which would minimize the surface area. What is the minimum surface area? Round your answers to two decimal places.

32. In Exercise 71 in Section 1.4, the population of Sasquatch in Portage County was modeled by the function $P(t) = \frac{150t}{t+15}$, where $t = 0$ represents the year 1803. When were there fewer than 100 Sasquatch in Portage County?

---

[13] 1.62 is a crude approximation of the so-called 'Golden Ratio' $\phi = \frac{1+\sqrt{5}}{2}$.

[14] According to www.dictionary.com, there are different values given for this conversion. We will stick with 33.6in^3 for this problem.

4.3 Rational Inequalities and Applications

In Exercises 33 - 38, translate the following into mathematical equations.

33. At a constant pressure, the temperature T of an ideal gas is directly proportional to its volume V. (This is Charles's Law)

34. The frequency of a wave f is inversely proportional to the wavelength of the wave λ.

35. The density d of a material is directly proportional to the mass of the object m and inversely proportional to its volume V.

36. The square of the orbital period of a planet P is directly proportional to the cube of the semi-major axis of its orbit a. (This is Kepler's Third Law of Planetary Motion)

37. The drag of an object traveling through a fluid D varies jointly with the density of the fluid ρ and the square of the velocity of the object ν.

38. Suppose two electric point charges, one with charge q and one with charge Q, are positioned r units apart. The electrostatic force F exerted on the charges varies directly with the product of the two charges and inversely with the square of the distance between the charges. (This is Coulomb's Law)

39. According to this webpage, the frequency f of a vibrating string is given by $f = \dfrac{1}{2L}\sqrt{\dfrac{T}{\mu}}$ where T is the tension, μ is the linear mass[15] of the string and L is the length of the vibrating part of the string. Express this relationship using the language of variation.

40. According to the Centers for Disease Control and Prevention www.cdc.gov, a person's Body Mass Index B is directly proportional to his weight W in pounds and inversely proportional to the square of his height h in inches.

    (a) Express this relationship as a mathematical equation.
    (b) If a person who was 5 feet, 10 inches tall weighed 235 pounds had a Body Mass Index of 33.7, what is the value of the constant of proportionality?
    (c) Rewrite the mathematical equation found in part 40a to include the value of the constant found in part 40b and then find your Body Mass Index.

41. We know that the circumference of a circle varies directly with its radius with 2π as the constant of proportionality. (That is, we know $C = 2\pi r$.) With the help of your classmates, compile a list of other basic geometric relationships which can be seen as variations.

---

[15] Also known as the linear density. It is simply a measure of mass per unit length.

4.3.3 Answers

1. $x = -\frac{6}{7}$
2. $x = 1$, $x = 2$
3. $x = -1$
4. $x = -6$, $x = 2$
5. No solution
6. $x = 0$, $x = \pm 2\sqrt{2}$
7. $(-2, \infty)$
8. $(-2, 3]$
9. $(-1, 0) \cup (1, \infty)$
10. $[0, \infty)$
11. $(-\infty, -3) \cup (-3, 2) \cup (4, \infty)$
12. $\left(-3, -\frac{1}{3}\right) \cup (2, 3)$
13. $(-1, 0] \cup (2, \infty)$
14. $(-\infty, -3) \cup (-2, -1) \cup (1, \infty)$
15. $(-\infty, 1] \cup [2, \infty)$
16. $(-\infty, -6) \cup (-1, 2)$
17. $(-\infty, -3) \cup \left[-2\sqrt{2}, 0\right] \cup \left[2\sqrt{2}, 3\right)$
18. No solution
19. $[-3, 0) \cup (0, 4) \cup [5, \infty)$
20. $\left(-1, -\frac{1}{2}\right] \cup (1, \infty)$
21. 4.5 miles per hour
22. 24 miles per hour
23. 3600 gallons
24. $\frac{40}{3} \approx 13.33$ minutes
25. 3 hours

26. The absolute minimum of $y = \overline{C}(x)$ occurs at $\approx (75.73, 59.57)$. Since x represents the number of game systems, we check $\overline{C}(75) \approx 59.58$ and $\overline{C}(76) \approx 59.57$. Hence, to minimize the average cost, 76 systems should be produced at an average cost of \$59.57 per system.

27. The width (and depth) should be 10.00 centimeters, the height should be 5.00 centimeters. The minimum surface area is 300.00 square centimeters.

28. The width of the base of the box should be ≈ 4.12 inches, the height of the box should be ≈ 6.67 inches, and the depth of the base of the box should be ≈ 5.09 inches; minimum surface area ≈ 164.91 square inches.

29. The dimensions are ≈ 7 feet by ≈ 14 feet; minimum amount of fencing required ≈ 28 feet.

30. (a) $V = \pi r^2 h$ (b) $S = 2\pi r^2 + 2\pi r h$
    (c) $S(r) = 2\pi r^2 + \frac{67.2}{r}$, Domain $r > 0$ (d) $r \approx 1.749$ in. and $h \approx 3.498$ in.

31. The radius of the drum should be ≈ 1.05 feet and the height of the drum should be ≈ 2.12 feet. The minimum surface area of the drum is ≈ 20.93 cubic feet.

32. $P(t) < 100$ on $(-15, 30)$, and the portion of this which lies in the applied domain is $[0, 30)$. Since $t = 0$ corresponds to the year 1803, from 1803 through the end of 1832, there were fewer than 100 Sasquatch in Portage County.

4.3 Rational Inequalities and Applications

33. $T = kV$

34.[16] $f = \dfrac{k}{\lambda}$

35. $d = \dfrac{km}{V}$

36. $P^2 = ka^3$

37.[17] $D = k\rho\nu^2$

38.[18] $F = \dfrac{kqQ}{r^2}$

39. Rewriting $f = \dfrac{1}{2L}\sqrt{\dfrac{T}{\mu}}$ as $f = \dfrac{\frac{1}{2}\sqrt{T}}{L\sqrt{\mu}}$ we see that the frequency f varies directly with the square root of the tension and varies inversely with the length and the square root of the linear mass.

40. (a) $B = \dfrac{kW}{h^2}$ \qquad (b)[19] $k = 702.68$ \qquad (c) $B = \dfrac{702.68W}{h^2}$

---

[16] The character λ is the lower case Greek letter 'lambda.'
[17] The characters ρ and ν are the lower case Greek letters 'rho' and 'nu,' respectively.
[18] Note the similarity to this formula and Newton's Law of Universal Gravitation as discussed in Example 5.
[19] The CDC uses 703.

Chapter 5

Further Topics in Functions

5.1 Function Composition

Before we embark upon any further adventures with functions, we need to take some time to gather our thoughts and gain some perspective. Chapter 1 first introduced us to functions in Section 1.3. At that time, functions were specific kinds of relations - sets of points in the plane which passed the Vertical Line Test, Theorem 1.1. In Section 1.4, we developed the idea that functions are processes - rules which match inputs to outputs - and this gave rise to the concepts of domain and range. We spoke about how functions could be combined in Section 1.5 using the four basic arithmetic operations, took a more detailed look at their graphs in Section 1.6 and studied how their graphs behaved under certain classes of transformations in Section 1.7. In Chapter 2, we took a closer look at three families of functions: linear functions (Section 2.1), absolute value functions[1] (Section 2.2), and quadratic functions (Section 2.3). Linear and quadratic functions were special cases of polynomial functions, which we studied in generality in Chapter 3. Chapter 3 culminated with the Real Factorization Theorem, Theorem 3.16, which says that all polynomial functions with real coefficients can be thought of as products of linear and quadratic functions. Our next step was to enlarge our field[2] of study to rational functions in Chapter 4. Being quotients of polynomials, we can ultimately view this family of functions as being built up of linear and quadratic functions as well. So in some sense, Chapters 2, 3, and 4 can be thought of as an exhaustive study of linear and quadratic[3] functions and their arithmetic combinations as described in Section 1.5. We now wish to study other algebraic functions, such as $f(x) = \sqrt{x}$ and $g(x) = x^{2/3}$, and the purpose of the first two sections of this chapter is to see how these kinds of functions arise from polynomial and rational functions. To that end, we first study a new way to combine functions as defined below.

---

[1] These were introduced, as you may recall, as piecewise-defined linear functions.

[2] This is a really bad math pun.

[3] If we broaden our concept of functions to allow for complex valued coefficients, the Complex Factorization Theorem, Theorem 3.14, tells us every function we have studied thus far is a combination of linear functions.

Definition 5.1. Suppose f and g are two functions. The **composite** of g with f, denoted $g \circ f$, is defined by the formula $(g \circ f)(x) = g(f(x))$, provided x is an element of the domain of f and $f(x)$ is an element of the domain of g.

The quantity $g \circ f$ is also read 'g composed with f' or, more simply 'g of f.' At its most basic level, Definition 5.1 tells us to obtain the formula for $(g \circ f)(x)$, we replace every occurrence of x in the formula for $g(x)$ with the formula we have for $f(x)$. If we take a step back and look at this from a procedural, 'inputs and outputs' perspective, Defintion 5.1 tells us the output from $g \circ f$ is found by taking the output from f, $f(x)$, and then making that the input to g. The result, $g(f(x))$, is the output from $g \circ f$. From this perspective, we see $g \circ f$ as a two step process taking an input x and first applying the procedure f then applying the procedure g. Abstractly, we have

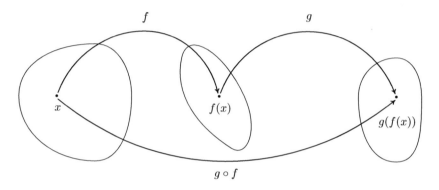

In the expression $g(f(x))$, the function f is often called the 'inside' function while g is often called the 'outside' function. There are two ways to go about evaluating composite functions - 'inside out' and 'outside in' - depending on which function we replace with its formula first. Both ways are demonstrated in the following example.

Example 5.1.1. Let $f(x) = x^2 - 4x$, $g(x) = 2 - \sqrt{x+3}$, and $h(x) = \dfrac{2x}{x+1}$.
In numbers 1 - 3, find the indicated function value.

1. $(g \circ f)(1)$ 2. $(f \circ g)(1)$ 3. $(g \circ g)(6)$

In numbers 4 - 10, find and simplify the indicated composite functions. State the domain of each.

4. $(g \circ f)(x)$ 5. $(f \circ g)(x)$ 6. $(g \circ h)(x)$ 7. $(h \circ g)(x)$

8. $(h \circ h)(x)$ 9. $(h \circ (g \circ f))(x)$ 10. $((h \circ g) \circ f)(x)$

Solution.

1. Using Definition 5.1, $(g \circ f)(1) = g(f(1))$. We find $f(1) = -3$, so
$$(g \circ f)(1) = g(f(1)) = g(-3) = 2$$

5.1 FUNCTION COMPOSITION

2. As before, we use Definition 5.1 to write $(f \circ g)(1) = f(g(1))$. We find $g(1) = 0$, so
$$(f \circ g)(1) = f(g(1)) = f(0) = 0$$

3. Once more, Definition 5.1 tells us $(g \circ g)(6) = g(g(6))$. That is, we evaluate g at 6, then plug that result back into g. Since $g(6) = -1$,
$$(g \circ g)(6) = g(g(6)) = g(-1) = 2 - \sqrt{2}$$

4. By definition, $(g \circ f)(x) = g(f(x))$. We now illustrate *two* ways to approach this problem.

    - *inside out*: We insert the expression $f(x)$ into g first to get
    $$(g \circ f)(x) = g(f(x)) = g\left(x^2 - 4x\right) = 2 - \sqrt{(x^2 - 4x) + 3} = 2 - \sqrt{x^2 - 4x + 3}$$
    Hence, $(g \circ f)(x) = 2 - \sqrt{x^2 - 4x + 3}$.
    - *outside in*: We use the formula for g first to get
    $$(g \circ f)(x) = g(f(x)) = 2 - \sqrt{f(x) + 3} = 2 - \sqrt{(x^2 - 4x) + 3} = 2 - \sqrt{x^2 - 4x + 3}$$
    We get the same answer as before, $(g \circ f)(x) = 2 - \sqrt{x^2 - 4x + 3}$.

    To find the domain of $g \circ f$, we need to find the elements in the domain of f whose outputs $f(x)$ are in the domain of g. We accomplish this by following the rule set forth in Section 1.4, that is, we find the domain *before* we simplify. To that end, we examine $(g \circ f)(x) = 2 - \sqrt{(x^2 - 4x) + 3}$. To keep the square root happy, we solve the inequality $x^2 - 4x + 3 \geq 0$ by creating a sign diagram. If we let $r(x) = x^2 - 4x + 3$, we find the zeros of r to be $x = 1$ and $x = 3$. We obtain

    $$\xleftarrow{\quad (+) \quad 0 \quad (-) \quad 0 \quad (+) \quad}_{ 1 3 } \rightarrow$$

    Our solution to $x^2 - 4x + 3 \geq 0$, and hence the domain of $g \circ f$, is $(-\infty, 1] \cup [3, \infty)$.

5. To find $(f \circ g)(x)$, we find $f(g(x))$.

    - *inside out*: We insert the expression $g(x)$ into f first to get
    $$\begin{aligned}(f \circ g)(x) &= f(g(x)) = f\left(2 - \sqrt{x+3}\right) \\ &= \left(2 - \sqrt{x+3}\right)^2 - 4\left(2 - \sqrt{x+3}\right) \\ &= 4 - 4\sqrt{x+3} + \left(\sqrt{x+3}\right)^2 - 8 + 4\sqrt{x+3} \\ &= 4 + x + 3 - 8 \\ &= x - 1\end{aligned}$$

- *outside in*: We use the formula for $f(x)$ first to get

$$\begin{aligned}(f \circ g)(x) &= f(g(x)) = (g(x))^2 - 4(g(x)) \\ &= \left(2 - \sqrt{x+3}\right)^2 - 4\left(2 - \sqrt{x+3}\right) \\ &= x - 1 \qquad \text{same algebra as before}\end{aligned}$$

Thus we get $(f \circ g)(x) = x - 1$. To find the domain of $(f \circ g)$, we look to the step before we did any simplification and find $(f \circ g)(x) = \left(2 - \sqrt{x+3}\right)^2 - 4\left(2 - \sqrt{x+3}\right)$. To keep the square root happy, we set $x + 3 \geq 0$ and find our domain to be $[-3, \infty)$.

6. To find $(g \circ h)(x)$, we compute $g(h(x))$.

    - *inside out*: We insert the expression $h(x)$ into g first to get

    $$\begin{aligned}(g \circ h)(x) &= g(h(x)) = g\left(\frac{2x}{x+1}\right) \\ &= 2 - \sqrt{\left(\frac{2x}{x+1}\right) + 3} \\ &= 2 - \sqrt{\frac{2x}{x+1} + \frac{3(x+1)}{x+1}} \qquad \text{get common denominators} \\ &= 2 - \sqrt{\frac{5x+3}{x+1}}\end{aligned}$$

    - *outside in*: We use the formula for $g(x)$ first to get

    $$\begin{aligned}(g \circ h)(x) &= g(h(x)) = 2 - \sqrt{h(x) + 3} \\ &= 2 - \sqrt{\left(\frac{2x}{x+1}\right) + 3} \\ &= 2 - \sqrt{\frac{5x+3}{x+1}} \qquad \text{get common denominators as before}\end{aligned}$$

To find the domain of $(g \circ h)$, we look to the step before we began to simplify:

$$(g \circ h)(x) = 2 - \sqrt{\left(\frac{2x}{x+1}\right) + 3}$$

To avoid division by zero, we need $x \neq -1$. To keep the radical happy, we need to solve

$$\frac{2x}{x+1} + 3 = \frac{5x+3}{x+1} \geq 0$$

Defining $r(x) = \frac{5x+3}{x+1}$, we see r is undefined at $x = -1$ and $r(x) = 0$ at $x = -\frac{3}{5}$. We get

5.1 FUNCTION COMPOSITION

$$
\begin{array}{c}
(+) \; ? \;\; (-) \;\; 0 \; (+) \\
\xleftarrow{\bullet\bullet}\rightarrow \\
-1 \qquad -\tfrac{3}{5}
\end{array}
$$

Our domain is $(-\infty, -1) \cup \left[-\tfrac{3}{5}, \infty\right)$.

7. We find $(h \circ g)(x)$ by finding $h(g(x))$.

   - *inside out*: We insert the expression $g(x)$ into h first to get

$$
\begin{aligned}
(h \circ g)(x) &= h(g(x)) = h\left(2 - \sqrt{x+3}\right) \\
&= \frac{2\left(2 - \sqrt{x+3}\right)}{\left(2 - \sqrt{x+3}\right) + 1} \\
&= \frac{4 - 2\sqrt{x+3}}{3 - \sqrt{x+3}}
\end{aligned}
$$

   - *outside in*: We use the formula for $h(x)$ first to get

$$
\begin{aligned}
(h \circ g)(x) &= h(g(x)) = \frac{2\,(g(x))}{(g(x)) + 1} \\
&= \frac{2\left(2 - \sqrt{x+3}\right)}{\left(2 - \sqrt{x+3}\right) + 1} \\
&= \frac{4 - 2\sqrt{x+3}}{3 - \sqrt{x+3}}
\end{aligned}
$$

To find the domain of $h \circ g$, we look to the step before any simplification:

$$
(h \circ g)(x) = \frac{2\left(2 - \sqrt{x+3}\right)}{\left(2 - \sqrt{x+3}\right) + 1}
$$

To keep the square root happy, we require $x + 3 \geq 0$ or $x \geq -3$. Setting the denominator equal to zero gives $\left(2 - \sqrt{x+3}\right) + 1 = 0$ or $\sqrt{x+3} = 3$. Squaring both sides gives us $x + 3 = 9$, or $x = 6$. Since $x = 6$ checks in the original equation, $\left(2 - \sqrt{x+3}\right) + 1 = 0$, we know $x = 6$ is the only zero of the denominator. Hence, the domain of $h \circ g$ is $[-3, 6) \cup (6, \infty)$.

8. To find $(h \circ h)(x)$, we substitute the function h into itself, $h(h(x))$.

   - *inside out*: We insert the expression $h(x)$ into h to get

$$
(h \circ h)(x) = h(h(x)) = h\left(\frac{2x}{x+1}\right)
$$

363

$$= \frac{2\left(\dfrac{2x}{x+1}\right)}{\left(\dfrac{2x}{x+1}\right)+1}$$

$$= \frac{\dfrac{4x}{x+1}}{\dfrac{2x}{x+1}+1} \cdot \frac{(x+1)}{(x+1)}$$

$$= \frac{\dfrac{4x}{x+1} \cdot (x+1)}{\left(\dfrac{2x}{x+1}\right) \cdot (x+1) + 1 \cdot (x+1)}$$

$$= \frac{\dfrac{4x}{\cancel{(x+1)}} \cdot \cancel{(x+1)}}{\dfrac{2x}{\cancel{(x+1)}} \cdot \cancel{(x+1)} + x + 1}$$

$$= \frac{4x}{3x+1}$$

- *outside in*: This approach yields

$$(h \circ h)(x) = h(h(x)) = \frac{2(h(x))}{h(x)+1}$$

$$= \frac{2\left(\dfrac{2x}{x+1}\right)}{\left(\dfrac{2x}{x+1}\right)+1}$$

$$= \frac{4x}{3x+1} \qquad \text{same algebra as before}$$

To find the domain of $h \circ h$, we analyze

$$(h \circ h)(x) = \frac{2\left(\dfrac{2x}{x+1}\right)}{\left(\dfrac{2x}{x+1}\right)+1}$$

To keep the denominator $x+1$ happy, we need $x \neq -1$. Setting the denominator

$$\frac{2x}{x+1} + 1 = 0$$

gives $x = -\frac{1}{3}$. Our domain is $(-\infty, -1) \cup \left(-1, -\frac{1}{3}\right) \cup \left(-\frac{1}{3}, \infty\right)$.

5.1 FUNCTION COMPOSITION

9. The expression $(h \circ (g \circ f))(x)$ indicates that we first find the composite, $g \circ f$ and compose the function h with the result. We know from number 1 that $(g \circ f)(x) = 2 - \sqrt{x^2 - 4x + 3}$. We now proceed as usual.

- *inside out*: We insert the expression $(g \circ f)(x)$ into h first to get

$$\begin{aligned}(h \circ (g \circ f))(x) &= h((g \circ f)(x)) = h\left(2 - \sqrt{x^2 - 4x + 3}\right) \\ &= \frac{2\left(2 - \sqrt{x^2 - 4x + 3}\right)}{\left(2 - \sqrt{x^2 - 4x + 3}\right) + 1} \\ &= \frac{4 - 2\sqrt{x^2 - 4x + 3}}{3 - \sqrt{x^2 - 4x + 3}}\end{aligned}$$

- *outside in*: We use the formula for $h(x)$ first to get

$$\begin{aligned}(h \circ (g \circ f))(x) &= h((g \circ f)(x)) = \frac{2((g \circ f)(x))}{((g \circ f)(x)) + 1} \\ &= \frac{2\left(2 - \sqrt{x^2 - 4x + 3}\right)}{\left(2 - \sqrt{x^2 - 4x + 3}\right) + 1} \\ &= \frac{4 - 2\sqrt{x^2 - 4x + 3}}{3 - \sqrt{x^2 - 4x + 3}}\end{aligned}$$

To find the domain of $(h \circ (g \circ f))$, we look at the step before we began to simplify,

$$(h \circ (g \circ f))(x) = \frac{2\left(2 - \sqrt{x^2 - 4x + 3}\right)}{\left(2 - \sqrt{x^2 - 4x + 3}\right) + 1}$$

For the square root, we need $x^2 - 4x + 3 \geq 0$, which we determined in number 1 to be $(-\infty, 1] \cup [3, \infty)$. Next, we set the denominator to zero and solve: $\left(2 - \sqrt{x^2 - 4x + 3}\right) + 1 = 0$. We get $\sqrt{x^2 - 4x + 3} = 3$, and, after squaring both sides, we have $x^2 - 4x + 3 = 9$. To solve $x^2 - 4x - 6 = 0$, we use the quadratic formula and get $x = 2 \pm \sqrt{10}$. The reader is encouraged to check that both of these numbers satisfy the original equation, $\left(2 - \sqrt{x^2 - 4x + 3}\right) + 1 = 0$. Hence we must exclude these numbers from the domain of $h \circ (g \circ f)$. Our final domain for $h \circ (f \circ g)$ is $(-\infty, 2 - \sqrt{10}) \cup (2 - \sqrt{10}, 1] \cup [3, 2 + \sqrt{10}) \cup (2 + \sqrt{10}, \infty)$.

10. The expression $((h \circ g) \circ f)(x)$ indicates that we first find the composite $h \circ g$ and then compose that with f. From number 4, we have

$$(h \circ g)(x) = \frac{4 - 2\sqrt{x + 3}}{3 - \sqrt{x + 3}}$$

We now proceed as before.

- *inside out*: We insert the expression $f(x)$ into $h \circ g$ first to get

$$\begin{aligned}((h \circ g) \circ f)(x) &= (h \circ g)(f(x)) = (h \circ g)\left(x^2 - 4x\right) \\ &= \frac{4 - 2\sqrt{(x^2 - 4x) + 3}}{3 - \sqrt{(x^2 - 4x) + 3}} \\ &= \frac{4 - 2\sqrt{x^2 - 4x + 3}}{3 - \sqrt{x^2 - 4x + 3}}\end{aligned}$$

- *outside in*: We use the formula for $(h \circ g)(x)$ first to get

$$\begin{aligned}((h \circ g) \circ f)(x) &= (h \circ g)(f(x)) = \frac{4 - 2\sqrt{(f(x)) + 3}}{3 - \sqrt{f(x)) + 3}} \\ &= \frac{4 - 2\sqrt{(x^2 - 4x) + 3}}{3 - \sqrt{(x^2 - 4x) + 3}} \\ &= \frac{4 - 2\sqrt{x^2 - 4x + 3}}{3 - \sqrt{x^2 - 4x + 3}}\end{aligned}$$

We note that the formula for $((h \circ g) \circ f)(x)$ before simplification is identical to that of $(h \circ (g \circ f))(x)$ before we simplified it. Hence, the two functions have the same domain, $h \circ (f \circ g)$ is $(-\infty, 2 - \sqrt{10}) \cup (2 - \sqrt{10}, 1] \cup [3, 2 + \sqrt{10}) \cup (2 + \sqrt{10}, \infty)$. □

It should be clear from Example 5.1.1 that, in general, when you compose two functions, such as f and g above, the order matters.[4] We found that the functions $f \circ g$ and $g \circ f$ were different as were $g \circ h$ and $h \circ g$. Thinking of functions as processes, this isn't all that surprising. If we think of one process as putting on our socks, and the other as putting on our shoes, the order in which we do these two tasks does matter.[5] Also note the importance of finding the domain of the composite function *before* simplifying. For instance, the domain of $f \circ g$ is much different than its simplified formula would indicate. Composing a function with itself, as in the case of finding $(g \circ g)(6)$ and $(h \circ h)(x)$, may seem odd. Looking at this from a procedural perspective, however, this merely indicates performing a task h and then doing it again - like setting the washing machine to do a 'double rinse'. Composing a function with itself is called 'iterating' the function, and we could easily spend an entire course on just that. The last two problems in Example 5.1.1 serve to demonstrate the **associative** property of functions. That is, when composing three (or more) functions, as long as we keep the order the same, it doesn't matter which two functions we compose first. This property as well as another important property are listed in the theorem below.

---

[4]This shows us function composition isn't **commutative**. An example of an operation we perform on two functions which is commutative is function addition, which we defined in Section 1.5. In other words, the functions $f + g$ and $g + f$ are always equal. Which of the remaining operations on functions we have discussed are commutative?

[5]A more mathematical example in which the order of two processes matters can be found in Section 1.7. In fact, all of the transformations in that section can be viewed in terms of composing functions with linear functions.

5.1 FUNCTION COMPOSITION

> **Theorem 5.1. Properties of Function Composition:** Suppose f, g, and h are functions.
>
> - $h \circ (g \circ f) = (h \circ g) \circ f$, provided the composite functions are defined.
> - If I is defined as $I(x) = x$ for all real numbers x, then $I \circ f = f \circ I = f$.

By repeated applications of Definition 5.1, we find $(h \circ (g \circ f))(x) = h((g \circ f)(x)) = h(g(f(x)))$. Similarly, $((h \circ g) \circ f)(x) = (h \circ g)(f(x)) = h(g(f(x)))$. This establishes that the formulas for the two functions are the same. We leave it to the reader to think about why the domains of these two functions are identical, too. These two facts establish the equality $h \circ (g \circ f) = (h \circ g) \circ f$. A consequence of the associativity of function composition is that there is no need for parentheses when we write $h \circ g \circ f$. The second property can also be verified using Definition 5.1. Recall that the function $I(x) = x$ is called the *identity function* and was introduced in Exercise 73 in Section 2.1. If we compose the function I with a function f, then we have $(I \circ f)(x) = I(f(x)) = f(x)$, and a similar computation shows $(f \circ I)(x) = f(x)$. This establishes that we have an identity for function composition much in the same way the real number 1 is an identity for real number multiplication. That is, just as for any real number x, $1 \cdot x = x \cdot 1 = x$, we have for any function f, $I \circ f = f \circ I = f$. We shall see the concept of an identity take on great significance in the next section. Out in the wild, function composition is often used to relate two quantities which may not be directly related, but have a variable in common, as illustrated in our next example.

Example 5.1.2. The surface area S of a sphere is a function of its radius r and is given by the formula $S(r) = 4\pi r^2$. Suppose the sphere is being inflated so that the radius of the sphere is increasing according to the formula $r(t) = 3t^2$, where t is measured in seconds, $t \geq 0$, and r is measured in inches. Find and interpret $(S \circ r)(t)$.

Solution. If we look at the functions $S(r)$ and $r(t)$ individually, we see the former gives the surface area of a sphere of a given radius while the latter gives the radius at a given time. So, given a specific time, t, we could find the radius at that time, $r(t)$ and feed that into $S(r)$ to find the surface area at that time. From this we see that the surface area S is ultimately a function of time t and we find $(S \circ r)(t) = S(r(t)) = 4\pi(r(t))^2 = 4\pi \left(3t^2\right)^2 = 36\pi t^4$. This formula allows us to compute the surface area directly given the time without going through the 'middle man' r. $\square$

A useful skill in Calculus is to be able to take a complicated function and break it down into a composition of easier functions which our last example illustrates.

Example 5.1.3. Write each of the following functions as a composition of two or more (non-identity) functions. Check your answer by performing the function composition.

1. $F(x) = |3x - 1|$
2. $G(x) = \dfrac{2}{x^2 + 1}$
3. $H(x) = \dfrac{\sqrt{x} + 1}{\sqrt{x} - 1}$

Solution. There are many approaches to this kind of problem, and we showcase a different methodology in each of the solutions below.

1. Our goal is to express the function F as $F = g \circ f$ for functions g and f. From Definition 5.1, we know $F(x) = g(f(x))$, and we can think of $f(x)$ as being the 'inside' function and g as being the 'outside' function. Looking at $F(x) = |3x - 1|$ from an 'inside versus outside' perspective, we can think of $3x - 1$ being inside the absolute value symbols. Taking this cue, we define $f(x) = 3x - 1$. At this point, we have $F(x) = |f(x)|$. What is the outside function? The function which takes the absolute value of its input, $g(x) = |x|$. Sure enough, $(g \circ f)(x) = g(f(x)) = |f(x)| = |3x - 1| = F(x)$, so we are done.

2. We attack deconstructing G from an operational approach. Given an input x, the first step is to square x, then add 1, then divide the result into 2. We will assign each of these steps a function so as to write G as a composite of three functions: f, g and h. Our first function, f, is the function that squares its input, $f(x) = x^2$. The next function is the function that adds 1 to its input, $g(x) = x + 1$. Our last function takes its input and divides it into 2, $h(x) = \frac{2}{x}$. The claim is that $G = h \circ g \circ f$. We find

$$(h \circ g \circ f)(x) = h(g(f(x))) = h(g(x^2)) = h(x^2 + 1) = \frac{2}{x^2 + 1} = G(x),$$

so we are done.

3. If we look $H(x) = \frac{\sqrt{x}+1}{\sqrt{x}-1}$ with an eye towards building a complicated function from simpler functions, we see the expression $\sqrt{x}$ is a simple piece of the larger function. If we define $f(x) = \sqrt{x}$, we have $H(x) = \frac{f(x)+1}{f(x)-1}$. If we want to decompose $H = g \circ f$, then we can glean the formula for $g(x)$ by looking at what is being done to $f(x)$. We take $g(x) = \frac{x+1}{x-1}$, so

$$(g \circ f)(x) = g(f(x)) = \frac{f(x)+1}{f(x)-1} = \frac{\sqrt{x}+1}{\sqrt{x}-1} = H(x),$$

as required. □

5.1 Function Composition

5.1.1 Exercises

In Exercises 1 - 12, use the given pair of functions to find the following values if they exist.

- $(g \circ f)(0)$
- $(f \circ g)(-1)$
- $(f \circ f)(2)$

- $(g \circ f)(-3)$
- $(f \circ g)\left(\frac{1}{2}\right)$
- $(f \circ f)(-2)$

1. $f(x) = x^2$, $g(x) = 2x + 1$
2. $f(x) = 4 - x$, $g(x) = 1 - x^2$
3. $f(x) = 4 - 3x$, $g(x) = |x|$
4. $f(x) = |x - 1|$, $g(x) = x^2 - 5$
5. $f(x) = 4x + 5$, $g(x) = \sqrt{x}$
6. $f(x) = \sqrt{3-x}$, $g(x) = x^2 + 1$
7. $f(x) = 6 - x - x^2$, $g(x) = x\sqrt{x+10}$
8. $f(x) = \sqrt[3]{x+1}$, $g(x) = 4x^2 - x$
9. $f(x) = \dfrac{3}{1-x}$, $g(x) = \dfrac{4x}{x^2+1}$
10. $f(x) = \dfrac{x}{x+5}$, $g(x) = \dfrac{2}{7-x^2}$
11. $f(x) = \dfrac{2x}{5-x^2}$, $g(x) = \sqrt{4x+1}$
12. $f(x) = \sqrt{2x+5}$, $g(x) = \dfrac{10x}{x^2+1}$

In Exercises 13 - 24, use the given pair of functions to find and simplify expressions for the following functions and state the domain of each using interval notation.

- $(g \circ f)(x)$
- $(f \circ g)(x)$
- $(f \circ f)(x)$

13. $f(x) = 2x + 3$, $g(x) = x^2 - 9$
14. $f(x) = x^2 - x + 1$, $g(x) = 3x - 5$
15. $f(x) = x^2 - 4$, $g(x) = |x|$
16. $f(x) = 3x - 5$, $g(x) = \sqrt{x}$
17. $f(x) = |x + 1|$, $g(x) = \sqrt{x}$
18. $f(x) = 3 - x^2$, $g(x) = \sqrt{x+1}$
19. $f(x) = |x|$, $g(x) = \sqrt{4-x}$
20. $f(x) = x^2 - x - 1$, $g(x) = \sqrt{x-5}$
21. $f(x) = 3x - 1$, $g(x) = \dfrac{1}{x+3}$
22. $f(x) = \dfrac{3x}{x-1}$, $g(x) = \dfrac{x}{x-3}$
23. $f(x) = \dfrac{x}{2x+1}$, $g(x) = \dfrac{2x+1}{x}$
24. $f(x) = \dfrac{2x}{x^2-4}$, $g(x) = \sqrt{1-x}$

In Exercises 25 - 30, use $f(x) = -2x$, $g(x) = \sqrt{x}$ and $h(x) = |x|$ to find and simplify expressions for the following functions and state the domain of each using interval notation.

25. $(h \circ g \circ f)(x)$
26. $(h \circ f \circ g)(x)$
27. $(g \circ f \circ h)(x)$
28. $(g \circ h \circ f)(x)$
29. $(f \circ h \circ g)(x)$
30. $(f \circ g \circ h)(x)$

In Exercises 31 - 40, write the given function as a composition of two or more non-identity functions. (There are several correct answers, so check your answer using function composition.)

31. $p(x) = (2x + 3)^3$
32. $P(x) = (x^2 - x + 1)^5$
33. $h(x) = \sqrt{2x - 1}$
34. $H(x) = |7 - 3x|$
35. $r(x) = \dfrac{2}{5x + 1}$
36. $R(x) = \dfrac{7}{x^2 - 1}$
37. $q(x) = \dfrac{|x| + 1}{|x| - 1}$
38. $Q(x) = \dfrac{2x^3 + 1}{x^3 - 1}$
39. $v(x) = \dfrac{2x + 1}{3 - 4x}$
40. $w(x) = \dfrac{x^2}{x^4 + 1}$

41. Write the function $F(x) = \sqrt{\dfrac{x^3 + 6}{x^3 - 9}}$ as a composition of three or more non-identity functions.

42. Let $g(x) = -x$, $h(x) = x+2$, $j(x) = 3x$ and $k(x) = x-4$. In what order must these functions be composed with $f(x) = \sqrt{x}$ to create $F(x) = 3\sqrt{-x+2} - 4$?

43. What linear functions could be used to transform $f(x) = x^3$ into $F(x) = -\frac{1}{2}(2x - 7)^3 + 1$? What is the proper order of composition?

In Exercises 44 - 55, let f be the function defined by

$$f = \{(-3,4), (-2,2), (-1,0), (0,1), (1,3), (2,4), (3,-1)\}$$

and let g be the function defined

$$g = \{(-3,-2), (-2,0), (-1,-4), (0,0), (1,-3), (2,1), (3,2)\}$$

. Find the value if it exists.

44. $(f \circ g)(3)$
45. $f(g(-1))$
46. $(f \circ f)(0)$
47. $(f \circ g)(-3)$
48. $(g \circ f)(3)$
49. $g(f(-3))$

5.1 FUNCTION COMPOSITION

50. $(g \circ g)(-2)$

51. $(g \circ f)(-2)$

52. $g(f(g(0)))$

53. $f(f(f(-1)))$

54. $f(f(f(f(f(1)))))$

55. $\underbrace{(g \circ g \circ \cdots \circ g)}_{n \text{ times}}(0)$

In Exercises 56 - 61, use the graphs of $y = f(x)$ and $y = g(x)$ below to find the function value.

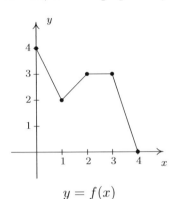

$y = f(x)$

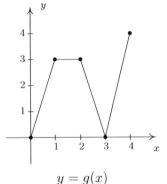

$y = g(x)$

56. $(g \circ f)(1)$

57. $(f \circ g)(3)$

58. $(g \circ f)(2)$

59. $(f \circ g)(0)$

60. $(f \circ f)(1)$

61. $(g \circ g)(1)$

62. The volume V of a cube is a function of its side length x. Let's assume that $x = t + 1$ is also a function of time t, where x is measured in inches and t is measured in minutes. Find a formula for V as a function of t.

63. Suppose a local vendor charges $2 per hot dog and that the number of hot dogs sold per hour x is given by $x(t) = -4t^2 + 20t + 92$, where t is the number of hours since 10 AM, $0 \leq t \leq 4$.

    (a) Find an expression for the revenue per hour R as a function of x.

    (b) Find and simplify $(R \circ x)(t)$. What does this represent?

    (c) What is the revenue per hour at noon?

64. Discuss with your classmates how 'real-world' processes such as filling out federal income tax forms or computing your final course grade could be viewed as a use of function composition. Find a process for which composition with itself (iteration) makes sense.

5.1.2 Answers

1. For $f(x) = x^2$ and $g(x) = 2x + 1$,

    - $(g \circ f)(0) = 1$
    - $(f \circ g)(-1) = 1$
    - $(f \circ f)(2) = 16$
    - $(g \circ f)(-3) = 19$
    - $(f \circ g)\left(\frac{1}{2}\right) = 4$
    - $(f \circ f)(-2) = 16$

2. For $f(x) = 4 - x$ and $g(x) = 1 - x^2$,

    - $(g \circ f)(0) = -15$
    - $(f \circ g)(-1) = 4$
    - $(f \circ f)(2) = 2$
    - $(g \circ f)(-3) = -48$
    - $(f \circ g)\left(\frac{1}{2}\right) = \frac{13}{4}$
    - $(f \circ f)(-2) = -2$

3. For $f(x) = 4 - 3x$ and $g(x) = |x|$,

    - $(g \circ f)(0) = 4$
    - $(f \circ g)(-1) = 1$
    - $(f \circ f)(2) = 10$
    - $(g \circ f)(-3) = 13$
    - $(f \circ g)\left(\frac{1}{2}\right) = \frac{5}{2}$
    - $(f \circ f)(-2) = -26$

4. For $f(x) = |x - 1|$ and $g(x) = x^2 - 5$,

    - $(g \circ f)(0) = -4$
    - $(f \circ g)(-1) = 5$
    - $(f \circ f)(2) = 0$
    - $(g \circ f)(-3) = 11$
    - $(f \circ g)\left(\frac{1}{2}\right) = \frac{23}{4}$
    - $(f \circ f)(-2) = 2$

5. For $f(x) = 4x + 5$ and $g(x) = \sqrt{x}$,

    - $(g \circ f)(0) = \sqrt{5}$
    - $(f \circ g)(-1)$ is not real
    - $(f \circ f)(2) = 57$
    - $(g \circ f)(-3)$ is not real
    - $(f \circ g)\left(\frac{1}{2}\right) = 5 + 2\sqrt{2}$
    - $(f \circ f)(-2) = -7$

6. For $f(x) = \sqrt{3 - x}$ and $g(x) = x^2 + 1$,

    - $(g \circ f)(0) = 4$
    - $(f \circ g)(-1) = 1$
    - $(f \circ f)(2) = \sqrt{2}$
    - $(g \circ f)(-3) = 7$
    - $(f \circ g)\left(\frac{1}{2}\right) = \frac{\sqrt{7}}{2}$
    - $(f \circ f)(-2) = \sqrt{3 - \sqrt{5}}$

5.1 Function Composition

7. For $f(x) = 6 - x - x^2$ and $g(x) = x\sqrt{x+10}$,

    - $(g \circ f)(0) = 24$
    - $(g \circ f)(-3) = 0$
    - $(f \circ g)(-1) = 0$
    - $(f \circ g)\left(\frac{1}{2}\right) = \frac{27-2\sqrt{42}}{8}$
    - $(f \circ f)(2) = 6$
    - $(f \circ f)(-2) = -14$

8. For $f(x) = \sqrt[3]{x+1}$ and $g(x) = 4x^2 - x$,

    - $(g \circ f)(0) = 3$
    - $(g \circ f)(-3) = 4\sqrt[3]{4} + \sqrt[3]{2}$
    - $(f \circ g)(-1) = \sqrt[3]{6}$
    - $(f \circ g)\left(\frac{1}{2}\right) = \frac{\sqrt[3]{12}}{2}$
    - $(f \circ f)(2) = \sqrt[3]{\sqrt[3]{3}+1}$
    - $(f \circ f)(-2) = 0$

9. For $f(x) = \frac{3}{1-x}$ and $g(x) = \frac{4x}{x^2+1}$,

    - $(g \circ f)(0) = \frac{6}{5}$
    - $(g \circ f)(-3) = \frac{48}{25}$
    - $(f \circ g)(-1) = 1$
    - $(f \circ g)\left(\frac{1}{2}\right) = -5$
    - $(f \circ f)(2) = \frac{3}{4}$
    - $(f \circ f)(-2)$ is undefined

10. For $f(x) = \frac{x}{x+5}$ and $g(x) = \frac{2}{7-x^2}$,

    - $(g \circ f)(0) = \frac{2}{7}$
    - $(g \circ f)(-3) = \frac{8}{19}$
    - $(f \circ g)(-1) = \frac{1}{16}$
    - $(f \circ g)\left(\frac{1}{2}\right) = \frac{8}{143}$
    - $(f \circ f)(2) = \frac{2}{37}$
    - $(f \circ f)(-2) = -\frac{2}{13}$

11. For $f(x) = \frac{2x}{5-x^2}$ and $g(x) = \sqrt{4x+1}$,

    - $(g \circ f)(0) = 1$
    - $(g \circ f)(-3) = \sqrt{7}$
    - $(f \circ g)(-1)$ is not real
    - $(f \circ g)\left(\frac{1}{2}\right) = \sqrt{3}$
    - $(f \circ f)(2) = -\frac{8}{11}$
    - $(f \circ f)(-2) = \frac{8}{11}$

12. For $f(x) = \sqrt{2x+5}$ and $g(x) = \frac{10x}{x^2+1}$,

    - $(g \circ f)(0) = \frac{5\sqrt{5}}{3}$
    - $(g \circ f)(-3)$ is not real
    - $(f \circ g)(-1)$ is not real
    - $(f \circ g)\left(\frac{1}{2}\right) = \sqrt{13}$
    - $(f \circ f)(2) = \sqrt{11}$
    - $(f \circ f)(-2) = \sqrt{7}$

13. For $f(x) = 2x + 3$ and $g(x) = x^2 - 9$

    - $(g \circ f)(x) = 4x^2 + 12x$, domain: $(-\infty, \infty)$
    - $(f \circ g)(x) = 2x^2 - 15$, domain: $(-\infty, \infty)$
    - $(f \circ f)(x) = 4x + 9$, domain: $(-\infty, \infty)$

14. For $f(x) = x^2 - x + 1$ and $g(x) = 3x - 5$

    - $(g \circ f)(x) = 3x^2 - 3x - 2$, domain: $(-\infty, \infty)$
    - $(f \circ g)(x) = 9x^2 - 33x + 31$, domain: $(-\infty, \infty)$
    - $(f \circ f)(x) = x^4 - 2x^3 + 2x^2 - x + 1$, domain: $(-\infty, \infty)$

15. For $f(x) = x^2 - 4$ and $g(x) = |x|$

    - $(g \circ f)(x) = |x^2 - 4|$, domain: $(-\infty, \infty)$
    - $(f \circ g)(x) = |x|^2 - 4 = x^2 - 4$, domain: $(-\infty, \infty)$
    - $(f \circ f)(x) = x^4 - 8x^2 + 12$, domain: $(-\infty, \infty)$

16. For $f(x) = 3x - 5$ and $g(x) = \sqrt{x}$

    - $(g \circ f)(x) = \sqrt{3x - 5}$, domain: $\left[\frac{5}{3}, \infty\right)$
    - $(f \circ g)(x) = 3\sqrt{x} - 5$, domain: $[0, \infty)$
    - $(f \circ f)(x) = 9x - 20$, domain: $(-\infty, \infty)$

17. For $f(x) = |x + 1|$ and $g(x) = \sqrt{x}$

    - $(g \circ f)(x) = \sqrt{|x + 1|}$, domain: $(-\infty, \infty)$
    - $(f \circ g)(x) = |\sqrt{x} + 1| = \sqrt{x} + 1$, domain: $[0, \infty)$
    - $(f \circ f)(x) = ||x + 1| + 1| = |x + 1| + 1$, domain: $(-\infty, \infty)$

18. For $f(x) = 3 - x^2$ and $g(x) = \sqrt{x + 1}$

    - $(g \circ f)(x) = \sqrt{4 - x^2}$, domain: $[-2, 2]$
    - $(f \circ g)(x) = 2 - x$, domain: $[-1, \infty)$
    - $(f \circ f)(x) = -x^4 + 6x^2 - 6$, domain: $(-\infty, \infty)$

19. For $f(x) = |x|$ and $g(x) = \sqrt{4 - x}$

    - $(g \circ f)(x) = \sqrt{4 - |x|}$, domain: $[-4, 4]$
    - $(f \circ g)(x) = |\sqrt{4 - x}| = \sqrt{4 - x}$, domain: $(-\infty, 4]$
    - $(f \circ f)(x) = ||x|| = |x|$, domain: $(-\infty, \infty)$

5.1 Function Composition

20. For $f(x) = x^2 - x - 1$ and $g(x) = \sqrt{x-5}$

    - $(g \circ f)(x) = \sqrt{x^2 - x - 6}$, domain: $(-\infty, -2] \cup [3, \infty)$
    - $(f \circ g)(x) = x - 6 - \sqrt{x-5}$, domain: $[5, \infty)$
    - $(f \circ f)(x) = x^4 - 2x^3 - 2x^2 + 3x + 1$, domain: $(-\infty, \infty)$

21. For $f(x) = 3x - 1$ and $g(x) = \frac{1}{x+3}$

    - $(g \circ f)(x) = \frac{1}{3x+2}$, domain: $\left(-\infty, -\frac{2}{3}\right) \cup \left(-\frac{2}{3}, \infty\right)$
    - $(f \circ g)(x) = -\frac{x}{x+3}$, domain: $(-\infty, -3) \cup (-3, \infty)$
    - $(f \circ f)(x) = 9x - 4$, domain: $(-\infty, \infty)$

22. For $f(x) = \frac{3x}{x-1}$ and $g(x) = \frac{x}{x-3}$

    - $(g \circ f)(x) = x$, domain: $(-\infty, 1) \cup (1, \infty)$
    - $(f \circ g)(x) = x$, domain: $(-\infty, 3) \cup (3, \infty)$
    - $(f \circ f)(x) = \frac{9x}{2x+1}$, domain: $\left(-\infty, -\frac{1}{2}\right) \cup \left(-\frac{1}{2}, 1\right) \cup (1, \infty)$

23. For $f(x) = \frac{x}{2x+1}$ and $g(x) = \frac{2x+1}{x}$

    - $(g \circ f)(x) = \frac{4x+1}{x}$, domain: $\left(-\infty, -\frac{1}{2}\right) \cup \left(-\frac{1}{2}, 0\right), \cup (0, \infty)$
    - $(f \circ g)(x) = \frac{2x+1}{5x+2}$, domain: $\left(-\infty, -\frac{2}{5}\right) \cup \left(-\frac{2}{5}, 0\right) \cup (0, \infty)$
    - $(f \circ f)(x) = \frac{x}{4x+1}$, domain: $\left(-\infty, -\frac{1}{2}\right) \cup \left(-\frac{1}{2}, -\frac{1}{4}\right) \cup \left(-\frac{1}{4}, \infty\right)$

24. For $f(x) = \frac{2x}{x^2-4}$ and $g(x) = \sqrt{1-x}$

    - $(g \circ f)(x) = \sqrt{\frac{x^2-2x-4}{x^2-4}}$, domain: $(-\infty, -2) \cup [1 - \sqrt{5}, 2) \cup [1 + \sqrt{5}, \infty)$
    - $(f \circ g)(x) = -\frac{2\sqrt{1-x}}{x+3}$, domain: $(-\infty, -3) \cup (-3, 1]$
    - $(f \circ f)(x) = \frac{4x-x^3}{x^4-9x^2+16}$, domain: $\left(-\infty, -\frac{1+\sqrt{17}}{2}\right) \cup \left(-\frac{1+\sqrt{17}}{2}, -2\right) \cup \left(-2, \frac{1-\sqrt{17}}{2}\right) \cup \left(\frac{1-\sqrt{17}}{2}, \frac{-1+\sqrt{17}}{2}\right) \cup \left(\frac{-1+\sqrt{17}}{2}, 2\right) \cup \left(2, \frac{1+\sqrt{17}}{2}\right) \cup \left(\frac{1+\sqrt{17}}{2}, \infty\right)$

25. $(h \circ g \circ f)(x) = |\sqrt{-2x}| = \sqrt{-2x}$, domain: $(-\infty, 0]$

26. $(h \circ f \circ g)(x) = |-2\sqrt{x}| = 2\sqrt{x}$, domain: $[0, \infty)$

27. $(g \circ f \circ h)(x) = \sqrt{-2|x|}$, domain: $\{0\}$

28. $(g \circ h \circ f)(x) = \sqrt{|-2x|} = \sqrt{2|x|}$, domain: $(-\infty, \infty)$

29. $(f \circ h \circ g)(x) = -2|\sqrt{x}| = -2\sqrt{x}$, domain: $[0, \infty)$

30. $(f \circ g \circ h)(x) = -2\sqrt{|x|}$, , domain: $(-\infty, \infty)$

31. Let $f(x) = 2x + 3$ and $g(x) = x^3$, then $p(x) = (g \circ f)(x)$.

32. Let $f(x) = x^2 - x + 1$ and $g(x) = x^5$, $P(x) = (g \circ f)(x)$.

33. Let $f(x) = 2x - 1$ and $g(x) = \sqrt{x}$, then $h(x) = (g \circ f)(x)$.

34. Let $f(x) = 7 - 3x$ and $g(x) = |x|$, then $H(x) = (g \circ f)(x)$.

35. Let $f(x) = 5x + 1$ and $g(x) = \frac{2}{x}$, then $r(x) = (g \circ f)(x)$.

36. Let $f(x) = x^2 - 1$ and $g(x) = \frac{7}{x}$, then $R(x) = (g \circ f)(x)$.

37. Let $f(x) = |x|$ and $g(x) = \frac{x+1}{x-1}$, then $q(x) = (g \circ f)(x)$.

38. Let $f(x) = x^3$ and $g(x) = \frac{2x+1}{x-1}$, then $Q(x) = (g \circ f)(x)$.

39. Let $f(x) = 2x$ and $g(x) = \frac{x+1}{3-2x}$, then $v(x) = (g \circ f)(x)$.

40. Let $f(x) = x^2$ and $g(x) = \frac{x}{x^2+1}$, then $w(x) = (g \circ f)(x)$.

41. $F(x) = \sqrt{\frac{x^3+6}{x^3-9}} = (h(g(f(x))))$ where $f(x) = x^3$, $g(x) = \frac{x+6}{x-9}$ and $h(x) = \sqrt{x}$.

42. $F(x) = 3\sqrt{-x+2} - 4 = k(j(f(h(g(x)))))$

43. One possible solution is $F(x) = -\frac{1}{2}(2x-7)^3 + 1 = k(j(f(h(g(x)))))$ where $g(x) = 2x$, $h(x) = x - 7$, $j(x) = -\frac{1}{2}x$ and $k(x) = x + 1$. You could also have $F(x) = H(f(G(x)))$ where $G(x) = 2x - 7$ and $H(x) = -\frac{1}{2}x + 1$.

44. $(f \circ g)(3) = f(g(3)) = f(2) = 4$

45. $f(g(-1)) = f(-4)$ which is undefined

46. $(f \circ f)(0) = f(f(0)) = f(1) = 3$

47. $(f \circ g)(-3) = f(g(-3)) = f(-2) = 2$

48. $(g \circ f)(3) = g(f(3)) = g(-1) = -4$

49. $g(f(-3)) = g(4)$ which is undefined

50. $(g \circ g)(-2) = g(g(-2)) = g(0) = 0$

51. $(g \circ f)(-2) = g(f(-2)) = g(2) = 1$

52. $g(f(g(0))) = g(f(0)) = g(1) = -3$

53. $f(f(f(-1))) = f(f(0)) = f(1) = 3$

54. $f(f(f(f(f(1))))) = f(f(f(f(3)))) = f(f(f(-1))) = f(f(0)) = f(1) = 3$

55. $\underbrace{(g \circ g \circ \cdots \circ g)}_{n \text{ times}}(0) = 0$

5.1 Function Composition

56. $(g \circ f)(1) = 3$

57. $(f \circ g)(3) = 4$

58. $(g \circ f)(2) = 0$

59. $(f \circ g)(0) = 4$

60. $(f \circ f)(1) = 3$

61. $(g \circ g)(1) = 0$

62. $V(x) = x^3$ so $V(x(t)) = (t+1)^3$

63. (a) $R(x) = 2x$

   (b) $(R \circ x)(t) = -8t^2 + 40t + 184$, $0 \leq t \leq 4$. This gives the revenue per hour as a function of time.

   (c) Noon corresponds to $t = 2$, so $(R \circ x)(2) = 232$. The hourly revenue at noon is $232 per hour.

5.2 Inverse Functions

Thinking of a function as a process like we did in Section 1.4, in this section we seek another function which might reverse that process. As in real life, we will find that some processes (like putting on socks and shoes) are reversible while some (like cooking a steak) are not. We start by discussing a very basic function which is reversible, $f(x) = 3x + 4$. Thinking of f as a process, we start with an input x and apply two steps, as we saw in Section 1.4

1. multiply by 3

2. add 4

To reverse this process, we seek a function g which will undo each of these steps and take the output from f, $3x + 4$, and return the input x. If we think of the real-world reversible two-step process of first putting on socks then putting on shoes, to reverse the process, we first take off the shoes, and then we take off the socks. In much the same way, the function g should undo the second step of f first. That is, the function g should

1. *subtract* 4

2. *divide* by 3

Following this procedure, we get $g(x) = \frac{x-4}{3}$. Let's check to see if the function g does the job. If $x = 5$, then $f(5) = 3(5) + 4 = 15 + 4 = 19$. Taking the output 19 from f, we substitute it into g to get $g(19) = \frac{19-4}{3} = \frac{15}{3} = 5$, which is our original input to f. To check that g does the job for all x in the domain of f, we take the generic output from f, $f(x) = 3x + 4$, and substitute that into g. That is, $g(f(x)) = g(3x + 4) = \frac{(3x+4)-4}{3} = \frac{3x}{3} = x$, which is our original input to f. If we carefully examine the arithmetic as we simplify $g(f(x))$, we actually see g first 'undoing' the addition of 4, and then 'undoing' the multiplication by 3. Not only does g undo f, but f also undoes g. That is, if we take the output from g, $g(x) = \frac{x-4}{3}$, and put that into f, we get $f(g(x)) = f\left(\frac{x-4}{3}\right) = 3\left(\frac{x-4}{3}\right) + 4 = (x - 4) + 4 = x$. Using the language of function composition developed in Section 5.1, the statements $g(f(x)) = x$ and $f(g(x)) = x$ can be written as $(g \circ f)(x) = x$ and $(f \circ g)(x) = x$, respectively. Abstractly, we can visualize the relationship between f and g in the diagram below.

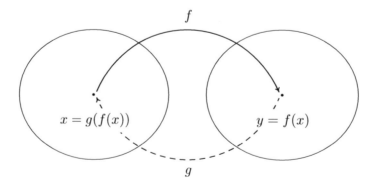

5.2 INVERSE FUNCTIONS

The main idea to get from the diagram is that g takes the outputs from f and returns them to their respective inputs, and conversely, f takes outputs from g and returns them to their respective inputs. We now have enough background to state the central definition of the section.

Definition 5.2. Suppose f and g are two functions such that

1. $(g \circ f)(x) = x$ for all x in the domain of f and

2. $(f \circ g)(x) = x$ for all x in the domain of g

then f and g are **inverses** of each other and the functions f and g are said to be **invertible**.

We now formalize the concept that inverse functions exchange inputs and outputs.

Theorem 5.2. Properties of Inverse Functions: Suppose f and g are inverse functions.

- The range[a] of f is the domain of g and the domain of f is the range of g

- $f(a) = b$ if and only if $g(b) = a$

- (a, b) is on the graph of f if and only if (b, a) is on the graph of g

---
[a]Recall this is the set of all outputs of a function.

Theorem 5.2 is a consequence of Definition 5.2 and the Fundamental Graphing Principle for Functions. We note the third property in Theorem 5.2 tells us that the graphs of inverse functions are reflections about the line $y = x$. For a proof of this, see Example 1.1.7 in Section 1.1 and Exercise 72 in Section 2.1. For example, we plot the inverse functions $f(x) = 3x + 4$ and $g(x) = \frac{x-4}{3}$ below.

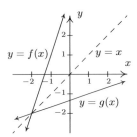

If we abstract one step further, we can express the sentiment in Definition 5.2 by saying that f and g are inverses if and only if $g \circ f = I_1$ and $f \circ g = I_2$ where I_1 is the identity function restricted[1] to the domain of f and I_2 is the identity function restricted to the domain of g. In other words, $I_1(x) = x$ for all x in the domain of f and $I_2(x) = x$ for all x in the domain of g. Using this description of inverses along with the properties of function composition listed in Theorem 5.1, we can show that function inverses are unique.[2] Suppose g and h are both inverses of a function

---
[1]The identity function I, which was introduced in Section 2.1 and mentioned in Theorem 5.1, has a domain of all real numbers. Since the domains of f and g may not be all real numbers, we need the restrictions listed here.
[2]In other words, invertible functions have exactly one inverse.

f. By Theorem 5.2, the domain of g is equal to the domain of h, since both are the range of f. This means the identity function I_2 applies both to the domain of h and the domain of g. Thus $h = h \circ I_2 = h \circ (f \circ g) = (h \circ f) \circ g = I_1 \circ g = g$, as required.[3] We summarize the discussion of the last two paragraphs in the following theorem.[4]

> **Theorem 5.3. Uniqueness of Inverse Functions and Their Graphs :** Suppose f is an invertible function.
>
> - There is exactly one inverse function for f, denoted f^{-1} (read f-inverse)
> - The graph of $y = f^{-1}(x)$ is the reflection of the graph of $y = f(x)$ across the line $y = x$.

The notation f^{-1} is an unfortunate choice since you've been programmed since Elementary Algebra to think of this as $\frac{1}{f}$. This is most definitely *not* the case since, for instance, $f(x) = 3x + 4$ has as its inverse $f^{-1}(x) = \frac{x-4}{3}$, which is certainly different than $\frac{1}{f(x)} = \frac{1}{3x+4}$. Why does this confusing notation persist? As we mentioned in Section 5.1, the identity function I is to function composition what the real number 1 is to real number multiplication. The choice of notation f^{-1} alludes to the property that $f^{-1} \circ f = I_1$ and $f \circ f^{-1} = I_2$, in much the same way as $3^{-1} \cdot 3 = 1$ and $3 \cdot 3^{-1} = 1$.

Let's turn our attention to the function $f(x) = x^2$. Is f invertible? A likely candidate for the inverse is the function $g(x) = \sqrt{x}$. Checking the composition yields $(g \circ f)(x) = g(f(x)) = \sqrt{x^2} = |x|$, which is not equal to x for all x in the domain $(-\infty, \infty)$. For example, when $x = -2$, $f(-2) = (-2)^2 = 4$, but $g(4) = \sqrt{4} = 2$, which means g failed to return the input -2 from its output 4. What g did, however, is match the output 4 to a *different* input, namely 2, which satisfies $f(2) = 4$. This issue is presented schematically in the picture below.

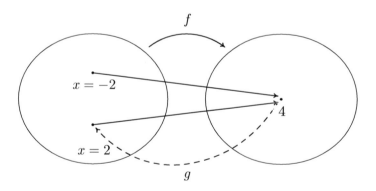

We see from the diagram that since both $f(-2)$ and $f(2)$ are 4, it is impossible to construct a *function* which takes 4 back to *both* $x = 2$ and $x = -2$. (By definition, a function matches a real number with exactly one other real number.) From a graphical standpoint, we know that if

---

[3] It is an excellent exercise to explain each step in this string of equalities.

[4] In the interests of full disclosure, the authors would like to admit that much of the discussion in the previous paragraphs could have easily been avoided had we appealed to the description of a function as a set of ordered pairs. We make no apology for our discussion from a function composition standpoint, however, since it exposes the reader to more abstract ways of thinking of functions and inverses. We will revisit this concept again in Chapter 8.

5.2 Inverse Functions

$y = f^{-1}(x)$ exists, its graph can be obtained by reflecting $y = x^2$ about the line $y = x$, in accordance with Theorem 5.3. Doing so produces

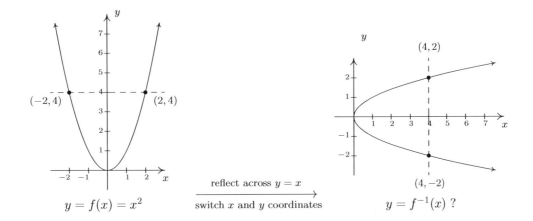

We see that the line $x = 4$ intersects the graph of the supposed inverse twice - meaning the graph fails the Vertical Line Test, Theorem 1.1, and as such, does not represent y as a function of x. The vertical line $x = 4$ on the graph on the right corresponds to the *horizontal line* $y = 4$ on the graph of $y = f(x)$. The fact that the horizontal line $y = 4$ intersects the graph of f twice means two *different* inputs, namely $x = -2$ and $x = 2$, are matched with the *same* output, 4, which is the cause of all of the trouble. In general, for a function to have an inverse, *different* inputs must go to *different* outputs, or else we will run into the same problem we did with $f(x) = x^2$. We give this property a name.

Definition 5.3. A function f is said to be **one-to-one** if f matches different inputs to different outputs. Equivalently, f is one-to-one if and only if whenever $f(c) = f(d)$, then $c = d$.

Graphically, we detect one-to-one functions using the test below.

Theorem 5.4. The Horizontal Line Test: A function f is one-to-one if and only if no horizontal line intersects the graph of f more than once.

We say that the graph of a function **passes** the Horizontal Line Test if no horizontal line intersects the graph more than once; otherwise, we say the graph of the function **fails** the Horizontal Line Test. We have argued that if f is invertible, then f must be one-to-one, otherwise the graph given by reflecting the graph of $y = f(x)$ about the line $y = x$ will fail the Vertical Line Test. It turns out that being one-to-one is also enough to guarantee invertibility. To see this, we think of f as the set of ordered pairs which constitute its graph. If switching the x- and y-coordinates of the points results in a function, then f is invertible and we have found f^{-1}. This is precisely what the Horizontal Line Test does for us: it checks to see whether or not a set of points describes x as a function of y. We summarize these results below.

> **Theorem 5.5. Equivalent Conditions for Invertibility:** Suppose f is a function. The following statements are equivalent.
>
> - f is invertible
>
> - f is one-to-one
>
> - The graph of f passes the Horizontal Line Test

We put this result to work in the next example.

Example 5.2.1. Determine if the following functions are one-to-one in two ways: (a) analytically using Definition 5.3 and (b) graphically using the Horizontal Line Test.

1. $f(x) = \dfrac{1-2x}{5}$

2. $g(x) = \dfrac{2x}{1-x}$

3. $h(x) = x^2 - 2x + 4$

4. $F = \{(-1,1), (0,2), (2,1)\}$

Solution.

1. (a) To determine if f is one-to-one analytically, we assume $f(c) = f(d)$ and attempt to deduce that $c = d$.

$$\begin{aligned} f(c) &= f(d) \\ \frac{1-2c}{5} &= \frac{1-2d}{5} \\ 1-2c &= 1-2d \\ -2c &= -2d \\ c &= d \checkmark \end{aligned}$$

   Hence, f is one-to-one.

   (b) To check if f is one-to-one graphically, we look to see if the graph of $y = f(x)$ passes the Horizontal Line Test. We have that f is a non-constant linear function, which means its graph is a non-horizontal line. Thus the graph of f passes the Horizontal Line Test.

2. (a) We begin with the assumption that $g(c) = g(d)$ and try to show $c = d$.

$$\begin{aligned} g(c) &= g(d) \\ \frac{2c}{1-c} &= \frac{2d}{1-d} \\ 2c(1-d) &= 2d(1-c) \\ 2c - 2cd &= 2d - 2dc \\ 2c &= 2d \\ c &= d \checkmark \end{aligned}$$

   We have shown that g is one-to-one.

5.2 INVERSE FUNCTIONS

(b) We can graph g using the six step procedure outlined in Section 4.2. We get the sole intercept at $(0,0)$, a vertical asymptote $x = 1$ and a horizontal asymptote (which the graph never crosses) $y = -2$. We see from that the graph of g passes the Horizontal Line Test.

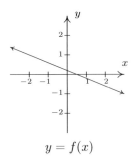

$y = f(x)$

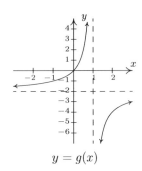

$y = g(x)$

3. (a) We begin with $h(c) = h(d)$. As we work our way through the problem, we encounter a nonlinear equation. We move the non-zero terms to the left, leave a 0 on the right and factor accordingly.

$$\begin{aligned} h(c) &= h(d) \\ c^2 - 2c + 4 &= d^2 - 2d + 4 \\ c^2 - 2c &= d^2 - 2d \\ c^2 - d^2 - 2c + 2d &= 0 \\ (c+d)(c-d) - 2(c-d) &= 0 \\ (c-d)((c+d) - 2) &= 0 \qquad \text{factor by grouping} \\ c - d = 0 \quad &\text{or} \quad c + d - 2 = 0 \\ c = d \quad &\text{or} \quad c = 2 - d \end{aligned}$$

We get $c = d$ as one possibility, but we also get the possibility that $c = 2 - d$. This suggests that f may not be one-to-one. Taking $d = 0$, we get $c = 0$ or $c = 2$. With $f(0) = 4$ and $f(2) = 4$, we have produced two different inputs with the same output meaning f is not one-to-one.

(b) We note that h is a quadratic function and we graph $y = h(x)$ using the techniques presented in Section 2.3. The vertex is $(1, 3)$ and the parabola opens upwards. We see immediately from the graph that h is not one-to-one, since there are several horizontal lines which cross the graph more than once.

4. (a) The function F is given to us as a set of ordered pairs. The condition $F(c) = F(d)$ means the outputs from the function (the y-coordinates of the ordered pairs) are the same. We see that the points $(-1, 1)$ and $(2, 1)$ are both elements of F with $F(-1) = 1$ and $F(2) = 1$. Since $-1 \neq 2$, we have established that F is *not* one-to-one.

(b) Graphically, we see the horizontal line $y = 1$ crosses the graph more than once. Hence, the graph of F fails the Horizontal Line Test.

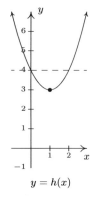

$y = h(x)$

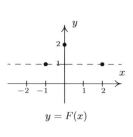

$y = F(x)$

We have shown that the functions f and g in Example 5.2.1 are one-to-one. This means they are invertible, so it is natural to wonder what $f^{-1}(x)$ and $g^{-1}(x)$ would be. For $f(x) = \frac{1-2x}{5}$, we can think our way through the inverse since there is only one occurrence of x. We can track step-by-step what is done to x and reverse those steps as we did at the beginning of the chapter. The function $g(x) = \frac{2x}{1-x}$ is a bit trickier since x occurs in two places. When one evaluates $g(x)$ for a specific value of x, which is first, the $2x$ or the $1-x$? We can imagine functions more complicated than these so we need to develop a general methodology to attack this problem. Theorem 5.2 tells us equation $y = f^{-1}(x)$ is equivalent to $f(y) = x$ and this is the basis of our algorithm.

Steps for finding the Inverse of a One-to-one Function

1. Write $y = f(x)$

2. Interchange x and y

3. Solve $x = f(y)$ for y to obtain $y = f^{-1}(x)$

Note that we could have simply written 'Solve $x = f(y)$ for y' and be done with it. The act of interchanging the x and y is there to remind us that we are finding the inverse function by switching the inputs and outputs.

Example 5.2.2. Find the inverse of the following one-to-one functions. Check your answers analytically using function composition and graphically.

1. $f(x) = \dfrac{1 - 2x}{5}$

2. $g(x) = \dfrac{2x}{1 - x}$

Solution.

1. As we mentioned earlier, it is possible to think our way through the inverse of f by recording the steps we apply to x and the order in which we apply them and then reversing those steps in the reverse order. We encourage the reader to do this. We, on the other hand, will practice the algorithm. We write $y = f(x)$ and proceed to switch x and y

5.2 Inverse Functions

$$\begin{aligned} y &= f(x) \\ y &= \frac{1-2x}{5} \\ x &= \frac{1-2y}{5} \quad \text{switch } x \text{ and } y \\ 5x &= 1-2y \\ 5x-1 &= -2y \\ \frac{5x-1}{-2} &= y \\ y &= -\frac{5}{2}x + \frac{1}{2} \end{aligned}$$

We have $f^{-1}(x) = -\frac{5}{2}x + \frac{1}{2}$. To check this answer analytically, we first check that $\left(f^{-1} \circ f\right)(x) = x$ for all x in the domain of f, which is all real numbers.

$$\begin{aligned} \left(f^{-1} \circ f\right)(x) &= f^{-1}(f(x)) \\ &= -\frac{5}{2}f(x) + \frac{1}{2} \\ &= -\frac{5}{2}\left(\frac{1-2x}{5}\right) + \frac{1}{2} \\ &= -\frac{1}{2}(1-2x) + \frac{1}{2} \\ &= -\frac{1}{2} + x + \frac{1}{2} \\ &= x \checkmark \end{aligned}$$

We now check that $\left(f \circ f^{-1}\right)(x) = x$ for all x in the range of f which is also all real numbers. (Recall that the domain of f^{-1}) is the range of f.)

$$\begin{aligned} \left(f \circ f^{-1}\right)(x) &= f(f^{-1}(x)) \\ &= \frac{1-2f^{-1}(x)}{5} \\ &= \frac{1-2\left(-\frac{5}{2}x + \frac{1}{2}\right)}{5} \\ &= \frac{1+5x-1}{5} \\ &= \frac{5x}{5} \\ &= x \checkmark \end{aligned}$$

To check our answer graphically, we graph $y = f(x)$ and $y = f^{-1}(x)$ on the same set of axes.[5] They appear to be reflections across the line $y = x$.

---

[5]Note that if you perform your check on a calculator for more sophisticated functions, you'll need to take advantage of the 'ZoomSquare' feature to get the correct geometric perspective.

386   FURTHER TOPICS IN FUNCTIONS

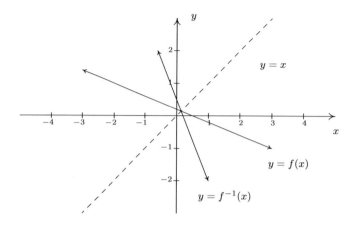

2. To find $g^{-1}(x)$, we start with $y = g(x)$. We note that the domain of g is $(-\infty, 1) \cup (1, \infty)$.

$$\begin{aligned}
y &= g(x) \\
y &= \frac{2x}{1-x} \\
x &= \frac{2y}{1-y} \qquad \text{switch } x \text{ and } y \\
x(1-y) &= 2y \\
x - xy &= 2y \\
x &= xy + 2y \\
x &= y(x+2) \qquad \text{factor} \\
y &= \frac{x}{x+2}
\end{aligned}$$

We obtain $g^{-1}(x) = \frac{x}{x+2}$. To check this analytically, we first check $\left(g^{-1} \circ g\right)(x) = x$ for all x in the domain of g, that is, for all $x \neq 1$.

$$\begin{aligned}
\left(g^{-1} \circ g\right)(x) &= g^{-1}(g(x)) \\
&= g^{-1}\left(\frac{2x}{1-x}\right) \\
&= \frac{\left(\frac{2x}{1-x}\right)}{\left(\frac{2x}{1-x}\right) + 2} \\
&= \frac{\left(\frac{2x}{1-x}\right)}{\left(\frac{2x}{1-x}\right) + 2} \cdot \frac{(1-x)}{(1-x)} \qquad \text{clear denominators}
\end{aligned}$$

5.2 Inverse Functions

$$= \frac{2x}{2x + 2(1-x)}$$

$$= \frac{2x}{2x + 2 - 2x}$$

$$= \frac{2x}{2}$$

$$= x \checkmark$$

Next, we check $g\left(g^{-1}(x)\right) = x$ for all x in the range of g. From the graph of g in Example 5.2.1, we have that the range of g is $(-\infty, -2) \cup (-2, \infty)$. This matches the domain we get from the formula $g^{-1}(x) = \frac{x}{x+2}$, as it should.

$$
\begin{aligned}
\left(g \circ g^{-1}\right)(x) &= g\left(g^{-1}(x)\right) \\
&= g\left(\frac{x}{x+2}\right) \\
&= \frac{2\left(\frac{x}{x+2}\right)}{1 - \left(\frac{x}{x+2}\right)} \\
&= \frac{2\left(\frac{x}{x+2}\right)}{1 - \left(\frac{x}{x+2}\right)} \cdot \frac{(x+2)}{(x+2)} \quad \text{clear denominators} \\
&= \frac{2x}{(x+2) - x} \\
&= \frac{2x}{2} \\
&= x \checkmark
\end{aligned}
$$

Graphing $y = g(x)$ and $y = g^{-1}(x)$ on the same set of axes is busy, but we can see the symmetric relationship if we thicken the curve for $y = g^{-1}(x)$. Note that the vertical asymptote $x = 1$ of the graph of g corresponds to the horizontal asymptote $y = 1$ of the graph of g^{-1}, as it should since x and y are switched. Similarly, the horizontal asymptote $y = -2$ of the graph of g corresponds to the vertical asymptote $x = -2$ of the graph of g^{-1}.

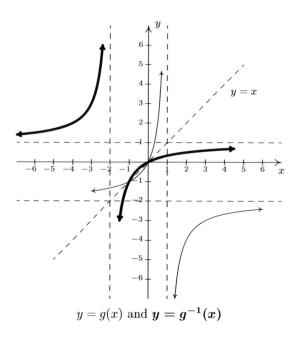

$$y = g(x) \text{ and } y = g^{-1}(x)$$

We now return to $f(x) = x^2$. We know that f is not one-to-one, and thus, is not invertible. However, if we restrict the domain of f, we can produce a new function g which is one-to-one. If we define $g(x) = x^2$, $x \geq 0$, then we have

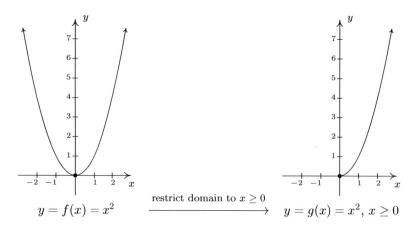

The graph of g passes the Horizontal Line Test. To find an inverse of g, we proceed as usual

$$\begin{aligned} y &= g(x) \\ y &= x^2, \ x \geq 0 \\ x &= y^2, \ y \geq 0 \quad \text{switch } x \text{ and } y \\ y &= \pm\sqrt{x} \\ y &= \sqrt{x} \quad\quad\quad\quad \text{since } y \geq 0 \end{aligned}$$

5.2 Inverse Functions

We get $g^{-1}(x) = \sqrt{x}$. At first it looks like we'll run into the same trouble as before, but when we check the composition, the domain restriction on g saves the day. We get $(g^{-1} \circ g)(x) = g^{-1}(g(x)) = g^{-1}(x^2) = \sqrt{x^2} = |x| = x$, since $x \geq 0$. Checking $(g \circ g^{-1})(x) = g(g^{-1}(x)) = g(\sqrt{x}) = (\sqrt{x})^2 = x$. Graphing[6] g and g^{-1} on the same set of axes shows that they are reflections about the line $y = x$.

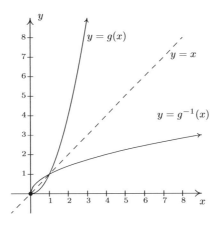

Our next example continues the theme of domain restriction.

Example 5.2.3. Graph the following functions to show they are one-to-one and find their inverses. Check your answers analytically using function composition and graphically.

1. $j(x) = x^2 - 2x + 4$, $x \leq 1$.
2. $k(x) = \sqrt{x+2} - 1$

Solution.

1. The function j is a restriction of the function h from Example 5.2.1. Since the domain of j is restricted to $x \leq 1$, we are selecting only the 'left half' of the parabola. We see that the graph of j passes the Horizontal Line Test and thus j is invertible.

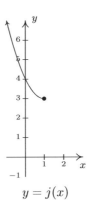

$y = j(x)$

---

[6] We graphed $y = \sqrt{x}$ in Section 1.7.

We now use our algorithm[7] to find $j^{-1}(x)$.

$$\begin{aligned}
y &= j(x) \\
y &= x^2 - 2x + 4, \ x \leq 1 \\
x &= y^2 - 2y + 4, \ y \leq 1 &&\text{switch } x \text{ and } y \\
0 &= y^2 - 2y + 4 - x \\
y &= \frac{2 \pm \sqrt{(-2)^2 - 4(1)(4-x)}}{2(1)} &&\text{quadratic formula, } c = 4 - x \\
y &= \frac{2 \pm \sqrt{4x - 12}}{2} \\
y &= \frac{2 \pm \sqrt{4(x-3)}}{2} \\
y &= \frac{2 \pm 2\sqrt{x-3}}{2} \\
y &= \frac{2\left(1 \pm \sqrt{x-3}\right)}{2} \\
y &= 1 \pm \sqrt{x-3} \\
y &= 1 - \sqrt{x-3} &&\text{since } y \leq 1.
\end{aligned}$$

We have $j^{-1}(x) = 1 - \sqrt{x-3}$. When we simplify $\left(j^{-1} \circ j\right)(x)$, we need to remember that the domain of j is $x \leq 1$.

$$\begin{aligned}
\left(j^{-1} \circ j\right)(x) &= j^{-1}(j(x)) \\
&= j^{-1}\left(x^2 - 2x + 4\right), \ x \leq 1 \\
&= 1 - \sqrt{(x^2 - 2x + 4) - 3} \\
&= 1 - \sqrt{x^2 - 2x + 1} \\
&= 1 - \sqrt{(x-1)^2} \\
&= 1 - |x-1| \\
&= 1 - (-(x-1)) &&\text{since } x \leq 1 \\
&= x \ \checkmark
\end{aligned}$$

Checking $j \circ j^{-1}$, we get

$$\begin{aligned}
\left(j \circ j^{-1}\right)(x) &= j\left(j^{-1}(x)\right) \\
&= j\left(1 - \sqrt{x-3}\right) \\
&= \left(1 - \sqrt{x-3}\right)^2 - 2\left(1 - \sqrt{x-3}\right) + 4 \\
&= 1 - 2\sqrt{x-3} + \left(\sqrt{x-3}\right)^2 - 2 + 2\sqrt{x-3} + 4 \\
&= 3 + x - 3 \\
&= x \ \checkmark
\end{aligned}$$

---

[7]Here, we use the Quadratic Formula to solve for y. For 'completeness,' we note you can (and should!) also consider solving for y by 'completing' the square.

5.2 Inverse Functions

Using what we know from Section 1.7, we graph $y = j^{-1}(x)$ and $y = j(x)$ below.

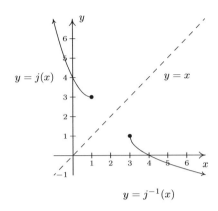

2. We graph $y = k(x) = \sqrt{x+2} - 1$ using what we learned in Section 1.7 and see k is one-to-one.

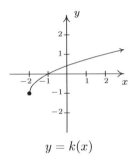

$y = k(x)$

We now try to find k^{-1}.

$$\begin{aligned} y &= k(x) \\ y &= \sqrt{x+2} - 1 \\ x &= \sqrt{y+2} - 1 \quad \text{switch } x \text{ and } y \\ x + 1 &= \sqrt{y+2} \\ (x+1)^2 &= \left(\sqrt{y+2}\right)^2 \\ x^2 + 2x + 1 &= y + 2 \\ y &= x^2 + 2x - 1 \end{aligned}$$

We have $k^{-1}(x) = x^2 + 2x - 1$. Based on our experience, we know something isn't quite right. We determined k^{-1} is a quadratic function, and we have seen several times in this section that these are not one-to-one unless their domains are suitably restricted. Theorem 5.2 tells us that the domain of k^{-1} is the range of k. From the graph of k, we see that the range is $[-1, \infty)$, which means we restrict the domain of k^{-1} to $x \geq -1$. We now check that this works in our compositions.

$$\begin{aligned}
\left(k^{-1} \circ k\right)(x) &= k^{-1}(k(x)) \\
&= k^{-1}\left(\sqrt{x+2} - 1\right), \ x \geq -2 \\
&= \left(\sqrt{x+2} - 1\right)^2 + 2\left(\sqrt{x+2} - 1\right) - 1 \\
&= \left(\sqrt{x+2}\right)^2 - 2\sqrt{x+2} + 1 + 2\sqrt{x+2} - 2 - 1 \\
&= x + 2 - 2 \\
&= x \ \checkmark
\end{aligned}$$

and

$$\begin{aligned}
\left(k \circ k^{-1}\right)(x) &= k\left(x^2 + 2x - 1\right) \ x \geq -1 \\
&= \sqrt{(x^2 + 2x - 1) + 2} - 1 \\
&= \sqrt{x^2 + 2x + 1} - 1 \\
&= \sqrt{(x+1)^2} - 1 \\
&= |x+1| - 1 \\
&= x + 1 - 1 \qquad \text{since } x \geq -1 \\
&= x \ \checkmark
\end{aligned}$$

Graphically, everything checks out as well, provided that we remember the domain restriction on k^{-1} means we take the right half of the parabola.

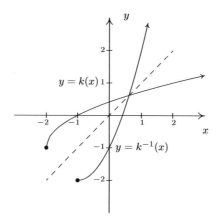

□

Our last example of the section gives an application of inverse functions.

Example 5.2.4. Recall from Section 2.1 that the price-demand equation for the PortaBoy game system is $p(x) = -1.5x + 250$ for $0 \leq x \leq 166$, where x represents the number of systems sold weekly and p is the price per system in dollars.

5.2 Inverse Functions

1. Explain why p is one-to-one and find a formula for $p^{-1}(x)$. State the restricted domain.

2. Find and interpret $p^{-1}(220)$.

3. Recall from Section 2.3 that the weekly profit P, in dollars, as a result of selling x systems is given by $P(x) = -1.5x^2 + 170x - 150$. Find and interpret $\left(P \circ p^{-1}\right)(x)$.

4. Use your answer to part 3 to determine the price per PortaBoy which would yield the maximum profit. Compare with Example 2.3.3.

Solution.

1. We leave to the reader to show the graph of $p(x) = -1.5x + 250$, $0 \leq x \leq 166$, is a line segment from $(0, 250)$ to $(166, 1)$, and as such passes the Horizontal Line Test. Hence, p is one-to-one. We find the expression for $p^{-1}(x)$ as usual and get $p^{-1}(x) = \frac{500-2x}{3}$. The domain of p^{-1} should match the range of p, which is $[1, 250]$, and as such, we restrict the domain of p^{-1} to $1 \leq x \leq 250$.

2. We find $p^{-1}(220) = \frac{500-2(220)}{3} = 20$. Since the function p took as inputs the weekly sales and furnished the price per system as the output, p^{-1} takes the price per system and returns the weekly sales as its output. Hence, $p^{-1}(220) = 20$ means 20 systems will be sold in a week if the price is set at $220 per system.

3. We compute $\left(P \circ p^{-1}\right)(x) = P\left(p^{-1}(x)\right) = P\left(\frac{500-2x}{3}\right) = -1.5\left(\frac{500-2x}{3}\right)^2 + 170\left(\frac{500-2x}{3}\right) - 150$. After a hefty amount of Elementary Algebra,[8] we obtain $\left(P \circ p^{-1}\right)(x) = -\frac{2}{3}x^2 + 220x - \frac{40450}{3}$. To understand what this means, recall that the original profit function P gave us the weekly profit as a function of the weekly sales. The function p^{-1} gives us the weekly sales as a function of the price. Hence, $P \circ p^{-1}$ takes as its input a price. The function p^{-1} returns the weekly sales, which in turn is fed into P to return the weekly profit. Hence, $\left(P \circ p^{-1}\right)(x)$ gives us the weekly profit (in dollars) as a function of the price per system, x, using the weekly sales $p^{-1}(x)$ as the 'middle man'.

4. We know from Section 2.3 that the graph of $y = \left(P \circ p^{-1}\right)(x)$ is a parabola opening downwards. The maximum profit is realized at the vertex. Since we are concerned only with the price per system, we need only find the x-coordinate of the vertex. Identifying $a = -\frac{2}{3}$ and $b = 220$, we get, by the Vertex Formula, Equation 2.4, $x = -\frac{b}{2a} = 165$. Hence, weekly profit is maximized if we set the price at $165 per system. Comparing this with our answer from Example 2.3.3, there is a slight discrepancy to the tune of $0.50. We leave it to the reader to balance the books appropriately. □

---

[8] It is good review to actually do this!

5.2.1 Exercises

In Exercises 1 - 20, show that the given function is one-to-one and find its inverse. Check your answers algebraically and graphically. Verify that the range of f is the domain of f^{-1} and vice-versa.

1. $f(x) = 6x - 2$

2. $f(x) = 42 - x$

3. $f(x) = \dfrac{x-2}{3} + 4$

4. $f(x) = 1 - \dfrac{4+3x}{5}$

5. $f(x) = \sqrt{3x-1} + 5$

6. $f(x) = 2 - \sqrt{x-5}$

7. $f(x) = 3\sqrt{x-1} - 4$

8. $f(x) = 1 - 2\sqrt{2x+5}$

9. $f(x) = \sqrt[5]{3x-1}$

10. $f(x) = 3 - \sqrt[3]{x-2}$

11. $f(x) = x^2 - 10x$, $x \geq 5$

12. $f(x) = 3(x+4)^2 - 5$, $x \leq -4$

13. $f(x) = x^2 - 6x + 5$, $x \leq 3$

14. $f(x) = 4x^2 + 4x + 1$, $x < -1$

15. $f(x) = \dfrac{3}{4-x}$

16. $f(x) = \dfrac{x}{1-3x}$

17. $f(x) = \dfrac{2x-1}{3x+4}$

18. $f(x) = \dfrac{4x+2}{3x-6}$

19. $f(x) = \dfrac{-3x-2}{x+3}$

20. $f(x) = \dfrac{x-2}{2x-1}$

With help from your classmates, find the inverses of the functions in Exercises 21 - 24.

21. $f(x) = ax + b$, $a \neq 0$

22. $f(x) = a\sqrt{x-h} + k$, $a \neq 0, x \geq h$

23. $f(x) = ax^2 + bx + c$ where $a \neq 0$, $x \geq -\dfrac{b}{2a}$.

24. $f(x) = \dfrac{ax+b}{cx+d}$, (See Exercise 33 below.)

25. In Example 1.5.3, the price of a dOpi media player, in dollars per dOpi, is given as a function of the weekly sales x according to the formula $p(x) = 450 - 15x$ for $0 \leq x \leq 30$.

    (a) Find $p^{-1}(x)$ and state its domain.

    (b) Find and interpret $p^{-1}(105)$.

    (c) In Example 1.5.3, we determined that the profit (in dollars) made from producing and selling x dOpis per week is $P(x) = -15x^2 + 350x - 2000$, for $0 \leq x \leq 30$. Find $\left(P \circ p^{-1}\right)(x)$ and determine what price per dOpi would yield the maximum profit. What is the maximum profit? How many dOpis need to be produced and sold to achieve the maximum profit?

5.2 INVERSE FUNCTIONS

26. Show that the Fahrenheit to Celsius conversion function found in Exercise 35 in Section 2.1 is invertible and that its inverse is the Celsius to Fahrenheit conversion function.

27. Analytically show that the function $f(x) = x^3 + 3x + 1$ is one-to-one. Since finding a formula for its inverse is beyond the scope of this textbook, use Theorem 5.2 to help you compute $f^{-1}(1)$, $f^{-1}(5)$, and $f^{-1}(-3)$.

28. Let $f(x) = \frac{2x}{x^2-1}$. Using the techniques in Section 4.2, graph $y = f(x)$. Verify that f is one-to-one on the interval $(-1, 1)$. Use the procedure outlined on Page 384 and your graphing calculator to find the formula for $f^{-1}(x)$. Note that since $f(0) = 0$, it should be the case that $f^{-1}(0) = 0$. What goes wrong when you attempt to substitute $x = 0$ into $f^{-1}(x)$? Discuss with your classmates how this problem arose and possible remedies.

29. With the help of your classmates, explain why a function which is either strictly increasing or strictly decreasing on its entire domain would have to be one-to-one, hence invertible.

30. If f is odd and invertible, prove that f^{-1} is also odd.

31. Let f and g be invertible functions. With the help of your classmates show that $(f \circ g)$ is one-to-one, hence invertible, and that $(f \circ g)^{-1}(x) = (g^{-1} \circ f^{-1})(x)$.

32. What graphical feature must a function f possess for it to be its own inverse?

33. What conditions must you place on the values of a, b, c and d in Exercise 24 in order to guarantee that the function is invertible?

5.2.2 Answers

1. $f^{-1}(x) = \dfrac{x+2}{6}$

2. $f^{-1}(x) = 42 - x$

3. $f^{-1}(x) = 3x - 10$

4. $f^{-1}(x) = -\frac{5}{3}x + \frac{1}{3}$

5. $f^{-1}(x) = \frac{1}{3}(x-5)^2 + \frac{1}{3}$, $x \geq 5$

6. $f^{-1}(x) = (x-2)^2 + 5$, $x \leq 2$

7. $f^{-1}(x) = \frac{1}{9}(x+4)^2 + 1$, $x \geq -4$

8. $f^{-1}(x) = \frac{1}{8}(x-1)^2 - \frac{5}{2}$, $x \leq 1$

9. $f^{-1}(x) = \frac{1}{3}x^5 + \frac{1}{3}$

10. $f^{-1}(x) = -(x-3)^3 + 2$

11. $f^{-1}(x) = 5 + \sqrt{x+25}$

12. $f^{-1}(x) = -\sqrt{\frac{x+5}{3}} - 4$

13. $f^{-1}(x) = 3 - \sqrt{x+4}$

14. $f^{-1}(x) = -\dfrac{\sqrt{x+1}}{2}$, $x > 1$

15. $f^{-1}(x) = \dfrac{4x-3}{x}$

16. $f^{-1}(x) = \dfrac{x}{3x+1}$

17. $f^{-1}(x) = \dfrac{4x+1}{2-3x}$

18. $f^{-1}(x) = \dfrac{6x+2}{3x-4}$

19. $f^{-1}(x) = \dfrac{-3x-2}{x+3}$

20. $f^{-1}(x) = \dfrac{x-2}{2x-1}$

25. (a) $p^{-1}(x) = \frac{450-x}{15}$. The domain of p^{-1} is the range of p which is $[0, 450]$

   (b) $p^{-1}(105) = 23$. This means that if the price is set to \$105 then 23 dOpis will be sold.

   (c) $(P \circ p^{-1})(x) = -\frac{1}{15}x^2 + \frac{110}{3}x - 5000$, $0 \leq x \leq 450$. The graph of $y = (P \circ p^{-1})(x)$ is a parabola opening downwards with vertex $\left(275, \frac{125}{3}\right) \approx (275, 41.67)$. This means that the maximum profit is a whopping \$41.67 when the price per dOpi is set to \$275. At this price, we can produce and sell $p^{-1}(275) = 11.\overline{6}$ dOpis. Since we cannot sell part of a system, we need to adjust the price to sell either 11 dOpis or 12 dOpis. We find $p(11) = 285$ and $p(12) = 270$, which means we set the price per dOpi at either \$285 or \$270, respectively. The profits at these prices are $(P \circ p^{-1})(285) = 35$ and $(P \circ p^{-1})(270) = 40$, so it looks as if the maximum profit is \$40 and it is made by producing and selling 12 dOpis a week at a price of \$270 per dOpi.

27. Given that $f(0) = 1$, we have $f^{-1}(1) = 0$. Similarly $f^{-1}(5) = 1$ and $f^{-1}(-3) = -1$

5.3 Other Algebraic Functions

This section serves as a watershed for functions which are combinations of polynomial, and more generally, rational functions, with the operations of radicals. It is business of Calculus to discuss these functions in all the detail they demand so our aim in this section is to help shore up the requisite skills needed so that the reader can answer Calculus's call when the time comes. We briefly recall the definition and some of the basic properties of radicals from Intermediate Algebra.[1]

> **Definition 5.4.** Let x be a real number and n a natural number.[a] If n is odd, the **principal n^{th} root** of x, denoted $\sqrt[n]{x}$ is the unique real number satisfying $\left(\sqrt[n]{x}\right)^n = x$. If n is even, $\sqrt[n]{x}$ is defined similarly[b] provided $x \geq 0$ and $\sqrt[n]{x} \geq 0$. The **index** is the number n and the **radicand** is the number x. For $n = 2$, we write $\sqrt{x}$ instead of $\sqrt[2]{x}$.
>
> [a]Recall this means $n = 1, 2, 3, \ldots$.
> [b]Recall both $x = -2$ and $x = 2$ satisfy $x^4 = 16$, but $\sqrt[4]{16} = 2$, not -2.

It is worth remarking that, in light of Section 5.2, we could define $f(x) = \sqrt[n]{x}$ functionally as the inverse of $g(x) = x^n$ with the stipulation that when n is even, the domain of g is restricted to $[0, \infty)$. From what we know about $g(x) = x^n$ from Section 3.1 along with Theorem 5.3, we can produce the graphs of $f(x) = \sqrt[n]{x}$ by reflecting the graphs of $g(x) = x^n$ across the line $y = x$. Below are the graphs of $y = \sqrt{x}$, $y = \sqrt[4]{x}$ and $y = \sqrt[6]{x}$. The point $(0, 0)$ is indicated as a reference. The axes are hidden so we can see the vertical steepening near $x = 0$ and the horizontal flattening as $x \to \infty$.

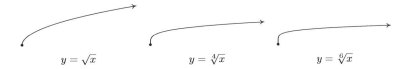

The odd-indexed radical functions also follow a predictable trend - steepening near $x = 0$ and flattening as $x \to \pm\infty$. In the exercises, you'll have a chance to graph some basic radical functions using the techniques presented in Section 1.7.

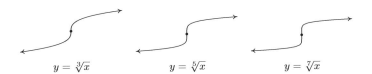

We have used all of the following properties at some point in the textbook for the case $n = 2$ (the square root), but we list them here in generality for completeness.

---

[1]Although we discussed imaginary numbers in Section 3.4, we restrict our attention to real numbers in this section. See the epilogue on page 294 for more details.

> **Theorem 5.6. Properties of Radicals:** Let x and y be real numbers and m and n be natural numbers. If $\sqrt[n]{x}$, $\sqrt[n]{y}$ are real numbers, then
>
> - **Product Rule:** $\sqrt[n]{xy} = \sqrt[n]{x}\sqrt[n]{y}$
>
> - **Powers of Radicals:** $\sqrt[n]{x^m} = (\sqrt[n]{x})^m$
>
> - **Quotient Rule:** $\sqrt[n]{\dfrac{x}{y}} = \dfrac{\sqrt[n]{x}}{\sqrt[n]{y}}$, provided $y \neq 0$.
>
> - If n is odd, $\sqrt[n]{x^n} = x$; if n is even, $\sqrt[n]{x^n} = |x|$.

The proof of Theorem 5.6 is based on the definition of the principal roots and properties of exponents. To establish the product rule, consider the following. If n is odd, then by definition $\sqrt[n]{xy}$ is the unique real number such that $(\sqrt[n]{xy})^n = xy$. Given that $(\sqrt[n]{x}\sqrt[n]{y})^n = (\sqrt[n]{x})^n(\sqrt[n]{y})^n = xy$, it must be the case that $\sqrt[n]{xy} = \sqrt[n]{x}\sqrt[n]{y}$. If n is even, then $\sqrt[n]{xy}$ is the unique non-negative real number such that $(\sqrt[n]{xy})^n = xy$. Also note that since n is even, $\sqrt[n]{x}$ and $\sqrt[n]{y}$ are also non-negative and hence so is $\sqrt[n]{x}\sqrt[n]{y}$. Proceeding as above, we find that $\sqrt[n]{xy} = \sqrt[n]{x}\sqrt[n]{y}$. The quotient rule is proved similarly and is left as an exercise. The power rule results from repeated application of the product rule, so long as $\sqrt[n]{x}$ is a real number to start with.[2] The last property is an application of the power rule when n is odd, and the occurrence of the absolute value when n is even is due to the requirement that $\sqrt[n]{x} \geq 0$ in Definition 5.4. For instance, $\sqrt[4]{(-2)^4} = \sqrt[4]{16} = 2 = |-2|$, not -2. It's this last property which makes compositions of roots and powers delicate. This is especially true when we use exponential notation for radicals. Recall the following definition.

> **Definition 5.5.** Let x be a real number, m an integer[a] and n a natural number.
>
> - $x^{\frac{1}{n}} = \sqrt[n]{x}$ and is defined whenever $\sqrt[n]{x}$ is defined.
>
> - $x^{\frac{m}{n}} = (\sqrt[n]{x})^m = \sqrt[n]{x^m}$, whenever $(\sqrt[n]{x})^m$ is defined.
>
> [a] Recall this means $m = 0, \pm 1, \pm 2, \ldots$

The rational exponents defined in Definition 5.5 behave very similarly to the usual integer exponents from Elementary Algebra with one critical exception. Consider the expression $(x^{2/3})^{3/2}$. Applying the usual laws of exponents, we'd be tempted to simplify this as $(x^{2/3})^{3/2} = x^{\frac{2}{3} \cdot \frac{3}{2}} = x^1 = x$. However, if we substitute $x = -1$ and apply Definition 5.5, we find $(-1)^{2/3} = (\sqrt[3]{-1})^2 = (-1)^2 = 1$ so that $((-1)^{2/3})^{3/2} = 1^{3/2} = (\sqrt{1})^3 = 1^3 = 1$. We see in this case that $(x^{2/3})^{3/2} \neq x$. If we take the time to rewrite $(x^{2/3})^{3/2}$ with radicals, we see

$$\left(x^{2/3}\right)^{3/2} = \left((\sqrt[3]{x})^2\right)^{3/2} = \left(\sqrt{(\sqrt[3]{x})^2}\right)^3 = (|\sqrt[3]{x}|)^3 = \left|(\sqrt[3]{x})^3\right| = |x|$$

---

[2] Otherwise we'd run into the same paradox we did in Section 3.4.

5.3 Other Algebraic Functions

In the play-by-play analysis, we see that when we canceled the 2's in multiplying $\frac{2}{3} \cdot \frac{3}{2}$, we were, in fact, attempting to cancel a square with a square root. The fact that $\sqrt{x^2} = |x|$ and not simply x is the root[3] of the trouble. It may amuse the reader to know that $\left(x^{3/2}\right)^{2/3} = x$, and this verification is left as an exercise. The moral of the story is that when simplifying fractional exponents, it's usually best to rewrite them as radicals.[4] The last major property we will state, and leave to Calculus to prove, is that radical functions are continuous on their domains, so the Intermediate Value Theorem, Theorem 3.1, applies. This means that if we take combinations of radical functions with polynomial and rational functions to form what the authors consider the **algebraic functions**,[5] we can make sign diagrams using the procedure set forth in Section 4.2.

Steps for Constructing a Sign Diagram for an Algebraic Function

Suppose f is an algebraic function.

1. Place any values excluded from the domain of f on the number line with an '?' above them.

2. Find the zeros of f and place them on the number line with the number 0 above them.

3. Choose a test value in each of the intervals determined in steps 1 and 2.

4. Determine the sign of $f(x)$ for each test value in step 3, and write that sign above the corresponding interval.

Our next example reviews quite a bit of Intermediate Algebra and demonstrates some of the new features of these graphs.

Example 5.3.1. For the following functions, state their domains and create sign diagrams. Check your answer graphically using your calculator.

1. $f(x) = 3x\sqrt[3]{2-x}$

2. $g(x) = \sqrt{2 - \sqrt[4]{x+3}}$

3. $h(x) = \sqrt[3]{\dfrac{8x}{x+1}}$

4. $k(x) = \dfrac{2x}{\sqrt{x^2 - 1}}$

Solution.

1. As far as domain is concerned, $f(x)$ has no denominators and no even roots, which means its domain is $(-\infty, \infty)$. To create the sign diagram, we find the zeros of f.

---

[3] Did you like that pun?
[4] In most other cases, though, rational exponents are preferred.
[5] As mentioned in Section 2.2, $f(x) = \sqrt{x^2} = |x|$ so that absolute value is also considered an algebraic function.

$$\begin{aligned} f(x) &= 0 \\ 3x\sqrt[3]{2-x} &= 0 \\ 3x = 0 \text{ or } \sqrt[3]{2-x} &= 0 \\ x = 0 \text{ or } \left(\sqrt[3]{2-x}\right)^3 &= 0^3 \\ x = 0 \text{ or } 2-x &= 0 \\ x = 0 \text{ or } x &= 2 \end{aligned}$$

The zeros 0 and 2 divide the real number line into three test intervals. The sign diagram and accompanying graph are below. Note that the intervals on which f is $(+)$ correspond to where the graph of f is above the x-axis, and where the graph of f is below the x-axis we have that f is $(-)$. The calculator suggests something mysterious happens near $x = 2$. Zooming in shows the graph becomes nearly vertical there. You'll have to wait until Calculus to fully understand this phenomenon.

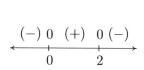

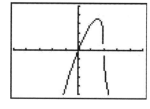

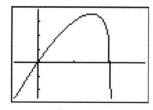

$y = f(x)$ $\qquad$ $y = f(x)$ near $x = 2$.

2. In $g(x) = \sqrt{2 - \sqrt[4]{x+3}}$, we have two radicals both of which are even indexed. To satisfy $\sqrt[4]{x+3}$, we require $x + 3 \geq 0$ or $x \geq -3$. To satisfy $\sqrt{2 - \sqrt[4]{x+3}}$, we need $2 - \sqrt[4]{x+3} \geq 0$. While it may be tempting to write this as $2 \geq \sqrt[4]{x+3}$ and take both sides to the fourth power, there are times when this technique will produce erroneous results.[6] Instead, we solve $2 - \sqrt[4]{x+3} \geq 0$ using a sign diagram. If we let $r(x) = 2 - \sqrt[4]{x+3}$, we know $x \geq -3$, so we concern ourselves with only this portion of the number line. To find the zeros of r we set $r(x) = 0$ and solve $2 - \sqrt[4]{x+3} = 0$. We get $\sqrt[4]{x+3} = 2$ so that $\left(\sqrt[4]{x+3}\right)^4 = 2^4$ from which we obtain $x + 3 = 16$ or $x = 13$. Since we raised both sides of an equation to an even power, we need to check to see if $x = 13$ is an extraneous solution.[7] We find $x = 13$ does check since $2 - \sqrt[4]{x+3} = 2 - \sqrt[4]{13+3} = 2 - \sqrt[4]{16} = 2 - 2 = 0$. Below is our sign diagram for r.

$$(+) \quad 0 \quad (-)$$
$$\xrightarrow{-313}$$

We find $2 - \sqrt[4]{x+3} \geq 0$ on $[-3, 13]$ so this is the domain of g. To find a sign diagram for g, we look for the zeros of g. Setting $g(x) = 0$ is equivalent to $\sqrt{2 - \sqrt[4]{x+3}} = 0$. After squaring

---

[6] For instance, $-2 \geq \sqrt[4]{x+3}$, which has no solution or $-2 \leq \sqrt[4]{x+3}$ whose solution is $[-3, \infty)$.

[7] Recall, this means we have produced a candidate which doesn't satisfy the original equation. Do you remember how raising both sides of an equation to an even power could cause this?

both sides, we get $2 - \sqrt[4]{x+3} = 0$, whose solution we have found to be $x = 13$. Since we squared both sides, we double check and find $g(13)$ is, in fact, 0. Our sign diagram and graph of g are below. Since the domain of g is $[-3, 13]$, what we have below is not just a *portion* of the graph of g, but the *complete* graph. It is always above or on the x-axis, which verifies our sign diagram.

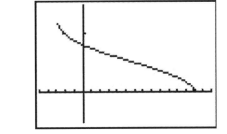

(+)
⊢————⊣
-3   13

The complete graph of $y = g(x)$.

3. The radical in $h(x)$ is odd, so our only concern is the denominator. Setting $x + 1 = 0$ gives $x = -1$, so our domain is $(-\infty, -1) \cup (-1, \infty)$. To find the zeros of h, we set $h(x) = 0$. To solve $\sqrt[3]{\frac{8x}{x+1}} = 0$, we cube both sides to get $\frac{8x}{x+1} = 0$. We get $8x = 0$, or $x = 0$. Below is the resulting sign diagram and corresponding graph. From the graph, it appears as though $x = -1$ is a vertical asymptote. Carrying out an analysis as $x \to -1$ as in Section 4.2 confirms this. (We leave the details to the reader.) Near $x = 0$, we have a situation similar to $x = 2$ in the graph of f in number 1 above. Finally, it appears as if the graph of h has a horizontal asymptote $y = 2$. Using techniques from Section 4.2, we find as $x \to \pm\infty$, $\frac{8x}{x+1} \to 8$. From this, it is hardly surprising that as $x \to \pm\infty$, $h(x) = \sqrt[3]{\frac{8x}{x+1}} \approx \sqrt[3]{8} = 2$.

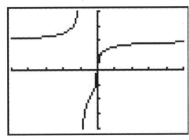

(+) ? (−) 0 (+)
←——+——+——→
   −1   0

$y = h(x)$

4. To find the domain of k, we have both an even root and a denominator to concern ourselves with. To satisfy the square root, $x^2 - 1 \geq 0$. Setting $r(x) = x^2 - 1$, we find the zeros of r to be $x = \pm 1$, and we find the sign diagram of r to be

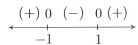

(+) 0 (−) 0 (+)
←——+——+——→
   −1   1

We find $x^2 - 1 \geq 0$ for $(-\infty, -1] \cup [1, \infty)$. To keep the denominator of $k(x)$ away from zero, we set $\sqrt{x^2 - 1} = 0$. We leave it to the reader to verify the solutions are $x = \pm 1$, both of which must be excluded from the domain. Hence, the domain of k is $(-\infty, -1) \cup (1, \infty)$. To build the sign diagram for k, we need the zeros of k. Setting $k(x) = 0$ results in $\frac{2x}{\sqrt{x^2-1}} = 0$. We get $2x = 0$ or $x = 0$. However, $x = 0$ isn't in the domain of k, which means k has no zeros. We construct our sign diagram on the domain of k below alongside the graph of k. It appears that the graph of k has two vertical asymptotes, one at $x = -1$ and one at $x = 1$. The gap in the graph between the asymptotes is because of the gap in the domain of k. Concerning end behavior, there appear to be two horizontal asymptotes, $y = 2$ and $y = -2$. To see why this is the case, we think of $x \to \pm\infty$. The radicand of the denominator $x^2 - 1 \approx x^2$, and as such, $k(x) = \frac{2x}{\sqrt{x^2-1}} \approx \frac{2x}{\sqrt{x^2}} = \frac{2x}{|x|}$. As $x \to \infty$, we have $|x| = x$ so $k(x) \approx \frac{2x}{x} = 2$. On the other hand, as $x \to -\infty$, $|x| = -x$, and as such $k(x) \approx \frac{2x}{-x} = -2$. Finally, it appears as though the graph of k passes the Horizontal Line Test which means k is one to one and k^{-1} exists. Computing k^{-1} is left as an exercise.

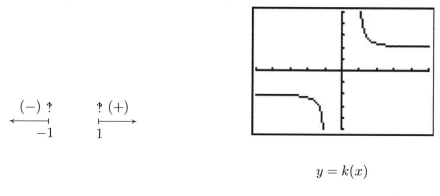

$y = k(x)$

As the previous example illustrates, the graphs of general algebraic functions can have features we've seen before, like vertical and horizontal asymptotes, but they can occur in new and exciting ways. For example, $k(x) = \frac{2x}{\sqrt{x^2-1}}$ had two distinct horizontal asymptotes. You'll recall that rational functions could have at most one horizontal asymptote. Also some new characteristics like 'unusual steepness'[8] and cusps[9] can appear in the graphs of arbitrary algebraic functions. Our next example first demonstrates how we can use sign diagrams to solve nonlinear inequalities. (Don't panic. The technique is very similar to the ones used in Chapters 2, 3 and 4.) We then check our answers graphically with a calculator and see some of the new graphical features of the functions in this extended family.

Example 5.3.2. Solve the following inequalities. Check your answers graphically with a calculator.

---

[8]The proper Calculus term for this is 'vertical tangent', but for now we'll be okay calling it 'unusual steepness'.
[9]See page 241 for the first reference to this feature.

5.3 OTHER ALGEBRAIC FUNCTIONS

1. $x^{2/3} < x^{4/3} - 6$
2. $3(2-x)^{1/3} \leq x(2-x)^{-2/3}$

Solution.

1. To solve $x^{2/3} < x^{4/3} - 6$, we get 0 on one side and attempt to solve $x^{4/3} - x^{2/3} - 6 > 0$. We set $r(x) = x^{4/3} - x^{2/3} - 6$ and note that since the denominators in the exponents are 3, they correspond to cube roots, which means the domain of r is $(-\infty, \infty)$. To find the zeros for the sign diagram, we set $r(x) = 0$ and attempt to solve $x^{4/3} - x^{2/3} - 6 = 0$. At this point, it may be unclear how to proceed. We could always try as a last resort converting back to radical notation, but in this case we can take a cue from Example 3.3.4. Since there are three terms, and the exponent on one of the variable terms, $x^{4/3}$, is exactly twice that of the other, $x^{2/3}$, we have ourselves a 'quadratic in disguise' and we can rewrite $x^{4/3} - x^{2/3} - 6 = 0$ as $\left(x^{2/3}\right)^2 - x^{2/3} - 6 = 0$. If we let $u = x^{2/3}$, then in terms of u, we get $u^2 - u - 6 = 0$. Solving for u, we obtain $u = -2$ or $u = 3$. Replacing $x^{2/3}$ back in for u, we get $x^{2/3} = -2$ or $x^{2/3} = 3$. To avoid the trouble we encountered in the discussion following Definition 5.5, we now convert back to radical notation. By interpreting $x^{2/3}$ as $\sqrt[3]{x^2}$ we have $\sqrt[3]{x^2} = -2$ or $\sqrt[3]{x^2} = 3$. Cubing both sides of these equations results in $x^2 = -8$, which admits no real solution, or $x^2 = 27$, which gives $x = \pm 3\sqrt{3}$. We construct a sign diagram and find $x^{4/3} - x^{2/3} - 6 > 0$ on $\left(-\infty, -3\sqrt{3}\right) \cup \left(3\sqrt{3}, \infty\right)$. To check our answer graphically, we set $f(x) = x^{2/3}$ and $g(x) = x^{4/3} - 6$. The solution to $x^{2/3} < x^{4/3} - 6$ corresponds to the inequality $f(x) < g(x)$, which means we are looking for the x values for which the graph of f is below the graph of g. Using the 'Intersect' command we confirm[10] that the graphs cross at $x = \pm 3\sqrt{3}$. We see that the graph of f is below the graph of g (the thicker curve) on $\left(-\infty, -3\sqrt{3}\right) \cup \left(3\sqrt{3}, \infty\right)$.

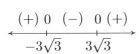

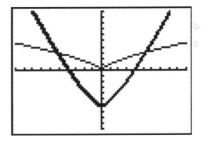

$y = f(x)$ and $y = g(x)$

As a point of interest, if we take a closer look at the graphs of f and g near $x = 0$ with the axes off, we see that despite the fact they both involve cube roots, they exhibit different behavior near $x = 0$. The graph of f has a sharp turn, or cusp, while g does not.[11]

---

[10] Or at least confirm to several decimal places

[11] Again, we introduced this feature on page 241 as a feature which makes the graph of a function 'not smooth'.

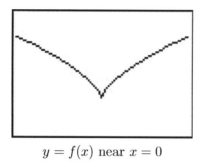
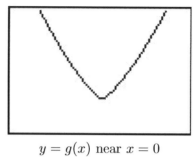

$y = f(x)$ near $x = 0$ $y = g(x)$ near $x = 0$

2. To solve $3(2-x)^{1/3} \leq x(2-x)^{-2/3}$, we gather all the nonzero terms on one side and obtain $3(2-x)^{1/3} - x(2-x)^{-2/3} \leq 0$. We set $r(x) = 3(2-x)^{1/3} - x(2-x)^{-2/3}$. As in number 1, the denominators of the rational exponents are odd, which means there are no domain concerns there. However, the negative exponent on the second term indicates a denominator. Rewriting $r(x)$ with positive exponents, we obtain

$$r(x) = 3(2-x)^{1/3} - \frac{x}{(2-x)^{2/3}}$$

Setting the denominator equal to zero we get $(2-x)^{2/3} = 0$, or $\sqrt[3]{(2-x)^2} = 0$. After cubing both sides, and subsequently taking square roots, we get $2 - x = 0$, or $x = 2$. Hence, the domain of r is $(-\infty, 2) \cup (2, \infty)$. To find the zeros of r, we set $r(x) = 0$. There are two school of thought on how to proceed and we demonstrate both.

- *Factoring Approach.* From $r(x) = 3(2-x)^{1/3} - x(2-x)^{-2/3}$, we note that the quantity $(2-x)$ is common to both terms. When we factor out common factors, we factor out the quantity with the *smaller* exponent. In this case, since $-\frac{2}{3} < \frac{1}{3}$, we factor $(2-x)^{-2/3}$ from both quantities. While it may seem odd to do so, we need to factor $(2-x)^{-2/3}$ *from* $(2-x)^{1/3}$, which results in subtracting the exponent $-\frac{2}{3}$ from $\frac{1}{3}$. We proceed using the usual properties of exponents.[12]

$$\begin{aligned} r(x) &= 3(2-x)^{1/3} - x(2-x)^{-2/3} \\ &= (2-x)^{-2/3} \left[3(2-x)^{\frac{1}{3} - \left(-\frac{2}{3}\right)} - x \right] \\ &= (2-x)^{-2/3} \left[3(2-x)^{3/3} - x \right] \\ &= (2-x)^{-2/3} \left[3(2-x)^{1} - x \right] \quad \text{since } \sqrt[3]{u^3} = \left(\sqrt[3]{u}\right)^3 = u \\ &= (2-x)^{-2/3} (6 - 4x) \\ &= (2-x)^{-2/3} (6 - 4x) \end{aligned}$$

To solve $r(x) = 0$, we set $(2-x)^{-2/3}(6-4x) = 0$, or $\frac{6-4x}{(2-x)^{2/3}} = 0$. We have $6 - 4x = 0$ or $x = \frac{3}{2}$.

---

[12] And we exercise special care when reducing the $\frac{3}{3}$ power to 1.

- *Common Denominator Approach.* We rewrite

$$\begin{aligned}
r(x) &= 3(2-x)^{1/3} - x(2-x)^{-2/3} \\
&= 3(2-x)^{1/3} - \frac{x}{(2-x)^{2/3}} \\
&= \frac{3(2-x)^{1/3}(2-x)^{2/3}}{(2-x)^{2/3}} - \frac{x}{(2-x)^{2/3}} \qquad \text{common denominator} \\
&= \frac{3(2-x)^{\frac{1}{3}+\frac{2}{3}}}{(2-x)^{2/3}} - \frac{x}{(2-x)^{2/3}} \\
&= \frac{3(2-x)^{3/3}}{(2-x)^{2/3}} - \frac{x}{(2-x)^{2/3}} \\
&= \frac{3(2-x)^{1}}{(2-x)^{2/3}} - \frac{x}{(2-x)^{2/3}} \qquad \text{since } \sqrt[3]{u^3} = (\sqrt[3]{u})^3 = u \\
&= \frac{3(2-x)-x}{(2-x)^{2/3}} \\
&= \frac{6-4x}{(2-x)^{2/3}}
\end{aligned}$$

As before, when we set $r(x) = 0$ we obtain $x = \frac{3}{2}$.

We now create our sign diagram and find $3(2-x)^{1/3} - x(2-x)^{-2/3} \leq 0$ on $\left[\frac{3}{2}, 2\right) \cup (2, \infty)$. To check this graphically, we set $f(x) = 3(2-x)^{1/3}$ and $g(x) = x(2-x)^{-2/3}$ (the thicker curve). We confirm that the graphs intersect at $x = \frac{3}{2}$ and the graph of f is below the graph of g for $x \geq \frac{3}{2}$, with the exception of $x = 2$ where it appears the graph of g has a vertical asymptote.

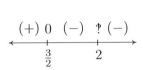

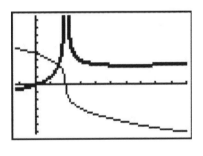

$y = f(x)$ and $y = g(x)$ □

One application of algebraic functions was given in Example 1.6.6 in Section 1.1. Our last example is a more sophisticated application of distance.

Example 5.3.3. Carl wishes to get high speed internet service installed in his remote Sasquatch observation post located 30 miles from Route 117. The nearest junction box is located 50 miles downroad from the post, as indicated in the diagram below. Suppose it costs $15 per mile to run cable along the road and $20 per mile to run cable off of the road.

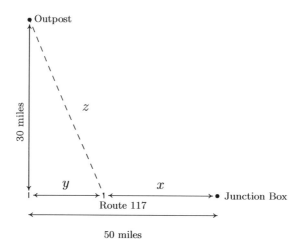

1. Express the total cost C of connecting the Junction Box to the Outpost as a function of x, the number of miles the cable is run along Route 117 before heading off road directly towards the Outpost. Determine a reasonable applied domain for the problem.

2. Use your calculator to graph $y = C(x)$ on its domain. What is the minimum cost? How far along Route 117 should the cable be run before turning off of the road?

Solution.

1. The cost is broken into two parts: the cost to run cable along Route 117 at $15 per mile, and the cost to run it off road at $20 per mile. Since x represents the miles of cable run along Route 117, the cost for that portion is $15x$. From the diagram, we see that the number of miles the cable is run off road is z, so the cost of that portion is $20z$. Hence, the total cost is $C = 15x + 20z$. Our next goal is to determine z as a function of x. The diagram suggests we can use the Pythagorean Theorem to get $y^2 + 30^2 = z^2$. But we also see $x + y = 50$ so that $y = 50 - x$. Hence, $z^2 = (50 - x)^2 + 900$. Solving for z, we obtain $z = \pm\sqrt{(50-x)^2 + 900}$. Since z represents a distance, we choose $z = \sqrt{(50-x)^2 + 900}$ so that our cost as a function of x only is given by
$$C(x) = 15x + 20\sqrt{(50-x)^2 + 900}$$
From the context of the problem, we have $0 \leq x \leq 50$.

2. Graphing $y = C(x)$ on a calculator and using the 'Minimum' feature, we find the relative minimum (which is also the absolute minimum in this case) to two decimal places to be $(15.98, 1146.86)$. Here the x-coordinate tells us that in order to minimize cost, we should run 15.98 miles of cable along Route 117 and then turn off of the road and head towards the outpost. The y-coordinate tells us that the minimum cost, in dollars, to do so is $1146.86. The ability to stream live SasquatchCasts? Priceless. □

5.3 Other Algebraic Functions

5.3.1 Exercises

For each function in Exercises 1 - 10 below

- Find its domain.
- Create a sign diagram.
- Use your calculator to help you sketch its graph and identify any vertical or horizontal asymptotes, 'unusual steepness' or cusps.

1. $f(x) = \sqrt{1-x^2}$
2. $f(x) = \sqrt{x^2-1}$
3. $f(x) = x\sqrt{1-x^2}$
4. $f(x) = x\sqrt{x^2-1}$
5. $f(x) = \sqrt[4]{\dfrac{16x}{x^2-9}}$
6. $f(x) = \dfrac{5x}{\sqrt[3]{x^3+8}}$
7. $f(x) = x^{\frac{2}{3}}(x-7)^{\frac{1}{3}}$
8. $f(x) = x^{\frac{3}{2}}(x-7)^{\frac{1}{3}}$
9. $f(x) = \sqrt{x(x+5)(x-4)}$
10. $f(x) = \sqrt[3]{x^3+3x^2-6x-8}$

In Exercises 11 - 16, sketch the graph of $y = g(x)$ by starting with the graph of $y = f(x)$ and using the transformations presented in Section 1.7.

11. $f(x) = \sqrt[3]{x}$, $g(x) = \sqrt[3]{x-1} - 2$
12. $f(x) = \sqrt[3]{x}$, $g(x) = -2\sqrt[3]{x+1} + 4$
13. $f(x) = \sqrt[4]{x}$, $g(x) = \sqrt[4]{x-1} - 2$
14. $f(x) = \sqrt[4]{x}$, $g(x) = 3\sqrt[4]{x-7} - 1$
15. $f(x) = \sqrt[5]{x}$, $g(x) = \sqrt[5]{x+2} + 3$
16. $f(x) = \sqrt[8]{x}$, $g(x) = \sqrt[8]{-x} - 2$

In Exercises 17 - 35, solve the equation or inequality.

17. $x + 1 = \sqrt{3x+7}$
18. $2x + 1 = \sqrt{3-3x}$
19. $x + \sqrt{3x+10} = -2$
20. $3x + \sqrt{6-9x} = 2$
21. $2x - 1 = \sqrt{x+3}$
22. $x^{\frac{3}{2}} = 8$
23. $x^{\frac{2}{3}} = 4$
24. $\sqrt{x-2} + \sqrt{x-5} = 3$
25. $\sqrt{2x+1} = 3 + \sqrt{4-x}$
26. $5 - (4-2x)^{\frac{2}{3}} = 1$
27. $10 - \sqrt{x-2} \leq 11$
28. $\sqrt[3]{x} \leq x$

29. $2(x-2)^{-\frac{1}{3}} - \frac{2}{3}x(x-2)^{-\frac{4}{3}} \leq 0$

30. $-\frac{4}{3}(x-2)^{-\frac{4}{3}} + \frac{8}{9}x(x-2)^{-\frac{7}{3}} \geq 0$

31. $2x^{-\frac{1}{3}}(x-3)^{\frac{1}{3}} + x^{\frac{2}{3}}(x-3)^{-\frac{2}{3}} \geq 0$

32. $\sqrt[3]{x^3 + 3x^2 - 6x - 8} > x + 1$

33. $\frac{1}{3}x^{\frac{3}{4}}(x-3)^{-\frac{2}{3}} + \frac{3}{4}x^{-\frac{1}{4}}(x-3)^{\frac{1}{3}} < 0$

34. $x^{-\frac{1}{3}}(x-3)^{-\frac{2}{3}} - x^{-\frac{4}{3}}(x-3)^{-\frac{5}{3}}(x^2 - 3x + 2) \geq 0$

35. $\frac{2}{3}(x+4)^{\frac{3}{5}}(x-2)^{-\frac{1}{3}} + \frac{3}{5}(x+4)^{-\frac{2}{5}}(x-2)^{\frac{2}{3}} \geq 0$

36. Rework Example 5.3.3 so that the outpost is 10 miles from Route 117 and the nearest junction box is 30 miles down the road for the post.

37. The volume V of a right cylindrical cone depends on the radius of its base r and its height h and is given by the formula $V = \frac{1}{3}\pi r^2 h$. The surface area S of a right cylindrical cone also depends on r and h according to the formula $S = \pi r \sqrt{r^2 + h^2}$. Suppose a cone is to have a volume of 100 cubic centimeters.

    (a) Use the formula for volume to find the height h as a function of r.

    (b) Use the formula for surface area and your answer to 37a to find the surface area S as a function of r.

    (c) Use your calculator to find the values of r and h which minimize the surface area. What is the minimum surface area? Round your answers to two decimal places.

38. The National Weather Service uses the following formula to calculate the wind chill:

    $$W = 35.74 + 0.6215\, T_a - 35.75\, V^{0.16} + 0.4275\, T_a\, V^{0.16}$$

    where W is the wind chill temperature in °F, T_a is the air temperature in °F, and V is the wind speed in miles per hour. Note that W is defined only for air temperatures at or lower than 50°F and wind speeds above 3 miles per hour.

    (a) Suppose the air temperature is 42° and the wind speed is 7 miles per hour. Find the wind chill temperature. Round your answer to two decimal places.

    (b) Suppose the air temperature is 37°F and the wind chill temperature is 30°F. Find the wind speed. Round your answer to two decimal places.

39. As a follow-up to Exercise 38, suppose the air temperature is 28°F.

    (a) Use the formula from Exercise 38 to find an expression for the wind chill temperature as a function of the wind speed, $W(V)$.

    (b) Solve $W(V) = 0$, round your answer to two decimal places, and interpret.

    (c) Graph the function W using your calculator and check your answer to part 39b.

5.3 OTHER ALGEBRAIC FUNCTIONS

40. The period of a pendulum in seconds is given by

$$T = 2\pi\sqrt{\frac{L}{g}}$$

(for small displacements) where L is the length of the pendulum in meters and $g = 9.8$ meters per second per second is the acceleration due to gravity. My Seth-Thomas antique schoolhouse clock needs $T = \frac{1}{2}$ second and I can adjust the length of the pendulum via a small dial on the bottom of the bob. At what length should I set the pendulum?

41. The Cobb-Douglas production model states that the yearly total dollar value of the production output P in an economy is a function of labor x (the total number of hours worked in a year) and capital y (the total dollar value of all of the stuff purchased in order to make things). Specifically, $P = ax^b y^{1-b}$. By fixing P, we create what's known as an 'isoquant' and we can then solve for y as a function of x. Let's assume that the Cobb-Douglas production model for the country of Sasquatchia is $P = 1.23 x^{0.4} y^{0.6}$.

    (a) Let $P = 300$ and solve for y in terms of x. If $x = 100$, what is y?

    (b) Graph the isoquant $300 = 1.23 x^{0.4} y^{0.6}$. What information does an ordered pair (x, y) which makes $P = 300$ give you? With the help of your classmates, find several different combinations of labor and capital all of which yield $P = 300$. Discuss any patterns you may see.

42. According to Einstein's Theory of Special Relativity, the observed mass m of an object is a function of how fast the object is traveling. Specifically,

$$m(x) = \frac{m_r}{\sqrt{1 - \frac{x^2}{c^2}}}$$

where $m(0) = m_r$ is the mass of the object at rest, x is the speed of the object and c is the speed of light.

    (a) Find the applied domain of the function.

    (b) Compute $m(.1c)$, $m(.5c)$, $m(.9c)$ and $m(.999c)$.

    (c) As $x \to c^-$, what happens to $m(x)$?

    (d) How slowly must the object be traveling so that the observed mass is no greater than 100 times its mass at rest?

43. Find the inverse of $k(x) = \dfrac{2x}{\sqrt{x^2 - 1}}$.

44. Suppose Fritzy the Fox, positioned at a point (x, y) in the first quadrant, spots Chewbacca the Bunny at $(0,0)$. Chewbacca begins to run along a fence (the positive y-axis) towards his warren. Fritzy, of course, takes chase and constantly adjusts his direction so that he is always running directly at Chewbacca. If Chewbacca's speed is v_1 and Fritzy's speed is v_2, the path Fritzy will take to intercept Chewbacca, provided v_2 is directly proportional to, but not equal to, v_1 is modeled by

$$y = \frac{1}{2}\left(\frac{x^{1+v_1/v_2}}{1+v_1/v_2} - \frac{x^{1-v_1/v_2}}{1-v_1/v_2}\right) + \frac{v_1 v_2}{v_2^2 - v_1^2}$$

(a) Determine the path that Fritzy will take if he runs exactly twice as fast as Chewbacca; that is, $v_2 = 2v_1$. Use your calculator to graph this path for $x \geq 0$. What is the significance of the y-intercept of the graph?

(b) Determine the path Fritzy will take if Chewbacca runs exactly twice as fast as he does; that is, $v_1 = 2v_2$. Use your calculator to graph this path for $x > 0$. Describe the behavior of y as $x \to 0^+$ and interpret this physically.

(c) With the help of your classmates, generalize parts (a) and (b) to two cases: $v_2 > v_1$ and $v_2 < v_1$. We will discuss the case of $v_1 = v_2$ in Exercise 32 in Section 6.5.

45. Verify the Quotient Rule for Radicals in Theorem 5.6.

46. Show that $\left(x^{\frac{3}{2}}\right)^{\frac{2}{3}} = x$ for all $x \geq 0$.

47. Show that $\sqrt[3]{2}$ is an irrational number by first showing that it is a zero of $p(x) = x^3 - 2$ and then showing p has no rational zeros. (You'll need the Rational Zeros Theorem, Theorem 3.9, in order to show this last part.)

48. With the help of your classmates, generalize Exercise 47 to show that $\sqrt[n]{c}$ is an irrational number for any natural numbers $c \geq 2$ and $n \geq 2$ provided that $c \neq p^n$ for some natural number p.

5.3 OTHER ALGEBRAIC FUNCTIONS

5.3.2 ANSWERS

1. $f(x) = \sqrt{1-x^2}$
   Domain: $[-1, 1]$

   $$\begin{array}{ccc} 0 & (+) & 0 \\ \vdash\!\!\!\!\!\!\!-\!\!\!\!\!\!\!-\!\!\!\!\!\!\!-\!\!\!\!\!\!\!\dashv \\ -1 & & 1 \end{array}$$

   No asymptotes
   Unusual steepness at $x = -1$ and $x = 1$
   No cusps

   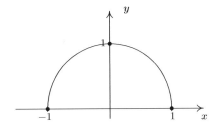

2. $f(x) = \sqrt{x^2 - 1}$
   Domain: $(-\infty, -1] \cup [1, \infty)$

   $$\begin{array}{cccc} (+) & 0 & 0 & (+) \\ \leftarrow\!\!\!\!\!\!\!-\!\!\!\!\!\!\!\dashv & & \vdash\!\!\!\!\!\!\!-\!\!\!\!\!\!\!\rightarrow \\ & -1 & 1 & \end{array}$$

   No asymptotes
   Unusual steepness at $x = -1$ and $x = 1$
   No cusps

   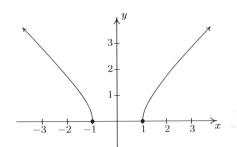

3. $f(x) = x\sqrt{1-x^2}$
   Domain: $[-1, 1]$

   $$\begin{array}{ccccc} 0 & (-) & 0 & (+) & 0 \\ \vdash\!\!\!\!\!\!\!-\!\!\!\!\!\!\!\!-\!\!\!\!\!\!\!\!-\!\!\!\!\!\!\!\!\dashv \\ -1 & & 0 & & 1 \end{array}$$

   No asymptotes
   Unusual steepness at $x = -1$ and $x = 1$
   No cusps

   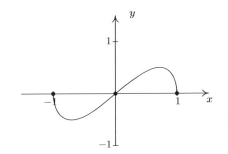

4. $f(x) = x\sqrt{x^2 - 1}$
   Domain: $(-\infty, -1] \cup [1, \infty)$

   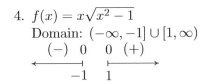

   No asymptotes
   Unusual steepness at $x = -1$ and $x = 1$
   No cusps

   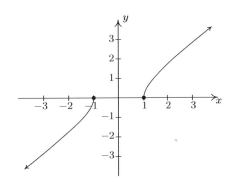

5. $f(x) = \sqrt[4]{\dfrac{16x}{x^2-9}}$
   Domain: $(-3, 0] \cup (3, \infty)$

   ```
     ?  (+)  0       ?  (+)
   ├────────┤      ├─────────→
   -3       0      3
   ```

   Vertical asymptotes: $x = -3$ and $x = 3$
   Horizontal asymptote: $y = 0$
   Unusual steepness at $x = 0$
   No cusps

   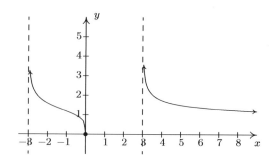

6. $f(x) = \dfrac{5x}{\sqrt[3]{x^3+8}}$
   Domain: $(-\infty, -2) \cup (-2, \infty)$

   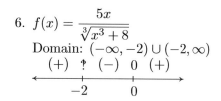

   Vertical asymptote $x = -2$
   Horizontal asymptote $y = 5$
   No unusual steepness or cusps

   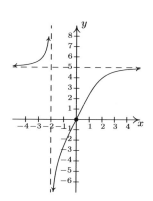

7. $f(x) = x^{\frac{2}{3}}(x-7)^{\frac{1}{3}}$
   Domain: $(-\infty, \infty)$

   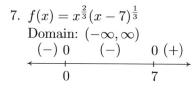

   No vertical or horizontal asymptotes[13]
   Unusual steepness at $x = 7$
   Cusp at $x = 0$

   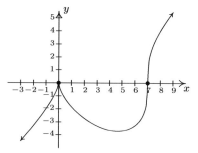

8. $f(x) = x^{\frac{3}{2}}(x-7)^{\frac{1}{3}}$
   Domain: $[0, \infty)$

   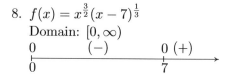

   No asymptotes
   Unusual steepness at $x = 7$
   No cusps

   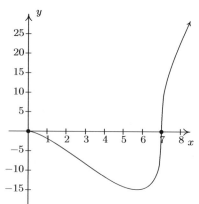

---

[13]Using Calculus it can be shown that $y = x - \frac{7}{3}$ is a slant asymptote of this graph.

5.3 OTHER ALGEBRAIC FUNCTIONS

9. $f(x) = \sqrt{x(x+5)(x-4)}$
Domain: $[-5, 0] \cup [4, \infty)$

```
   0  (+)  0       0  (+)
   ├────────┤       ├──────→
  -5       0       4
```

No asymptotes
Unusual steepness at $x = -5, x = 0$ and $x = 4$
No cusps

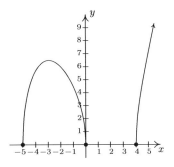

10. $f(x) = \sqrt[3]{x^3 + 3x^2 - 6x - 8}$
Domain: $(-\infty, \infty)$

```
   (-)  0 (+) 0 (-) 0  (+)
   ←────┼─────┼────┼─────→
       -4    -1    2
```

No vertical or horizontal asymptotes[14]
Unusual steepness at $x = -4, x = -1$ and $x = 2$
No cusps

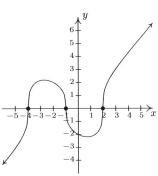

11. $g(x) = \sqrt[3]{x-1} - 2$

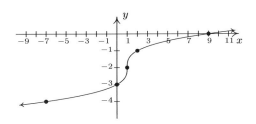

12. $g(x) = -2\sqrt[3]{x+1} + 4$

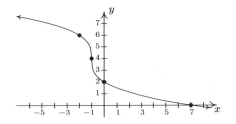

13. $g(x) = \sqrt[4]{x-1} - 2$

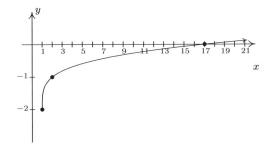

14. $g(x) = 3\sqrt[4]{x-7} - 1$

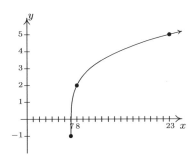

---

[14]Using Calculus it can be shown that $y = x + 1$ is a slant asymptote of this graph.

15. $g(x) = \sqrt[5]{x+2} + 3$

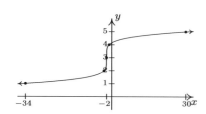

16. $g(x) = \sqrt[8]{-x} - 2$

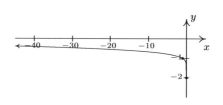

17. $x = 3$

18. $x = \frac{1}{4}$

19. $x = -3$

20. $x = -\frac{1}{3}, \frac{2}{3}$

21. $x = \frac{5+\sqrt{57}}{8}$

22. $x = 4$

23. $x = \pm 8$

24. $x = 6$

25. $x = 4$

26. $x = -2, 6$

27. $[2, \infty)$

28. $[-1, 0] \cup [1, \infty)$

29. $(-\infty, 2) \cup (2, 3]$

30. $(2, 6]$

31. $(-\infty, 0) \cup [2, 3) \cup (3, \infty)$

32. $(-\infty, -1)$

33. $\left(0, \frac{27}{13}\right)$

34. $(-\infty, 0) \cup (0, 3)$

35. $(-\infty, -4) \cup \left(-4, -\frac{22}{19}\right] \cup (2, \infty)$

36. $C(x) = 15x + 20\sqrt{100 + (30-x)^2}$, $0 \leq x \leq 30$. The calculator gives the absolute minimum at $\approx (18.66, 582.29)$. This means to minimize the cost, approximately 18.66 miles of cable should be run along Route 117 before turning off the road and heading towards the outpost. The minimum cost to run the cable is approximately \$582.29.

37. (a) $h(r) = \frac{300}{\pi r^2}$, $r > 0$.

   (b) $S(r) = \pi r \sqrt{r^2 + \left(\frac{300}{\pi r^2}\right)^2} = \frac{\sqrt{\pi^2 r^6 + 90000}}{r}$, $r > 0$

   (c) The calculator gives the absolute minimum at the point $\approx (4.07, 90.23)$. This means the radius should be (approximately) 4.07 centimeters and the height should be 5.76 centimeters to give a minimum surface area of 90.23 square centimeters.

38. (a) $W \approx 37.55°F$.

   (b) $V \approx 9.84$ miles per hour.

39. (a) $W(V) = 53.142 - 23.78V^{0.16}$. Since we are told in Exercise 38 that wind chill is only effect for wind speeds of more than 3 miles per hour, we restrict the domain to $V > 3$.

   (b) $W(V) = 0$ when $V \approx 152.29$. This means, according to the model, for the wind chill temperature to be $0°F$, the wind speed needs to be 152.29 miles per hour.

5.3 OTHER ALGEBRAIC FUNCTIONS 415

(c) The graph is below.

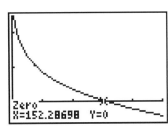

40. $9.8\left(\dfrac{1}{4\pi}\right)^2 \approx 0.062$ meters or 6.2 centimeters

41. (a) First rewrite the model as $P = 1.23x^{\frac{2}{5}}y^{\frac{3}{5}}$. Then $300 = 1.23x^{\frac{2}{5}}y^{\frac{3}{5}}$ yields $y = \left(\dfrac{300}{1.23x^{\frac{2}{5}}}\right)^{\frac{5}{3}}$.
    If $x = 100$ then $y \approx 441.93687$.

42. (a) $[0, c)$

    (b)
    $$m(.1c) = \dfrac{m_r}{\sqrt{.99}} \approx 1.005 m_r \quad m(.5c) = \dfrac{m_r}{\sqrt{.75}} \approx 1.155 m_r$$
    $$m(.9c) = \dfrac{m_r}{\sqrt{.19}} \approx 2.294 m_r \quad m(.999c) = \dfrac{m_r}{\sqrt{.0.001999}} \approx 22.366 m_r$$

    (c) As $x \to c^-$, $m(x) \to \infty$

    (d) If the object is traveling no faster than approximately 0.99995 times the speed of light, then its observed mass will be no greater than $100 m_r$.

43. $k^{-1}(x) = \dfrac{x}{\sqrt{x^2 - 4}}$

44. (a) $y = \frac{1}{3}x^{3/2} - \sqrt{x} + \frac{2}{3}$. The point $\left(0, \frac{2}{3}\right)$ is when Fritzy's path crosses Chewbacca's path - in other words, where Fritzy catches Chewbacca.

    (b) $y = \frac{1}{6}x^3 + \frac{1}{2x} - \frac{2}{3}$. Using the techniques from Chapter 4, we find as $x \to 0^+$, $y \to \infty$ which means, in this case, Fritzy's pursuit never ends; he never catches Chewbacca. This makes sense since Chewbacca has a head start and is running faster than Fritzy.

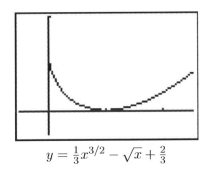

$y = \frac{1}{3}x^{3/2} - \sqrt{x} + \frac{2}{3}$

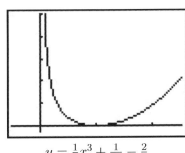

$y = \frac{1}{6}x^3 + \frac{1}{2x} - \frac{2}{3}$

Chapter 6

Exponential and Logarithmic Functions

6.1 Introduction to Exponential and Logarithmic Functions

Of all of the functions we study in this text, exponential and logarithmic functions are possibly the ones which impact everyday life the most.[1] This section introduces us to these functions while the rest of the chapter will more thoroughly explore their properties. Up to this point, we have dealt with functions which involve terms like x^2 or $x^{2/3}$, in other words, terms of the form x^p where the base of the term, x, varies but the exponent of each term, p, remains constant. In this chapter, we study functions of the form $f(x) = b^x$ where the base b is a constant and the exponent x is the variable. We start our exploration of these functions with $f(x) = 2^x$. (Apparently this is a tradition. Every College Algebra book we have ever read starts with $f(x) = 2^x$.) We make a table of values, plot the points and connect the dots in a pleasing fashion.

x	$f(x)$	$(x, f(x))$
-3	$2^{-3} = \frac{1}{8}$	$\left(-3, \frac{1}{8}\right)$
-2	$2^{-2} = \frac{1}{4}$	$\left(-2, \frac{1}{4}\right)$
-1	$2^{-1} = \frac{1}{2}$	$\left(-1, \frac{1}{2}\right)$
0	$2^0 = 1$	$(0, 1)$
1	$2^1 = 2$	$(1, 2)$
2	$2^2 = 4$	$(2, 4)$
3	$2^3 = 8$	$(3, 8)$

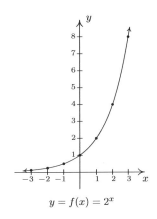

$y = f(x) = 2^x$

A few remarks about the graph of $f(x) = 2^x$ which we have constructed are in order. As $x \to -\infty$

---

[1] Take a class in Differential Equations and you'll see why.

and attains values like $x = -100$ or $x = -1000$, the function $f(x) = 2^x$ takes on values like $f(-100) = 2^{-100} = \frac{1}{2^{100}}$ or $f(-1000) = 2^{-1000} = \frac{1}{2^{1000}}$. In other words, as $x \to -\infty$,

$$2^x \approx \frac{1}{\text{very big } (+)} \approx \text{very small } (+)$$

So as $x \to -\infty$, $2^x \to 0^+$. This is represented graphically using the x-axis (the line $y = 0$) as a horizontal asymptote. On the flip side, as $x \to \infty$, we find $f(100) = 2^{100}$, $f(1000) = 2^{1000}$, and so on, thus $2^x \to \infty$. As a result, our graph suggests the range of f is $(0, \infty)$. The graph of f passes the Horizontal Line Test which means f is one-to-one and hence invertible. We also note that when we 'connected the dots in a pleasing fashion', we have made the implicit assumption that $f(x) = 2^x$ is continuous[2] and has a domain of all real numbers. In particular, we have suggested that things like $2^{\sqrt{3}}$ exist as real numbers. We should take a moment to discuss what something like $2^{\sqrt{3}}$ might mean, and refer the interested reader to a solid course in Calculus for a more rigorous explanation. The number $\sqrt{3} = 1.73205\ldots$ is an irrational number[3] and as such, its decimal representation neither repeats nor terminates. We can, however, approximate $\sqrt{3}$ by terminating decimals, and it stands to reason[4] we can use these to approximate $2^{\sqrt{3}}$. For example, if we approximate $\sqrt{3}$ by 1.73, we can approximate $2^{\sqrt{3}} \approx 2^{1.73} = 2^{\frac{173}{100}} = \sqrt[100]{2^{173}}$. It is not, by any means, a pleasant number, but it is at least a number that we understand in terms of powers and roots. It also stands to reason that better and better approximations of $\sqrt{3}$ yield better and better approximations of $2^{\sqrt{3}}$, so the value of $2^{\sqrt{3}}$ should be the result of this sequence of approximations.[5]

Suppose we wish to study the family of functions $f(x) = b^x$. Which bases b make sense to study? We find that we run into difficulty if $b < 0$. For example, if $b = -2$, then the function $f(x) = (-2)^x$ has trouble, for instance, at $x = \frac{1}{2}$ since $(-2)^{1/2} = \sqrt{-2}$ is not a real number. In general, if x is any rational number with an even denominator, then $(-2)^x$ is not defined, so we must restrict our attention to bases $b \geq 0$. What about $b = 0$? The function $f(x) = 0^x$ is undefined for $x \leq 0$ because we cannot divide by 0 and 0^0 is an indeterminant form. For $x > 0$, $0^x = 0$ so the function $f(x) = 0^x$ is the same as the function $f(x) = 0$, $x > 0$. We know everything we can possibly know about this function, so we exclude it from our investigations. The only other base we exclude is $b = 1$, since the function $f(x) = 1^x = 1$ is, once again, a function we have already studied. We are now ready for our definition of exponential functions.

Definition 6.1. A function of the form $f(x) = b^x$ where b is a fixed real number, $b > 0$, $b \neq 1$ is called a **base b exponential function**.

We leave it to the reader to verify[6] that if $b > 1$, then the exponential function $f(x) = b^x$ will share the same basic shape and characteristics as $f(x) = 2^x$. What if $0 < b < 1$? Consider $g(x) = \left(\frac{1}{2}\right)^x$. We could certainly build a table of values and connect the points, or we could take a step back and

---

[2]Recall that this means there are no holes or other kinds of breaks in the graph.
[3]You can actually prove this by considering the polynomial $p(x) = x^2 - 3$ and showing it has no rational zeros by applying Theorem 3.9.
[4]This is where Calculus and continuity come into play.
[5]Want more information? Look up "convergent sequences" on the Internet.
[6]Meaning, graph some more examples on your own.

note that $g(x) = \left(\frac{1}{2}\right)^x = \left(2^{-1}\right)^x = 2^{-x} = f(-x)$, where $f(x) = 2^x$. Thinking back to Section 1.7, the graph of $f(-x)$ is obtained from the graph of $f(x)$ by reflecting it across the y-axis. We get

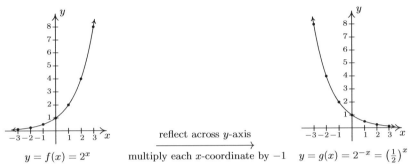

We see that the domain and range of g match that of f, namely $(-\infty, \infty)$ and $(0, \infty)$, respectively. Like f, g is also one-to-one. Whereas f is always increasing, g is always decreasing. As a result, as $x \to -\infty$, $g(x) \to \infty$, and on the flip side, as $x \to \infty$, $g(x) \to 0^+$. It shouldn't be too surprising that for all choices of the base $0 < b < 1$, the graph of $y = b^x$ behaves similarly to the graph of g. We summarize the basic properties of exponential functions in the following theorem.[7]

Theorem 6.1. Properties of Exponential Functions: Suppose $f(x) = b^x$.

- The domain of f is $(-\infty, \infty)$ and the range of f is $(0, \infty)$.

- $(0, 1)$ is on the graph of f and $y = 0$ is a horizontal asymptote to the graph of f.

- f is one-to-one, continuous and smooth[a]

- If $b > 1$:
  - f is always increasing
  - As $x \to -\infty$, $f(x) \to 0^+$
  - As $x \to \infty$, $f(x) \to \infty$
  - The graph of f resembles:

    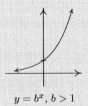

    $y = b^x, b > 1$

- If $0 < b < 1$:
  - f is always decreasing
  - As $x \to -\infty$, $f(x) \to \infty$
  - As $x \to \infty$, $f(x) \to 0^+$
  - The graph of f resembles:

    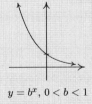

    $y = b^x, 0 < b < 1$

[a]Recall that this means the graph of f has no sharp turns or corners.

---

[7]The proof of which, like many things discussed in the text, requires Calculus.

Of all of the bases for exponential functions, two occur the most often in scientific circles. The first, base 10, is often called the **common base**. The second base is an irrational number, $e \approx 2.718$, called the **natural base**. We will more formally discuss the origins of this number in Section 6.5. For now, it is enough to know that since $e > 1$, $f(x) = e^x$ is an increasing exponential function. The following examples give us an idea how these functions are used in the wild.

Example 6.1.1. The value of a car can be modeled by $V(x) = 25\left(\frac{4}{5}\right)^x$, where $x \geq 0$ is age of the car in years and $V(x)$ is the value in thousands of dollars.

1. Find and interpret $V(0)$.

2. Sketch the graph of $y = V(x)$ using transformations.

3. Find and interpret the horizontal asymptote of the graph you found in 2.

Solution.

1. To find $V(0)$, we replace x with 0 to obtain $V(0) = 25\left(\frac{4}{5}\right)^0 = 25$. Since x represents the age of the car in years, $x = 0$ corresponds to the car being brand new. Since $V(x)$ is measured in thousands of dollars, $V(0) = 25$ corresponds to a value of \$25,000. Putting it all together, we interpret $V(0) = 25$ to mean the purchase price of the car was \$25,000.

2. To graph $y = 25\left(\frac{4}{5}\right)^x$, we start with the basic exponential function $f(x) = \left(\frac{4}{5}\right)^x$. Since the base $b = \frac{4}{5}$ is between 0 and 1, the graph of $y = f(x)$ is decreasing. We plot the y-intercept $(0, 1)$ and two other points, $\left(-1, \frac{5}{4}\right)$ and $\left(1, \frac{4}{5}\right)$, and label the horizontal asymptote $y = 0$. To obtain $V(x) = 25\left(\frac{4}{5}\right)^x$, $x \geq 0$, we multiply the output from f by 25, in other words, $V(x) = 25f(x)$. In accordance with Theorem 1.5, this results in a vertical stretch by a factor of 25. We multiply all of the y values in the graph by 25 (including the y value of the horizontal asymptote) and obtain the points $\left(-1, \frac{125}{4}\right)$, $(0, 25)$ and $(1, 20)$. The horizontal asymptote remains $y = 0$. Finally, we restrict the domain to $[0, \infty)$ to fit with the applied domain given to us. We have the result below.

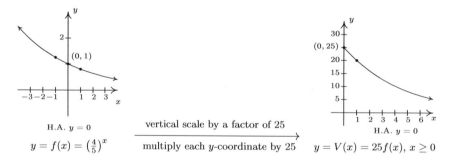

3. We see from the graph of V that its horizontal asymptote is $y = 0$. (We leave it to reader to verify this analytically by thinking about what happens as we take larger and larger powers of $\frac{4}{5}$.) This means as the car gets older, its value diminishes to 0. □

6.1 INTRODUCTION TO EXPONENTIAL AND LOGARITHMIC FUNCTIONS

The function in the previous example is often called a 'decay curve'. Increasing exponential functions are used to model 'growth curves' and we shall see several different examples of those in Section 6.5. For now, we present another common decay curve which will serve as the basis for further study of exponential functions. Although it may look more complicated than the previous example, it is actually just a basic exponential function which has been modified by a few transformations from Section 1.7.

Example 6.1.2. According to Newton's Law of Cooling[8] the temperature of coffee T (in degrees Fahrenheit) t minutes after it is served can be modeled by $T(t) = 70 + 90e^{-0.1t}$.

1. Find and interpret $T(0)$.

2. Sketch the graph of $y = T(t)$ using transformations.

3. Find and interpret the horizontal asymptote of the graph.

Solution.

1. To find $T(0)$, we replace every occurrence of the independent variable t with 0 to obtain $T(0) = 70 + 90e^{-0.1(0)} = 160$. This means that the coffee was served at 160°F.

2. To graph $y = T(t)$ using transformations, we start with the basic function, $f(t) = e^t$. As we have already remarked, $e \approx 2.718 > 1$ so the graph of f is an increasing exponential with y-intercept $(0, 1)$ and horizontal asymptote $y = 0$. The points $(-1, e^{-1}) \approx (-1, 0.37)$ and $(1, e) \approx (1, 2.72)$ are also on the graph. Since the formula $T(t)$ looks rather complicated, we rewrite $T(t)$ in the form presented in Theorem 1.7 and use that result to track the changes to our three points and the horizontal asymptote. We have

$$T(t) = 70 + 90e^{-0.1t} = 90e^{-0.1t} + 70 = 90f(-0.1t) + 70$$

Multiplication of the input to f, t, by -0.1 results in a horizontal expansion by a factor of 10 as well as a reflection about the y-axis. We divide each of the x values of our points by -0.1 (which amounts to multiplying them by -10) to obtain $(10, e^{-1})$, $(0, 1)$, and $(-10, e)$. Since none of these changes affected the y values, the horizontal asymptote remains $y = 0$. Next, we see that the output from f is being multiplied by 90. This results in a vertical stretch by a factor of 90. We multiply the y-coordinates by 90 to obtain $(10, 90e^{-1})$, $(0, 90)$, and $(-10, 90e)$. We also multiply the y value of the horizontal asymptote $y = 0$ by 90, and it remains $y = 0$. Finally, we add 70 to all of the y-coordinates, which shifts the graph upwards to obtain $(10, 90e^{-1} + 70) \approx (10, 103.11)$, $(0, 160)$, and $(-10, 90e + 70) \approx (-10, 314.64)$. Adding 70 to the horizontal asymptote shifts it upwards as well to $y = 70$. We connect these three points using the same shape in the same direction as in the graph of f and, last but not least, we restrict the domain to match the applied domain $[0, \infty)$. The result is below.

---

[8] We will discuss this in greater detail in Section 6.5.

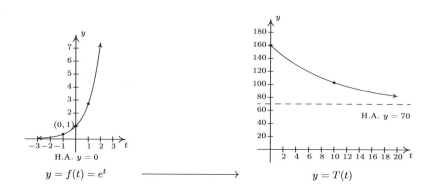

3. From the graph, we see that the horizontal asymptote is $y = 70$. It is worth a moment or two of our time to see how this happens analytically and to review some of the 'number sense' developed in Chapter 4. As $t \to \infty$, We get $T(t) = 70 + 90e^{-0.1t} \approx 70 + 90e^{\text{very big } (-)}$. Since $e > 1$,

$$e^{\text{very big } (-)} = \frac{1}{e^{\text{very big } (+)}} \approx \frac{1}{\text{very big } (+)} \approx \text{very small } (+)$$

The larger t becomes, the smaller $e^{-0.1t}$ becomes, so the term $90e^{-0.1t} \approx$ very small $(+)$. Hence, $T(t) \approx 70 +$ very small $(+)$ which means the graph is approaching the horizontal line $y = 70$ from above. This means that as time goes by, the temperature of the coffee is cooling to $70°F$, presumably room temperature. □

As we have already remarked, the graphs of $f(x) = b^x$ all pass the Horizontal Line Test. Thus the exponential functions are invertible. We now turn our attention to these inverses, the logarithmic functions, which are called 'logs' for short.

Definition 6.2. The inverse of the exponential function $f(x) = b^x$ is called the **base b logarithm function**, and is denoted $f^{-1}(x) = \log_b(x)$ We read '$\log_b(x)$' as 'log base b of x.'

We have special notations for the common base, $b = 10$, and the natural base, $b = e$.

Definition 6.3. The **common logarithm** of a real number x is $\log_{10}(x)$ and is usually written $\log(x)$. The **natural logarithm** of a real number x is $\log_e(x)$ and is usually written $\ln(x)$.

Since logs are defined as the inverses of exponential functions, we can use Theorems 5.2 and 5.3 to tell us about logarithmic functions. For example, we know that the domain of a log function is the range of an exponential function, namely $(0, \infty)$, and that the range of a log function is the domain of an exponential function, namely $(-\infty, \infty)$. Since we know the basic shapes of $y = f(x) = b^x$ for the different cases of b, we can obtain the graph of $y = f^{-1}(x) = \log_b(x)$ by reflecting the graph of f across the line $y = x$ as shown below. The y-intercept $(0, 1)$ on the graph of f corresponds to an x-intercept of $(1, 0)$ on the graph of f^{-1}. The horizontal asymptotes $y = 0$ on the graphs of the exponential functions become vertical asymptotes $x = 0$ on the log graphs.

6.1 Introduction to Exponential and Logarithmic Functions

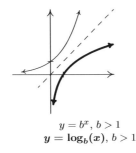
$y = b^x$, $b > 1$
$y = \log_b(x)$, $b > 1$

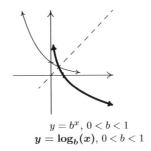
$y = b^x$, $0 < b < 1$
$y = \log_b(x)$, $0 < b < 1$

On a procedural level, logs undo the exponentials. Consider the function $f(x) = 2^x$. When we evaluate $f(3) = 2^3 = 8$, the input 3 becomes the exponent on the base 2 to produce the real number 8. The function $f^{-1}(x) = \log_2(x)$ then takes the number 8 as its input and returns the exponent 3 as its output. In symbols, $\log_2(8) = 3$. More generally, $\log_2(x)$ is the exponent you put on 2 to get x. Thus, $\log_2(16) = 4$, because $2^4 = 16$. The following theorem summarizes the basic properties of logarithmic functions, all of which come from the fact that they are inverses of exponential functions.

Theorem 6.2. Properties of Logarithmic Functions: Suppose $f(x) = \log_b(x)$.

- The domain of f is $(0, \infty)$ and the range of f is $(-\infty, \infty)$.

- $(1, 0)$ is on the graph of f and $x = 0$ is a vertical asymptote of the graph of f.

- f is one-to-one, continuous and smooth

- $b^a = c$ if and only if $\log_b(c) = a$. That is, $\log_b(c)$ is the exponent you put on b to obtain c.

- $\log_b(b^x) = x$ for all x and $b^{\log_b(x)} = x$ for all $x > 0$

- If $b > 1$:
  - f is always increasing
  - As $x \to 0^+$, $f(x) \to -\infty$
  - As $x \to \infty$, $f(x) \to \infty$
  - The graph of f resembles:

  $y = \log_b(x)$, $b > 1$

- If $0 < b < 1$:
  - f is always decreasing
  - As $x \to 0^+$, $f(x) \to \infty$
  - As $x \to \infty$, $f(x) \to -\infty$
  - The graph of f resembles:

  $y = \log_b(x)$, $0 < b < 1$

As we have mentioned, Theorem 6.2 is a consequence of Theorems 5.2 and 5.3. However, it is worth the reader's time to understand Theorem 6.2 from an exponential perspective. For instance, we know that the domain of $g(x) = \log_2(x)$ is $(0, \infty)$. Why? Because the range of $f(x) = 2^x$ is $(0, \infty)$. In a way, this says everything, but at the same time, it doesn't. For example, if we try to find $\log_2(-1)$, we are trying to find the exponent we put on 2 to give us -1. In other words, we are looking for x that satisfies $2^x = -1$. There is no such real number, since all powers of 2 are positive. While what we have said is exactly the same thing as saying 'the domain of $g(x) = \log_2(x)$ is $(0, \infty)$ because the range of $f(x) = 2^x$ is $(0, \infty)$', we feel it is in a student's best interest to understand the statements in Theorem 6.2 at this level instead of just merely memorizing the facts.

Example 6.1.3. Simplify the following.

1. $\log_3(81)$
2. $\log_2\left(\frac{1}{8}\right)$
3. $\log_{\sqrt{5}}(25)$
4. $\ln\left(\sqrt[3]{e^2}\right)$
5. $\log(0.001)$
6. $2^{\log_2(8)}$
7. $117^{-\log_{117}(6)}$

Solution.

1. The number $\log_3(81)$ is the exponent we put on 3 to get 81. As such, we want to write 81 as a power of 3. We find $81 = 3^4$, so that $\log_3(81) = 4$.

2. To find $\log_2\left(\frac{1}{8}\right)$, we need rewrite $\frac{1}{8}$ as a power of 2. We find $\frac{1}{8} = \frac{1}{2^3} = 2^{-3}$, so $\log_2\left(\frac{1}{8}\right) = -3$.

3. To determine $\log_{\sqrt{5}}(25)$, we need to express 25 as a power of $\sqrt{5}$. We know $25 = 5^2$, and $5 = \left(\sqrt{5}\right)^2$, so we have $25 = \left(\left(\sqrt{5}\right)^2\right)^2 = \left(\sqrt{5}\right)^4$. We get $\log_{\sqrt{5}}(25) = 4$.

4. First, recall that the notation $\ln\left(\sqrt[3]{e^2}\right)$ means $\log_e\left(\sqrt[3]{e^2}\right)$, so we are looking for the exponent to put on e to obtain $\sqrt[3]{e^2}$. Rewriting $\sqrt[3]{e^2} = e^{2/3}$, we find $\ln\left(\sqrt[3]{e^2}\right) = \ln\left(e^{2/3}\right) = \frac{2}{3}$.

5. Rewriting $\log(0.001)$ as $\log_{10}(0.001)$, we see that we need to write 0.001 as a power of 10. We have $0.001 = \frac{1}{1000} = \frac{1}{10^3} = 10^{-3}$. Hence, $\log(0.001) = \log\left(10^{-3}\right) = -3$.

6. We can use Theorem 6.2 directly to simplify $2^{\log_2(8)} = 8$. We can also understand this problem by first finding $\log_2(8)$. By definition, $\log_2(8)$ is the exponent we put on 2 to get 8. Since $8 = 2^3$, we have $\log_2(8) = 3$. We now substitute to find $2^{\log_2(8)} = 2^3 = 8$.

7. From Theorem 6.2, we know $117^{\log_{117}(6)} = 6$, but we cannot directly apply this formula to the expression $117^{-\log_{117}(6)}$. (Can you see why?) At this point, we use a property of exponents followed by Theorem 6.2 to get[9]

$$117^{-\log_{117}(6)} = \frac{1}{117^{\log_{117}(6)}} = \frac{1}{6}$$

□

---

[9]It is worth a moment of your time to think your way through why $117^{\log_{117}(6)} = 6$. By definition, $\log_{117}(6)$ is the exponent we put on 117 to get 6. What are we doing with this exponent? We are putting it on 117. By definition we get 6. In other words, the exponential function $f(x) = 117^x$ undoes the logarithmic function $g(x) = \log_{117}(x)$.

6.1 Introduction to Exponential and Logarithmic Functions

Up until this point, restrictions on the domains of functions came from avoiding division by zero and keeping negative numbers from beneath even radicals. With the introduction of logs, we now have another restriction. Since the domain of $f(x) = \log_b(x)$ is $(0, \infty)$, the argument[10] of the log must be strictly positive.

Example 6.1.4. Find the domain of the following functions. Check your answers graphically using the calculator.

1. $f(x) = 2\log(3 - x) - 1$

2. $g(x) = \ln\left(\dfrac{x}{x-1}\right)$

Solution.

1. We set $3 - x > 0$ to obtain $x < 3$, or $(-\infty, 3)$. The graph from the calculator below verifies this. Note that we could have graphed f using transformations. Taking a cue from Theorem 1.7, we rewrite $f(x) = 2\log_{10}(-x+3) - 1$ and find the main function involved is $y = h(x) = \log_{10}(x)$. We select three points to track, $\left(\frac{1}{10}, -1\right)$, $(1, 0)$ and $(10, 1)$, along with the vertical asymptote $x = 0$. Since $f(x) = 2h(-x + 3) - 1$, Theorem 1.7 tells us that to obtain the destinations of these points, we first subtract 3 from the x-coordinates (shifting the graph left 3 units), then divide (multiply) by the x-coordinates by -1 (causing a reflection across the y-axis). These transformations apply to the vertical asymptote $x = 0$ as well. Subtracting 3 gives us $x = -3$ as our asymptote, then multplying by -1 gives us the vertical asymptote $x = 3$. Next, we multiply the y-coordinates by 2 which results in a vertical stretch by a factor of 2, then we finish by subtracting 1 from the y-coordinates which shifts the graph down 1 unit. We leave it to the reader to perform the indicated arithmetic on the points themselves and to verify the graph produced by the calculator below.

2. To find the domain of g, we need to solve the inequality $\frac{x}{x-1} > 0$. As usual, we proceed using a sign diagram. If we define $r(x) = \frac{x}{x-1}$, we find r is undefined at $x = 1$ and $r(x) = 0$ when $x = 0$. Choosing some test values, we generate the sign diagram below.

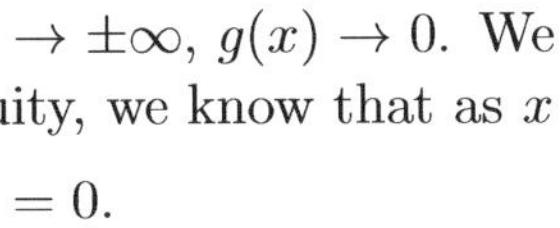

$$(+) \;\; 0 \;\; (-) \;\; ? \;\; (+)$$
$$ 0 1$$

We find $\frac{x}{x-1} > 0$ on $(-\infty, 0) \cup (1, \infty)$ to get the domain of g. The graph of $y = g(x)$ confirms this. We can tell from the graph of g that it is not the result of Section 1.7 transformations being applied to the graph $y = \ln(x)$, so barring a more detailed analysis using Calculus, the calculator graph is the best we can do. One thing worthy of note, however, is the end behavior of g. The graph suggests that as $x \to \pm\infty$, $g(x) \to 0$. We can verify this analytically. Using results from Chapter 4 and continuity, we know that as $x \to \pm\infty$, $\frac{x}{x-1} \approx 1$. Hence, it makes sense that $g(x) = \ln\left(\frac{x}{x-1}\right) \approx \ln(1) = 0$.

---

[10] See page 55 if you've forgotten what this term means.

426   EXPONENTIAL AND LOGARITHMIC FUNCTIONS

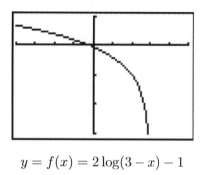
$y = f(x) = 2\log(3-x) - 1$

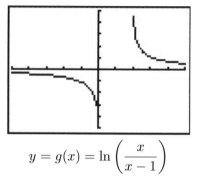
$y = g(x) = \ln\left(\dfrac{x}{x-1}\right)$

While logarithms have some interesting applications of their own which you'll explore in the exercises, their primary use to us will be to undo exponential functions. (This is, after all, how they were defined.) Our last example solidifies this and reviews all of the material in the section.

Example 6.1.5. Let $f(x) = 2^{x-1} - 3$.

1. Graph f using transformations and state the domain and range of f.

2. Explain why f is invertible and find a formula for $f^{-1}(x)$.

3. Graph f^{-1} using transformations and state the domain and range of f^{-1}.

4. Verify $\left(f^{-1} \circ f\right)(x) = x$ for all x in the domain of f and $\left(f \circ f^{-1}\right)(x) = x$ for all x in the domain of f^{-1}.

5. Graph f and f^{-1} on the same set of axes and check the symmetry about the line $y = x$.

Solution.

1. If we identify $g(x) = 2^x$, we see $f(x) = g(x-1) - 3$. We pick the points $\left(-1, \frac{1}{2}\right)$, $(0,1)$ and $(1,2)$ on the graph of g along with the horizontal asymptote $y = 0$ to track through the transformations. By Theorem 1.7 we first add 1 to the x-coordinates of the points on the graph of g (shifting g to the right 1 unit) to get $\left(0, \frac{1}{2}\right)$, $(1,1)$ and $(2,2)$. The horizontal asymptote remains $y = 0$. Next, we subtract 3 from the y-coordinates, shifting the graph down 3 units. We get the points $\left(0, -\frac{5}{2}\right)$, $(1, -2)$ and $(2, -1)$ with the horizontal asymptote now at $y = -3$. Connecting the dots in the order and manner as they were on the graph of g, we get the graph below. We see that the domain of f is the same as g, namely $(-\infty, \infty)$, but that the range of f is $(-3, \infty)$.

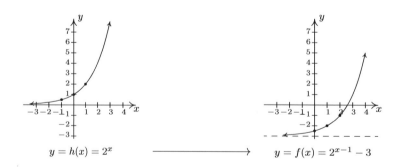

2. The graph of f passes the Horizontal Line Test so f is one-to-one, hence invertible. To find a formula for $f^{-1}(x)$, we normally set $y = f(x)$, interchange the x and y, then proceed to solve for y. Doing so in this situation leads us to the equation $x = 2^{y-1} - 3$. We have yet to discuss how to solve this kind of equation, so we will attempt to find the formula for f^{-1} from a procedural perspective. If we break $f(x) = 2^{x-1} - 3$ into a series of steps, we find f takes an input x and applies the steps

   (a) subtract 1
   (b) put as an exponent on 2
   (c) subtract 3

   Clearly, to undo subtracting 1, we will add 1, and similarly we undo subtracting 3 by adding 3. How do we undo the second step? The answer is we use the logarithm. By definition, $\log_2(x)$ undoes exponentiation by 2. Hence, f^{-1} should

   (a) add 3
   (b) take the logarithm base 2
   (c) add 1

   In symbols, $f^{-1}(x) = \log_2(x+3) + 1$.

3. To graph $f^{-1}(x) = \log_2(x+3) + 1$ using transformations, we start with $j(x) = \log_2(x)$. We track the points $\left(\frac{1}{2}, -1\right)$, $(1, 0)$ and $(2, 1)$ on the graph of j along with the vertical asymptote $x = 0$ through the transformations using Theorem 1.7. Since $f^{-1}(x) = j(x+3) + 1$, we first subtract 3 from each of the x values (including the vertical asymptote) to obtain $\left(-\frac{5}{2}, -1\right)$, $(-2, 0)$ and $(-1, 1)$ with a vertical asymptote $x = -3$. Next, we add 1 to the y values on the graph and get $\left(-\frac{5}{2}, 0\right)$, $(-2, 1)$ and $(-1, 2)$. If you are experiencing *déjà vu*, there is a good reason for it but we leave it to the reader to determine the source of this uncanny familiarity. We obtain the graph below. The domain of f^{-1} is $(-3, \infty)$, which matches the range of f, and the range of f^{-1} is $(-\infty, \infty)$, which matches the domain of f.

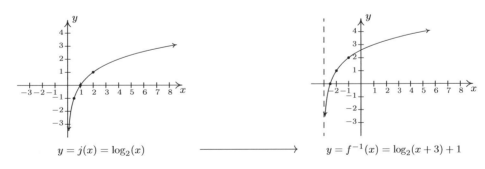

4. We now verify that $f(x) = 2^{x-1} - 3$ and $f^{-1}(x) = \log_2(x+3) + 1$ satisfy the composition requirement for inverses. For all real numbers x,

$$\begin{aligned}
\left(f^{-1} \circ f\right)(x) &= f^{-1}(f(x)) \\
&= f^{-1}\left(2^{x-1} - 3\right) \\
&= \log_2\left(\left[2^{x-1} - 3\right] + 3\right) + 1 \\
&= \log_2\left(2^{x-1}\right) + 1 \\
&= (x-1) + 1 \qquad \text{Since } \log_2\left(2^u\right) = u \text{ for all real numbers } u \\
&= x \checkmark
\end{aligned}$$

For all real numbers $x > -3$, we have[11]

$$\begin{aligned}
\left(f \circ f^{-1}\right)(x) &= f\left(f^{-1}(x)\right) \\
&= f\left(\log_2(x+3) + 1\right) \\
&= 2^{(\log_2(x+3)+1)-1} - 3 \\
&= 2^{\log_2(x+3)} - 3 \\
&= (x+3) - 3 \qquad \text{Since } 2^{\log_2(u)} = u \text{ for all real numbers } u > 0 \\
&= x \checkmark
\end{aligned}$$

5. Last, but certainly not least, we graph $y = f(x)$ and $y = f^{-1}(x)$ on the same set of axes and see the symmetry about the line $y = x$.

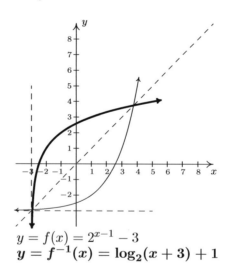

$y = f(x) = 2^{x-1} - 3$
$y = f^{-1}(x) = \log_2(x+3) + 1$

□

---

[11] Pay attention - can you spot in which step below we need $x > -3$?

6.1 Introduction to Exponential and Logarithmic Functions

6.1.1 Exercises

In Exercises 1 - 15, use the property: $b^a = c$ if and only if $\log_b(c) = a$ from Theorem 6.2 to rewrite the given equation in the other form. That is, rewrite the exponential equations as logarithmic equations and rewrite the logarithmic equations as exponential equations.

1. $2^3 = 8$
2. $5^{-3} = \frac{1}{125}$
3. $4^{5/2} = 32$
4. $\left(\frac{1}{3}\right)^{-2} = 9$
5. $\left(\frac{4}{25}\right)^{-1/2} = \frac{5}{2}$
6. $10^{-3} = 0.001$
7. $e^0 = 1$
8. $\log_5(25) = 2$
9. $\log_{25}(5) = \frac{1}{2}$
10. $\log_3\left(\frac{1}{81}\right) = -4$
11. $\log_{\frac{4}{3}}\left(\frac{3}{4}\right) = -1$
12. $\log(100) = 2$
13. $\log(0.1) = -1$
14. $\ln(e) = 1$
15. $\ln\left(\frac{1}{\sqrt{e}}\right) = -\frac{1}{2}$

In Exercises 16 - 42, evaluate the expression.

16. $\log_3(27)$
17. $\log_6(216)$
18. $\log_2(32)$
19. $\log_6\left(\frac{1}{36}\right)$
20. $\log_8(4)$
21. $\log_{36}(216)$
22. $\log_{\frac{1}{5}}(625)$
23. $\log_{\frac{1}{6}}(216)$
24. $\log_{36}(36)$
25. $\log\left(\frac{1}{1000000}\right)$
26. $\log(0.01)$
27. $\ln\left(e^3\right)$
28. $\log_4(8)$
29. $\log_6(1)$
30. $\log_{13}\left(\sqrt{13}\right)$
31. $\log_{36}\left(\sqrt[4]{36}\right)$
32. $7^{\log_7(3)}$
33. $36^{\log_{36}(216)}$
34. $\log_{36}\left(36^{216}\right)$
35. $\ln\left(e^5\right)$
36. $\log\left(\sqrt[9]{10^{11}}\right)$
37. $\log\left(\sqrt[3]{10^5}\right)$
38. $\ln\left(\frac{1}{\sqrt{e}}\right)$
39. $\log_5\left(3^{\log_3(5)}\right)$
40. $\log\left(e^{\ln(100)}\right)$
41. $\log_2\left(3^{-\log_3(2)}\right)$
42. $\ln\left(42^{6\log(1)}\right)$

In Exercises 43 - 57, find the domain of the function.

43. $f(x) = \ln(x^2 + 1)$
44. $f(x) = \log_7(4x + 8)$
45. $f(x) = \ln(4x - 20)$
46. $f(x) = \log\left(x^2 + 9x + 18\right)$

47. $f(x) = \log\left(\dfrac{x+2}{x^2-1}\right)$

48. $f(x) = \log\left(\dfrac{x^2+9x+18}{4x-20}\right)$

49. $f(x) = \ln(7-x) + \ln(x-4)$

50. $f(x) = \ln(4x-20) + \ln\left(x^2+9x+18\right)$

51. $f(x) = \log\left(x^2+x+1\right)$

52. $f(x) = \sqrt[4]{\log_4(x)}$

53. $f(x) = \log_9(|x+3|-4)$

54. $f(x) = \ln(\sqrt{x-4}-3)$

55. $f(x) = \dfrac{1}{3-\log_5(x)}$

56. $f(x) = \dfrac{\sqrt{-1-x}}{\log_{\frac{1}{2}}(x)}$

57. $f(x) = \ln(-2x^3 - x^2 + 13x - 6)$

In Exercises 58 - 63, sketch the graph of $y = g(x)$ by starting with the graph of $y = f(x)$ and using transformations. Track at least three points of your choice and the horizontal asymptote through the transformations. State the domain and range of g.

58. $f(x) = 2^x$, $g(x) = 2^x - 1$

59. $f(x) = \left(\frac{1}{3}\right)^x$, $g(x) = \left(\frac{1}{3}\right)^{x-1}$

60. $f(x) = 3^x$, $g(x) = 3^{-x} + 2$

61. $f(x) = 10^x$, $g(x) = 10^{\frac{x+1}{2}} - 20$

62. $f(x) = e^x$, $g(x) = 8 - e^{-x}$

63. $f(x) = e^x$, $g(x) = 10e^{-0.1x}$

In Exercises 64 - 69, sketch the graph of $y = g(x)$ by starting with the graph of $y = f(x)$ and using transformations. Track at least three points of your choice and the vertical asymptote through the transformations. State the domain and range of g.

64. $f(x) = \log_2(x)$, $g(x) = \log_2(x+1)$

65. $f(x) = \log_{\frac{1}{3}}(x)$, $g(x) = \log_{\frac{1}{3}}(x) + 1$

66. $f(x) = \log_3(x)$, $g(x) = -\log_3(x-2)$

67. $f(x) = \log(x)$, $g(x) = 2\log(x+20) - 1$

68. $f(x) = \ln(x)$, $g(x) = -\ln(8-x)$

69. $f(x) = \ln(x)$, $g(x) = -10\ln\left(\frac{x}{10}\right)$

70. Verify that each function in Exercises 64 - 69 is the inverse of the corresponding function in Exercises 58 - 63. (Match up #58 and #64, and so on.)

In Exercises 71 - 74, find the inverse of the function from the 'procedural perspective' discussed in Example 6.1.5 and graph the function and its inverse on the same set of axes.

71. $f(x) = 3^{x+2} - 4$

72. $f(x) = \log_4(x-1)$

73. $f(x) = -2^{-x} + 1$

74. $f(x) = 5\log(x) - 2$

6.1 Introduction to Exponential and Logarithmic Functions

(Logarithmic Scales) In Exercises 75 - 77, we introduce three widely used measurement scales which involve common logarithms: the Richter scale, the decibel scale and the pH scale. The computations involved in all three scales are nearly identical so pay attention to the subtle differences.

75. Earthquakes are complicated events and it is not our intent to provide a complete discussion of the science involved in them. Instead, we refer the interested reader to a solid course in Geology[12] or the U.S. Geological Survey's Earthquake Hazards Program found here and present only a simplified version of the Richter scale. The Richter scale measures the magnitude of an earthquake by comparing the amplitude of the seismic waves of the given earthquake to those of a "magnitude 0 event", which was chosen to be a seismograph reading of 0.001 millimeters recorded on a seismometer 100 kilometers from the earthquake's epicenter. Specifically, the magnitude of an earthquake is given by

$$M(x) = \log\left(\frac{x}{0.001}\right)$$

    where x is the seismograph reading in millimeters of the earthquake recorded 100 kilometers from the epicenter.

    (a) Show that $M(0.001) = 0$.
    (b) Compute $M(80,000)$.
    (c) Show that an earthquake which registered 6.7 on the Richter scale had a seismograph reading ten times larger than one which measured 5.7.
    (d) Find two news stories about recent earthquakes which give their magnitudes on the Richter scale. How many times larger was the seismograph reading of the earthquake with larger magnitude?

76. While the decibel scale can be used in many disciplines,[13] we shall restrict our attention to its use in acoustics, specifically its use in measuring the intensity level of sound.[14] The Sound Intensity Level L (measured in decibels) of a sound intensity I (measured in watts per square meter) is given by

$$L(I) = 10\log\left(\frac{I}{10^{-12}}\right).$$

    Like the Richter scale, this scale compares I to baseline: $10^{-12}\frac{W}{m^2}$ is the threshold of human hearing.

    (a) Compute $L(10^{-6})$.

---

[12] Rock-solid, perhaps?

[13] See this webpage for more information.

[14] As of the writing of this exercise, the Wikipedia page given here states that it may not meet the "general notability guideline" nor does it cite any references or sources. I find this odd because it is this very usage of the decibel scale which shows up in every College Algebra book I have read. Perhaps those other books have been wrong all along and we're just blindly following tradition.

(b) Damage to your hearing can start with short term exposure to sound levels around 115 decibels. What intensity I is needed to produce this level?

(c) Compute $L(1)$. How does this compare with the threshold of pain which is around 140 decibels?

77. The pH of a solution is a measure of its acidity or alkalinity. Specifically, pH $= -\log[\text{H}^+]$ where $[\text{H}^+]$ is the hydrogen ion concentration in moles per liter. A solution with a pH less than 7 is an acid, one with a pH greater than 7 is a base (alkaline) and a pH of 7 is regarded as neutral.

    (a) The hydrogen ion concentration of pure water is $[\text{H}^+] = 10^{-7}$. Find its pH.
    (b) Find the pH of a solution with $[\text{H}^+] = 6.3 \times 10^{-13}$.
    (c) The pH of gastric acid (the acid in your stomach) is about 0.7. What is the corresponding hydrogen ion concentration?

78. Show that $\log_b 1 = 0$ and $\log_b b = 1$ for every $b > 0$, $b \neq 1$.

79. (Crazy bonus question) Without using your calculator, determine which is larger: e^π or π^e.

6.1 Introduction to Exponential and Logarithmic Functions

6.1.2 Answers

1. $\log_2(8) = 3$
2. $\log_5\left(\frac{1}{125}\right) = -3$
3. $\log_4(32) = \frac{5}{2}$
4. $\log_{\frac{1}{3}}(9) = -2$
5. $\log_{\frac{4}{25}}\left(\frac{5}{2}\right) = -\frac{1}{2}$
6. $\log(0.001) = -3$
7. $\ln(1) = 0$
8. $5^2 = 25$
9. $(25)^{\frac{1}{2}} = 5$
10. $3^{-4} = \frac{1}{81}$
11. $\left(\frac{4}{3}\right)^{-1} = \frac{3}{4}$
12. $10^2 = 100$
13. $10^{-1} = 0.1$
14. $e^1 = e$
15. $e^{-\frac{1}{2}} = \frac{1}{\sqrt{e}}$
16. $\log_3(27) = 3$
17. $\log_6(216) = 3$
18. $\log_2(32) = 5$
19. $\log_6\left(\frac{1}{36}\right) = -2$
20. $\log_8(4) = \frac{2}{3}$
21. $\log_{36}(216) = \frac{3}{2}$
22. $\log_{\frac{1}{5}}(625) = -4$
23. $\log_{\frac{1}{6}}(216) = -3$
24. $\log_{36}(36) = 1$
25. $\log\frac{1}{1000000} = -6$
26. $\log(0.01) = -2$
27. $\ln\left(e^3\right) = 3$
28. $\log_4(8) = \frac{3}{2}$
29. $\log_6(1) = 0$
30. $\log_{13}\left(\sqrt{13}\right) = \frac{1}{2}$
31. $\log_{36}\left(\sqrt[4]{36}\right) = \frac{1}{4}$
32. $7^{\log_7(3)} = 3$
33. $36^{\log_{36}(216)} = 216$
34. $\log_{36}\left(36^{216}\right) = 216$
35. $\ln(e^5) = 5$
36. $\log\left(\sqrt[9]{10^{11}}\right) = \frac{11}{9}$
37. $\log\left(\sqrt[3]{10^5}\right) = \frac{5}{3}$
38. $\ln\left(\frac{1}{\sqrt{e}}\right) = -\frac{1}{2}$
39. $\log_5\left(3^{\log_3 5}\right) = 1$
40. $\log\left(e^{\ln(100)}\right) = 2$
41. $\log_2\left(3^{-\log_3(2)}\right) = -1$
42. $\ln\left(42^{6\log(1)}\right) = 0$
43. $(-\infty, \infty)$
44. $(-2, \infty)$
45. $(5, \infty)$
46. $(-\infty, -6) \cup (-3, \infty)$
47. $(-2, -1) \cup (1, \infty)$
48. $(-6, -3) \cup (5, \infty)$
49. $(4, 7)$
50. $(5, \infty)$
51. $(-\infty, \infty)$
52. $[1, \infty)$
53. $(-\infty, -7) \cup (1, \infty)$
54. $(13, \infty)$
55. $(0, 125) \cup (125, \infty)$
56. No domain
57. $(-\infty, -3) \cup \left(\frac{1}{2}, 2\right)$

58. Domain of g: $(-\infty, \infty)$
    Range of g: $(-1, \infty)$

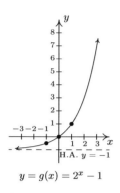

$y = g(x) = 2^x - 1$

59. Domain of g: $(-\infty, \infty)$
    Range of g: $(0, \infty)$

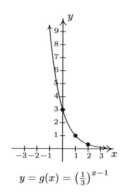

$y = g(x) = \left(\frac{1}{3}\right)^{x-1}$

60. Domain of g: $(-\infty, \infty)$
    Range of g: $(2, \infty)$

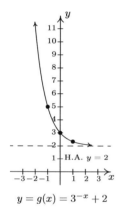

$y = g(x) = 3^{-x} + 2$

61. Domain of g: $(-\infty, \infty)$
    Range of g: $(-20, \infty)$

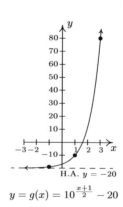

$y = g(x) = 10^{\frac{x+1}{2}} - 20$

62. Domain of g: $(-\infty, \infty)$
    Range of g: $(-\infty, 8)$

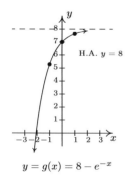

$y = g(x) = 8 - e^{-x}$

63. Domain of g: $(-\infty, \infty)$
    Range of g: $(0, \infty)$

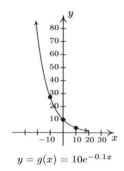

$y = g(x) = 10e^{-0.1x}$

64. Domain of g: $(-1, \infty)$
    Range of g: $(-\infty, \infty)$

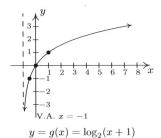

$y = g(x) = \log_2(x+1)$

65. Domain of g: $(0, \infty)$
    Range of g: $(-\infty, \infty)$

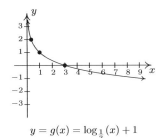

$y = g(x) = \log_{\frac{1}{3}}(x) + 1$

66. Domain of g: $(2, \infty)$
    Range of g: $(-\infty, \infty)$

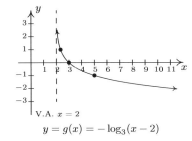

$y = g(x) = -\log_3(x-2)$

67. Domain of g: $(-20, \infty)$
    Range of g: $(-\infty, \infty)$

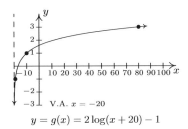

$y = g(x) = 2\log(x+20) - 1$

68. Domain of g: $(-\infty, 8)$
    Range of g: $(-\infty, \infty)$

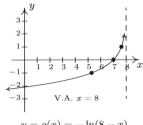

$y = g(x) = -\ln(8-x)$

69. Domain of g: $(0, \infty)$
    Range of g: $(-\infty, \infty)$

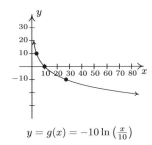

$y = g(x) = -10\ln\left(\frac{x}{10}\right)$

71. $f(x) = 3^{x+2} - 4$
    $f^{-1}(x) = \log_3(x+4) - 2$

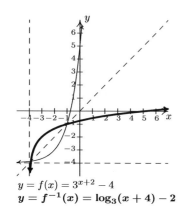

$y = f(x) = 3^{x+2} - 4$
$y = f^{-1}(x) = \log_3(x+4) - 2$

72. $f(x) = \log_4(x-1)$
    $f^{-1}(x) = 4^x + 1$

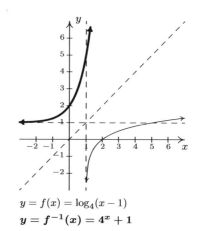

$y = f(x) = \log_4(x-1)$
$y = f^{-1}(x) = 4^x + 1$

73. $f(x) = -2^{-x} + 1$
    $f^{-1}(x) = -\log_2(1-x)$

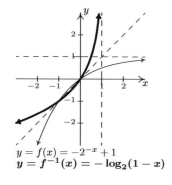

$y = f(x) = -2^{-x} + 1$
$y = f^{-1}(x) = -\log_2(1-x)$

74. $f(x) = 5\log(x) - 2$
    $f^{-1}(x) = 10^{\frac{x+2}{5}}$

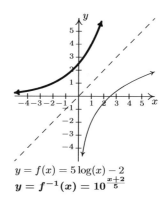

$y = f(x) = 5\log(x) - 2$
$y = f^{-1}(x) = 10^{\frac{x+2}{5}}$

75. (a) $M(0.001) = \log\left(\frac{0.001}{0.001}\right) = \log(1) = 0$.

    (b) $M(80,000) = \log\left(\frac{80,000}{0.001}\right) = \log(80,000,000) \approx 7.9$.

76. (a) $L(10^{-6}) = 60$ decibels.

    (b) $I = 10^{-.5} \approx 0.316$ watts per square meter.

    (c) Since $L(1) = 120$ decibels and $L(100) = 140$ decibels, a sound with intensity level 140 decibels has an intensity 100 times greater than a sound with intensity level 120 decibels.

77. (a) The pH of pure water is 7.

    (b) If $[\text{H}^+] = 6.3 \times 10^{-13}$ then the solution has a pH of 12.2.

    (c) $[\text{H}^+] = 10^{-0.7} \approx .1995$ moles per liter.

6.2 Properties of Logarithms

In Section 6.1, we introduced the logarithmic functions as inverses of exponential functions and discussed a few of their functional properties from that perspective. In this section, we explore the algebraic properties of logarithms. Historically, these have played a huge role in the scientific development of our society since, among other things, they were used to develop analog computing devices called slide rules which enabled scientists and engineers to perform accurate calculations leading to such things as space travel and the moon landing. As we shall see shortly, logs inherit analogs of all of the properties of exponents you learned in Elementary and Intermediate Algebra. We first extract two properties from Theorem 6.2 to remind us of the definition of a logarithm as the inverse of an exponential function.

Theorem 6.3. (Inverse Properties of Exponential and Logarithmic Functions)
Let $b > 0$, $b \neq 1$.

- $b^a = c$ if and only if $\log_b(c) = a$

- $\log_b(b^x) = x$ for all x and $b^{\log_b(x)} = x$ for all $x > 0$

Next, we spell out what it means for exponential and logarithmic functions to be one-to-one.

Theorem 6.4. (One-to-one Properties of Exponential and Logarithmic Functions)
Let $f(x) = b^x$ and $g(x) = \log_b(x)$ where $b > 0$, $b \neq 1$. Then f and g are one-to-one and

- $b^u = b^w$ if and only if $u = w$ for all real numbers u and w.

- $\log_b(u) = \log_b(w)$ if and only if $u = w$ for all real numbers $u > 0$, $w > 0$.

We now state the algebraic properties of exponential functions which will serve as a basis for the properties of logarithms. While these properties may look identical to the ones you learned in Elementary and Intermediate Algebra, they apply to real number exponents, not just rational exponents. Note that in the theorem that follows, we are interested in the properties of exponential functions, so the base b is restricted to $b > 0$, $b \neq 1$. An added benefit of this restriction is that it eliminates the pathologies discussed in Section 5.3 when, for example, we simplified $\left(x^{2/3}\right)^{3/2}$ and obtained $|x|$ instead of what we had expected from the arithmetic in the exponents, $x^1 = x$.

Theorem 6.5. (Algebraic Properties of Exponential Functions) Let $f(x) = b^x$ be an exponential function ($b > 0$, $b \neq 1$) and let u and w be real numbers.

- **Product Rule:** $f(u + w) = f(u)f(w)$. In other words, $b^{u+w} = b^u b^w$

- **Quotient Rule:** $f(u - w) = \dfrac{f(u)}{f(w)}$. In other words, $b^{u-w} = \dfrac{b^u}{b^w}$

- **Power Rule:** $(f(u))^w = f(uw)$. In other words, $(b^u)^w = b^{uw}$

While the properties listed in Theorem 6.5 are certainly believable based on similar properties of integer and rational exponents, the full proofs require Calculus. To each of these properties of exponential functions corresponds an analogous property of logarithmic functions. We list these below in our next theorem.

> **Theorem 6.6. (Algebraic Properties of Logarithmic Functions)** Let $g(x) = \log_b(x)$ be a logarithmic function ($b > 0$, $b \neq 1$) and let $u > 0$ and $w > 0$ be real numbers.
>
> - **Product Rule:** $g(uw) = g(u) + g(w)$. In other words, $\log_b(uw) = \log_b(u) + \log_b(w)$
> - **Quotient Rule:** $g\left(\dfrac{u}{w}\right) = g(u) - g(w)$. In other words, $\log_b\left(\dfrac{u}{w}\right) = \log_b(u) - \log_b(w)$
> - **Power Rule:** $g\left(u^w\right) = w g(u)$. In other words, $\log_b\left(u^w\right) = w \log_b(u)$

There are a couple of different ways to understand why Theorem 6.6 is true. Consider the product rule: $\log_b(uw) = \log_b(u) + \log_b(w)$. Let $a = \log_b(uw)$, $c = \log_b(u)$, and $d = \log_b(w)$. Then, by definition, $b^a = uw$, $b^c = u$ and $b^d = w$. Hence, $b^a = uw = b^c b^d = b^{c+d}$, so that $b^a = b^{c+d}$. By the one-to-one property of b^x, we have $a = c + d$. In other words, $\log_b(uw) = \log_b(u) + \log_b(w)$. The remaining properties are proved similarly. From a purely functional approach, we can see the properties in Theorem 6.6 as an example of how inverse functions interchange the roles of inputs in outputs. For instance, the Product Rule for exponential functions given in Theorem 6.5, $f(u+w) = f(u)f(w)$, says that adding inputs results in multiplying outputs. Hence, whatever f^{-1} is, it must take the products of outputs from f and return them to the sum of their respective inputs. Since the outputs from f are the inputs to f^{-1} and vice-versa, we have that that f^{-1} must take products of its inputs to the sum of their respective outputs. This is precisely what the Product Rule for Logarithmic functions states in Theorem 6.6: $g(uw) = g(u) + g(w)$. The reader is encouraged to view the remaining properties listed in Theorem 6.6 similarly. The following examples help build familiarity with these properties. In our first example, we are asked to 'expand' the logarithms. This means that we read the properties in Theorem 6.6 from left to right and rewrite products inside the log as sums outside the log, quotients inside the log as differences outside the log, and powers inside the log as factors outside the log.[1]

Example 6.2.1. Expand the following using the properties of logarithms and simplify. Assume when necessary that all quantities represent positive real numbers.

1. $\log_2\left(\dfrac{8}{x}\right)$
2. $\log_{0.1}\left(10x^2\right)$
3. $\ln\left(\dfrac{3}{ex}\right)^2$

4. $\log \sqrt[3]{\dfrac{100x^2}{yz^5}}$
5. $\log_{117}\left(x^2 - 4\right)$

Solution.

1. To expand $\log_2\left(\dfrac{8}{x}\right)$, we use the Quotient Rule identifying $u = 8$ and $w = x$ and simplify.

---

[1] Interestingly enough, it is the exact *opposite* process (which we will practice later) that is most useful in Algebra, the utility of expanding logarithms becomes apparent in Calculus.

6.2 Properties of Logarithms

$$\begin{aligned}
\log_2\left(\frac{8}{x}\right) &= \log_2(8) - \log_2(x) &\text{Quotient Rule} \\
&= 3 - \log_2(x) &\text{Since } 2^3 = 8 \\
&= -\log_2(x) + 3
\end{aligned}$$

2. In the expression $\log_{0.1}\left(10x^2\right)$, we have a power (the x^2) and a product. In order to use the Product Rule, the *entire* quantity inside the logarithm must be raised to the same exponent. Since the exponent 2 applies only to the x, we first apply the Product Rule with $u = 10$ and $w = x^2$. Once we get the x^2 by itself inside the log, we may apply the Power Rule with $u = x$ and $w = 2$ and simplify.

$$\begin{aligned}
\log_{0.1}\left(10x^2\right) &= \log_{0.1}(10) + \log_{0.1}\left(x^2\right) &\text{Product Rule} \\
&= \log_{0.1}(10) + 2\log_{0.1}(x) &\text{Power Rule} \\
&= -1 + 2\log_{0.1}(x) &\text{Since } (0.1)^{-1} = 10 \\
&= 2\log_{0.1}(x) - 1
\end{aligned}$$

3. We have a power, quotient and product occurring in $\ln\left(\frac{3}{ex}\right)^2$. Since the exponent 2 applies to the entire quantity inside the logarithm, we begin with the Power Rule with $u = \frac{3}{ex}$ and $w = 2$. Next, we see the Quotient Rule is applicable, with $u = 3$ and $w = ex$, so we replace $\ln\left(\frac{3}{ex}\right)$ with the quantity $\ln(3) - \ln(ex)$. Since $\ln\left(\frac{3}{ex}\right)$ is being multiplied by 2, the entire quantity $\ln(3) - \ln(ex)$ is multiplied by 2. Finally, we apply the Product Rule with $u = e$ and $w = x$, and replace $\ln(ex)$ with the quantity $\ln(e) + \ln(x)$, and simplify, keeping in mind that the natural log is log base e.

$$\begin{aligned}
\ln\left(\frac{3}{ex}\right)^2 &= 2\ln\left(\frac{3}{ex}\right) &\text{Power Rule} \\
&= 2\left[\ln(3) - \ln(ex)\right] &\text{Quotient Rule} \\
&= 2\ln(3) - 2\ln(ex) \\
&= 2\ln(3) - 2\left[\ln(e) + \ln(x)\right] &\text{Product Rule} \\
&= 2\ln(3) - 2\ln(e) - 2\ln(x) \\
&= 2\ln(3) - 2 - 2\ln(x) &\text{Since } e^1 = e \\
&= -2\ln(x) + 2\ln(3) - 2
\end{aligned}$$

4. In Theorem 6.6, there is no mention of how to deal with radicals. However, thinking back to Definition 5.5, we can rewrite the cube root as a $\frac{1}{3}$ exponent. We begin by using the Power

Rule[2], and we keep in mind that the common log is log base 10.

$$\begin{aligned}
\log\sqrt[3]{\frac{100x^2}{yz^5}} &= \log\left(\frac{100x^2}{yz^5}\right)^{1/3} \\
&= \tfrac{1}{3}\log\left(\frac{100x^2}{yz^5}\right) & \text{Power Rule} \\
&= \tfrac{1}{3}\left[\log\left(100x^2\right) - \log\left(yz^5\right)\right] & \text{Quotient Rule} \\
&= \tfrac{1}{3}\log\left(100x^2\right) - \tfrac{1}{3}\log\left(yz^5\right) \\
&= \tfrac{1}{3}\left[\log(100) + \log\left(x^2\right)\right] - \tfrac{1}{3}\left[\log(y) + \log\left(z^5\right)\right] & \text{Product Rule} \\
&= \tfrac{1}{3}\log(100) + \tfrac{1}{3}\log\left(x^2\right) - \tfrac{1}{3}\log(y) - \tfrac{1}{3}\log\left(z^5\right) \\
&= \tfrac{1}{3}\log(100) + \tfrac{2}{3}\log(x) - \tfrac{1}{3}\log(y) - \tfrac{5}{3}\log(z) & \text{Power Rule} \\
&= \tfrac{2}{3} + \tfrac{2}{3}\log(x) - \tfrac{1}{3}\log(y) - \tfrac{5}{3}\log(z) & \text{Since } 10^2 = 100 \\
&= \tfrac{2}{3}\log(x) - \tfrac{1}{3}\log(y) - \tfrac{5}{3}\log(z) + \tfrac{2}{3}
\end{aligned}$$

5. At first it seems as if we have no means of simplifying $\log_{117}\left(x^2 - 4\right)$, since none of the properties of logs addresses the issue of expanding a difference *inside* the logarithm. However, we may factor $x^2 - 4 = (x+2)(x-2)$ thereby introducing a product which gives us license to use the Product Rule.

$$\begin{aligned}
\log_{117}\left(x^2 - 4\right) &= \log_{117}\left[(x+2)(x-2)\right] & \text{Factor} \\
&= \log_{117}(x+2) + \log_{117}(x-2) & \text{Product Rule}
\end{aligned}$$

□

A couple of remarks about Example 6.2.1 are in order. First, while not explicitly stated in the above example, a general rule of thumb to determine which log property to apply first to a complicated problem is 'reverse order of operations.' For example, if we were to substitute a number for x into the expression $\log_{0.1}\left(10x^2\right)$, we would first square the x, then multiply by 10. The last step is the multiplication, which tells us the first log property to apply is the Product Rule. In a multi-step problem, this rule can give the required guidance on which log property to apply at each step. The reader is encouraged to look through the solutions to Example 6.2.1 to see this rule in action. Second, while we were instructed to assume when necessary that all quantities represented positive real numbers, the authors would be committing a sin of omission if we failed to point out that, for instance, the functions $f(x) = \log_{117}\left(x^2 - 4\right)$ and $g(x) = \log_{117}(x+2) + \log_{117}(x-2)$ have different domains, and, hence, are different functions. We leave it to the reader to verify the domain of f is $(-\infty, -2) \cup (2, \infty)$ whereas the domain of g is $(2, \infty)$. In general, when using log properties to

---

[2]At this point in the text, the reader is encouraged to carefully read through each step and think of which quantity is playing the role of u and which is playing the role of w as we apply each property.

6.2 Properties of Logarithms

expand a logarithm, we may very well be restricting the domain as we do so. One last comment before we move to reassembling logs from their various bits and pieces. The authors are well aware of the propensity for some students to become overexcited and invent their own properties of logs like $\log_{117}(x^2 - 4) = \log_{117}(x^2) - \log_{117}(4)$, which simply isn't true, in general. The unwritten[3] property of logarithms is that if it isn't written in a textbook, it probably isn't true.

Example 6.2.2. Use the properties of logarithms to write the following as a single logarithm.

1. $\log_3(x-1) - \log_3(x+1)$
2. $\log(x) + 2\log(y) - \log(z)$
3. $4\log_2(x) + 3$
4. $-\ln(x) - \frac{1}{2}$

Solution. Whereas in Example 6.2.1 we read the properties in Theorem 6.6 from left to right to expand logarithms, in this example we read them from right to left.

1. The difference of logarithms requires the Quotient Rule: $\log_3(x-1) - \log_3(x+1) = \log_3\left(\frac{x-1}{x+1}\right)$.

2. In the expression, $\log(x) + 2\log(y) - \log(z)$, we have both a sum and difference of logarithms. However, before we use the product rule to combine $\log(x) + 2\log(y)$, we note that we need to somehow deal with the coefficient 2 on $\log(y)$. This can be handled using the Power Rule. We can then apply the Product and Quotient Rules as we move from left to right. Putting it all together, we have

$$\begin{aligned} \log(x) + 2\log(y) - \log(z) &= \log(x) + \log(y^2) - \log(z) & \text{Power Rule} \\ &= \log(xy^2) - \log(z) & \text{Product Rule} \\ &= \log\left(\frac{xy^2}{z}\right) & \text{Quotient Rule} \end{aligned}$$

3. We can certainly get started rewriting $4\log_2(x) + 3$ by applying the Power Rule to $4\log_2(x)$ to obtain $\log_2(x^4)$, but in order to use the Product Rule to handle the addition, we need to rewrite 3 as a logarithm base 2. From Theorem 6.3, we know $3 = \log_2(2^3)$, so we get

$$\begin{aligned} 4\log_2(x) + 3 &= \log_2(x^4) + 3 & \text{Power Rule} \\ &= \log_2(x^4) + \log_2(2^3) & \text{Since } 3 = \log_2(2^3) \\ &= \log_2(x^4) + \log_2(8) \\ &= \log_2(8x^4) & \text{Product Rule} \end{aligned}$$

---

[3]The authors relish the irony involved in writing what follows.

4. To get started with $-\ln(x) - \frac{1}{2}$, we rewrite $-\ln(x)$ as $(-1)\ln(x)$. We can then use the Power Rule to obtain $(-1)\ln(x) = \ln\left(x^{-1}\right)$. In order to use the Quotient Rule, we need to write $\frac{1}{2}$ as a natural logarithm. Theorem 6.3 gives us $\frac{1}{2} = \ln\left(e^{1/2}\right) = \ln\left(\sqrt{e}\right)$. We have

$$\begin{aligned}
-\ln(x) - \tfrac{1}{2} &= (-1)\ln(x) - \tfrac{1}{2} \\
&= \ln\left(x^{-1}\right) - \tfrac{1}{2} && \text{Power Rule} \\
&= \ln\left(x^{-1}\right) - \ln\left(e^{1/2}\right) && \text{Since } \tfrac{1}{2} = \ln\left(e^{1/2}\right) \\
&= \ln\left(x^{-1}\right) - \ln\left(\sqrt{e}\right) \\
&= \ln\left(\frac{x^{-1}}{\sqrt{e}}\right) && \text{Quotient Rule} \\
&= \ln\left(\frac{1}{x\sqrt{e}}\right)
\end{aligned}$$

$\square$

As we would expect, the rule of thumb for re-assembling logarithms is the opposite of what it was for dismantling them. That is, if we are interested in rewriting an expression as a single logarithm, we apply log properties following the usual order of operations: deal with multiples of logs first with the Power Rule, then deal with addition and subtraction using the Product and Quotient Rules, respectively. Additionally, we find that using log properties in this fashion can increase the domain of the expression. For example, we leave it to the reader to verify the domain of $f(x) = \log_3(x-1) - \log_3(x+1)$ is $(1, \infty)$ but the domain of $g(x) = \log_3\left(\frac{x-1}{x+1}\right)$ is $(-\infty, -1) \cup (1, \infty)$. We will need to keep this in mind when we solve equations involving logarithms in Section 6.4 - it is precisely for this reason we will have to check for extraneous solutions.

The two logarithm buttons commonly found on calculators are the 'LOG' and 'LN' buttons which correspond to the common and natural logs, respectively. Suppose we wanted an approximation to $\log_2(7)$. The answer should be a little less than 3, (Can you explain why?) but how do we coerce the calculator into telling us a more accurate answer? We need the following theorem.

Theorem 6.7. (Change of Base Formulas) Let $a, b > 0$, $a, b \neq 1$.

- $a^x = b^{x \log_b(a)}$ for all real numbers x.

- $\log_a(x) = \dfrac{\log_b(x)}{\log_b(a)}$ for all real numbers $x > 0$.

The proofs of the Change of Base formulas are a result of the other properties studied in this section. If we start with $b^{x \log_b(a)}$ and use the Power Rule in the exponent to rewrite $x \log_b(a)$ as $\log_b(a^x)$ and then apply one of the Inverse Properties in Theorem 6.3, we get

$$b^{x \log_b(a)} = b^{\log_b(a^x)} = a^x,$$

6.2 Properties of Logarithms

as required. To verify the logarithmic form of the property, we also use the Power Rule and an Inverse Property. We note that

$$\log_a(x) \cdot \log_b(a) = \log_b\left(a^{\log_a(x)}\right) = \log_b(x),$$

and we get the result by dividing through by $\log_b(a)$. Of course, the authors can't help but point out the inverse relationship between these two change of base formulas. To change the base of an exponential expression, we *multiply* the *input* by the factor $\log_b(a)$. To change the base of a logarithmic expression, we *divide* the *output* by the factor $\log_b(a)$. While, in the grand scheme of things, both change of base formulas are really saying the same thing, the logarithmic form is the one usually encountered in Algebra while the exponential form isn't usually introduced until Calculus.[4] What Theorem 6.7 really tells us is that all exponential and logarithmic functions are just scalings of one another. Not only does this explain why their graphs have similar shapes, but it also tells us that we could do all of mathematics with a single base - be it 10, e, 42, or 117. Your Calculus teacher will have more to say about this when the time comes.

Example 6.2.3. Use an appropriate change of base formula to convert the following expressions to ones with the indicated base. Verify your answers using a calculator, as appropriate.

1. 3^2 to base 10
2. 2^x to base e
3. $\log_4(5)$ to base e
4. $\ln(x)$ to base 10

Solution.

1. We apply the Change of Base formula with $a = 3$ and $b = 10$ to obtain $3^2 = 10^{2\log(3)}$. Typing the latter in the calculator produces an answer of 9 as required.

2. Here, $a = 2$ and $b = e$ so we have $2^x = e^{x\ln(2)}$. To verify this on our calculator, we can graph $f(x) = 2^x$ and $g(x) = e^{x\ln(2)}$. Their graphs are indistinguishable which provides evidence that they are the same function.

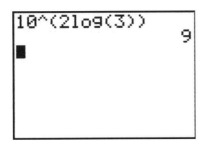

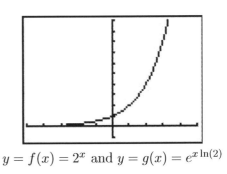

$y = f(x) = 2^x$ and $y = g(x) = e^{x\ln(2)}$

---

[4] The authors feel so strongly about showing students that every property of logarithms comes from and corresponds to a property of exponents that we have broken tradition with the vast majority of other authors in this field. This isn't the first time this happened, and it certainly won't be the last.

3. Applying the change of base with $a = 4$ and $b = e$ leads us to write $\log_4(5) = \frac{\ln(5)}{\ln(4)}$. Evaluating this in the calculator gives $\frac{\ln(5)}{\ln(4)} \approx 1.16$. How do we check this really is the value of $\log_4(5)$? By definition, $\log_4(5)$ is the exponent we put on 4 to get 5. The calculator confirms this.[5]

4. We write $\ln(x) = \log_e(x) = \frac{\log(x)}{\log(e)}$. We graph both $f(x) = \ln(x)$ and $g(x) = \frac{\log(x)}{\log(e)}$ and find both graphs appear to be identical.

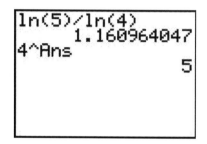
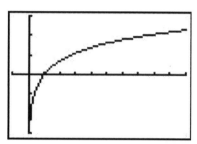

$y = f(x) = \ln(x)$ and $y = g(x) = \frac{\log(x)}{\log(e)}$

□

---

[5]Which means if it is lying to us about the first answer it gave us, at least it is being consistent.

6.2 Properties of Logarithms

6.2.1 Exercises

In Exercises 1 - 15, expand the given logarithm and simplify. Assume when necessary that all quantities represent positive real numbers.

1. $\ln(x^3 y^2)$

2. $\log_2 \left(\dfrac{128}{x^2 + 4} \right)$

3. $\log_5 \left(\dfrac{z}{25} \right)^3$

4. $\log(1.23 \times 10^{37})$

5. $\ln \left(\dfrac{\sqrt{z}}{xy} \right)$

6. $\log_5 \left(x^2 - 25 \right)$

7. $\log_{\sqrt{2}} \left(4x^3 \right)$

8. $\log_{\frac{1}{3}} (9x(y^3 - 8))$

9. $\log \left(1000 x^3 y^5 \right)$

10. $\log_3 \left(\dfrac{x^2}{81 y^4} \right)$

11. $\ln \left(\sqrt[4]{\dfrac{xy}{ez}} \right)$

12. $\log_6 \left(\dfrac{216}{x^3 y} \right)^4$

13. $\log \left(\dfrac{100 x \sqrt{y}}{\sqrt[3]{10}} \right)$

14. $\log_{\frac{1}{2}} \left(\dfrac{4 \sqrt[3]{x^2}}{y \sqrt{z}} \right)$

15. $\ln \left(\dfrac{\sqrt[3]{x}}{10 \sqrt{yz}} \right)$

In Exercises 16 - 29, use the properties of logarithms to write the expression as a single logarithm.

16. $4\ln(x) + 2\ln(y)$

17. $\log_2(x) + \log_2(y) - \log_2(z)$

18. $\log_3(x) - 2\log_3(y)$

19. $\frac{1}{2}\log_3(x) - 2\log_3(y) - \log_3(z)$

20. $2\ln(x) - 3\ln(y) - 4\ln(z)$

21. $\log(x) - \frac{1}{3}\log(z) + \frac{1}{2}\log(y)$

22. $-\frac{1}{3}\ln(x) - \frac{1}{3}\ln(y) + \frac{1}{3}\ln(z)$

23. $\log_5(x) - 3$

24. $3 - \log(x)$

25. $\log_7(x) + \log_7(x - 3) - 2$

26. $\ln(x) + \frac{1}{2}$

27. $\log_2(x) + \log_4(x)$

28. $\log_2(x) + \log_4(x - 1)$

29. $\log_2(x) + \log_{\frac{1}{2}}(x - 1)$

In Exercises 30 - 33, use the appropriate change of base formula to convert the given expression to an expression with the indicated base.

30. 7^{x-1} to base e

31. $\log_3(x+2)$ to base 10

32. $\left(\dfrac{2}{3}\right)^x$ to base e

33. $\log(x^2+1)$ to base e

In Exercises 34 - 39, use the appropriate change of base formula to approximate the logarithm.

34. $\log_3(12)$

35. $\log_5(80)$

36. $\log_6(72)$

37. $\log_4\left(\dfrac{1}{10}\right)$

38. $\log_{\frac{3}{5}}(1000)$

39. $\log_{\frac{2}{3}}(50)$

40. Compare and contrast the graphs of $y = \ln(x^2)$ and $y = 2\ln(x)$.

41. Prove the Quotient Rule and Power Rule for Logarithms.

42. Give numerical examples to show that, in general,

    (a) $\log_b(x+y) \neq \log_b(x) + \log_b(y)$
    (b) $\log_b(x-y) \neq \log_b(x) - \log_b(y)$
    (c) $\log_b\left(\dfrac{x}{y}\right) \neq \dfrac{\log_b(x)}{\log_b(y)}$

43. The Henderson-Hasselbalch Equation: Suppose HA represents a weak acid. Then we have a reversible chemical reaction
    $$HA \rightleftharpoons H^+ + A^-.$$
    The acid disassociation constant, K_a, is given by
    $$K_a = \dfrac{[H^+][A^-]}{[HA]} = [H^+]\dfrac{[A^-]}{[HA]},$$
    where the square brackets denote the concentrations just as they did in Exercise 77 in Section 6.1. The symbol pK_a is defined similarly to pH in that $pK_a = -\log(K_a)$. Using the definition of pH from Exercise 77 and the properties of logarithms, derive the Henderson-Hasselbalch Equation which states
    $$\text{pH} = pK_a + \log\dfrac{[A^-]}{[HA]}$$

44. Research the history of logarithms including the origin of the word 'logarithm' itself. Why is the abbreviation of natural log 'ln' and not 'nl'?

45. There is a scene in the movie 'Apollo 13' in which several people at Mission Control use slide rules to verify a computation. Was that scene accurate? Look for other pop culture references to logarithms and slide rules.

6.2 Properties of Logarithms

6.2.2 Answers

1. $3\ln(x) + 2\ln(y)$
2. $7 - \log_2(x^2 + 4)$
3. $3\log_5(z) - 6$
4. $\log(1.23) + 37$
5. $\frac{1}{2}\ln(z) - \ln(x) - \ln(y)$
6. $\log_5(x-5) + \log_5(x+5)$
7. $3\log_{\sqrt{2}}(x) + 4$
8. $-2 + \log_{\frac{1}{3}}(x) + \log_{\frac{1}{3}}(y-2) + \log_{\frac{1}{3}}(y^2 + 2y + 4)$
9. $3 + 3\log(x) + 5\log(y)$
10. $2\log_3(x) - 4 - 4\log_3(y)$
11. $\frac{1}{4}\ln(x) + \frac{1}{4}\ln(y) - \frac{1}{4} - \frac{1}{4}\ln(z)$
12. $12 - 12\log_6(x) - 4\log_6(y)$
13. $\frac{5}{3} + \log(x) + \frac{1}{2}\log(y)$
14. $-2 + \frac{2}{3}\log_{\frac{1}{2}}(x) - \log_{\frac{1}{2}}(y) - \frac{1}{2}\log_{\frac{1}{2}}(z)$
15. $\frac{1}{3}\ln(x) - \ln(10) - \frac{1}{2}\ln(y) - \frac{1}{2}\ln(z)$
16. $\ln(x^4 y^2)$
17. $\log_2\left(\frac{xy}{z}\right)$
18. $\log_3\left(\frac{x}{y^2}\right)$
19. $\log_3\left(\frac{\sqrt{x}}{y^2 z}\right)$
20. $\ln\left(\frac{x^2}{y^3 z^4}\right)$
21. $\log\left(\frac{x\sqrt{y}}{\sqrt[3]{z}}\right)$
22. $\ln\left(\sqrt[3]{\frac{z}{xy}}\right)$
23. $\log_5\left(\frac{x}{125}\right)$
24. $\log\left(\frac{1000}{x}\right)$
25. $\log_7\left(\frac{x(x-3)}{49}\right)$
26. $\ln\left(x\sqrt{e}\right)$
27. $\log_2\left(x^{3/2}\right)$
28. $\log_2\left(x\sqrt{x-1}\right)$
29. $\log_2\left(\frac{x}{x-1}\right)$
30. $7^{x-1} = e^{(x-1)\ln(7)}$
31. $\log_3(x+2) = \frac{\log(x+2)}{\log(3)}$
32. $\left(\frac{2}{3}\right)^x = e^{x\ln\left(\frac{2}{3}\right)}$
33. $\log(x^2 + 1) = \frac{\ln(x^2+1)}{\ln(10)}$
34. $\log_3(12) \approx 2.26186$
35. $\log_5(80) \approx 2.72271$
36. $\log_6(72) \approx 2.38685$
37. $\log_4\left(\frac{1}{10}\right) \approx -1.66096$
38. $\log_{\frac{3}{5}}(1000) \approx -13.52273$
39. $\log_{\frac{2}{3}}(50) \approx -9.64824$

6.3 Exponential Equations and Inequalities

In this section we will develop techniques for solving equations involving exponential functions. Suppose, for instance, we wanted to solve the equation $2^x = 128$. After a moment's calculation, we find $128 = 2^7$, so we have $2^x = 2^7$. The one-to-one property of exponential functions, detailed in Theorem 6.4, tells us that $2^x = 2^7$ if and only if $x = 7$. This means that not only is $x = 7$ a solution to $2^x = 2^7$, it is the *only* solution. Now suppose we change the problem ever so slightly to $2^x = 129$. We could use one of the inverse properties of exponentials and logarithms listed in Theorem 6.3 to write $129 = 2^{\log_2(129)}$. We'd then have $2^x = 2^{\log_2(129)}$, which means our solution is $x = \log_2(129)$. This makes sense because, after all, the definition of $\log_2(129)$ is 'the exponent we put on 2 to get 129.' Indeed we could have obtained this solution directly by rewriting the equation $2^x = 129$ in its logarithmic form $\log_2(129) = x$. Either way, in order to get a reasonable decimal approximation to this number, we'd use the change of base formula, Theorem 6.7, to give us something more calculator friendly,[1] say $\log_2(129) = \frac{\ln(129)}{\ln(2)}$. Another way to arrive at this answer is as follows

$$\begin{aligned} 2^x &= 129 \\ \ln\left(2^x\right) &= \ln(129) \quad \text{Take the natural log of both sides.} \\ x\ln(2) &= \ln(129) \quad \text{Power Rule} \\ x &= \frac{\ln(129)}{\ln(2)} \end{aligned}$$

'Taking the natural log' of both sides is akin to squaring both sides: since $f(x) = \ln(x)$ is a *function*, as long as two quantities are equal, their natural logs are equal.[2] Also note that we treat $\ln(2)$ as any other non-zero real number and divide it through[3] to isolate the variable x. We summarize below the two common ways to solve exponential equations, motivated by our examples.

Steps for Solving an Equation involving Exponential Functions

1. Isolate the exponential function.

2. (a) If convenient, express both sides with a common base and equate the exponents.

   (b) Otherwise, take the natural log of both sides of the equation and use the Power Rule.

Example 6.3.1. Solve the following equations. Check your answer graphically using a calculator.

1. $2^{3x} = 16^{1-x}$
2. $2000 = 1000 \cdot 3^{-0.1t}$
3. $9 \cdot 3^x = 7^{2x}$
4. $75 = \frac{100}{1+3e^{-2t}}$
5. $25^x = 5^x + 6$
6. $\frac{e^x - e^{-x}}{2} = 5$

Solution.

---
[1] You can use natural logs or common logs. We choose natural logs. (In Calculus, you'll learn these are the most 'mathy' of the logarithms.)

[2] This is also the 'if' part of the statement $\log_b(u) = \log_b(w)$ if and only if $u = w$ in Theorem 6.4.

[3] Please resist the temptation to divide both sides by 'ln' instead of $\ln(2)$. Just like it wouldn't make sense to divide both sides by the square root symbol '$\sqrt{}$' when solving $x\sqrt{2} = 5$, it makes no sense to divide by 'ln'.

1. Since 16 is a power of 2, we can rewrite $2^{3x} = 16^{1-x}$ as $2^{3x} = (2^4)^{1-x}$. Using properties of exponents, we get $2^{3x} = 2^{4(1-x)}$. Using the one-to-one property of exponential functions, we get $3x = 4(1-x)$ which gives $x = \frac{4}{7}$. To check graphically, we set $f(x) = 2^{3x}$ and $g(x) = 16^{1-x}$ and see that they intersect at $x = \frac{4}{7} \approx 0.5714$.

2. We begin solving $2000 = 1000 \cdot 3^{-0.1t}$ by dividing both sides by 1000 to isolate the exponential which yields $3^{-0.1t} = 2$. Since it is inconvenient to write 2 as a power of 3, we use the natural log to get $\ln\left(3^{-0.1t}\right) = \ln(2)$. Using the Power Rule, we get $-0.1t \ln(3) = \ln(2)$, so we divide both sides by $-0.1 \ln(3)$ to get $t = -\frac{\ln(2)}{0.1 \ln(3)} = -\frac{10 \ln(2)}{\ln(3)}$. On the calculator, we graph $f(x) = 2000$ and $g(x) = 1000 \cdot 3^{-0.1x}$ and find that they intersect at $x = -\frac{10 \ln(2)}{\ln(3)} \approx -6.3093$.

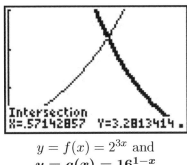
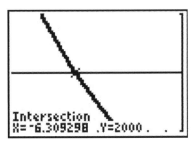

$y = f(x) = 2^{3x}$ and
$y = g(x) = 16^{1-x}$

$y = f(x) = 2000$ and
$y = g(x) = 1000 \cdot 3^{-0.1x}$

3. We first note that we can rewrite the equation $9 \cdot 3^x = 7^{2x}$ as $3^2 \cdot 3^x = 7^{2x}$ to obtain $3^{x+2} = 7^{2x}$. Since it is not convenient to express both sides as a power of 3 (or 7 for that matter) we use the natural log: $\ln\left(3^{x+2}\right) = \ln\left(7^{2x}\right)$. The power rule gives $(x+2) \ln(3) = 2x \ln(7)$. Even though this equation appears very complicated, keep in mind that $\ln(3)$ and $\ln(7)$ are just constants. The equation $(x+2) \ln(3) = 2x \ln(7)$ is actually a linear equation and as such we gather all of the terms with x on one side, and the constants on the other. We then divide both sides by the coefficient of x, which we obtain by factoring.

$$\begin{aligned} (x+2) \ln(3) &= 2x \ln(7) \\ x \ln(3) + 2 \ln(3) &= 2x \ln(7) \\ 2 \ln(3) &= 2x \ln(7) - x \ln(3) \\ 2 \ln(3) &= x(2 \ln(7) - \ln(3)) \quad \text{Factor.} \\ x &= \frac{2 \ln(3)}{2 \ln(7) - \ln(3)} \end{aligned}$$

Graphing $f(x) = 9 \cdot 3^x$ and $g(x) = 7^{2x}$ on the calculator, we see that these two graphs intersect at $x = \frac{2 \ln(3)}{2 \ln(7) - \ln(3)} \approx 0.7866$.

4. Our objective in solving $75 = \frac{100}{1 + 3e^{-2t}}$ is to first isolate the exponential. To that end, we clear denominators and get $75\left(1 + 3e^{-2t}\right) = 100$. From this we get $75 + 225e^{-2t} = 100$, which leads to $225e^{-2t} = 25$, and finally, $e^{-2t} = \frac{1}{9}$. Taking the natural log of both sides

gives $\ln\left(e^{-2t}\right) = \ln\left(\frac{1}{9}\right)$. Since natural log is log base e, $\ln\left(e^{-2t}\right) = -2t$. We can also use the Power Rule to write $\ln\left(\frac{1}{9}\right) = -\ln(9)$. Putting these two steps together, we simplify $\ln\left(e^{-2t}\right) = \ln\left(\frac{1}{9}\right)$ to $-2t = -\ln(9)$. We arrive at our solution, $t = \frac{\ln(9)}{2}$ which simplifies to $t = \ln(3)$. (Can you explain why?) The calculator confirms the graphs of $f(x) = 75$ and $g(x) = \frac{100}{1+3e^{-2x}}$ intersect at $x = \ln(3) \approx 1.099$.

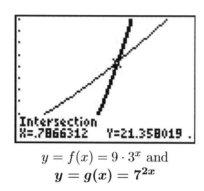
$y = f(x) = 9 \cdot 3^x$ and
$y = g(x) = 7^{2x}$

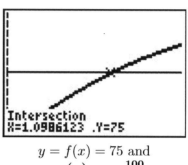
$y = f(x) = 75$ and
$y = g(x) = \frac{100}{1+3e^{-2x}}$

5. We start solving $25^x = 5^x + 6$ by rewriting $25 = 5^2$ so that we have $\left(5^2\right)^x = 5^x + 6$, or $5^{2x} = 5^x + 6$. Even though we have a common base, having two terms on the right hand side of the equation foils our plan of equating exponents or taking logs. If we stare at this long enough, we notice that we have three terms with the exponent on one term exactly twice that of another. To our surprise and delight, we have a 'quadratic in disguise'. Letting $u = 5^x$, we have $u^2 = (5^x)^2 = 5^{2x}$ so the equation $5^{2x} = 5^x + 6$ becomes $u^2 = u + 6$. Solving this as $u^2 - u - 6 = 0$ gives $u = -2$ or $u = 3$. Since $u = 5^x$, we have $5^x = -2$ or $5^x = 3$. Since $5^x = -2$ has no real solution, (Why not?) we focus on $5^x = 3$. Since it isn't convenient to express 3 as a power of 5, we take natural logs and get $\ln(5^x) = \ln(3)$ so that $x\ln(5) = \ln(3)$ or $x = \frac{\ln(3)}{\ln(5)}$. On the calculator, we see the graphs of $f(x) = 25^x$ and $g(x) = 5^x + 6$ intersect at $x = \frac{\ln(3)}{\ln(5)} \approx 0.6826$.

6. At first, it's unclear how to proceed with $\frac{e^x - e^{-x}}{2} = 5$, besides clearing the denominator to obtain $e^x - e^{-x} = 10$. Of course, if we rewrite $e^{-x} = \frac{1}{e^x}$, we see we have another denominator lurking in the problem: $e^x - \frac{1}{e^x} = 10$. Clearing this denominator gives us $e^{2x} - 1 = 10e^x$, and once again, we have an equation with three terms where the exponent on one term is exactly twice that of another - a 'quadratic in disguise.' If we let $u = e^x$, then $u^2 = e^{2x}$ so the equation $e^{2x} - 1 = 10e^x$ can be viewed as $u^2 - 1 = 10u$. Solving $u^2 - 10u - 1 = 0$, we obtain by the quadratic formula $u = 5 \pm \sqrt{26}$. From this, we have $e^x = 5 \pm \sqrt{26}$. Since $5 - \sqrt{26} < 0$, we get no real solution to $e^x = 5 - \sqrt{26}$, but for $e^x = 5 + \sqrt{26}$, we take natural logs to obtain $x = \ln\left(5 + \sqrt{26}\right)$. If we graph $f(x) = \frac{e^x - e^{-x}}{2}$ and $g(x) = 5$, we see that the graphs intersect at $x = \ln\left(5 + \sqrt{26}\right) \approx 2.312$

6.3 Exponential Equations and Inequalities

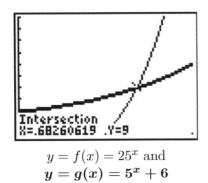
$y = f(x) = 25^x$ and
$y = g(x) = 5^x + 6$

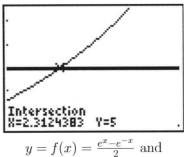
$y = f(x) = \frac{e^x - e^{-x}}{2}$ and
$y = g(x) = 5$

□

The authors would be remiss not to mention that Example 6.3.1 still holds great educational value. Much can be learned about logarithms and exponentials by verifying the solutions obtained in Example 6.3.1 analytically. For example, to verify our solution to $2000 = 1000 \cdot 3^{-0.1t}$, we substitute $t = -\frac{10 \ln(2)}{\ln(3)}$ and obtain

$$
\begin{aligned}
2000 &\stackrel{?}{=} 1000 \cdot 3^{-0.1\left(-\frac{10\ln(2)}{\ln(3)}\right)} \\
2000 &\stackrel{?}{=} 1000 \cdot 3^{\frac{\ln(2)}{\ln(3)}} \\
2000 &\stackrel{?}{=} 1000 \cdot 3^{\log_3(2)} \qquad \text{Change of Base} \\
2000 &\stackrel{?}{=} 1000 \cdot 2 \qquad \text{Inverse Property} \\
2000 &\stackrel{\checkmark}{=} 2000
\end{aligned}
$$

The other solutions can be verified by using a combination of log and inverse properties. Some fall out quite quickly, while others are more involved. We leave them to the reader.

Since exponential functions are continuous on their domains, the Intermediate Value Theorem 3.1 applies. As with the algebraic functions in Section 5.3, this allows us to solve inequalities using sign diagrams as demonstrated below.

Example 6.3.2. Solve the following inequalities. Check your answer graphically using a calculator.

1. $2^{x^2 - 3x} - 16 \geq 0$
2. $\dfrac{e^x}{e^x - 4} \leq 3$
3. $xe^{2x} < 4x$

Solution.

1. Since we already have 0 on one side of the inequality, we set $r(x) = 2^{x^2-3x} - 16$. The domain of r is all real numbers, so in order to construct our sign diagram, we need to find the zeros of r. Setting $r(x) = 0$ gives $2^{x^2-3x} - 16 = 0$ or $2^{x^2-3x} = 16$. Since $16 = 2^4$ we have $2^{x^2-3x} = 2^4$, so by the one-to-one property of exponential functions, $x^2 - 3x = 4$. Solving $x^2 - 3x - 4 = 0$ gives $x = 4$ and $x = -1$. From the sign diagram, we see $r(x) \geq 0$ on $(-\infty, -1] \cup [4, \infty)$, which corresponds to where the graph of $y = r(x) = 2^{x^2-3x} - 16$, is on or above the x-axis.

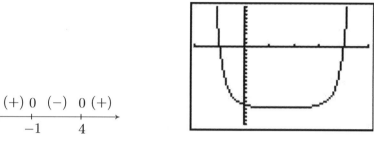

$$y = r(x) = 2^{x^2-3x} - 16$$

2. The first step we need to take to solve $\frac{e^x}{e^x-4} \leq 3$ is to get 0 on one side of the inequality. To that end, we subtract 3 from both sides and get a common denominator

$$\frac{e^x}{e^x - 4} \leq 3$$

$$\frac{e^x}{e^x - 4} - 3 \leq 0$$

$$\frac{e^x}{e^x - 4} - \frac{3(e^x - 4)}{e^x - 4} \leq 0 \quad \text{Common denomintors.}$$

$$\frac{12 - 2e^x}{e^x - 4} \leq 0$$

We set $r(x) = \frac{12-2e^x}{e^x-4}$ and we note that r is undefined when its denominator $e^x - 4 = 0$, or when $e^x = 4$. Solving this gives $x = \ln(4)$, so the domain of r is $(-\infty, \ln(4)) \cup (\ln(4), \infty)$. To find the zeros of r, we solve $r(x) = 0$ and obtain $12 - 2e^x = 0$. Solving for e^x, we find $e^x = 6$, or $x = \ln(6)$. When we build our sign diagram, finding test values may be a little tricky since we need to check values around $\ln(4)$ and $\ln(6)$. Recall that the function $\ln(x)$ is increasing[4] which means $\ln(3) < \ln(4) < \ln(5) < \ln(6) < \ln(7)$. While the prospect of determining the sign of $r(\ln(3))$ may be very unsettling, remember that $e^{\ln(3)} = 3$, so

$$r(\ln(3)) = \frac{12 - 2e^{\ln(3)}}{e^{\ln(3)} - 4} = \frac{12 - 2(3)}{3 - 4} = -6$$

We determine the signs of $r(\ln(5))$ and $r(\ln(7))$ similarly.[5] From the sign diagram, we find our answer to be $(-\infty, \ln(4)) \cup [\ln(6), \infty)$. Using the calculator, we see the graph of $f(x) = \frac{e^x}{e^x-4}$ is below the graph of $g(x) = 3$ on $(-\infty, \ln(4)) \cup (\ln(6), \infty)$, and they intersect at $x = \ln(6) \approx 1.792$.

---

[4]This is because the base of $\ln(x)$ is $e > 1$. If the base b were in the interval $0 < b < 1$, then $\log_b(x)$ would decreasing.
[5]We could, of course, use the calculator, but what fun would that be?

6.3 Exponential Equations and Inequalities

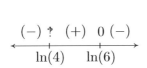

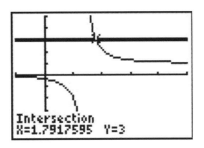

$$y = f(x) = \frac{e^x}{e^x - 4}$$
$$y = g(x) = 3$$

3. As before, we start solving $xe^{2x} < 4x$ by getting 0 on one side of the inequality, $xe^{2x} - 4x < 0$. We set $r(x) = xe^{2x} - 4x$ and since there are no denominators, even-indexed radicals, or logs, the domain of r is all real numbers. Setting $r(x) = 0$ produces $xe^{2x} - 4x = 0$. We factor to get $x\left(e^{2x} - 4\right) = 0$ which gives $x = 0$ or $e^{2x} - 4 = 0$. To solve the latter, we isolate the exponential and take logs to get $2x = \ln(4)$, or $x = \frac{\ln(4)}{2} = \ln(2)$. (Can you explain the last equality using properties of logs?) As in the previous example, we need to be careful about choosing test values. Since $\ln(1) = 0$, we choose $\ln\left(\frac{1}{2}\right)$, $\ln\left(\frac{3}{2}\right)$ and $\ln(3)$. Evaluating,[6] we get

$$\begin{aligned} r\left(\ln\left(\tfrac{1}{2}\right)\right) &= \ln\left(\tfrac{1}{2}\right) e^{2\ln\left(\tfrac{1}{2}\right)} - 4\ln\left(\tfrac{1}{2}\right) \\ &= \ln\left(\tfrac{1}{2}\right) e^{\ln\left(\tfrac{1}{2}\right)^2} - 4\ln\left(\tfrac{1}{2}\right) \qquad \text{Power Rule} \\ &= \ln\left(\tfrac{1}{2}\right) e^{\ln\left(\tfrac{1}{4}\right)} - 4\ln\left(\tfrac{1}{2}\right) \\ &= \tfrac{1}{4}\ln\left(\tfrac{1}{2}\right) - 4\ln\left(\tfrac{1}{2}\right) = -\tfrac{15}{4}\ln\left(\tfrac{1}{2}\right) \end{aligned}$$

Since $\frac{1}{2} < 1$, $\ln\left(\frac{1}{2}\right) < 0$ and we get $r(\ln\left(\frac{1}{2}\right))$ is $(+)$, so $r(x) < 0$ on $(0, \ln(2))$. The calculator confirms that the graph of $f(x) = xe^{2x}$ is below the graph of $g(x) = 4x$ on these intervals.[7]

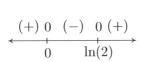

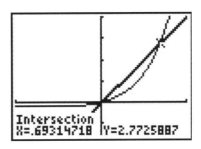

$$y = f(x) = xe^{2x} \text{ and } y = g(x) = 4x$$

□

---

[6] A calculator can be used at this point. As usual, we proceed without apologies, with the analytical method.
[7] Note: $\ln(2) \approx 0.693$.

Example 6.3.3. Recall from Example 6.1.2 that the temperature of coffee T (in degrees Fahrenheit) t minutes after it is served can be modeled by $T(t) = 70 + 90e^{-0.1t}$. When will the coffee be warmer than $100°F$?

Solution. We need to find when $T(t) > 100$, or in other words, we need to solve the inequality $70 + 90e^{-0.1t} > 100$. Getting 0 on one side of the inequality, we have $90e^{-0.1t} - 30 > 0$, and we set $r(t) = 90e^{-0.1t} - 30$. The domain of r is artificially restricted due to the context of the problem to $[0, \infty)$, so we proceed to find the zeros of r. Solving $90e^{-0.1t} - 30 = 0$ results in $e^{-0.1t} = \frac{1}{3}$ so that $t = -10 \ln\left(\frac{1}{3}\right)$ which, after a quick application of the Power Rule leaves us with $t = 10\ln(3)$. If we wish to avoid using the calculator to choose test values, we note that since $1 < 3$, $0 = \ln(1) < \ln(3)$ so that $10\ln(3) > 0$. So we choose $t = 0$ as a test value in $[0, 10\ln(3))$. Since $3 < 4$, $10\ln(3) < 10\ln(4)$, so the latter is our choice of a test value for the interval $(10\ln(3), \infty)$. Our sign diagram is below, and next to it is our graph of $y = T(t)$ from Example 6.1.2 with the horizontal line $y = 100$.

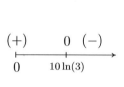
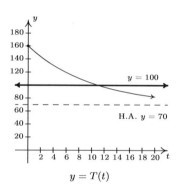

$y = T(t)$

In order to interpret what this means in the context of the real world, we need a reasonable approximation of the number $10 \ln(3) \approx 10.986$. This means it takes approximately 11 minutes for the coffee to cool to $100°F$. Until then, the coffee is warmer than that.[8] □

We close this section by finding the inverse of a function which is a composition of a rational function with an exponential function.

Example 6.3.4. The function $f(x) = \dfrac{5e^x}{e^x + 1}$ is one-to-one. Find a formula for $f^{-1}(x)$ and check your answer graphically using your calculator.

Solution. We start by writing $y = f(x)$, and interchange the roles of x and y. To solve for y, we first clear denominators and then isolate the exponential function.

---

[8]Critics may point out that since we needed to use the calculator to interpret our answer anyway, why not use it earlier to simplify the computations? It is a fair question which we answer unfairly: it's our book.

6.3 Exponential Equations and Inequalities

$$y = \frac{5e^x}{e^x + 1}$$

$$x = \frac{5e^y}{e^y + 1} \qquad \text{Switch } x \text{ and } y$$

$$x(e^y + 1) = 5e^y$$

$$xe^y + x = 5e^y$$

$$x = 5e^y - xe^y$$

$$x = e^y(5 - x)$$

$$e^y = \frac{x}{5 - x}$$

$$\ln(e^y) = \ln\left(\frac{x}{5 - x}\right)$$

$$y = \ln\left(\frac{x}{5 - x}\right)$$

We claim $f^{-1}(x) = \ln\left(\frac{x}{5-x}\right)$. To verify this analytically, we would need to verify the compositions $\left(f^{-1} \circ f\right)(x) = x$ for all x in the domain of f and that $\left(f \circ f^{-1}\right)(x) = x$ for all x in the domain of f^{-1}. We leave this to the reader. To verify our solution graphically, we graph $y = f(x) = \frac{5e^x}{e^x+1}$ and $y = g(x) = \ln\left(\frac{x}{5-x}\right)$ on the same set of axes and observe the symmetry about the line $y = x$. Note the domain of f is the range of g and vice-versa.

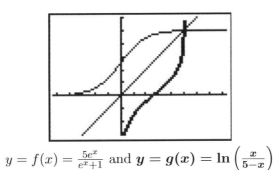

$y = f(x) = \frac{5e^x}{e^x+1}$ and $y = g(x) = \ln\left(\frac{x}{5-x}\right)$

□

6.3.1 Exercises

In Exercises 1 - 33, solve the equation analytically.

1. $2^{4x} = 8$
2. $3^{(x-1)} = 27$
3. $5^{2x-1} = 125$
4. $4^{2x} = \frac{1}{2}$
5. $8^x = \frac{1}{128}$
6. $2^{(x^3-x)} = 1$
7. $3^{7x} = 81^{4-2x}$
8. $9 \cdot 3^{7x} = \left(\frac{1}{9}\right)^{2x}$
9. $3^{2x} = 5$
10. $5^{-x} = 2$
11. $5^x = -2$
12. $3^{(x-1)} = 29$
13. $(1.005)^{12x} = 3$
14. $e^{-5730k} = \frac{1}{2}$
15. $2000e^{0.1t} = 4000$
16. $500\left(1 - e^{2x}\right) = 250$
17. $70 + 90e^{-0.1t} = 75$
18. $30 - 6e^{-0.1x} = 20$
19. $\dfrac{100e^x}{e^x + 2} = 50$
20. $\dfrac{5000}{1 + 2e^{-3t}} = 2500$
21. $\dfrac{150}{1 + 29e^{-0.8t}} = 75$
22. $25\left(\frac{4}{5}\right)^x = 10$
23. $e^{2x} = 2e^x$
24. $7e^{2x} = 28e^{-6x}$
25. $3^{(x-1)} = 2^x$
26. $3^{(x-1)} = \left(\frac{1}{2}\right)^{(x+5)}$
27. $7^{3+7x} = 3^{4-2x}$
28. $e^{2x} - 3e^x - 10 = 0$
29. $e^{2x} = e^x + 6$
30. $4^x + 2^x = 12$
31. $e^x - 3e^{-x} = 2$
32. $e^x + 15e^{-x} = 8$
33. $3^x + 25 \cdot 3^{-x} = 10$

In Exercises 34 - 39, solve the inequality analytically.

34. $e^x > 53$
35. $1000(1.005)^{12t} \geq 3000$
36. $2^{(x^3-x)} < 1$
37. $25\left(\frac{4}{5}\right)^x \geq 10$
38. $\dfrac{150}{1 + 29e^{-0.8t}} \leq 130$
39. $70 + 90e^{-0.1t} \leq 75$

In Exercises 40 - 45, use your calculator to help you solve the equation or inequality.

40. $2^x = x^2$
41. $e^x = \ln(x) + 5$
42. $e^{\sqrt{x}} = x + 1$
43. $e^{-x} - xe^{-x} \geq 0$
44. $3^{(x-1)} < 2^x$
45. $e^x < x^3 - x$

46. Since $f(x) = \ln(x)$ is a strictly increasing function, if $0 < a < b$ then $\ln(a) < \ln(b)$. Use this fact to solve the inequality $e^{(3x-1)} > 6$ without a sign diagram. Use this technique to solve the inequalities in Exercises 34 - 39. (NOTE: Isolate the exponential function first!)

47. Compute the inverse of $f(x) = \dfrac{e^x - e^{-x}}{2}$. State the domain and range of both f and f^{-1}.

6.3 Exponential Equations and Inequalities

48. In Example 6.3.4, we found that the inverse of $f(x) = \dfrac{5e^x}{e^x + 1}$ was $f^{-1}(x) = \ln\left(\dfrac{x}{5-x}\right)$ but we left a few loose ends for you to tie up.

    (a) Show that $\left(f^{-1} \circ f\right)(x) = x$ for all x in the domain of f and that $\left(f \circ f^{-1}\right)(x) = x$ for all x in the domain of f^{-1}.

    (b) Find the range of f by finding the domain of f^{-1}.

    (c) Let $g(x) = \dfrac{5x}{x+1}$ and $h(x) = e^x$. Show that $f = g \circ h$ and that $(g \circ h)^{-1} = h^{-1} \circ g^{-1}$.
    (We know this is true in general by Exercise 31 in Section 5.2, but it's nice to see a specific example of the property.)

49. With the help of your classmates, solve the inequality $e^x > x^n$ for a variety of natural numbers n. What might you conjecture about the "speed" at which $f(x) = e^x$ grows versus any polynomial?

6.3.2 Answers

1. $x = \frac{3}{4}$
2. $x = 4$
3. $x = 2$
4. $x = -\frac{1}{4}$
5. $x = -\frac{7}{3}$
6. $x = -1, 0, 1$
7. $x = \frac{16}{15}$
8. $x = -\frac{2}{11}$
9. $x = \frac{\ln(5)}{2\ln(3)}$
10. $x = -\frac{\ln(2)}{\ln(5)}$
11. No solution.
12. $x = \frac{\ln(29)+\ln(3)}{\ln(3)}$
13. $x = \frac{\ln(3)}{12\ln(1.005)}$
14. $k = \frac{\ln\left(\frac{1}{2}\right)}{-5730} = \frac{\ln(2)}{5730}$
15. $t = \frac{\ln(2)}{0.1} = 10\ln(2)$
16. $x = \frac{1}{2}\ln\left(\frac{1}{2}\right) = -\frac{1}{2}\ln(2)$
17. $t = \frac{\ln\left(\frac{1}{18}\right)}{-0.1} = 10\ln(18)$
18. $x = -10\ln\left(\frac{5}{3}\right) = 10\ln\left(\frac{3}{5}\right)$
19. $x = \ln(2)$
20. $t = \frac{1}{3}\ln(2)$
21. $t = \frac{\ln\left(\frac{1}{29}\right)}{-0.8} = \frac{5}{4}\ln(29)$
22. $x = \frac{\ln\left(\frac{2}{5}\right)}{\ln\left(\frac{4}{5}\right)} = \frac{\ln(2)-\ln(5)}{\ln(4)-\ln(5)}$
23. $x = \ln(2)$
24. $x = -\frac{1}{8}\ln\left(\frac{1}{4}\right) = \frac{1}{4}\ln(2)$
25. $x = \frac{\ln(3)}{\ln(3)-\ln(2)}$
26. $x = \frac{\ln(3)+5\ln\left(\frac{1}{2}\right)}{\ln(3)-\ln\left(\frac{1}{2}\right)} = \frac{\ln(3)-5\ln(2)}{\ln(3)+\ln(2)}$
27. $x = \frac{4\ln(3)-3\ln(7)}{7\ln(7)+2\ln(3)}$
28. $x = \ln(5)$
29. $x = \ln(3)$
30. $x = \frac{\ln(3)}{\ln(2)}$
31. $x = \ln(3)$
32. $x = \ln(3), \ln(5)$
33. $x = \frac{\ln(5)}{\ln(3)}$
34. $(\ln(53), \infty)$
35. $\left[\frac{\ln(3)}{12\ln(1.005)}, \infty\right)$
36. $(-\infty, -1) \cup (0, 1)$
37. $\left(-\infty, \frac{\ln\left(\frac{2}{5}\right)}{\ln\left(\frac{4}{5}\right)}\right] = \left(-\infty, \frac{\ln(2)-\ln(5)}{\ln(4)-\ln(5)}\right]$
38. $\left(-\infty, \frac{\ln\left(\frac{2}{377}\right)}{-0.8}\right] = \left(-\infty, \frac{5}{4}\ln\left(\frac{377}{2}\right)\right]$
39. $\left[\frac{\ln\left(\frac{1}{18}\right)}{-0.1}, \infty\right) = [10\ln(18), \infty)$
40. $x \approx -0.76666, x = 2, x = 4$
41. $x \approx 0.01866, x \approx 1.7115$
42. $x = 0$
43. $(-\infty, 1]$
44. $\approx (-\infty, 2.7095)$
45. $\approx (2.3217, 4.3717)$
46. $x > \frac{1}{3}(\ln(6) + 1)$
47. $f^{-1} = \ln\left(x + \sqrt{x^2 + 1}\right)$. Both f and f^{-1} have domain $(-\infty, \infty)$ and range $(-\infty, \infty)$.

6.4 Logarithmic Equations and Inequalities

In Section 6.3 we solved equations and inequalities involving exponential functions using one of two basic strategies. We now turn our attention to equations and inequalities involving logarithmic functions, and not surprisingly, there are two basic strategies to choose from. For example, suppose we wish to solve $\log_2(x) = \log_2(5)$. Theorem 6.4 tells us that the *only* solution to this equation is $x = 5$. Now suppose we wish to solve $\log_2(x) = 3$. If we want to use Theorem 6.4, we need to rewrite 3 as a logarithm base 2. We can use Theorem 6.3 to do just that: $3 = \log_2\left(2^3\right) = \log_2(8)$. Our equation then becomes $\log_2(x) = \log_2(8)$ so that $x = 8$. However, we could have arrived at the same answer, in fewer steps, by using Theorem 6.3 to rewrite the equation $\log_2(x) = 3$ as $2^3 = x$, or $x = 8$. We summarize the two common ways to solve log equations below.

Steps for Solving an Equation involving Logarithmic Functions

1. Isolate the logarithmic function.

2. (a) If convenient, express both sides as logs with the same base and equate the arguments of the log functions.

   (b) Otherwise, rewrite the log equation as an exponential equation.

Example 6.4.1. Solve the following equations. Check your solutions graphically using a calculator.

1. $\log_{117}(1 - 3x) = \log_{117}\left(x^2 - 3\right)$

2. $2 - \ln(x - 3) = 1$

3. $\log_6(x + 4) + \log_6(3 - x) = 1$

4. $\log_7(1 - 2x) = 1 - \log_7(3 - x)$

5. $\log_2(x + 3) = \log_2(6 - x) + 3$

6. $1 + 2\log_4(x + 1) = 2\log_2(x)$

Solution.

1. Since we have the same base on both sides of the equation $\log_{117}(1 - 3x) = \log_{117}\left(x^2 - 3\right)$, we equate what's inside the logs to get $1 - 3x = x^2 - 3$. Solving $x^2 + 3x - 4 = 0$ gives $x = -4$ and $x = 1$. To check these answers using the calculator, we make use of the change of base formula and graph $f(x) = \frac{\ln(1-3x)}{\ln(117)}$ and $g(x) = \frac{\ln\left(x^2-3\right)}{\ln(117)}$ and we see they intersect only at $x = -4$. To see what happened to the solution $x = 1$, we substitute it into our original equation to obtain $\log_{117}(-2) = \log_{117}(-2)$. While these expressions look identical, neither is a real number,[1] which means $x = 1$ is not in the domain of the original equation, and is not a solution.

2. Our first objective in solving $2 - \ln(x - 3) = 1$ is to isolate the logarithm. We get $\ln(x - 3) = 1$, which, as an exponential equation, is $e^1 = x - 3$. We get our solution $x = e + 3$. On the calculator, we see the graph of $f(x) = 2 - \ln(x - 3)$ intersects the graph of $g(x) = 1$ at $x = e + 3 \approx 5.718$.

---

[1]They do, however, represent the same **family** of complex numbers. We stop ourselves at this point and refer the reader to a good course in Complex Variables.

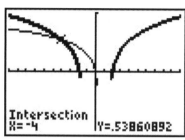
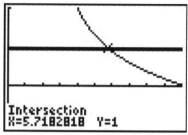

$y = f(x) = \log_{117}(1 - 3x)$ and
$y = g(x) = \log_{117}\left(x^2 - 3\right)$

$y = f(x) = 2 - \ln(x - 3)$ and
$y = g(x) = 1$

3. We can start solving $\log_6(x+4) + \log_6(3-x) = 1$ by using the Product Rule for logarithms to rewrite the equation as $\log_6\left[(x+4)(3-x)\right] = 1$. Rewriting this as an exponential equation, we get $6^1 = (x+4)(3-x)$. This reduces to $x^2 + x - 6 = 0$, which gives $x = -3$ and $x = 2$. Graphing $y = f(x) = \frac{\ln(x+4)}{\ln(6)} + \frac{\ln(3-x)}{\ln(6)}$ and $y = g(x) = 1$, we see they intersect twice, at $x = -3$ and $x = 2$.

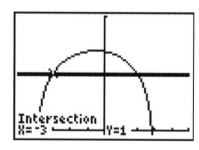
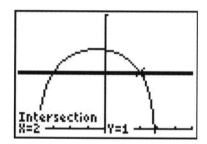

$y = f(x) = \log_6(x + 4) + \log_6(3 - x)$ and $y = g(x) = 1$

4. Taking a cue from the previous problem, we begin solving $\log_7(1 - 2x) = 1 - \log_7(3 - x)$ by first collecting the logarithms on the same side, $\log_7(1 - 2x) + \log_7(3 - x) = 1$, and then using the Product Rule to get $\log_7[(1 - 2x)(3 - x)] = 1$. Rewriting this as an exponential equation gives $7^1 = (1 - 2x)(3 - x)$ which gives the quadratic equation $2x^2 - 7x - 4 = 0$. Solving, we find $x = -\frac{1}{2}$ and $x = 4$. Graphing, we find $y = f(x) = \frac{\ln(1-2x)}{\ln(7)}$ and $y = g(x) = 1 - \frac{\ln(3-x)}{\ln(7)}$ intersect only at $x = -\frac{1}{2}$. Checking $x = 4$ in the original equation produces $\log_7(-7) = 1 - \log_7(-1)$, which is a clear domain violation.

5. Starting with $\log_2(x + 3) = \log_2(6 - x) + 3$, we gather the logarithms to one side and get $\log_2(x + 3) - \log_2(6 - x) = 3$. We then use the Quotient Rule and convert to an exponential equation

$$\log_2\left(\frac{x+3}{6-x}\right) = 3 \iff 2^3 = \frac{x+3}{6-x}$$

This reduces to the linear equation $8(6 - x) = x + 3$, which gives us $x = 5$. When we graph $f(x) = \frac{\ln(x+3)}{\ln(2)}$ and $g(x) = \frac{\ln(6-x)}{\ln(2)} + 3$, we find they intersect at $x = 5$.

6.4 Logarithmic Equations and Inequalities

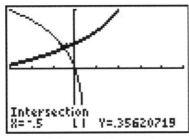

$y = f(x) = \log_7(1 - 2x)$ and
$y = g(x) = 1 - \log_7(3 - x)$

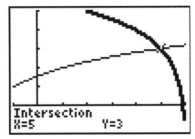

$y = f(x) = \log_2(x + 3)$ and
$y = g(x) = \log_2(6 - x) + 3$

6. Starting with $1 + 2\log_4(x+1) = 2\log_2(x)$, we gather the logs to one side to get the equation $1 = 2\log_2(x) - 2\log_4(x+1)$. Before we can combine the logarithms, however, we need a common base. Since 4 is a power of 2, we use change of base to convert

$$\log_4(x+1) = \frac{\log_2(x+1)}{\log_2(4)} = \frac{1}{2}\log_2(x+1)$$

Hence, our original equation becomes

$$\begin{array}{rll} 1 &=& 2\log_2(x) - 2\left(\frac{1}{2}\log_2(x+1)\right) \\ 1 &=& 2\log_2(x) - \log_2(x+1) \\ 1 &=& \log_2\left(x^2\right) - \log_2(x+1) \qquad \text{Power Rule} \\ 1 &=& \log_2\left(\dfrac{x^2}{x+1}\right) \qquad \text{Quotient Rule} \end{array}$$

Rewriting this in exponential form, we get $\frac{x^2}{x+1} = 2$ or $x^2 - 2x - 2 = 0$. Using the quadratic formula, we get $x = 1 \pm \sqrt{3}$. Graphing $f(x) = 1 + \frac{2\ln(x+1)}{\ln(4)}$ and $g(x) = \frac{2\ln(x)}{\ln(2)}$, we see the graphs intersect only at $x = 1 + \sqrt{3} \approx 2.732$. The solution $x = 1 - \sqrt{3} < 0$, which means if substituted into the original equation, the term $2\log_2\left(1 - \sqrt{3}\right)$ is undefined.

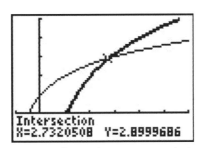

$y = f(x) = 1 + 2\log_4(x + 1)$ and $y = g(x) = 2\log_2(x)$

□

If nothing else, Example 6.4.1 demonstrates the importance of checking for extraneous solutions[2] when solving equations involving logarithms. Even though we checked our answers graphically, extraneous solutions are easy to spot - any supposed solution which causes a negative number inside a logarithm needs to be discarded. As with the equations in Example 6.3.1, much can be learned from checking all of the answers in Example 6.4.1 analytically. We leave this to the reader and turn our attention to inequalities involving logarithmic functions. Since logarithmic functions are continuous on their domains, we can use sign diagrams.

Example 6.4.2. Solve the following inequalities. Check your answer graphically using a calculator.

1. $\dfrac{1}{\ln(x) + 1} \leq 1$
2. $(\log_2(x))^2 < 2\log_2(x) + 3$
3. $x\log(x+1) \geq x$

Solution.

1. We start solving $\frac{1}{\ln(x)+1} \leq 1$ by getting 0 on one side of the inequality: $\frac{1}{\ln(x)+1} - 1 \leq 0$. Getting a common denominator yields $\frac{1}{\ln(x)+1} - \frac{\ln(x)+1}{\ln(x)+1} \leq 0$ which reduces to $\frac{-\ln(x)}{\ln(x)+1} \leq 0$, or $\frac{\ln(x)}{\ln(x)+1} \geq 0$. We define $r(x) = \frac{\ln(x)}{\ln(x)+1}$ and set about finding the domain and the zeros of r. Due to the appearance of the term $\ln(x)$, we require $x > 0$. In order to keep the denominator away from zero, we solve $\ln(x) + 1 = 0$ so $\ln(x) = -1$, so $x = e^{-1} = \frac{1}{e}$. Hence, the domain of r is $\left(0, \frac{1}{e}\right) \cup \left(\frac{1}{e}, \infty\right)$. To find the zeros of r, we set $r(x) = \frac{\ln(x)}{\ln(x)+1} = 0$ so that $\ln(x) = 0$, and we find $x = e^0 = 1$. In order to determine test values for r without resorting to the calculator, we need to find numbers between 0, $\frac{1}{e}$, and 1 which have a base of e. Since $e \approx 2.718 > 1$, $0 < \frac{1}{e^2} < \frac{1}{e} < \frac{1}{\sqrt{e}} < 1 < e$. To determine the sign of $r\left(\frac{1}{e^2}\right)$, we use the fact that $\ln\left(\frac{1}{e^2}\right) = \ln\left(e^{-2}\right) = -2$, and find $r\left(\frac{1}{e^2}\right) = \frac{-2}{-2+1} = 2$, which is $(+)$. The rest of the test values are determined similarly. From our sign diagram, we find the solution to be $\left(0, \frac{1}{e}\right) \cup [1, \infty)$. Graphing $f(x) = \frac{1}{\ln(x)+1}$ and $g(x) = 1$, we see the graph of f is below the graph of g on the solution intervals, and that the graphs intersect at $x = 1$.

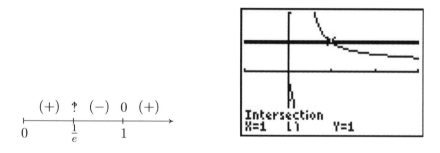

$y = f(x) = \frac{1}{\ln(x)+1}$ and $y = g(x) = 1$

---

[2]Recall that an extraneous solution is an answer obtained analytically which does not satisfy the original equation.

6.4 Logarithmic Equations and Inequalities

2. Moving all of the nonzero terms of $(\log_2(x))^2 < 2\log_2(x) + 3$ to one side of the inequality, we have $(\log_2(x))^2 - 2\log_2(x) - 3 < 0$. Defining $r(x) = (\log_2(x))^2 - 2\log_2(x) - 3$, we get the domain of r is $(0, \infty)$, due to the presence of the logarithm. To find the zeros of r, we set $r(x) = (\log_2(x))^2 - 2\log_2(x) - 3 = 0$ which results in a 'quadratic in disguise.' We set $u = \log_2(x)$ so our equation becomes $u^2 - 2u - 3 = 0$ which gives us $u = -1$ and $u = 3$. Since $u = \log_2(x)$, we get $\log_2(x) = -1$, which gives us $x = 2^{-1} = \frac{1}{2}$, and $\log_2(x) = 3$, which yields $x = 2^3 = 8$. We use test values which are powers of 2: $0 < \frac{1}{4} < \frac{1}{2} < 1 < 8 < 16$, and from our sign diagram, we see $r(x) < 0$ on $\left(\frac{1}{2}, 8\right)$. Geometrically, we see the graph of $f(x) = \left(\frac{\ln(x)}{\ln(2)}\right)^2$ is below the graph of $y = g(x) = \frac{2\ln(x)}{\ln(2)} + 3$ on the solution interval.

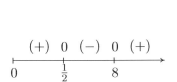

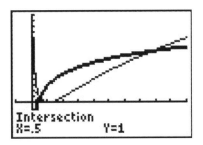

$y = f(x) = (\log_2(x))^2$ and $y = g(x) = 2\log_2(x) + 3$

3. We begin to solve $x\log(x+1) \geq x$ by subtracting x from both sides to get $x\log(x+1) - x \geq 0$. We define $r(x) = x\log(x+1) - x$ and due to the presence of the logarithm, we require $x+1 > 0$, or $x > -1$. To find the zeros of r, we set $r(x) = x\log(x+1) - x = 0$. Factoring, we get $x(\log(x+1) - 1) = 0$, which gives $x = 0$ or $\log(x+1) - 1 = 0$. The latter gives $\log(x+1) = 1$, or $x + 1 = 10^1$, which admits $x = 9$. We select test values x so that $x+1$ is a power of 10, and we obtain $-1 < -0.9 < 0 < \sqrt{10} - 1 < 9 < 99$. Our sign diagram gives the solution to be $(-1, 0] \cup [9, \infty)$. The calculator indicates the graph of $y = f(x) = x\log(x+1)$ is above $y = g(x) = x$ on the solution intervals, and the graphs intersect at $x = 0$ and $x = 9$.

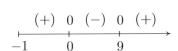

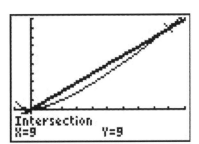

$y = f(x) = x\log(x+1)$ and $y = g(x) = x$

□

Our next example revisits the concept of pH first seen in Exercise 77 in Section 6.1.

Example 6.4.3. In order to successfully breed Ippizuti fish the pH of a freshwater tank must be at least 7.8 but can be no more than 8.5. Determine the corresponding range of hydrogen ion concentration, and check your answer using a calculator.

Solution. Recall from Exercise 77 in Section 6.1 that pH $= -\log[\text{H}^+]$ where $[\text{H}^+]$ is the hydrogen ion concentration in moles per liter. We require $7.8 \leq -\log[\text{H}^+] \leq 8.5$ or $-7.8 \geq \log[\text{H}^+] \geq -8.5$. To solve this compound inequality we solve $-7.8 \geq \log[\text{H}^+]$ and $\log[\text{H}^+] \geq -8.5$ and take the intersection of the solution sets.[3] The former inequality yields $0 < [\text{H}^+] \leq 10^{-7.8}$ and the latter yields $[\text{H}^+] \geq 10^{-8.5}$. Taking the intersection gives us our final answer $10^{-8.5} \leq [\text{H}^+] \leq 10^{-7.8}$. (Your Chemistry professor may want the answer written as $3.16 \times 10^{-9} \leq [\text{H}^+] \leq 1.58 \times 10^{-8}$.) After carefully adjusting the viewing window on the graphing calculator we see that the graph of $f(x) = -\log(x)$ lies between the lines $y = 7.8$ and $y = 8.5$ on the interval $[3.16 \times 10^{-9}, 1.58 \times 10^{-8}]$.

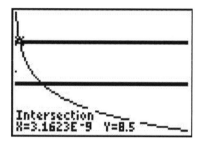 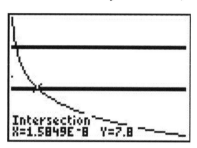

The graphs of $y = f(x) = -\log(x)$, $y = 7.8$ and $y = 8.5$

□

We close this section by finding an inverse of a one-to-one function which involves logarithms.

Example 6.4.4. The function $f(x) = \dfrac{\log(x)}{1 - \log(x)}$ is one-to-one. Find a formula for $f^{-1}(x)$ and check your answer graphically using your calculator.

Solution. We first write $y = f(x)$ then interchange the x and y and solve for y.

$$
\begin{aligned}
y &= f(x) \\
y &= \frac{\log(x)}{1 - \log(x)} \\
x &= \frac{\log(y)}{1 - \log(y)} \qquad &&\text{Interchange } x \text{ and } y. \\
x(1 - \log(y)) &= \log(y) \\
x - x\log(y) &= \log(y) \\
x &= x\log(y) + \log(y) \\
x &= (x+1)\log(y) \\
\frac{x}{x+1} &= \log(y) \\
y &= 10^{\frac{x}{x+1}} \qquad &&\text{Rewrite as an exponential equation.}
\end{aligned}
$$

---

[3] Refer to page 4 for a discussion of what this means.

We have $f^{-1}(x) = 10^{\frac{x}{x+1}}$. Graphing f and f^{-1} on the same viewing window yields

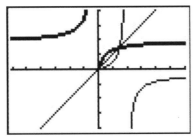

$$y = f(x) = \frac{\log(x)}{1 - \log(x)} \text{ and } y = g(x) = 10^{\frac{x}{x+1}}$$

□

6.4.1 Exercises

In Exercises 1 - 24, solve the equation analytically.

1. $\log(3x - 1) = \log(4 - x)$

2. $\log_2(x^3) = \log_2(x)$

3. $\ln(8 - x^2) = \ln(2 - x)$

4. $\log_5(18 - x^2) = \log_5(6 - x)$

5. $\log_3(7 - 2x) = 2$

6. $\log_{\frac{1}{2}}(2x - 1) = -3$

7. $\ln(x^2 - 99) = 0$

8. $\log(x^2 - 3x) = 1$

9. $\log_{125}\left(\dfrac{3x - 2}{2x + 3}\right) = \dfrac{1}{3}$

10. $\log\left(\dfrac{x}{10^{-3}}\right) = 4.7$

11. $-\log(x) = 5.4$

12. $10\log\left(\dfrac{x}{10^{-12}}\right) = 150$

13. $6 - 3\log_5(2x) = 0$

14. $3\ln(x) - 2 = 1 - \ln(x)$

15. $\log_3(x - 4) + \log_3(x + 4) = 2$

16. $\log_5(2x + 1) + \log_5(x + 2) = 1$

17. $\log_{169}(3x + 7) - \log_{169}(5x - 9) = \dfrac{1}{2}$

18. $\ln(x + 1) - \ln(x) = 3$

19. $2\log_7(x) = \log_7(2) + \log_7(x + 12)$

20. $\log(x) - \log(2) = \log(x + 8) - \log(x + 2)$

21. $\log_3(x) = \log_{\frac{1}{3}}(x) + 8$

22. $\ln(\ln(x)) = 3$

23. $(\log(x))^2 = 2\log(x) + 15$

24. $\ln(x^2) = (\ln(x))^2$

In Exercises 25 - 30, solve the inequality analytically.

25. $\dfrac{1 - \ln(x)}{x^2} < 0$

26. $x\ln(x) - x > 0$

27. $10\log\left(\dfrac{x}{10^{-12}}\right) \geq 90$

28. $5.6 \leq \log\left(\dfrac{x}{10^{-3}}\right) \leq 7.1$

29. $2.3 < -\log(x) < 5.4$

30. $\ln(x^2) \leq (\ln(x))^2$

In Exercises 31 - 34, use your calculator to help you solve the equation or inequality.

31. $\ln(x) = e^{-x}$

32. $\ln(x) = \sqrt[4]{x}$

33. $\ln(x^2 + 1) \geq 5$

34. $\ln(-2x^3 - x^2 + 13x - 6) < 0$

6.4 LOGARITHMIC EQUATIONS AND INEQUALITIES

35. Since $f(x) = e^x$ is a strictly increasing function, if $a < b$ then $e^a < e^b$. Use this fact to solve the inequality $\ln(2x + 1) < 3$ without a sign diagram. Use this technique to solve the inequalities in Exercises 27 - 29. (Compare this to Exercise 46 in Section 6.3.)

36. Solve $\ln(3 - y) - \ln(y) = 2x + \ln(5)$ for y.

37. In Example 6.4.4 we found the inverse of $f(x) = \dfrac{\log(x)}{1 - \log(x)}$ to be $f^{-1}(x) = 10^{\frac{x}{x+1}}$.

    (a) Show that $\left(f^{-1} \circ f\right)(x) = x$ for all x in the domain of f and that $\left(f \circ f^{-1}\right)(x) = x$ for all x in the domain of f^{-1}.

    (b) Find the range of f by finding the domain of f^{-1}.

    (c) Let $g(x) = \dfrac{x}{1 - x}$ and $h(x) = \log(x)$. Show that $f = g \circ h$ and $(g \circ h)^{-1} = h^{-1} \circ g^{-1}$.
    (We know this is true in general by Exercise 31 in Section 5.2, but it's nice to see a specific example of the property.)

38. Let $f(x) = \dfrac{1}{2} \ln\left(\dfrac{1+x}{1-x}\right)$. Compute $f^{-1}(x)$ and find its domain and range.

39. Explain the equation in Exercise 10 and the inequality in Exercise 28 above in terms of the Richter scale for earthquake magnitude. (See Exercise 75 in Section 6.1.)

40. Explain the equation in Exercise 12 and the inequality in Exercise 27 above in terms of sound intensity level as measured in decibels. (See Exercise 76 in Section 6.1.)

41. Explain the equation in Exercise 11 and the inequality in Exercise 29 above in terms of the pH of a solution. (See Exercise 77 in Section 6.1.)

42. With the help of your classmates, solve the inequality $\sqrt[n]{x} > \ln(x)$ for a variety of natural numbers n. What might you conjecture about the "speed" at which $f(x) = \ln(x)$ grows versus any principal n^{th} root function?

6.4.2 Answers

1. $x = \frac{5}{4}$
2. $x = 1$
3. $x = -2$
4. $x = -3, 4$
5. $x = -1$
6. $x = \frac{9}{2}$
7. $x = \pm 10$
8. $x = -2, 5$
9. $x = -\frac{17}{7}$
10. $x = 10^{1.7}$
11. $x = 10^{-5.4}$
12. $x = 10^3$
13. $x = \frac{25}{2}$
14. $x = e^{3/4}$
15. $x = 5$
16. $x = \frac{1}{2}$
17. $x = 2$
18. $x = \frac{1}{e^3-1}$
19. $x = 6$
20. $x = 4$
21. $x = 81$
22. $x = e^{e^3}$
23. $x = 10^{-3}, 10^5$
24. $x = 1, x = e^2$
25. (e, ∞)
26. (e, ∞)
27. $[10^{-3}, \infty)$
28. $[10^{2.6}, 10^{4.1}]$
29. $(10^{-5.4}, 10^{-2.3})$
30. $(0, 1] \cup [e^2, \infty)$
31. $x \approx 1.3098$
32. $x \approx 4.177, x \approx 5503.665$
33. $\approx (-\infty, -12.1414) \cup (12.1414, \infty)$
34. $\approx (-3.0281, -3) \cup (0.5, 0.5991) \cup (1.9299, 2)$
35. $-\frac{1}{2} < x < \frac{e^3-1}{2}$
36. $y = \dfrac{3}{5e^{2x}+1}$
38. $f^{-1}(x) = \dfrac{e^{2x}-1}{e^{2x}+1} = \dfrac{e^x - e^{-x}}{e^x + e^{-x}}$. (To see why we rewrite this in this form, see Exercise 51 in Section 11.10.) The domain of f^{-1} is $(-\infty, \infty)$ and its range is the same as the domain of f, namely $(-1, 1)$.

6.5 Applications of Exponential and Logarithmic Functions

As we mentioned in Section 6.1, exponential and logarithmic functions are used to model a wide variety of behaviors in the real world. In the examples that follow, note that while the applications are drawn from many different disciplines, the mathematics remains essentially the same. Due to the applied nature of the problems we will examine in this section, the calculator is often used to express our answers as decimal approximations.

6.5.1 Applications of Exponential Functions

Perhaps the most well-known application of exponential functions comes from the financial world. Suppose you have $100 to invest at your local bank and they are offering a whopping 5% annual percentage interest rate. This means that after one year, the bank will pay *you* 5% of that $100, or $100(0.05) = \$5$ in interest, so you now have $105.[1] This is in accordance with the formula for *simple interest* which you have undoubtedly run across at some point before.

Equation 6.1. Simple Interest The amount of interest I accrued at an annual rate r on an investment[a] P after t years is

$$I = Prt$$

The amount A in the account after t years is given by

$$A = P + I = P + Prt = P(1 + rt)$$

---
[a] Called the **principal**

Suppose, however, that six months into the year, you hear of a better deal at a rival bank.[2] Naturally, you withdraw your money and try to invest it at the higher rate there. Since six months is one half of a year, that initial $100 yields $100(0.05)\left(\frac{1}{2}\right) = \$2.50$ in interest. You take your $102.50 off to the competitor and find out that those restrictions which *may* apply actually <u>do</u> apply to you, and you return to your bank which happily accepts your $102.50 for the remaining six months of the year. To your surprise and delight, at the end of the year your statement reads $105.06, not $105 as you had expected.[3] Where did those extra six cents come from? For the first six months of the year, interest was earned on the original principal of $100, but for the second six months, interest was earned on $102.50, that is, you earned interest on your interest. This is the basic concept behind **compound interest**. In the previous discussion, we would say that the interest was compounded twice, or semiannually.[4] If more money can be earned by earning interest on interest already earned, a natural question to ask is what happens if the interest is compounded more often, say 4 times a year, which is every three months, or 'quarterly.' In this case, the money is in the account for three months, or $\frac{1}{4}$ of a year, at a time. After the first quarter, we have $A = P(1 + rt) = \$100\left(1 + 0.05 \cdot \frac{1}{4}\right) = \$101.25$. We now invest the $101.25 for the next three

---
[1] How generous of them!
[2] Some restrictions may apply.
[3] Actually, the final balance should be $105.0625.
[4] Using this convention, simple interest after one year is the same as compounding the interest only once.

months and find that at the end of the second quarter, we have $A = \$101.25 \left(1 + 0.05 \cdot \frac{1}{4}\right) \approx \$102.51$. Continuing in this manner, the balance at the end of the third quarter is $\$103.79$, and, at last, we obtain $\$105.08$. The extra two cents hardly seems worth it, but we see that we do in fact get more money the more often we compound. In order to develop a formula for this phenomenon, we need to do some abstract calculations. Suppose we wish to invest our principal P at an annual rate r and compound the interest n times per year. This means the money sits in the account $\frac{1}{n}^{\text{th}}$ of a year between compoundings. Let A_k denote the amount in the account after the kth compounding. Then $A_1 = P\left(1 + r\left(\frac{1}{n}\right)\right)$ which simplifies to $A_1 = P\left(1 + \frac{r}{n}\right)$. After the second compounding, we use A_1 as our new principal and get $A_2 = A_1\left(1 + \frac{r}{n}\right) = \left[P\left(1 + \frac{r}{n}\right)\right]\left(1 + \frac{r}{n}\right) = P\left(1 + \frac{r}{n}\right)^2$. Continuing in this fashion, we get $A_3 = P\left(1 + \frac{r}{n}\right)^3$, $A_4 = P\left(1 + \frac{r}{n}\right)^4$, and so on, so that $A_k = P\left(1 + \frac{r}{n}\right)^k$. Since we compound the interest n times per year, after t years, we have nt compoundings. We have just derived the general formula for compound interest below.

> **Equation 6.2. Compounded Interest:** If an initial principal P is invested at an annual rate r and the interest is compounded n times per year, the amount A in the account after t years is
>
> $$A(t) = P\left(1 + \frac{r}{n}\right)^{nt}$$

If we take $P = 100$, $r = 0.05$, and $n = 4$, Equation 6.2 becomes $A(t) = 100\left(1 + \frac{0.05}{4}\right)^{4t}$ which reduces to $A(t) = 100(1.0125)^{4t}$. To check this new formula against our previous calculations, we find $A\left(\frac{1}{4}\right) = 100(1.0125)^{4\left(\frac{1}{4}\right)} = 101.25$, $A\left(\frac{1}{2}\right) \approx \$102.51$, $A\left(\frac{3}{4}\right) \approx \$103.79$, and $A(1) \approx \$105.08$.

Example 6.5.1. Suppose $\$2000$ is invested in an account which offers 7.125% compounded monthly.

1. Express the amount A in the account as a function of the term of the investment t in years.

2. How much is in the account after 5 years?

3. How long will it take for the initial investment to double?

4. Find and interpret the average rate of change[5] of the amount in the account from the end of the fourth year to the end of the fifth year, and from the end of the thirty-fourth year to the end of the thirty-fifth year.

Solution.

1. Substituting $P = 2000$, $r = 0.07125$, and $n = 12$ (since interest is compounded *monthly*) into Equation 6.2 yields $A(t) = 2000\left(1 + \frac{0.07125}{12}\right)^{12t} = 2000(1.0059375)^{12t}$.

2. Since t represents the length of the investment in years, we substitute $t = 5$ into $A(t)$ to find $A(5) = 2000(1.0059375)^{12(5)} \approx 2852.92$. After 5 years, we have approximately $\$2852.92$.

---

[5]See Definition 2.3 in Section 2.1.

6.5 Applications of Exponential and Logarithmic Functions

3. Our initial investment is $2000, so to find the time it takes this to double, we need to find t when $A(t) = 4000$. We get $2000(1.0059375)^{12t} = 4000$, or $(1.0059375)^{12t} = 2$. Taking natural logs as in Section 6.3, we get $t = \frac{\ln(2)}{12\ln(1.0059375)} \approx 9.75$. Hence, it takes approximately 9 years 9 months for the investment to double.

4. To find the average rate of change of A from the end of the fourth year to the end of the fifth year, we compute $\frac{A(5)-A(4)}{5-4} \approx 195.63$. Similarly, the average rate of change of A from the end of the thirty-fourth year to the end of the thirty-fifth year is $\frac{A(35)-A(34)}{35-34} \approx 1648.21$. This means that the value of the investment is increasing at a rate of approximately \$195.63 per year between the end of the fourth and fifth years, while that rate jumps to \$1648.21 per year between the end of the thirty-fourth and thirty-fifth years. So, not only is it true that the longer you wait, the more money you have, but also the longer you wait, the faster the money increases.[6]   □

We have observed that the more times you compound the interest per year, the more money you will earn in a year. Let's push this notion to the limit.[7] Consider an investment of \$1 invested at 100% interest for 1 year compounded n times a year. Equation 6.2 tells us that the amount of money in the account after 1 year is $A = \left(1 + \frac{1}{n}\right)^n$. Below is a table of values relating n and A.

n	A
1	2
2	2.25
4	≈ 2.4414
12	≈ 2.6130
360	≈ 2.7145
1000	≈ 2.7169
10000	≈ 2.7181
100000	≈ 2.7182

As promised, the more compoundings per year, the more money there is in the account, but we also observe that the increase in money is greatly diminishing. We are witnessing a mathematical 'tug of war'. While we are compounding more times per year, and hence getting interest on our interest more often, the amount of time between compoundings is getting smaller and smaller, so there is less time to build up additional interest. With Calculus, we can show[8] that as $n \to \infty$, $A = \left(1 + \frac{1}{n}\right)^n \to e$, where e is the natural base first presented in Section 6.1. Taking the number of compoundings per year to infinity results in what is called **continuously** compounded interest.

Theorem 6.8. If you invest \$1 at 100% interest compounded continuously, then you will have \$$e$ at the end of one year.

---

[6] In fact, the rate of increase of the amount in the account is exponential as well. This is the quality that really defines exponential functions and we refer the reader to a course in Calculus.
[7] Once you've had a semester of Calculus, you'll be able to fully appreciate this very lame pun.
[8] Or define, depending on your point of view.

Using this definition of e and a little Calculus, we can take Equation 6.2 and produce a formula for continuously compounded interest.

> **Equation 6.3. Continuously Compounded Interest:** If an initial principal P is invested at an annual rate r and the interest is compounded continuously, the amount A in the account after t years is
> $$A(t) = Pe^{rt}$$

If we take the scenario of Example 6.5.1 and compare monthly compounding to continuous compounding over 35 years, we find that monthly compounding yields $A(35) = 2000(1.0059375)^{12(35)}$ which is about \$24,035.28, whereas continuously compounding gives $A(35) = 2000e^{0.07125(35)}$ which is about \$24,213.18 - a difference of less than 1%.

Equations 6.2 and 6.3 both use exponential functions to describe the growth of an investment. Curiously enough, the same principles which govern compound interest are also used to model short term growth of populations. In Biology, **The Law of Uninhibited Growth** states as its premise that the *instantaneous* rate at which a population increases at any time is directly proportional to the population at that time.[9] In other words, the more organisms there are at a given moment, the faster they reproduce. Formulating the law as stated results in a differential equation, which requires Calculus to solve. Its solution is stated below.

> **Equation 6.4. Uninhibited Growth:** If a population increases according to The Law of Uninhibited Growth, the number of organisms N at time t is given by the formula
> $$N(t) = N_0 e^{kt},$$
> where $N(0) = N_0$ (read 'N nought') is the initial number of organisms and $k > 0$ is the constant of proportionality which satisfies the equation
>
> (instantaneous rate of change of $N(t)$ at time t) = $k\,N(t)$

It is worth taking some time to compare Equations 6.3 and 6.4. In Equation 6.3, we use P to denote the initial investment; in Equation 6.4, we use N_0 to denote the initial population. In Equation 6.3, r denotes the annual interest rate, and so it shouldn't be too surprising that the k in Equation 6.4 corresponds to a growth rate as well. While Equations 6.3 and 6.4 look entirely different, they both represent the same mathematical concept.

Example 6.5.2. In order to perform arthrosclerosis research, epithelial cells are harvested from discarded umbilical tissue and grown in the laboratory. A technician observes that a culture of twelve thousand cells grows to five million cells in one week. Assuming that the cells follow The Law of Uninhibited Growth, find a formula for the number of cells, N, in thousands, after t days.

Solution. We begin with $N(t) = N_0 e^{kt}$. Since N is to give the number of cells *in thousands*, we have $N_0 = 12$, so $N(t) = 12e^{kt}$. In order to complete the formula, we need to determine the

---

[9]The average rate of change of a function over an interval was first introduced in Section 2.1. *Instantaneous* rates of change are the business of Calculus, as is mentioned on Page 161.

6.5 Applications of Exponential and Logarithmic Functions

growth rate k. We know that after one week, the number of cells has grown to five million. Since t measures days and the units of N are in thousands, this translates mathematically to $N(7) = 5000$. We get the equation $12e^{7k} = 5000$ which gives $k = \frac{1}{7}\ln\left(\frac{1250}{3}\right)$. Hence, $N(t) = 12e^{\frac{t}{7}\ln\left(\frac{1250}{3}\right)}$. Of course, in practice, we would approximate k to some desired accuracy, say $k \approx 0.8618$, which we can interpret as an 86.18% daily growth rate for the cells. □

Whereas Equations 6.3 and 6.4 model the growth of quantities, we can use equations like them to describe the decline of quantities. One example we've seen already is Example 6.1.1 in Section 6.1. There, the value of a car declined from its purchase price of $25,000 to nothing at all. Another real world phenomenon which follows suit is radioactive decay. There are elements which are unstable and emit energy spontaneously. In doing so, the amount of the element itself diminishes. The assumption behind this model is that the rate of decay of an element at a particular time is directly proportional to the amount of the element present at that time. In other words, the more of the element there is, the faster the element decays. This is precisely the same kind of hypothesis which drives The Law of Uninhibited Growth, and as such, the equation governing radioactive decay is hauntingly similar to Equation 6.4 with the exception that the rate constant k is negative.

Equation 6.5. Radioactive Decay The amount of a radioactive element A at time t is given by the formula

$$A(t) = A_0 e^{kt},$$

where $A(0) = A_0$ is the initial amount of the element and $k < 0$ is the constant of proportionality which satisfies the equation

(instantaneous rate of change of $A(t)$ at time t) $= k\, A(t)$

Example 6.5.3. Iodine-131 is a commonly used radioactive isotope used to help detect how well the thyroid is functioning. Suppose the decay of Iodine-131 follows the model given in Equation 6.5, and that the half-life[10] of Iodine-131 is approximately 8 days. If 5 grams of Iodine-131 is present initially, find a function which gives the amount of Iodine-131, A, in grams, t days later.

Solution. Since we start with 5 grams initially, Equation 6.5 gives $A(t) = 5e^{kt}$. Since the half-life is 8 days, it takes 8 days for half of the Iodine-131 to decay, leaving half of it behind. Hence, $A(8) = 2.5$ which means $5e^{8k} = 2.5$. Solving, we get $k = \frac{1}{8}\ln\left(\frac{1}{2}\right) = -\frac{\ln(2)}{8} \approx -0.08664$, which we can interpret as a loss of material at a rate of 8.664% daily. Hence, $A(t) = 5e^{-\frac{t\ln(2)}{8}} \approx 5e^{-0.08664t}$. □

We now turn our attention to some more mathematically sophisticated models. One such model is Newton's Law of Cooling, which we first encountered in Example 6.1.2 of Section 6.1. In that example we had a cup of coffee cooling from 160°F to room temperature 70°F according to the formula $T(t) = 70 + 90e^{-0.1t}$, where t was measured in minutes. In this situation, we know the physical limit of the temperature of the coffee is room temperature,[11] and the differential equation

---

[10] The time it takes for half of the substance to decay.

[11] The Second Law of Thermodynamics states that heat can spontaneously flow from a hotter object to a colder one, but not the other way around. Thus, the coffee could not continue to release heat into the air so as to cool below room temperature.

which gives rise to our formula for $T(t)$ takes this into account. Whereas the radioactive decay model had a rate of decay at time t directly proportional to the amount of the element which remained at time t, Newton's Law of Cooling states that the rate of cooling of the coffee at a given time t is directly proportional to how much of a temperature <u>gap</u> exists between the coffee at time t and room temperature, not the temperature of the coffee itself. In other words, the coffee cools faster when it is first served, and as its temperature nears room temperature, the coffee cools ever more slowly. Of course, if we take an item from the refrigerator and let it sit out in the kitchen, the object's temperature will rise to room temperature, and since the physics behind warming and cooling is the same, we combine both cases in the equation below.

Equation 6.6. Newton's Law of Cooling (Warming): The temperature T of an object at time t is given by the formula

$$T(t) = T_a + (T_0 - T_a) e^{-kt},$$

where $T(0) = T_0$ is the initial temperature of the object, T_a is the ambient temperature[a] and $k > 0$ is the constant of proportionality which satisfies the equation

$$\text{(instantaneous rate of change of } T(t) \text{ at time } t) = k \left(T(t) - T_a \right)$$

[a]That is, the temperature of the surroundings.

If we re-examine the situation in Example 6.1.2 with $T_0 = 160$, $T_a = 70$, and $k = 0.1$, we get, according to Equation 6.6, $T(t) = 70 + (160 - 70)e^{-0.1t}$ which reduces to the original formula given. The rate constant $k = 0.1$ indicates the coffee is cooling at a rate equal to 10% of the difference between the temperature of the coffee and its surroundings. Note in Equation 6.6 that the constant k is positive for both the cooling and warming scenarios. What determines if the function $T(t)$ is increasing or decreasing is if T_0 (the initial temperature of the object) is greater than T_a (the ambient temperature) or vice-versa, as we see in our next example.

Example 6.5.4. A 40°F roast is cooked in a 350°F oven. After 2 hours, the temperature of the roast is 125°F.

1. Assuming the temperature of the roast follows Newton's Law of Warming, find a formula for the temperature of the roast T as a function of its time in the oven, t, in hours.

2. The roast is done when the internal temperature reaches 165°F. When will the roast be done?

Solution.

1. The initial temperature of the roast is 40°F, so $T_0 = 40$. The environment in which we are placing the roast is the 350°F oven, so $T_a = 350$. Newton's Law of Warming tells us $T(t) = 350 + (40 - 350)e^{-kt}$, or $T(t) = 350 - 310e^{-kt}$. To determine k, we use the fact that after 2 hours, the roast is 125°F, which means $T(2) = 125$. This gives rise to the equation $350 - 310e^{-2k} = 125$ which yields $k = -\frac{1}{2} \ln\left(\frac{45}{62}\right) \approx 0.1602$. The temperature function is

$$T(t) = 350 - 310 e^{\frac{t}{2} \ln\left(\frac{45}{62}\right)} \approx 350 - 310 e^{-0.1602t}.$$

6.5 Applications of Exponential and Logarithmic Functions

2. To determine when the roast is done, we set $T(t) = 165$. This gives $350 - 310e^{-0.1602t} = 165$ whose solution is $t = -\frac{1}{0.1602} \ln\left(\frac{37}{62}\right) \approx 3.22$. It takes roughly 3 hours and 15 minutes to cook the roast completely. □

If we had taken the time to graph $y = T(t)$ in Example 6.5.4, we would have found the horizontal asymptote to be $y = 350$, which corresponds to the temperature of the oven. We can also arrive at this conclusion by applying a bit of 'number sense'. As $t \to \infty$, $-0.1602t \approx$ very big $(-)$ so that $e^{-0.1602t} \approx$ very small $(+)$. The larger the value of t, the smaller $e^{-0.1602t}$ becomes so that $T(t) \approx 350 -$ very small $(+)$, which indicates the graph of $y = T(t)$ is approaching its horizontal asymptote $y = 350$ from below. Physically, this means the roast will eventually warm up to $350°F$.[12] The function T is sometimes called a **limited** growth model, since the function T remains bounded as $t \to \infty$. If we apply the principles behind Newton's Law of Cooling to a biological example, it says the growth rate of a population is directly proportional to how much room the population has to grow. In other words, the more room for expansion, the faster the growth rate. The **logistic** growth model combines The Law of Uninhibited Growth with limited growth and states that the rate of growth of a population varies jointly with the population itself as well as the room the population has to grow.

Equation 6.7. Logistic Growth: If a population behaves according to the assumptions of logistic growth, the number of organisms N at time t is given by the equation

$$N(t) = \frac{L}{1 + Ce^{-kLt}},$$

where $N(0) = N_0$ is the initial population, L is the limiting population,[a] C is a measure of how much room there is to grow given by

$$C = \frac{L}{N_0} - 1.$$

and $k > 0$ is the constant of proportionality which satisfies the equation

(instantaneous rate of change of $N(t)$ at time t) $= k\, N(t)\, (L - N(t))$

---
[a]That is, as $t \to \infty$, $N(t) \to L$.

The logistic function is used not only to model the growth of organisms, but is also often used to model the spread of disease and rumors.[13]

Example 6.5.5. The number of people N, in hundreds, at a local community college who have heard the rumor 'Carl is afraid of Virginia Woolf' can be modeled using the logistic equation

$$N(t) = \frac{84}{1 + 2799e^{-t}},$$

---
[12]at which point it would be more toast than roast.
[13]Which can be just as damaging as diseases.

where $t \geq 0$ is the number of days after April 1, 2009.

1. Find and interpret $N(0)$.

2. Find and interpret the end behavior of $N(t)$.

3. How long until 4200 people have heard the rumor?

4. Check your answers to 2 and 3 using your calculator.

Solution.

1. We find $N(0) = \frac{84}{1+2799e^0} = \frac{84}{2800} = \frac{3}{100}$. Since $N(t)$ measures the number of people who have heard the rumor in hundreds, $N(0)$ corresponds to 3 people. Since $t = 0$ corresponds to April 1, 2009, we may conclude that on that day, 3 people have heard the rumor.[14]

2. We could simply note that $N(t)$ is written in the form of Equation 6.7, and identify $L = 84$. However, to see why the answer is 84, we proceed analytically. Since the domain of N is restricted to $t \geq 0$, the only end behavior of significance is $t \to \infty$. As we've seen before,[15] as $t \to \infty$, we have $1997e^{-t} \to 0^+$ and so $N(t) \approx \frac{84}{1+\text{very small }(+)} \approx 84$. Hence, as $t \to \infty$, $N(t) \to 84$. This means that as time goes by, the number of people who will have heard the rumor approaches 8400.

3. To find how long it takes until 4200 people have heard the rumor, we set $N(t) = 42$. Solving $\frac{84}{1+2799e^{-t}} = 42$ gives $t = \ln(2799) \approx 7.937$. It takes around 8 days until 4200 people have heard the rumor.

4. We graph $y = N(x)$ using the calculator and see that the line $y = 84$ is the horizontal asymptote of the graph, confirming our answer to part 2, and the graph intersects the line $y = 42$ at $x = \ln(2799) \approx 7.937$, which confirms our answer to part 3.

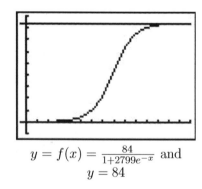

$y = f(x) = \frac{84}{1+2799e^{-x}}$ and $y = 84$

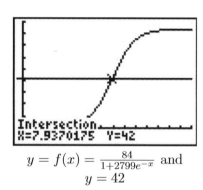

$y = f(x) = \frac{84}{1+2799e^{-x}}$ and $y = 42$

---

[14]Or, more likely, three people started the rumor. I'd wager Jeff, Jamie, and Jason started it. So much for telling your best friends something in confidence!

[15]See, for example, Example 6.1.2.

6.5 Applications of Exponential and Logarithmic Functions

If we take the time to analyze the graph of $y = N(x)$ above, we can see graphically how logistic growth combines features of uninhibited and limited growth. The curve seems to rise steeply, then at some point, begins to level off. The point at which this happens is called an **inflection point** or is sometimes called the 'point of diminishing returns'. At this point, even though the function is still increasing, the rate at which it does so begins to decline. It turns out the point of diminishing returns always occurs at half the limiting population. (In our case, when $y = 42$.) While these concepts are more precisely quantified using Calculus, below are two views of the graph of $y = N(x)$, one on the interval $[0, 8]$, the other on $[8, 15]$. The former looks strikingly like uninhibited growth; the latter like limited growth.

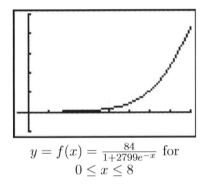

$y = f(x) = \frac{84}{1+2799e^{-x}}$ for $0 \leq x \leq 8$

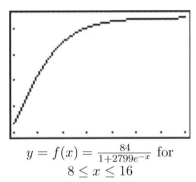

$y = f(x) = \frac{84}{1+2799e^{-x}}$ for $8 \leq x \leq 16$

6.5.2 Applications of Logarithms

Just as many physical phenomena can be modeled by exponential functions, the same is true of logarithmic functions. In Exercises 75, 76 and 77 of Section 6.1, we showed that logarithms are useful in measuring the intensities of earthquakes (the Richter scale), sound (decibels) and acids and bases (pH). We now present yet a different use of the a basic logarithm function, password strength.

Example 6.5.6. The information entropy H, in bits, of a randomly generated password consisting of L characters is given by $H = L \log_2(N)$, where N is the number of possible symbols for each character in the password. In general, the higher the entropy, the stronger the password.

1. If a 7 character case-sensitive[16] password is comprised of letters and numbers only, find the associated information entropy.

2. How many possible symbol options per character is required to produce a 7 character password with an information entropy of 50 bits?

Solution.

1. There are 26 letters in the alphabet, 52 if upper and lower case letters are counted as different. There are 10 digits (0 through 9) for a total of $N = 62$ symbols. Since the password is to be 7 characters long, $L = 7$. Thus, $H = 7 \log_2(62) = \frac{7 \ln(62)}{\ln(2)} \approx 41.68$.

---
[16]That is, upper and lower case letters are treated as different characters.

2. We have $L = 7$ and $H = 50$ and we need to find N. Solving the equation $50 = 7\log_2(N)$ gives $N = 2^{50/7} \approx 141.323$, so we would need 142 different symbols to choose from.[17] □

Chemical systems known as buffer solutions have the ability to adjust to small changes in acidity to maintain a range of pH values. Buffer solutions have a wide variety of applications from maintaining a healthy fish tank to regulating the pH levels in blood. Our next example shows how the pH in a buffer solution is a little more complicated than the pH we first encountered in Exercise 77 in Section 6.1.

Example 6.5.7. Blood is a buffer solution. When carbon dioxide is absorbed into the bloodstream it produces carbonic acid and lowers the pH. The body compensates by producing bicarbonate, a weak base to partially neutralize the acid. The equation[18] which models blood pH in this situation is pH $= 6.1 + \log\left(\frac{800}{x}\right)$, where x is the partial pressure of carbon dioxide in arterial blood, measured in torr. Find the partial pressure of carbon dioxide in arterial blood if the pH is 7.4.

Solution. We set pH $= 7.4$ and get $7.4 = 6.1 + \log\left(\frac{800}{x}\right)$, or $\log\left(\frac{800}{x}\right) = 1.3$. Solving, we find $x = \frac{800}{10^{1.3}} \approx 40.09$. Hence, the partial pressure of carbon dioxide in the blood is about 40 torr. □

Another place logarithms are used is in data analysis. Suppose, for instance, we wish to model the spread of influenza A (H1N1), the so-called 'Swine Flu'. Below is data taken from the World Health Organization (WHO) where t represents the number of days since April 28, 2009, and N represents the number of confirmed cases of H1N1 virus worldwide.

t	1	2	3	4	5	6	7	8	9	10	11	12	13
N	148	257	367	658	898	1085	1490	1893	2371	2500	3440	4379	4694

t	14	15	16	17	18	19	20
N	5251	5728	6497	7520	8451	8480	8829

Making a scatter plot of the data treating t as the independent variable and N as the dependent variable gives

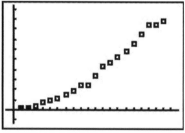

Which models are suggested by the shape of the data? Thinking back Section 2.5, we try a Quadratic Regression, with pretty good results.

---

[17]Since there are only 94 distinct ASCII keyboard characters, to achieve this strength, the number of characters in the password should be increased.

[18]Derived from the Henderson-Hasselbalch Equation. See Exercise 43 in Section 6.2. Hasselbalch himself was studying carbon dioxide dissolving in blood - a process called metabolic acidosis.

6.5 Applications of Exponential and Logarithmic Functions

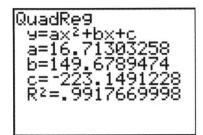 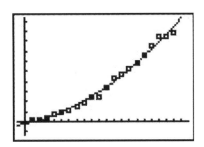

However, is there any scientific reason for the data to be quadratic? Are there other models which fit the data equally well, or better? Scientists often use logarithms in an attempt to 'linearize' data sets - in other words, transform the data sets to produce ones which result in straight lines. To see how this could work, suppose we guessed the relationship between N and t was some kind of power function, not necessarily quadratic, say $N = Bt^A$. To try to determine the A and B, we can take the natural log of both sides and get $\ln(N) = \ln(Bt^A)$. Using properties of logs to expand the right hand side of this equation, we get $\ln(N) = A\ln(t) + \ln(B)$. If we set $X = \ln(t)$ and $Y = \ln(N)$, this equation becomes $Y = AX + \ln(B)$. In other words, we have a line with slope A and Y-intercept $\ln(B)$. So, instead of plotting N versus t, we plot $\ln(N)$ versus $\ln(t)$.

$\ln(t)$	0	0.693	1.099	1.386	1.609	1.792	1.946	2.079	2.197	2.302	2.398	2.485	2.565
$\ln(N)$	4.997	5.549	5.905	6.489	6.800	6.989	7.306	7.546	7.771	7.824	8.143	8.385	8.454

$\ln(t)$	2.639	2.708	2.773	2.833	2.890	2.944	2.996
$\ln(N)$	8.566	8.653	8.779	8.925	9.042	9.045	9.086

Running a linear regression on the data gives

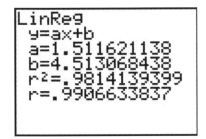 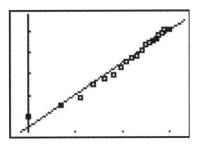

The slope of the regression line is $a \approx 1.512$ which corresponds to our exponent A. The y-intercept $b \approx 4.513$ corresponds to $\ln(B)$, so that $B \approx 91.201$. Hence, we get the model $N = 91.201t^{1.512}$, something from Section 5.3. Of course, the calculator has a built-in 'Power Regression' feature. If we apply this to our original data set, we get the same model we arrived at before.[19]

---

[19]Critics may question why the authors of the book have chosen to even discuss linearization of data when the calculator has a Power Regression built-in and ready to go. Our response: talk to your science faculty.

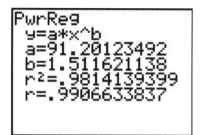 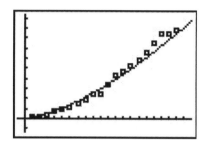

This is all well and good, but the quadratic model appears to fit the data better, and we've yet to mention any scientific principle which would lead us to believe the actual spread of the flu follows any kind of power function at all. If we are to attack this data from a scientific perspective, it does seem to make sense that, at least in the early stages of the outbreak, the more people who have the flu, the faster it will spread, which leads us to proposing an uninhibited growth model. If we assume $N = Be^{At}$ then, taking logs as before, we get $\ln(N) = At + \ln(B)$. If we set $X = t$ and $Y = \ln(N)$, then, once again, we get $Y = AX + \ln(B)$, a line with slope A and Y-intercept $\ln(B)$. Plotting $\ln(N)$ versus t gives the following linear regression.

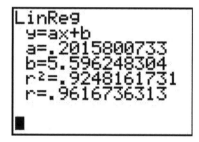 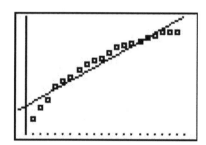

We see the slope is $a \approx 0.202$ and which corresponds to A in our model, and the y-intercept is $b \approx 5.596$ which corresponds to $\ln(B)$. We get $B \approx 269.414$, so that our model is $N = 269.414e^{0.202t}$. Of course, the calculator has a built-in 'Exponential Regression' feature which produces what appears to be a different model $N = 269.414(1.22333419)^t$. Using properties of exponents, we write $e^{0.202t} = \left(e^{0.202}\right)^t \approx (1.223848)^t$, which, had we carried more decimal places, would have matched the base of the calculator model exactly.

 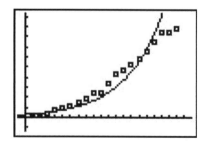

The exponential model didn't fit the data as well as the quadratic or power function model, but it stands to reason that, perhaps, the spread of the flu is not unlike that of the spread of a rumor

6.5 Applications of Exponential and Logarithmic Functions

and that a logistic model can be used to model the data. The calculator does have a 'Logistic Regression' feature, and using it produces the model $N = \frac{10739.147}{1+42.416e^{0.268t}}$.

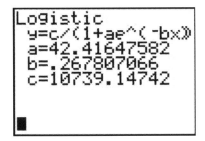 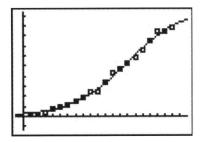

This appears to be an excellent fit, but there is no friendly coefficient of determination, R^2, by which to judge this numerically. There are good reasons for this, but they are far beyond the scope of the text. Which of the models, quadratic, power, exponential, or logistic is the 'best model'? If by 'best' we mean 'fits closest to the data,' then the quadratic and logistic models are arguably the winners with the power function model a close second. However, if we think about the science behind the spread of the flu, the logistic model gets an edge. For one thing, it takes into account that only a finite number of people will ever get the flu (according to our model, 10,739), whereas the quadratic model predicts no limit to the number of cases. As we have stated several times before in the text, mathematical models, regardless of their sophistication, are just that: models, and they all have their limitations.[20]

---

[20]Speaking of limitations, as of June 3, 2009, there were 19,273 confirmed cases of influenza A (H1N1). This is well above our prediction of 10,739. Each time a new report is issued, the data set increases and the model must be recalculated. We leave this recalculation to the reader.

6.5.3 Exercises

For each of the scenarios given in Exercises 1 - 6,

- Find the amount A in the account as a function of the term of the investment t in years.

- Determine how much is in the account after 5 years, 10 years, 30 years and 35 years. Round your answers to the nearest cent.

- Determine how long will it take for the initial investment to double. Round your answer to the nearest year.

- Find and interpret the average rate of change of the amount in the account from the end of the fourth year to the end of the fifth year, and from the end of the thirty-fourth year to the end of the thirty-fifth year. Round your answer to two decimal places.

1. $500 is invested in an account which offers 0.75%, compounded monthly.

2. $500 is invested in an account which offers 0.75%, compounded continuously.

3. $1000 is invested in an account which offers 1.25%, compounded monthly.

4. $1000 is invested in an account which offers 1.25%, compounded continuously.

5. $5000 is invested in an account which offers 2.125%, compounded monthly.

6. $5000 is invested in an account which offers 2.125%, compounded continuously.

7. Look back at your answers to Exercises 1 - 6. What can be said about the difference between monthly compounding and continuously compounding the interest in those situations? With the help of your classmates, discuss scenarios where the difference between monthly and continuously compounded interest would be more dramatic. Try varying the interest rate, the term of the investment and the principal. Use computations to support your answer.

8. How much money needs to be invested now to obtain $2000 in 3 years if the interest rate in a savings account is 0.25%, compounded continuously? Round your answer to the nearest cent.

9. How much money needs to be invested now to obtain $5000 in 10 years if the interest rate in a CD is 2.25%, compounded monthly? Round your answer to the nearest cent.

10. On May, 31, 2009, the Annual Percentage Rate listed at Jeff's bank for regular savings accounts was 0.25% compounded monthly. Use Equation 6.2 to answer the following.

    (a) If $P = 2000$ what is $A(8)$?
    (b) Solve the equation $A(t) = 4000$ for t.
    (c) What principal P should be invested so that the account balance is $2000 is three years?

6.5 Applications of Exponential and Logarithmic Functions

11. Jeff's bank also offers a 36-month Certificate of Deposit (CD) with an APR of 2.25%.

    (a) If $P = 2000$ what is $A(8)$?
    (b) Solve the equation $A(t) = 4000$ for t.
    (c) What principal P should be invested so that the account balance is $2000 in three years?
    (d) The Annual Percentage Yield is the <u>simple</u> interest rate that returns the same amount of interest after one year as the compound interest does. With the help of your classmates, compute the APY for this investment.

12. A finance company offers a promotion on $5000 loans. The borrower does not have to make any payments for the first three years, however interest will continue to be charged to the loan at 29.9% compounded continuously. What amount will be due at the end of the three year period, assuming no payments are made? If the promotion is extended an additional three years, and no payments are made, what amount would be due?

13. Use Equation 6.2 to show that the time it takes for an investment to double in value does <u>not</u> depend on the principal P, but rather, depends only on the APR and the number of compoundings per year. Let $n = 12$ and with the help of your classmates compute the doubling time for a variety of rates r. Then look up the Rule of 72 and compare your answers to what that rule says. If you're really interested[21] in Financial Mathematics, you could also compare and contrast the Rule of 72 with the Rule of 70 and the Rule of 69.

In Exercises 14 - 18, we list some radioactive isotopes and their associated half-lives. Assume that each decays according to the formula $A(t) = A_0 e^{kt}$ where A_0 is the initial amount of the material and k is the decay constant. For each isotope:

- Find the decay constant k. Round your answer to four decimal places.

- Find a function which gives the amount of isotope A which remains after time t. (Keep the units of A and t the same as the given data.)

- Determine how long it takes for 90% of the material to decay. Round your answer to two decimal places. (HINT: If 90% of the material decays, how much is left?)

14. Cobalt 60, used in food irradiation, initial amount 50 grams, half-life of 5.27 years.

15. Phosphorus 32, used in agriculture, initial amount 2 milligrams, half-life 14 days.

16. Chromium 51, used to track red blood cells, initial amount 75 milligrams, half-life 27.7 days.

17. Americium 241, used in smoke detectors, initial amount 0.29 micrograms, half-life 432.7 years.

18. Uranium 235, used for nuclear power, initial amount 1 kg grams, half-life 704 million years.

---
[21] Awesome pun!

19. With the help of your classmates, show that the time it takes for 90% of each isotope listed in Exercises 14 - 18 to decay does not depend on the initial amount of the substance, but rather, on only the decay constant k. Find a formula, in terms of k only, to determine how long it takes for 90% of a radioactive isotope to decay.

20. In Example 6.1.1 in Section 6.1, the exponential function $V(x) = 25 \left(\frac{4}{5}\right)^x$ was used to model the value of a car over time. Use the properties of logs and/or exponents to rewrite the model in the form $V(t) = 25e^{kt}$.

21. The Gross Domestic Product (GDP) of the US (in billions of dollars) t years after the year 2000 can be modeled by:
$$G(t) = 9743.77e^{0.0514t}$$

    (a) Find and interpret $G(0)$.

    (b) According to the model, what should have been the GDP in 2007? In 2010? (According to the US Department of Commerce, the 2007 GDP was \$14,369.1 billion and the 2010 GDP was \$14,657.8 billion.)

22. The diameter D of a tumor, in millimeters, t days after it is detected is given by:
$$D(t) = 15e^{0.0277t}$$

    (a) What was the diameter of the tumor when it was originally detected?

    (b) How long until the diameter of the tumor doubles?

23. Under optimal conditions, the growth of a certain strain of E. Coli is modeled by the Law of Uninhibited Growth $N(t) = N_0 e^{kt}$ where N_0 is the initial number of bacteria and t is the elapsed time, measured in minutes. From numerous experiments, it has been determined that the doubling time of this organism is 20 minutes. Suppose 1000 bacteria are present initially.

    (a) Find the growth constant k. Round your answer to four decimal places.

    (b) Find a function which gives the number of bacteria $N(t)$ after t minutes.

    (c) How long until there are 9000 bacteria? Round your answer to the nearest minute.

24. Yeast is often used in biological experiments. A research technician estimates that a sample of yeast suspension contains 2.5 million organisms per cubic centimeter (cc). Two hours later, she estimates the population density to be 6 million organisms per cc. Let t be the time elapsed since the first observation, measured in hours. Assume that the yeast growth follows the Law of Uninhibited Growth $N(t) = N_0 e^{kt}$.

    (a) Find the growth constant k. Round your answer to four decimal places.

    (b) Find a function which gives the number of yeast (in millions) per cc $N(t)$ after t hours.

    (c) What is the doubling time for this strain of yeast?

6.5 APPLICATIONS OF EXPONENTIAL AND LOGARITHMIC FUNCTIONS 485

25. The Law of Uninhibited Growth also applies to situations where an animal is re-introduced into a suitable environment. Such a case is the reintroduction of wolves to Yellowstone National Park. According to the National Park Service, the wolf population in Yellowstone National Park was 52 in 1996 and 118 in 1999. Using these data, find a function of the form $N(t) = N_0 e^{kt}$ which models the number of wolves t years after 1996. (Use $t = 0$ to represent the year 1996. Also, round your value of k to four decimal places.) According to the model, how many wolves were in Yellowstone in 2002? (The recorded number is 272.)

26. During the early years of a community, it is not uncommon for the population to grow according to the Law of Uninhibited Growth. According to the Painesville Wikipedia entry, in 1860, the Village of Painesville had a population of 2649. In 1920, the population was 7272. Use these two data points to fit a model of the form $N(t) = N_0 e^{kt}$ were $N(t)$ is the number of Painesville Residents t years after 1860. (Use $t = 0$ to represent the year 1860. Also, round the value of k to four decimal places.) According to this model, what was the population of Painesville in 2010? (The 2010 census gave the population as 19,563) What could be some causes for such a vast discrepancy? For more on this, see Exercise 37.

27. The population of Sasquatch in Bigfoot county is modeled by

$$P(t) = \frac{120}{1 + 3.167 e^{-0.05t}}$$

where $P(t)$ is the population of Sasquatch t years after 2010.

(a) Find and interpret $P(0)$.

(b) Find the population of Sasquatch in Bigfoot county in 2013. Round your answer to the nearest Sasquatch.

(c) When will the population of Sasquatch in Bigfoot county reach 60? Round your answer to the nearest year.

(d) Find and interpret the end behavior of the graph of $y = P(t)$. Check your answer using a graphing utility.

28. The half-life of the radioactive isotope Carbon-14 is about 5730 years.

(a) Use Equation 6.5 to express the amount of Carbon-14 left from an initial N milligrams as a function of time t in years.

(b) What percentage of the original amount of Carbon-14 is left after 20,000 years?

(c) If an old wooden tool is found in a cave and the amount of Carbon-14 present in it is estimated to be only 42% of the original amount, approximately how old is the tool?

(d) Radiocarbon dating is not as easy as these exercises might lead you to believe. With the help of your classmates, research radiocarbon dating and discuss why our model is somewhat over-simplified.

29. Carbon-14 cannot be used to date inorganic material such as rocks, but there are many other methods of radiometric dating which estimate the age of rocks. One of them, Rubidium-Strontium dating, uses Rubidium-87 which decays to Strontium-87 with a half-life of 50 billion years. Use Equation 6.5 to express the amount of Rubidium-87 left from an initial 2.3 micrograms as a function of time t in *billions* of years. Research this and other radiometric techniques and discuss the margins of error for various methods with your classmates.

30. Use Equation 6.5 to show that $k = -\dfrac{\ln(2)}{h}$ where h is the half-life of the radioactive isotope.

31. A pork roast[22] was taken out of a hardwood smoker when its internal temperature had reached 180°F and it was allowed to rest in a 75°F house for 20 minutes after which its internal temperature had dropped to 170°F. Assuming that the temperature of the roast follows Newton's Law of Cooling (Equation 6.6),

    (a) Express the temperature T (in °F) as a function of time t (in minutes).

    (b) Find the time at which the roast would have dropped to 140°F had it not been carved and eaten.

32. In reference to Exercise 44 in Section 5.3, if Fritzy the Fox's speed is the same as Chewbacca the Bunny's speed, Fritzy's pursuit curve is given by

    $$y(x) = \frac{1}{4}x^2 - \frac{1}{4}\ln(x) - \frac{1}{4}$$

    Use your calculator to graph this path for $x > 0$. Describe the behavior of y as $x \to 0^+$ and interpret this physically.

33. The current i measured in amps in a certain electronic circuit with a constant impressed voltage of 120 volts is given by $i(t) = 2 - 2e^{-10t}$ where $t \geq 0$ is the number of seconds after the circuit is switched on. Determine the value of i as $t \to \infty$. (This is called the **steady state** current.)

34. If the voltage in the circuit in Exercise 33 above is switched off after 30 seconds, the current is given by the piecewise-defined function

    $$i(t) = \begin{cases} 2 - 2e^{-10t} & \text{if } 0 \leq t < 30 \\ \left(2 - 2e^{-300}\right)e^{-10t+300} & \text{if } t \geq 30 \end{cases}$$

    With the help of your calculator, graph $y = i(t)$ and discuss with your classmates the physical significance of the two parts of the graph $0 \leq t < 30$ and $t \geq 30$.

---

[22]This roast was enjoyed by Jeff and his family on June 10, 2009. This is real data, folks!

6.5 APPLICATIONS OF EXPONENTIAL AND LOGARITHMIC FUNCTIONS

35. In Exercise 26 in Section 2.3, we stated that the cable of a suspension bridge formed a parabola but that a free hanging cable did not. A free hanging cable forms a <u>catenary</u> and its basic shape is given by $y = \frac{1}{2}(e^x + e^{-x})$. Use your calculator to graph this function. What are its domain and range? What is its end behavior? Is it invertible? How do you think it is related to the function given in Exercise 47 in Section 6.3 and the one given in the answer to Exercise 38 in Section 6.4? When flipped upside down, the catenary makes an arch. The Gateway Arch in St. Louis, Missouri has the shape

$$y = 757.7 - \frac{127.7}{2}\left(e^{\frac{x}{127.7}} + e^{-\frac{x}{127.7}}\right)$$

where x and y are measured in feet and $-315 \leq x \leq 315$. Find the highest point on the arch.

36. In Exercise 6a in Section 2.5, we examined the data set given below which showed how two cats and their surviving offspring can produce over 80 million cats in just ten years. It is virtually impossible to see this data plotted on your calculator, so plot x versus $\ln(x)$ as was done on page 480. Find a linear model for this new data and comment on its goodness of fit. Find an exponential model for the original data and comment on its goodness of fit.

Year x	1	2	3	4	5	6	7	8	9	10
Number of Cats $N(x)$	12	66	382	2201	12680	73041	420715	2423316	13968290	80399780

37. This exercise is a follow-up to Exercise 26 which more thoroughly explores the population growth of Painesville, Ohio. According to <u>Wikipedia</u>, the population of Painesville, Ohio is given by

Year t	1860	1870	1880	1890	1900	1910	1920	1930	1940	1950
Population	2649	3728	3841	4755	5024	5501	7272	10944	12235	14432

Year t	1960	1970	1980	1990	2000
Population	16116	16536	16351	15699	17503

(a) Use a graphing utility to perform an exponential regression on the data from 1860 through 1920 only, letting $t = 0$ represent the year 1860 as before. How does this calculator model compare with the model you found in Exercise 26? Use the calculator's exponential model to predict the population in 2010. (The 2010 census gave the population as 19,563)

(b) The logistic model fit to *all* of the given data points for the population of Painesville t years after 1860 (again, using $t = 0$ as 1860) is

$$P(t) = \frac{18691}{1 + 9.8505e^{-0.03617t}}$$

According to this model, what should the population of Painesville have been in 2010? (The 2010 census gave the population as 19,563.) What is the population limit of Painesville?

38. According to OhioBiz, the census data for Lake County, Ohio is as follows:

Year t	1860	1870	1880	1890	1900	1910	1920	1930	1940	1950
Population	15576	15935	16326	18235	21680	22927	28667	41674	50020	75979

Year t	1960	1970	1980	1990	2000
Population	148700	197200	212801	215499	227511

(a) Use your calculator to fit a logistic model to these data, using $x = 0$ to represent the year 1860.

(b) Graph these data and your logistic function on your calculator to judge the reasonableness of the fit.

(c) Use this model to estimate the population of Lake County in 2010. (The 2010 census gave the population to be 230,041.)

(d) According to your model, what is the population limit of Lake County, Ohio?

39. According to facebook, the number of active users of facebook has grown significantly since its initial launch from a Harvard dorm room in February 2004. The chart below has the approximate number $U(x)$ of active users, in millions, x months after February 2004. For example, the first entry $(10, 1)$ means that there were 1 million active users in December 2004 and the last entry $(77, 500)$ means that there were 500 million active users in July 2010.

Month x	10	22	34	38	44	54	59	60	62	65	67	70	72	77
Active Users in Millions $U(x)$	1	5.5	12	20	50	100	150	175	200	250	300	350	400	500

With the help of your classmates, find a model for this data.

40. Each Monday during the registration period before the Fall Semester at LCCC, the Enrollment Planning Council gets a report prepared by the data analysts in Institutional Effectiveness and Planning.[23] While the ongoing enrollment data is analyzed in many different ways, we shall focus only on the overall headcount. Below is a chart of the enrollment data for Fall Semester 2008. It starts 21 weeks before "Opening Day" and ends on "Day 15" of the semester, but we have relabeled the top row to be $x = 1$ through $x = 24$ so that the math is easier. (Thus, $x = 22$ is Opening Day.)

Week x	1	2	3	4	5	6	7	8
Total Headcount	1194	1564	2001	2475	2802	3141	3527	3790

Week x	9	10	11	12	13	14	15	16
Total Headcount	4065	4371	4611	4945	5300	5657	6056	6478

---

[23]The authors thank Dr. Wendy Marley and her staff for this data and Dr. Marcia Ballinger for the permission to use it in this problem.

6.5 APPLICATIONS OF EXPONENTIAL AND LOGARITHMIC FUNCTIONS

Week x	17	18	19	20	21	22	23	24
Total Headcount	7161	7772	8505	9256	10201	10743	11102	11181

With the help of your classmates, find a model for this data. Unlike most of the phenomena we have studied in this section, there is no single differential equation which governs the enrollment growth. Thus there is no scientific reason to rely on a logistic function even though the data plot may lead us to that model. What are some factors which influence enrollment at a community college and how can you take those into account mathematically?

41. When we wrote this exercise, the Enrollment Planning Report for Fall Semester 2009 had only 10 data points for the first 10 weeks of the registration period. Those numbers are given below.

Week x	1	2	3	4	5	6	7	8	9	10
Total Headcount	1380	2000	2639	3153	3499	3831	4283	4742	5123	5398

With the help of your classmates, find a model for this data and make a prediction for the Opening Day enrollment as well as the Day 15 enrollment. (WARNING: The registration period for 2009 was one week shorter than it was in 2008 so Opening Day would be $x = 21$ and Day 15 is $x = 23$.)

6.5.4 Answers

1. - $A(t) = 500\left(1 + \frac{0.0075}{12}\right)^{12t}$
   - $A(5) \approx \$519.10$, $A(10) \approx \$538.93$, $A(30) \approx \$626.12$, $A(35) \approx \$650.03$
   - It will take approximately 92 years for the investment to double.
   - The average rate of change from the end of the fourth year to the end of the fifth year is approximately 3.88. This means that the investment is growing at an average rate of \$3.88 per year at this point. The average rate of change from the end of the thirty-fourth year to the end of the thirty-fifth year is approximately 4.85. This means that the investment is growing at an average rate of \$4.85 per year at this point.

2. - $A(t) = 500e^{0.0075t}$
   - $A(5) \approx \$519.11$, $A(10) \approx \$538.94$, $A(30) \approx \$626.16$, $A(35) \approx \$650.09$
   - It will take approximately 92 years for the investment to double.
   - The average rate of change from the end of the fourth year to the end of the fifth year is approximately 3.88. This means that the investment is growing at an average rate of \$3.88 per year at this point. The average rate of change from the end of the thirty-fourth year to the end of the thirty-fifth year is approximately 4.86. This means that the investment is growing at an average rate of \$4.86 per year at this point.

3. - $A(t) = 1000\left(1 + \frac{0.0125}{12}\right)^{12t}$
   - $A(5) \approx \$1064.46$, $A(10) \approx \$1133.07$, $A(30) \approx \$1454.71$, $A(35) \approx \$1548.48$
   - It will take approximately 55 years for the investment to double.
   - The average rate of change from the end of the fourth year to the end of the fifth year is approximately 13.22. This means that the investment is growing at an average rate of \$13.22 per year at this point. The average rate of change from the end of the thirty-fourth year to the end of the thirty-fifth year is approximately 19.23. This means that the investment is growing at an average rate of \$19.23 per year at this point.

4. - $A(t) = 1000e^{0.0125t}$
   - $A(5) \approx \$1064.49$, $A(10) \approx \$1133.15$, $A(30) \approx \$1454.99$, $A(35) \approx \$1548.83$
   - It will take approximately 55 years for the investment to double.
   - The average rate of change from the end of the fourth year to the end of the fifth year is approximately 13.22. This means that the investment is growing at an average rate of \$13.22 per year at this point. The average rate of change from the end of the thirty-fourth year to the end of the thirty-fifth year is approximately 19.24. This means that the investment is growing at an average rate of \$19.24 per year at this point.

5. - $A(t) = 5000\left(1 + \frac{0.02125}{12}\right)^{12t}$
   - $A(5) \approx \$5559.98$, $A(10) \approx \$6182.67$, $A(30) \approx \$9453.40$, $A(35) \approx \$10512.13$
   - It will take approximately 33 years for the investment to double.

6.5 Applications of Exponential and Logarithmic Functions

- The average rate of change from the end of the fourth year to the end of the fifth year is approximately 116.80. This means that the investment is growing at an average rate of $116.80 per year at this point. The average rate of change from the end of the thirty-fourth year to the end of the thirty-fifth year is approximately 220.83. This means that the investment is growing at an average rate of $220.83 per year at this point.

6.
- $A(t) = 5000e^{0.02125t}$
- $A(5) \approx \$5560.50$, $A(10) \approx \$6183.83$, $A(30) \approx \$9458.73$, $A(35) \approx \$10519.05$
- It will take approximately 33 years for the investment to double.
- The average rate of change from the end of the fourth year to the end of the fifth year is approximately 116.91. This means that the investment is growing at an average rate of $116.91 per year at this point. The average rate of change from the end of the thirty-fourth year to the end of the thirty-fifth year is approximately 221.17. This means that the investment is growing at an average rate of $221.17 per year at this point.

8. $P = \dfrac{2000}{e^{0.0025 \cdot 3}} \approx \$1985.06$

9. $P = \dfrac{5000}{\left(1+\frac{0.0225}{12}\right)^{12 \cdot 10}} \approx \$3993.42$

10. (a) $A(8) = 2000\left(1 + \frac{0.0025}{12}\right)^{12 \cdot 8} \approx \$2040.40$

    (b) $t = \dfrac{\ln(2)}{12\ln\left(1 + \frac{0.0025}{12}\right)} \approx 277.29$ years

    (c) $P = \dfrac{2000}{\left(1 + \frac{0.0025}{12}\right)^{36}} \approx \$1985.06$

11. (a) $A(8) = 2000\left(1 + \frac{0.0225}{12}\right)^{12 \cdot 8} \approx \$2394.03$

    (b) $t = \dfrac{\ln(2)}{12\ln\left(1 + \frac{0.0225}{12}\right)} \approx 30.83$ years

    (c) $P = \dfrac{2000}{\left(1 + \frac{0.0225}{12}\right)^{36}} \approx \$1869.57$

    (d) $\left(1 + \frac{0.0225}{12}\right)^{12} \approx 1.0227$ so the APY is 2.27%

12. $A(3) = 5000e^{0.299 \cdot 3} \approx \$12,226.18$, $A(6) = 5000e^{0.299 \cdot 6} \approx \$30,067.29$

14.
- $k = \dfrac{\ln(1/2)}{5.27} \approx -0.1315$
- $A(t) = 50e^{-0.1315t}$
- $t = \dfrac{\ln(0.1)}{-0.1315} \approx 17.51$ years.

15.
- $k = \dfrac{\ln(1/2)}{14} \approx -0.0495$
- $A(t) = 2e^{-0.0495t}$
- $t = \dfrac{\ln(0.1)}{-0.0495} \approx 46.52$ days.

16.
- $k = \frac{\ln(1/2)}{27.7} \approx -0.0250$
- $A(t) = 75e^{-0.0250t}$
- $t = \frac{\ln(0.1)}{-0.025} \approx 92.10$ days.

17.
- $k = \frac{\ln(1/2)}{432.7} \approx -0.0016$
- $A(t) = 0.29e^{-0.0016t}$
- $t = \frac{\ln(0.1)}{-0.0016} \approx 1439.11$ years.

18.
- $k = \frac{\ln(1/2)}{704} \approx -0.0010$
- $A(t) = e^{-0.0010t}$
- $t = \frac{\ln(0.1)}{-0.0010} \approx 2302.58$ million years, or 2.30 billion years.

19. $t = \frac{\ln(0.1)}{k} = -\frac{\ln(10)}{k}$

20. $V(t) = 25e^{\ln\left(\frac{4}{5}\right)t} \approx 25e^{-0.22314355t}$

21. (a) $G(0) = 9743.77$ This means that the GDP of the US in 2000 was \$9743.77 billion dollars.

(b) $G(7) = 13963.24$ and $G(10) = 16291.25$, so the model predicted a GDP of \$13,963.24 billion in 2007 and \$16,291.25 billion in 2010.

22. (a) $D(0) = 15$, so the tumor was 15 millimeters in diameter when it was first detected.

(b) $t = \frac{\ln(2)}{0.0277} \approx 25$ days.

23. (a) $k = \frac{\ln(2)}{20} \approx 0.0346$

(b) $N(t) = 1000e^{0.0346t}$

(c) $t = \frac{\ln(9)}{0.0346} \approx 63$ minutes

24. (a) $k = \frac{1}{2}\frac{\ln(6)}{2.5} \approx 0.4377$

(b) $N(t) = 2.5e^{0.4377t}$

(c) $t = \frac{\ln(2)}{0.4377} \approx 1.58$ hours

25. $N_0 = 52$, $k = \frac{1}{3}\ln\left(\frac{118}{52}\right) \approx 0.2731$, $N(t) = 52e^{0.2731t}$. $N(6) \approx 268$.

26. $N_0 = 2649$, $k = \frac{1}{60}\ln\left(\frac{7272}{2649}\right) \approx 0.0168$, $N(t) = 2649e^{0.0168t}$. $N(150) \approx 32923$, so the population of Painesville in 2010 based on this model would have been 32,923.

27. (a) $P(0) = \frac{120}{4.167} \approx 29$. There are 29 Sasquatch in Bigfoot County in 2010.

(b) $P(3) = \frac{120}{1+3.167e^{-0.05(3)}} \approx 32$ Sasquatch.

(c) $t = 20\ln(3.167) \approx 23$ years.

(d) As $t \to \infty$, $P(t) \to 120$. As time goes by, the Sasquatch Population in Bigfoot County will approach 120. Graphically, $y = P(x)$ has a horizontal asymptote $y = 120$.

28. (a) $A(t) = Ne^{-\left(\frac{\ln(2)}{5730}\right)t} \approx Ne^{-0.00012097t}$

(b) $A(20000) \approx 0.088978 \cdot N$ so about 8.9% remains

(c) $t \approx \frac{\ln(.42)}{-0.00012097} \approx 7171$ years old

29. $A(t) = 2.3e^{-0.0138629t}$

6.5 Applications of Exponential and Logarithmic Functions

31. (a) $T(t) = 75 + 105e^{-0.005005t}$

    (b) The roast would have cooled to 140°F in about 95 minutes.

32. From the graph, it appears that as $x \to 0^+$, $y \to \infty$. This is due to the presence of the $\ln(x)$ term in the function. This means that Fritzy will never catch Chewbacca, which makes sense since Chewbacca has a head start and Fritzy only runs as fast as he does.

    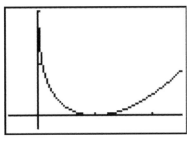

    $y(x) = \frac{1}{4}x^2 - \frac{1}{4}\ln(x) - \frac{1}{4}$

33. The steady state current is 2 amps.

36. The linear regression on the data below is $y = 1.74899x + 0.70739$ with $r^2 \approx 0.999995$. This is an excellent fit.

x	1	2	3	4	5	6	7	8	9	10
$\ln(N(x))$	2.4849	4.1897	5.9454	7.6967	9.4478	11.1988	12.9497	14.7006	16.4523	18.2025

    $N(x) = 2.02869(5.74879)^x = 2.02869e^{1.74899x}$ with $r^2 \approx 0.999995$. This is also an excellent fit and corresponds to our linearized model because $\ln(2.02869) \approx 0.70739$.

37. (a) The calculator gives: $y = 2895.06(1.0147)^x$. Graphing this along with our answer from Exercise 26 over the interval $[0, 60]$ shows that they are pretty close. From this model, $y(150) \approx 25840$ which once again overshoots the actual data value.

    (b) $P(150) \approx 18717$, so this model predicts 17,914 people in Painesville in 2010, a more conservative number than was recorded in the 2010 census. As $t \to \infty$, $P(t) \to 18691$. So the limiting population of Painesville based on this model is 18,691 people.

38. (a) $y = \dfrac{242526}{1 + 874.62e^{-0.07113x}}$, where x is the number of years since 1860.

    (b) The plot of the data and the curve is below.

    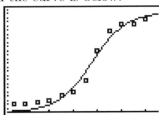

    (c) $y(140) \approx 232889$, so this model predicts 232,889 people in Lake County in 2010.

    (d) As $x \to \infty$, $y \to 242526$, so the limiting population of Lake County based on this model is 242,526 people.

Chapter 7

Hooked on Conics

7.1 Introduction to Conics

In this chapter, we study the **Conic Sections** - literally 'sections of a cone'. Imagine a double-napped cone as seen below being 'sliced' by a plane.

If we slice the cone with a horizontal plane the resulting curve is a **circle**.

Tilting the plane ever so slightly produces an **ellipse**.

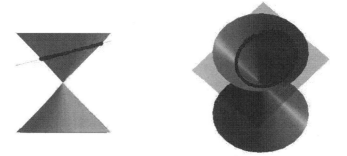

If the plane cuts parallel to the cone, we get a **parabola**.

If we slice the cone with a vertical plane, we get a **hyperbola**.

For a wonderful animation describing the conics as intersections of planes and cones, see Dr. Louis Talman's Mathematics Animated Website.

7.1 Introduction to Conics

If the slicing plane contains the vertex of the cone, we get the so-called 'degenerate' conics: a point, a line, or two intersecting lines.

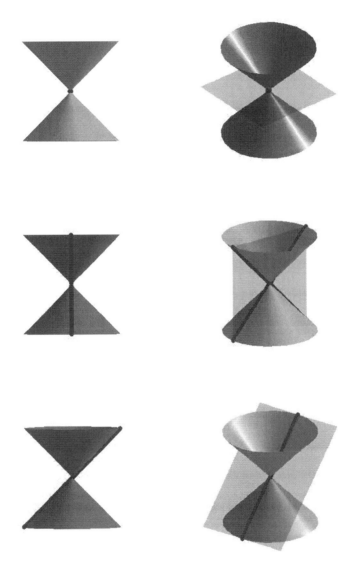

We will focus the discussion on the non-degenerate cases: circles, parabolas, ellipses, and hyperbolas, in that order. To determine equations which describe these curves, we will make use of their definitions in terms of distances.

7.2 Circles

Recall from Geometry that a circle can be determined by fixing a point (called the center) and a positive number (called the radius) as follows.

Definition 7.1. A **circle** with center (h, k) and radius $r > 0$ is the set of all points (x, y) in the plane whose distance to (h, k) is r.

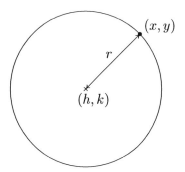

From the picture, we see that a point (x, y) is on the circle if and only if its distance to (h, k) is r. We express this relationship algebraically using the Distance Formula, Equation 1.1, as

$$r = \sqrt{(x-h)^2 + (y-k)^2}$$

By squaring both sides of this equation, we get an equivalent equation (since $r > 0$) which gives us the standard equation of a circle.

Equation 7.1. The Standard Equation of a Circle: The equation of a circle with center (h, k) and radius $r > 0$ is $(x - h)^2 + (y - k)^2 = r^2$.

Example 7.2.1. Write the standard equation of the circle with center $(-2, 3)$ and radius 5.

Solution. Here, $(h, k) = (-2, 3)$ and $r = 5$, so we get

$$\begin{aligned} (x - (-2))^2 + (y - 3)^2 &= (5)^2 \\ (x + 2)^2 + (y - 3)^2 &= 25 \end{aligned}$$

□

Example 7.2.2. Graph $(x + 2)^2 + (y - 1)^2 = 4$. Find the center and radius.

Solution. From the standard form of a circle, Equation 7.1, we have that $x + 2$ is $x - h$, so $h = -2$ and $y - 1$ is $y - k$ so $k = 1$. This tells us that our center is $(-2, 1)$. Furthermore, $r^2 = 4$, so $r = 2$. Thus we have a circle centered at $(-2, 1)$ with a radius of 2. Graphing gives us

7.2 CIRCLES

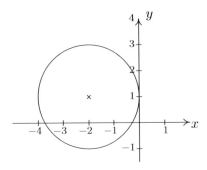

If we were to expand the equation in the previous example and gather up like terms, instead of the easily recognizable $(x+2)^2 + (y-1)^2 = 4$, we'd be contending with $x^2 + 4x + y^2 - 2y + 1 = 0$. If we're given such an equation, we can complete the square in each of the variables to see if it fits the form given in Equation 7.1 by following the steps given below.

To Write the Equation of a Circle in Standard Form

1. Group the same variables together on one side of the equation and position the constant on the other side.

2. Complete the square on both variables as needed.

3. Divide both sides by the coefficient of the squares. (For circles, they will be the same.)

Example 7.2.3. Complete the square to find the center and radius of $3x^2 - 6x + 3y^2 + 4y - 4 = 0$.

Solution.

$$\begin{aligned}
3x^2 - 6x + 3y^2 + 4y - 4 &= 0 \\
3x^2 - 6x + 3y^2 + 4y &= 4 & \text{add 4 to both sides} \\
3\left(x^2 - 2x\right) + 3\left(y^2 + \frac{4}{3}y\right) &= 4 & \text{factor out leading coefficients} \\
3\left(x^2 - 2x + \underline{1}\right) + 3\left(y^2 + \frac{4}{3}y + \underline{\frac{4}{9}}\right) &= 4 + 3(\underline{1}) + 3\left(\underline{\frac{4}{9}}\right) & \text{complete the square in } x, y \\
3(x-1)^2 + 3\left(y + \frac{2}{3}\right)^2 &= \frac{25}{3} & \text{factor} \\
(x-1)^2 + \left(y + \frac{2}{3}\right)^2 &= \frac{25}{9} & \text{divide both sides by 3}
\end{aligned}$$

From Equation 7.1, we identify $x - 1$ as $x - h$, so $h = 1$, and $y + \frac{2}{3}$ as $y - k$, so $k = -\frac{2}{3}$. Hence, the center is $(h, k) = \left(1, -\frac{2}{3}\right)$. Furthermore, we see that $r^2 = \frac{25}{9}$ so the radius is $r = \frac{5}{3}$. □

It is possible to obtain equations like $(x-3)^2 + (y+1)^2 = 0$ or $(x-3)^2 + (y+1)^2 = -1$, neither of which describes a circle. (Do you see why not?) The reader is encouraged to think about what, if any, points lie on the graphs of these two equations. The next example uses the Midpoint Formula, Equation 1.2, in conjunction with the ideas presented so far in this section.

Example 7.2.4. Write the standard equation of the circle which has $(-1, 3)$ and $(2, 4)$ as the endpoints of a diameter.

Solution. We recall that a diameter of a circle is a line segment containing the center and two points on the circle. Plotting the given data yields

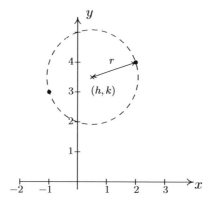

Since the given points are endpoints of a diameter, we know their midpoint (h, k) is the center of the circle. Equation 1.2 gives us

$$\begin{aligned}(h, k) &= \left(\frac{x_1 + x_2}{2}, \frac{y_1 + y_2}{2}\right) \\ &= \left(\frac{-1 + 2}{2}, \frac{3 + 4}{2}\right) \\ &= \left(\frac{1}{2}, \frac{7}{2}\right)\end{aligned}$$

The diameter of the circle is the distance between the given points, so we know that half of the distance is the radius. Thus,

$$\begin{aligned}r &= \frac{1}{2}\sqrt{(x_2 - x_1)^2 + (y_2 - y_1)^2} \\ &= \frac{1}{2}\sqrt{(2 - (-1))^2 + (4 - 3)^2} \\ &= \frac{1}{2}\sqrt{3^2 + 1^2} \\ &= \frac{\sqrt{10}}{2}\end{aligned}$$

Finally, since $\left(\frac{\sqrt{10}}{2}\right)^2 = \frac{10}{4}$, our answer becomes $\left(x - \frac{1}{2}\right)^2 + \left(y - \frac{7}{2}\right)^2 = \frac{10}{4}$ □

7.2 Circles

We close this section with the most important[1] circle in all of mathematics: the **Unit Circle**.

Definition 7.2. The **Unit Circle** is the circle centered at $(0,0)$ with a radius of 1. The standard equation of the Unit Circle is $x^2 + y^2 = 1$.

Example 7.2.5. Find the points on the unit circle with y-coordinate $\frac{\sqrt{3}}{2}$.

Solution. We replace y with $\frac{\sqrt{3}}{2}$ in the equation $x^2 + y^2 = 1$ to get

$$\begin{aligned} x^2 + y^2 &= 1 \\ x^2 + \left(\frac{\sqrt{3}}{2}\right)^2 &= 1 \\ \frac{3}{4} + x^2 &= 1 \\ x^2 &= \frac{1}{4} \\ x &= \pm\sqrt{\frac{1}{4}} \\ x &= \pm\frac{1}{2} \end{aligned}$$

Our final answers are $\left(\frac{1}{2}, \frac{\sqrt{3}}{2}\right)$ and $\left(-\frac{1}{2}, \frac{\sqrt{3}}{2}\right)$. □

---

[1] While this may seem like an opinion, it is indeed a fact. See Chapters 10 and 11 for details.

7.2.1 Exercises

In Exercises 1 - 6, find the standard equation of the circle and then graph it.

1. Center $(-1, -5)$, radius 10
2. Center $(4, -2)$, radius 3
3. Center $\left(-3, \frac{7}{13}\right)$, radius $\frac{1}{2}$
4. Center $(5, -9)$, radius $\ln(8)$
5. Center $(-e, \sqrt{2})$, radius π
6. Center (π, e^2), radius $\sqrt[3]{91}$

In Exercises 7 - 12, complete the square in order to put the equation into standard form. Identify the center and the radius or explain why the equation does not represent a circle.

7. $x^2 - 4x + y^2 + 10y = -25$
8. $-2x^2 - 36x - 2y^2 - 112 = 0$
9. $x^2 + y^2 + 8x - 10y - 1 = 0$
10. $x^2 + y^2 + 5x - y - 1 = 0$
11. $4x^2 + 4y^2 - 24y + 36 = 0$
12. $x^2 + x + y^2 - \frac{6}{5}y = 1$

In Exercises 13 - 16, find the standard equation of the circle which satisfies the given criteria.

13. center $(3, 5)$, passes through $(-1, -2)$
14. center $(3, 6)$, passes through $(-1, 4)$
15. endpoints of a diameter: $(3, 6)$ and $(-1, 4)$
16. endpoints of a diameter: $\left(\frac{1}{2}, 4\right)$, $\left(\frac{3}{2}, -1\right)$

17. The Giant Wheel at Cedar Point is a circle with diameter 128 feet which sits on an 8 foot tall platform making its overall height is 136 feet.[2] Find an equation for the wheel assuming that its center lies on the y-axis and that the ground is the x-axis.

18. Verify that the following points lie on the Unit Circle: $(\pm 1, 0)$, $(0, \pm 1)$, $\left(\pm\frac{\sqrt{2}}{2}, \pm\frac{\sqrt{2}}{2}\right)$, $\left(\pm\frac{1}{2}, \pm\frac{\sqrt{3}}{2}\right)$ and $\left(\pm\frac{\sqrt{3}}{2}, \pm\frac{1}{2}\right)$

19. Discuss with your classmates how to obtain the standard equation of a circle, Equation 7.1, from the equation of the Unit Circle, $x^2 + y^2 = 1$ using the transformations discussed in Section 1.7. (Thus every circle is just a few transformations away from the Unit Circle.)

20. Find an equation for the function represented graphically by the top half of the Unit Circle. Explain how the transformations is Section 1.7 can be used to produce a function whose graph is either the top or bottom of an arbitrary circle.

21. Find a one-to-one function whose graph is half of a circle. (Hint: Think piecewise.)

---

[2] Source: Cedar Point's webpage.

7.2.2 Answers

1. $(x+1)^2 + (y+5)^2 = 100$

2. $(x-4)^2 + (y+2)^2 = 9$

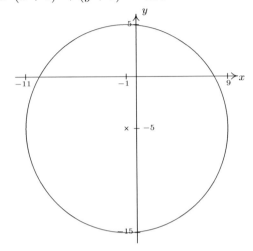
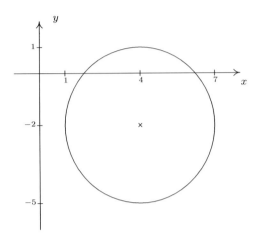

3. $(x+3)^2 + \left(y - \frac{7}{13}\right)^2 = \frac{1}{4}$

4. $(x-5)^2 + (y+9)^2 = (\ln(8))^2$

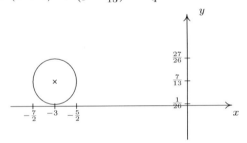
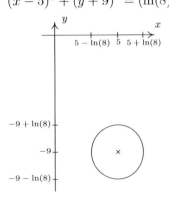

5. $(x+e)^2 + \left(y - \sqrt{2}\right)^2 = \pi^2$

6. $(x-\pi)^2 + \left(y - e^2\right)^2 = 91^{\frac{2}{3}}$

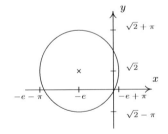
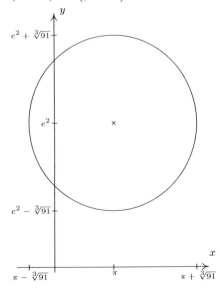

7. $(x-2)^2 + (y+5)^2 = 4$
   Center $(2,-5)$, radius $r = 2$

8. $(x+9)^2 + y^2 = 25$
   Center $(-9,0)$, radius $r = 5$

9. $(x+4)^2 + (y-5)^2 = 42$
   Center $(-4,5)$, radius $r = \sqrt{42}$

10. $\left(x + \frac{5}{2}\right)^2 + \left(y - \frac{1}{2}\right)^2 = \frac{30}{4}$
    Center $\left(-\frac{5}{2}, \frac{1}{2}\right)$, radius $r = \frac{\sqrt{30}}{2}$

11. $x^2 + (y-3)^2 = 0$
    This is not a circle.

12. $\left(x + \frac{1}{2}\right)^2 + \left(y - \frac{3}{5}\right)^2 = \frac{161}{100}$
    Center $\left(-\frac{1}{2}, \frac{3}{5}\right)$, radius $r = \frac{\sqrt{161}}{10}$

13. $(x-3)^2 + (y-5)^2 = 65$

14. $(x-3)^2 + (y-6)^2 = 20$

15. $(x-1)^2 + (y-5)^2 = 5$

16. $(x-1)^2 + \left(y - \frac{3}{2}\right)^2 = \frac{13}{2}$

17. $x^2 + (y-72)^2 = 4096$

7.3 Parabolas

We have already learned that the graph of a quadratic function $f(x) = ax^2 + bx + c$ ($a \neq 0$) is called a **parabola**. To our surprise and delight, we may also define parabolas in terms of distance.

> **Definition 7.3.** Let F be a point in the plane and D be a line not containing F. A **parabola** is the set of all points equidistant from F and D. The point F is called the **focus** of the parabola and the line D is called the **directrix** of the parabola.

Schematically, we have the following.

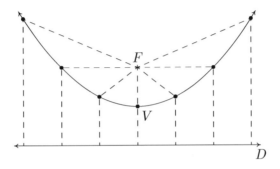

Each dashed line from the point F to a point on the curve has the same length as the dashed line from the point on the curve to the line D. The point suggestively labeled V is, as you should expect, the **vertex**. The vertex is the point on the parabola closest to the focus.

We want to use only the distance definition of parabola to derive the equation of a parabola and, if all is right with the universe, we should get an expression much like those studied in Section 2.3. Let p denote the directed[1] distance from the vertex to the focus, which by definition is the same as the distance from the vertex to the directrix. For simplicity, assume that the vertex is $(0,0)$ and that the parabola opens upwards. Hence, the focus is $(0, p)$ and the directrix is the line $y = -p$. Our picture becomes

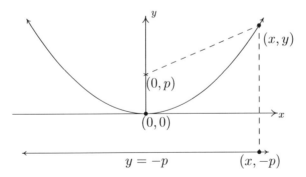

From the definition of parabola, we know the distance from $(0, p)$ to (x, y) is the same as the distance from $(x, -p)$ to (x, y). Using the Distance Formula, Equation 1.1, we get

---
[1] We'll talk more about what 'directed' means later.

$$\begin{aligned}\sqrt{(x-0)^2+(y-p)^2} &= \sqrt{(x-x)^2+(y-(-p))^2}\\ \sqrt{x^2+(y-p)^2} &= \sqrt{(y+p)^2}\\ x^2+(y-p)^2 &= (y+p)^2 \qquad \text{square both sides}\\ x^2+y^2-2py+p^2 &= y^2+2py+p^2 \qquad \text{expand quantities}\\ x^2 &= 4py \qquad \text{gather like terms}\end{aligned}$$

Solving for y yields $y = \frac{x^2}{4p}$, which is a quadratic function of the form found in Equation 2.4 with $a = \frac{1}{4p}$ and vertex $(0,0)$.

We know from previous experience that if the coefficient of x^2 is negative, the parabola opens downwards. In the equation $y = \frac{x^2}{4p}$ this happens when $p < 0$. In our formulation, we say that p is a 'directed distance' from the vertex to the focus: if $p > 0$, the focus is above the vertex; if $p < 0$, the focus is below the vertex. The **focal length** of a parabola is $|p|$.

If we choose to place the vertex at an arbitrary point (h, k), we arrive at the following formula using either transformations from Section 1.7 or re-deriving the formula from Definition 7.3.

> **Equation 7.2. The Standard Equation of a Vertical[a] Parabola:** The equation of a (vertical) parabola with vertex (h, k) and focal length $|p|$ is
>
> $$(x-h)^2 = 4p(y-k)$$
>
> If $p > 0$, the parabola opens upwards; if $p < 0$, it opens downwards.
>
> ---
> [a]That is, a parabola which opens either upwards or downwards.

Notice that in the standard equation of the parabola above, only one of the variables, x, is squared. This is a quick way to distinguish an equation of a parabola from that of a circle because in the equation of a circle, both variables are squared.

Example 7.3.1. Graph $(x+1)^2 = -8(y-3)$. Find the vertex, focus, and directrix.

Solution. We recognize this as the form given in Equation 7.2. Here, $x - h$ is $x + 1$ so $h = -1$, and $y - k$ is $y - 3$ so $k = 3$. Hence, the vertex is $(-1, 3)$. We also see that $4p = -8$ so $p = -2$. Since $p < 0$, the focus will be below the vertex and the parabola will open downwards.

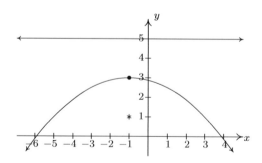

7.3 Parabolas

The distance from the vertex to the focus is $|p| = 2$, which means the focus is 2 units below the vertex. From $(-1, 3)$, we move down 2 units and find the focus at $(-1, 1)$. The directrix, then, is 2 units above the vertex, so it is the line $y = 5$. □

Of all of the information requested in the previous example, only the vertex is part of the graph of the parabola. So in order to get a sense of the actual shape of the graph, we need some more information. While we could plot a few points randomly, a more useful measure of how wide a parabola opens is the length of the parabola's latus rectum.[2] The **latus rectum** of a parabola is the line segment parallel to the directrix which contains the focus. The endpoints of the latus rectum are, then, two points on 'opposite' sides of the parabola. Graphically, we have the following.

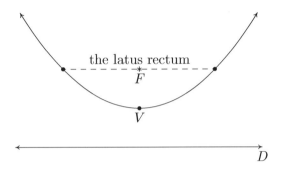

It turns out[3] that the length of the latus rectum, called the **focal diameter** of the parabola is $|4p|$, which, in light of Equation 7.2, is easy to find. In our last example, for instance, when graphing $(x + 1)^2 = -8(y - 3)$, we can use the fact that the focal diameter is $|-8| = 8$, which means the parabola is 8 units wide at the focus, to help generate a more accurate graph by plotting points 4 units to the left and right of the focus.

Example 7.3.2. Find the standard form of the parabola with focus $(2, 1)$ and directrix $y = -4$.

Solution. Sketching the data yields,

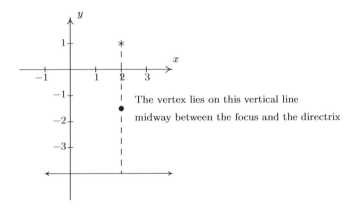

---
[2]No, I'm not making this up.
[3]Consider this an exercise to show what follows.

From the diagram, we see the parabola opens upwards. (Take a moment to think about it if you don't see that immediately.) Hence, the vertex lies below the focus and has an x-coordinate of 2. To find the y-coordinate, we note that the distance from the focus to the directrix is $1 - (-4) = 5$, which means the vertex lies $\frac{5}{2}$ units (halfway) below the focus. Starting at $(2, 1)$ and moving down $5/2$ units leaves us at $(2, -3/2)$, which is our vertex. Since the parabola opens upwards, we know p is positive. Thus $p = 5/2$. Plugging all of this data into Equation 7.2 give us

$$(x - 2)^2 = 4\left(\frac{5}{2}\right)\left(y - \left(-\frac{3}{2}\right)\right)$$
$$(x - 2)^2 = 10\left(y + \frac{3}{2}\right)$$

$\square$

If we interchange the roles of x and y, we can produce 'horizontal' parabolas: parabolas which open to the left or to the right. The directrices[4] of such animals would be vertical lines and the focus would either lie to the left or to the right of the vertex, as seen below.

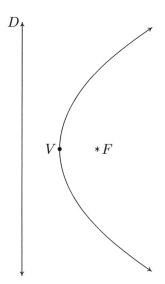

> **Equation 7.3. The Standard Equation of a Horizontal Parabola:** The equation of a (horizontal) parabola with vertex (h, k) and focal length $|p|$ is
>
> $$(y - k)^2 = 4p(x - h)$$
>
> If $p > 0$, the parabola opens to the right; if $p < 0$, it opens to the left.

---

[4] plural of 'directrix'

7.3 PARABOLAS

Example 7.3.3. Graph $(y-2)^2 = 12(x+1)$. Find the vertex, focus, and directrix.

Solution. We recognize this as the form given in Equation 7.3. Here, $x - h$ is $x + 1$ so $h = -1$, and $y - k$ is $y - 2$ so $k = 2$. Hence, the vertex is $(-1, 2)$. We also see that $4p = 12$ so $p = 3$. Since $p > 0$, the focus will be the right of the vertex and the parabola will open to the right. The distance from the vertex to the focus is $|p| = 3$, which means the focus is 3 units to the right. If we start at $(-1, 2)$ and move right 3 units, we arrive at the focus $(2, 2)$. The directrix, then, is 3 units to the left of the vertex and if we move left 3 units from $(-1, 2)$, we'd be on the vertical line $x = -4$. Since the focal diameter is $|4p| = 12$, the parabola is 12 units wide at the focus, and thus there are points 6 units above and below the focus on the parabola.

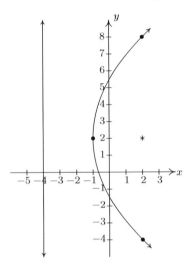

□

As with circles, not all parabolas will come to us in the forms in Equations 7.2 or 7.3. If we encounter an equation with two variables in which exactly one variable is squared, we can attempt to put the equation into a standard form using the following steps.

To Write the Equation of a Parabola in Standard Form

1. Group the variable which is squared on one side of the equation and position the non-squared variable and the constant on the other side.

2. Complete the square if necessary and divide by the coefficient of the perfect square.

3. Factor out the coefficient of the non-squared variable from it and the constant.

Example 7.3.4. Consider the equation $y^2 + 4y + 8x = 4$. Put this equation into standard form and graph the parabola. Find the vertex, focus, and directrix.

Solution. We need a perfect square (in this case, using y) on the left-hand side of the equation and factor out the coefficient of the non-squared variable (in this case, the x) on the other.

$$
\begin{aligned}
y^2 + 4y + 8x &= 4 \\
y^2 + 4y &= -8x + 4 \\
y^2 + 4y + 4 &= -8x + 4 + 4 \quad \text{complete the square in } y \text{ only} \\
(y+2)^2 &= -8x + 8 \quad\quad\quad\quad\quad \text{factor} \\
(y+2)^2 &= -8(x-1)
\end{aligned}
$$

Now that the equation is in the form given in Equation 7.3, we see that $x - h$ is $x - 1$ so $h = 1$, and $y - k$ is $y + 2$ so $k = -2$. Hence, the vertex is $(1, -2)$. We also see that $4p = -8$ so that $p = -2$. Since $p < 0$, the focus will be the left of the vertex and the parabola will open to the left. The distance from the vertex to the focus is $|p| = 2$, which means the focus is 2 units to the left of 1, so if we start at $(1, -2)$ and move left 2 units, we arrive at the focus $(-1, -2)$. The directrix, then, is 2 units to the right of the vertex, so if we move right 2 units from $(1, -2)$, we'd be on the vertical line $x = 3$. Since the focal diameter is $|4p|$ is 8, the parabola is 8 units wide at the focus, so there are points 4 units above and below the focus on the parabola.

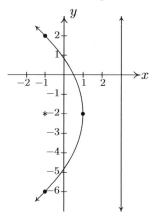

In studying quadratic functions, we have seen parabolas used to model physical phenomena such as the trajectories of projectiles. Other applications of the parabola concern its 'reflective property' which necessitates knowing about the focus of a parabola. For example, many satellite dishes are formed in the shape of a **paraboloid of revolution** as depicted below.

7.3 Parabolas

Every cross section through the vertex of the paraboloid is a parabola with the same focus. To see why this is important, imagine the dashed lines below as electromagnetic waves heading towards a parabolic dish. It turns out that the waves reflect off the parabola and concentrate at the focus which then becomes the optimal place for the receiver. If, on the other hand, we imagine the dashed lines as emanating from the focus, we see that the waves are reflected off the parabola in a coherent fashion as in the case in a flashlight. Here, the bulb is placed at the focus and the light rays are reflected off a parabolic mirror to give directional light.

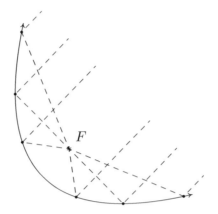

Example 7.3.5. A satellite dish is to be constructed in the shape of a paraboloid of revolution. If the receiver placed at the focus is located 2 ft above the vertex of the dish, and the dish is to be 12 feet wide, how deep will the dish be?

Solution. One way to approach this problem is to determine the equation of the parabola suggested to us by this data. For simplicity, we'll assume the vertex is $(0,0)$ and the parabola opens upwards. Our standard form for such a parabola is $x^2 = 4py$. Since the focus is 2 units above the vertex, we know $p = 2$, so we have $x^2 = 8y$. Visually,

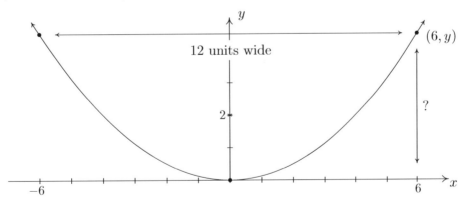

Since the parabola is 12 feet wide, we know the edge is 6 feet from the vertex. To find the depth, we are looking for the y value when $x = 6$. Substituting $x = 6$ into the equation of the parabola yields $6^2 = 8y$ or $y = \frac{36}{8} = \frac{9}{2} = 4.5$. Hence, the dish will be 4.5 feet deep. $\square$

7.3.1 Exercises

In Exercises 1 - 8, sketch the graph of the given parabola. Find the vertex, focus and directrix. Include the endpoints of the latus rectum in your sketch.

1. $(x-3)^2 = -16y$
2. $\left(x+\frac{7}{3}\right)^2 = 2\left(y+\frac{5}{2}\right)$
3. $(y-2)^2 = -12(x+3)$
4. $(y+4)^2 = 4x$
5. $(x-1)^2 = 4(y+3)$
6. $(x+2)^2 = -20(y-5)$
7. $(y-4)^2 = 18(x-2)$
8. $\left(y+\frac{3}{2}\right)^2 = -7\left(x+\frac{9}{2}\right)$

In Exercises 9 - 14, put the equation into standard form and identify the vertex, focus and directrix.

9. $y^2 - 10y - 27x + 133 = 0$
10. $25x^2 + 20x + 5y - 1 = 0$
11. $x^2 + 2x - 8y + 49 = 0$
12. $2y^2 + 4y + x - 8 = 0$
13. $x^2 - 10x + 12y + 1 = 0$
14. $3y^2 - 27y + 4x + \frac{211}{4} = 0$

In Exercises 15 - 18, find an equation for the parabola which fits the given criteria.

15. Vertex $(7, 0)$, focus $(0, 0)$
16. Focus $(10, 1)$, directrix $x = 5$
17. Vertex $(-8, -9)$; $(0, 0)$ and $(-16, 0)$ are points on the curve
18. The endpoints of latus rectum are $(-2, -7)$ and $(4, -7)$

19. The mirror in Carl's flashlight is a paraboloid of revolution. If the mirror is 5 centimeters in diameter and 2.5 centimeters deep, where should the light bulb be placed so it is at the focus of the mirror?

20. A parabolic Wi-Fi antenna is constructed by taking a flat sheet of metal and bending it into a parabolic shape.[5] If the cross section of the antenna is a parabola which is 45 centimeters wide and 25 centimeters deep, where should the receiver be placed to maximize reception?

21. A parabolic arch is constructed which is 6 feet wide at the base and 9 feet tall in the middle. Find the height of the arch exactly 1 foot in from the base of the arch.

22. A popular novelty item is the 'mirage bowl.' Follow this link to see another startling application of the reflective property of the parabola.

23. With the help of your classmates, research spinning liquid mirrors. To get you started, check out this website.

---

[5]This shape is called a 'parabolic cylinder.'

7.3 Parabolas

7.3.2 Answers

1. $(x-3)^2 = -16y$
   Vertex $(3, 0)$
   Focus $(3, -4)$
   Directrix $y = 4$
   Endpoints of latus rectum $(-5, -4)$, $(11, -4)$

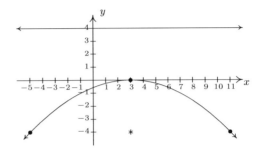

2. $\left(x + \frac{7}{3}\right)^2 = 2\left(y + \frac{5}{2}\right)$
   Vertex $\left(-\frac{7}{3}, -\frac{5}{2}\right)$
   Focus $\left(-\frac{7}{3}, -2\right)$
   Directrix $y = -3$
   Endpoints of latus rectum $\left(-\frac{10}{3}, -2\right)$, $\left(-\frac{4}{3}, -2\right)$

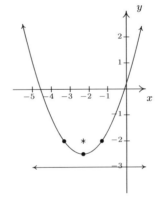

3. $(y-2)^2 = -12(x+3)$
   Vertex $(-3, 2)$
   Focus $(-6, 2)$
   Directrix $x = 0$
   Endpoints of latus rectum $(-6, 8)$, $(-6, -4)$

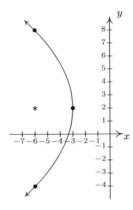

4. $(y+4)^2 = 4x$
   Vertex $(0, -4)$
   Focus $(1, -4)$
   Directrix $x = -1$
   Endpoints of latus rectum $(1, -2)$, $(1, -6)$

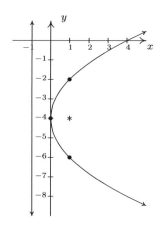

5. $(x-1)^2 = 4(y+3)$
   Vertex $(1, -3)$
   Focus $(1, -2)$
   Directrix $y = -4$
   Endpoints of latus rectum $(3, -2)$, $(-1, -2)$

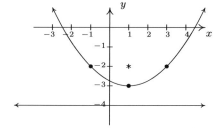

6. $(x+2)^2 = -20(y-5)$
   Vertex $(-2, 5)$
   Focus $(-2, 0)$
   Directrix $y = 10$
   Endpoints of latus rectum $(-12, 0)$, $(8, 0)$

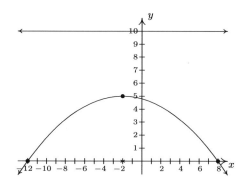

7. $(y-4)^2 = 18(x-2)$
   Vertex $(2, 4)$
   Focus $\left(\frac{13}{2}, 4\right)$
   Directrix $x = -\frac{5}{2}$
   Endpoints of latus rectum $\left(\frac{13}{2}, -5\right)$, $\left(\frac{13}{2}, 13\right)$

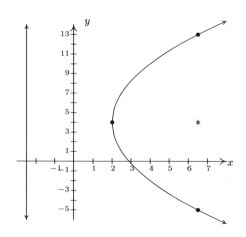

7.3 Parabolas

8. $\left(y+\frac{3}{2}\right)^2 = -7\left(x+\frac{9}{2}\right)$
   Vertex $\left(-\frac{9}{2}, -\frac{3}{2}\right)$
   Focus $\left(-\frac{25}{4}, -\frac{3}{2}\right)$
   Directrix $x = -\frac{11}{4}$
   Endpoints of latus rectum $\left(-\frac{25}{4}, 2\right)$, $\left(-\frac{25}{4}, -5\right)$

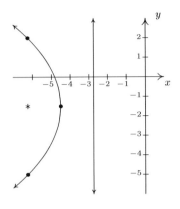

9. $(y-5)^2 = 27(x-4)$
   Vertex $(4, 5)$
   Focus $\left(\frac{43}{4}, 5\right)$
   Directrix $x = -\frac{11}{4}$

10. $\left(x+\frac{2}{5}\right)^2 = -\frac{1}{5}(y-1)$
    Vertex $\left(-\frac{2}{5}, 1\right)$
    Focus $\left(-\frac{2}{5}, \frac{19}{20}\right)$
    Directrix $y = \frac{21}{20}$

11. $(x+1)^2 = 8(y-6)$
    Vertex $(-1, 6)$
    Focus $(-1, 8)$
    Directrix $y = 4$

12. $(y+1)^2 = -\frac{1}{2}(x-10)$
    Vertex $(10, -1)$
    Focus $\left(\frac{79}{8}, -1\right)$
    Directrix $x = \frac{81}{8}$

13. $(x-5)^2 = -12(y-2)$
    Vertex $(5, 2)$
    Focus $(5, -1)$
    Directrix $y = 5$

14. $\left(y-\frac{9}{2}\right)^2 = -\frac{4}{3}(x-2)$
    Vertex $\left(2, \frac{9}{2}\right)$
    Focus $\left(\frac{5}{3}, \frac{9}{2}\right)$
    Directrix $x = \frac{7}{3}$

15. $y^2 = -28(x-7)$

16. $(y-1)^2 = 10\left(x-\frac{15}{2}\right)$

17. $(x+8)^2 = \frac{64}{9}(y+9)$

18. $(x-1)^2 = 6\left(y+\frac{17}{2}\right)$ or
    $(x-1)^2 = -6\left(y+\frac{11}{2}\right)$

19. The bulb should be placed 0.625 centimeters above the vertex of the mirror. (As verified by Carl himself!)

20. The receiver should be placed 5.0625 centimeters from the vertex of the cross section of the antenna.

21. The arch can be modeled by $x^2 = -(y-9)$ or $y = 9 - x^2$. One foot in from the base of the arch corresponds to either $x = \pm 2$, so the height is $y = 9 - (\pm 2)^2 = 5$ feet.

7.4 Ellipses

In the definition of a circle, Definition 7.1, we fixed a point called the **center** and considered all of the points which were a fixed distance r from that one point. For our next conic section, the ellipse, we fix two distinct points and a distance d to use in our definition.

> **Definition 7.4.** Given two distinct points F_1 and F_2 in the plane and a fixed distance d, an **ellipse** is the set of all points (x, y) in the plane such that the sum of each of the distances from F_1 and F_2 to (x, y) is d. The points F_1 and F_2 are called the **foci**[a] of the ellipse.
>
> [a] the plural of 'focus'

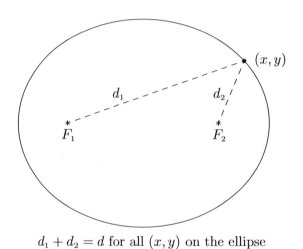

$d_1 + d_2 = d$ for all (x, y) on the ellipse

We may imagine taking a length of string and anchoring it to two points on a piece of paper. The curve traced out by taking a pencil and moving it so the string is always taut is an ellipse.

The **center** of the ellipse is the midpoint of the line segment connecting the two foci. The **major axis** of the ellipse is the line segment connecting two opposite ends of the ellipse which also contains the center and foci. The **minor axis** of the ellipse is the line segment connecting two opposite ends of the ellipse which contains the center but is perpendicular to the major axis. The **vertices** of an ellipse are the points of the ellipse which lie on the major axis. Notice that the center is also the midpoint of the major axis, hence it is the midpoint of the vertices. In pictures we have,

7.4 Ellipses

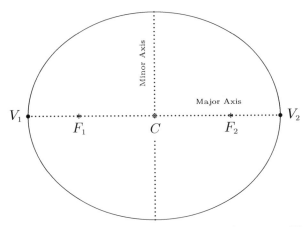

An ellipse with center C; foci F_1, F_2; and vertices V_1, V_2

Note that the major axis is the longer of the two axes through the center, and likewise, the minor axis is the shorter of the two. In order to derive the standard equation of an ellipse, we assume that the ellipse has its center at $(0,0)$, its major axis along the x-axis, and has foci $(c, 0)$ and $(-c, 0)$ and vertices $(-a, 0)$ and $(a, 0)$. We will label the y-intercepts of the ellipse as $(0, b)$ and $(0, -b)$ (We assume a, b, and c are all positive numbers.) Schematically,

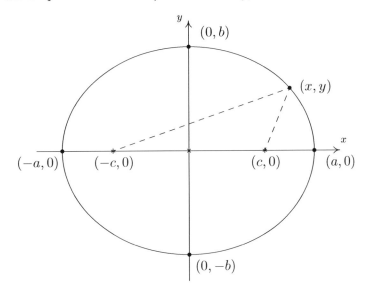

Note that since $(a, 0)$ is on the ellipse, it must satisfy the conditions of Definition 7.4. That is, the distance from $(-c, 0)$ to $(a, 0)$ plus the distance from $(c, 0)$ to $(a, 0)$ must equal the fixed distance d. Since all of these points lie on the x-axis, we get

$$\begin{aligned} \text{distance from } (-c, 0) \text{ to } (a, 0) + \text{distance from } (c, 0) \text{ to } (a, 0) &= d \\ (a + c) + (a - c) &= d \\ 2a &= d \end{aligned}$$

In other words, the fixed distance d mentioned in the definition of the ellipse is none other than the length of the major axis. We now use that fact $(0, b)$ is on the ellipse, along with the fact that $d = 2a$ to get

$$\begin{aligned} \text{distance from } (-c, 0) \text{ to } (0, b) + \text{distance from } (c, 0) \text{ to } (0, b) &= 2a \\ \sqrt{(0 - (-c))^2 + (b - 0)^2} + \sqrt{(0 - c)^2 + (b - 0)^2} &= 2a \\ \sqrt{b^2 + c^2} + \sqrt{b^2 + c^2} &= 2a \\ 2\sqrt{b^2 + c^2} &= 2a \\ \sqrt{b^2 + c^2} &= a \end{aligned}$$

From this, we get $a^2 = b^2 + c^2$, or $b^2 = a^2 - c^2$, which will prove useful later. Now consider a point (x, y) on the ellipse. Applying Definition 7.4, we get

$$\begin{aligned} \text{distance from } (-c, 0) \text{ to } (x, y) + \text{distance from } (c, 0) \text{ to } (x, y) &= 2a \\ \sqrt{(x - (-c))^2 + (y - 0)^2} + \sqrt{(x - c)^2 + (y - 0)^2} &= 2a \\ \sqrt{(x + c)^2 + y^2} + \sqrt{(x - c)^2 + y^2} &= 2a \end{aligned}$$

In order to make sense of this situation, we need to make good use of Intermediate Algebra.

$$\begin{aligned} \sqrt{(x + c)^2 + y^2} + \sqrt{(x - c)^2 + y^2} &= 2a \\ \sqrt{(x + c)^2 + y^2} &= 2a - \sqrt{(x - c)^2 + y^2} \\ \left(\sqrt{(x + c)^2 + y^2}\right)^2 &= \left(2a - \sqrt{(x - c)^2 + y^2}\right)^2 \\ (x + c)^2 + y^2 &= 4a^2 - 4a\sqrt{(x - c)^2 + y^2} + (x - c)^2 + y^2 \\ 4a\sqrt{(x - c)^2 + y^2} &= 4a^2 + (x - c)^2 - (x + c)^2 \\ 4a\sqrt{(x - c)^2 + y^2} &= 4a^2 - 4cx \\ a\sqrt{(x - c)^2 + y^2} &= a^2 - cx \\ \left(a\sqrt{(x - c)^2 + y^2}\right)^2 &= \left(a^2 - cx\right)^2 \\ a^2 \left((x - c)^2 + y^2\right) &= a^4 - 2a^2 cx + c^2 x^2 \\ a^2 x^2 - 2a^2 cx + a^2 c^2 + a^2 y^2 &= a^4 - 2a^2 cx + c^2 x^2 \\ a^2 x^2 - c^2 x^2 + a^2 y^2 &= a^4 - a^2 c^2 \\ \left(a^2 - c^2\right) x^2 + a^2 y^2 &= a^2 \left(a^2 - c^2\right) \end{aligned}$$

We are nearly finished. Recall that $b^2 = a^2 - c^2$ so that

$$\begin{aligned} \left(a^2 - c^2\right) x^2 + a^2 y^2 &= a^2 \left(a^2 - c^2\right) \\ b^2 x^2 + a^2 y^2 &= a^2 b^2 \\ \frac{x^2}{a^2} + \frac{y^2}{b^2} &= 1 \end{aligned}$$

7.4 Ellipses

This equation is for an ellipse centered at the origin. To get the formula for the ellipse centered at (h, k), we could use the transformations from Section 1.7 or re-derive the equation using Definition 7.4 and the distance formula to obtain the formula below.

> **Equation 7.4. The Standard Equation of an Ellipse:** For positive unequal numbers a and b, the equation of an ellipse with center (h, k) is
> $$\frac{(x-h)^2}{a^2} + \frac{(y-k)^2}{b^2} = 1$$

Some remarks about Equation 7.4 are in order. First note that the values a and b determine how far in the x and y directions, respectively, one counts from the center to arrive at points on the ellipse. Also take note that if $a > b$, then we have an ellipse whose major axis is horizontal, and hence, the foci lie to the left and right of the center. In this case, as we've seen in the derivation, the distance from the center to the focus, c, can be found by $c = \sqrt{a^2 - b^2}$. If $b > a$, the roles of the major and minor axes are reversed, and the foci lie above and below the center. In this case, $c = \sqrt{b^2 - a^2}$. In either case, c is the distance from the center to each focus, and $c = \sqrt{\text{bigger denominator} - \text{smaller denominator}}$. Finally, it is worth mentioning that if we take the standard equation of a circle, Equation 7.1, and divide both sides by r^2, we get

> **Equation 7.5. The Alternate Standard Equation of a Circle:** The equation of a circle with center (h, k) and radius $r > 0$ is
> $$\frac{(x-h)^2}{r^2} + \frac{(y-k)^2}{r^2} = 1$$

Notice the similarity between Equation 7.4 and Equation 7.5. Both equations involve a sum of squares equal to 1; the difference is that with a circle, the denominators are the same, and with an ellipse, they are different. If we take a transformational approach, we can consider both Equations 7.4 and 7.5 as shifts and stretches of the Unit Circle $x^2 + y^2 = 1$ in Definition 7.2. Replacing x with $(x-h)$ and y with $(y-k)$ causes the usual horizontal and vertical shifts. Replacing x with $\frac{x}{a}$ and y with $\frac{y}{b}$ causes the usual vertical and horizontal stretches. In other words, it is perfectly fine to think of an ellipse as the deformation of a circle in which the circle is stretched farther in one direction than the other.[1]

Example 7.4.1. Graph $\frac{(x+1)^2}{9} + \frac{(y-2)^2}{25} = 1$. Find the center, the lines which contain the major and minor axes, the vertices, the endpoints of the minor axis, and the foci.

Solution. We see that this equation is in the standard form of Equation 7.4. Here $x - h$ is $x + 1$ so $h = -1$, and $y - k$ is $y - 2$ so $k = 2$. Hence, our ellipse is centered at $(-1, 2)$. We see that $a^2 = 9$ so $a = 3$, and $b^2 = 25$ so $b = 5$. This means that we move 3 units left and right from the center and 5 units up and down from the center to arrive at points on the ellipse. As an aid to sketching, we draw a rectangle matching this description, called a **guide rectangle**, and sketch the ellipse inside this rectangle as seen below on the left.

---
[1] This was foreshadowed in Exercise 19 in Section 7.2.

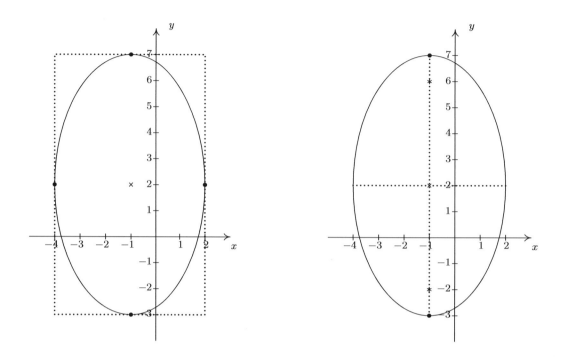

Since we moved farther in the y direction than in the x direction, the major axis will lie along the vertical line $x = -1$, which means the minor axis lies along the horizontal line, $y = 2$. The vertices are the points on the ellipse which lie along the major axis so in this case, they are the points $(-1, 7)$ and $(-1, -3)$, and the endpoints of the minor axis are $(-4, 2)$ and $(2, 2)$. (Notice these points are the four points we used to draw the guide rectangle.) To find the foci, we find $c = \sqrt{25 - 9} = \sqrt{16} = 4$, which means the foci lie 4 units from the center. Since the major axis is vertical, the foci lie 4 units above and below the center, at $(-1, -2)$ and $(-1, 6)$. Plotting all this information gives the graph seen above on the right. □

Example 7.4.2. Find the equation of the ellipse with foci $(2, 1)$ and $(4, 1)$ and vertex $(0, 1)$.

Solution. Plotting the data given to us, we have

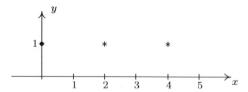

From this sketch, we know that the major axis is horizontal, meaning $a > b$. Since the center is the midpoint of the foci, we know it is $(3, 1)$. Since one vertex is $(0, 1)$ we have that $a = 3$, so $a^2 = 9$. All that remains is to find b^2. Since the foci are 1 unit away from the center, we know $c = 1$. Since $a > b$, we have $c = \sqrt{a^2 - b^2}$, or $1 = \sqrt{9 - b^2}$, so $b^2 = 8$. Substituting all of our findings into the equation $\frac{(x-h)^2}{a^2} + \frac{(y-k)^2}{b^2} = 1$, we get our final answer to be $\frac{(x-3)^2}{9} + \frac{(y-1)^2}{8} = 1$. □

7.4 Ellipses

As with circles and parabolas, an equation may be given which is an ellipse, but isn't in the standard form of Equation 7.4. In those cases, as with circles and parabolas before, we will need to massage the given equation into the standard form.

> **To Write the Equation of an Ellipse in Standard Form**
>
> 1. Group the same variables together on one side of the equation and position the constant on the other side.
>
> 2. Complete the square in both variables as needed.
>
> 3. Divide both sides by the constant term so that the constant on the other side of the equation becomes 1.

Example 7.4.3. Graph $x^2 + 4y^2 - 2x + 24y + 33 = 0$. Find the center, the lines which contain the major and minor axes, the vertices, the endpoints of the minor axis, and the foci.

Solution. Since we have a sum of squares and the squared terms have unequal coefficients, it's a good bet we have an ellipse on our hands.[2] We need to complete both squares, and then divide, if necessary, to get the right-hand side equal to 1.

$$
\begin{aligned}
x^2 + 4y^2 - 2x + 24y + 33 &= 0 \\
x^2 - 2x + 4y^2 + 24y &= -33 \\
x^2 - 2x + 4\left(y^2 + 6y\right) &= -33 \\
\left(x^2 - 2x + 1\right) + 4\left(y^2 + 6y + 9\right) &= -33 + 1 + 4(9) \\
(x-1)^2 + 4(y+3)^2 &= 4 \\
\frac{(x-1)^2 + 4(y+3)^2}{4} &= \frac{4}{4} \\
\frac{(x-1)^2}{4} + (y+3)^2 &= 1 \\
\frac{(x-1)^2}{4} + \frac{(y+3)^2}{1} &= 1
\end{aligned}
$$

Now that this equation is in the standard form of Equation 7.4, we see that $x - h$ is $x - 1$ so $h = 1$, and $y - k$ is $y + 3$ so $k = -3$. Hence, our ellipse is centered at $(1, -3)$. We see that $a^2 = 4$ so $a = 2$, and $b^2 = 1$ so $b = 1$. This means we move 2 units left and right from the center and 1 unit up and down from the center to arrive at points on the ellipse. Since we moved farther in the x direction than in the y direction, the major axis will lie along the horizontal line $y = -3$, which means the minor axis lies along the vertical line $x = 1$. The vertices are the points on the ellipse which lie along the major axis so in this case, they are the points $(-1, -3)$ and $(3, -3)$, and the endpoints of the minor axis are $(1, -2)$ and $(1, -4)$. To find the foci, we find $c = \sqrt{4-1} = \sqrt{3}$, which means

---

[2] The equation of a parabola has only one squared variable and the equation of a circle has two squared variables with *identical* coefficients.

the foci lie $\sqrt{3}$ units from the center. Since the major axis is horizontal, the foci lie $\sqrt{3}$ units to the left and right of the center, at $(1-\sqrt{3}, -3)$ and $(1+\sqrt{3}, -3)$. Plotting all of this information gives

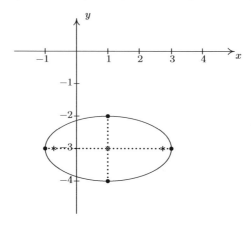

□

As you come across ellipses in the homework exercises and in the wild, you'll notice they come in all shapes in sizes. Compare the two ellipses below.

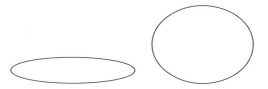

Certainly, one ellipse is more round than the other. This notion of 'roundness' is quantified below.

Definition 7.5. The **eccentricity** of an ellipse, denoted e, is the following ratio:
$$e = \frac{\text{distance from the center to a focus}}{\text{distance from the center to a vertex}}$$

In an ellipse, the foci are closer to the center than the vertices, so $0 < e < 1$. The ellipse above on the left has eccentricity $e \approx 0.98$; for the ellipse above on the right, $e \approx 0.66$. In general, the closer the eccentricity is to 0, the more 'circular' the ellipse; the closer the eccentricity is to 1, the more 'eccentric' the ellipse.

Example 7.4.4. Find the equation of the ellipse whose vertices are $(\pm 5, 0)$ with eccentricity $e = \frac{1}{4}$.

Solution. As before, we plot the data given to us

From this sketch, we know that the major axis is horizontal, meaning $a > b$. With the vertices located at $(\pm 5, 0)$, we get $a = 5$ so $a^2 = 25$. We also know that the center is $(0,0)$ because the center is the midpoint of the vertices. All that remains is to find b^2. To that end, we use the fact that the eccentricity $e = \frac{1}{4}$ which means

$$e = \frac{\text{distance from the center to a focus}}{\text{distance from the center to a vertex}} = \frac{c}{a} = \frac{c}{5} = \frac{1}{4}$$

from which we get $c = \frac{5}{4}$. To get b^2, we use the fact that $c = \sqrt{a^2 - b^2}$, so $\frac{5}{4} = \sqrt{25 - b^2}$ from which we get $b^2 = \frac{375}{16}$. Substituting all of our findings into the equation $\frac{(x-h)^2}{a^2} + \frac{(y-k)^2}{b^2} = 1$, yields our final answer $\frac{x^2}{25} + \frac{16y^2}{375} = 1$. $\square$

As with parabolas, ellipses have a reflective property. If we imagine the dashed lines below representing sound waves, then the waves emanating from one focus reflect off the top of the ellipse and head towards the other focus.

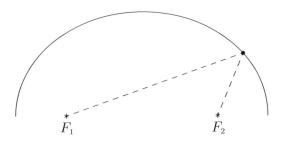

Such geometry is exploited in the construction of so-called 'Whispering Galleries'. If a person whispers at one focus, a person standing at the other focus will hear the first person as if they were standing right next to them. We explore the Whispering Galleries in our last example.

Example 7.4.5. Jamie and Jason want to exchange secrets (terrible secrets) from across a crowded whispering gallery. Recall that a whispering gallery is a room which, in cross section, is half of an ellipse. If the room is 40 feet high at the center and 100 feet wide at the floor, how far from the outer wall should each of them stand so that they will be positioned at the foci of the ellipse?

Solution. Graphing the data yields

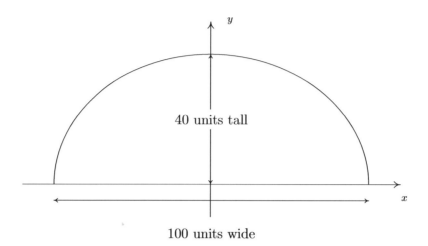

It's most convenient to imagine this ellipse centered at $(0,0)$. Since the ellipse is 100 units wide and 40 units tall, we get $a = 50$ and $b = 40$. Hence, our ellipse has the equation $\frac{x^2}{50^2} + \frac{y^2}{40^2} = 1$. We're looking for the foci, and we get $c = \sqrt{50^2 - 40^2} = \sqrt{900} = 30$, so that the foci are 30 units from the center. That means they are $50 - 30 = 20$ units from the vertices. Hence, Jason and Jamie should stand 20 feet from opposite ends of the gallery. □

7.4 Ellipses

7.4.1 Exercises

In Exercises 1 - 8, graph the ellipse. Find the center, the lines which contain the major and minor axes, the vertices, the endpoints of the minor axis, the foci and the eccentricity.

1. $\dfrac{x^2}{169} + \dfrac{y^2}{25} = 1$

2. $\dfrac{x^2}{9} + \dfrac{y^2}{25} = 1$

3. $\dfrac{(x-2)^2}{4} + \dfrac{(y+3)^2}{9} = 1$

4. $\dfrac{(x+5)^2}{16} + \dfrac{(y-4)^2}{1} = 1$

5. $\dfrac{(x-1)^2}{10} + \dfrac{(y-3)^2}{11} = 1$

6. $\dfrac{(x-1)^2}{9} + \dfrac{(y+3)^2}{4} = 1$

7. $\dfrac{(x+2)^2}{16} + \dfrac{(y-5)^2}{20} = 1$

8. $\dfrac{(x-4)^2}{8} + \dfrac{(y-2)^2}{18} = 1$

In Exercises 9 - 14, put the equation in standard form. Find the center, the lines which contain the major and minor axes, the vertices, the endpoints of the minor axis, the foci and the eccentricity.

9. $9x^2 + 25y^2 - 54x - 50y - 119 = 0$

10. $12x^2 + 3y^2 - 30y + 39 = 0$

11. $5x^2 + 18y^2 - 30x + 72y + 27 = 0$

12. $x^2 - 2x + 2y^2 - 12y + 3 = 0$

13. $9x^2 + 4y^2 - 4y - 8 = 0$

14. $6x^2 + 5y^2 - 24x + 20y + 14 = 0$

In Exercises 15 - 20, find the standard form of the equation of the ellipse which has the given properties.

15. Center $(3, 7)$, Vertex $(3, 2)$, Focus $(3, 3)$

16. Foci $(0, \pm 5)$, Vertices $(0, \pm 8)$.

17. Foci $(\pm 3, 0)$, length of the Minor Axis 10

18. Vertices $(3, 2)$, $(13, 2)$; Endpoints of the Minor Axis $(8, 4)$, $(8, 0)$

19. Center $(5, 2)$, Vertex $(0, 2)$, eccentricity $\frac{1}{2}$

20. All points on the ellipse are in Quadrant IV except $(0, -9)$ and $(8, 0)$. (One might also say that the ellipse is "tangent to the axes" at those two points.)

21. Repeat Example 7.4.5 for a whispering gallery 200 feet wide and 75 feet tall.

22. An elliptical arch is constructed which is 6 feet wide at the base and 9 feet tall in the middle. Find the height of the arch exactly 1 foot in from the base of the arch. Compare your result with your answer to Exercise 21 in Section 7.3.

23. The Earth's orbit around the sun is an ellipse with the sun at one focus and eccentricity $e \approx 0.0167$. The length of the semimajor axis (that is, half of the major axis) is defined to be 1 astronomical unit (AU). The vertices of the elliptical orbit are given special names: 'aphelion' is the vertex farthest from the sun, and 'perihelion' is the vertex closest to the sun. Find the distance in AU between the sun and aphelion and the distance in AU between the sun and perihelion.

24. The graph of an ellipse clearly fails the Vertical Line Test, Theorem 1.1, so the equation of an ellipse does not define y as a function of x. However, much like with circles and horizontal parabolas, we can split an ellipse into a top half and a bottom half, each of which would indeed represent y as a function of x. With the help of your classmates, use your calculator to graph the ellipses given in Exercises 1 - 8 above. What difficulties arise when you plot them on the calculator?

25. Some famous examples of whispering galleries include St. Paul's Cathedral in London, England, National Statuary Hall in Washington, D.C., and The Cincinnati Museum Center. With the help of your classmates, research these whispering galleries. How does the whispering effect compare and contrast with the scenario in Example 7.4.5?

26. With the help of your classmates, research "extracorporeal shock-wave lithotripsy". It uses the reflective property of the ellipsoid to dissolve kidney stones.

7.4 Ellipses

7.4.2 Answers

1. $\dfrac{x^2}{169} + \dfrac{y^2}{25} = 1$
   Center $(0,0)$
   Major axis along $y = 0$
   Minor axis along $x = 0$
   Vertices $(13, 0)$, $(-13, 0)$
   Endpoints of Minor Axis $(0, -5)$, $(0, 5)$
   Foci $(12, 0)$, $(-12, 0)$
   $e = \dfrac{12}{13}$

   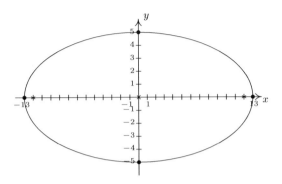

2. $\dfrac{x^2}{9} + \dfrac{y^2}{25} = 1$
   Center $(0,0)$
   Major axis along $x = 0$
   Minor axis along $y = 0$
   Vertices $(0, 5)$, $(0, -5)$
   Endpoints of Minor Axis $(-3, 0)$, $(3, 0)$
   Foci $(0, -4)$, $(0, 4)$
   $e = \dfrac{4}{5}$

   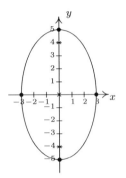

3. $\dfrac{(x-2)^2}{4} + \dfrac{(y+3)^2}{9} = 1$
   Center $(2, -3)$
   Major axis along $x = 2$
   Minor axis along $y = -3$
   Vertices $(2, 0)$, $(2, -6)$
   Endpoints of Minor Axis $(0, -3)$, $(4, -3)$
   Foci $(2, -3 + \sqrt{5})$, $(2, -3 - \sqrt{5})$
   $e = \dfrac{\sqrt{5}}{3}$

   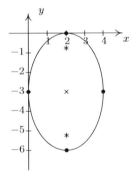

4. $\dfrac{(x+5)^2}{16} + \dfrac{(y-4)^2}{1} = 1$
   Center $(-5, 4)$
   Major axis along $y = 4$
   Minor axis along $x = -5$
   Vertices $(-9, 4)$, $(-1, 4)$
   Endpoints of Minor Axis $(-5, 3)$, $(-5, 5)$
   Foci $(-5 + \sqrt{15}, 4)$, $(-5 - \sqrt{15}, 4)$
   $e = \dfrac{\sqrt{15}}{4}$

   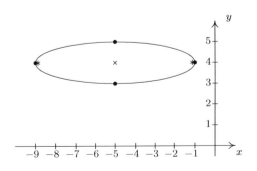

5. $\dfrac{(x-1)^2}{10} + \dfrac{(y-3)^2}{11} = 1$
   Center $(1,3)$
   Major axis along $x = 1$
   Minor axis along $y = 3$
   Vertices $(1, 3+\sqrt{11})$, $(1, 3-\sqrt{11})$
   Endpoints of the Minor Axis
   $(1-\sqrt{10}, 3)$, $(1+\sqrt{10}, 3)$
   Foci $(1,2)$, $(1,4)$
   $e = \dfrac{\sqrt{11}}{11}$

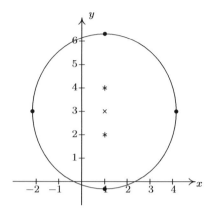

6. $\dfrac{(x-1)^2}{9} + \dfrac{(y+3)^2}{4} = 1$
   Center $(1,-3)$
   Major axis along $y = -3$
   Minor axis along $x = 1$
   Vertices $(4,-3)$, $(-2,-3)$
   Endpoints of the Minor Axis $(1,-1)$, $(1,-5)$
   Foci $(1+\sqrt{5}, -3)$, $(1-\sqrt{5}, -3)$
   $e = \dfrac{\sqrt{5}}{3}$

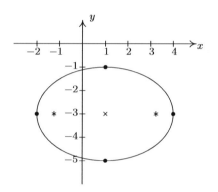

7. $\dfrac{(x+2)^2}{16} + \dfrac{(y-5)^2}{20} = 1$
   Center $(-2, 5)$
   Major axis along $x = -2$
   Minor axis along $y = 5$
   Vertices $(-2, 5+2\sqrt{5})$, $(-2, 5-2\sqrt{5})$
   Endpoints of the Minor Axis $(-6, 5)$, $(2, 5)$
   Foci $(-2, 7)$, $(-2, 3)$
   $e = \dfrac{\sqrt{5}}{5}$

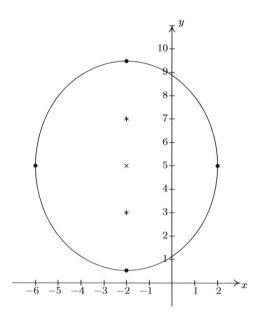

7.4 Ellipses

8. $\dfrac{(x-4)^2}{8} + \dfrac{(y-2)^2}{18} = 1$

Center $(4,2)$
Major axis along $x = 4$
Minor axis along $y = 2$
Vertices $(4, 2 + 3\sqrt{2}), (4, 2 - 3\sqrt{2})$
Endpoints of the Minor Axis
$(4 - 2\sqrt{2}, 2), (4 + 2\sqrt{2}, 2)$
Foci $(4, 2 + \sqrt{10}), (4, 2 - \sqrt{10})$
$e = \dfrac{\sqrt{5}}{3}$

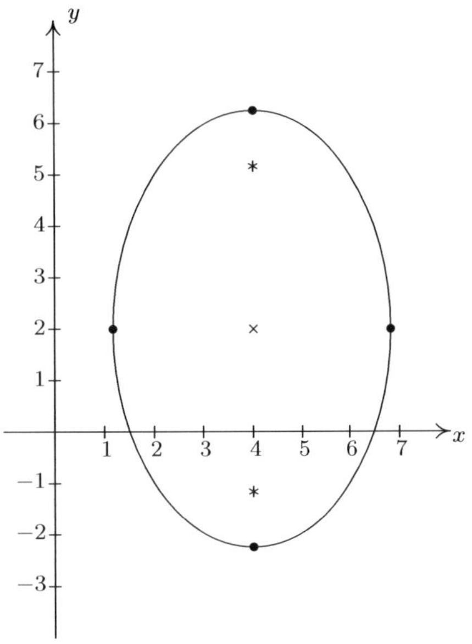

9. $\dfrac{(x-3)^2}{25} + \dfrac{(y-1)^2}{9} = 1$

Center $(3,1)$
Major Axis along $y = 1$
Minor Axis along $x = 3$
Vertices $(8,1), (-2,1)$
Endpoints of Minor Axis $(3, 4), (3, -2)$
Foci $(7, 1), (-1, 1)$
$e = \dfrac{4}{5}$

10. $\dfrac{x^2}{3} + \dfrac{(y-5)^2}{12} = 1$

Center $(0, 5)$
Major axis along $x = 0$
Minor axis along $y = 5$
Vertices $(0, 5 - 2\sqrt{3}), (0, 5 + 2\sqrt{3})$
Endpoints of Minor Axis $(-\sqrt{3}, 5), (\sqrt{3}, 5)$
Foci $(0, 2), (0, 8)$
$e = \dfrac{\sqrt{3}}{2}$

11. $\dfrac{(x-3)^2}{18} + \dfrac{(y+2)^2}{5} = 1$

Center $(3, -2)$
Major axis along $y = -2$
Minor axis along $x = 3$
Vertices $(3 - 3\sqrt{2}, -2), (3 + 3\sqrt{2}, -2)$
Endpoints of Minor Axis $(3, -2 + \sqrt{5})$, $(3, -2 - \sqrt{5})$
Foci $(3 - \sqrt{13}, -2), (3 + \sqrt{13}, -2)$
$e = \dfrac{\sqrt{26}}{6}$

12. $\dfrac{(x-1)^2}{16} + \dfrac{(y-3)^2}{8} = 1$

Center $(1, 3)$
Major Axis along $y = 3$
Minor Axis along $x = 1$
Vertices $(5, 3), (-3, 3)$
Endpoints of Minor Axis $(1, 3 + 2\sqrt{2})$, $(1, 3 - 2\sqrt{2})$
Foci $(1 + 2\sqrt{2}, 3), (1 - 2\sqrt{2}, 3)$
$e = \dfrac{\sqrt{2}}{2}$

13. $\dfrac{x^2}{1} + \dfrac{4\left(y-\frac{1}{2}\right)^2}{9} = 1$
    Center $\left(0, \frac{1}{2}\right)$
    Major Axis along $x = 0$ (the y-axis)
    Minor Axis along $y = \frac{1}{2}$
    Vertices $(0, 2)$, $(0, -1)$
    Endpoints of Minor Axis $\left(-1, \frac{1}{2}\right)$, $\left(1, \frac{1}{2}\right)$
    Foci $\left(0, \frac{1+\sqrt{5}}{2}\right)$, $\left(0, \frac{1-\sqrt{5}}{2}\right)$
    $e = \frac{\sqrt{5}}{3}$

14. $\dfrac{(x-2)^2}{5} + \dfrac{(y+2)^2}{6} = 1$
    Center $(2, -2)$
    Major Axis along $x = 2$
    Minor Axis along $y = -2$
    Vertices $\left(2, -2+\sqrt{6}\right)$, $\left(2, -2-\sqrt{6}\right)$
    Endpoints of Minor Axis $\left(2-\sqrt{5}, -2\right)$, $\left(2+\sqrt{5}, -2\right)$
    Foci $(2, -1)$, $(2, -3)$
    $e = \frac{\sqrt{6}}{6}$

15. $\dfrac{(x-3)^2}{9} + \dfrac{(y-7)^2}{25} = 1$

16. $\dfrac{x^2}{39} + \dfrac{y^2}{64} = 1$

17. $\dfrac{x^2}{34} + \dfrac{y^2}{25} = 1$

18. $\dfrac{(x-8)^2}{25} + \dfrac{(y-2)^2}{4} = 1$

19. $\dfrac{(x-5)^2}{25} + \dfrac{4(y-2)^2}{75} = 1$

20. $\dfrac{(x-8)^2}{64} + \dfrac{(y+9)^2}{81} = 1$

21. Jamie and Jason should stand $100 - 25\sqrt{7} \approx 33.86$ feet from opposite ends of the gallery.

22. The arch can be modeled by the top half of $\frac{x^2}{9} + \frac{y^2}{81} = 1$. One foot in from the base of the arch corresponds to either $x = \pm 2$. Plugging in $x = \pm 2$ gives $y = \pm 3\sqrt{5}$ and since y represents a height, we choose $y = 3\sqrt{5} \approx 6.71$ feet.

23. Distance from the sun to aphelion ≈ 1.0167 AU.
    Distance from the sun to perihelion ≈ 0.9833 AU.

7.5 Hyperbolas

In the definition of an ellipse, Definition 7.4, we fixed two points called foci and looked at points whose distances to the foci always **added** to a constant distance d. Those prone to syntactical tinkering may wonder what, if any, curve we'd generate if we replaced **added** with **subtracted**. The answer is a hyperbola.

> **Definition 7.6.** Given two distinct points F_1 and F_2 in the plane and a fixed distance d, a **hyperbola** is the set of all points (x, y) in the plane such that the absolute value of the difference of each of the distances from F_1 and F_2 to (x, y) is d. The points F_1 and F_2 are called the **foci** of the hyperbola.

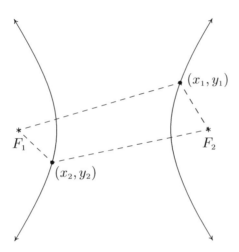

In the figure above:

$$\text{the distance from } F_1 \text{ to } (x_1, y_1) - \text{the distance from } F_2 \text{ to } (x_1, y_1) \;=\; d$$

and

$$\text{the distance from } F_2 \text{ to } (x_2, y_2) - \text{the distance from } F_1 \text{ to } (x_2, y_2) \;=\; d$$

Note that the hyperbola has two parts, called **branches**. The **center** of the hyperbola is the midpoint of the line segment connecting the two foci. The **transverse axis** of the hyperbola is the line segment connecting two opposite ends of the hyperbola which also contains the center and foci. The **vertices** of a hyperbola are the points of the hyperbola which lie on the transverse axis. In addition, we will show momentarily that there are lines called **asymptotes** which the branches of the hyperbola approach for large x and y values. They serve as guides to the graph. In pictures,

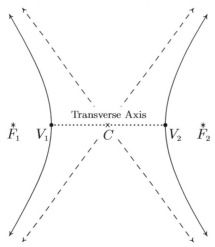

A hyperbola with center C; foci F_1, F_2; and vertices V_1, V_2 and asymptotes (dashed)

Before we derive the standard equation of the hyperbola, we need to discuss one further parameter, the **conjugate axis** of the hyperbola. The conjugate axis of a hyperbola is the line segment through the center which is perpendicular to the transverse axis and has the same length as the line segment through a vertex which connects the asymptotes. In pictures we have

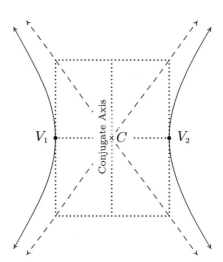

Note that in the diagram, we can construct a rectangle using line segments with lengths equal to the lengths of the transverse and conjugate axes whose center is the center of the hyperbola and whose diagonals are contained in the asymptotes. This **guide rectangle**, much akin to the one we saw Section 7.4 to help us graph ellipses, will aid us in graphing hyperbolas.

Suppose we wish to derive the equation of a hyperbola. For simplicity, we shall assume that the center is $(0,0)$, the vertices are $(a,0)$ and $(-a,0)$ and the foci are $(c,0)$ and $(-c,0)$. We label the

7.5 Hyperbolas

endpoints of the conjugate axis $(0, b)$ and $(0, -b)$. (Although b does not enter into our derivation, we will have to justify this choice as you shall see later.) As before, we assume a, b, and c are all positive numbers. Schematically we have

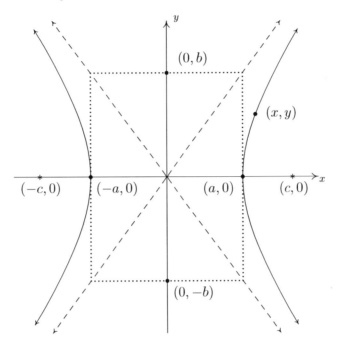

Since $(a, 0)$ is on the hyperbola, it must satisfy the conditions of Definition 7.6. That is, the distance from $(-c, 0)$ to $(a, 0)$ minus the distance from $(c, 0)$ to $(a, 0)$ must equal the fixed distance d. Since all these points lie on the x-axis, we get

$$\begin{aligned}
\text{distance from } (-c, 0) \text{ to } (a, 0) - \text{distance from } (c, 0) \text{ to } (a, 0) &= d \\
(a + c) - (c - a) &= d \\
2a &= d
\end{aligned}$$

In other words, the fixed distance d from the definition of the hyperbola is actually the length of the transverse axis! (Where have we seen that type of coincidence before?) Now consider a point (x, y) on the hyperbola. Applying Definition 7.6, we get

$$\begin{aligned}
\text{distance from } (-c, 0) \text{ to } (x, y) - \text{distance from } (c, 0) \text{ to } (x, y) &= 2a \\
\sqrt{(x - (-c))^2 + (y - 0)^2} - \sqrt{(x - c)^2 + (y - 0)^2} &= 2a \\
\sqrt{(x + c)^2 + y^2} - \sqrt{(x - c)^2 + y^2} &= 2a
\end{aligned}$$

Using the same arsenal of Intermediate Algebra weaponry we used in deriving the standard formula of an ellipse, Equation 7.4, we arrive at the following.[1]

---
[1] It is a good exercise to actually work this out.

$$(a^2 - c^2) x^2 + a^2 y^2 = a^2 (a^2 - c^2)$$

What remains is to determine the relationship between a, b and c. To that end, we note that since a and c are both positive numbers with $a < c$, we get $a^2 < c^2$ so that $a^2 - c^2$ is a negative number. Hence, $c^2 - a^2$ is a positive number. For reasons which will become clear soon, we re-write the equation by solving for y^2/x^2 to get

$$
\begin{aligned}
(a^2 - c^2) x^2 + a^2 y^2 &= a^2 (a^2 - c^2) \\
-(c^2 - a^2) x^2 + a^2 y^2 &= -a^2 (c^2 - a^2) \\
a^2 y^2 &= (c^2 - a^2) x^2 - a^2 (c^2 - a^2) \\
\frac{y^2}{x^2} &= \frac{(c^2 - a^2)}{a^2} - \frac{(c^2 - a^2)}{x^2}
\end{aligned}
$$

As x and y attain very large values, the quantity $\frac{(c^2-a^2)}{x^2} \to 0$ so that $\frac{y^2}{x^2} \to \frac{(c^2-a^2)}{a^2}$. By setting $b^2 = c^2 - a^2$ we get $\frac{y^2}{x^2} \to \frac{b^2}{a^2}$. This shows that $y \to \pm \frac{b}{a} x$ as $|x|$ grows large. Thus $y = \pm \frac{b}{a} x$ are the asymptotes to the graph as predicted and our choice of labels for the endpoints of the conjugate axis is justified. In our equation of the hyperbola we can substitute $a^2 - c^2 = -b^2$ which yields

$$
\begin{aligned}
(a^2 - c^2) x^2 + a^2 y^2 &= a^2 (a^2 - c^2) \\
-b^2 x^2 + a^2 y^2 &= -a^2 b^2 \\
\frac{x^2}{a^2} - \frac{y^2}{b^2} &= 1
\end{aligned}
$$

The equation above is for a hyperbola whose center is the origin and which opens to the left and right. If the hyperbola were centered at a point (h, k), we would get the following.

> **Equation 7.6. The Standard Equation of a Horizontal[a] Hyperbola** For positive numbers a and b, the equation of a horizontal hyperbola with center (h, k) is
>
> $$\frac{(x-h)^2}{a^2} - \frac{(y-k)^2}{b^2} = 1$$
>
> ---
> [a]That is, a hyperbola whose branches open to the left and right

If the roles of x and y were interchanged, then the hyperbola's branches would open upwards and downwards and we would get a 'vertical' hyperbola.

> **Equation 7.7. The Standard Equation of a Vertical Hyperbola** For positive numbers a and b, the equation of a vertical hyperbola with center (h, k) is:
>
> $$\frac{(y-k)^2}{b^2} - \frac{(x-h)^2}{a^2} = 1$$

The values of a and b determine how far in the x and y directions, respectively, one counts from the center to determine the rectangle through which the asymptotes pass. In both cases, the distance

7.5 Hyperbolas

from the center to the foci, c, as seen in the derivation, can be found by the formula $c = \sqrt{a^2 + b^2}$. Lastly, note that we can quickly distinguish the equation of a hyperbola from that of a circle or ellipse because the hyperbola formula involves a **difference** of squares where the circle and ellipse formulas both involve the **sum** of squares.

Example 7.5.1. Graph the equation $\frac{(x-2)^2}{4} - \frac{y^2}{25} = 1$. Find the center, the lines which contain the transverse and conjugate axes, the vertices, the foci and the equations of the asymptotes.

Solution. We first see that this equation is given to us in the standard form of Equation 7.6. Here $x - h$ is $x - 2$ so $h = 2$, and $y - k$ is y so $k = 0$. Hence, our hyperbola is centered at $(2, 0)$. We see that $a^2 = 4$ so $a = 2$, and $b^2 = 25$ so $b = 5$. This means we move 2 units to the left and right of the center and 5 units up and down from the center to arrive at points on the guide rectangle. The asymptotes pass through the center of the hyperbola as well as the corners of the rectangle. This yields the following set up.

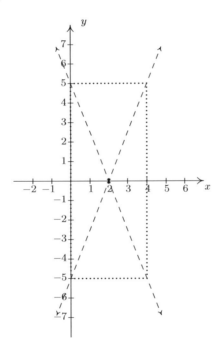

Since the y^2 term is being subtracted from the x^2 term, we know that the branches of the hyperbola open to the left and right. This means that the transverse axis lies along the x-axis. Hence, the conjugate axis lies along the vertical line $x = 2$. Since the vertices of the hyperbola are where the hyperbola intersects the transverse axis, we get that the vertices are 2 units to the left and right of $(2, 0)$ at $(0, 0)$ and $(4, 0)$. To find the foci, we need $c = \sqrt{a^2 + b^2} = \sqrt{4 + 25} = \sqrt{29}$. Since the foci lie on the transverse axis, we move $\sqrt{29}$ units to the left and right of $(2, 0)$ to arrive at $(2 - \sqrt{29}, 0)$ (approximately $(-3.39, 0)$) and $(2 + \sqrt{29}, 0)$ (approximately $(7.39, 0)$). To determine the equations of the asymptotes, recall that the asymptotes go through the center of the hyperbola, $(2, 0)$, as well as the corners of guide rectangle, so they have slopes of $\pm\frac{b}{a} = \pm\frac{5}{2}$. Using the point-slope equation of a line, Equation 2.2, yields $y - 0 = \pm \frac{5}{2}(x - 2)$, so we get $y = \frac{5}{2}x - 5$ and $y = -\frac{5}{2}x + 5$. Putting it all together, we get

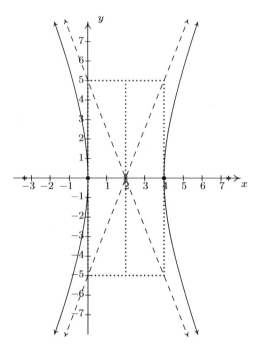

□

Example 7.5.2. Find the equation of the hyperbola with asymptotes $y = \pm 2x$ and vertices $(\pm 5, 0)$.

Solution. Plotting the data given to us, we have

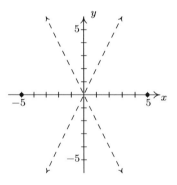

This graph not only tells us that the branches of the hyperbola open to the left and to the right, it also tells us that the center is $(0, 0)$. Hence, our standard form is $\frac{x^2}{a^2} - \frac{y^2}{b^2} = 1$. Since the vertices are $(\pm 5, 0)$, we have $a = 5$ so $a^2 = 25$. In order to determine b^2, we recall that the slopes of the asymptotes are $\pm \frac{b}{a}$. Since $a = 5$ and the slope of the line $y = 2x$ is 2, we have that $\frac{b}{5} = 2$, so $b = 10$. Hence, $b^2 = 100$ and our final answer is $\frac{x^2}{25} - \frac{y^2}{100} = 1$. □

7.5 HYPERBOLAS

As with the other conic sections, an equation whose graph is a hyperbola may not be given in either of the standard forms. To rectify that, we have the following.

> **To Write the Equation of a Hyperbola in Standard Form**
> 1. Group the same variables together on one side of the equation and position the constant on the other side
> 2. Complete the square in both variables as needed
> 3. Divide both sides by the constant term so that the constant on the other side of the equation becomes 1

Example 7.5.3. Consider the equation $9y^2 - x^2 - 6x = 10$. Put this equation in to standard form and graph. Find the center, the lines which contain the transverse and conjugate axes, the vertices, the foci, and the equations of the asymptotes.

Solution. We need only complete the square on x:

$$\begin{aligned} 9y^2 - x^2 - 6x &= 10 \\ 9y^2 - 1\left(x^2 + 6x\right) &= 10 \\ 9y^2 - \left(x^2 + 6x + 9\right) &= 10 - 1(9) \\ 9y^2 - (x+3)^2 &= 1 \\ \frac{y^2}{\frac{1}{9}} - \frac{(x+3)^2}{1} &= 1 \end{aligned}$$

Now that this equation is in the standard form of Equation 7.7, we see that $x - h$ is $x + 3$ so $h = -3$, and $y - k$ is y so $k = 0$. Hence, our hyperbola is centered at $(-3, 0)$. We find that $a^2 = 1$ so $a = 1$, and $b^2 = \frac{1}{9}$ so $b = \frac{1}{3}$. This means that we move 1 unit to the left and right of the center and $\frac{1}{3}$ units up and down from the center to arrive at points on the guide rectangle. Since the x^2 term is being subtracted from the y^2 term, we know the branches of the hyperbola open upwards and downwards. This means the transverse axis lies along the vertical line $x = -3$ and the conjugate axis lies along the x-axis. Since the vertices of the hyperbola are where the hyperbola intersects the transverse axis, we get that the vertices are $\frac{1}{3}$ of a unit above and below $(-3, 0)$ at $\left(-3, \frac{1}{3}\right)$ and $\left(-3, -\frac{1}{3}\right)$. To find the foci, we use

$$c = \sqrt{a^2 + b^2} = \sqrt{\frac{1}{9} + 1} = \frac{\sqrt{10}}{3}$$

Since the foci lie on the transverse axis, we move $\frac{\sqrt{10}}{3}$ units above and below $(-3, 0)$ to arrive at $\left(-3, \frac{\sqrt{10}}{3}\right)$ and $\left(-3, -\frac{\sqrt{10}}{3}\right)$. To determine the asymptotes, recall that the asymptotes go through the center of the hyperbola, $(-3, 0)$, as well as the corners of guide rectangle, so they have slopes of $\pm \frac{b}{a} = \pm \frac{1}{3}$. Using the point-slope equation of a line, Equation 2.2, we get $y = \frac{1}{3}x + 1$ and $y = -\frac{1}{3}x - 1$. Putting it all together, we get

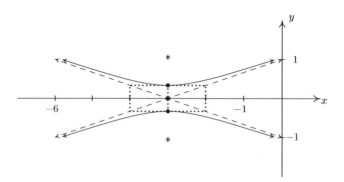

□

Hyperbolas can be used in so-called 'trilateration,' or 'positioning' problems. The procedure outlined in the next example is the basis of the (now virtually defunct) LOng Range Aid to Navigation (LORAN for short) system.[2]

Example 7.5.4. Jeff is stationed 10 miles due west of Carl in an otherwise empty forest in an attempt to locate an elusive Sasquatch. At the stroke of midnight, Jeff records a Sasquatch call 9 seconds earlier than Carl. If the speed of sound that night is 760 miles per hour, determine a hyperbolic path along which Sasquatch must be located.

Solution. Since Jeff hears Sasquatch sooner, it is closer to Jeff than it is to Carl. Since the speed of sound is 760 miles per hour, we can determine how much closer Sasquatch is to Jeff by multiplying

$$760 \frac{\text{miles}}{\text{hour}} \times \frac{1\,\text{hour}}{3600\,\text{seconds}} \times 9\,\text{seconds} = 1.9\,\text{miles}$$

This means that Sasquatch is 1.9 miles closer to Jeff than it is to Carl. In other words, Sasquatch must lie on a path where

$$(\text{the distance to Carl}) - (\text{the distance to Jeff}) = 1.9$$

This is exactly the situation in the definition of a hyperbola, Definition 7.6. In this case, Jeff and Carl are located at the foci,[3] and our fixed distance d is 1.9. For simplicity, we assume the hyperbola is centered at $(0,0)$ with its foci at $(-5,0)$ and $(5,0)$. Schematically, we have

---

[2] GPS now rules the positioning kingdom. Is there still a place for LORAN and other land-based systems? Do satellites ever malfunction?

[3] We usually like to be the *center* of attention, but being the *focus* of attention works equally well.

7.5 Hyperbolas

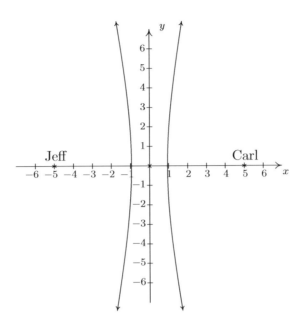

We are seeking a curve of the form $\frac{x^2}{a^2} - \frac{y^2}{b^2} = 1$ in which the distance from the center to each focus is $c = 5$. As we saw in the derivation of the standard equation of the hyperbola, Equation 7.6, $d = 2a$, so that $2a = 1.9$, or $a = 0.95$ and $a^2 = 0.9025$. All that remains is to find b^2. To that end, we recall that $a^2 + b^2 = c^2$ so $b^2 = c^2 - a^2 = 25 - 0.9025 = 24.0975$. Since Sasquatch is closer to Jeff than it is to Carl, it must be on the western (left hand) branch of $\frac{x^2}{0.9025} - \frac{y^2}{24.0975} = 1$. $\square$

In our previous example, we did not have enough information to pin down the exact location of Sasquatch. To accomplish this, we would need a third observer.

Example 7.5.5. By a stroke of luck, Kai was also camping in the woods during the events of the previous example. He was located 6 miles due north of Jeff and heard the Sasquatch call 18 seconds after Jeff did. Use this added information to locate Sasquatch.

Solution. Kai and Jeff are now the foci of a second hyperbola where the fixed distance d can be determined as before

$$760 \, \frac{\text{miles}}{\text{hour}} \times \frac{1 \, \text{hour}}{3600 \, \text{seconds}} \times 18 \, \text{seconds} = 3.8 \, \text{miles}$$

Since Jeff was positioned at $(-5, 0)$, we place Kai at $(-5, 6)$. This puts the center of the new hyperbola at $(-5, 3)$. Plotting Kai's position and the new center gives us the diagram below on the left. The second hyperbola is vertical, so it must be of the form $\frac{(y-3)^2}{b^2} - \frac{(x+5)^2}{a^2} = 1$. As before, the distance d is the length of the major axis, which in this case is $2b$. We get $2b = 3.8$ so that $b = 1.9$ and $b^2 = 3.61$. With Kai 6 miles due North of Jeff, we have that the distance from the center to the focus is $c = 3$. Since $a^2 + b^2 = c^2$, we get $a^2 = c^2 - b^2 = 9 - 3.61 = 5.39$. Kai heard the Sasquatch call after Jeff, so Kai is farther from Sasquatch than Jeff. Thus Sasquatch must lie on the southern branch of the hyperbola $\frac{(y-3)^2}{3.61} - \frac{(x+5)^2}{5.39} = 1$. Looking at the western branch of the hyperbola determined by Jeff and Carl along with the southern branch of the hyperbola determined by Kai and Jeff, we see that there is exactly one point in common, and this is where Sasquatch must have been when it called.

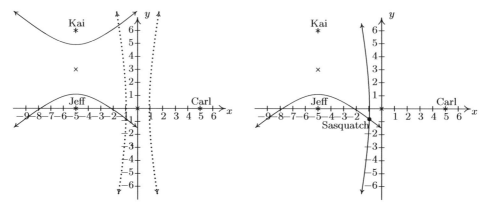

To determine the coordinates of this point of intersection exactly, we would need techniques for solving systems of non-linear equations (which we won't see until Section 8.7), so we use the calculator[4] Doing so, we get Sasquatch is approximately at $(-0.9629, -0.8113)$. □

Each of the conic sections we have studied in this chapter result from graphing equations of the form $Ax^2 + Cy^2 + Dx + Ey + F = 0$ for different choices of A, C, D, E, and[5] F. While we've seen examples[6] demonstrate *how* to convert an equation from this general form to one of the standard forms, we close this chapter with some advice about *which* standard form to choose.[7]

Strategies for Identifying Conic Sections

Suppose the graph of equation $Ax^2 + Cy^2 + Dx + Ey + F = 0$ is a non-degenerate conic section.[a]

- If just *one* variable is squared, the graph is a parabola. Put the equation in the form of Equation 7.2 (if x is squared) or Equation 7.3 (if y is squared).

If *both* variables are squared, look at the coefficients of x^2 and y^2, A and B.

- If $A = B$, the graph is a circle. Put the equation in the form of Equation 7.1.

- If $A \neq B$ but A and B have the *same sign*, the graph is an ellipse. Put the equation in the form of Equation 7.4.

- If A and B have the *different signs*, the graph is a hyperbola. Put the equation in the form of either Equation 7.6 or Equation 7.7.

---
[a]That is, a parabola, circle, ellipse, or hyperbola – see Section 7.1.

---

[4]First solve each hyperbola for y, and choose the correct equation (branch) before proceeding.
[5]See Section 11.6 to see why we skip B.
[6]Examples 7.2.3, 7.3.4, 7.4.3, and 7.5.3, in particular.
[7]We formalize this in Exercise 34.

7.5 Hyperbolas

7.5.1 Exercises

In Exercises 1 - 8, graph the hyperbola. Find the center, the lines which contain the transverse and conjugate axes, the vertices, the foci and the equations of the asymptotes.

1. $\dfrac{x^2}{16} - \dfrac{y^2}{9} = 1$

2. $\dfrac{y^2}{9} - \dfrac{x^2}{16} = 1$

3. $\dfrac{(x-2)^2}{4} - \dfrac{(y+3)^2}{9} = 1$

4. $\dfrac{(y-3)^2}{11} - \dfrac{(x-1)^2}{10} = 1$

5. $\dfrac{(x+4)^2}{16} - \dfrac{(y-4)^2}{1} = 1$

6. $\dfrac{(x+1)^2}{9} - \dfrac{(y-3)^2}{4} = 1$

7. $\dfrac{(y+2)^2}{16} - \dfrac{(x-5)^2}{20} = 1$

8. $\dfrac{(x-4)^2}{8} - \dfrac{(y-2)^2}{18} = 1$

In Exercises 9 - 12, put the equation in standard form. Find the center, the lines which contain the transverse and conjugate axes, the vertices, the foci and the equations of the asymptotes.

9. $12x^2 - 3y^2 + 30y - 111 = 0$

10. $18y^2 - 5x^2 + 72y + 30x - 63 = 0$

11. $9x^2 - 25y^2 - 54x - 50y - 169 = 0$

12. $-6x^2 + 5y^2 - 24x + 40y + 26 = 0$

In Exercises 13 - 18, find the standard form of the equation of the hyperbola which has the given properties.

13. Center $(3,7)$, Vertex $(3,3)$, Focus $(3,2)$

14. Vertex $(0,1)$, Vertex $(8,1)$, Focus $(-3,1)$

15. Foci $(0, \pm 8)$, Vertices $(0, \pm 5)$.

16. Foci $(\pm 5, 0)$, length of the Conjugate Axis 6

17. Vertices $(3,2)$, $(13,2)$; Endpoints of the Conjugate Axis $(8,4)$, $(8,0)$

18. Vertex $(-10, 5)$, Asymptotes $y = \pm \frac{1}{2}(x-6) + 5$

In Exercises 19 - 28, find the standard form of the equation using the guidelines on page 540 and then graph the conic section.

19. $x^2 - 2x - 4y - 11 = 0$

20. $x^2 + y^2 - 8x + 4y + 11 = 0$

21. $9x^2 + 4y^2 - 36x + 24y + 36 = 0$

22. $9x^2 - 4y^2 - 36x - 24y - 36 = 0$

23. $y^2 + 8y - 4x + 16 = 0$

24. $4x^2 + y^2 - 8x + 4 = 0$

25. $4x^2 + 9y^2 - 8x + 54y + 49 = 0$

26. $x^2 + y^2 - 6x + 4y + 14 = 0$

27. $2x^2 + 4y^2 + 12x - 8y + 25 = 0$

28. $4x^2 - 5y^2 - 40x - 20y + 160 = 0$

29. The graph of a vertical or horizontal hyperbola clearly fails the Vertical Line Test, Theorem 1.1, so the equation of a vertical of horizontal hyperbola does not define y as a function of x.[8] However, much like with circles, horizontal parabolas and ellipses, we can split a hyperbola into pieces, each of which would indeed represent y as a function of x. With the help of your classmates, use your calculator to graph the hyperbolas given in Exercises 1 - 8 above. How many pieces do you need for a vertical hyperbola? How many for a horizontal hyperbola?

30. The location of an earthquake's epicenter – the point on the surface of the Earth directly above where the earthquake actually occurred – can be determined by a process similar to how we located Sasquatch in Example 7.5.5. (As we said back in Exercise 75 in Section 6.1, earthquakes are complicated events and it is not our intent to provide a complete discussion of the science involved in them. Instead, we refer the interested reader to a course in Geology or the U.S. Geological Survey's Earthquake Hazards Program found here.) Our technique works only for relatively small distances because we need to assume that the Earth is flat in order to use hyperbolas in the plane.[9] The P-waves ("P" stands for Primary) of an earthquake in Sasquatchia travel at 6 kilometers per second.[10] Station A records the waves first. Then Station B, which is 100 kilometers due north of Station A, records the waves 2 seconds later. Station C, which is 150 kilometers due west of Station A records the waves 3 seconds after that (a total of 5 seconds after Station A). Where is the epicenter?

31. The notion of eccentricity introduced for ellipses in Definition 7.5 in Section 7.4 is the same for hyperbolas in that we can define the eccentricity e of a hyperbola as

$$e = \frac{\text{distance from the center to a focus}}{\text{distance from the center to a vertex}}$$

   (a) With the help of your classmates, explain why $e > 1$ for any hyperbola.

   (b) Find the equation of the hyperbola with vertices $(\pm 3, 0)$ and eccentricity $e = 2$.

   (c) With the help of your classmates, find the eccentricity of each of the hyperbolas in Exercises 1 - 8. What role does eccentricity play in the shape of the graphs?

32. On page 510 in Section 7.3, we discussed paraboloids of revolution when studying the design of satellite dishes and parabolic mirrors. In much the same way, 'natural draft' cooling towers are often shaped as **hyperboloids of revolution**. Each vertical cross section of these towers

---

[8]We will see later in the text that the graphs of certain rotated hyperbolas pass the Vertical Line Test.

[9]Back in the Exercises in Section 1.1 you were asked to research people who believe the world is flat. What did you discover?

[10]Depending on the composition of the crust at a specific location, P-waves can travel between 5 kps and 8 kps.

7.5 Hyperbolas

is a hyperbola. Suppose the a natural draft cooling tower has the cross section below. Suppose the tower is 450 feet wide at the base, 275 feet wide at the top, and 220 feet at its narrowest point (which occurs 330 feet above the ground.) Determine the height of the tower to the nearest foot.

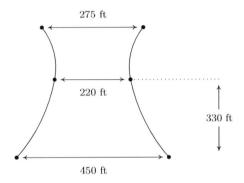

33. With the help of your classmates, research the Cassegrain Telescope. It uses the reflective property of the hyperbola as well as that of the parabola to make an ingenious telescope.

34. With the help of your classmates show that if $Ax^2 + Cy^2 + Dx + Ey + F = 0$ determines a non-degenerate conic[11] then

    - $AC < 0$ means that the graph is a hyperbola
    - $AC = 0$ means that the graph is a parabola
    - $AC > 0$ means that the graph is an ellipse or circle

    NOTE: This result will be generalized in Theorem 11.11 in Section 11.6.1.

---

[11] Recall that this means its graph is either a circle, parabola, ellipse or hyperbola.

7.5.2 Answers

1. $\dfrac{x^2}{16} - \dfrac{y^2}{9} = 1$
   Center $(0, 0)$
   Transverse axis on $y = 0$
   Conjugate axis on $x = 0$
   Vertices $(4, 0), (-4, 0)$
   Foci $(5, 0), (-5, 0)$
   Asymptotes $y = \pm \frac{3}{4} x$

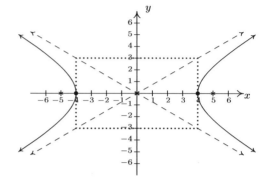

2. $\dfrac{y^2}{9} - \dfrac{x^2}{16} = 1$
   Center $(0, 0)$
   Transverse axis on $x = 0$
   Conjugate axis on $y = 0$
   Vertices $(0, 3), (0, -3)$
   Foci $(0, 5), (0, -5)$
   Asymptotes $y = \pm \frac{3}{4} x$

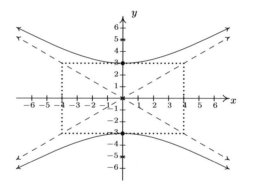

3. $\dfrac{(x-2)^2}{4} - \dfrac{(y+3)^2}{9} = 1$
   Center $(2, -3)$
   Transverse axis on $y = -3$
   Conjugate axis on $x = 2$
   Vertices $(0, -3), (4, -3)$
   Foci $(2 + \sqrt{13}, -3), (2 - \sqrt{13}, -3)$
   Asymptotes $y = \pm \frac{3}{2}(x - 2) - 3$

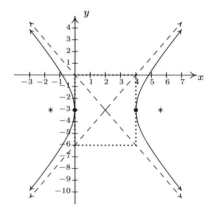

7.5 Hyperbolas

4. $\dfrac{(y-3)^2}{11} - \dfrac{(x-1)^2}{10} = 1$
   Center $(1,3)$
   Transverse axis on $x = 1$
   Conjugate axis on $y = 3$
   Vertices $(1, 3+\sqrt{11}), (1, 3-\sqrt{11})$
   Foci $(1, 3+\sqrt{21}), (1, 3-\sqrt{21})$
   Asymptotes $y = \pm\dfrac{\sqrt{110}}{10}(x-1) + 3$

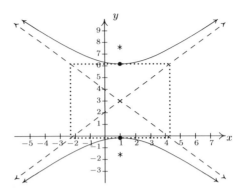

5. $\dfrac{(x+4)^2}{16} - \dfrac{(y-4)^2}{1} = 1$
   Center $(-4, 4)$
   Transverse axis on $y = 4$
   Conjugate axis on $x = -4$
   Vertices $(-8, 4), (0, 4)$
   Foci $(-4+\sqrt{17}, 4), (-4-\sqrt{17}, 4)$
   Asymptotes $y = \pm\dfrac{1}{4}(x+4) + 4$

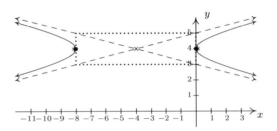

6. $\dfrac{(x+1)^2}{9} - \dfrac{(y-3)^2}{4} = 1$
   Center $(-1, 3)$
   Transverse axis on $y = 3$
   Conjugate axis on $x = -1$
   Vertices $(2, 3), (-4, 3)$
   Foci $\left(-1+\sqrt{13}, 3\right), \left(-1-\sqrt{13}, 3\right)$
   Asymptotes $y = \pm\dfrac{2}{3}(x+1) + 3$

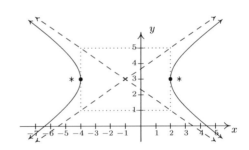

7. $\dfrac{(y+2)^2}{16} - \dfrac{(x-5)^2}{20} = 1$
   Center $(5, -2)$
   Transverse axis on $x = 5$
   Conjugate axis on $y = -2$
   Vertices $(5, 2), (5, -6)$
   Foci $(5, 4), (5, -8)$
   Asymptotes $y = \pm\dfrac{2\sqrt{5}}{5}(x-5) - 2$

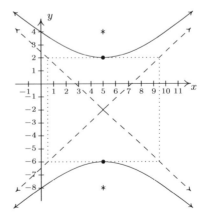

8. $\dfrac{(x-4)^2}{8} - \dfrac{(y-2)^2}{18} = 1$

Center $(4, 2)$
Transverse axis on $y = 2$
Conjugate axis on $x = 4$
Vertices $(4 + 2\sqrt{2}, 2), (4 - 2\sqrt{2}, 2)$
Foci $(4 + \sqrt{26}, 2), (4 - \sqrt{26}, 2)$
Asymptotes $y = \pm\frac{3}{2}(x - 4) + 2$

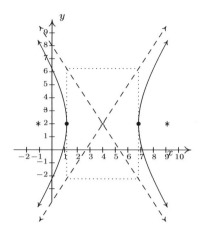

9. $\dfrac{x^2}{3} - \dfrac{(y-5)^2}{12} = 1$

Center $(0, 5)$
Transverse axis on $y = 5$
Conjugate axis on $x = 0$
Vertices $(\sqrt{3}, 5), (-\sqrt{3}, 5)$
Foci $(\sqrt{15}, 5), (-\sqrt{15}, 5)$
Asymptotes $y = \pm 2x + 5$

10. $\dfrac{(y+2)^2}{5} - \dfrac{(x-3)^2}{18} = 1$

Center $(3, -2)$
Transverse axis on $x = 3$
Conjugate axis on $y = -2$
Vertices $(3, -2 + \sqrt{5}), (3, -2 - \sqrt{5})$
Foci $(3, -2 + \sqrt{23}), (3, -2 - \sqrt{23})$
Asymptotes $y = \pm\frac{\sqrt{10}}{6}(x - 3) - 2$

11. $\dfrac{(x-3)^2}{25} - \dfrac{(y+1)^2}{9} = 1$

Center $(3, -1)$
Transverse axis on $y = -1$
Conjugate axis on $x = 3$
Vertices $(8, -1), (-2, -1)$
Foci $(3 + \sqrt{34}, -1), (3 - \sqrt{34}, -1)$
Asymptotes $y = \pm\frac{3}{5}(x - 3) - 1$

12. $\dfrac{(y+4)^2}{6} - \dfrac{(x+2)^2}{5} = 1$

Center $(-2, -4)$
Transverse axis on $x = -2$
Conjugate axis on $y = -4$
Vertices $(-2, -4 + \sqrt{6}), (-2, -4 - \sqrt{6})$
Foci $(-2, -4 + \sqrt{11}), (-2, -4 - \sqrt{11})$
Asymptotes $y = \pm\frac{\sqrt{30}}{5}(x + 2) - 4$

13. $\dfrac{(y-7)^2}{16} - \dfrac{(x-3)^2}{9} = 1$

14. $\dfrac{(x-4)^2}{16} - \dfrac{(y-1)^2}{33} = 1$

15. $\dfrac{y^2}{25} - \dfrac{x^2}{39} = 1$

16. $\dfrac{x^2}{16} - \dfrac{y^2}{9} = 1$

17. $\dfrac{(x-8)^2}{25} - \dfrac{(y-2)^2}{4} = 1$

18. $\dfrac{(x-6)^2}{256} - \dfrac{(y-5)^2}{64} = 1$

7.5 Hyperbolas

19. $(x-1)^2 = 4(y+3)$

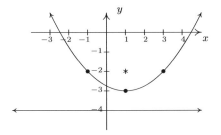

20. $(x-4)^2 + (y+2)^2 = 9$

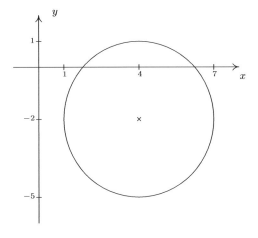

21. $\dfrac{(x-2)^2}{4} + \dfrac{(y+3)^2}{9} = 1$

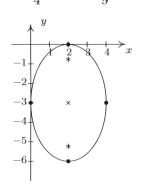

22. $\dfrac{(x-2)^2}{4} - \dfrac{(y+3)^2}{9} = 1$

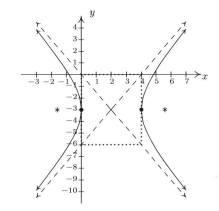

23. $(y+4)^2 = 4x$

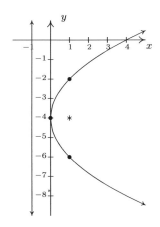

24. $\dfrac{(x-1)^2}{1} + \dfrac{y^2}{4} = 0$

The graph is the point $(1,0)$ only.

25. $\dfrac{(x-1)^2}{9} + \dfrac{(y+3)^2}{4} = 1$

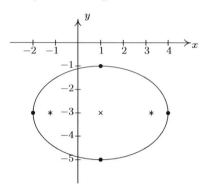

26. $(x-3)^2 + (y+2)^2 = -1$
There is no graph.

27. $\dfrac{(x+3)^2}{2} + \dfrac{(y-1)^2}{1} = -\dfrac{3}{4}$
There is no graph.

28. $\dfrac{(y+2)^2}{16} - \dfrac{(x-5)^2}{20} = 1$

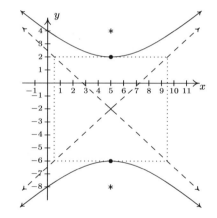

30. By placing Station A at $(0,-50)$ and Station B at $(0,50)$, the two second time difference yields the hyperbola $\dfrac{y^2}{36} - \dfrac{x^2}{2464} = 1$ with foci A and B and center $(0,0)$. Placing Station C at $(-150,-50)$ and using foci A and C gives us a center of $(-75,-50)$ and the hyperbola $\dfrac{(x+75)^2}{225} - \dfrac{(y+50)^2}{5400} = 1$. The point of intersection of these two hyperbolas which is closer to A than B and closer to A than C is $(-57.8444, -9.21336)$ so that is the epicenter.

31. (b) $\dfrac{x^2}{9} - \dfrac{y^2}{27} = 1$.

32. The tower may be modeled (approximately)[12] by $\dfrac{x^2}{12100} - \dfrac{(y-330)^2}{34203} = 1$. To find the height, we plug in $x = 137.5$ which yields $y \approx 191$ or $y \approx 469$. Since the top of the tower is above the narrowest point, we get the tower is approximately 469 feet tall.

---

[12] The exact value underneath $(y-330)^2$ is $\dfrac{52707600}{1541}$ in case you need more precision.

Chapter 8

Systems of Equations and Matrices

8.1 Systems of Linear Equations: Gaussian Elimination

Up until now, when we concerned ourselves with solving different types of equations there was only one equation to solve at a time. Given an equation $f(x) = g(x)$, we could check our solutions geometrically by finding where the graphs of $y = f(x)$ and $y = g(x)$ intersect. The x-coordinates of these intersection points correspond to the solutions to the equation $f(x) = g(x)$, and the y-coordinates were largely ignored. If we modify the problem and ask for the intersection points of the graphs of $y = f(x)$ and $y = g(x)$, where both the solution to x and y are of interest, we have what is known as a **system of equations**, usually written as

$$\begin{cases} y &= f(x) \\ y &= g(x) \end{cases}$$

The 'curly bracket' notation means we are to find all **pairs** of points (x, y) which satisfy **both** equations. We begin our study of systems of equations by reviewing some basic notions from Intermediate Algebra.

> **Definition 8.1.** A **linear equation in two variables** is an equation of the form $a_1 x + a_2 y = c$ where a_1, a_2 and c are real numbers and at least one of a_1 and a_2 is nonzero.

For reasons which will become clear later in the section, we are using subscripts in Definition 8.1 to indicate different, but fixed, real numbers and those subscripts have no mathematical meaning beyond that. For example, $3x - \frac{y}{2} = 0.1$ is a linear equation in two variables with $a_1 = 3$, $a_2 = -\frac{1}{2}$ and $c = 0.1$. We can also consider $x = 5$ to be a linear equation in two variables[1] by identifying $a_1 = 1$, $a_2 = 0$, and $c = 5$. If a_1 and a_2 are both 0, then depending on c, we get either an equation which is *always* true, called an **identity**, or an equation which is *never* true, called a **contradiction**. (If $c = 0$, then we get $0 = 0$, which is always true. If $c \neq 0$, then we'd have $0 \neq 0$, which is never true.) Even though identities and contradictions have a large role to play

---

[1] Critics may argue that $x = 5$ is clearly an equation in one variable. It can also be considered an equation in 117 variables with the coefficients of 116 variables set to 0. As with many conventions in Mathematics, the context will clarify the situation.

in the upcoming sections, we do not consider them linear equations. The key to identifying linear equations is to note that the variables involved are to the first power and that the coefficients of the variables are numbers. Some examples of equations which are non-linear are $x^2 + y = 1$, $xy = 5$ and $e^{2x} + \ln(y) = 1$. We leave it to the reader to explain why these do not satisfy Definition 8.1. From what we know from Sections 1.2 and 2.1, the graphs of linear equations are lines. If we couple two or more linear equations together, in effect to find the points of intersection of two or more lines, we obtain a **system of linear equations in two variables**. Our first example reviews some of the basic techniques first learned in Intermediate Algebra.

Example 8.1.1. Solve the following systems of equations. Check your answer algebraically and graphically.

1. $\begin{cases} 2x - y = 1 \\ y = 3 \end{cases}$

2. $\begin{cases} 3x + 4y = -2 \\ -3x - y = 5 \end{cases}$

3. $\begin{cases} \frac{x}{3} - \frac{4y}{5} = \frac{7}{5} \\ \frac{2x}{9} + \frac{y}{3} = \frac{1}{2} \end{cases}$

4. $\begin{cases} 2x - 4y = 6 \\ 3x - 6y = 9 \end{cases}$

5. $\begin{cases} 6x + 3y = 9 \\ 4x + 2y = 12 \end{cases}$

6. $\begin{cases} x - y = 0 \\ x + y = 2 \\ -2x + y = -2 \end{cases}$

Solution.

1. Our first system is nearly solved for us. The second equation tells us that $y = 3$. To find the corresponding value of x, we **substitute** this value for y into the the first equation to obtain $2x - 3 = 1$, so that $x = 2$. Our solution to the system is $(2, 3)$. To check this algebraically, we substitute $x = 2$ and $y = 3$ into each equation and see that they are satisfied. We see $2(2) - 3 = 1$, and $3 = 3$, as required. To check our answer graphically, we graph the lines $2x - y = 1$ and $y = 3$ and verify that they intersect at $(2, 3)$.

2. To solve the second system, we use the **addition** method to **eliminate** the variable x. We take the two equations as given and 'add equals to equals' to obtain

$$\begin{array}{rcl} 3x + 4y &=& -2 \\ + \;\; (-3x - y &=& 5) \\ \hline 3y &=& 3 \end{array}$$

This gives us $y = 1$. We now substitute $y = 1$ into either of the two equations, say $-3x - y = 5$, to get $-3x - 1 = 5$ so that $x = -2$. Our solution is $(-2, 1)$. Substituting $x = -2$ and $y = 1$ into the first equation gives $3(-2) + 4(1) = -2$, which is true, and, likewise, when we check $(-2, 1)$ in the second equation, we get $-3(-2) - 1 = 5$, which is also true. Geometrically, the lines $3x + 4y = -2$ and $-3x - y = 5$ intersect at $(-2, 1)$.

8.1 Systems of Linear Equations: Gaussian Elimination

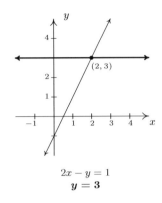

$2x - y = 1$
$y = 3$

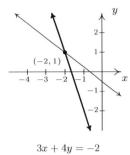

$3x + 4y = -2$
$-3x - y = 5$

3. The equations in the third system are more approachable if we clear denominators. We multiply both sides of the first equation by 15 and both sides of the second equation by 18 to obtain the kinder, gentler system

$$\begin{cases} 5x - 12y &= 21 \\ 4x + 6y &= 9 \end{cases}$$

Adding these two equations directly fails to eliminate either of the variables, but we note that if we multiply the first equation by 4 and the second by -5, we will be in a position to eliminate the x term

$$\begin{array}{rrr} & 20x - 48y = & 84 \\ + & (-20x - 30y = & -45) \\ \hline & -78y = & 39 \end{array}$$

From this we get $y = -\frac{1}{2}$. We can temporarily avoid too much unpleasantness by choosing to substitute $y = -\frac{1}{2}$ into one of the equivalent equations we found by clearing denominators, say into $5x - 12y = 21$. We get $5x + 6 = 21$ which gives $x = 3$. Our answer is $\left(3, -\frac{1}{2}\right)$. At this point, we have no choice — in order to check an answer algebraically, we must see if the answer satisfies both of the *original* equations, so we substitute $x = 3$ and $y = -\frac{1}{2}$ into both $\frac{x}{3} - \frac{4y}{5} = \frac{7}{5}$ and $\frac{2x}{9} + \frac{y}{3} = \frac{1}{2}$. We leave it to the reader to verify that the solution is correct. Graphing both of the lines involved with considerable care yields an intersection point of $\left(3, -\frac{1}{2}\right)$.

4. An eerie calm settles over us as we cautiously approach our fourth system. Do its friendly integer coefficients belie something more sinister? We note that if we multiply both sides of the first equation by 3 and both sides of the second equation by -2, we are ready to eliminate the x

$$\begin{array}{rcr} 6x - 12y &=& 18 \\ +\ (-6x + 12y &=& -18) \\ \hline 0 &=& 0 \end{array}$$

We eliminated not only the x, but the y as well and we are left with the identity $0 = 0$. This means that these two different linear equations are, in fact, equivalent. In other words, if an ordered pair (x, y) satisfies the equation $2x - 4y = 6$, it *automatically* satisfies the equation $3x - 6y = 9$. One way to describe the solution set to this system is to use the roster method[2] and write $\{(x, y) \mid 2x - 4y = 6\}$. While this is correct (and corresponds exactly to what's happening graphically, as we shall see shortly), we take this opportunity to introduce the notion of a **parametric solution to a system**. Our first step is to solve $2x - 4y = 6$ for one of the variables, say $y = \frac{1}{2}x - \frac{3}{2}$. For each value of x, the formula $y = \frac{1}{2}x - \frac{3}{2}$ determines the corresponding y-value of a solution. Since we have no restriction on x, it is called a **free variable**. We let $x = t$, a so-called 'parameter', and get $y = \frac{1}{2}t - \frac{3}{2}$. Our set of solutions can then be described as $\left\{ \left(t, \frac{1}{2}t - \frac{3}{2}\right) \mid -\infty < t < \infty \right\}$.[3] For specific values of t, we can generate solutions. For example, $t = 0$ gives us the solution $\left(0, -\frac{3}{2}\right)$; $t = 117$ gives us $(117, 57)$, and while we can readily check each of these particular solutions satisfy both equations, the question is how do we check our general answer algebraically? Same as always. We claim that for any real number t, the pair $\left(t, \frac{1}{2}t - \frac{3}{2}\right)$ satisfies both equations. Substituting $x = t$ and $y = \frac{1}{2}t - \frac{3}{2}$ into $2x - 4y = 6$ gives $2t - 4\left(\frac{1}{2}t - \frac{3}{2}\right) = 6$. Simplifying, we get $2t - 2t + 6 = 6$, which is always true. Similarly, when we make these substitutions in the equation $3x - 6y = 9$, we get $3t - 6\left(\frac{1}{2}t - \frac{3}{2}\right) = 9$ which reduces to $3t - 3t + 9 = 9$, so it checks out, too. Geometrically, $2x - 4y = 6$ and $3x - 6y = 9$ are the same line, which means that they intersect at every point on their graphs. The reader is encouraged to think about how our parametric solution says exactly that.

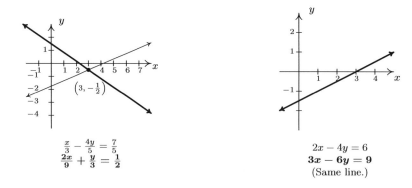

---

[2] See Section 1.2 for a review of this.

[3] Note that we could have just as easily chosen to solve $2x - 4y = 6$ for x to obtain $x = 2y + 3$. Letting y be the parameter t, we have that for any value of t, $x = 2t + 3$, which gives $\{(2t + 3, t) \mid -\infty < t < \infty\}$. There is no one correct way to parameterize the solution set, which is why it is always best to check your answer.

5. Multiplying both sides of the first equation by 2 and the both sides of the second equation by -3, we set the stage to eliminate x

$$\begin{array}{rcr} 12x + 6y & = & 18 \\ +\ (-12x - 6y & = & -36) \\ \hline 0 & = & -18 \end{array}$$

As in the previous example, both x and y dropped out of the equation, but we are left with an irrevocable contradiction, $0 = -18$. This tells us that it is impossible to find a pair (x, y) which satisfies both equations; in other words, the system has no solution. Graphically, the lines $6x + 3y = 9$ and $4x + 2y = 12$ are distinct and parallel, so they do not intersect.

6. We can begin to solve our last system by adding the first two equations

$$\begin{array}{rcr} x - y & = & 0 \\ +\ (x + y & = & 2) \\ \hline 2x & = & 2 \end{array}$$

which gives $x = 1$. Substituting this into the first equation gives $1 - y = 0$ so that $y = 1$. We seem to have determined a solution to our system, $(1, 1)$. While this checks in the first two equations, when we substitute $x = 1$ and $y = 1$ into the third equation, we get $-2(1)+(1) = -2$ which simplifies to the contradiction $-1 = -2$. Graphing the lines $x - y = 0$, $x + y = 2$, and $-2x + y = -2$, we see that the first two lines do, in fact, intersect at $(1, 1)$, however, all three lines never intersect at the same point simultaneously, which is what is required if a solution to the system is to be found.

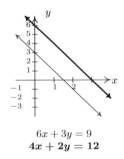

$6x + 3y = 9$
$4x + 2y = 12$

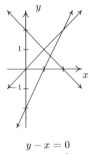

$y - x = 0$
$y + x = 2$
$-2x + y = -2$

□

A few remarks about Example 8.1.1 are in order. It is clear that some systems of equations have solutions, and some do not. Those which have solutions are called **consistent**, those with no solution are called **inconsistent**. We also distinguish the two different types of behavior among consistent systems. Those which admit free variables are called **dependent**; those with no free variables are called **independent**.[4] Using this new vocabulary, we classify numbers 1, 2 and 3 in Example 8.1.1 as consistent independent systems, number 4 is consistent dependent, and numbers 5 and 6 are inconsistent.[5] The system in 6 above is called **overdetermined**, since we have more equations than variables.[6] Not surprisingly, a system with more variables than equations is called **underdetermined**. While the system in number 6 above is overdetermined and inconsistent, there exist overdetermined consistent systems (both dependent and independent) and we leave it to the reader to think about what is happening algebraically and geometrically in these cases. Likewise, there are both consistent and inconsistent underdetermined systems,[7] but a consistent underdetermined system of linear equations is necessarily dependent.[8]

In order to move this section beyond a review of Intermediate Algebra, we now define what is meant by a linear equation in n variables.

> **Definition 8.2.** A **linear equation in n variables**, $x_1, x_2, \ldots, x_n$, is an equation of the form $a_1 x_1 + a_2 x_2 + \ldots + a_n x_n = c$ where $a_1, a_2, \ldots a_n$ and c are real numbers and at least one of $a_1, a_2, \ldots, a_n$ is nonzero.

Instead of using more familiar variables like x, y, and even z and/or w in Definition 8.2, we use subscripts to distinguish the different variables. We have no idea how many variables may be involved, so we use numbers to distinguish them instead of letters. (There is an endless supply of distinct numbers.) As an example, the linear equation $3x_1 - x_2 = 4$ represents the same relationship between the variables x_1 and x_2 as the equation $3x - y = 4$ does between the variables x and y. In addition, just as we cannot combine the terms in the expression $3x - y$, we cannot combine the terms in the expression $3x_1 - x_2$. Coupling more than one linear equation in n variables results in a **system of linear equations in n variables**. When solving these systems, it becomes increasingly important to keep track of what operations are performed to which equations and to develop a strategy based on the kind of manipulations we've already employed. To this end, we first remind ourselves of the maneuvers which can be applied to a system of linear equations that result in an equivalent system.[9]

---

[4] In the case of systems of linear equations, regardless of the number of equations or variables, consistent independent systems have exactly one solution. The reader is encouraged to think about why this is the case for linear equations in two variables. Hint: think geometrically.

[5] The adjectives 'dependent' and 'independent' apply only to *consistent* systems – they describe the *type* of solutions. Is there a free variable (dependent) or not (independent)?

[6] If we think if each variable being an unknown quantity, then ostensibly, to recover two unknown quantities, we need two pieces of information - i.e., two equations. Having more than two equations suggests we have more information than necessary to determine the values of the unknowns. While this is not necessarily the case, it does explain the choice of terminology 'overdetermined'.

[7] We need more than two variables to give an example of the latter.

[8] Again, experience with systems with more variables helps to see this here, as does a solid course in Linear Algebra.

[9] That is, a system with the same solution set.

8.1 Systems of Linear Equations: Gaussian Elimination

> **Theorem 8.1.** Given a system of equations, the following moves will result in an equivalent system of equations.
>
> - Interchange the position of any two equations.
> - Replace an equation with a nonzero multiple of itself.[a]
> - Replace an equation with itself plus a nonzero multiple of another equation.
>
> ---
> [a]That is, an equation which results from multiplying both sides of the equation by the same nonzero number.

We have seen plenty of instances of the second and third moves in Theorem 8.1 when we solved the systems in Example 8.1.1. The first move, while it obviously admits an equivalent system, seems silly. Our perception will change as we consider more equations and more variables in this, and later sections.

Consider the system of equations

$$\begin{cases} x - \frac{1}{3}y + \frac{1}{2}z = 1 \\ y - \frac{1}{2}z = 4 \\ z = -1 \end{cases}$$

Clearly $z = -1$, and we substitute this into the second equation $y - \frac{1}{2}(-1) = 4$ to obtain $y = \frac{7}{2}$. Finally, we substitute $y = \frac{7}{2}$ and $z = -1$ into the first equation to get $x - \frac{1}{3}\left(\frac{7}{2}\right) + \frac{1}{2}(-1) = 1$, so that $x = \frac{8}{3}$. The reader can verify that these values of x, y and z satisfy all three original equations. It is tempting for us to write the solution to this system by extending the usual (x, y) notation to (x, y, z) and list our solution as $\left(\frac{8}{3}, \frac{7}{2}, -1\right)$. The question quickly becomes what does an 'ordered triple' like $\left(\frac{8}{3}, \frac{7}{2}, -1\right)$ represent? Just as ordered pairs are used to locate points on the two-dimensional plane, ordered triples can be used to locate points in space.[10] Moreover, just as equations involving the variables x and y describe graphs of one-dimensional lines and curves in the two-dimensional plane, equations involving variables x, y, and z describe objects called **surfaces** in three-dimensional space. Each of the equations in the above system can be visualized as a plane situated in three-space. Geometrically, the system is trying to find the intersection, or common point, of all three planes. If you imagine three sheets of notebook paper each representing a portion of these planes, you will start to see the complexities involved in how three such planes can intersect. Below is a sketch of the three planes. It turns out that any two of these planes intersect in a line,[11] so our intersection point is where all three of these lines meet.

---
[10]You were asked to think about this in Exercise 40 in Section 1.1.
[11]In fact, these lines are described by the parametric solutions to the systems formed by taking any two of these equations by themselves.

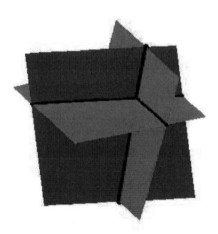

Since the geometry for equations involving more than two variables is complicated, we will focus our efforts on the algebra. Returning to the system

$$\begin{cases} x - \frac{1}{3}y + \frac{1}{2}z = 1 \\ y - \frac{1}{2}z = 4 \\ z = -1 \end{cases}$$

we note the reason it was so easy to solve is that the third equation is solved for z, the second equation involves only y and z, and since the coefficient of y is 1, it makes it easy to solve for y using our known value for z. Lastly, the coefficient of x in the first equation is 1 making it easy to substitute the known values of y and z and then solve for x. We formalize this pattern below for the most general systems of linear equations. Again, we use subscripted variables to describe the general case. The variable with the smallest subscript in a given equation is typically called the **leading variable** of that equation.

Definition 8.3. A system of linear equations with variables x_1, x_2, ... x_n is said to be in **triangular form** provided all of the following conditions hold:

1. The subscripts of the variables in each equation are always increasing from left to right.

2. The leading variable in each equation has coefficient 1.

3. The subscript on the leading variable in a given equation is greater than the subscript on the leading variable in the equation above it.

4. Any equation without variables[a] cannot be placed above an equation with variables.

---
[a] necessarily an identity or contradiction

8.1 Systems of Linear Equations: Gaussian Elimination

In our previous system, if we make the obvious choices $x = x_1$, $y = x_2$, and $z = x_3$, we see that the system is in triangular form.[12] An example of a more complicated system in triangular form is

$$\begin{cases} x_1 - 4x_3 + x_4 - x_6 = 6 \\ x_2 + 2x_3 = 1 \\ x_4 + 3x_5 - x_6 = 8 \\ x_5 + 9x_6 = 10 \end{cases}$$

Our goal henceforth will be to transform a given system of linear equations into triangular form using the moves in Theorem 8.1.

Example 8.1.2. Use Theorem 8.1 to put the following systems into triangular form and then solve the system if possible. Classify each system as consistent independent, consistent dependent, or inconsistent.

1. $\begin{cases} 3x - y + z = 3 \\ 2x - 4y + 3z = 16 \\ x - y + z = 5 \end{cases}$
2. $\begin{cases} 2x + 3y - z = 1 \\ 10x - z = 2 \\ 4x - 9y + 2z = 5 \end{cases}$
3. $\begin{cases} 3x_1 + x_2 + x_4 = 6 \\ 2x_1 + x_2 - x_3 = 4 \\ x_2 - 3x_3 - 2x_4 = 0 \end{cases}$

Solution.

1. For definitiveness, we label the topmost equation in the system $E1$, the equation beneath that $E2$, and so forth. We now attempt to put the system in triangular form using an algorithm known as **Gaussian Elimination**. What this means is that, starting with x, we transform the system so that conditions 2 and 3 in Definition 8.3 are satisfied. Then we move on to the next variable, in this case y, and repeat. Since the variables in all of the equations have a consistent ordering from left to right, our first move is to get an x in $E1$'s spot with a coefficient of 1. While there are many ways to do this, the easiest is to apply the first move listed in Theorem 8.1 and interchange $E1$ and $E3$.

$$\begin{cases} (E1) & 3x - y + z = 3 \\ (E2) & 2x - 4y + 3z = 16 \\ (E3) & x - y + z = 5 \end{cases} \xrightarrow{\text{Switch } E1 \text{ and } E3} \begin{cases} (E1) & x - y + z = 5 \\ (E2) & 2x - 4y + 3z = 16 \\ (E3) & 3x - y + z = 3 \end{cases}$$

To satisfy Definition 8.3, we need to eliminate the x's from $E2$ and $E3$. We accomplish this by replacing each of them with a sum of themselves and a multiple of $E1$. To eliminate the x from $E2$, we need to multiply $E1$ by -2 then add; to eliminate the x from $E3$, we need to multiply $E1$ by -3 then add. Applying the third move listed in Theorem 8.1 twice, we get

$$\begin{cases} (E1) & x - y + z = 5 \\ (E2) & 2x - 4y + 3z = 16 \\ (E3) & 3x - y + z = 3 \end{cases} \xrightarrow[\text{Replace } E3 \text{ with } -3E1 + E3]{\text{Replace } E2 \text{ with } -2E1 + E2} \begin{cases} (E1) & x - y + z = 5 \\ (E2) & -2y + z = 6 \\ (E3) & 2y - 2z = -12 \end{cases}$$

---

[12] If letters are used instead of subscripted variables, Definition 8.3 can be suitably modified using alphabetical order of the variables instead of numerical order on the subscripts of the variables.

Now we enforce the conditions stated in Definition 8.3 for the variable y. To that end we need to get the coefficient of y in $E2$ equal to 1. We apply the second move listed in Theorem 8.1 and replace $E2$ with itself times $-\frac{1}{2}$.

$$\begin{cases} (E1) & x - y + z = 5 \\ (E2) & -2y + z = 6 \\ (E3) & 2y - 2z = -12 \end{cases} \xrightarrow{\text{Replace } E2 \text{ with } -\frac{1}{2}E2} \begin{cases} (E1) & x - y + z = 5 \\ (E2) & y - \frac{1}{2}z = -3 \\ (E3) & 2y - 2z = -12 \end{cases}$$

To eliminate the y in $E3$, we add $-2E2$ to it.

$$\begin{cases} (E1) & x - y + z = 5 \\ (E2) & y - \frac{1}{2}z = -3 \\ (E3) & 2y - 2z = -12 \end{cases} \xrightarrow{\text{Replace } E3 \text{ with } -2E2 + E3} \begin{cases} (E1) & x - y + z = 5 \\ (E2) & y - \frac{1}{2}z = -3 \\ (E3) & -z = -6 \end{cases}$$

Finally, we apply the second move from Theorem 8.1 one last time and multiply $E3$ by -1 to satisfy the conditions of Definition 8.3 for the variable z.

$$\begin{cases} (E1) & x - y + z = 5 \\ (E2) & y - \frac{1}{2}z = -3 \\ (E3) & -z = -6 \end{cases} \xrightarrow{\text{Replace } E3 \text{ with } -1E3} \begin{cases} (E1) & x - y + z = 5 \\ (E2) & y - \frac{1}{2}z = -3 \\ (E3) & z = 6 \end{cases}$$

Now we proceed to substitute. Plugging in $z = 6$ into $E2$ gives $y - 3 = -3$ so that $y = 0$. With $y = 0$ and $z = 6$, $E1$ becomes $x - 0 + 6 = 5$, or $x = -1$. Our solution is $(-1, 0, 6)$. We leave it to the reader to check that substituting the respective values for x, y, and z into the original system results in three identities. Since we have found a solution, the system is consistent; since there are no free variables, it is independent.

2. Proceeding as we did in 1, our first step is to get an equation with x in the $E1$ position with 1 as its coefficient. Since there is no easy fix, we multiply $E1$ by $\frac{1}{2}$.

$$\begin{cases} (E1) & 2x + 3y - z = 1 \\ (E2) & 10x - z = 2 \\ (E3) & 4x - 9y + 2z = 5 \end{cases} \xrightarrow{\text{Replace } E1 \text{ with } \frac{1}{2}E1} \begin{cases} (E1) & x + \frac{3}{2}y - \frac{1}{2}z = \frac{1}{2} \\ (E2) & 10x - z = 2 \\ (E3) & 4x - 9y + 2z = 5 \end{cases}$$

Now it's time to take care of the x's in $E2$ and $E3$.

$$\begin{cases} (E1) & x + \frac{3}{2}y - \frac{1}{2}z = \frac{1}{2} \\ (E2) & 10x - z = 2 \\ (E3) & 4x - 9y + 2z = 5 \end{cases} \xrightarrow[\text{Replace } E3 \text{ with } -4E1 + E3]{\text{Replace } E2 \text{ with } -10E1 + E2} \begin{cases} (E1) & x + \frac{3}{2}y - \frac{1}{2}z = \frac{1}{2} \\ (E2) & -15y + 4z = -3 \\ (E3) & -15y + 4z = 3 \end{cases}$$

8.1 Systems of Linear Equations: Gaussian Elimination

Our next step is to get the coefficient of y in $E2$ equal to 1. To that end, we have

$$\begin{cases} (E1) & x + \frac{3}{2}y - \frac{1}{2}z = \frac{1}{2} \\ (E2) & -15y + 4z = -3 \\ (E3) & -15y + 4z = 3 \end{cases} \xrightarrow{\text{Replace } E2 \text{ with } -\frac{1}{15}E2} \begin{cases} (E1) & x + \frac{3}{2}y - \frac{1}{2}z = \frac{1}{2} \\ (E2) & y - \frac{4}{15}z = \frac{1}{5} \\ (E3) & -15y + 4z = 3 \end{cases}$$

Finally, we rid $E3$ of y.

$$\begin{cases} (E1) & x + \frac{3}{2}y - \frac{1}{2}z = \frac{1}{2} \\ (E2) & y - \frac{4}{15}z = \frac{1}{5} \\ (E3) & -15y + 4z = 3 \end{cases} \xrightarrow{\text{Replace } E3 \text{ with } 15E2 + E3} \begin{cases} (E1) & x - y + z = 5 \\ (E2) & y - \frac{1}{2}z = -3 \\ (E3) & 0 = 6 \end{cases}$$

The last equation, $0 = 6$, is a contradiction so the system has no solution. According to Theorem 8.1, since this system has no solutions, neither does the original, thus we have an inconsistent system.

3. For our last system, we begin by multiplying $E1$ by $\frac{1}{3}$ to get a coefficient of 1 on x_1.

$$\begin{cases} (E1) & 3x_1 + x_2 + x_4 = 6 \\ (E2) & 2x_1 + x_2 - x_3 = 4 \\ (E3) & x_2 - 3x_3 - 2x_4 = 0 \end{cases} \xrightarrow{\text{Replace } E1 \text{ with } \frac{1}{3}E1} \begin{cases} (E1) & x_1 + \frac{1}{3}x_2 + \frac{1}{3}x_4 = 2 \\ (E2) & 2x_1 + x_2 - x_3 = 4 \\ (E3) & x_2 - 3x_3 - 2x_4 = 0 \end{cases}$$

Next we eliminate x_1 from $E2$

$$\begin{cases} (E1) & x_1 + \frac{1}{3}x_2 + \frac{1}{3}x_4 = 2 \\ (E2) & 2x_1 + x_2 - x_3 = 4 \\ (E3) & x_2 - 3x_3 - 2x_4 = 0 \end{cases} \xrightarrow[\text{with } -2E1 + E2]{\text{Replace } E2} \begin{cases} (E1) & x_1 + \frac{1}{3}x_2 + \frac{1}{3}x_4 = 2 \\ (E2) & \frac{1}{3}x_2 - x_3 - \frac{2}{3}x_4 = 0 \\ (E3) & x_2 - 3x_3 - 2x_4 = 0 \end{cases}$$

We switch $E2$ and $E3$ to get a coefficient of 1 for x_2.

$$\begin{cases} (E1) & x_1 + \frac{1}{3}x_2 + \frac{1}{3}x_4 = 2 \\ (E2) & \frac{1}{3}x_2 - x_3 - \frac{2}{3}x_4 = 0 \\ (E3) & x_2 - 3x_3 - 2x_4 = 0 \end{cases} \xrightarrow{\text{Switch } E2 \text{ and } E3} \begin{cases} (E1) & x_1 + \frac{1}{3}x_2 + \frac{1}{3}x_4 = 2 \\ (E2) & x_2 - 3x_3 - 2x_4 = 0 \\ (E3) & \frac{1}{3}x_2 - x_3 - \frac{2}{3}x_4 = 0 \end{cases}$$

Finally, we eliminate x_2 in $E3$.

$$\begin{cases} (E1) & x_1 + \frac{1}{3}x_2 + \frac{1}{3}x_4 = 2 \\ (E2) & x_2 - 3x_3 - 2x_4 = 0 \\ (E3) & \frac{1}{3}x_2 - x_3 - \frac{2}{3}x_4 = 0 \end{cases} \xrightarrow[\text{with } -\frac{1}{3}E2 + E3]{\text{Replace } E3} \begin{cases} (E1) & x_1 + \frac{1}{3}x_2 + \frac{1}{3}x_4 = 2 \\ (E2) & x_2 - 3x_3 - 2x_4 = 0 \\ (E3) & 0 = 0 \end{cases}$$

Equation $E3$ reduces to $0 = 0$, which is always true. Since we have no equations with x_3 or x_4 as leading variables, they are both free, which means we have a consistent dependent system. We parametrize the solution set by letting $x_3 = s$ and $x_4 = t$ and obtain from $E2$ that $x_2 = 3s + 2t$. Substituting this and $x_4 = t$ into $E1$, we have $x_1 + \frac{1}{3}(3s + 2t) + \frac{1}{3}t = 2$ which gives $x_1 = 2 - s - t$. Our solution is the set $\{(2-s-t, 2s+3t, s, t) \mid -\infty < s, t < \infty\}$.[13] We leave it to the reader to verify that the substitutions $x_1 = 2-s-t$, $x_2 = 3s+2t$, $x_3 = s$ and $x_4 = t$ satisfy the equations in the original system. □

Like all algorithms, Gaussian Elimination has the advantage of always producing what we need, but it can also be inefficient at times. For example, when solving 2 above, it is clear after we eliminated the x's in the second step to get the system

$$\begin{cases} (E1) & x + \frac{3}{2}y - \frac{1}{2}z = \frac{1}{2} \\ (E2) & -15y + 4z = -3 \\ (E3) & -15y + 4z = 3 \end{cases}$$

that equations $E2$ and $E3$ when taken together form a contradiction since we have identical left hand sides and different right hand sides. The algorithm takes two more steps to reach this contradiction. We also note that substitution in Gaussian Elimination is delayed until all the elimination is done, thus it gets called **back-substitution**. This may also be inefficient in many cases. Rest assured, the technique of substitution as you may have learned it in Intermediate Algebra will once again take center stage in Section 8.7. Lastly, we note that the system in 3 above is underdetermined, and as it is consistent, we have free variables in our answer. We close this section with a standard 'mixture' type application of systems of linear equations.

Example 8.1.3. Lucas needs to create a 500 milliliters (mL) of a 40% acid solution. He has stock solutions of 30% and 90% acid as well as all of the distilled water he wants. Set-up and solve a system of linear equations which determines all of the possible combinations of the stock solutions and water which would produce the required solution.

Solution. We are after three unknowns, the amount (in mL) of the 30% stock solution (which we'll call x), the amount (in mL) of the 90% stock solution (which we'll call y) and the amount (in mL) of water (which we'll call w). We now need to determine some relationships between these variables. Our goal is to produce 500 milliliters of a 40% acid solution. This product has two defining characteristics. First, it must be 500 mL; second, it must be 40% acid. We take each

---

[13] Here, any choice of s and t will determine a solution which is a point in 4-dimensional space. Yeah, we have trouble visualizing that, too.

8.1 SYSTEMS OF LINEAR EQUATIONS: GAUSSIAN ELIMINATION

of these qualities in turn. First, the total volume of 500 mL must be the sum of the contributed volumes of the two stock solutions and the water. That is

amount of 30% stock solution + amount of 90% stock solution + amount of water = 500 mL

Using our defined variables, this reduces to $x + y + w = 500$. Next, we need to make sure the final solution is 40% acid. Since water contains no acid, the acid will come from the stock solutions only. We find 40% of 500 mL to be 200 mL which means the final solution must contain 200 mL of acid. We have

amount of acid in 30% stock solution + amount of acid 90% stock solution = 200 mL

The amount of acid in x mL of 30% stock is $0.30x$ and the amount of acid in y mL of 90% solution is $0.90y$. We have $0.30x + 0.90y = 200$. Converting to fractions,[14] our system of equations becomes

$$\begin{cases} x + y + w = 500 \\ \frac{3}{10}x + \frac{9}{10}y = 200 \end{cases}$$

We first eliminate the x from the second equation

$$\begin{cases} (E1) & x + y + w = 500 \\ (E2) & \frac{3}{10}x + \frac{9}{10}y = 200 \end{cases} \xrightarrow{\text{Replace } E2 \text{ with } -\frac{3}{10}E1 + E2} \begin{cases} (E1) & x + y + w = 500 \\ (E2) & \frac{3}{5}y - \frac{3}{10}w = 50 \end{cases}$$

Next, we get a coefficient of 1 on the leading variable in $E2$

$$\begin{cases} (E1) & x + y + w = 500 \\ (E2) & \frac{3}{5}y - \frac{3}{10}w = 50 \end{cases} \xrightarrow{\text{Replace } E2 \text{ with } \frac{5}{3}E2} \begin{cases} (E1) & x + y + w = 500 \\ (E2) & y - \frac{1}{2}w = \frac{250}{3} \end{cases}$$

Notice that we have no equation to determine w, and as such, w is free. We set $w = t$ and from $E2$ get $y = \frac{1}{2}t + \frac{250}{3}$. Substituting into $E1$ gives $x + \left(\frac{1}{2}t + \frac{250}{3}\right) + t = 500$ so that $x = -\frac{3}{2}t + \frac{1250}{3}$. This system is consistent, dependent and its solution set is $\{\left(-\frac{3}{2}t + \frac{1250}{3}, \frac{1}{2}t + \frac{250}{3}, t\right) \mid -\infty < t < \infty\}$. While this answer checks algebraically, we have neglected to take into account that x, y and w, being amounts of acid and water, need to be nonnegative. That is, $x \geq 0$, $y \geq 0$ and $w \geq 0$. The constraint $x \geq 0$ gives us $-\frac{3}{2}t + \frac{1250}{3} \geq 0$, or $t \leq \frac{2500}{9}$. From $y \geq 0$, we get $\frac{1}{2}t + \frac{250}{3} \geq 0$ or $t \geq -\frac{500}{3}$. The condition $z \geq 0$ yields $t \geq 0$, and we see that when we take the set theoretic intersection of these intervals, we get $0 \leq t \leq \frac{2500}{9}$. Our final answer is $\{\left(-\frac{3}{2}t + \frac{1250}{3}, \frac{1}{2}t + \frac{250}{3}, t\right) \mid 0 \leq t \leq \frac{2500}{9}\}$. Of what practical use is our answer? Suppose there is only 100 mL of the 90% solution remaining and it is due to expire. Can we use all of it to make our required solution? We would have $y = 100$ so that $\frac{1}{2}t + \frac{250}{3} = 100$, and we get $t = \frac{100}{3}$. This means the amount of 30% solution required is $x = -\frac{3}{2}t + \frac{1250}{3} = -\frac{3}{2}\left(\frac{100}{3}\right) + \frac{1250}{3} = \frac{1100}{3}$ mL, and for the water, $w = t = \frac{100}{3}$ mL. The reader is invited to check that mixing these three amounts of our constituent solutions produces the required 40% acid mix. □

---

[14] We do this only because we believe students can use all of the practice with fractions they can get!

8.1.1 Exercises

(Review Exercises) In Exercises 1 - 8, take a trip down memory lane and solve the given system using substitution and/or elimination. Classify each system as consistent independent, consistent dependent, or inconsistent. Check your answers both algebraically and graphically.

1. $\begin{cases} x + 2y = 5 \\ x = 6 \end{cases}$

2. $\begin{cases} 2y - 3x = 1 \\ y = -3 \end{cases}$

3. $\begin{cases} \frac{x+2y}{4} = -5 \\ \frac{3x-y}{2} = 1 \end{cases}$

4. $\begin{cases} \frac{2}{3}x - \frac{1}{5}y = 3 \\ \frac{1}{2}x + \frac{3}{4}y = 1 \end{cases}$

5. $\begin{cases} \frac{1}{2}x - \frac{1}{3}y = -1 \\ 2y - 3x = 6 \end{cases}$

6. $\begin{cases} x + 4y = 6 \\ \frac{1}{12}x + \frac{1}{3}y = \frac{1}{2} \end{cases}$

7. $\begin{cases} 3y - \frac{3}{2}x = -\frac{15}{2} \\ \frac{1}{2}x - y = \frac{3}{2} \end{cases}$

8. $\begin{cases} \frac{5}{6}x + \frac{5}{3}y = -\frac{7}{3} \\ -\frac{10}{3}x - \frac{20}{3}y = 10 \end{cases}$

In Exercises 9 - 26, put each system of linear equations into triangular form and solve the system if possible. Classify each system as consistent independent, consistent dependent, or inconsistent.

9. $\begin{cases} -5x + y = 17 \\ x + y = 5 \end{cases}$

10. $\begin{cases} x + y + z = 3 \\ 2x - y + z = 0 \\ -3x + 5y + 7z = 7 \end{cases}$

11. $\begin{cases} 4x - y + z = 5 \\ 2y + 6z = 30 \\ x + z = 5 \end{cases}$

12. $\begin{cases} 4x - y + z = 5 \\ 2y + 6z = 30 \\ x + z = 6 \end{cases}$

13. $\begin{cases} x + y + z = -17 \\ y - 3z = 0 \end{cases}$

14. $\begin{cases} x - 2y + 3z = 7 \\ -3x + y + 2z = -5 \\ 2x + 2y + z = 3 \end{cases}$

15. $\begin{cases} 3x - 2y + z = -5 \\ x + 3y - z = 12 \\ x + y + 2z = 0 \end{cases}$

16. $\begin{cases} 2x - y + z = -1 \\ 4x + 3y + 5z = 1 \\ 5y + 3z = 4 \end{cases}$

17. $\begin{cases} x - y + z = -4 \\ -3x + 2y + 4z = -5 \\ x - 5y + 2z = -18 \end{cases}$

18. $\begin{cases} 2x - 4y + z = -7 \\ x - 2y + 2z = -2 \\ -x + 4y - 2z = 3 \end{cases}$

19. $\begin{cases} 2x - y + z = 1 \\ 2x + 2y - z = 1 \\ 3x + 6y + 4z = 9 \end{cases}$

20. $\begin{cases} x - 3y - 4z = 3 \\ 3x + 4y - z = 13 \\ 2x - 19y - 19z = 2 \end{cases}$

8.1 SYSTEMS OF LINEAR EQUATIONS: GAUSSIAN ELIMINATION

21. $\begin{cases} x + y + z = 4 \\ 2x - 4y - z = -1 \\ x - y = 2 \end{cases}$

22. $\begin{cases} x - y + z = 8 \\ 3x + 3y - 9z = -6 \\ 7x - 2y + 5z = 39 \end{cases}$

23. $\begin{cases} 2x - 3y + z = -1 \\ 4x - 4y + 4z = -13 \\ 6x - 5y + 7z = -25 \end{cases}$

24. $\begin{cases} 2x_1 + x_2 - 12x_3 - x_4 = 16 \\ -x_1 + x_2 + 12x_3 - 4x_4 = -5 \\ 3x_1 + 2x_2 - 16x_3 - 3x_4 = 25 \\ x_1 + 2x_2 - 5x_4 = 11 \end{cases}$

25. $\begin{cases} x_1 - x_3 = -2 \\ 2x_2 - x_4 = 0 \\ x_1 - 2x_2 + x_3 = 0 \\ -x_3 + x_4 = 1 \end{cases}$

26. $\begin{cases} x_1 - x_2 - 5x_3 + 3x_4 = -1 \\ x_1 + x_2 + 5x_3 - 3x_4 = 0 \\ x_2 + 5x_3 - 3x_4 = 1 \\ x_1 - 2x_2 - 10x_3 + 6x_4 = -1 \end{cases}$

27. Find two other forms of the parametric solution to Exercise 11 above by reorganizing the equations so that x or y can be the free variable.

28. A local buffet charges $7.50 per person for the basic buffet and $9.25 for the deluxe buffet (which includes crab legs.) If 27 diners went out to eat and the total bill was $227.00 before taxes, how many chose the basic buffet and how many chose the deluxe buffet?

29. At The Old Home Fill'er Up and Keep on a-Truckin' Cafe, Mavis mixes two different types of coffee beans to produce a house blend. The first type costs $3 per pound and the second costs $8 per pound. How much of each type does Mavis use to make 50 pounds of a blend which costs $6 per pound?

30. Skippy has a total of $10,000 to split between two investments. One account offers 3% simple interest, and the other account offers 8% simple interest. For tax reasons, he can only earn $500 in interest the entire year. How much money should Skippy invest in each account to earn $500 in interest for the year?

31. A 10% salt solution is to be mixed with pure water to produce 75 gallons of a 3% salt solution. How much of each are needed?

32. At The Crispy Critter's Head Shop and Patchouli Emporium along with their dried up weeds, sunflower seeds and astrological postcards they sell an herbal tea blend. By weight, Type I herbal tea is 30% peppermint, 40% rose hips and 30% chamomile, Type II has percents 40%, 20% and 40%, respectively, and Type III has percents 35%, 30% and 35%, respectively. How much of each Type of tea is needed to make 2 pounds of a new blend of tea that is equal parts peppermint, rose hips and chamomile?

33. Discuss with your classmates how you would approach Exercise 32 above if they needed to use up a pound of Type I tea to make room on the shelf for a new canister.

34. If you were to try to make 100 mL of a 60% acid solution using stock solutions at 20% and 40%, respectively, what would the triangular form of the resulting system look like? Explain.

8.1.2 Answers

1. Consistent independent
   Solution $\left(6, -\frac{1}{2}\right)$

2. Consistent independent
   Solution $\left(-\frac{7}{3}, -3\right)$

3. Consistent independent
   Solution $\left(-\frac{16}{7}, -\frac{62}{7}\right)$

4. Consistent independent
   Solution $\left(\frac{49}{12}, -\frac{25}{18}\right)$

5. Consistent dependent
   Solution $\left(t, \frac{3}{2}t + 3\right)$
   for all real numbers t

6. Consistent dependent
   Solution $(6 - 4t, t)$
   for all real numbers t

7. Inconsistent
   No solution

8. Inconsistent
   No solution

Because triangular form is not unique, we give only one possible answer to that part of the question. Yours may be different and still be correct.

9. $\begin{cases} x + y = 5 \\ y = 7 \end{cases}$
   Consistent independent
   Solution $(-2, 7)$

10. $\begin{cases} x - \frac{5}{3}y - \frac{7}{3}z = -\frac{7}{3} \\ y + \frac{5}{4}z = 2 \\ z = 0 \end{cases}$
    Consistent independent
    Solution $(1, 2, 0)$

11. $\begin{cases} x - \frac{1}{4}y + \frac{1}{4}z = \frac{5}{4} \\ y + 3z = 15 \\ 0 = 0 \end{cases}$
    Consistent dependent
    Solution $(-t + 5, -3t + 15, t)$
    for all real numbers t

12. $\begin{cases} x - \frac{1}{4}y + \frac{1}{4}z = \frac{5}{4} \\ y + 3z = 15 \\ 0 = 1 \end{cases}$
    Inconsistent
    No solution

13. $\begin{cases} x + y + z = -17 \\ y - 3z = 0 \end{cases}$
    Consistent dependent
    Solution $(-4t - 17, 3t, t)$
    for all real numbers t

14. $\begin{cases} x - 2y + 3z = 7 \\ y - \frac{11}{5}z = -\frac{16}{5} \\ z = 1 \end{cases}$
    Consistent independent
    Solution $(2, -1, 1)$

8.1 SYSTEMS OF LINEAR EQUATIONS: GAUSSIAN ELIMINATION

15. $\begin{cases} x + y + 2z = 0 \\ y - \frac{3}{2}z = 6 \\ z = -2 \end{cases}$
Consistent independent
Solution $(1, 3, -2)$

16. $\begin{cases} x - \frac{1}{2}y + \frac{1}{2}z = -\frac{1}{2} \\ y + \frac{3}{5}z = \frac{3}{5} \\ 0 = 1 \end{cases}$
Inconsistent
no solution

17. $\begin{cases} x - y + z = -4 \\ y - 7z = 17 \\ z = -2 \end{cases}$
Consistent independent
Solution $(1, 3, -2)$

18. $\begin{cases} x - 2y + 2z = -2 \\ y = \frac{1}{2} \\ z = 1 \end{cases}$
Consistent independent
Solution $\left(-3, \frac{1}{2}, 1\right)$

19. $\begin{cases} x - \frac{1}{2}y + \frac{1}{2}z = \frac{1}{2} \\ y - \frac{2}{3}z = 0 \\ z = 1 \end{cases}$
Consistent independent
Solution $\left(\frac{1}{3}, \frac{2}{3}, 1\right)$

20. $\begin{cases} x - 3y - 4z = 3 \\ y + \frac{11}{13}z = \frac{4}{13} \\ 0 = 0 \end{cases}$
Consistent dependent
Solution $\left(\frac{19}{13}t + \frac{51}{13}, -\frac{11}{13}t + \frac{4}{13}, t\right)$
for all real numbers t

21. $\begin{cases} x + y + z = 4 \\ y + \frac{1}{2}z = \frac{3}{2} \\ 0 = 1 \end{cases}$
Inconsistent
no solution

22. $\begin{cases} x - y + z = 8 \\ y - 2z = -5 \\ z = 1 \end{cases}$
Consistent independent
Solution $(4, -3, 1)$

23. $\begin{cases} x - \frac{3}{2}y + \frac{1}{2}z = -\frac{1}{2} \\ y + z = -\frac{11}{2} \\ 0 = 0 \end{cases}$
Consistent dependent
Solution $\left(-2t - \frac{35}{4}, -t - \frac{11}{2}, t\right)$
for all real numbers t

24. $\begin{cases} x_1 + \frac{2}{3}x_2 - \frac{16}{3}x_3 - x_4 = \frac{25}{3} \\ x_2 + 4x_3 - 3x_4 = 2 \\ 0 = 0 \\ 0 = 0 \end{cases}$
Consistent dependent
Solution $(8s - t + 7, -4s + 3t + 2, s, t)$
for all real numbers s and t

25. $\begin{cases} x_1 - x_3 = -2 \\ x_2 - \frac{1}{2}x_4 = 0 \\ x_3 - \frac{1}{2}x_4 = 1 \\ x_4 = 4 \end{cases}$
Consistent independent
Solution $(1, 2, 3, 4)$

26. $\begin{cases} x_1 - x_2 - 5x_3 + 3x_4 &= -1 \\ x_2 + 5x_3 - 3x_4 &= \frac{1}{2} \\ 0 &= 1 \\ 0 &= 0 \end{cases}$ Inconsistent
   No solution

27. If x is the free variable then the solution is $(t, 3t, -t+5)$ and if y is the free variable then the solution is $\left(\frac{1}{3}t, t, -\frac{1}{3}t + 5\right)$.

28. 13 chose the basic buffet and 14 chose the deluxe buffet.

29. Mavis needs 20 pounds of \$3 per pound coffee and 30 pounds of \$8 per pound coffee.

30. Skippy needs to invest \$6000 in the 3% account and \$4000 in the 8% account.

31. 22.5 gallons of the 10% solution and 52.5 gallons of pure water.

32. $\frac{4}{3} - \frac{1}{2}t$ pounds of Type I, $\frac{2}{3} - \frac{1}{2}t$ pounds of Type II and t pounds of Type III where $0 \leq t \leq \frac{4}{3}$.

8.2 Systems of Linear Equations: Augmented Matrices

In Section 8.1 we introduced Gaussian Elimination as a means of transforming a system of linear equations into triangular form with the ultimate goal of producing an equivalent system of linear equations which is easier to solve. If we take a step back and study the process, we see that all of our moves are determined entirely by the *coefficients* of the variables involved, and not the variables themselves. Much the same thing happened when we studied long division in Section 3.2. Just as we developed synthetic division to streamline that process, in this section, we introduce a similar bookkeeping device to help us solve systems of linear equations. To that end, we define a **matrix** as a rectangular array of real numbers. We typically enclose matrices with square brackets, '[' and ']', and we size matrices by the number of rows and columns they have. For example, the **size** (sometimes called the **dimension**) of

$$\begin{bmatrix} 3 & 0 & -1 \\ 2 & -5 & 10 \end{bmatrix}$$

is 2×3 because it has 2 rows and 3 columns. The individual numbers in a matrix are called its **entries** and are usually labeled with double subscripts: the first tells which row the element is in and the second tells which column it is in. The rows are numbered from top to bottom and the columns are numbered from left to right. Matrices themselves are usually denoted by uppercase letters (A, B, C, etc.) while their entries are usually denoted by the corresponding letter. So, for instance, if we have

$$A = \begin{bmatrix} 3 & 0 & -1 \\ 2 & -5 & 10 \end{bmatrix}$$

then $a_{11} = 3$, $a_{12} = 0$, $a_{13} = -1$, $a_{21} = 2$, $a_{22} = -5$, and $a_{23} = 10$. We shall explore matrices as mathematical objects with their own algebra in Section 8.3 and introduce them here solely as a bookkeeping device. Consider the system of linear equations from number 2 in Example 8.1.2

$$\begin{cases} (E1) & 2x + 3y - z = 1 \\ (E2) & 10x - z = 2 \\ (E3) & 4x - 9y + 2z = 5 \end{cases}$$

We encode this system into a matrix by assigning each equation to a corresponding row. Within that row, each variable and the constant gets its own column, and to separate the variables on the left hand side of the equation from the constants on the right hand side, we use a vertical bar, |. Note that in $E2$, since y is not present, we record its coefficient as 0. The matrix associated with this system is

$$\begin{array}{c} (E1) \to \\ (E2) \to \\ (E3) \to \end{array} \begin{bmatrix} x & y & z & c \\ 2 & 3 & -1 & 1 \\ 10 & 0 & -1 & 2 \\ 4 & -9 & 2 & 5 \end{bmatrix}$$

This matrix is called an **augmented matrix** because the column containing the constants is appended to the matrix containing the coefficients.[1] To solve this system, we can use the same kind operations on the *rows* of the matrix that we performed on the *equations* of the system. More specifically, we have the following analog of Theorem 8.1 below.

> **Theorem 8.2. Row Operations:** Given an augmented matrix for a system of linear equations, the following row operations produce an augmented matrix which corresponds to an equivalent system of linear equations.
>
> - Interchange any two rows.
>
> - Replace a row with a nonzero multiple of itself.[a]
>
> - Replace a row with itself plus a nonzero multiple of another row.[b]
>
> ---
> [a]That is, the row obtained by multiplying each entry in the row by the same nonzero number.
> [b]Where we add entries in corresponding columns.

As a demonstration of the moves in Theorem 8.2, we revisit some of the steps that were used in solving the systems of linear equations in Example 8.1.2 of Section 8.1. The reader is encouraged to perform the indicated operations on the rows of the augmented matrix to see that the machinations are identical to what is done to the coefficients of the variables in the equations. We first see a demonstration of switching two rows using the first step of part 1 in Example 8.1.2.

$$\begin{cases} (E1) & 3x - y + z = 3 \\ (E2) & 2x - 4y + 3z = 16 \\ (E3) & x - y + z = 5 \end{cases} \xrightarrow{\text{Switch } E1 \text{ and } E3} \begin{cases} (E1) & x - y + z = 5 \\ (E2) & 2x - 4y + 3z = 16 \\ (E3) & 3x - y + z = 3 \end{cases}$$

$$\begin{bmatrix} 3 & -1 & 1 & | & 3 \\ 2 & -4 & 3 & | & 16 \\ 1 & -1 & 1 & | & 5 \end{bmatrix} \xrightarrow{\text{Switch } R1 \text{ and } R3} \begin{bmatrix} 1 & -1 & 1 & | & 5 \\ 2 & -4 & 3 & | & 16 \\ 3 & -1 & 1 & | & 3 \end{bmatrix}$$

Next, we have a demonstration of replacing a row with a nonzero multiple of itself using the first step of part 3 in Example 8.1.2.

$$\begin{cases} (E1) & 3x_1 + x_2 + x_4 = 6 \\ (E2) & 2x_1 + x_2 - x_3 = 4 \\ (E3) & x_2 - 3x_3 - 2x_4 = 0 \end{cases} \xrightarrow{\text{Replace } E1 \text{ with } \frac{1}{3}E1} \begin{cases} (E1) & x_1 + \frac{1}{3}x_2 + \frac{1}{3}x_4 = 2 \\ (E2) & 2x_1 + x_2 - x_3 = 4 \\ (E3) & x_2 - 3x_3 - 2x_4 = 0 \end{cases}$$

$$\begin{bmatrix} 3 & 1 & 0 & 1 & | & 6 \\ 2 & 1 & -1 & 0 & | & 4 \\ 0 & 1 & -3 & -2 & | & 0 \end{bmatrix} \xrightarrow{\text{Replace } R1 \text{ with } \frac{1}{3}R1} \begin{bmatrix} 1 & \frac{1}{3} & 0 & \frac{1}{3} & | & 2 \\ 2 & 1 & -1 & 0 & | & 4 \\ 0 & 1 & -3 & -2 & | & 0 \end{bmatrix}$$

Finally, we have an example of replacing a row with itself plus a multiple of another row using the second step from part 2 in Example 8.1.2.

---
[1]We shall study the coefficient and constant matrices separately in Section 8.3.

8.2 Systems of Linear Equations: Augmented Matrices

$$\begin{cases} (E1) & x + \frac{3}{2}y - \frac{1}{2}z = \frac{1}{2} \\ (E2) & 10x - z = 2 \\ (E3) & 4x - 9y + 2z = 5 \end{cases} \xrightarrow[\text{Replace } E3 \text{ with } -4E1 + E3]{\text{Replace } E2 \text{ with } -10E1 + E2} \begin{cases} (E1) & x + \frac{3}{2}y - \frac{1}{2}z = \frac{1}{2} \\ (E2) & -15y + 4z = -3 \\ (E3) & -15y + 4z = 3 \end{cases}$$

$$\begin{bmatrix} 1 & \frac{3}{2} & -\frac{1}{2} & | & \frac{1}{2} \\ 10 & 0 & -1 & | & 2 \\ 4 & -9 & 2 & | & 5 \end{bmatrix} \xrightarrow[\text{Replace } R3 \text{ with } -4R1 + R3]{\text{Replace } R2 \text{ with } -10R1 + R2} \begin{bmatrix} 1 & \frac{3}{2} & -\frac{1}{2} & | & \frac{1}{2} \\ 0 & -15 & 4 & | & -3 \\ 0 & -15 & 4 & | & 3 \end{bmatrix}$$

The matrix equivalent of 'triangular form' is **row echelon form**. The reader is encouraged to refer to Definition 8.3 for comparison. Note that the analog of 'leading variable' of an equation is 'leading entry' of a row. Specifically, the first nonzero entry (if it exists) in a row is called the **leading entry** of that row.

> **Definition 8.4.** A matrix is said to be in **row echelon form** provided all of the following conditions hold:
>
> 1. The first nonzero entry in each row is 1.
>
> 2. The leading 1 of a given row must be to the right of the leading 1 of the row above it.
>
> 3. Any row of all zeros cannot be placed above a row with nonzero entries.

To solve a system of a linear equations using an augmented matrix, we encode the system into an augmented matrix and apply Gaussian Elimination to the rows to get the matrix into row-echelon form. We then decode the matrix and back substitute. The next example illustrates this nicely.

Example 8.2.1. Use an augmented matrix to transform the following system of linear equations into triangular form. Solve the system.

$$\begin{cases} 3x - y + z = 8 \\ x + 2y - z = 4 \\ 2x + 3y - 4z = 10 \end{cases}$$

Solution. We first encode the system into an augmented matrix.

$$\begin{cases} 3x - y + z = 8 \\ x + 2y - z = 4 \\ 2x + 3y - 4z = 10 \end{cases} \xrightarrow{\text{Encode into the matrix}} \begin{bmatrix} 3 & -1 & 1 & | & 8 \\ 1 & 2 & -1 & | & 4 \\ 2 & 3 & -4 & | & 10 \end{bmatrix}$$

Thinking back to Gaussian Elimination at an equations level, our first order of business is to get x in $E1$ with a coefficient of 1. At the matrix level, this means getting a leading 1 in $R1$. This is in accordance with the first criteria in Definition 8.4. To that end, we interchange $R1$ and $R2$.

$$\begin{bmatrix} 3 & -1 & 1 & | & 8 \\ 1 & 2 & -1 & | & 4 \\ 2 & 3 & -4 & | & 10 \end{bmatrix} \xrightarrow{\text{Switch } R1 \text{ and } R2} \begin{bmatrix} 1 & 2 & -1 & | & 4 \\ 3 & -1 & 1 & | & 8 \\ 2 & 3 & -4 & | & 10 \end{bmatrix}$$

Our next step is to eliminate the x's from $E2$ and $E3$. From a matrix standpoint, this means we need 0's below the leading 1 in $R1$. This guarantees the leading 1 in $R2$ will be to the right of the leading 1 in $R1$ in accordance with the second requirement of Definition 8.4.

$$\begin{bmatrix} 1 & 2 & -1 & | & 4 \\ 3 & -1 & 1 & | & 8 \\ 2 & 3 & -4 & | & 10 \end{bmatrix} \xrightarrow[\text{Replace } R3 \text{ with } -2R1 + R3]{\text{Replace } R2 \text{ with } -3R1 + R2} \begin{bmatrix} 1 & 2 & -1 & | & 4 \\ 0 & -7 & 4 & | & -4 \\ 0 & -1 & -2 & | & 2 \end{bmatrix}$$

Now we repeat the above process for the variable y which means we need to get the leading entry in $R2$ to be 1.

$$\begin{bmatrix} 1 & 2 & -1 & | & 4 \\ 0 & -7 & 4 & | & -4 \\ 0 & -1 & -2 & | & 2 \end{bmatrix} \xrightarrow{\text{Replace } R2 \text{ with } -\frac{1}{7}R2} \begin{bmatrix} 1 & 2 & -1 & | & 4 \\ 0 & 1 & -\frac{4}{7} & | & \frac{4}{7} \\ 0 & -1 & -2 & | & 2 \end{bmatrix}$$

To guarantee the leading 1 in $R3$ is to the right of the leading 1 in $R2$, we get a 0 in the second column of $R3$.

$$\begin{bmatrix} 1 & 2 & -1 & | & 4 \\ 0 & 1 & -\frac{4}{7} & | & \frac{4}{7} \\ 0 & -1 & -2 & | & 2 \end{bmatrix} \xrightarrow{\text{Replace } R3 \text{ with } R2 + R3} \begin{bmatrix} 1 & 2 & -1 & | & 4 \\ 0 & 1 & -\frac{4}{7} & | & \frac{4}{7} \\ 0 & 0 & -\frac{18}{7} & | & \frac{18}{7} \end{bmatrix}$$

Finally, we get the leading entry in $R3$ to be 1.

$$\begin{bmatrix} 1 & 2 & -1 & | & 4 \\ 0 & 1 & -\frac{4}{7} & | & \frac{4}{7} \\ 0 & 0 & -\frac{18}{7} & | & \frac{18}{7} \end{bmatrix} \xrightarrow{\text{Replace } R3 \text{ with } -\frac{7}{18}R3} \begin{bmatrix} 1 & 2 & -1 & | & 4 \\ 0 & 1 & -\frac{4}{7} & | & \frac{4}{7} \\ 0 & 0 & 1 & | & -1 \end{bmatrix}$$

Decoding from the matrix gives a system in triangular form

$$\begin{bmatrix} 1 & 2 & -1 & | & 4 \\ 0 & 1 & -\frac{4}{7} & | & \frac{4}{7} \\ 0 & 0 & 1 & | & -1 \end{bmatrix} \xrightarrow{\text{Decode from the matrix}} \begin{cases} x + 2y - z = 4 \\ y - \frac{4}{7}z = \frac{4}{7} \\ z = -1 \end{cases}$$

We get $z = -1$, $y = \frac{4}{7}z + \frac{4}{7} = \frac{4}{7}(-1) + \frac{4}{7} = 0$ and $x = -2y + z + 4 = -2(0) + (-1) + 4 = 3$ for a final answer of $(3, 0, -1)$. We leave it to the reader to check. $\square$

As part of Gaussian Elimination, we used row operations to obtain 0's beneath each leading 1 to put the matrix into row echelon form. If we also require that 0's are the only numbers above a leading 1, we have what is known as the **reduced row echelon form** of the matrix.

> **Definition 8.5.** A matrix is said to be in **reduced row echelon form** provided both of the following conditions hold:
>
> 1. The matrix is in row echelon form.
>
> 2. The leading 1s are the only nonzero entry in their respective columns.

8.2 Systems of Linear Equations: Augmented Matrices

Of what significance is the reduced row echelon form of a matrix? To illustrate, let's take the row echelon form from Example 8.2.1 and perform the necessary steps to put into reduced row echelon form. We start by using the leading 1 in $R3$ to zero out the numbers in the rows above it.

$$\begin{bmatrix} 1 & 2 & -1 & | & 4 \\ 0 & 1 & -\frac{4}{7} & | & \frac{4}{7} \\ 0 & 0 & 1 & | & -1 \end{bmatrix} \xrightarrow[\text{Replace } R2 \text{ with } \frac{4}{7}R3 + R2]{\text{Replace } R1 \text{ with } R3 + R1} \begin{bmatrix} 1 & 2 & 0 & | & 3 \\ 0 & 1 & 0 & | & 0 \\ 0 & 0 & 1 & | & -1 \end{bmatrix}$$

Finally, we take care of the 2 in $R1$ above the leading 1 in $R2$.

$$\begin{bmatrix} 1 & 2 & 0 & | & 3 \\ 0 & 1 & 0 & | & 0 \\ 0 & 0 & 1 & | & -1 \end{bmatrix} \xrightarrow{\text{Replace } R1 \text{ with } -2R2 + R1} \begin{bmatrix} 1 & 0 & 0 & | & 3 \\ 0 & 1 & 0 & | & 0 \\ 0 & 0 & 1 & | & -1 \end{bmatrix}$$

To our surprise and delight, when we decode this matrix, we obtain the solution instantly without having to deal with any back-substitution at all.

$$\begin{bmatrix} 1 & 0 & 0 & | & 3 \\ 0 & 1 & 0 & | & 0 \\ 0 & 0 & 1 & | & -1 \end{bmatrix} \xrightarrow{\text{Decode from the matrix}} \begin{cases} x & = & 3 \\ y & = & 0 \\ z & = & -1 \end{cases}$$

Note that in the previous discussion, we could have started with $R2$ and used it to get a zero above its leading 1 and then done the same for the leading 1 in $R3$. By starting with $R3$, however, we get more zeros first, and the more zeros there are, the faster the remaining calculations will be.[2] It is also worth noting that while a matrix has several[3] row echelon forms, it has only one reduced row echelon form. The process by which we have put a matrix into reduced row echelon form is called **Gauss-Jordan Elimination**.

Example 8.2.2. Solve the following system using an augmented matrix. Use Gauss-Jordan Elimination to put the augmented matrix into reduced row echelon form.

$$\begin{cases} x_2 - 3x_1 + x_4 & = & 2 \\ 2x_1 + 4x_3 & = & 5 \\ 4x_2 - x_4 & = & 3 \end{cases}$$

Solution. We first encode the system into a matrix. (Pay attention to the subscripts!)

$$\begin{cases} x_2 - 3x_1 + x_4 & = & 2 \\ 2x_1 + 4x_3 & = & 5 \\ 4x_2 - x_4 & = & 3 \end{cases} \xrightarrow{\text{Encode into the matrix}} \begin{bmatrix} -3 & 1 & 0 & 1 & | & 2 \\ 2 & 0 & 4 & 0 & | & 5 \\ 0 & 4 & 0 & -1 & | & 3 \end{bmatrix}$$

Next, we get a leading 1 in the first column of $R1$.

$$\begin{bmatrix} -3 & 1 & 0 & 1 & | & 2 \\ 2 & 0 & 4 & 0 & | & 5 \\ 0 & 4 & 0 & -1 & | & 3 \end{bmatrix} \xrightarrow{\text{Replace } R1 \text{ with } -\frac{1}{3}R1} \begin{bmatrix} 1 & -\frac{1}{3} & 0 & -\frac{1}{3} & | & -\frac{2}{3} \\ 2 & 0 & 4 & 0 & | & 5 \\ 0 & 4 & 0 & -1 & | & 3 \end{bmatrix}$$

---
[2] Carl also finds starting with $R3$ to be more symmetric, in a purely poetic way.
[3] infinite, in fact

Now we eliminate the nonzero entry below our leading 1.

$$\begin{bmatrix} 1 & -\frac{1}{3} & 0 & -\frac{1}{3} & -\frac{2}{3} \\ 2 & 0 & 4 & 0 & 5 \\ 0 & 4 & 0 & -1 & 3 \end{bmatrix} \xrightarrow{\text{Replace } R2 \text{ with } -2R1 + R2} \begin{bmatrix} 1 & -\frac{1}{3} & 0 & -\frac{1}{3} & -\frac{2}{3} \\ 0 & \frac{2}{3} & 4 & \frac{2}{3} & \frac{19}{3} \\ 0 & 4 & 0 & -1 & 3 \end{bmatrix}$$

We proceed to get a leading 1 in $R2$.

$$\begin{bmatrix} 1 & -\frac{1}{3} & 0 & -\frac{1}{3} & -\frac{2}{3} \\ 0 & \frac{2}{3} & 4 & \frac{2}{3} & \frac{19}{3} \\ 0 & 4 & 0 & -1 & 3 \end{bmatrix} \xrightarrow{\text{Replace } R2 \text{ with } \frac{3}{2}R2} \begin{bmatrix} 1 & -\frac{1}{3} & 0 & -\frac{1}{3} & -\frac{2}{3} \\ 0 & 1 & 6 & 1 & \frac{19}{2} \\ 0 & 4 & 0 & -1 & 3 \end{bmatrix}$$

We now zero out the entry below the leading 1 in $R2$.

$$\begin{bmatrix} 1 & -\frac{1}{3} & 0 & -\frac{1}{3} & -\frac{2}{3} \\ 0 & 1 & 6 & 1 & \frac{19}{2} \\ 0 & 4 & 0 & -1 & 3 \end{bmatrix} \xrightarrow{\text{Replace } R3 \text{ with } -4R2 + R3} \begin{bmatrix} 1 & -\frac{1}{3} & 0 & -\frac{1}{3} & -\frac{2}{3} \\ 0 & 1 & 6 & 1 & \frac{19}{2} \\ 0 & 0 & -24 & -5 & -35 \end{bmatrix}$$

Next, it's time for a leading 1 in $R3$.

$$\begin{bmatrix} 1 & -\frac{1}{3} & 0 & -\frac{1}{3} & -\frac{2}{3} \\ 0 & 1 & 6 & 1 & \frac{19}{2} \\ 0 & 0 & -24 & -5 & -35 \end{bmatrix} \xrightarrow{\text{Replace } R3 \text{ with } -\frac{1}{24}R3} \begin{bmatrix} 1 & -\frac{1}{3} & 0 & -\frac{1}{3} & -\frac{2}{3} \\ 0 & 1 & 6 & 1 & \frac{19}{2} \\ 0 & 0 & 1 & \frac{5}{24} & \frac{35}{24} \end{bmatrix}$$

The matrix is now in row echelon form. To get the reduced row echelon form, we start with the last leading 1 we produced and work to get 0's above it.

$$\begin{bmatrix} 1 & -\frac{1}{3} & 0 & -\frac{1}{3} & -\frac{2}{3} \\ 0 & 1 & 6 & 1 & \frac{19}{2} \\ 0 & 0 & 1 & \frac{5}{24} & \frac{35}{24} \end{bmatrix} \xrightarrow{\text{Replace } R2 \text{ with } -6R3 + R2} \begin{bmatrix} 1 & -\frac{1}{3} & 0 & -\frac{1}{3} & -\frac{2}{3} \\ 0 & 1 & 0 & -\frac{1}{4} & \frac{3}{4} \\ 0 & 0 & 1 & \frac{5}{24} & \frac{35}{24} \end{bmatrix}$$

Lastly, we get a 0 above the leading 1 of $R2$.

$$\begin{bmatrix} 1 & -\frac{1}{3} & 0 & -\frac{1}{3} & -\frac{2}{3} \\ 0 & 1 & 0 & -\frac{1}{4} & \frac{3}{4} \\ 0 & 0 & 1 & \frac{5}{24} & \frac{35}{24} \end{bmatrix} \xrightarrow{\text{Replace } R1 \text{ with } \frac{1}{3}R2 + R1} \begin{bmatrix} 1 & 0 & 0 & -\frac{5}{12} & -\frac{5}{12} \\ 0 & 1 & 0 & -\frac{1}{4} & \frac{3}{4} \\ 0 & 0 & 1 & \frac{5}{24} & \frac{35}{24} \end{bmatrix}$$

At last, we decode to get

$$\begin{bmatrix} 1 & 0 & 0 & -\frac{5}{12} & -\frac{5}{12} \\ 0 & 1 & 0 & -\frac{1}{4} & \frac{3}{4} \\ 0 & 0 & 1 & \frac{5}{24} & \frac{35}{24} \end{bmatrix} \xrightarrow{\text{Decode from the matrix}} \begin{cases} x_1 - \frac{5}{12}x_4 = -\frac{5}{12} \\ x_2 - \frac{1}{4}x_4 = \frac{3}{4} \\ x_3 + \frac{5}{24}x_4 = \frac{35}{24} \end{cases}$$

We have that x_4 is free and we assign it the parameter t. We obtain $x_3 = -\frac{5}{24}t + \frac{35}{24}$, $x_2 = \frac{1}{4}t + \frac{3}{4}$, and $x_1 = \frac{5}{12}t - \frac{5}{12}$. Our solution is $\left\{\left(\frac{5}{12}t - \frac{5}{12}, \frac{1}{4}t + \frac{3}{4}, -\frac{5}{24}t + \frac{35}{24}, t\right) : -\infty < t < \infty\right\}$ and leave it to the reader to check. $\square$

8.2 Systems of Linear Equations: Augmented Matrices

Like all good algorithms, putting a matrix in row echelon or reduced row echelon form can easily be programmed into a calculator, and, doubtless, your graphing calculator has such a feature. We use this in our next example.

Example 8.2.3. Find the quadratic function passing through the points $(-1, 3)$, $(2, 4)$, $(5, -2)$.

Solution. According to Definition 2.5, a quadratic function has the form $f(x) = ax^2 + bx + c$ where $a \neq 0$. Our goal is to find a, b and c so that the three given points are on the graph of f. If $(-1, 3)$ is on the graph of f, then $f(-1) = 3$, or $a(-1)^2 + b(-1) + c = 3$ which reduces to $a - b + c = 3$, an honest-to-goodness linear equation with the variables a, b and c. Since the point $(2, 4)$ is also on the graph of f, then $f(2) = 4$ which gives us the equation $4a + 2b + c = 4$. Lastly, the point $(5, -2)$ is on the graph of f gives us $25a + 5b + c = -2$. Putting these together, we obtain a system of three linear equations. Encoding this into an augmented matrix produces

$$\left\{ \begin{array}{rcl} a - b + c & = & 3 \\ 4a + 2b + c & = & 4 \\ 25a + 5b + c & = & -2 \end{array} \right. \quad \xrightarrow{\text{Encode into the matrix}} \quad \left[\begin{array}{rrr|r} 1 & -1 & 1 & 3 \\ 4 & 2 & 1 & 4 \\ 25 & 5 & 1 & -2 \end{array} \right]$$

Using a calculator,[4] we find $a = -\frac{7}{18}$, $b = \frac{13}{18}$ and $c = \frac{37}{9}$. Hence, the one and only quadratic which fits the bill is $f(x) = -\frac{7}{18}x^2 + \frac{13}{18}x + \frac{37}{9}$. To verify this analytically, we see that $f(-1) = 3$, $f(2) = 4$, and $f(5) = -2$. We can use the calculator to check our solution as well by plotting the three data points and the function f.

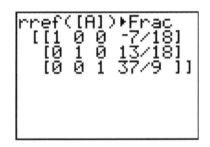
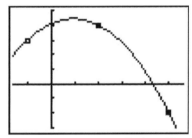

The graph of $f(x) = -\frac{7}{18}x^2 + \frac{13}{18}x + \frac{37}{9}$ with the points $(-1, 3)$, $(2, 4)$ and $(5, -2)$

□

---

[4]We've tortured you enough already with fractions in this exposition!

8.2.1 Exercises

In Exercises 1 - 6, state whether the given matrix is in reduced row echelon form, row echelon form only or in neither of those forms.

1. $\begin{bmatrix} 1 & 0 & | & 3 \\ 0 & 1 & | & 3 \end{bmatrix}$

2. $\begin{bmatrix} 3 & -1 & 1 & | & 3 \\ 2 & -4 & 3 & | & 16 \\ 1 & -1 & 1 & | & 5 \end{bmatrix}$

3. $\begin{bmatrix} 1 & 1 & 4 & | & 3 \\ 0 & 1 & 3 & | & 6 \\ 0 & 0 & 0 & | & 1 \end{bmatrix}$

4. $\begin{bmatrix} 1 & 0 & 0 & | & 0 \\ 0 & 1 & 0 & | & 0 \\ 0 & 0 & 0 & | & 1 \end{bmatrix}$

5. $\begin{bmatrix} 1 & 0 & 4 & 3 & | & 0 \\ 0 & 1 & 3 & 6 & | & 0 \\ 0 & 0 & 0 & 0 & | & 0 \end{bmatrix}$

6. $\begin{bmatrix} 1 & 1 & 4 & | & 3 \\ 0 & 1 & 3 & | & 6 \end{bmatrix}$

In Exercises 7 - 12, the following matrices are in reduced row echelon form. Determine the solution of the corresponding system of linear equations or state that the system is inconsistent.

7. $\begin{bmatrix} 1 & 0 & | & -2 \\ 0 & 1 & | & 7 \end{bmatrix}$

8. $\begin{bmatrix} 1 & 0 & 0 & | & -3 \\ 0 & 1 & 0 & | & 20 \\ 0 & 0 & 1 & | & 19 \end{bmatrix}$

9. $\begin{bmatrix} 1 & 0 & 0 & 3 & | & 4 \\ 0 & 1 & 0 & 6 & | & -6 \\ 0 & 0 & 1 & 0 & | & 2 \end{bmatrix}$

10. $\begin{bmatrix} 1 & 0 & 0 & 3 & | & 0 \\ 0 & 1 & 2 & 6 & | & 0 \\ 0 & 0 & 0 & 0 & | & 1 \end{bmatrix}$

11. $\begin{bmatrix} 1 & 0 & -8 & 1 & | & 7 \\ 0 & 1 & 4 & -3 & | & 2 \\ 0 & 0 & 0 & 0 & | & 0 \\ 0 & 0 & 0 & 0 & | & 0 \end{bmatrix}$

12. $\begin{bmatrix} 1 & 0 & 9 & | & -3 \\ 0 & 1 & -4 & | & 20 \\ 0 & 0 & 0 & | & 0 \end{bmatrix}$

In Exercises 13 - 26, solve the following systems of linear equations using the techniques discussed in this section. Compare and contrast these techniques with those you used to solve the systems in the Exercises in Section 8.1.

13. $\begin{cases} -5x + y = 17 \\ x + y = 5 \end{cases}$

14. $\begin{cases} x + y + z = 3 \\ 2x - y + z = 0 \\ -3x + 5y + 7z = 7 \end{cases}$

15. $\begin{cases} 4x - y + z = 5 \\ 2y + 6z = 30 \\ x + z = 5 \end{cases}$

16. $\begin{cases} x - 2y + 3z = 7 \\ -3x + y + 2z = -5 \\ 2x + 2y + z = 3 \end{cases}$

17. $\begin{cases} 3x - 2y + z = -5 \\ x + 3y - z = 12 \\ x + y + 2z = 0 \end{cases}$

18. $\begin{cases} 2x - y + z = -1 \\ 4x + 3y + 5z = 1 \\ 5y + 3z = 4 \end{cases}$

19. $\begin{cases} x - y + z = -4 \\ -3x + 2y + 4z = -5 \\ x - 5y + 2z = -18 \end{cases}$

20. $\begin{cases} 2x - 4y + z = -7 \\ x - 2y + 2z = -2 \\ -x + 4y - 2z = 3 \end{cases}$

21. $\begin{cases} 2x - y + z = 1 \\ 2x + 2y - z = 1 \\ 3x + 6y + 4z = 9 \end{cases}$

22. $\begin{cases} x - 3y - 4z = 3 \\ 3x + 4y - z = 13 \\ 2x - 19y - 19z = 2 \end{cases}$

23. $\begin{cases} x + y + z = 4 \\ 2x - 4y - z = -1 \\ x - y = 2 \end{cases}$

24. $\begin{cases} x - y + z = 8 \\ 3x + 3y - 9z = -6 \\ 7x - 2y + 5z = 39 \end{cases}$

25. $\begin{cases} 2x - 3y + z = -1 \\ 4x - 4y + 4z = -13 \\ 6x - 5y + 7z = -25 \end{cases}$

26. $\begin{cases} x_1 - x_3 = -2 \\ 2x_2 - x_4 = 0 \\ x_1 - 2x_2 + x_3 = 0 \\ -x_3 + x_4 = 1 \end{cases}$

27. It's time for another meal at our local buffet. This time, 22 diners (5 of whom were children) feasted for $162.25, before taxes. If the kids buffet is $4.50, the basic buffet is $7.50, and the deluxe buffet (with crab legs) is $9.25, find out how many diners chose the deluxe buffet.

28. Carl wants to make a party mix consisting of almonds (which cost $7 per pound), cashews (which cost $5 per pound), and peanuts (which cost $2 per pound.) If he wants to make a 10 pound mix with a budget of $35, what are the possible combinations almonds, cashews, and peanuts? (You may find it helpful to review Example 8.1.3 in Section 8.1.)

29. Find the quadratic function passing through the points $(-2, 1)$, $(1, 4)$, $(3, -2)$

30. At 9 PM, the temperature was 60°F; at midnight, the temperature was 50°F; and at 6 AM, the temperature was 70°F. Use the technique in Example 8.2.3 to fit a quadratic function to these data with the temperature, T, measured in degrees Fahrenheit, as the dependent variable, and the number of hours after 9 PM, t, measured in hours, as the independent variable. What was the coldest temperature of the night? When did it occur?

31. The price for admission into the Stitz-Zeager Sasquatch Museum and Research Station is $15 for adults and $8 for kids 13 years old and younger. When the Zahlenreich family visits the museum their bill is $38 and when the Nullsatz family visits their bill is $39. One day both families went together and took an adult babysitter along to watch the kids and the total admission charge was $92. Later that summer, the adults from both families went without the kids and the bill was $45. Is that enough information to determine how many adults and children are in each family? If not, state whether the resulting system is inconsistent or consistent dependent. In the latter case, give at least two plausible solutions.

32. Use the technique in Example 8.2.3 to find the line between the points $(-3, 4)$ and $(6, 1)$. How does your answer compare to the slope-intercept form of the line in Equation 2.3?

33. With the help of your classmates, find at least two different row echelon forms for the matrix

$$\begin{bmatrix} 1 & 2 & | & 3 \\ 4 & 12 & | & 8 \end{bmatrix}$$

8.2.2 Answers

1. Reduced row echelon form
2. Neither
3. Row echelon form only
4. Reduced row echelon form
5. Reduced row echelon form
6. Row echelon form only
7. $(-2, 7)$
8. $(-3, 20, 19)$
9. $(-3t + 4, -6t - 6, 2, t)$
   for all real numbers t
10. Inconsistent
11. $(8s - t + 7, -4s + 3t + 2, s, t)$
    for all real numbers s and t
12. $(-9t - 3, 4t + 20, t)$
    for all real numbers t
13. $(-2, 7)$
14. $(1, 2, 0)$
15. $(-t + 5, -3t + 15, t)$
    for all real numbers t
16. $(2, -1, 1)$
17. $(1, 3, -2)$
18. Inconsistent
19. $(1, 3, -2)$
20. $\left(-3, \frac{1}{2}, 1\right)$
21. $\left(\frac{1}{3}, \frac{2}{3}, 1\right)$
22. $\left(\frac{19}{13}t + \frac{51}{13}, -\frac{11}{13}t + \frac{4}{13}, t\right)$
    for all real numbers t
23. Inconsistent
24. $(4, -3, 1)$
25. $\left(-2t - \frac{35}{4}, -t - \frac{11}{2}, t\right)$
    for all real numbers t
26. $(1, 2, 3, 4)$

27. This time, 7 diners chose the deluxe buffet.

28. If t represents the amount (in pounds) of peanuts, then we need $1.5t - 7.5$ pounds of almonds and $17.5 - 2.5t$ pounds of cashews. Since we can't have a negative amount of nuts, $5 \leq t \leq 7$.

29. $f(x) = -\frac{4}{5}x^2 + \frac{1}{5}x + \frac{23}{5}$

30. $T(t) = \frac{20}{27}t^2 - \frac{50}{9}t + 60$. Lowest temperature of the evening $\frac{595}{12} \approx 49.58°F$ at 12:45 AM.

8.2 SYSTEMS OF LINEAR EQUATIONS: AUGMENTED MATRICES

31. Let x_1 and x_2 be the numbers of adults and children, respectively, in the Zahlenreich family and let x_3 and x_4 be the numbers of adults and children, respectively, in the Nullsatz family. The system of equations determined by the given information is

$$\begin{cases} 15x_1 + 8x_2 &= 38 \\ 15x_3 + 8x_4 &= 39 \\ 15x_1 + 8x_2 + 15x_3 + 8x_4 &= 77 \\ 15x_1 + 15x_3 &= 45 \end{cases}$$

We subtracted the cost of the babysitter in E3 so the constant is 77, not 92. This system is consistent dependent and its solution is $\left(\frac{8}{15}t + \frac{2}{5}, -t + 4, -\frac{8}{15}t + \frac{13}{5}, t\right)$. Our variables represent numbers of adults and children so they must be whole numbers. Running through the values $t = 0, 1, 2, 3, 4$ yields only one solution where all four variables are whole numbers; $t = 3$ gives us $(2, 1, 1, 3)$. Thus there are 2 adults and 1 child in the Zahlenreichs and 1 adult and 3 kids in the Nullsatzs.

8.3 Matrix Arithmetic

In Section 8.2, we used a special class of matrices, the augmented matrices, to assist us in solving systems of linear equations. In this section, we study matrices as mathematical objects of their own accord, temporarily divorced from systems of linear equations. To do so conveniently requires some more notation. When we write $A = [a_{ij}]_{m \times n}$, we mean A is an m by n matrix[1] and a_{ij} is the entry found in the ith row and jth column. Schematically, we have

$$A = \begin{bmatrix} a_{11} & a_{12} & \cdots & a_{1n} \\ a_{21} & a_{22} & \cdots & a_{2n} \\ \vdots & \vdots & & \vdots \\ a_{m1} & a_{m2} & \cdots & a_{mn} \end{bmatrix}$$

with j counting columns from left to right, and i counting rows from top to bottom.

With this new notation we can define what it means for two matrices to be equal.

> **Definition 8.6. Matrix Equality:** Two matrices are said to be **equal** if they are the same size and their corresponding entries are equal. More specifically, if $A = [a_{ij}]_{m \times n}$ and $B = [b_{ij}]_{p \times r}$, we write $A = B$ provided
>
> 1. $m = p$ and $n = r$
> 2. $a_{ij} = b_{ij}$ for all $1 \leq i \leq m$ and all $1 \leq j \leq n$.

Essentially, two matrices are equal if they are the same size and they have the same numbers in the same spots.[2] For example, the two 2×3 matrices below are, despite appearances, equal.

$$\begin{bmatrix} 0 & -2 & 9 \\ 25 & 117 & -3 \end{bmatrix} = \begin{bmatrix} \ln(1) & \sqrt[3]{-8} & e^{2\ln(3)} \\ 125^{2/3} & 3^2 \cdot 13 & \log(0.001) \end{bmatrix}$$

Now that we have an agreed upon understanding of what it means for two matrices to equal each other, we may begin defining arithmetic operations on matrices. Our first operation is addition.

> **Definition 8.7. Matrix Addition:** Given two matrices of the same size, the matrix obtained by adding the corresponding entries of the two matrices is called the **sum** of the two matrices. More specifically, if $A = [a_{ij}]_{m \times n}$ and $B = [b_{ij}]_{m \times n}$, we define
>
> $$A + B = [a_{ij}]_{m \times n} + [b_{ij}]_{m \times n} = [a_{ij} + b_{ij}]_{m \times n}$$

As an example, consider the sum below.

---

[1] Recall that means A has m rows and n columns.
[2] Critics may well ask: Why not leave it at that? Why the need for all the notation in Definition 8.6? It is the authors' attempt to expose you to the wonderful world of mathematical precision.

8.3 Matrix Arithmetic

$$\begin{bmatrix} 2 & 3 \\ 4 & -1 \\ 0 & -7 \end{bmatrix} + \begin{bmatrix} -1 & 4 \\ -5 & -3 \\ 8 & 1 \end{bmatrix} = \begin{bmatrix} 2+(-1) & 3+4 \\ 4+(-5) & (-1)+(-3) \\ 0+8 & (-7)+1 \end{bmatrix} = \begin{bmatrix} 1 & 7 \\ -1 & -4 \\ 8 & -6 \end{bmatrix}$$

It is worth the reader's time to think what would have happened had we reversed the order of the summands above. As we would expect, we arrive at the same answer. In general, $A + B = B + A$ for matrices A and B, provided they are the same size so that the sum is defined in the first place. This is the **commutative property** of matrix addition. To see why this is true in general, we appeal to the definition of matrix addition. Given $A = [a_{ij}]_{m \times n}$ and $B = [b_{ij}]_{m \times n}$,

$$A + B = [a_{ij}]_{m \times n} + [b_{ij}]_{m \times n} = [a_{ij} + b_{ij}]_{m \times n} = [b_{ij} + a_{ij}]_{m \times n} = [b_{ij}]_{m \times n} + [a_{ij}]_{m \times n} = B + A$$

where the second equality is the definition of $A + B$, the third equality holds by the commutative law of real number addition, and the fourth equality is the definition of $B + A$. In other words, matrix addition is commutative because real number addition is. A similar argument shows the **associative property** of matrix addition also holds, inherited in turn from the associative law of real number addition. Specifically, for matrices A, B, and C of the same size, $(A + B) + C = A + (B + C)$. In other words, when adding more than two matrices, it doesn't matter how they are grouped. This means that we can write $A + B + C$ without parentheses and there is no ambiguity as to what this means.[3] These properties and more are summarized in the following theorem.

Theorem 8.3. Properties of Matrix Addition

- **Commutative Property:** For all $m \times n$ matrices, $A + B = B + A$

- **Associative Property:** For all $m \times n$ matrices, $(A + B) + C = A + (B + C)$

- **Identity Property:** If $0_{m \times n}$ is the $m \times n$ matrix whose entries are all 0, then $0_{m \times n}$ is called the $m \times n$ **additive identity** and for all $m \times n$ matrices A

$$A + 0_{m \times n} = 0_{m \times n} + A = A$$

- **Inverse Property:** For every given $m \times n$ matrix A, there is a unique matrix denoted $-A$ called the **additive inverse of A** such that

$$A + (-A) = (-A) + A = 0_{m \times n}$$

The identity property is easily verified by resorting to the definition of matrix addition; just as the number 0 is the additive identity for real numbers, the matrix comprised of all 0's does the same job for matrices. To establish the inverse property, given a matrix $A = [a_{ij}]_{m \times n}$, we are looking for a matrix $B = [b_{ij}]_{m \times n}$ so that $A + B = 0_{m \times n}$. By the definition of matrix addition, we must have that $a_{ij} + b_{ij} = 0$ for all i and j. Solving, we get $b_{ij} = -a_{ij}$. Hence, given a matrix A, its additive inverse, which we call $-A$, does exist and is unique and, moreover, is given by the formula: $-A = [-a_{ij}]_{m \times n}$. The long and short of this is: to get the additive inverse of a matrix,

---
[3]A technical detail which is sadly lost on most readers.

take additive inverses of each of its entries. With the concept of additive inverse well in hand, we may now discuss what is meant by subtracting matrices. You may remember from arithmetic that $a - b = a + (-b)$; that is, subtraction is defined as 'adding the opposite (inverse).' We extend this concept to matrices. For two matrices A and B of the same size, we define $A - B = A + (-B)$. At the level of entries, this amounts to

$$A - B = A + (-B) = [a_{ij}]_{m \times n} + [-b_{ij}]_{m \times n} = [a_{ij} + (-b_{ij})]_{m \times n} = [a_{ij} - b_{ij}]_{m \times n}$$

Thus to subtract two matrices of equal size, we subtract their corresponding entries. Surprised?

Our next task is to define what it means to multiply a matrix by a real number. Thinking back to arithmetic, you may recall that multiplication, at least by a natural number, can be thought of as 'rapid addition.' For example, $2 + 2 + 2 = 3 \cdot 2$. We know from algebra[4] that $3x = x + x + x$, so it seems natural that given a matrix A, we define $3A = A + A + A$. If $A = [a_{ij}]_{m \times n}$, we have

$$3A = A + A + A = [a_{ij}]_{m \times n} + [a_{ij}]_{m \times n} + [a_{ij}]_{m \times n} = [a_{ij} + a_{ij} + a_{ij}]_{m \times n} = [3a_{ij}]_{m \times n}$$

In other words, multiplying the *matrix* in this fashion by 3 is the same as multiplying *each entry* by 3. This leads us to the following definition.

Definition 8.8. Scalar[a] Multiplication: We define the product of a real number and a matrix to be the matrix obtained by multiplying each of its entries by said real number. More specifically, if k is a real number and $A = [a_{ij}]_{m \times n}$, we define

$$kA = k[a_{ij}]_{m \times n} = [ka_{ij}]_{m \times n}$$

[a]The word 'scalar' here refers to real numbers. 'Scalar multiplication' in this context means we are multiplying a matrix by a real number (a scalar).

One may well wonder why the word 'scalar' is used for 'real number.' It has everything to do with 'scaling' factors.[5] A point $P(x, y)$ in the plane can be represented by its position matrix, P:

$$(x, y) \leftrightarrow P = \begin{bmatrix} x \\ y \end{bmatrix}$$

Suppose we take the point $(-2, 1)$ and multiply its position matrix by 3. We have

$$3P = 3 \begin{bmatrix} -2 \\ 1 \end{bmatrix} = \begin{bmatrix} 3(-2) \\ 3(1) \end{bmatrix} = \begin{bmatrix} -6 \\ 3 \end{bmatrix}$$

which corresponds to the point $(-6, 3)$. We can imagine taking $(-2, 1)$ to $(-6, 3)$ in this fashion as a dilation by a factor of 3 in both the horizontal and vertical directions. Doing this to all points (x, y) in the plane, therefore, has the effect of magnifying (scaling) the plane by a factor of 3.

---

[4]The Distributive Property, in particular.
[5]See Section 1.7.

8.3 Matrix Arithmetic

As did matrix addition, scalar multiplication inherits many properties from real number arithmetic. Below we summarize these properties.

Theorem 8.4. Properties of Scalar Multiplication

- **Associative Property:** For every $m \times n$ matrix A and scalars k and r, $(kr)A = k(rA)$.

- **Identity Property:** For all $m \times n$ matrices A, $1A = A$.

- **Additive Inverse Property:** For all $m \times n$ matrices A, $-A = (-1)A$.

- **Distributive Property of Scalar Multiplication over Scalar Addition:** For every $m \times n$ matrix A and scalars k and r,

$$(k+r)A = kA + rA$$

- **Distributive Property of Scalar Multiplication over Matrix Addition:** For all $m \times n$ matrices A and B scalars k,

$$k(A+B) = kA + kB$$

- **Zero Product Property:** If A is an $m \times n$ matrix and k is a scalar, then

$$kA = 0_{m \times n} \quad \text{if and only if} \quad k = 0 \quad \text{or} \quad A = 0_{m \times n}$$

As with the other results in this section, Theorem 8.4 can be proved using the definitions of scalar multiplication and matrix addition. For example, to prove that $k(A+B) = kA + kB$ for a scalar k and $m \times n$ matrices A and B, we start by adding A and B, then multiplying by k and seeing how that compares with the sum of kA and kB.

$$k(A+B) = k\left([a_{ij}]_{m \times n} + [b_{ij}]_{m \times n}\right) = k\left[a_{ij} + b_{ij}\right]_{m \times n} = \left[k\left(a_{ij} + b_{ij}\right)\right]_{m \times n} = \left[ka_{ij} + kb_{ij}\right]_{m \times n}$$

As for $kA + kB$, we have

$$kA + kB = k\left[a_{ij}\right]_{m \times n} + k\left[b_{ij}\right]_{m \times n} = \left[ka_{ij}\right]_{m \times n} + \left[kb_{ij}\right]_{m \times n} = \left[ka_{ij} + kb_{ij}\right]_{m \times n} \checkmark$$

which establishes the property. The remaining properties are left to the reader. The properties in Theorems 8.3 and 8.4 establish an algebraic system that lets us treat matrices and scalars more or less as we would real numbers and variables, as the next example illustrates.

Example 8.3.1. Solve for the matrix A: $3A - \left(\begin{bmatrix} 2 & -1 \\ 3 & 5 \end{bmatrix} + 5A\right) = \begin{bmatrix} -4 & 2 \\ 6 & -2 \end{bmatrix} + \dfrac{1}{3}\begin{bmatrix} 9 & 12 \\ -3 & 39 \end{bmatrix}$ using the definitions and properties of matrix arithmetic.

Solution.

$$3A - \left(\begin{bmatrix} 2 & -1 \\ 3 & 5 \end{bmatrix} + 5A\right) = \begin{bmatrix} -4 & 2 \\ 6 & -2 \end{bmatrix} + \frac{1}{3}\begin{bmatrix} 9 & 12 \\ -3 & 39 \end{bmatrix}$$

$$3A + \left\{-\left(\begin{bmatrix} 2 & -1 \\ 3 & 5 \end{bmatrix} + 5A\right)\right\} = \begin{bmatrix} -4 & 2 \\ 6 & -2 \end{bmatrix} + \begin{bmatrix} \left(\frac{1}{3}\right)(9) & \left(\frac{1}{3}\right)(12) \\ \left(\frac{1}{3}\right)(-3) & \left(\frac{1}{3}\right)(39) \end{bmatrix}$$

$$3A + (-1)\left(\begin{bmatrix} 2 & -1 \\ 3 & 5 \end{bmatrix} + 5A\right) = \begin{bmatrix} -4 & 2 \\ 6 & -2 \end{bmatrix} + \begin{bmatrix} 3 & 4 \\ -1 & 13 \end{bmatrix}$$

$$3A + \left\{(-1)\begin{bmatrix} 2 & -1 \\ 3 & 5 \end{bmatrix} + (-1)(5A)\right\} = \begin{bmatrix} -1 & 6 \\ 5 & 11 \end{bmatrix}$$

$$3A + (-1)\begin{bmatrix} 2 & -1 \\ 3 & 5 \end{bmatrix} + (-1)(5A) = \begin{bmatrix} -1 & 6 \\ 5 & 11 \end{bmatrix}$$

$$3A + \begin{bmatrix} (-1)(2) & (-1)(-1) \\ (-1)(3) & (-1)(5) \end{bmatrix} + ((-1)(5))A = \begin{bmatrix} -1 & 6 \\ 5 & 11 \end{bmatrix}$$

$$3A + \begin{bmatrix} -2 & 1 \\ -3 & -5 \end{bmatrix} + (-5)A = \begin{bmatrix} -1 & 6 \\ 5 & 11 \end{bmatrix}$$

$$3A + (-5)A + \begin{bmatrix} -2 & 1 \\ -3 & -5 \end{bmatrix} = \begin{bmatrix} -1 & 6 \\ 5 & 11 \end{bmatrix}$$

$$(3+(-5))A + \begin{bmatrix} -2 & 1 \\ -3 & -5 \end{bmatrix} + \left(-\begin{bmatrix} -2 & 1 \\ -3 & -5 \end{bmatrix}\right) = \begin{bmatrix} -1 & 6 \\ 5 & 11 \end{bmatrix} + \left(-\begin{bmatrix} -2 & 1 \\ -3 & -5 \end{bmatrix}\right)$$

$$(-2)A + 0_{2\times 2} = \begin{bmatrix} -1 & 6 \\ 5 & 11 \end{bmatrix} - \begin{bmatrix} -2 & 1 \\ -3 & -5 \end{bmatrix}$$

$$(-2)A = \begin{bmatrix} -1-(-2) & 6-1 \\ 5-(-3) & 11-(-5) \end{bmatrix}$$

$$(-2)A = \begin{bmatrix} 1 & 5 \\ 8 & 16 \end{bmatrix}$$

$$\left(-\tfrac{1}{2}\right)((-2)A) = -\tfrac{1}{2}\begin{bmatrix} 1 & 5 \\ 8 & 16 \end{bmatrix}$$

$$\left(\left(-\tfrac{1}{2}\right)(-2)\right)A = \begin{bmatrix} \left(-\tfrac{1}{2}\right)(1) & \left(-\tfrac{1}{2}\right)(5) \\ \left(-\tfrac{1}{2}\right)(8) & \left(-\tfrac{1}{2}\right)(16) \end{bmatrix}$$

$$1A = \begin{bmatrix} -\tfrac{1}{2} & -\tfrac{5}{2} \\ -4 & -\tfrac{16}{2} \end{bmatrix}$$

$$A = \begin{bmatrix} -\tfrac{1}{2} & -\tfrac{5}{2} \\ -4 & -8 \end{bmatrix}$$

The reader is encouraged to check our answer in the original equation. □

8.3 Matrix Arithmetic

While the solution to the previous example is written in excruciating detail, in practice many of the steps above are omitted. We have spelled out each step in this example to encourage the reader to justify each step using the definitions and properties we have established thus far for matrix arithmetic. The reader is encouraged to solve the equation in Example 8.3.1 as they would any other linear equation, for example: $3a - (2 + 5a) = -4 + \frac{1}{3}(9)$.

We now turn our attention to **matrix multiplication** - that is, multiplying a matrix by another matrix. Based on the 'no surprises' trend so far in the section, you may expect that in order to multiply two matrices, they must be of the same size and you find the product by multiplying the corresponding entries. While this kind of product is used in other areas of mathematics,[6] we define matrix multiplication to serve us in solving systems of linear equations. To that end, we begin by defining the product of a row and a column. We motivate the general definition with an example. Consider the two matrices A and B below.

$$A = \begin{bmatrix} 2 & 0 & -1 \\ -10 & 3 & 5 \end{bmatrix} \quad B = \begin{bmatrix} 3 & 1 & 2 & -8 \\ 4 & 8 & -5 & 9 \\ 5 & 0 & -2 & -12 \end{bmatrix}$$

Let $R1$ denote the first row of A and $C1$ denote the first column of B. To find the 'product' of $R1$ with $C1$, denoted $R1 \cdot C1$, we first find the product of the first entry in $R1$ and the first entry in $C1$. Next, we add to that the product of the second entry in $R1$ and the second entry in $C1$. Finally, we take that sum and we add to that the product of the last entry in $R1$ and the last entry in $C1$. Using entry notation, $R1 \cdot C1 = a_{11}b_{11} + a_{12}b_{21} + a_{13}b_{31} = (2)(3) + (0)(4) + (-1)(5) = 6 + 0 + (-5) = 1$. We can visualize this schematically as follows

To find $R2 \cdot C3$ where $R2$ denotes the second row of A and $C3$ denotes the third column of B, we proceed similarly. We start with finding the product of the first entry of $R2$ with the first entry in $C3$ then add to it the product of the second entry in $R2$ with the second entry in $C3$, and so forth. Using entry notation, we have $R2 \cdot C3 = a_{21}b_{13} + a_{22}b_{23} + a_{23}b_{33} = (-10)(2) + (3)(-5) + (5)(-2) = -45$. Schematically,

---

[6]See this article on the Hadamard Product.

$$\underbrace{\xrightarrow{} \begin{array}{cccc} \boxed{-10} & 3 & 5 & \boxed{2} \\ & & & -5 \\ & & & -2 \end{array} \Bigg\downarrow}_{a_{21}b_{13} = (-10)(2) = -20} \;+\; \underbrace{\xrightarrow{} \begin{array}{cccc} -10 & \boxed{3} & 5 & 2 \\ & & & \boxed{-5} \\ & & & -2 \end{array} \Bigg\downarrow}_{a_{22}b_{23} = (3)(-5) = -15} \;+\; \underbrace{\xrightarrow{} \begin{array}{cccc} -10 & 3 & \boxed{5} & 2 \\ & & & -5 \\ & & & \boxed{-2} \end{array} \Bigg\downarrow}_{a_{23}b_{33} = (5)(-2) = -10}$$

Generalizing this process, we have the following definition.

Definition 8.9. Product of a Row and a Column: Suppose $A = [a_{ij}]_{m \times n}$ and $B = [b_{ij}]_{n \times r}$. Let Ri denote the ith row of A and let Cj denote the jth column of B. The **product of R_i and C_j**, denoted $R_i \cdot C_j$ is the real number defined by

$$Ri \cdot Cj = a_{i1}b_{1j} + a_{i2}b_{2j} + \ldots a_{in}b_{nj}$$

Note that in order to multiply a row by a column, the number of entries in the row must match the number of entries in the column. We are now in the position to define matrix multiplication.

Definition 8.10. Matrix Multiplication: Suppose $A = [a_{ij}]_{m \times n}$ and $B = [b_{ij}]_{n \times r}$. Let Ri denote the ith row of A and let Cj denote the jth column of B. The **product of A and B**, denoted AB, is the matrix defined by

$$AB = [Ri \cdot Cj]_{m \times r}$$

that is

$$AB = \begin{bmatrix} R1 \cdot C1 & R1 \cdot C2 & \ldots & R1 \cdot Cr \\ R2 \cdot C1 & R2 \cdot C2 & \ldots & R2 \cdot Cr \\ \vdots & \vdots & & \vdots \\ Rm \cdot C1 & Rm \cdot C2 & \ldots & Rm \cdot Cr \end{bmatrix}$$

There are a number of subtleties in Definition 8.10 which warrant closer inspection. First and foremost, Definition 8.10 tells us that the ij-entry of a matrix product AB is the ith row of A times the jth column of B. In order for this to be defined, the number of entries in the rows of A must match the number of entries in the columns of B. This means that the number of columns of A must match[7] the number of rows of B. In other words, to multiply A times B, the second dimension of A must match the first dimension of B, which is why in Definition 8.10, $A_{m \times \underline{n}}$ is being multiplied by a matrix $B_{\underline{n} \times r}$. Furthermore, the product matrix AB has as many rows as A and as many columns of B. As a result, when multiplying a matrix $A_{\underline{m} \times n}$ by a matrix $B_{n \times \underline{r}}$, the result is the matrix $AB_{\underline{m} \times \underline{r}}$. Returning to our example matrices below, we see that A is a $2 \times \underline{3}$ matrix and B is a $\underline{3} \times 4$ matrix. This means that the product matrix AB is defined and will be a 2×4 matrix.

$$A = \begin{bmatrix} 2 & 0 & -1 \\ -10 & 3 & 5 \end{bmatrix} \quad B = \begin{bmatrix} 3 & 1 & 2 & -8 \\ 4 & 8 & -5 & 9 \\ 5 & 0 & -2 & -12 \end{bmatrix}$$

---

[7]The reader is encouraged to think this through carefully.

8.3 MATRIX ARITHMETIC

Using Ri to denote the ith row of A and Cj to denote the jth column of B, we form AB according to Definition 8.10.

$$AB = \begin{bmatrix} R1 \cdot C1 & R1 \cdot C2 & R1 \cdot C3 & R1 \cdot C4 \\ R2 \cdot C1 & R2 \cdot C2 & R2 \cdot C3 & R2 \cdot C4 \end{bmatrix} = \begin{bmatrix} 1 & 2 & 6 & -4 \\ 7 & 14 & -45 & 47 \end{bmatrix}$$

Note that the product BA is not defined, since B is a $3 \times \underline{4}$ matrix while A is a $\underline{2} \times 3$ matrix; B has more columns than A has rows, and so it is not possible to multiply a row of B by a column of A. Even when the dimensions of A and B are compatible such that AB and BA are both defined, the product AB and BA aren't necessarily equal.[8] In other words, AB may not equal BA. Although there is no commutative property of matrix multiplication in general, several other real number properties are inherited by matrix multiplication, as illustrated in our next theorem.

Theorem 8.5. Properties of Matrix Multiplication Let A, B and C be matrices such that all of the matrix products below are defined and let k be a real number.

- **Associative Property of Matrix Multiplication:** $(AB)C = A(BC)$

- **Associative Property with Scalar Multiplication:** $k(AB) = (kA)B = A(kB)$

- **Identity Property:** For a natural number k, the $k \times k$ **identity matrix**, denoted I_k, is defined by $I_k = [d_{ij}]_{k \times k}$ where

$$d_{ij} = \begin{cases} 1, & \text{if } i = j \\ 0, & \text{otherwise} \end{cases}$$

For all $m \times n$ matrices, $I_m A = A I_n = A$.

- **Distributive Property of Matrix Multiplication over Matrix Addition:**

$$A(B \pm C) = AB \pm AC \text{ and } (A \pm B)C = AC \pm BC$$

The one property in Theorem 8.5 which begs further investigation is, without doubt, the multiplicative identity. The entries in a matrix where $i = j$ comprise what is called the **main diagonal** of the matrix. The identity matrix has 1's along its main diagonal and 0's everywhere else. A few examples of the matrix I_k mentioned in Theorem 8.5 are given below. The reader is encouraged to see how they match the definition of the identity matrix presented there.

$$[1] \quad \begin{bmatrix} 1 & 0 \\ 0 & 1 \end{bmatrix} \quad \begin{bmatrix} 1 & 0 & 0 \\ 0 & 1 & 0 \\ 0 & 0 & 1 \end{bmatrix} \quad \begin{bmatrix} 1 & 0 & 0 & 0 \\ 0 & 1 & 0 & 0 \\ 0 & 0 & 1 & 0 \\ 0 & 0 & 0 & 1 \end{bmatrix}$$

$$I_1 \qquad I_2 \qquad\quad I_3 \qquad\qquad I_4$$

---

[8]And may not even have the same dimensions. For example, if A is a 2×3 matrix and B is a 3×2 matrix, then AB is defined and is a 2×2 matrix while BA is also defined... but is a 3×3 matrix!

The identity matrix is an example of what is called a **square matrix** as it has the same number of rows as columns. Note that to in order to verify that the identity matrix acts as a multiplicative identity, some care must be taken depending on the order of the multiplication. For example, take the matrix 2×3 matrix A from earlier

$$A = \begin{bmatrix} 2 & 0 & -1 \\ -10 & 3 & 5 \end{bmatrix}$$

In order for the product $I_k A$ to be defined, $k = 2$; similarly, for AI_k to be defined, $k = 3$. We leave it to the reader to show $I_2 A = A$ and $AI_3 = A$. In other words,

$$\begin{bmatrix} 1 & 0 \\ 0 & 1 \end{bmatrix} \begin{bmatrix} 2 & 0 & -1 \\ -10 & 3 & 5 \end{bmatrix} = \begin{bmatrix} 2 & 0 & -1 \\ -10 & 3 & 5 \end{bmatrix}$$

and

$$\begin{bmatrix} 2 & 0 & -1 \\ -10 & 3 & 5 \end{bmatrix} \begin{bmatrix} 1 & 0 & 0 \\ 0 & 1 & 0 \\ 0 & 0 & 1 \end{bmatrix} = \begin{bmatrix} 2 & 0 & -1 \\ -10 & 3 & 5 \end{bmatrix}$$

While the proofs of the properties in Theorem 8.5 are computational in nature, the notation becomes quite involved very quickly, so they are left to a course in Linear Algebra. The following example provides some practice with matrix multiplication and its properties. As usual, some valuable lessons are to be learned.

Example 8.3.2.

1. Find AB for $A = \begin{bmatrix} -23 & -1 & 17 \\ 46 & 2 & -34 \end{bmatrix}$ and $B = \begin{bmatrix} -3 & 2 \\ 1 & 5 \\ -4 & 3 \end{bmatrix}$

2. Find $C^2 - 5C + 10 I_2$ for $C = \begin{bmatrix} 1 & -2 \\ 3 & 4 \end{bmatrix}$

3. Suppose M is a 4×4 matrix. Use Theorem 8.5 to expand $(M - 2I_4)(M + 3I_4)$.

Solution.

1. We have $AB = \begin{bmatrix} -23 & -1 & 17 \\ 46 & 2 & -34 \end{bmatrix} \begin{bmatrix} -3 & 2 \\ 1 & 5 \\ -4 & 3 \end{bmatrix} = \begin{bmatrix} 0 & 0 \\ 0 & 0 \end{bmatrix}$

2. Just as x^2 means x times itself, C^2 denotes the matrix C times itself. We get

8.3 MATRIX ARITHMETIC

$$\begin{aligned} C^2 - 5C + 10I_2 &= \begin{bmatrix} 1 & -2 \\ 3 & 4 \end{bmatrix}^2 - 5 \begin{bmatrix} 1 & -2 \\ 3 & 4 \end{bmatrix} + 10 \begin{bmatrix} 1 & 0 \\ 0 & 1 \end{bmatrix} \\ &= \begin{bmatrix} 1 & -2 \\ 3 & 4 \end{bmatrix} \begin{bmatrix} 1 & -2 \\ 3 & 4 \end{bmatrix} + \begin{bmatrix} -5 & 10 \\ -15 & -20 \end{bmatrix} + \begin{bmatrix} 10 & 0 \\ 0 & 10 \end{bmatrix} \\ &= \begin{bmatrix} -5 & -10 \\ 15 & 10 \end{bmatrix} + \begin{bmatrix} 5 & 10 \\ -15 & -10 \end{bmatrix} \\ &= \begin{bmatrix} 0 & 0 \\ 0 & 0 \end{bmatrix} \end{aligned}$$

3. We expand $(M - 2I_4)(M + 3I_4)$ with the same pedantic zeal we showed in Example 8.3.1. The reader is encouraged to determine which property of matrix arithmetic is used as we proceed from one step to the next.

$$\begin{aligned} (M - 2I_4)(M + 3I_4) &= (M - 2I_4) M + (M - 2I_4) (3I_4) \\ &= MM - (2I_4) M + M (3I_4) - (2I_4)(3I_4) \\ &= M^2 - 2(I_4 M) + 3(MI_4) - 2(I_4 (3I_4)) \\ &= M^2 - 2M + 3M - 2(3(I_4 I_4)) \\ &= M^2 + M - 6I_4 \end{aligned}$$

□

Example 8.3.2 illustrates some interesting features of matrix multiplication. First note that in part 1, neither A nor B is the zero matrix, yet the product AB is the zero matrix. Hence, the the zero product property enjoyed by real numbers and scalar multiplication does not hold for matrix multiplication. Parts 2 and 3 introduce us to polynomials involving matrices. The reader is encouraged to step back and compare our expansion of the matrix product $(M - 2I_4)(M + 3I_4)$ in part 3 with the product $(x - 2)(x + 3)$ from real number algebra. The exercises explore this kind of parallel further.

As we mentioned earlier, a point $P(x, y)$ in the xy-plane can be represented as a 2×1 position matrix. We now show that matrix multiplication can be used to rotate these points, and hence graphs of equations.

Example 8.3.3. Let $R = \begin{bmatrix} \frac{\sqrt{2}}{2} & -\frac{\sqrt{2}}{2} \\ \frac{\sqrt{2}}{2} & \frac{\sqrt{2}}{2} \end{bmatrix}$.

1. Plot $P(2, -2)$, $Q(4, 0)$, $S(0, 3)$, and $T(-3, -3)$ in the plane as well as the points RP, RQ, RS, and RT. Plot the lines $y = x$ and $y = -x$ as guides. What does R appear to be doing to these points?

2. If a point P is on the hyperbola $x^2 - y^2 = 4$, show that the point RP is on the curve $y = \frac{2}{x}$.

Solution. For $P(2, -2)$, the position matrix is $P = \begin{bmatrix} 2 \\ -2 \end{bmatrix}$, and

$$RP = \begin{bmatrix} \frac{\sqrt{2}}{2} & -\frac{\sqrt{2}}{2} \\ \frac{\sqrt{2}}{2} & \frac{\sqrt{2}}{2} \end{bmatrix} \begin{bmatrix} 2 \\ -2 \end{bmatrix}$$
$$= \begin{bmatrix} 2\sqrt{2} \\ 0 \end{bmatrix}$$

We have that R takes $(2, -2)$ to $(2\sqrt{2}, 0)$. Similarly, we find $(4, 0)$ is moved to $(2\sqrt{2}, 2\sqrt{2})$, $(0, 3)$ is moved to $\left(-\frac{3\sqrt{2}}{2}, \frac{3\sqrt{2}}{2}\right)$, and $(-3, -3)$ is moved to $(0, -3\sqrt{2})$. Plotting these in the coordinate plane along with the lines $y = x$ and $y = -x$, we see that the matrix R is rotating these points counterclockwise by $45°$.

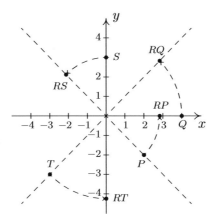

For a generic point $P(x, y)$ on the hyperbola $x^2 - y^2 = 4$, we have

$$RP = \begin{bmatrix} \frac{\sqrt{2}}{2} & -\frac{\sqrt{2}}{2} \\ \frac{\sqrt{2}}{2} & \frac{\sqrt{2}}{2} \end{bmatrix} \begin{bmatrix} x \\ y \end{bmatrix}$$
$$= \begin{bmatrix} \frac{\sqrt{2}}{2}x - \frac{\sqrt{2}}{2}y \\ \frac{\sqrt{2}}{2}x + \frac{\sqrt{2}}{2}y \end{bmatrix}$$

which means R takes (x, y) to $\left(\frac{\sqrt{2}}{2}x - \frac{\sqrt{2}}{2}y, \frac{\sqrt{2}}{2}x + \frac{\sqrt{2}}{2}y\right)$. To show that this point is on the curve $y = \frac{2}{x}$, we replace x with $\frac{\sqrt{2}}{2}x - \frac{\sqrt{2}}{2}y$ and y with $\frac{\sqrt{2}}{2}x + \frac{\sqrt{2}}{2}y$ and simplify.

8.3 Matrix Arithmetic

$$y = \frac{2}{x}$$
$$\frac{\sqrt{2}}{2}x + \frac{\sqrt{2}}{2}y \stackrel{?}{=} \frac{2}{\frac{\sqrt{2}}{2}x - \frac{\sqrt{2}}{2}y}$$
$$\left(\frac{\sqrt{2}}{2}x - \frac{\sqrt{2}}{2}y\right)\left(\frac{\sqrt{2}}{2}x + \frac{\sqrt{2}}{2}y\right) \stackrel{?}{=} \left(\frac{2}{\frac{\sqrt{2}}{2}x - \frac{\sqrt{2}}{2}y}\right)\left(\frac{\sqrt{2}}{2}x - \frac{\sqrt{2}}{2}y\right)$$
$$\left(\frac{\sqrt{2}}{2}x\right)^2 - \left(\frac{\sqrt{2}}{2}y\right)^2 \stackrel{?}{=} 2$$
$$\frac{x^2}{2} - \frac{y^2}{2} \stackrel{?}{=} 2$$
$$x^2 - y^2 \stackrel{\checkmark}{=} 4$$

Since (x, y) is on the hyperbola $x^2 - y^2 = 4$, we know that this last equation is true. Since all of our steps are reversible, this last equation is equivalent to our original equation, which establishes the point is, indeed, on the graph of $y = \frac{2}{x}$. This means the graph of $y = \frac{2}{x}$ is a hyperbola, and it is none other than the hyperbola $x^2 - y^2 = 4$ rotated counterclockwise by $45°$.[9] Below we have the graph of $x^2 - y^2 = 4$ (solid line) and $y = \frac{2}{x}$ (dashed line) for comparison.

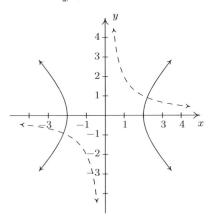

□

When we started this section, we mentioned that we would temporarily consider matrices as their own entities, but that the algebra developed here would ultimately allow us to solve systems of linear equations. To that end, consider the system

$$\begin{cases} 3x - y + z = 8 \\ x + 2y - z = 4 \\ 2x + 3y - 4z = 10 \end{cases}$$

In Section 8.2, we encoded this system into the augmented matrix

$$\begin{bmatrix} 3 & -1 & 1 & | & 8 \\ 1 & 2 & -1 & | & 4 \\ 2 & 3 & -4 & | & 10 \end{bmatrix}$$

---
[9]See Section 7.5 for more details.

Recall that the entries to the left of the vertical line come from the coefficients of the variables in the system, while those on the right comprise the associated constants. For that reason, we may form the **coefficient matrix** A, the **unknowns matrix** X and the **constant matrix** B as below

$$A = \begin{bmatrix} 3 & -1 & 1 \\ 1 & 2 & -1 \\ 2 & 3 & -4 \end{bmatrix} \quad X = \begin{bmatrix} x \\ y \\ z \end{bmatrix} \quad B = \begin{bmatrix} 8 \\ 4 \\ 10 \end{bmatrix}$$

We now consider the matrix equation $AX = B$.

$$AX = B$$

$$\begin{bmatrix} 3 & -1 & 1 \\ 1 & 2 & -1 \\ 2 & 3 & -4 \end{bmatrix} \begin{bmatrix} x \\ y \\ z \end{bmatrix} = \begin{bmatrix} 8 \\ 4 \\ 10 \end{bmatrix}$$

$$\begin{bmatrix} 3x - y + z \\ x + 2y - z \\ 2x + 3y - 4z \end{bmatrix} = \begin{bmatrix} 8 \\ 4 \\ 10 \end{bmatrix}$$

We see that finding a solution (x, y, z) to the original system corresponds to finding a solution X for the matrix equation $AX = B$. If we think about solving the real number equation $ax = b$, we would simply 'divide' both sides by a. Is it possible to 'divide' both sides of the matrix equation $AX = B$ by the matrix A? This is the central topic of Section 8.4.

8.3 Matrix Arithmetic

8.3.1 Exercises

For each pair of matrices A and B in Exercises 1 - 7, find the following, if defined

- $3A$
- $-B$
- A^2
- $A - 2B$
- AB
- BA

1. $A = \begin{bmatrix} 2 & -3 \\ 1 & 4 \end{bmatrix}$, $B = \begin{bmatrix} 5 & -2 \\ 4 & 8 \end{bmatrix}$

2. $A = \begin{bmatrix} -1 & 5 \\ -3 & 6 \end{bmatrix}$, $B = \begin{bmatrix} 2 & 10 \\ -7 & 1 \end{bmatrix}$

3. $A = \begin{bmatrix} -1 & 3 \\ 5 & 2 \end{bmatrix}$, $B = \begin{bmatrix} 7 & 0 & 8 \\ -3 & 1 & 4 \end{bmatrix}$

4. $A = \begin{bmatrix} 2 & 4 \\ 6 & 8 \end{bmatrix}$, $B = \begin{bmatrix} -1 & 3 & -5 \\ 7 & -9 & 11 \end{bmatrix}$

5. $A = \begin{bmatrix} 7 \\ 8 \\ 9 \end{bmatrix}$, $B = \begin{bmatrix} 1 & 2 & 3 \end{bmatrix}$

6. $A = \begin{bmatrix} 1 & -2 \\ -3 & 4 \\ 5 & -6 \end{bmatrix}$, $B = \begin{bmatrix} -5 & 1 & 8 \end{bmatrix}$

7. $A = \begin{bmatrix} 2 & -3 & 5 \\ 3 & 1 & -2 \\ -7 & 1 & -1 \end{bmatrix}$, $B = \begin{bmatrix} 1 & 2 & 1 \\ 17 & 33 & 19 \\ 10 & 19 & 11 \end{bmatrix}$

In Exercises 8 - 21, use the matrices

$$A = \begin{bmatrix} 1 & 2 \\ 3 & 4 \end{bmatrix} \quad B = \begin{bmatrix} 0 & -3 \\ -5 & 2 \end{bmatrix} \quad C = \begin{bmatrix} 10 & -\frac{11}{2} & 0 \\ \frac{3}{5} & 5 & 9 \end{bmatrix}$$

$$D = \begin{bmatrix} 7 & -13 \\ -\frac{4}{3} & 0 \\ 6 & 8 \end{bmatrix} \quad E = \begin{bmatrix} 1 & 2 & 3 \\ 0 & 4 & -9 \\ 0 & 0 & -5 \end{bmatrix}$$

to compute the following or state that the indicated operation is undefined.

8. $7B - 4A$
9. AB
10. BA
11. $E + D$
12. ED
13. $CD + 2I_2 A$
14. $A - 4I_2$
15. $A^2 - B^2$
16. $(A+B)(A-B)$
17. $A^2 - 5A - 2I_2$
18. $E^2 + 5E - 36I_3$
19. EDC
20. CDE
21. $ABCEDI_2$

22. Let $A = \begin{bmatrix} a & b & c \\ d & e & f \end{bmatrix}$ $E_1 = \begin{bmatrix} 0 & 1 \\ 1 & 0 \end{bmatrix}$ $E_2 = \begin{bmatrix} 5 & 0 \\ 0 & 1 \end{bmatrix}$ $E_3 = \begin{bmatrix} 1 & -2 \\ 0 & 1 \end{bmatrix}$

Compute $E_1 A$, $E_2 A$ and $E_3 A$. What effect did each of the E_i matrices have on the rows of A? Create E_4 so that its effect on A is to multiply the bottom row by -6. How would you extend this idea to matrices with more than two rows?

In Exercises 23 - 29, consider the following scenario. In the small village of Pedimaxus in the country of Sasquatchia, all 150 residents get one of the two local newspapers. Market research has shown that in any given week, 90% of those who subscribe to the Pedimaxus Tribune want to keep getting it, but 10% want to switch to the Sasquatchia Picayune. Of those who receive the Picayune, 80% want to continue with it and 20% want switch to the Tribune. We can express this situation using matrices. Specifically, let X be the 'state matrix' given by

$$X = \begin{bmatrix} T \\ P \end{bmatrix}$$

where T is the number of people who get the Tribune and P is the number of people who get the Picayune in a given week. Let Q be the 'transition matrix' given by

$$Q = \begin{bmatrix} 0.90 & 0.20 \\ 0.10 & 0.80 \end{bmatrix}$$

such that QX will be the state matrix for the next week.

23. Let's assume that when Pedimaxus was founded, all 150 residents got the Tribune. (Let's call this Week 0.) This would mean

$$X = \begin{bmatrix} 150 \\ 0 \end{bmatrix}$$

Since 10% of that 150 want to switch to the Picayune, we should have that for Week 1, 135 people get the Tribune and 15 people get the Picayune. Show that QX in this situation is indeed

$$QX = \begin{bmatrix} 135 \\ 15 \end{bmatrix}$$

24. Assuming that the percentages stay the same, we can get to the subscription numbers for Week 2 by computing $Q^2 X$. How many people get each paper in Week 2?

25. Explain why the transition matrix does what we want it to do.

26. If the conditions do not change from week to week, then Q remains the same and we have what's known as a **Stochastic Process**[10] because Week n's numbers are found by computing $Q^n X$. Choose a few values of n and, with the help of your classmates and calculator, find out how many people get each paper for that week. You should start to see a pattern as $n \to \infty$.

27. If you didn't see the pattern, we'll help you out. Let

$$X_s = \begin{bmatrix} 100 \\ 50 \end{bmatrix}.$$

Show that $QX_s = X_s$ This is called the **steady state** because the number of people who get each paper didn't change for the next week. Show that $Q^n X \to X_s$ as $n \to \infty$.

---
[10]More specifically, we have a Markov Chain, which is a special type of stochastic process.

28. Now let
$$S = \begin{bmatrix} \frac{2}{3} & \frac{2}{3} \\ \frac{1}{3} & \frac{1}{3} \end{bmatrix}$$

Show that $Q^n \to S$ as $n \to \infty$.

29. Show that $SY = X_s$ for any matrix Y of the form
$$Y = \begin{bmatrix} y \\ 150 - y \end{bmatrix}$$

This means that no matter how the distribution starts in Pedimaxus, if Q is applied often enough, we always end up with 100 people getting the Tribune and 50 people getting the Picayune.

30. Let $z = a + bi$ and $w = c + di$ be arbitrary complex numbers. Associate z and w with the matrices
$$Z = \begin{bmatrix} a & b \\ -b & a \end{bmatrix} \text{ and } W = \begin{bmatrix} c & d \\ -d & c \end{bmatrix}$$

Show that complex number addition, subtraction and multiplication are mirrored by the associated *matrix* arithmetic. That is, show that $Z + W$, $Z - W$ and ZW produce matrices which can be associated with the complex numbers $z + w$, $z - w$ and zw, respectively.

31. Let
$$A = \begin{bmatrix} 1 & 2 \\ 3 & 4 \end{bmatrix} \text{ and } B = \begin{bmatrix} 0 & -3 \\ -5 & 2 \end{bmatrix}$$

Compare $(A + B)^2$ to $A^2 + 2AB + B^2$. Discuss with your classmates what constraints must be placed on two arbitrary matrices A and B so that both $(A+B)^2$ and $A^2 + 2AB + B^2$ exist. When will $(A + B)^2 = A^2 + 2AB + B^2$? In general, what is the correct formula for $(A + B)^2$?

In Exercises 32 - 36, consider the following definitions. A square matrix is said to be an **upper triangular matrix** if all of its entries below the main diagonal are zero and it is said to be a **lower triangular matrix** if all of its entries above the main diagonal are zero. For example,
$$E = \begin{bmatrix} 1 & 2 & 3 \\ 0 & 4 & -9 \\ 0 & 0 & -5 \end{bmatrix}$$

from Exercises 8 - 21 above is an upper triangular matrix whereas
$$F = \begin{bmatrix} 1 & 0 \\ 3 & 0 \end{bmatrix}$$

is a lower triangular matrix. (Zeros are allowed on the main diagonal.) Discuss the following questions with your classmates.

32. Give an example of a matrix which is neither upper triangular nor lower triangular.

33. Is the product of two $n \times n$ upper triangular matrices always upper triangular?

34. Is the product of two $n \times n$ lower triangular matrices always lower triangular?

35. Given the matrix
$$A = \begin{bmatrix} 1 & 2 \\ 3 & 4 \end{bmatrix}$$
write A as LU where L is a lower triangular matrix and U is an upper triangular matrix?

36. Are there any matrices which are simultaneously upper and lower triangular?

8.3.2 Answers

1. For $A = \begin{bmatrix} 2 & -3 \\ 1 & 4 \end{bmatrix}$ and $B = \begin{bmatrix} 5 & -2 \\ 4 & 8 \end{bmatrix}$

   - $3A = \begin{bmatrix} 6 & -9 \\ 3 & 12 \end{bmatrix}$
   - $-B = \begin{bmatrix} -5 & 2 \\ -4 & -8 \end{bmatrix}$
   - $A^2 = \begin{bmatrix} 1 & -18 \\ 6 & 13 \end{bmatrix}$

   - $A - 2B = \begin{bmatrix} -8 & 1 \\ -7 & -12 \end{bmatrix}$
   - $AB = \begin{bmatrix} -2 & -28 \\ 21 & 30 \end{bmatrix}$
   - $BA = \begin{bmatrix} 8 & -23 \\ 16 & 20 \end{bmatrix}$

2. For $A = \begin{bmatrix} -1 & 5 \\ -3 & 6 \end{bmatrix}$ and $B = \begin{bmatrix} 2 & 10 \\ -7 & 1 \end{bmatrix}$

   - $3A = \begin{bmatrix} -3 & 15 \\ -9 & 18 \end{bmatrix}$
   - $-B = \begin{bmatrix} -2 & -10 \\ 7 & -1 \end{bmatrix}$
   - $A^2 = \begin{bmatrix} -14 & 25 \\ -15 & 21 \end{bmatrix}$

   - $A - 2B = \begin{bmatrix} -5 & -15 \\ 11 & 4 \end{bmatrix}$
   - $AB = \begin{bmatrix} -37 & -5 \\ -48 & -24 \end{bmatrix}$
   - $BA = \begin{bmatrix} -32 & 70 \\ 4 & -29 \end{bmatrix}$

3. For $A = \begin{bmatrix} -1 & 3 \\ 5 & 2 \end{bmatrix}$ and $B = \begin{bmatrix} 7 & 0 & 8 \\ -3 & 1 & 4 \end{bmatrix}$

   - $3A = \begin{bmatrix} -3 & 9 \\ 15 & 6 \end{bmatrix}$
   - $-B = \begin{bmatrix} -7 & 0 & -8 \\ 3 & -1 & -4 \end{bmatrix}$
   - $A^2 = \begin{bmatrix} 16 & 3 \\ 5 & 19 \end{bmatrix}$

   - $A - 2B$ is not defined
   - $AB = \begin{bmatrix} -16 & 3 & 4 \\ 29 & 2 & 48 \end{bmatrix}$
   - BA is not defined

4. For $A = \begin{bmatrix} 2 & 4 \\ 6 & 8 \end{bmatrix}$ and $B = \begin{bmatrix} -1 & 3 & -5 \\ 7 & -9 & 11 \end{bmatrix}$

   - $3A = \begin{bmatrix} 6 & 12 \\ 18 & 24 \end{bmatrix}$
   - $-B = \begin{bmatrix} 1 & -3 & 5 \\ -7 & 9 & -11 \end{bmatrix}$
   - $A^2 = \begin{bmatrix} 28 & 40 \\ 60 & 88 \end{bmatrix}$

   - $A - 2B$ is not defined
   - $AB = \begin{bmatrix} 26 & -30 & 34 \\ 50 & -54 & 58 \end{bmatrix}$
   - BA is not defined

5. For $A = \begin{bmatrix} 7 \\ 8 \\ 9 \end{bmatrix}$ and $B = \begin{bmatrix} 1 & 2 & 3 \end{bmatrix}$

- $3A = \begin{bmatrix} 21 \\ 24 \\ 27 \end{bmatrix}$
- $-B = \begin{bmatrix} -1 & -2 & -3 \end{bmatrix}$
- A^2 is not defined
- $A - 2B$ is not defined
- $AB = \begin{bmatrix} 7 & 14 & 21 \\ 8 & 16 & 24 \\ 9 & 18 & 27 \end{bmatrix}$
- $BA = [50]$

6. For $A = \begin{bmatrix} 1 & -2 \\ -3 & 4 \\ 5 & -6 \end{bmatrix}$ and $B = \begin{bmatrix} -5 & 1 & 8 \end{bmatrix}$

- $3A = \begin{bmatrix} 3 & -6 \\ -9 & 12 \\ 15 & -18 \end{bmatrix}$
- $-B = \begin{bmatrix} 5 & -1 & -8 \end{bmatrix}$
- A^2 is not defined
- $A - 2B$ is not defined
- AB is not defined
- $BA = \begin{bmatrix} 32 & -34 \end{bmatrix}$

7. For $A = \begin{bmatrix} 2 & -3 & 5 \\ 3 & 1 & -2 \\ -7 & 1 & -1 \end{bmatrix}$ and $B = \begin{bmatrix} 1 & 2 & 1 \\ 17 & 33 & 19 \\ 10 & 19 & 11 \end{bmatrix}$

- $3A = \begin{bmatrix} 6 & -9 & 15 \\ 9 & 3 & -6 \\ -21 & 3 & -3 \end{bmatrix}$
- $-B = \begin{bmatrix} -1 & -2 & -1 \\ -17 & -33 & -19 \\ -10 & -19 & -11 \end{bmatrix}$
- $A^2 = \begin{bmatrix} -40 & -4 & 11 \\ 23 & -10 & 15 \\ -4 & 21 & -36 \end{bmatrix}$
- $A - 2B = \begin{bmatrix} 0 & -7 & 3 \\ -31 & -65 & -40 \\ -27 & -37 & -23 \end{bmatrix}$
- $AB = \begin{bmatrix} 1 & 0 & 0 \\ 0 & 1 & 0 \\ 0 & 0 & 1 \end{bmatrix}$
- $BA = \begin{bmatrix} 1 & 0 & 0 \\ 0 & 1 & 0 \\ 0 & 0 & 1 \end{bmatrix}$

8. $7B - 4A = \begin{bmatrix} -4 & -29 \\ -47 & -2 \end{bmatrix}$

9. $AB = \begin{bmatrix} -10 & 1 \\ -20 & -1 \end{bmatrix}$

10. $BA = \begin{bmatrix} -9 & -12 \\ 1 & -2 \end{bmatrix}$

11. $E + D$ is undefined

12. $ED = \begin{bmatrix} \frac{67}{3} & 11 \\ -\frac{178}{3} & -72 \\ -30 & -40 \end{bmatrix}$

13. $CD + 2I_2 A = \begin{bmatrix} \frac{238}{3} & -126 \\ \frac{863}{15} & \frac{361}{5} \end{bmatrix}$

14. $A - 4I_2 = \begin{bmatrix} -3 & 2 \\ 3 & 0 \end{bmatrix}$

15. $A^2 - B^2 = \begin{bmatrix} -8 & 16 \\ 25 & 3 \end{bmatrix}$

16. $(A+B)(A-B) = \begin{bmatrix} -7 & 3 \\ 46 & 2 \end{bmatrix}$

17. $A^2 - 5A - 2I_2 = \begin{bmatrix} 0 & 0 \\ 0 & 0 \end{bmatrix}$

18. $E^2 + 5E - 36I_3 = \begin{bmatrix} -30 & 20 & -15 \\ 0 & 0 & -36 \\ 0 & 0 & -36 \end{bmatrix}$

19. $EDC = \begin{bmatrix} \frac{3449}{15} & -\frac{407}{6} & 99 \\ -\frac{9548}{15} & -\frac{101}{3} & -648 \\ -324 & -35 & -360 \end{bmatrix}$

20. CDE is undefined

21. $ABCEDI_2 = \begin{bmatrix} -\frac{90749}{15} & -\frac{28867}{5} \\ -\frac{156601}{15} & -\frac{47033}{5} \end{bmatrix}$

22. $E_1 A = \begin{bmatrix} d & e & f \\ a & b & c \end{bmatrix}$ E_1 interchanged $R1$ and $R2$ of A.

$E_2 A = \begin{bmatrix} 5a & 5b & 5c \\ d & e & f \end{bmatrix}$ E_2 multiplied $R1$ of A by 5.

$E_3 A = \begin{bmatrix} a-2d & b-2e & c-2f \\ d & e & f \end{bmatrix}$ E_3 replaced $R1$ in A with $R1 - 2R2$.

$E_4 = \begin{bmatrix} 1 & 0 \\ 0 & -6 \end{bmatrix}$

8.4 Systems of Linear Equations: Matrix Inverses

We concluded Section 8.3 by showing how we can rewrite a system of linear equations as the matrix equation $AX = B$ where A and B are known matrices and the solution matrix X of the equation corresponds to the solution of the system. In this section, we develop the method for solving such an equation. To that end, consider the system

$$\begin{cases} 2x - 3y = 16 \\ 3x + 4y = 7 \end{cases}$$

To write this as a matrix equation, we follow the procedure outlined on page 590. We find the coefficient matrix A, the unknowns matrix X and constant matrix B to be

$$A = \begin{bmatrix} 2 & -3 \\ 3 & 4 \end{bmatrix} \quad X = \begin{bmatrix} x \\ y \end{bmatrix} \quad B = \begin{bmatrix} 16 \\ 7 \end{bmatrix}$$

In order to motivate how we solve a matrix equation like $AX = B$, we revisit solving a similar equation involving real numbers. Consider the equation $3x = 5$. To solve, we simply divide both sides by 3 and obtain $x = \frac{5}{3}$. How can we go about defining an analogous process for matrices? To answer this question, we solve $3x = 5$ again, but this time, we pay attention to the properties of real numbers being used at each step. Recall that dividing by 3 is the same as multiplying by $\frac{1}{3} = 3^{-1}$, the so-called *multiplicative inverse*[1] of 3.

$$\begin{aligned} 3x &= 5 \\ 3^{-1}(3x) &= 3^{-1}(5) \quad &\text{Multiply by the (multiplicative) inverse of 3} \\ \left(3^{-1} \cdot 3\right)x &= 3^{-1}(5) \quad &\text{Associative property of multiplication} \\ 1 \cdot x &= 3^{-1}(5) \quad &\text{Inverse property} \\ x &= 3^{-1}(5) \quad &\text{Multiplicative Identity} \end{aligned}$$

If we wish to check our answer, we substitute $x = 3^{-1}(5)$ into the original equation

$$\begin{aligned} 3x &\stackrel{?}{=} 5 \\ 3\left(3^{-1}(5)\right) &\stackrel{?}{=} 5 \\ \left(3 \cdot 3^{-1}\right)(5) &\stackrel{?}{=} 5 \quad &\text{Associative property of multiplication} \\ 1 \cdot 5 &\stackrel{?}{=} 5 \quad &\text{Inverse property} \\ 5 &\stackrel{\checkmark}{=} 5 \quad &\text{Multiplicative Identity} \end{aligned}$$

Thinking back to Theorem 8.5, we know that matrix multiplication enjoys both an associative property and a multiplicative identity. What's missing from the mix is a multiplicative inverse for the coefficient matrix A. Assuming we can find such a beast, we can mimic our solution (and check) to $3x = 5$ as follows

---

[1] Every nonzero real number a has a multiplicative inverse, denoted a^{-1}, such that $a^{-1} \cdot a = a \cdot a^{-1} = 1$.

8.4 Systems of Linear Equations: Matrix Inverses

Solving $AX = B$

$$\begin{aligned} AX &= B \\ A^{-1}(AX) &= A^{-1}B \\ (A^{-1}A)X &= A^{-1}B \\ I_2 X &= A^{-1}B \\ X &= A^{-1}B \end{aligned}$$

Checking our answer

$$\begin{aligned} AX &\stackrel{?}{=} B \\ A\left(A^{-1}B\right) &\stackrel{?}{=} B \\ \left(AA^{-1}\right)B &\stackrel{?}{=} B \\ I_2 B &\stackrel{?}{=} B \\ B &\stackrel{\checkmark}{=} B \end{aligned}$$

The matrix A^{-1} is read 'A-inverse' and we will define it formally later in the section. At this stage, we have no idea if such a matrix A^{-1} exists, but that won't deter us from trying to find it.[2] We want A^{-1} to satisfy two equations, $A^{-1}A = I_2$ and $AA^{-1} = I_2$, making A^{-1} necessarily a 2×2 matrix.[3] Hence, we assume A^{-1} has the form

$$A^{-1} = \begin{bmatrix} x_1 & x_2 \\ x_3 & x_4 \end{bmatrix}$$

for real numbers x_1, x_2, x_3 and x_4. For reasons which will become clear later, we focus our attention on the equation $AA^{-1} = I_2$. We have

$$AA^{-1} = I_2$$

$$\begin{bmatrix} 2 & -3 \\ 3 & 4 \end{bmatrix} \begin{bmatrix} x_1 & x_2 \\ x_3 & x_4 \end{bmatrix} = \begin{bmatrix} 1 & 0 \\ 0 & 1 \end{bmatrix}$$

$$\begin{bmatrix} 2x_1 - 3x_3 & 2x_2 - 3x_4 \\ 3x_1 + 4x_3 & 3x_2 + 4x_4 \end{bmatrix} = \begin{bmatrix} 1 & 0 \\ 0 & 1 \end{bmatrix}$$

This gives rise to *two* more systems of equations

$$\begin{cases} 2x_1 - 3x_3 = 1 \\ 3x_1 + 4x_3 = 0 \end{cases} \quad \begin{cases} 2x_2 - 3x_4 = 0 \\ 3x_2 + 4x_4 = 1 \end{cases}$$

At this point, it may seem absurd to continue with this venture. After all, the intent was to solve *one* system of equations, and in doing so, we have produced *two* more to solve. Remember, the objective of this discussion is to develop a general *method* which, when used in the correct scenarios, allows us to do far more than just solve a system of equations. If we set about to solve these systems using augmented matrices using the techniques in Section 8.2, we see that not only do both systems have the same coefficient matrix, this coefficient matrix is none other than the matrix A itself. (We will come back to this observation in a moment.)

---

[2] Much like Carl's quest to find Sasquatch.
[3] Since matrix multiplication isn't necessarily commutative, at this stage, these are two different equations.

$$\begin{cases} 2x_1 - 3x_3 = 1 \\ 3x_1 + 4x_3 = 0 \end{cases} \xrightarrow{\text{Encode into a matrix}} \begin{bmatrix} 2 & -3 & | & 1 \\ 3 & 4 & | & 0 \end{bmatrix}$$

$$\begin{cases} 2x_2 - 3x_4 = 0 \\ 3x_2 + 4x_4 = 1 \end{cases} \xrightarrow{\text{Encode into a matrix}} \begin{bmatrix} 2 & -3 & | & 0 \\ 3 & 4 & | & 1 \end{bmatrix}$$

To solve these two systems, we use Gauss-Jordan Elimination to put the augmented matrices into reduced row echelon form. (We leave the details to the reader.) For the first system, we get

$$\begin{bmatrix} 2 & -3 & | & 1 \\ 3 & 4 & | & 0 \end{bmatrix} \xrightarrow{\text{Gauss Jordan Elimination}} \begin{bmatrix} 1 & 0 & | & \frac{4}{17} \\ 0 & 1 & | & -\frac{3}{17} \end{bmatrix}$$

which gives $x_1 = \frac{4}{17}$ and $x_3 = -\frac{3}{17}$. To solve the second system, we use the exact same row operations, in the same order, to put its augmented matrix into reduced row echelon form (Think about why that works.) and we obtain

$$\begin{bmatrix} 2 & -3 & | & 0 \\ 3 & 4 & | & 1 \end{bmatrix} \xrightarrow{\text{Gauss Jordan Elimination}} \begin{bmatrix} 1 & 0 & | & \frac{3}{17} \\ 0 & 1 & | & \frac{2}{17} \end{bmatrix}$$

which means $x_2 = \frac{3}{17}$ and $x_4 = \frac{2}{17}$. Hence,

$$A^{-1} = \begin{bmatrix} x_1 & x_2 \\ x_3 & x_4 \end{bmatrix} = \begin{bmatrix} \frac{4}{17} & \frac{3}{17} \\ -\frac{3}{17} & \frac{2}{17} \end{bmatrix}$$

We can check to see that A^{-1} behaves as it should by computing AA^{-1}

$$AA^{-1} = \begin{bmatrix} 2 & -3 \\ 3 & 4 \end{bmatrix} \begin{bmatrix} \frac{4}{17} & \frac{3}{17} \\ -\frac{3}{17} & \frac{2}{17} \end{bmatrix} = \begin{bmatrix} 1 & 0 \\ 0 & 1 \end{bmatrix} = I_2 \checkmark$$

As an added bonus,

$$A^{-1}A = \begin{bmatrix} \frac{4}{17} & \frac{3}{17} \\ -\frac{3}{17} & \frac{2}{17} \end{bmatrix} \begin{bmatrix} 2 & -3 \\ 3 & 4 \end{bmatrix} = \begin{bmatrix} 1 & 0 \\ 0 & 1 \end{bmatrix} = I_2 \checkmark$$

We can now return to the problem at hand. From our discussion at the beginning of the section on page 599, we know

$$X = A^{-1}B = \begin{bmatrix} \frac{4}{17} & \frac{3}{17} \\ -\frac{3}{17} & \frac{2}{17} \end{bmatrix} \begin{bmatrix} 16 \\ 7 \end{bmatrix} = \begin{bmatrix} 5 \\ -2 \end{bmatrix}$$

so that our final solution to the system is $(x, y) = (5, -2)$.

As we mentioned, the point of this exercise was not just to solve the system of linear equations, but to develop a general method for finding A^{-1}. We now take a step back and analyze the foregoing discussion in a more general context. In solving for A^{-1}, we used two augmented matrices, both of which contained the same entries as A

8.4 Systems of Linear Equations: Matrix Inverses

$$\begin{bmatrix} 2 & -3 & | & 1 \\ 3 & 4 & | & 0 \end{bmatrix} = \begin{bmatrix} A & | & 1 \\ & & 0 \end{bmatrix}$$

$$\begin{bmatrix} 2 & -3 & | & 0 \\ 3 & 4 & | & 1 \end{bmatrix} = \begin{bmatrix} A & | & 0 \\ & & 1 \end{bmatrix}$$

We also note that the reduced row echelon forms of these augmented matrices can be written as

$$\begin{bmatrix} 1 & 0 & | & \frac{4}{17} \\ 0 & 1 & | & -\frac{3}{17} \end{bmatrix} = \begin{bmatrix} I_2 & | & x_1 \\ & & x_3 \end{bmatrix}$$

$$\begin{bmatrix} 1 & 0 & | & \frac{3}{17} \\ 0 & 1 & | & \frac{2}{17} \end{bmatrix} = \begin{bmatrix} I_2 & | & x_2 \\ & & x_4 \end{bmatrix}$$

where we have identified the entries to the left of the vertical bar as the identity I_2 and the entries to the right of the vertical bar as the solutions to our systems. The long and short of the solution process can be summarized as

$$\begin{bmatrix} A & | & 1 \\ & & 0 \end{bmatrix} \xrightarrow{\text{Gauss Jordan Elimination}} \begin{bmatrix} I_2 & | & x_1 \\ & & x_3 \end{bmatrix}$$

$$\begin{bmatrix} A & | & 0 \\ & & 1 \end{bmatrix} \xrightarrow{\text{Gauss Jordan Elimination}} \begin{bmatrix} I_2 & | & x_2 \\ & & x_4 \end{bmatrix}$$

Since the row operations for both processes are the same, all of the arithmetic on the left hand side of the vertical bar is identical in both problems. The only difference between the two processes is what happens to the constants to the right of the vertical bar. As long as we keep these separated into columns, we can combine our efforts into one 'super-sized' augmented matrix and describe the above process as

$$\begin{bmatrix} A & | & 1 & 0 \\ & & 0 & 1 \end{bmatrix} \xrightarrow{\text{Gauss Jordan Elimination}} \begin{bmatrix} I_2 & | & x_1 & x_2 \\ & & x_3 & x_4 \end{bmatrix}$$

We have the identity matrix I_2 appearing as the right hand side of the first super-sized augmented matrix and the left hand side of the second super-sized augmented matrix. To our surprise and delight, the elements on the right hand side of the second super-sized augmented matrix are none other than those which comprise A^{-1}. Hence, we have

$$\begin{bmatrix} A & | & I_2 \end{bmatrix} \xrightarrow{\text{Gauss Jordan Elimination}} \begin{bmatrix} I_2 & | & A^{-1} \end{bmatrix}$$

In other words, the process of finding A^{-1} for a matrix A can be viewed as performing a series of row operations which transform A into the identity matrix of the same dimension. We can view this process as follows. In trying to find A^{-1}, we are trying to 'undo' multiplication by the matrix A. The identity matrix in the super-sized augmented matrix $[A|I]$ keeps a running memory of all of the moves required to 'undo' A. This results in exactly what we want, A^{-1}. We are now ready to formalize and generalize the foregoing discussion. We begin with the formal definition of an invertible matrix.

> **Definition 8.11.** An $n \times n$ matrix A is said to be **invertible** if there exists a matrix A^{-1}, read 'A inverse', such that $A^{-1}A = AA^{-1} = I_n$.

Note that, as a consequence of our definition, invertible matrices are square, and as such, the conditions in Definition 8.11 force the matrix A^{-1} to be same dimensions as A, that is, $n \times n$. Since not all matrices are square, not all matrices are invertible. However, just because a matrix is square doesn't guarantee it is invertible. (See the exercises.) Our first result summarizes some of the important characteristics of invertible matrices and their inverses.

> **Theorem 8.6.** Suppose A is an $n \times n$ matrix.
> 1. If A is invertible then A^{-1} is unique.
> 2. A is invertible if and only if $AX = B$ has a unique solution for every $n \times r$ matrix B.

The proofs of the properties in Theorem 8.6 rely on a healthy mix of definition and matrix arithmetic. To establish the first property, we assume that A is invertible and suppose the matrices B and C act as inverses for A. That is, $BA = AB = I_n$ and $CA = AC = I_n$. We need to show that B and C are, in fact, the same matrix. To see this, we note that $B = I_nB = (CA)B = C(AB) = CI_n = C$. Hence, any two matrices that act like A^{-1} are, in fact, the same matrix.[4] To prove the second property of Theorem 8.6, we note that if A is invertible then the discussion on page 599 shows the solution to $AX = B$ to be $X = A^{-1}B$, and since A^{-1} is unique, so is $A^{-1}B$. Conversely, if $AX = B$ has a unique solution for every $n \times r$ matrix B, then, in particular, there is a unique solution X_0 to the equation $AX = I_n$. The solution matrix X_0 is our candidate for A^{-1}. We have $AX_0 = I_n$ by definition, but we need to also show $X_0A = I_n$. To that end, we note that $A(X_0A) = (AX_0)A = I_nA = A$. In other words, the matrix X_0A is a solution to the equation $AX = A$. Clearly, $X = I_n$ is also a solution to the equation $AX = A$, and since we are assuming every such equation as a *unique* solution, we must have $X_0A = I_n$. Hence, we have $X_0A = AX_0 = I_n$, so that $X_0 = A^{-1}$ and A is invertible. The foregoing discussion justifies our quest to find A^{-1} using our super-sized augmented matrix approach

$$[\, A \mid I_n \,] \xrightarrow{\text{Gauss Jordan Elimination}} [\, I_n \mid A^{-1} \,]$$

We are, in essence, trying to find the unique solution to the equation $AX = I_n$ using row operations.

What does all of this mean for a system of linear equations? Theorem 8.6 tells us that if we write the system in the form $AX = B$, then if the coefficient matrix A is invertible, there is only one solution to the system — that is, if A is invertible, the system is consistent and independent.[5] We also know that the process by which we find A^{-1} is determined completely by A, and not by the

---

[4]If this proof sounds familiar, it should. See the discussion following Theorem 5.2 on page 380.

[5]It can be shown that a matrix is invertible if and only if when it serves as a coefficient matrix for a system of equations, the system is always consistent independent. It amounts to the second property in Theorem 8.6 where the matrices B are restricted to being $n \times 1$ matrices. We note that, owing to how matrix multiplication is defined, being able to find unique solutions to $AX = B$ for $n \times 1$ matrices B gives you the same statement about solving such equations for $n \times r$ matrices — since we can find a unique solution to them one column at a time.

8.4 Systems of Linear Equations: Matrix Inverses

constants in B. This answers the question as to why we would bother doing row operations on a super-sized augmented matrix to find A^{-1} instead of an ordinary augmented matrix to solve a system; by finding A^{-1} we have done all of the row operations we ever need to do, once and for all, since we can quickly solve *any* equation $AX = B$ using *one* multiplication, $A^{-1}B$.

Example 8.4.1. Let $A = \begin{bmatrix} 3 & 1 & 2 \\ 0 & -1 & 5 \\ 2 & 1 & 4 \end{bmatrix}$

1. Use row operations to find A^{-1}. Check your answer by finding $A^{-1}A$ and AA^{-1}.

2. Use A^{-1} to solve the following systems of equations

(a) $\begin{cases} 3x + y + 2z = 26 \\ -y + 5z = 39 \\ 2x + y + 4z = 117 \end{cases}$ (b) $\begin{cases} 3x + y + 2z = 4 \\ -y + 5z = 2 \\ 2x + y + 4z = 5 \end{cases}$ (c) $\begin{cases} 3x + y + 2z = 1 \\ -y + 5z = 0 \\ 2x + y + 4z = 0 \end{cases}$

Solution.

1. We begin with a super-sized augmented matrix and proceed with Gauss-Jordan elimination.

$$\begin{bmatrix} 3 & 1 & 2 & | & 1 & 0 & 0 \\ 0 & -1 & 5 & | & 0 & 1 & 0 \\ 2 & 1 & 4 & | & 0 & 0 & 1 \end{bmatrix} \xrightarrow{\text{Replace } R1 \text{ with } \frac{1}{3}R1} \begin{bmatrix} 1 & \frac{1}{3} & \frac{2}{3} & | & \frac{1}{3} & 0 & 0 \\ 0 & -1 & 5 & | & 0 & 1 & 0 \\ 2 & 1 & 4 & | & 0 & 0 & 1 \end{bmatrix}$$

$$\begin{bmatrix} 1 & \frac{1}{3} & \frac{2}{3} & | & \frac{1}{3} & 0 & 0 \\ 0 & -1 & 5 & | & 0 & 1 & 0 \\ 2 & 1 & 4 & | & 0 & 0 & 1 \end{bmatrix} \xrightarrow{\text{Replace } R3 \text{ with } -2R1 + R3} \begin{bmatrix} 1 & \frac{1}{3} & \frac{2}{3} & | & \frac{1}{3} & 0 & 0 \\ 0 & -1 & 5 & | & 0 & 1 & 0 \\ 0 & \frac{1}{3} & \frac{8}{3} & | & -\frac{2}{3} & 0 & 1 \end{bmatrix}$$

$$\begin{bmatrix} 1 & \frac{1}{3} & \frac{2}{3} & | & \frac{1}{3} & 0 & 0 \\ 0 & -1 & 5 & | & 0 & 1 & 0 \\ 0 & \frac{1}{3} & \frac{8}{3} & | & -\frac{2}{3} & 0 & 1 \end{bmatrix} \xrightarrow{\text{Replace } R2 \text{ with } (-1)R2} \begin{bmatrix} 1 & \frac{1}{3} & \frac{2}{3} & | & \frac{1}{3} & 0 & 0 \\ 0 & 1 & -5 & | & 0 & -1 & 0 \\ 0 & \frac{1}{3} & \frac{8}{3} & | & -\frac{2}{3} & 0 & 1 \end{bmatrix}$$

$$\begin{bmatrix} 1 & \frac{1}{3} & \frac{2}{3} & | & \frac{1}{3} & 0 & 0 \\ 0 & 1 & -5 & | & 0 & -1 & 0 \\ 0 & \frac{1}{3} & \frac{8}{3} & | & -\frac{2}{3} & 0 & 1 \end{bmatrix} \xrightarrow{\text{Replace } R3 \text{ with } -\frac{1}{3}R2 + R3} \begin{bmatrix} 1 & \frac{1}{3} & \frac{2}{3} & | & \frac{1}{3} & 0 & 0 \\ 0 & 1 & -5 & | & 0 & -1 & 0 \\ 0 & 0 & \frac{13}{3} & | & -\frac{2}{3} & \frac{1}{3} & 1 \end{bmatrix}$$

$$\begin{bmatrix} 1 & \frac{1}{3} & \frac{2}{3} & | & \frac{1}{3} & 0 & 0 \\ 0 & 1 & -5 & | & 0 & -1 & 0 \\ 0 & 0 & \frac{13}{3} & | & -\frac{2}{3} & \frac{1}{3} & 1 \end{bmatrix} \xrightarrow{\text{Replace } R3 \text{ with } \frac{3}{13}R3} \begin{bmatrix} 1 & \frac{1}{3} & \frac{2}{3} & | & \frac{1}{3} & 0 & 0 \\ 0 & 1 & -5 & | & 0 & -1 & 0 \\ 0 & 0 & 1 & | & -\frac{2}{13} & \frac{1}{13} & \frac{3}{13} \end{bmatrix}$$

$$\begin{bmatrix} 1 & \frac{1}{3} & \frac{2}{3} & | & \frac{1}{3} & 0 & 0 \\ 0 & 1 & -5 & | & 0 & -1 & 0 \\ 0 & 0 & 1 & | & -\frac{2}{13} & \frac{1}{13} & \frac{3}{13} \end{bmatrix} \xrightarrow[\text{Replace } R2 \text{ with } 5R3 + R2]{\text{Replace } R1 \text{ with } -\frac{2}{3}R3 + R1} \begin{bmatrix} 1 & \frac{1}{3} & 0 & | & \frac{17}{39} & -\frac{2}{39} & -\frac{2}{13} \\ 0 & 1 & 0 & | & -\frac{10}{13} & -\frac{8}{13} & \frac{15}{13} \\ 0 & 0 & 1 & | & -\frac{2}{13} & \frac{1}{13} & \frac{3}{13} \end{bmatrix}$$

$$\begin{bmatrix} 1 & \frac{1}{3} & 0 & \vline & \frac{17}{39} & -\frac{2}{39} & -\frac{2}{13} \\ 0 & 1 & 0 & \vline & -\frac{10}{13} & -\frac{8}{13} & \frac{15}{13} \\ 0 & 0 & 1 & \vline & -\frac{2}{13} & \frac{1}{13} & \frac{3}{13} \end{bmatrix} \xrightarrow[-\frac{1}{3}R2 + R1]{\text{Replace } R1 \text{ with}} \begin{bmatrix} 1 & 0 & 0 & \vline & \frac{9}{13} & \frac{2}{13} & -\frac{7}{13} \\ 0 & 1 & 0 & \vline & -\frac{10}{13} & -\frac{8}{13} & \frac{15}{13} \\ 0 & 0 & 1 & \vline & -\frac{2}{13} & \frac{1}{13} & \frac{3}{13} \end{bmatrix}$$

We find $A^{-1} = \begin{bmatrix} \frac{9}{13} & \frac{2}{13} & -\frac{7}{13} \\ -\frac{10}{13} & -\frac{8}{13} & \frac{15}{13} \\ -\frac{2}{13} & \frac{1}{13} & \frac{3}{13} \end{bmatrix}$. To check our answer, we compute

$$A^{-1}A = \begin{bmatrix} \frac{9}{13} & \frac{2}{13} & -\frac{7}{13} \\ -\frac{10}{13} & -\frac{8}{13} & \frac{15}{13} \\ -\frac{2}{13} & \frac{1}{13} & \frac{3}{13} \end{bmatrix} \begin{bmatrix} 3 & 1 & 2 \\ 0 & -1 & 5 \\ 2 & 1 & 4 \end{bmatrix} = \begin{bmatrix} 1 & 0 & 0 \\ 0 & 1 & 0 \\ 0 & 0 & 1 \end{bmatrix} = I_3 \checkmark$$

and

$$AA^{-1} = \begin{bmatrix} 3 & 1 & 2 \\ 0 & -1 & 5 \\ 2 & 1 & 4 \end{bmatrix} \begin{bmatrix} \frac{9}{13} & \frac{2}{13} & -\frac{7}{13} \\ -\frac{10}{13} & -\frac{8}{13} & \frac{15}{13} \\ -\frac{2}{13} & \frac{1}{13} & \frac{3}{13} \end{bmatrix} = \begin{bmatrix} 1 & 0 & 0 \\ 0 & 1 & 0 \\ 0 & 0 & 1 \end{bmatrix} = I_3 \checkmark$$

2. Each of the systems in this part has A as its coefficient matrix. The only difference between the systems is the constants which is the matrix B in the associated matrix equation $AX = B$. We solve each of them using the formula $X = A^{-1}B$.

(a) $X = A^{-1}B = \begin{bmatrix} \frac{9}{13} & \frac{2}{13} & -\frac{7}{13} \\ -\frac{10}{13} & -\frac{8}{13} & \frac{15}{13} \\ -\frac{2}{13} & \frac{1}{13} & \frac{3}{13} \end{bmatrix} \begin{bmatrix} 26 \\ 39 \\ 117 \end{bmatrix} = \begin{bmatrix} -39 \\ 91 \\ 26 \end{bmatrix}$. Our solution is $(-39, 91, 26)$.

(b) $X = A^{-1}B = \begin{bmatrix} \frac{9}{13} & \frac{2}{13} & -\frac{7}{13} \\ -\frac{10}{13} & -\frac{8}{13} & \frac{15}{13} \\ -\frac{2}{13} & \frac{1}{13} & \frac{3}{13} \end{bmatrix} \begin{bmatrix} 4 \\ 2 \\ 5 \end{bmatrix} = \begin{bmatrix} \frac{5}{13} \\ \frac{19}{13} \\ \frac{9}{13} \end{bmatrix}$. We get $\left(\frac{5}{13}, \frac{19}{13}, \frac{9}{13}\right)$.

(c) $X = A^{-1}B = \begin{bmatrix} \frac{9}{13} & \frac{2}{13} & -\frac{7}{13} \\ -\frac{10}{13} & -\frac{8}{13} & \frac{15}{13} \\ -\frac{2}{13} & \frac{1}{13} & \frac{3}{13} \end{bmatrix} \begin{bmatrix} 1 \\ 0 \\ 0 \end{bmatrix} = \begin{bmatrix} \frac{9}{13} \\ -\frac{10}{13} \\ -\frac{2}{13} \end{bmatrix}$. We find $\left(\frac{9}{13}, -\frac{10}{13}, -\frac{2}{13}\right)$.[6]

□

In Example 8.4.1, we see that finding one inverse matrix can enable us to solve an entire family of systems of linear equations. There are many examples of where this comes in handy 'in the wild', and we chose our example for this section from the field of electronics. We also take this opportunity to introduce the student to how we can compute inverse matrices using the calculator.

---

[6] Note that the solution is the first column of the A^{-1}. The reader is encouraged to meditate on this 'coincidence'.

8.4 Systems of Linear Equations: Matrix Inverses

Example 8.4.2. Consider the circuit diagram below.[7] We have two batteries with source voltages VB_1 and VB_2, measured in volts V, along with six resistors with resistances R_1 through R_6, measured in kiloohms, $k\Omega$. Using Ohm's Law and Kirchhoff's Voltage Law, we can relate the voltage supplied to the circuit by the two batteries to the voltage drops across the six resistors in order to find the four 'mesh' currents: i_1, i_2, i_3 and i_4, measured in milliamps, mA. If we think of electrons flowing through the circuit, we can think of the voltage sources as providing the 'push' which makes the electrons move, the resistors as obstacles for the electrons to overcome, and the mesh current as a net rate of flow of electrons around the indicated loops.

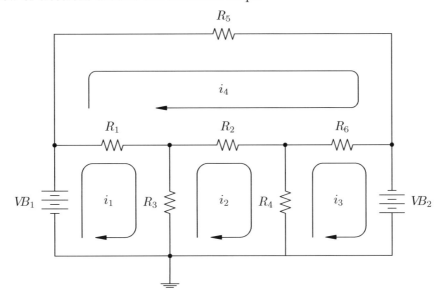

The system of linear equations associated with this circuit is

$$\begin{cases} (R_1 + R_3)\,i_1 - R_3 i_2 - R_1 i_4 = VB_1 \\ -R_3 i_1 + (R_2 + R_3 + R_4)\,i_2 - R_4 i_3 - R_2 i_4 = 0 \\ -R_4 i_2 + (R_4 + R_6)\,i_3 - R_6 i_4 = -VB_2 \\ -R_1 i_1 - R_2 i_2 - R_6 i_3 + (R_1 + R_2 + R_5 + R_6)\,i_4 = 0 \end{cases}$$

1. Assuming the resistances are all $1k\Omega$, find the mesh currents if the battery voltages are

    (a) $VB_1 = 10V$ and $VB_2 = 5V$

    (b) $VB_1 = 10V$ and $VB_2 = 0V$

    (c) $VB_1 = 0V$ and $VB_2 = 10V$

    (d) $VB_1 = 10V$ and $VB_2 = 10V$

2. Assuming $VB_1 = 10V$ and $VB_2 = 5V$, find the possible combinations of resistances which would yield the mesh currents you found in 1(a).

---

[7] The authors wish to thank Don Anthan of Lakeland Community College for the design of this example.

Solution.

1. Substituting the resistance values into our system of equations, we get

$$\begin{cases} 2i_1 - i_2 - i_4 = VB_1 \\ -i_1 + 3i_2 - i_3 - i_4 = 0 \\ -i_2 + 2i_3 - i_4 = -VB_2 \\ -i_1 - i_2 - i_3 + 4i_4 = 0 \end{cases}$$

This corresponds to the matrix equation $AX = B$ where

$$A = \begin{bmatrix} 2 & -1 & 0 & -1 \\ -1 & 3 & -1 & -1 \\ 0 & -1 & 2 & -1 \\ -1 & -1 & -1 & 4 \end{bmatrix} \quad X = \begin{bmatrix} i_1 \\ i_2 \\ i_3 \\ i_4 \end{bmatrix} \quad B = \begin{bmatrix} VB_1 \\ 0 \\ -VB_2 \\ 0 \end{bmatrix}$$

When we input the matrix A into the calculator, we find

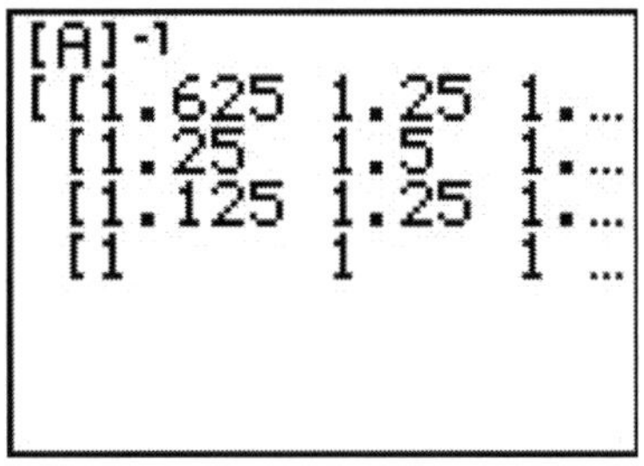

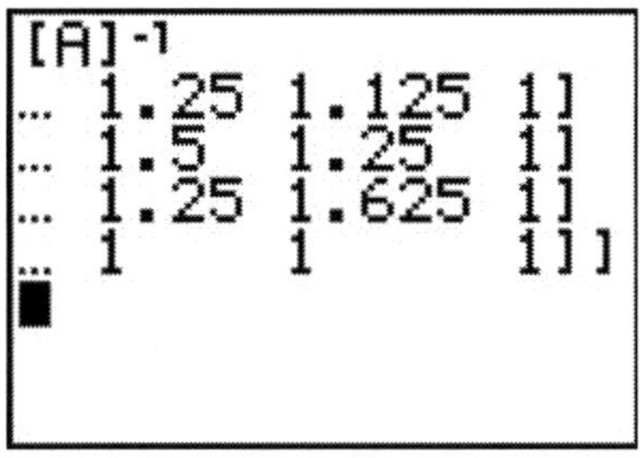

from which we have $A^{-1} = \begin{bmatrix} 1.625 & 1.25 & 1.125 & 1 \\ 1.25 & 1.5 & 1.25 & 1 \\ 1.125 & 1.25 & 1.625 & 1 \\ 1 & 1 & 1 & 1 \end{bmatrix}$.

To solve the four systems given to us, we find $X = A^{-1}B$ where the value of B is determined by the given values of VB_1 and VB_2

$$1\ (a) \quad B = \begin{bmatrix} 10 \\ 0 \\ -5 \\ 0 \end{bmatrix}, \quad 1\ (b) \quad B = \begin{bmatrix} 10 \\ 0 \\ 0 \\ 0 \end{bmatrix}, \quad 1\ (c) \quad B = \begin{bmatrix} 0 \\ 0 \\ -10 \\ 0 \end{bmatrix}, \quad 1\ (d) \quad B = \begin{bmatrix} 10 \\ 0 \\ 10 \\ 0 \end{bmatrix}$$

(a) For $VB_1 = 10V$ and $VB_2 = 5V$, the calculator gives $i_1 = 10.625\ mA$, $i_2 = 6.25\ mA$, $i_3 = 3.125\ mA$, and $i_4 = 5\ mA$. We include a calculator screenshot below for this part (and this part only!) for reference.

8.4 SYSTEMS OF LINEAR EQUATIONS: MATRIX INVERSES

```
[A]⁻¹[B]
      [[10.625]
       [6.25  ]
       [3.125 ]
       [5     ]]
```

(b) By keeping $VB_1 = 10V$ and setting $VB_2 = 0V$, we are removing the effect of the second battery. We get $i_1 = 16.25 \ mA$, $i_2 = 12.5 \ mA$, $i_3 = 11.25 \ mA$, and $i_4 = 10 \ mA$.

(c) Part (c) is a symmetric situation to part (b) in so much as we are zeroing out VB_1 and making $VB_2 = 10$. We find $i_1 = -11.25 \ mA$, $i_2 = -12.5 \ mA$, $i_3 = -16.25 \ mA$, and $i_4 = -10 \ mA$, where the negatives indicate that the current is flowing in the opposite direction as is indicated on the diagram. The reader is encouraged to study the symmetry here, and if need be, hold up a mirror to the diagram to literally 'see' what is happening.

(d) For $VB_1 = 10V$ and $VB_2 = 10V$, we get $i_1 = 5 \ mA$, $i_2 = 0 \ mA$, $i_3 = -5 \ mA$, and $i_4 = 0 \ mA$. The mesh currents i_2 and i_4 being zero is a consequence of both batteries 'pushing' in equal but opposite directions, causing the net flow of electrons in these two regions to cancel out.

2. We now turn the tables and are given $VB_1 = 10V$, $VB_2 = 5V$, $i_1 = 10.625 \ mA$, $i_2 = 6.25 \ mA$, $i_3 = 3.125 \ mA$ and $i_4 = 5 \ mA$ and our unknowns are the resistance values. Rewriting our system of equations, we get

$$\begin{cases} 5.625R_1 + 4.375R_3 = 10 \\ 1.25R_2 - 4.375R_3 + 3.125R_4 = 0 \\ -3.125R_4 - 1.875R_6 = -5 \\ -5.625R_1 - 1.25R_2 + 5R_5 + 1.875R_6 = 0 \end{cases}$$

The coefficient matrix for this system is 4×6 (4 equations with 6 unknowns) and is therefore not invertible. We do know, however, this system is consistent, since setting all the resistance values equal to 1 corresponds to our situation in problem 1a. This means we have an underdetermined consistent system which is necessarily dependent. To solve this system, we encode it into an augmented matrix

$$\left[\begin{array}{cccccc|c} 5.25 & 0 & 4.375 & 0 & 0 & 0 & 10 \\ 0 & 1.25 & -4.375 & 3.125 & 0 & 0 & 0 \\ 0 & 0 & 0 & -3.125 & 0 & -1.875 & -5 \\ -5.625 & -1.25 & 0 & 0 & 5 & 1.875 & 0 \end{array} \right]$$

and use the calculator to write in reduced row echelon form

$$\begin{bmatrix} 1 & 0 & 0.\overline{7} & 0 & 0 & 0 & | & 1.\overline{7} \\ 0 & 1 & -3.5 & 0 & 0 & -1.5 & | & -4 \\ 0 & 0 & 0 & 1 & 0 & 0.6 & | & 1.6 \\ 0 & 0 & 0 & 0 & 1 & 0 & | & 1 \end{bmatrix}$$

Decoding this system from the matrix, we get

$$\begin{cases} R_1 + 0.\overline{7}R_3 &= 1.\overline{7} \\ R_2 - 3.5R_3 - 1.5R_6 &= -4 \\ R_4 + 0.6R_6 &= 1.6 \\ R_5 &= 1 \end{cases}$$

We can solve for R_1, R_2, R_4 and R_5 leaving R_3 and R_6 as free variables. Labeling $R_3 = s$ and $R_6 = t$, we have $R_1 = -0.\overline{7}s + 1.\overline{7}$, $R_2 = 3.5s + 1.5t - 4$, $R_4 = -0.6t + 1.6$ and $R_5 = 1$. Since resistance values are always positive, we need to restrict our values of s and t. We know $R_3 = s > 0$ and when we combine that with $R_1 = -0.\overline{7}s + 1.\overline{7} > 0$, we get $0 < s < \frac{16}{7}$. Similarly, $R_6 = t > 0$ and with $R_4 = -0.6t + 1.6 > 0$, we find $0 < t < \frac{8}{3}$. In order visualize the inequality $R_2 = 3.5s + 1.5t - 4 > 0$, we graph the line $3.5s + 1.5t - 4 = 0$ on the st-plane and shade accordingly.[8] Imposing the additional conditions $0 < s < \frac{16}{7}$ and $0 < t < \frac{8}{3}$, we find our values of s and t restricted to the region depicted on the right. Using the roster method, the values of s and t are pulled from the region $\left\{ (s,t) : 0 < s < \frac{16}{7},\ 0 < t < \frac{8}{3},\ 3.5s + 1.5t - 4 > 0 \right\}$. The reader is encouraged to check that the solution presented in 1(a), namely all resistance values equal to 1, corresponds to a pair (s,t) in the region.

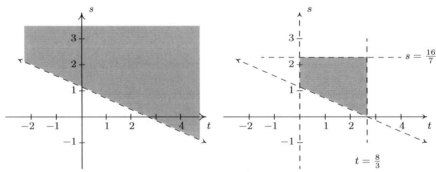

The region where $3.5s + 1.5t - 4 > 0$ The region for our parameters s and t.

□

---

[8]See Section 2.4 for a review of this procedure.

8.4.1 Exercises

In Exercises 1 - 8, find the inverse of the matrix or state that the matrix is not invertible.

1. $A = \begin{bmatrix} 1 & 2 \\ 3 & 4 \end{bmatrix}$

2. $B = \begin{bmatrix} 12 & -7 \\ -5 & 3 \end{bmatrix}$

3. $C = \begin{bmatrix} 6 & 15 \\ 14 & 35 \end{bmatrix}$

4. $D = \begin{bmatrix} 2 & -1 \\ 16 & -9 \end{bmatrix}$

5. $E = \begin{bmatrix} 3 & 0 & 4 \\ 2 & -1 & 3 \\ -3 & 2 & -5 \end{bmatrix}$

6. $F = \begin{bmatrix} 4 & 6 & -3 \\ 3 & 4 & -3 \\ 1 & 2 & 6 \end{bmatrix}$

7. $G = \begin{bmatrix} 1 & 2 & 3 \\ 2 & 3 & 11 \\ 3 & 4 & 19 \end{bmatrix}$

8. $H = \begin{bmatrix} 1 & 0 & -3 & 0 \\ 2 & -2 & 8 & 7 \\ -5 & 0 & 16 & 0 \\ 1 & 0 & 4 & 1 \end{bmatrix}$

In Exercises 9 - 11, use one matrix inverse to solve the following systems of linear equations.

9. $\begin{cases} 3x + 7y = 26 \\ 5x + 12y = 39 \end{cases}$

10. $\begin{cases} 3x + 7y = 0 \\ 5x + 12y = -1 \end{cases}$

11. $\begin{cases} 3x + 7y = -7 \\ 5x + 12y = 5 \end{cases}$

In Exercises 12 - 14, use the inverse of E from Exercise 5 above to solve the following systems of linear equations.

12. $\begin{cases} 3x + 4z = 1 \\ 2x - y + 3z = 0 \\ -3x + 2y - 5z = 0 \end{cases}$

13. $\begin{cases} 3x + 4z = 0 \\ 2x - y + 3z = 1 \\ -3x + 2y - 5z = 0 \end{cases}$

14. $\begin{cases} 3x + 4z = 0 \\ 2x - y + 3z = 0 \\ -3x + 2y - 5z = 1 \end{cases}$

15. This exercise is a continuation of Example 8.3.3 in Section 8.3 and gives another application of matrix inverses. Recall that given the position matrix P for a point in the plane, the matrix RP corresponds to a point rotated $45°$ counterclockwise from P where

$$R = \begin{bmatrix} \frac{\sqrt{2}}{2} & -\frac{\sqrt{2}}{2} \\ \frac{\sqrt{2}}{2} & \frac{\sqrt{2}}{2} \end{bmatrix}$$

(a) Find R^{-1}.

(b) If RP rotates a point counterclockwise $45°$, what should $R^{-1}P$ do? Check your answer by finding $R^{-1}P$ for various points on the coordinate axes and the lines $y = \pm x$.

(c) Find $R^{-1}P$ where P corresponds to a generic point $P(x, y)$. Verify that this takes points on the curve $y = \frac{2}{x}$ to points on the curve $x^2 - y^2 = 4$.

16. A Sasquatch's diet consists of three primary foods: Ippizuti Fish, Misty Mushrooms, and Sun Berries. Each serving of Ippizuti Fish is 500 calories, contains 40 grams of protein, and has no Vitamin X. Each serving of Misty Mushrooms is 50 calories, contains 1 gram of protein, and 5 milligrams of Vitamin X. Finally, each serving of Sun Berries is 80 calories, contains no protein, but has 15 milligrams of Vitamin X.[9]

    (a) If an adult male Sasquatch requires 3200 calories, 130 grams of protein, and 275 milligrams of Vitamin X daily, use a matrix inverse to find how many servings each of Ippizuti Fish, Misty Mushrooms, and Sun Berries he needs to eat each day.

    (b) An adult female Sasquatch requires 3100 calories, 120 grams of protein, and 300 milligrams of Vitamin X daily. Use the matrix inverse you found in part (a) to find how many servings each of Ippizuti Fish, Misty Mushrooms, and Sun Berries she needs to eat each day.

    (c) An adolescent Sasquatch requires 5000 calories, 400 grams of protein daily, but no Vitamin X daily.[10] Use the matrix inverse you found in part (a) to find how many servings each of Ippizuti Fish, Misty Mushrooms, and Sun Berries she needs to eat each day.

17. Matrices can be used in cryptography. Suppose we wish to encode the message 'BIGFOOT LIVES'. We start by assigning a number to each letter of the alphabet, say $A = 1$, $B = 2$ and so on. We reserve 0 to act as a space. Hence, our message 'BIGFOOT LIVES' corresponds to the string of numbers '2, 9, 7, 6, 15, 15, 20, 0, 12, 9, 22, 5, 19.' To encode this message, we use an invertible matrix. Any invertible matrix will do, but for this exercise, we choose

$$A = \begin{bmatrix} 2 & -3 & 5 \\ 3 & 1 & -2 \\ -7 & 1 & -1 \end{bmatrix}$$

Since A is 3×3 matrix, we encode our message string into a matrix M with 3 rows. To do this, we take the first three numbers, 2 9 7, and make them our first column, the next three numbers, 6 15 15, and make them our second column, and so on. We put 0's to round out the matrix.

$$M = \begin{bmatrix} 2 & 6 & 20 & 9 & 19 \\ 9 & 15 & 0 & 22 & 0 \\ 7 & 15 & 12 & 5 & 0 \end{bmatrix}$$

To encode the message, we find the product AM

$$AM = \begin{bmatrix} 2 & -3 & 5 \\ 3 & 1 & -2 \\ -7 & 1 & -1 \end{bmatrix} \begin{bmatrix} 2 & 6 & 20 & 9 & 19 \\ 9 & 15 & 0 & 22 & 0 \\ 7 & 15 & 12 & 5 & 0 \end{bmatrix} = \begin{bmatrix} 12 & 42 & 100 & -23 & 38 \\ 1 & 3 & 36 & 39 & 57 \\ -12 & -42 & -152 & -46 & -133 \end{bmatrix}$$

---

[9] Misty Mushrooms and Sun Berries are the only known fictional sources of Vitamin X.
[10] Vitamin X is needed to sustain Sasquatch longevity only.

8.4 SYSTEMS OF LINEAR EQUATIONS: MATRIX INVERSES 611

So our coded message is '12, 1, −12, 42, 3, −42, 100, 36, −152, −23, 39, −46, 38, 57, −133.'
To decode this message, we start with this string of numbers, construct a message matrix as we did earlier (we should get the matrix AM again) and then multiply by A^{-1}.

(a) Find A^{-1}.

(b) Use A^{-1} to decode the message and check this method actually works.

(c) Decode the message '14, 37, −76, 128, 21, −151, 31, 65, −140'

(d) Choose another invertible matrix and encode and decode your own messages.

18. Using the matrices A from Exercise 1, B from Exercise 2 and D from Exercise 4, show $AB = D$ and $D^{-1} = B^{-1}A^{-1}$. That is, show that $(AB)^{-1} = B^{-1}A^{-1}$.

19. Let M and N be invertible $n \times n$ matrices. Show that $(MN)^{-1} = N^{-1}M^{-1}$ and compare your work to Exercise 31 in Section 5.2.

8.4.2 Answers

1. $A^{-1} = \begin{bmatrix} -2 & 1 \\ \frac{3}{2} & -\frac{1}{2} \end{bmatrix}$

2. $B^{-1} = \begin{bmatrix} 3 & 7 \\ 5 & 12 \end{bmatrix}$

3. C is not invertible

4. $D^{-1} = \begin{bmatrix} \frac{9}{2} & -\frac{1}{2} \\ 8 & -1 \end{bmatrix}$

5. $E^{-1} = \begin{bmatrix} -1 & 8 & 4 \\ 1 & -3 & -1 \\ 1 & -6 & -3 \end{bmatrix}$

6. $F^{-1} = \begin{bmatrix} -\frac{5}{2} & \frac{7}{2} & \frac{1}{2} \\ \frac{7}{4} & -\frac{9}{4} & -\frac{1}{4} \\ -\frac{1}{6} & \frac{1}{6} & \frac{1}{6} \end{bmatrix}$

7. G is not invertible

8. $H^{-1} = \begin{bmatrix} 16 & 0 & 3 & 0 \\ -90 & -\frac{1}{2} & -\frac{35}{2} & \frac{7}{2} \\ 5 & 0 & 1 & 0 \\ -36 & 0 & -7 & 1 \end{bmatrix}$

The coefficient matrix is B^{-1} from Exercise 2 above so the inverse we need is $(B^{-1})^{-1} = B$.

9. $\begin{bmatrix} 12 & -7 \\ -5 & 3 \end{bmatrix} \begin{bmatrix} 26 \\ 39 \end{bmatrix} = \begin{bmatrix} 39 \\ -13 \end{bmatrix}$ So $x = 39$ and $y = -13$.

10. $\begin{bmatrix} 12 & -7 \\ -5 & 3 \end{bmatrix} \begin{bmatrix} 0 \\ -1 \end{bmatrix} = \begin{bmatrix} 7 \\ -3 \end{bmatrix}$ So $x = 7$ and $y = -3$.

11. $\begin{bmatrix} 12 & -7 \\ -5 & 3 \end{bmatrix} \begin{bmatrix} -7 \\ 5 \end{bmatrix} = \begin{bmatrix} -119 \\ 50 \end{bmatrix}$ So $x = -119$ and $y = 50$.

The coefficient matrix is $E = \begin{bmatrix} 3 & 0 & 4 \\ 2 & -1 & 3 \\ -3 & 2 & -5 \end{bmatrix}$ from Exercise 5, so $E^{-1} = \begin{bmatrix} -1 & 8 & 4 \\ 1 & -3 & -1 \\ 1 & -6 & -3 \end{bmatrix}$

12. $\begin{bmatrix} -1 & 8 & 4 \\ 1 & -3 & -1 \\ 1 & -6 & -3 \end{bmatrix} \begin{bmatrix} 1 \\ 0 \\ 0 \end{bmatrix} = \begin{bmatrix} -1 \\ 1 \\ 1 \end{bmatrix}$ So $x = -1$, $y = 1$ and $z = 1$.

13. $\begin{bmatrix} -1 & 8 & 4 \\ 1 & -3 & -1 \\ 1 & -6 & -3 \end{bmatrix} \begin{bmatrix} 0 \\ 1 \\ 0 \end{bmatrix} = \begin{bmatrix} 8 \\ -3 \\ -6 \end{bmatrix}$ So $x = 8$, $y = -3$ and $z = -6$.

14. $\begin{bmatrix} -1 & 8 & 4 \\ 1 & -3 & -1 \\ 1 & -6 & -3 \end{bmatrix} \begin{bmatrix} 0 \\ 0 \\ 1 \end{bmatrix} = \begin{bmatrix} 4 \\ -1 \\ -3 \end{bmatrix}$ So $x = 4$, $y = -1$ and $z = -3$.

8.4 Systems of Linear Equations: Matrix Inverses   613

16. (a) The adult male Sasquatch needs: 3 servings of Ippizuti Fish, 10 servings of Misty Mushrooms, and 15 servings of Sun Berries daily.

   (b) The adult female Sasquatch needs: 3 servings of Ippizuti Fish and 20 servings of Sun Berries daily. (No Misty Mushrooms are needed!)

   (c) The adolescent Sasquatch requires 10 servings of Ippizuti Fish daily. (No Misty Mushrooms or Sun Berries are needed!)

17. (a) $A^{-1} = \begin{bmatrix} 1 & 2 & 1 \\ 17 & 33 & 19 \\ 10 & 19 & 11 \end{bmatrix}$

   (b) $\begin{bmatrix} 1 & 2 & 1 \\ 17 & 33 & 19 \\ 10 & 19 & 11 \end{bmatrix} \begin{bmatrix} 12 & 42 & 100 & -23 & 38 \\ 1 & 3 & 36 & 39 & 57 \\ -12 & -42 & -152 & -46 & -133 \end{bmatrix} = \begin{bmatrix} 2 & 6 & 20 & 9 & 19 \\ 9 & 15 & 0 & 22 & 0 \\ 7 & 15 & 12 & 5 & 0 \end{bmatrix}$ ✓

   (c) 'LOGS RULE'

8.5 Determinants and Cramer's Rule

8.5.1 Definition and Properties of the Determinant

In this section we assign to each square matrix A a real number, called the **determinant** of A, which will eventually lead us to yet another technique for solving consistent independent systems of linear equations. The determinant is defined recursively, that is, we define it for 1×1 matrices and give a rule by which we can reduce determinants of $n \times n$ matrices to a sum of determinants of $(n-1) \times (n-1)$ matrices.[1] This means we will be able to evaluate the determinant of a 2×2 matrix as a sum of the determinants of 1×1 matrices; the determinant of a 3×3 matrix as a sum of the determinants of 2×2 matrices, and so forth. To explain how we will take an $n \times n$ matrix and distill from it an $(n-1) \times (n-1)$, we use the following notation.

> **Definition 8.12.** Given an $n \times n$ matrix A where $n > 1$, the matrix A_{ij} is the $(n-1) \times (n-1)$ matrix formed by deleting the ith row of A and the jth column of A.

For example, using the matrix A below, we find the matrix A_{23} by deleting the second row and third column of A.

$$A = \begin{bmatrix} 3 & 1 & 2 \\ 0 & -1 & 5 \\ 2 & 1 & 4 \end{bmatrix} \xrightarrow{\text{Delete } R2 \text{ and } C3} A_{23} = \begin{bmatrix} 3 & 1 \\ 2 & 1 \end{bmatrix}$$

We are now in the position to define the determinant of a matrix.

> **Definition 8.13.** Given an $n \times n$ matrix A the **determinant of A**, denoted $\det(A)$, is defined as follows
>
> - If $n = 1$, then $A = [a_{11}]$ and $\det(A) = \det([a_{11}]) = a_{11}$.
>
> - If $n > 1$, then $A = [a_{ij}]_{n \times n}$ and
>
> $$\det(A) = \det\left([a_{ij}]_{n \times n}\right) = a_{11}\det(A_{11}) - a_{12}\det(A_{12}) + - \ldots + (-1)^{1+n}a_{1n}\det(A_{1n})$$

There are two commonly used notations for the determinant of a matrix A: '$\det(A)$' and '$|A|$' We have chosen to use the notation $\det(A)$ as opposed to $|A|$ because we find that the latter is often confused with absolute value, especially in the context of a 1×1 matrix. In the expansion $a_{11}\det(A_{11}) - a_{12}\det(A_{12}) + - \ldots + (-1)^{1+n}a_{1n}\det(A_{1n})$, the notation '$+ - \ldots + (-1)^{1+n}a_{1n}$' means that the signs alternate and the final sign is dictated by the sign of the quantity $(-1)^{1+n}$. Since the entries a_{11}, a_{12} and so forth up through a_{1n} comprise the first row of A, we say we are finding the determinant of A by 'expanding along the first row'. Later in the section, we will develop a formula for $\det(A)$ which allows us to find it by expanding along any row.

Applying Definition 8.13 to the matrix $A = \begin{bmatrix} 4 & -3 \\ 2 & 1 \end{bmatrix}$ we get

---

[1] We will talk more about the term 'recursively' in Section 9.1.

8.5 Determinants and Cramer's Rule

$$\begin{aligned}\det(A) &= \det\left(\begin{bmatrix} 4 & -3 \\ 2 & 1 \end{bmatrix}\right) \\ &= 4\det(A_{11}) - (-3)\det(A_{12}) \\ &= 4\det([1]) + 3\det([2]) \\ &= 4(1) + 3(2) \\ &= 10 \end{aligned}$$

For a generic 2×2 matrix $A = \begin{bmatrix} a & b \\ c & d \end{bmatrix}$ we get

$$\begin{aligned}\det(A) &= \det\left(\begin{bmatrix} a & b \\ c & d \end{bmatrix}\right) \\ &= a\det(A_{11}) - b\det(A_{12}) \\ &= a\det([d]) - b\det([c]) \\ &= ad - bc \end{aligned}$$

This formula is worth remembering

Equation 8.1. For a 2×2 matrix,

$$\det\left(\begin{bmatrix} a & b \\ c & d \end{bmatrix}\right) = ad - bc$$

Applying Definition 8.13 to the 3×3 matrix $A = \begin{bmatrix} 3 & 1 & 2 \\ 0 & -1 & 5 \\ 2 & 1 & 4 \end{bmatrix}$ we obtain

$$\begin{aligned}\det(A) &= \det\left(\begin{bmatrix} 3 & 1 & 2 \\ 0 & -1 & 5 \\ 2 & 1 & 4 \end{bmatrix}\right) \\ &= 3\det(A_{11}) - 1\det(A_{12}) + 2\det(A_{13}) \\ &= 3\det\left(\begin{bmatrix} -1 & 5 \\ 1 & 4 \end{bmatrix}\right) - \det\left(\begin{bmatrix} 0 & 5 \\ 2 & 4 \end{bmatrix}\right) + 2\det\left(\begin{bmatrix} 0 & -1 \\ 2 & 1 \end{bmatrix}\right) \\ &= 3((-1)(4) - (5)(1)) - ((0)(4) - (5)(2)) + 2((0)(1) - (-1)(2)) \\ &= 3(-9) - (-10) + 2(2) \\ &= -13 \end{aligned}$$

To evaluate the determinant of a 4×4 matrix, we would have to evaluate the determinants of *four* 3×3 matrices, each of which involves the finding the determinants of *three* 2×2 matrices. As you can see, our method of evaluating determinants quickly gets out of hand and many of you may be reaching for the calculator. There is some mathematical machinery which can assist us in calculating determinants and we present that here. Before we state the theorem, we need some more terminology.

Definition 8.14. Let A be an $n \times n$ matrix and A_{ij} be defined as in Definition 8.12. The ij **minor** of A, denoted M_{ij} is defined by $M_{ij} = \det(A_{ij})$. The ij **cofactor** of A, denoted C_{ij} is defined by $C_{ij} = (-1)^{i+j} M_{ij} = (-1)^{i+j} \det(A_{ij})$.

We note that in Definition 8.13, the sum

$$a_{11} \det(A_{11}) - a_{12} \det(A_{12}) + - \ldots + (-1)^{1+n} a_{1n} \det(A_{1n})$$

can be rewritten as

$$a_{11}(-1)^{1+1} \det(A_{11}) + a_{12}(-1)^{1+2} \det(A_{12}) + \ldots + a_{1n}(-1)^{1+n} \det(A_{1n})$$

which, in the language of cofactors is

$$a_{11} C_{11} + a_{12} C_{12} + \ldots + a_{1n} C_{1n}$$

We are now ready to state our main theorem concerning determinants.

Theorem 8.7. Properties of the Determinant: Let $A = [a_{ij}]_{n \times n}$.

- We may find the determinant by expanding along any row. That is, for any $1 \leq k \leq n$,

$$\det(A) = a_{k1} C_{k1} + a_{k2} C_{k2} + \ldots + a_{kn} C_{kn}$$

- If A' is the matrix obtained from A by:

    - interchanging any two rows, then $\det(A') = -\det(A)$.
    - replacing a row with a nonzero multiple (say c) of itself, then $\det(A') = c \det(A)$
    - replacing a row with itself plus a multiple of another row, then $\det(A') = \det(A)$

- If A has two identical rows, or a row consisting of all 0's, then $\det(A) = 0$.

- If A is upper or lower triangular,[a] then $\det(A)$ is the product of the entries on the main diagonal.[b]

- If B is an $n \times n$ matrix, then $\det(AB) = \det(A) \det(B)$.

- $\det(A^n) = \det(A)^n$ for all natural numbers n.

- A is invertible if and only if $\det(A) \neq 0$. In this case, $\det(A^{-1}) = \dfrac{1}{\det(A)}$.

---
[a]See Exercise 8.3.1 in 8.3.
[b]See page 585 in Section 8.3.

Unfortunately, while we can easily *demonstrate* the results in Theorem 8.7, the proofs of most of these properties are beyond the scope of this text. We could prove these properties for generic 2×2

8.5 Determinants and Cramer's Rule

or even 3×3 matrices by brute force computation, but this manner of proof belies the elegance and symmetry of the determinant. We will prove what few properties we can after we have developed some more tools such as the Principle of Mathematical Induction in Section 9.3.[2] For the moment, let us demonstrate some of the properties listed in Theorem 8.7 on the matrix A below. (Others will be discussed in the Exercises.)

$$A = \begin{bmatrix} 3 & 1 & 2 \\ 0 & -1 & 5 \\ 2 & 1 & 4 \end{bmatrix}$$

We found $\det(A) = -13$ by expanding along the first row. To take advantage of the 0 in the second row, we use Theorem 8.7 to find $\det(A) = -13$ by expanding along that row.

$$\begin{aligned}
\det\left(\begin{bmatrix} 3 & 1 & 2 \\ 0 & -1 & 5 \\ 2 & 1 & 4 \end{bmatrix}\right) &= 0C_{21} + (-1)C_{22} + 5C_{23} \\
&= (-1)(-1)^{2+2}\det(A_{22}) + 5(-1)^{2+3}\det(A_{23}) \\
&= -\det\left(\begin{bmatrix} 3 & 2 \\ 2 & 4 \end{bmatrix}\right) - 5\det\left(\begin{bmatrix} 3 & 1 \\ 2 & 1 \end{bmatrix}\right) \\
&= -((3)(4) - (2)(2)) - 5((3)(1) - (2)(1)) \\
&= -8 - 5 \\
&= -13 \checkmark
\end{aligned}$$

In general, the sign of $(-1)^{i+j}$ in front of the minor in the expansion of the determinant follows an alternating pattern. Below is the pattern for 2×2, 3×3 and 4×4 matrices, and it extends naturally to higher dimensions.

$$\begin{bmatrix} + & - \\ - & + \end{bmatrix} \quad \begin{bmatrix} + & - & + \\ - & + & - \\ + & - & + \end{bmatrix} \quad \begin{bmatrix} + & - & + & - \\ - & + & - & + \\ + & - & + & - \\ - & + & - & + \end{bmatrix}$$

The reader is cautioned, however, against reading too much into these sign patterns. In the example above, we expanded the 3×3 matrix A by its second row and the term which corresponds to the second entry ended up being negative even though the sign attached to the minor is $(+)$. These signs represent only the signs of the $(-1)^{i+j}$ in the formula; the sign of the corresponding entry as well as the minor itself determine the ultimate sign of the term in the expansion of the determinant.

To illustrate some of the other properties in Theorem 8.7, we use row operations to transform our 3×3 matrix A into an upper triangular matrix, keeping track of the row operations, and labeling

---

[2] For a very elegant treatment, take a course in Linear Algebra. There, you will most likely see the treatment of determinants logically reversed than what is presented here. Specifically, the determinant is defined as a function which takes a square matrix to a real number and satisfies some of the properties in Theorem 8.7. From that function, a formula for the determinant is developed.

each successive matrix.[3]

$$\begin{bmatrix} 3 & 1 & 2 \\ 0 & -1 & 5 \\ 2 & 1 & 4 \end{bmatrix} \xrightarrow[\text{with } -\frac{2}{3}R1 + R3]{\text{Replace } R3} \begin{bmatrix} 3 & 1 & 2 \\ 0 & -1 & 5 \\ 0 & \frac{1}{3} & \frac{8}{3} \end{bmatrix} \xrightarrow[\frac{1}{3}R2 + R3]{\text{Replace } R3 \text{ with}} \begin{bmatrix} 3 & 1 & 2 \\ 0 & -1 & 5 \\ 0 & 0 & \frac{13}{3} \end{bmatrix}$$
$$\quad A \qquad\qquad\qquad\qquad\qquad\qquad B \qquad\qquad\qquad\qquad\qquad\qquad C$$

Theorem 8.7 guarantees us that $\det(A) = \det(B) = \det(C)$ since we are replacing a row with itself plus a multiple of another row moving from one matrix to the next. Furthermore, since C is upper triangular, $\det(C)$ is the product of the entries on the main diagonal, in this case $\det(C) = (3)(-1)\left(\frac{13}{3}\right) = -13$. This demonstrates the utility of using row operations to assist in calculating determinants. This also sheds some light on the connection between a determinant and invertibility. Recall from Section 8.4 that in order to find A^{-1}, we attempt to transform A to I_n using row operations

$$[\, A \mid I_n \,] \xrightarrow{\text{Gauss Jordan Elimination}} [\, I_n \mid A^{-1} \,]$$

As we apply our allowable row operations on A to put it into reduced row echelon form, the determinant of the intermediate matrices can vary from the determinant of A by at most a *nonzero* multiple. This means that if $\det(A) \neq 0$, then the determinant of A's reduced row echelon form must also be nonzero, which, according to Definition 8.4 means that all the main diagonal entries on A's reduced row echelon form must be 1. That is, A's reduced row echelon form is I_n, and A is invertible. Conversely, if A is invertible, then A can be transformed into I_n using row operations. Since $\det(I_n) = 1 \neq 0$, our same logic implies $\det(A) \neq 0$. Basically, we have established that the determinant *determines* whether or not the matrix A is invertible.[4]

It is worth noting that when we first introduced the notion of a matrix inverse, it was in the context of solving a linear matrix equation. In effect, we were trying to 'divide' both sides of the matrix equation $AX = B$ by the matrix A. Just like we cannot divide a real number by 0, Theorem 8.7 tells us we cannot 'divide' by a matrix whose *determinant* is 0. We also know that if the coefficient matrix of a system of linear equations is invertible, then system is consistent and independent. It follows, then, that if the determinant of said coefficient is not zero, the system is consistent and independent.

8.5.2 Cramer's Rule and Matrix Adjoints

In this section, we introduce a theorem which enables us to solve a system of linear equations by means of determinants only. As usual, the theorem is stated in full generality, using numbered unknowns x_1, x_2, etc., instead of the more familiar letters x, y, z, etc. The proof of the general case is best left to a course in Linear Algebra.

---

[3]Essentially, we follow the Gauss Jordan algorithm but we don't care about getting leading 1's.

[4]In Section 8.5.2, we learn determinants (specifically cofactors) are deeply connected with the inverse of a matrix.

8.5 Determinants and Cramer's Rule

Theorem 8.8. Cramer's Rule: Suppose $AX = B$ is the matrix form of a system of n linear equations in n unknowns where A is the coefficient matrix, X is the unknowns matrix, and B is the constant matrix. If $\det(A) \neq 0$, then the corresponding system is consistent and independent and the solution for unknowns $x_1, x_2, \ldots x_n$ is given by:

$$x_j = \frac{\det(A_j)}{\det(A)},$$

where A_j is the matrix A whose jth column has been replaced by the constants in B.

In words, Cramer's Rule tells us we can solve for each unknown, one at a time, by finding the ratio of the determinant of A_j to that of the determinant of the coefficient matrix. The matrix A_j is found by replacing the column in the coefficient matrix which holds the coefficients of x_j with the constants of the system. The following example fleshes out this method.

Example 8.5.1. Use Cramer's Rule to solve for the indicated unknowns.

1. Solve $\begin{cases} 2x_1 - 3x_2 = 4 \\ 5x_1 + x_2 = -2 \end{cases}$ for x_1 and x_2

2. Solve $\begin{cases} 2x - 3y + z = -1 \\ x - y + z = 1 \\ 3x - 4z = 0 \end{cases}$ for z.

Solution.

1. Writing this system in matrix form, we find

$$A = \begin{bmatrix} 2 & -3 \\ 5 & 1 \end{bmatrix} \qquad X = \begin{bmatrix} x_1 \\ x_2 \end{bmatrix} \qquad B = \begin{bmatrix} 4 \\ -2 \end{bmatrix}$$

To find the matrix A_1, we remove the column of the coefficient matrix A which holds the coefficients of x_1 and replace it with the corresponding entries in B. Likewise, we replace the column of A which corresponds to the coefficients of x_2 with the constants to form the matrix A_2. This yields

$$A_1 = \begin{bmatrix} 4 & -3 \\ -2 & 1 \end{bmatrix} \qquad A_2 = \begin{bmatrix} 2 & 4 \\ 5 & -2 \end{bmatrix}$$

Computing determinants, we get $\det(A) = 17$, $\det(A_1) = -2$ and $\det(A_2) = -24$, so that

$$x_1 = \frac{\det(A_1)}{\det(A)} = -\frac{2}{17} \qquad x_2 = \frac{\det(A_2)}{\det(A)} = -\frac{24}{17}$$

The reader can check that the solution to the system is $\left(-\frac{2}{17}, -\frac{24}{17}\right)$.

2. To use Cramer's Rule to find z, we identify x_3 as z. We have

$$A = \begin{bmatrix} 2 & -3 & 1 \\ 1 & -1 & 1 \\ 3 & 0 & -4 \end{bmatrix} \quad X = \begin{bmatrix} x \\ y \\ z \end{bmatrix} \quad B = \begin{bmatrix} -1 \\ 1 \\ 0 \end{bmatrix} \quad A_3 = A_z = \begin{bmatrix} 2 & -3 & -1 \\ 1 & -1 & 1 \\ 3 & 0 & 0 \end{bmatrix}$$

Expanding both det(A) and det (A_z) along the third rows (to take advantage of the 0's) gives

$$z = \frac{\det(A_z)}{\det(A)} = \frac{-12}{-10} = \frac{6}{5}$$

The reader is encouraged to solve this system for x and y similarly and check the answer. □

Our last application of determinants is to develop an alternative method for finding the inverse of a matrix.[5] Let us consider the 3×3 matrix A which we so extensively studied in Section 8.5.1

$$A = \begin{bmatrix} 3 & 1 & 2 \\ 0 & -1 & 5 \\ 2 & 1 & 4 \end{bmatrix}$$

We found through a variety of methods that det$(A) = -13$. To our surprise and delight, its inverse below has a remarkable number of 13's in the denominators of its entries. This is no coincidence.

$$A^{-1} = \begin{bmatrix} \frac{9}{13} & \frac{2}{13} & -\frac{7}{13} \\ -\frac{10}{13} & -\frac{8}{13} & \frac{15}{13} \\ -\frac{2}{13} & \frac{1}{13} & \frac{3}{13} \end{bmatrix}$$

Recall that to find A^{-1}, we are essentially solving the matrix equation $AX = I_3$, where $X = [x_{ij}]_{3 \times 3}$ is a 3×3 matrix. Because of how matrix multiplication is defined, the first column of I_3 is the product of A with the first column of X, the second column of I_3 is the product of A with the second column of X and the third column of I_3 is the product of A with the third column of X. In other words, we are solving three equations[6]

$$A \begin{bmatrix} x_{11} \\ x_{21} \\ x_{31} \end{bmatrix} = \begin{bmatrix} 1 \\ 0 \\ 0 \end{bmatrix} \quad A \begin{bmatrix} x_{12} \\ x_{22} \\ x_{32} \end{bmatrix} = \begin{bmatrix} 0 \\ 1 \\ 0 \end{bmatrix} \quad A \begin{bmatrix} x_{13} \\ x_{23} \\ x_{33} \end{bmatrix} = \begin{bmatrix} 0 \\ 0 \\ 1 \end{bmatrix}$$

We can solve each of these systems using Cramer's Rule. Focusing on the first system, we have

$$A_1 = \begin{bmatrix} 1 & 1 & 2 \\ 0 & -1 & 5 \\ 0 & 1 & 4 \end{bmatrix} \quad A_2 = \begin{bmatrix} 3 & 1 & 2 \\ 0 & 0 & 5 \\ 2 & 0 & 4 \end{bmatrix} \quad A_3 = \begin{bmatrix} 3 & 1 & 1 \\ 0 & -1 & 0 \\ 2 & 1 & 0 \end{bmatrix}$$

---

[5] We are developing a *method* in the forthcoming discussion. As with the discussion in Section 8.4 when we developed the first algorithm to find matrix inverses, we ask that you indulge us.

[6] The reader is encouraged to stop and think this through.

8.5 Determinants and Cramer's Rule

If we expand $\det(A_1)$ along the first row, we get

$$\det(A_1) = \det\left(\begin{bmatrix} -1 & 5 \\ 1 & 4 \end{bmatrix}\right) - \det\left(\begin{bmatrix} 0 & 5 \\ 0 & 4 \end{bmatrix}\right) + 2\det\left(\begin{bmatrix} 0 & -1 \\ 0 & 1 \end{bmatrix}\right)$$

$$= \det\left(\begin{bmatrix} -1 & 5 \\ 1 & 4 \end{bmatrix}\right)$$

Amazingly, this is none other than the C_{11} cofactor of A. The reader is invited to check this, as well as the claims that $\det(A_2) = C_{12}$ and $\det(A_3) = C_{13}$.[7] (To see this, though it seems unnatural to do so, expand along the first row.) Cramer's Rule tells us

$$x_{11} = \frac{\det(A_1)}{\det(A)} = \frac{C_{11}}{\det(A)}, \quad x_{21} = \frac{\det(A_2)}{\det(A)} = \frac{C_{12}}{\det(A)}, \quad x_{31} = \frac{\det(A_3)}{\det(A)} = \frac{C_{13}}{\det(A)}$$

So the first column of the inverse matrix X is:

$$\begin{bmatrix} x_{11} \\ x_{21} \\ x_{31} \end{bmatrix} = \begin{bmatrix} \dfrac{C_{11}}{\det(A)} \\ \dfrac{C_{12}}{\det(A)} \\ \dfrac{C_{13}}{\det(A)} \end{bmatrix} = \frac{1}{\det(A)} \begin{bmatrix} C_{11} \\ C_{12} \\ C_{13} \end{bmatrix}$$

Notice the reversal of the subscripts going from the unknown to the corresponding cofactor of A. This trend continues and we get

$$\begin{bmatrix} x_{12} \\ x_{22} \\ x_{32} \end{bmatrix} = \frac{1}{\det(A)} \begin{bmatrix} C_{21} \\ C_{22} \\ C_{23} \end{bmatrix} \qquad \begin{bmatrix} x_{13} \\ x_{23} \\ x_{33} \end{bmatrix} = \frac{1}{\det(A)} \begin{bmatrix} C_{31} \\ C_{32} \\ C_{33} \end{bmatrix}$$

Putting all of these together, we have obtained a new and surprising formula for A^{-1}, namely

$$A^{-1} = \frac{1}{\det(A)} \begin{bmatrix} C_{11} & C_{21} & C_{31} \\ C_{12} & C_{22} & C_{32} \\ C_{13} & C_{23} & C_{33} \end{bmatrix}$$

To see that this does indeed yield A^{-1}, we find all of the cofactors of A

$$\begin{array}{lll} C_{11} = -9, & C_{21} = -2, & C_{31} = 7 \\ C_{12} = 10, & C_{22} = 8, & C_{32} = -15 \\ C_{13} = 2, & C_{23} = -1, & C_{33} = -3 \end{array}$$

And, as promised,

---

[7] In a solid Linear Algebra course you will learn that the properties in Theorem 8.7 hold equally well if the word 'row' is replaced by the word 'column'. We're not going to get into column operations in this text, but they do make some of what we're trying to say easier to follow.

$$A^{-1} = \frac{1}{\det(A)} \begin{bmatrix} C_{11} & C_{21} & C_{31} \\ C_{12} & C_{22} & C_{32} \\ C_{13} & C_{23} & C_{33} \end{bmatrix} = -\frac{1}{13} \begin{bmatrix} -9 & -2 & 7 \\ 10 & 8 & -15 \\ 2 & -1 & -3 \end{bmatrix} = \begin{bmatrix} \frac{9}{13} & \frac{2}{13} & -\frac{7}{13} \\ -\frac{10}{13} & -\frac{8}{13} & \frac{15}{13} \\ -\frac{2}{13} & \frac{1}{13} & \frac{3}{13} \end{bmatrix}$$

To generalize this to invertible $n \times n$ matrices, we need another definition and a theorem. Our definition gives a special name to the cofactor matrix, and the theorem tells us how to use it along with $\det(A)$ to find the inverse of a matrix.

Definition 8.15. Let A be an $n \times n$ matrix, and C_{ij} denote the ij cofactor of A. The **adjoint** of A, denoted $\text{adj}(A)$ is the matrix whose ij-entry is the ji cofactor of A, C_{ji}. That is

$$\text{adj}(A) = \begin{bmatrix} C_{11} & C_{21} & \cdots & C_{n1} \\ C_{12} & C_{22} & \cdots & C_{n2} \\ \vdots & \vdots & & \vdots \\ C_{1n} & C_{2n} & \cdots & C_{nn} \end{bmatrix}$$

This new notation greatly shortens the statement of the formula for the inverse of a matrix.

Theorem 8.9. Let A be an invertible $n \times n$ matrix. Then

$$A^{-1} = \frac{1}{\det(A)} \text{adj}(A)$$

For 2×2 matrices, Theorem 8.9 reduces to a fairly simple formula.

Equation 8.2. For an invertible 2×2 matrix,

$$\begin{bmatrix} a & b \\ c & d \end{bmatrix}^{-1} = \frac{1}{ad - bc} \begin{bmatrix} d & -b \\ -c & a \end{bmatrix}$$

The proof of Theorem 8.9 is, like so many of the results in this section, best left to a course in Linear Algebra. In such a course, not only do you gain some more sophisticated proof techniques, you also gain a larger perspective. The authors assure you that persistence pays off. If you stick around a few semesters and take a course in Linear Algebra, you'll see just how pretty all things matrix really are - in spite of the tedious notation and sea of subscripts. Within the scope of this text, we will prove a few results involving determinants in Section 9.3 once we have the Principle of Mathematical Induction well in hand. Until then, make sure you have a handle on the *mechanics* of matrices and the theory will come eventually.

8.5.3 Exercises

In Exercises 1 - 8, compute the determinant of the given matrix. (Some of these matrices appeared in Exercises 1 - 8 in Section 8.4.)

1. $B = \begin{bmatrix} 12 & -7 \\ -5 & 3 \end{bmatrix}$

2. $C = \begin{bmatrix} 6 & 15 \\ 14 & 35 \end{bmatrix}$

3. $Q = \begin{bmatrix} x & x^2 \\ 1 & 2x \end{bmatrix}$

4. $L = \begin{bmatrix} \dfrac{1}{x^3} & \dfrac{\ln(x)}{x^3} \\ -\dfrac{3}{x^4} & \dfrac{1-3\ln(x)}{x^4} \end{bmatrix}$

5. $F = \begin{bmatrix} 4 & 6 & -3 \\ 3 & 4 & -3 \\ 1 & 2 & 6 \end{bmatrix}$

6. $G = \begin{bmatrix} 1 & 2 & 3 \\ 2 & 3 & 11 \\ 3 & 4 & 19 \end{bmatrix}$

7. $V = \begin{bmatrix} i & j & k \\ -1 & 0 & 5 \\ 9 & -4 & -2 \end{bmatrix}$

8. $H = \begin{bmatrix} 1 & 0 & -3 & 0 \\ 2 & -2 & 8 & 7 \\ -5 & 0 & 16 & 0 \\ 1 & 0 & 4 & 1 \end{bmatrix}$

In Exercises 9 - 14, use Cramer's Rule to solve the system of linear equations.

9. $\begin{cases} 3x + 7y = 26 \\ 5x + 12y = 39 \end{cases}$

10. $\begin{cases} 2x - 4y = 5 \\ 10x + 13y = -6 \end{cases}$

11. $\begin{cases} x + y = 8000 \\ 0.03x + 0.05y = 250 \end{cases}$

12. $\begin{cases} \frac{1}{2}x - \frac{1}{5}y = 1 \\ 6x + 7y = 3 \end{cases}$

13. $\begin{cases} x + y + z = 3 \\ 2x - y + z = 0 \\ -3x + 5y + 7z = 7 \end{cases}$

14. $\begin{cases} 3x + y - 2z = 10 \\ 4x - y + z = 5 \\ x - 3y - 4z = -1 \end{cases}$

In Exercises 15 - 16, use Cramer's Rule to solve for x_4.

15. $\begin{cases} x_1 - x_3 = -2 \\ 2x_2 - x_4 = 0 \\ x_1 - 2x_2 + x_3 = 0 \\ -x_3 + x_4 = 1 \end{cases}$

16. $\begin{cases} 4x_1 + x_2 = 4 \\ x_2 - 3x_3 = 1 \\ 10x_1 + x_3 + x_4 = 0 \\ -x_2 + x_3 = -3 \end{cases}$

In Exercises 17 - 18, find the inverse of the given matrix using their determinants and adjoints.

17. $B = \begin{bmatrix} 12 & -7 \\ -5 & 3 \end{bmatrix}$

18. $F = \begin{bmatrix} 4 & 6 & -3 \\ 3 & 4 & -3 \\ 1 & 2 & 6 \end{bmatrix}$

19. Carl's Sasquatch Attack! Game Card Collection is a mixture of common and rare cards. Each common card is worth $0.25 while each rare card is worth $0.75. If his entire 117 card collection is worth $48.75, how many of each kind of card does he own?

20. How much of a 5 gallon 40% salt solution should be replaced with pure water to obtain 5 gallons of a 15% solution?

21. How much of a 10 liter 30% acid solution must be replaced with pure acid to obtain 10 liters of a 50% solution?

22. Daniel's Exotic Animal Rescue houses snakes, tarantulas and scorpions. When asked how many animals of each kind he boards, Daniel answered: 'We board 49 total animals, and I am responsible for each of their 272 legs and 28 tails.' How many of each animal does the Rescue board? (Recall: tarantulas have 8 legs and no tails, scorpions have 8 legs and one tail, and snakes have no legs and one tail.)

23. This exercise is a continuation of Exercise 16 in Section 8.4. Just because a system is consistent independent doesn't mean it will admit a solution that makes sense in an applied setting. Using the nutrient values given for Ippizuti Fish, Misty Mushrooms, and Sun Berries, use Cramer's Rule to determine the number of servings of Ippizuti Fish needed to meet the needs of a daily diet which requires 2500 calories, 1000 grams of protein, and 400 milligrams of Vitamin X. Now use Cramer's Rule to find the number of servings of Misty Mushrooms required. Does a solution to this diet problem exist?

24. Let $R = \begin{bmatrix} -7 & 3 \\ 11 & 2 \end{bmatrix}$, $S = \begin{bmatrix} 1 & -5 \\ 6 & 9 \end{bmatrix}$, $T = \begin{bmatrix} 11 & 2 \\ -7 & 3 \end{bmatrix}$, and $U = \begin{bmatrix} -3 & 15 \\ 6 & 9 \end{bmatrix}$

   (a) Show that $\det(RS) = \det(R)\det(S)$

   (b) Show that $\det(T) = -\det(R)$

   (c) Show that $\det(U) = -3\det(S)$

25. For M, N, and P below, show that $\det(M) = 0$, $\det(N) = 0$ and $\det(P) = 0$.

$$M = \begin{bmatrix} 1 & 2 & 3 \\ 0 & 0 & 0 \\ 7 & 8 & 9 \end{bmatrix}, \quad N = \begin{bmatrix} 1 & 2 & 3 \\ 1 & 2 & 3 \\ 4 & 5 & 6 \end{bmatrix}, \quad P = \begin{bmatrix} 1 & 2 & 3 \\ -2 & -4 & -6 \\ 7 & 8 & 9 \end{bmatrix}$$

26. Let A be an arbitrary invertible 3×3 matrix.

   (a) Show that $\det(I_3) = 1$. (See footnote[8] below.)
   (b) Using the facts that $AA^{-1} = I_3$ and $\det(AA^{-1}) = \det(A)\det(A^{-1})$, show that

   $$\det(A^{-1}) = \frac{1}{\det(A)}$$

The purpose of Exercises 27 - 30 is to introduce you to the eigenvalues and eigenvectors of a matrix.[9] We begin with an example using a 2×2 matrix and then guide you through some exercises using a 3×3 matrix. Consider the matrix

$$C = \begin{bmatrix} 6 & 15 \\ 14 & 35 \end{bmatrix}$$

from Exercise 2. We know that $\det(C) = 0$ which means that $CX = 0_{2 \times 2}$ does not have a unique solution. So there is a nonzero matrix Y with $CY = 0_{2 \times 2}$. In fact, every matrix of the form

$$Y = \begin{bmatrix} -\frac{5}{2}t \\ t \end{bmatrix}$$

is a solution to $CX = 0_{2 \times 2}$, so there are infinitely many matrices such that $CX = 0_{2 \times 2}$. But consider the matrix

$$X_{41} = \begin{bmatrix} 3 \\ 7 \end{bmatrix}$$

It is NOT a solution to $CX = 0_{2 \times 2}$, but rather,

$$CX_{41} = \begin{bmatrix} 6 & 15 \\ 14 & 35 \end{bmatrix} \begin{bmatrix} 3 \\ 7 \end{bmatrix} = \begin{bmatrix} 123 \\ 287 \end{bmatrix} = 41 \begin{bmatrix} 3 \\ 7 \end{bmatrix}$$

In fact, if Z is of the form

$$Z = \begin{bmatrix} \frac{3}{7}t \\ t \end{bmatrix}$$

then

$$CZ = \begin{bmatrix} 6 & 15 \\ 14 & 35 \end{bmatrix} \begin{bmatrix} \frac{3}{7}t \\ t \end{bmatrix} = \begin{bmatrix} \frac{123}{7}t \\ 41t \end{bmatrix} = 41 \begin{bmatrix} \frac{3}{7}t \\ t \end{bmatrix} = 41Z$$

for all t. The big question is "How did we know to use 41?"

We need a number λ such that $CX = \lambda X$ has nonzero solutions. We have demonstrated that $\lambda = 0$ and $\lambda = 41$ both worked. Are there others? If we look at the matrix equation more closely, what

---

[8] If you think about it for just a moment, you'll see that $\det(I_n) = 1$ for any natural number n. The formal proof of this fact requires the Principle of Mathematical Induction (Section 9.3) so we'll stick with $n = 3$ for the time being.

[9] This material is usually given its own chapter in a Linear Algebra book so clearly we're not able to tell you everything you need to know about eigenvalues and eigenvectors. They are a nice application of determinants, though, so we're going to give you enough background so that you can start playing around with them.

we *really* wanted was a nonzero solution to $(C - \lambda I_2)X = 0_{2\times 2}$ which we know exists if and only if the determinant of $C - \lambda I_2$ is zero.[10] So we computed

$$\det(C - \lambda I_2) = \det\left(\begin{bmatrix} 6 - \lambda & 15 \\ 14 & 35 - \lambda \end{bmatrix}\right) = (6-\lambda)(35-\lambda) - 14 \cdot 15 = \lambda^2 - 41\lambda$$

This is called the **characteristic polynomial** of the matrix C and it has two zeros: $\lambda = 0$ and $\lambda = 41$. That's how we knew to use 41 in our work above. The fact that $\lambda = 0$ showed up as one of the zeros of the characteristic polynomial just means that C itself had determinant zero which we already knew. Those two numbers are called the **eigenvalues** of C. The corresponding matrix solutions to $CX = \lambda X$ are called the **eigenvectors** of C and the 'vector' portion of the name will make more sense after you've studied vectors.

Now it's your turn. In the following exercises, you'll be using the matrix G from Exercise 6.

$$G = \begin{bmatrix} 1 & 2 & 3 \\ 2 & 3 & 11 \\ 3 & 4 & 19 \end{bmatrix}$$

27. Show that the characteristic polynomial of G is $p(\lambda) = -\lambda(\lambda - 1)(\lambda - 22)$. That is, compute $\det(G - \lambda I_3)$.

28. Let $G_0 = G$. Find the parametric description of the solution to the system of linear equations given by $GX = 0_{3\times 3}$.

29. Let $G_1 = G - I_3$. Find the parametric description of the solution to the system of linear equations given by $G_1 X = 0_{3\times 3}$. Show that any solution to $G_1 X = 0_{3\times 3}$ also has the property that $GX = 1X$.

30. Let $G_{22} = G - 22 I_3$. Find the parametric description of the solution to the system of linear equations given by $G_{22} X = 0_{3\times 3}$. Show that any solution to $G_{22} X = 0_{3\times 3}$ also has the property that $GX = 22X$.

---

[10] Think about this.

8.5.4 Answers

1. $\det(B) = 1$
2. $\det(C) = 0$
3. $\det(Q) = x^2$
4. $\det(L) = \dfrac{1}{x^7}$
5. $\det(F) = -12$
6. $\det(G) = 0$
7. $\det(V) = 20i + 43j + 4k$
8. $\det(H) = -2$
9. $x = 39$, $y = -13$
10. $x = \frac{41}{66}$, $y = -\frac{31}{33}$
11. $x = 7500$, $y = 500$
12. $x = \frac{76}{47}$, $y = -\frac{45}{47}$
13. $x = 1$, $y = 2$, $z = 0$
14. $x = \frac{121}{60}$, $y = \frac{131}{60}$, $z = -\frac{53}{60}$
15. $x_4 = 4$
16. $x_4 = -1$

17. $B^{-1} = \begin{bmatrix} 3 & 7 \\ 5 & 12 \end{bmatrix}$

18. $F^{-1} = \begin{bmatrix} -\frac{5}{2} & \frac{7}{2} & \frac{1}{2} \\ \frac{7}{4} & -\frac{9}{4} & -\frac{1}{4} \\ -\frac{1}{6} & \frac{1}{6} & \frac{1}{6} \end{bmatrix}$

19. Carl owns 78 common cards and 39 rare cards.

20. 3.125 gallons.

21. $\frac{20}{7} \approx 2.85$ liters.

22. The rescue houses 15 snakes, 21 tarantulas and 13 scorpions.

23. Using Cramer's Rule, we find we need 53 servings of Ippizuti Fish to satisfy the dietary requirements. The number of servings of Misty Mushrooms required, however, is -1120. Since it's impossible to have a negative number of servings, there is no solution to the applied problem, despite there being a solution to the mathematical problem. A cautionary tale about using Cramer's Rule: just because you are guaranteed a mathematical answer for each variable doesn't mean the solution will make sense in the 'real' world.

8.6 Partial Fraction Decomposition

This section uses systems of linear equations to rewrite rational functions in a form more palatable to Calculus students. In College Algebra, the function

$$f(x) = \frac{x^2 - x - 6}{x^4 + x^2} \tag{1}$$

is written in the best form possible to construct a sign diagram and to find zeros and asymptotes, but certain applications in Calculus require us to rewrite $f(x)$ as

$$f(x) = \frac{x+7}{x^2+1} - \frac{1}{x} - \frac{6}{x^2} \tag{2}$$

If we are given the form of $f(x)$ in (2), it is a matter of Intermediate Algebra to determine a common denominator to obtain the form of $f(x)$ given in (1). The focus of this section is to develop a method by which we start with $f(x)$ in the form of (1) and 'resolve it into **partial fractions**' to obtain the form in (2). Essentially, we need to reverse the least common denominator process. Starting with the form of $f(x)$ in (1), we begin by factoring the denominator

$$\frac{x^2 - x - 6}{x^4 + x^2} = \frac{x^2 - x - 6}{x^2(x^2+1)}$$

We now think about which individual denominators could contribute to obtain $x^2(x^2+1)$ as the least common denominator. Certainly x^2 and x^2+1, but are there any other factors? Since x^2+1 is an irreducible quadratic[1] there are no factors of it that have real coefficients which can contribute to the denominator. The factor x^2, however, is not irreducible, since we can think of it as $x^2 = xx = (x-0)(x-0)$, a so-called 'repeated' linear factor.[2] This means it's possible that a term with a denominator of just x contributed to the expression as well. What about something like $x(x^2+1)$? This, too, could contribute, but we would then wish to break down that denominator into x and (x^2+1), so we leave out a term of that form. At this stage, we have guessed

$$\frac{x^2-x-6}{x^4+x^2} = \frac{x^2-x-6}{x^2(x^2+1)} = \frac{?}{x} + \frac{?}{x^2} + \frac{?}{x^2+1}$$

Our next task is to determine what form the unknown numerators take. It stands to reason that since the expression $\frac{x^2-x-6}{x^4+x^2}$ is 'proper' in the sense that the degree of the numerator is less than the degree of the denominator, we are safe to make the <u>ansatz</u> that all of the partial fraction resolvents are also. This means that the numerator of the fraction with x as its denominator is just a constant and the numerators on the terms involving the denominators x^2 and x^2+1 are at most linear polynomials. That is, we guess that there are real numbers A, B, C, D and E so that

$$\frac{x^2-x-6}{x^4+x^2} = \frac{x^2-x-6}{x^2(x^2+1)} = \frac{A}{x} + \frac{Bx+C}{x^2} + \frac{Dx+E}{x^2+1}$$

---
[1] Recall this means it has no real zeros; see Section 3.4.
[2] Recall this means $x = 0$ is a zero of multiplicity 2.

8.6 Partial Fraction Decomposition

However, if we look more closely at the term $\frac{Bx+C}{x^2}$, we see that $\frac{Bx+C}{x^2} = \frac{Bx}{x^2} + \frac{C}{x^2} = \frac{B}{x} + \frac{C}{x^2}$. The term $\frac{B}{x}$ has the same form as the term $\frac{A}{x}$ which means it contributes nothing new to our expansion. Hence, we drop it and, after re-labeling, we find ourselves with our new guess:

$$\frac{x^2 - x - 6}{x^4 + x^2} = \frac{x^2 - x - 6}{x^2(x^2 + 1)} = \frac{A}{x} + \frac{B}{x^2} + \frac{Cx + D}{x^2 + 1}$$

Our next task is to determine the values of our unknowns. Clearing denominators gives

$$x^2 - x - 6 = Ax(x^2 + 1) + B(x^2 + 1) + (Cx + D)x^2$$

Gathering the like powers of x we have

$$x^2 - x - 6 = (A + C)x^3 + (B + D)x^2 + Ax + B$$

In order for this to hold for all values of x in the domain of f, we equate the coefficients of corresponding powers of x on each side of the equation[3] and obtain the system of linear equations

$$\begin{cases} (E1) & A + C = 0 & \text{From equating coefficients of } x^3 \\ (E2) & B + D = 1 & \text{From equating coefficients of } x^2 \\ (E3) & A = -1 & \text{From equating coefficients of } x \\ (E4) & B = -6 & \text{From equating the constant terms} \end{cases}$$

To solve this system of equations, we could use any of the methods presented in Sections 8.1 through 8.5, but none of these methods are as efficient as the good old-fashioned substitution you learned in Intermediate Algebra. From $E3$, we have $A = -1$ and we substitute this into $E1$ to get $C = 1$. Similarly, since $E4$ gives us $B = -6$, we have from $E2$ that $D = 7$. We get

$$\frac{x^2 - x - 6}{x^4 + x^2} = \frac{x^2 - x - 6}{x^2(x^2 + 1)} = -\frac{1}{x} - \frac{6}{x^2} + \frac{x + 7}{x^2 + 1}$$

which matches the formula given in (2). As we have seen in this opening example, resolving a rational function into partial fractions takes two steps: first, we need to determine the *form* of the decomposition, and then we need to determine the unknown coefficients which appear in said form. Theorem 3.16 guarantees that any polynomial with real coefficients can be factored over the real numbers as a product of linear factors and irreducible quadratic factors. Once we have this factorization of the denominator of a rational function, the next theorem tells us the form the decomposition takes. The reader is encouraged to review the Factor Theorem (Theorem 3.6) and its connection to the role of multiplicity to fully appreciate the statement of the following theorem.

---

[3] We will justify this shortly.

> **Theorem 8.10.** Suppose $R(x) = \dfrac{N(x)}{D(x)}$ is a rational function where the degree of $N(x)$ less than the degree of $D(x)$ and $N(x)$ and $D(x)$ have no common factors.[a]
>
> - If α is a real zero of D of multiplicity m which corresponds to the linear factor $ax + b$, the partial fraction decomposition includes
>
> $$\frac{A_1}{ax+b} + \frac{A_2}{(ax+b)^2} + \cdots + \frac{A_m}{(ax+b)^m}$$
>
> for real numbers $A_1, A_2, \ldots A_m$.
>
> - If α is a non-real zero of D of multiplicity m which corresponds to the irreducible quadratic $ax^2 + bx + c$, the partial fraction decomposition includes
>
> $$\frac{B_1 x + C_1}{ax^2 + bx + c} + \frac{B_2 x + C_2}{(ax^2 + bx + c)^2} + \cdots + \frac{B_m x + C_m}{(ax^2 + bx + c)^m}$$
>
> for real numbers $B_1, B_2, \ldots B_m$ and $C_1, C_2, \ldots C_m$.
>
> [a] In other words, $R(x)$ is a proper rational function which has been fully reduced.

The proof of Theorem 8.10 is best left to a course in Abstract Algebra. Notice that the theorem provides for the general case, so we need to use subscripts, A_1, A_2, etc., to denote different unknown coefficients as opposed to the usual convention of A, B, etc.. The stress on multiplicities is to help us correctly group factors in the denominator. For example, consider the rational function

$$\frac{3x-1}{(x^2-1)(2-x-x^2)}$$

Factoring the denominator to find the zeros, we get $(x+1)(x-1)(1-x)(2+x)$. We find $x = -1$ and $x = -2$ are zeros of multiplicity one but that $x = 1$ is a zero of multiplicity two due to the two different factors $(x-1)$ and $(1-x)$. One way to handle this is to note that $(1-x) = -(x-1)$ so

$$\frac{3x-1}{(x+1)(x-1)(1-x)(2+x)} = \frac{3x-1}{-(x-1)^2(x+1)(x+2)} = \frac{1-3x}{(x-1)^2(x+1)(x+2)}$$

from which we proceed with the partial fraction decomposition

$$\frac{1-3x}{(x-1)^2(x+1)(x+2)} = \frac{A}{x-1} + \frac{B}{(x-1)^2} + \frac{C}{x+1} + \frac{D}{x+2}$$

Turning our attention to non-real zeros, we note that the tool of choice to determine the irreducibility of a quadratic $ax^2 + bx + c$ is the discriminant, $b^2 - 4ac$. If $b^2 - 4ac < 0$, the quadratic admits a *pair* of non-real complex conjugate zeros. Even though *one* irreducible quadratic gives *two* distinct non-real zeros, we list the terms with denominators involving a given irreducible quadratic only once to avoid duplication in the form of the decomposition. The trick, of course, is factoring the

8.6 PARTIAL FRACTION DECOMPOSITION

denominator or otherwise finding the zeros and their multiplicities in order to apply Theorem 8.10. We recommend that the reader review the techniques set forth in Sections 3.3 and 3.4. Next, we state a theorem that if two polynomials are equal, the corresponding coefficients of the like powers of x are equal. This is the principal by which we shall determine the unknown coefficients in our partial fraction decomposition.

Theorem 8.11. Suppose

$$a_n x^n + a_{n-1} x^{n-1} + \cdots + a_2 x^2 + a_1 x + a_0 = b_m x^m + m_{m-1} x^{m-1} + \cdots + b_2 x^2 + b_1 x + b_0$$

for all x in an open interval I. Then $n = m$ and $a_i = b_i$ for all $i = 1 \ldots n$.

Believe it or not, the proof of Theorem 8.11 is a consequence of Theorem 3.14. Define $p(x)$ to be the difference of the left hand side of the equation in Theorem 8.11 and the right hand side. Then $p(x) = 0$ for all x in the open interval I. If $p(x)$ were a nonzero polynomial of degree k, then, by Theorem 3.14, p could have at most k zeros in I, and k is a finite number. Since $p(x) = 0$ for all the x in I, p has infinitely many zeros, and hence, p is the zero polynomial. This means there can be no nonzero terms in $p(x)$ and the theorem follows. Arguably, the best way to make sense of either of the two preceding theorems is to work some examples.

Example 8.6.1. Resolve the following rational functions into partial fractions.

1. $R(x) = \dfrac{x+5}{2x^2 - x - 1}$
2. $R(x) = \dfrac{3}{x^3 - 2x^2 + x}$
3. $R(x) = \dfrac{3}{x^3 - x^2 + x}$
4. $R(x) = \dfrac{4x^3}{x^2 - 2}$
5. $R(x) = \dfrac{x^3 + 5x - 1}{x^4 + 6x^2 + 9}$
6. $R(x) = \dfrac{8x^2}{x^4 + 16}$

Solution.

1. We begin by factoring the denominator to find $2x^2 - x - 1 = (2x+1)(x-1)$. We get $x = -\frac{1}{2}$ and $x = 1$ are both zeros of multiplicity one and thus we know

$$\frac{x+5}{2x^2 - x - 1} = \frac{x+5}{(2x+1)(x-1)} = \frac{A}{2x+1} + \frac{B}{x-1}$$

Clearing denominators, we get $x + 5 = A(x-1) + B(2x+1)$ so that $x + 5 = (A+2B)x + B - A$. Equating coefficients, we get the system

$$\begin{cases} A + 2B = 1 \\ -A + B = 5 \end{cases}$$

This system is readily handled using the Addition Method from Section 8.1, and after adding both equations, we get $3B = 6$ so $B = 2$. Using back substitution, we find $A = -3$. Our answer is easily checked by getting a common denominator and adding the fractions.

$$\frac{x+5}{2x^2 - x - 1} = \frac{2}{x-1} - \frac{3}{2x+1}$$

2. Factoring the denominator gives $x^3 - 2x^2 + x = x\left(x^2 - 2x + 1\right) = x(x-1)^2$ which gives $x = 0$ as a zero of multiplicity one and $x = 1$ as a zero of multiplicity two. We have

$$\frac{3}{x^3 - 2x^2 + x} = \frac{3}{x(x-1)^2} = \frac{A}{x} + \frac{B}{x-1} + \frac{C}{(x-1)^2}$$

Clearing denominators, we get $3 = A(x-1)^2 + Bx(x-1) + Cx$, which, after gathering up the like terms becomes $3 = (A+B)x^2 + (-2A - B + C)x + A$. Our system is

$$\begin{cases} A + B & = & 0 \\ -2A - B + C & = & 0 \\ A & = & 3 \end{cases}$$

Substituting $A = 3$ into $A + B = 0$ gives $B = -3$, and substituting both for A and B in $-2A - B + C = 0$ gives $C = 3$. Our final answer is

$$\frac{3}{x^3 - 2x^2 + x} = \frac{3}{x} - \frac{3}{x-1} + \frac{3}{(x-1)^2}$$

3. The denominator factors as $x\left(x^2 - x + 1\right)$. We see immediately that $x = 0$ is a zero of multiplicity one, but the zeros of $x^2 - x + 1$ aren't as easy to discern. The quadratic doesn't factor easily, so we check the discriminant and find it to be $(-1)^2 - 4(1)(1) = -3 < 0$. We find its zeros are not real so it is an irreducible quadratic. The form of the partial fraction decomposition is then

$$\frac{3}{x^3 - x^2 + x} = \frac{3}{x\left(x^2 - x + 1\right)} = \frac{A}{x} + \frac{Bx + C}{x^2 - x + 1}$$

Proceeding as usual, we clear denominators and get $3 = A\left(x^2 - x + 1\right) + (Bx + C)x$ or $3 = (A+B)x^2 + (-A+C)x + A$. We get

$$\begin{cases} A + B & = & 0 \\ -A + C & = & 0 \\ A & = & 3 \end{cases}$$

From $A = 3$ and $A + B = 0$, we get $B = -3$. From $-A + C = 0$, we get $C = A = 3$. We get

$$\frac{3}{x^3 - x^2 + x} = \frac{3}{x} + \frac{3 - 3x}{x^2 - x + 1}$$

4. Since $\frac{4x^3}{x^2 - 2}$ isn't proper, we use long division and we get a quotient of $4x$ with a remainder of $8x$. That is, $\frac{4x^3}{x^2 - 2} = 4x + \frac{8x}{x^2 - 2}$ so we now work on resolving $\frac{8x}{x^2 - 2}$ into partial fractions. The quadratic $x^2 - 2$, though it doesn't factor nicely, is, nevertheless, reducible. Solving $x^2 - 2 = 0$

8.6 PARTIAL FRACTION DECOMPOSITION

gives us $x = \pm\sqrt{2}$, and each of these zeros must be of multiplicity one since Theorem 3.14 enables us to now factor $x^2 - 2 = (x - \sqrt{2})(x + \sqrt{2})$. Hence,

$$\frac{8x}{x^2 - 2} = \frac{8x}{(x - \sqrt{2})(x + \sqrt{2})} = \frac{A}{x - \sqrt{2}} + \frac{B}{x + \sqrt{2}}$$

Clearing fractions, we get $8x = A(x + \sqrt{2}) + B(x - \sqrt{2})$ or $8x = (A + B)x + (A - B)\sqrt{2}$. We get the system

$$\begin{cases} A + B = 8 \\ (A - B)\sqrt{2} = 0 \end{cases}$$

From $(A - B)\sqrt{2} = 0$, we get $A = B$, which, when substituted into $A + B = 8$ gives $B = 4$. Hence, $A = B = 4$ and we get

$$\frac{4x^3}{x^2 - 2} = 4x + \frac{8x}{x^2 - 2} = 4x + \frac{4}{x + \sqrt{2}} + \frac{4}{x - \sqrt{2}}$$

5. At first glance, the denominator $D(x) = x^4 + 6x^2 + 9$ appears irreducible. However, $D(x)$ has three terms, and the exponent on the first term is exactly twice that of the second. Rewriting $D(x) = (x^2)^2 + 6x^2 + 9$, we see it is a quadratic in disguise and factor $D(x) = (x^2 + 3)^2$. Since $x^2 + 3$ clearly has no real zeros, it is irreducible and the form of the decomposition is

$$\frac{x^3 + 5x - 1}{x^4 + 6x^2 + 9} = \frac{x^3 + 5x - 1}{(x^2 + 3)^2} = \frac{Ax + B}{x^2 + 3} + \frac{Cx + D}{(x^2 + 3)^2}$$

When we clear denominators, we find $x^3 + 5x - 1 = (Ax + B)(x^2 + 3) + Cx + D$ which yields $x^3 + 5x - 1 = Ax^3 + Bx^2 + (3A + C)x + 3B + D$. Our system is

$$\begin{cases} A = 1 \\ B = 0 \\ 3A + C = 5 \\ 3B + D = -1 \end{cases}$$

We have $A = 1$ and $B = 0$ from which we get $C = 2$ and $D = -1$. Our final answer is

$$\frac{x^3 + 5x - 1}{x^4 + 6x^2 + 9} = \frac{x}{x^2 + 3} + \frac{2x - 1}{(x^2 + 3)^2}$$

6. Once again, the difficulty in our last example is factoring the denominator. In an attempt to get a quadratic in disguise, we write

$$x^4 + 16 = (x^2)^2 + 4^2 = (x^2)^2 + 8x^2 + 4^2 - 8x^2 = (x^2 + 4)^2 - 8x^2$$

and obtain a difference of two squares: $\left(x^2+4\right)^2$ and $8x^2 = \left(2x\sqrt{2}\right)^2$. Hence,

$$x^4 + 16 = \left(x^2 + 4 - 2x\sqrt{2}\right)\left(x^2 + 4 + 2x\sqrt{2}\right) = \left(x^2 - 2x\sqrt{2} + 4\right)\left(x^2 + 2x\sqrt{2} + 4\right)$$

The discrimant of both of these quadratics works out to be $-8 < 0$, which means they are irreducible. We leave it to the reader to verify that, despite having the same discriminant, these quadratics have different zeros. The partial fraction decomposition takes the form

$$\frac{8x^2}{x^4+16} = \frac{8x^2}{\left(x^2 - 2x\sqrt{2} + 4\right)\left(x^2 + 2x\sqrt{2} + 4\right)} = \frac{Ax+B}{x^2 - 2x\sqrt{2} + 4} + \frac{Cx+D}{x^2 + 2x\sqrt{2} + 4}$$

We get $8x^2 = (Ax+B)\left(x^2 + 2x\sqrt{2} + 4\right) + (Cx+D)\left(x^2 - 2x\sqrt{2} + 4\right)$ or

$$8x^2 = (A+C)x^3 + (2A\sqrt{2} + B - 2C\sqrt{2} + D)x^2 + (4A + 2B\sqrt{2} + 4C - 2D\sqrt{2})x + 4B + 4D$$

which gives the system

$$\begin{cases} A + C &= 0 \\ 2A\sqrt{2} + B - 2C\sqrt{2} + D &= 8 \\ 4A + 2B\sqrt{2} + 4C - 2D\sqrt{2} &= 0 \\ 4B + 4D &= 0 \end{cases}$$

We choose substitution as the weapon of choice to solve this system. From $A + C = 0$, we get $A = -C$; from $4B + 4D = 0$, we get $B = -D$. Substituting these into the remaining two equations, we get

$$\begin{cases} -2C\sqrt{2} - D - 2C\sqrt{2} + D &= 8 \\ -4C - 2D\sqrt{2} + 4C - 2D\sqrt{2} &= 0 \end{cases}$$

or

$$\begin{cases} -4C\sqrt{2} &= 8 \\ -4D\sqrt{2} &= 0 \end{cases}$$

We get $C = -\sqrt{2}$ so that $A = -C = \sqrt{2}$ and $D = 0$ which means $B = -D = 0$. We get

$$\frac{8x^2}{x^4 + 16} = \frac{x\sqrt{2}}{x^2 - 2x\sqrt{2} + 4} - \frac{x\sqrt{2}}{x^2 + 2x\sqrt{2} + 4}$$

$\square$

8.6 Partial Fraction Decomposition

8.6.1 Exercises

In Exercises 1 - 6, find only the *form* needed to begin the process of partial fraction decomposition. Do not create the system of linear equations or attempt to find the actual decomposition.

1. $\dfrac{7}{(x-3)(x+5)}$

2. $\dfrac{5x+4}{x(x-2)(2-x)}$

3. $\dfrac{m}{(7x-6)(x^2+9)}$

4. $\dfrac{ax^2+bx+c}{x^3(5x+9)(3x^2+7x+9)}$

5. $\dfrac{\text{A polynomial of degree } < 9}{(x+4)^5(x^2+1)^2}$

6. $\dfrac{\text{A polynomial of degree } < 7}{x(4x-1)^2(x^2+5)(9x^2+16)}$

In Exercises 7 - 18, find the partial fraction decomposition of the following rational expressions.

7. $\dfrac{2x}{x^2-1}$

8. $\dfrac{-7x+43}{3x^2+19x-14}$

9. $\dfrac{11x^2-5x-10}{5x^3-5x^2}$

10. $\dfrac{-2x^2+20x-68}{x^3+4x^2+4x+16}$

11. $\dfrac{-x^2+15}{4x^4+40x^2+36}$

12. $\dfrac{-21x^2+x-16}{3x^3+4x^2-3x+2}$

13. $\dfrac{5x^4-34x^3+70x^2-33x-19}{(x-3)^2}$

14. $\dfrac{x^6+5x^5+16x^4+80x^3-2x^2+6x-43}{x^3+5x^2+16x+80}$

15. $\dfrac{-7x^2-76x-208}{x^3+18x^2+108x+216}$

16. $\dfrac{-10x^4+x^3-19x^2+x-10}{x^5+2x^3+x}$

17. $\dfrac{4x^3-9x^2+12x+12}{x^4-4x^3+8x^2-16x+16}$

18. $\dfrac{2x^2+3x+14}{(x^2+2x+9)(x^2+x+5)}$

19. As we stated at the beginning of this section, the technique of resolving a rational function into partial fractions is a skill needed for Calculus. However, we hope to have shown you that it is worth doing if, for no other reason, it reinforces a hefty amount of algebra. One of the common algebraic errors the authors find students make is something along the lines of

$$\dfrac{8}{x^2-9} \neq \dfrac{8}{x^2} - \dfrac{8}{9}$$

Think about why if the above were true, this section would have no need to exist.

8.6.2 Answers

1. $\dfrac{A}{x-3} + \dfrac{B}{x+5}$

2. $\dfrac{A}{x} + \dfrac{B}{x-2} + \dfrac{C}{(x-2)^2}$

3. $\dfrac{A}{7x-6} + \dfrac{Bx+C}{x^2+9}$

4. $\dfrac{A}{x} + \dfrac{B}{x^2} + \dfrac{C}{x^3} + \dfrac{D}{5x+9} + \dfrac{Ex+F}{3x^2+7x+9}$

5. $\dfrac{A}{x+4} + \dfrac{B}{(x+4)^2} + \dfrac{C}{(x+4)^3} + \dfrac{D}{(x+4)^4} + \dfrac{E}{(x+4)^5} + \dfrac{Fx+G}{x^2+1} + \dfrac{Hx+I}{(x^2+1)^2}$

6. $\dfrac{A}{x} + \dfrac{B}{4x-1} + \dfrac{C}{(4x-1)^2} + \dfrac{Dx+E}{x^2+5} + \dfrac{Fx+G}{9x^2+16}$

7. $\dfrac{2x}{x^2-1} = \dfrac{1}{x+1} + \dfrac{1}{x-1}$

8. $\dfrac{-7x+43}{3x^2+19x-14} = \dfrac{5}{3x-2} - \dfrac{4}{x+7}$

9. $\dfrac{11x^2-5x-10}{5x^3-5x^2} = \dfrac{3}{x} + \dfrac{2}{x^2} - \dfrac{4}{5(x-1)}$

10. $\dfrac{-2x^2+20x-68}{x^3+4x^2+4x+16} = -\dfrac{9}{x+4} + \dfrac{7x-8}{x^2+4}$

11. $\dfrac{-x^2+15}{4x^4+40x^2+36} = \dfrac{1}{2(x^2+1)} - \dfrac{3}{4(x^2+9)}$

12. $\dfrac{-21x^2+x-16}{3x^3+4x^2-3x+2} = -\dfrac{6}{x+2} - \dfrac{3x+5}{3x^2-2x+1}$

13. $\dfrac{5x^4-34x^3+70x^2-33x-19}{(x-3)^2} = 5x^2 - 4x + 1 + \dfrac{9}{x-3} - \dfrac{1}{(x-3)^2}$

14. $\dfrac{x^6+5x^5+16x^4+80x^3-2x^2+6x-43}{x^3+5x^2+16x+80} = x^3 + \dfrac{x+1}{x^2+16} - \dfrac{3}{x+5}$

15. $\dfrac{-7x^2-76x-208}{x^3+18x^2+108x+216} = -\dfrac{7}{x+6} + \dfrac{8}{(x+6)^2} - \dfrac{4}{(x+6)^3}$

16. $\dfrac{-10x^4+x^3-19x^2+x-10}{x^5+2x^3+x} = -\dfrac{10}{x} + \dfrac{1}{x^2+1} + \dfrac{x}{(x^2+1)^2}$

17. $\dfrac{4x^3-9x^2+12x+12}{x^4-4x^3+8x^2-16x+16} = \dfrac{1}{x-2} + \dfrac{4}{(x-2)^2} + \dfrac{3x+1}{x^2+4}$

18. $\dfrac{2x^2+3x+14}{(x^2+2x+9)(x^2+x+5)} = \dfrac{1}{x^2+2x+9} + \dfrac{1}{x^2+x+5}$

8.7 Systems of Non-Linear Equations and Inequalities

In this section, we study systems of non-linear equations and inequalities. Unlike the systems of linear equations for which we have developed several algorithmic solution techniques, there is no general algorithm to solve systems of non-linear equations. Moreover, all of the usual hazards of non-linear equations like extraneous solutions and unusual function domains are once again present. Along with the tried and true techniques of substitution and elimination, we shall often need equal parts tenacity and ingenuity to see a problem through to the end. You may find it necessary to review topics throughout the text which pertain to solving equations involving the various functions we have studied thus far. To get the section rolling we begin with a fairly routine example.

Example 8.7.1. Solve the following systems of equations. Verify your answers algebraically and graphically.

1. $\begin{cases} x^2 + y^2 = 4 \\ 4x^2 + 9y^2 = 36 \end{cases}$

2. $\begin{cases} x^2 + y^2 = 4 \\ 4x^2 - 9y^2 = 36 \end{cases}$

3. $\begin{cases} x^2 + y^2 = 4 \\ y - 2x = 0 \end{cases}$

4. $\begin{cases} x^2 + y^2 = 4 \\ y - x^2 = 0 \end{cases}$

SOLUTION:

1. Since both equations contain x^2 and y^2 only, we can eliminate one of the variables as we did in Section 8.1.

$$\begin{cases} (E1) & x^2 + y^2 = 4 \\ (E2) & 4x^2 + 9y^2 = 36 \end{cases} \xrightarrow[-4E1 + E2]{\text{Replace } E2 \text{ with}} \begin{cases} (E1) & x^2 + y^2 = 4 \\ (E2) & 5y^2 = 20 \end{cases}$$

From $5y^2 = 20$, we get $y^2 = 4$ or $y = \pm 2$. To find the associated x values, we substitute each value of y into one of the equations to find the resulting value of x. Choosing $x^2 + y^2 = 4$, we find that for both $y = -2$ and $y = 2$, we get $x = 0$. Our solution is thus $\{(0, 2), (0, -2)\}$. To check this algebraically, we need to show that both points satisfy both of the original equations. We leave it to the reader to verify this. To check our answer graphically, we sketch both equations and look for their points of intersection. The graph of $x^2 + y^2 = 4$ is a circle centered at $(0, 0)$ with a radius of 2, whereas the graph of $4x^2 + 9y^2 = 36$, when written in the standard form $\frac{x^2}{9} + \frac{y^2}{4} = 1$ is easily recognized as an ellipse centered at $(0, 0)$ with a major axis along the x-axis of length 6 and a minor axis along the y-axis of length 4. We see from the graph that the two curves intersect at their y-intercepts only, $(0, \pm 2)$.

2. We proceed as before to eliminate one of the variables

$$\begin{cases} (E1) & x^2 + y^2 = 4 \\ (E2) & 4x^2 - 9y^2 = 36 \end{cases} \xrightarrow[-4E1 + E2]{\text{Replace } E2 \text{ with}} \begin{cases} (E1) & x^2 + y^2 = 4 \\ (E2) & -13y^2 = 20 \end{cases}$$

Since the equation $-13y^2 = 20$ admits no real solution, the system is inconsistent. To verify this graphically, we note that $x^2 + y^2 = 4$ is the same circle as before, but when writing the second equation in standard form, $\frac{x^2}{9} - \frac{y^2}{4} = 1$, we find a hyperbola centered at $(0,0)$ opening to the left and right with a transverse axis of length 6 and a conjugate axis of length 4. We see that the circle and the hyperbola have no points in common.

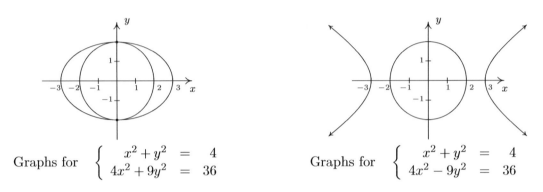

Graphs for $\begin{cases} x^2 + y^2 = 4 \\ 4x^2 + 9y^2 = 36 \end{cases}$ Graphs for $\begin{cases} x^2 + y^2 = 4 \\ 4x^2 - 9y^2 = 36 \end{cases}$

3. Since there are no like terms among the two equations, elimination won't do us any good. We turn to substitution and from the equation $y - 2x = 0$, we get $y = 2x$. Substituting this into $x^2 + y^2 = 4$ gives $x^2 + (2x)^2 = 4$. Solving, we find $5x^2 = 4$ or $x = \pm \frac{2\sqrt{5}}{5}$. Returning to the equation we used for the substitution, $y = 2x$, we find $y = \frac{4\sqrt{5}}{5}$ when $x = \frac{2\sqrt{5}}{5}$, so one solution is $\left(\frac{2\sqrt{5}}{5}, \frac{4\sqrt{5}}{5}\right)$. Similarly, we find the other solution to be $\left(-\frac{2\sqrt{5}}{5}, -\frac{4\sqrt{5}}{5}\right)$. We leave it to the reader that both points satisfy both equations, so that our final answer is $\left\{\left(\frac{2\sqrt{5}}{5}, \frac{4\sqrt{5}}{5}\right), \left(-\frac{2\sqrt{5}}{5}, -\frac{4\sqrt{5}}{5}\right)\right\}$. The graph of $x^2 + y^2 = 4$ is our circle from before and the graph of $y - 2x = 0$ is a line through the origin with slope 2. Though we cannot verify the numerical values of the points of intersection from our sketch, we do see that we have two solutions: one in Quadrant I and one in Quadrant III as required.

4. While it may be tempting to solve $y - x^2 = 0$ as $y = x^2$ and substitute, we note that this system is set up for elimination.[1]

$$\begin{cases} (E1) & x^2 + y^2 = 4 \\ (E2) & y - x^2 = 0 \end{cases} \xrightarrow{\text{Replace } E2 \text{ with } E1 + E2} \begin{cases} (E1) & x^2 + y^2 = 4 \\ (E2) & y^2 + y = 4 \end{cases}$$

From $y^2 + y = 4$ we get $y^2 + y - 4 = 0$ which gives $y = \frac{-1 \pm \sqrt{17}}{2}$. Due to the complicated nature of these answers, it is worth our time to make a quick sketch of both equations to head off any extraneous solutions we may encounter. We see that the circle $x^2 + y^2 = 4$ intersects the parabola $y = x^2$ exactly twice, and both of these points have a positive y value. Of the two solutions for y, only $y = \frac{-1+\sqrt{17}}{2}$ is positive, so to get our solution, we substitute this

---

[1]We encourage the reader to solve the system using substitution to see that you get the same solution.

8.7 SYSTEMS OF NON-LINEAR EQUATIONS AND INEQUALITIES

into $y - x^2 = 0$ and solve for x. We get $x = \pm\sqrt{\frac{-1+\sqrt{17}}{2}} = \pm\frac{\sqrt{-2+2\sqrt{17}}}{2}$. Our solution is $\left\{\left(\frac{\sqrt{-2+2\sqrt{17}}}{2}, \frac{-1+\sqrt{17}}{2}\right), \left(-\frac{\sqrt{-2+2\sqrt{17}}}{2}, \frac{-1+\sqrt{17}}{2}\right)\right\}$, which we leave to the reader to verify.

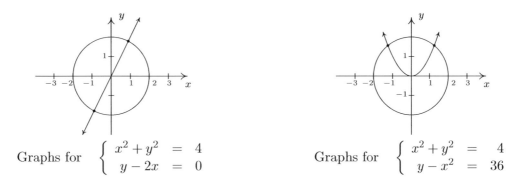

Graphs for $\begin{cases} x^2 + y^2 = 4 \\ y - 2x = 0 \end{cases}$ Graphs for $\begin{cases} x^2 + y^2 = 4 \\ y - x^2 = 36 \end{cases}$

□

A couple of remarks about Example 8.7.1 are in order. First note that, unlike systems of linear equations, it is possible for a system of non-linear equations to have more than one solution without having infinitely many solutions. In fact, while we characterize systems of nonlinear equations as being 'consistent' or 'inconsistent,' we generally don't use the labels 'dependent' or 'independent'. Secondly, as we saw with number 4, sometimes making a quick sketch of the problem situation can save a lot of time and effort. While in general the curves in a system of non-linear equations may not be easily visualized, it sometimes pays to take advantage when they are. Our next example provides some considerable review of many of the topics introduced in this text.

Example 8.7.2. Solve the following systems of equations. Verify your answers algebraically and graphically, as appropriate.

1. $\begin{cases} x^2 + 2xy - 16 = 0 \\ y^2 + 2xy - 16 = 0 \end{cases}$ 2. $\begin{cases} y + 4e^{2x} = 1 \\ y^2 + 2e^x = 1 \end{cases}$ 3. $\begin{cases} z(x - 2) = x \\ yz = y \\ (x - 2)^2 + y^2 = 1 \end{cases}$

Solution.

1. At first glance, it doesn't appear as though elimination will do us any good since it's clear that we cannot completely eliminate one of the variables. The alternative, solving one of the equations for one variable and substituting it into the other, is full of unpleasantness. Returning to elimination, we note that it is possible to eliminate the troublesome xy term, and the constant term as well, by elimination and doing so we get a more tractable relationship between x and y

$\begin{cases} (E1) & x^2 + 2xy - 16 = 0 \\ (E2) & y^2 + 2xy - 16 = 0 \end{cases} \xrightarrow[-E1+E2]{\text{Replace } E2 \text{ with}} \begin{cases} (E1) & x^2 + 2xy - 16 = 0 \\ (E2) & y^2 - x^2 = 0 \end{cases}$

We get $y^2 - x^2 = 0$ or $y = \pm x$. Substituting $y = x$ into $E1$ we get $x^2 + 2x^2 - 16 = 0$ so that $x^2 = \frac{16}{3}$ or $x = \pm \frac{4\sqrt{3}}{3}$. On the other hand, when we substitute $y = -x$ into $E1$, we get $x^2 - 2x^2 - 16 = 0$ or $x^2 = -16$ which gives no real solutions. Substituting each of $x = \pm \frac{4\sqrt{3}}{3}$ into the substitution equation $y = x$ yields the solution $\left\{ \left(\frac{4\sqrt{3}}{3}, \frac{4\sqrt{3}}{3} \right), \left(-\frac{4\sqrt{3}}{3}, -\frac{4\sqrt{3}}{3} \right) \right\}$. We leave it to the reader to show that both points satisfy both equations and now turn to verifying our solution graphically. We begin by solving $x^2 + 2xy - 16 = 0$ for y to obtain $y = \frac{16-x^2}{2x}$. This function is easily graphed using the techniques of Section 4.2. Solving the second equation, $y^2 + 2xy - 16 = 0$, for y, however, is more complicated. We use the quadratic formula to obtain $y = -x \pm \sqrt{x^2 + 16}$ which would require the use of Calculus or a calculator to graph. Believe it or not, we don't need either because the equation $y^2 + 2xy - 16 = 0$ can be obtained from the equation $x^2 + 2xy - 16 = 0$ by interchanging y and x. Thinking back to Section 5.2, this means we can obtain the graph of $y^2 + 2xy - 16 = 0$ by reflecting the graph of $x^2 + 2xy - 16 = 0$ across the line $y = x$. Doing so confirms that the two graphs intersect twice: once in Quadrant I, and once in Quadrant III as required.

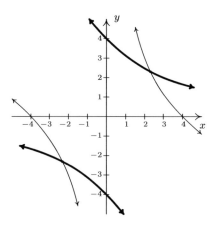

The graphs of $x^2 + 2xy - 16 = 0$ and $\mathbf{y^2 + 2xy - 16 = 0}$

2. Unlike the previous problem, there seems to be no avoiding substitution and a bit of algebraic unpleasantness. Solving $y + 4e^{2x} = 1$ for y, we get $y = 1 - 4e^{2x}$ which, when substituted into the second equation, yields $\left(1 - 4e^{2x} \right)^2 + 2e^x = 1$. After expanding and gathering like terms, we get $16e^{4x} - 8e^{2x} + 2e^x = 0$. Factoring gives us $2e^x \left(8e^{3x} - 4e^x + 1 \right) = 0$, and since $2e^x \neq 0$ for any real x, we are left with solving $8e^{3x} - 4e^x + 1 = 0$. We have three terms, and even though this is not a 'quadratic in disguise', we can benefit from the substitution $u = e^x$. The equation becomes $8u^3 - 4u + 1 = 0$. Using the techniques set forth in Section 3.3, we find $u = \frac{1}{2}$ is a zero and use synthetic division to factor the left hand side as $\left(u - \frac{1}{2} \right) \left(8u^2 + 4u - 2 \right)$. We use the quadratic formula to solve $8u^2 + 4u - 2 = 0$ and find $u = \frac{-1 \pm \sqrt{5}}{4}$. Since $u = e^x$, we now must solve $e^x = \frac{1}{2}$ and $e^x = \frac{-1 \pm \sqrt{5}}{4}$. From $e^x = \frac{1}{2}$, we get $x = \ln\left(\frac{1}{2}\right) = -\ln(2)$. As for $e^x = \frac{-1 \pm \sqrt{5}}{4}$, we first note that $\frac{-1 - \sqrt{5}}{4} < 0$, so $e^x = \frac{-1 - \sqrt{5}}{4}$ has no real solutions. We are

8.7 Systems of Non-Linear Equations and Inequalities

left with $e^x = \frac{-1+\sqrt{5}}{4}$, so that $x = \ln\left(\frac{-1+\sqrt{5}}{4}\right)$. We now return to $y = 1 - 4e^{2x}$ to find the accompanying y values for each of our solutions for x. For $x = -\ln(2)$, we get

$$\begin{aligned} y &= 1 - 4e^{2x} \\ &= 1 - 4e^{-2\ln(2)} \\ &= 1 - 4e^{\ln\left(\frac{1}{4}\right)} \\ &= 1 - 4\left(\frac{1}{4}\right) \\ &= 0 \end{aligned}$$

For $x = \ln\left(\frac{-1+\sqrt{5}}{4}\right)$, we have

$$\begin{aligned} y &= 1 - 4e^{2x} \\ &= 1 - 4e^{2\ln\left(\frac{-1+\sqrt{5}}{4}\right)} \\ &= 1 - 4e^{\ln\left(\frac{-1+\sqrt{5}}{4}\right)^2} \\ &= 1 - 4\left(\frac{-1+\sqrt{5}}{4}\right)^2 \\ &= 1 - 4\left(\frac{3-\sqrt{5}}{8}\right) \\ &= \frac{-1+\sqrt{5}}{2} \end{aligned}$$

We get two solutions, $\left\{(0, -\ln(2)), \left(\ln\left(\frac{-1+\sqrt{5}}{4}\right), \frac{-1+\sqrt{5}}{2}\right)\right\}$. It is a good review of the properties of logarithms to verify both solutions, so we leave that to the reader. We are able to sketch $y = 1 - 4e^{2x}$ using transformations, but the second equation is more difficult and we resort to the calculator. We note that to graph $y^2 + 2e^x = 1$, we need to graph both the positive and negative roots, $y = \pm\sqrt{1 - 2e^x}$. After some careful zooming,[2] we get

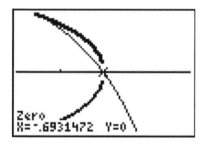 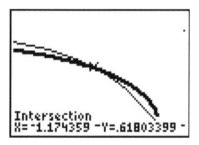

The graphs of $y = 1 - 4e^{2x}$ and $y = \pm\sqrt{1 - 2e^x}$.

3. Our last system involves three variables and gives some insight on how to keep such systems organized. Labeling the equations as before, we have

---

[2] The calculator has trouble confirming the solution $(-\ln(2), 0)$ due to its issues in graphing square root functions. If we mentally connect the two branches of the thicker curve, we see the intersection.

$$\begin{cases} E1 & z(x-2) = x \\ E2 & yz = y \\ E3 & (x-2)^2 + y^2 = 1 \end{cases}$$

The easiest equation to start with appears to be $E2$. While it may be tempting to divide both sides of $E2$ by y, we caution against this practice because it presupposes $y \neq 0$. Instead, we take $E2$ and rewrite it as $yz - y = 0$ so $y(z-1) = 0$. From this, we get two cases: $y = 0$ or $z = 1$. We take each case in turn.

CASE 1: $y = 0$. Substituting $y = 0$ into $E1$ and $E3$, we get

$$\begin{cases} E1 & z(x-2) = x \\ E3 & (x-2)^2 = 1 \end{cases}$$

Solving $E3$ for x gives $x = 1$ or $x = 3$. Substituting these values into $E1$ gives $z = -1$ when $x = 1$ and $z = 3$ when $x = 3$. We obtain two solutions, $(1, 0, -1)$ and $(3, 0, 3)$.

CASE 2: $z = 1$. Substituting $z = 1$ into $E1$ and $E3$ gives us

$$\begin{cases} E1 & (1)(x-2) = x \\ E3 & (1-2)^2 + y^2 = 1 \end{cases}$$

Equation $E1$ gives us $x - 2 = x$ or $-2 = 0$, which is a contradiction. This means we have no solution to the system in this case, even though $E3$ is solvable and gives $y = 0$. Hence, our final answer is $\{(1, 0, -1), (3, 0, 3)\}$. These points are easy enough to check algebraically in our three original equations, so that is left to the reader. As for verifying these solutions graphically, they require plotting surfaces in three dimensions and looking for intersection points. While this is beyond the scope of this book, we provide a snapshot of the graphs of our three equations near one of the solution points, $(1, 0, -1)$.

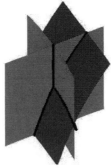

□

Example 8.7.2 showcases some of the ingenuity and tenacity mentioned at the beginning of the section. Sometimes you just have to look at a system the right way to find the most efficient method to solve it. Sometimes you just have to try something.

8.7 Systems of Non-Linear Equations and Inequalities

We close this section discussing how non-linear inequalities can be used to describe regions in the plane which we first introduced in Section 2.4. Before we embark on some examples, a little motivation is in order. Suppose we wish to solve $x^2 < 4-y^2$. If we mimic the algorithms for solving nonlinear inequalities in one variable, we would gather all of the terms on one side and leave a 0 on the other to obtain $x^2 + y^2 - 4 < 0$. Then we would find the zeros of the left hand side, that is, where is $x^2 + y^2 - 4 = 0$, or $x^2 + y^2 = 4$. Instead of obtaining a few *numbers* which divide the real number *line* into *intervals*, we get an equation of a *curve*, in this case, a circle, which divides the *plane* into two *regions* - the 'inside' and 'outside' of the circle - with the circle itself as the boundary between the two. Just like we used test *values* to determine whether or not an interval belongs to the solution of the inequality, we use test *points* in the each of the regions to see which of these belong to our solution set.[3] We choose $(0,0)$ to represent the region inside the circle and $(0,3)$ to represent the points outside of the circle. When we substitute $(0,0)$ into $x^2 + y^2 - 4 < 0$, we get $-4 < 4$ which is true. This means $(0,0)$ and all the other points inside the circle are part of the solution. On the other hand, when we substitute $(0,3)$ into the same inequality, we get $5 < 0$ which is false. This means $(0,3)$ along with all other points outside the circle are not part of the solution. What about points on the circle itself? Choosing a point on the circle, say $(0,2)$, we get $0 < 0$, which means the circle itself does not satisfy the inequality.[4] As a result, we leave the circle dashed in the final diagram.

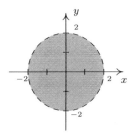

The solution to $x^2 < 4 - y^2$

We put this technique to good use in the following example.

Example 8.7.3. Sketch the solution to the following nonlinear inequalities in the plane.

1. $y^2 - 4 \leq x < y + 2$

2. $\begin{cases} x^2 + y^2 \geq 4 \\ x^2 - 2x + y^2 - 2y \leq 0 \end{cases}$

Solution.

1. The inequality $y^2 - 4 \leq x < y + 2$ is a compound inequality. It translates as $y^2 - 4 \leq x$ and $x < y + 2$. As usual, we solve each inequality and take the set theoretic intersection to determine the region which satisfies both inequalities. To solve $y^2 - 4 \leq x$, we write

---

[3]The theory behind why all this works is, surprisingly, the same theory which guarantees that sign diagrams work the way they do - continuity and the Intermediate Value Theorem - but in this case, applied to functions of more than one variable.

[4]Another way to see this is that points on the circle satisfy $x^2 + y^2 - 4 = 0$, so they do not satisfy $x^2 + y^2 - 4 < 0$.

$y^2 - x - 4 \leq 0$. The curve $y^2 - x - 4 = 0$ describes a parabola since exactly one of the variables is squared. Rewriting this in standard form, we get $y^2 = x + 4$ and we see that the vertex is $(-4, 0)$ and the parabola opens to the right. Using the test points $(-5, 0)$ and $(0, 0)$, we find that the solution to the inequality includes the region to the right of, or 'inside', the parabola. The points on the parabola itself are also part of the solution, since the vertex $(-4, 0)$ satisfies the inequality. We now turn our attention to $x < y + 2$. Proceeding as before, we write $x - y - 2 < 0$ and focus our attention on $x - y - 2 = 0$, which is the line $y = x - 2$. Using the test points $(0, 0)$ and $(0, -4)$, we find points in the region above the line $y = x - 2$ satisfy the inequality. The points on the line $y = x - 2$ do not satisfy the inequality, since the y-intercept $(0, -2)$ does not. We see that these two regions do overlap, and to make the graph more precise, we seek the intersection of these two curves. That is, we need to solve the system of nonlinear equations

$$\begin{cases} (E1) & y^2 = x + 4 \\ (E2) & y = x - 2 \end{cases}$$

Solving $E1$ for x, we get $x = y^2 - 4$. Substituting this into $E2$ gives $y = y^2 - 4 - 2$, or $y^2 - y - 6 = 0$. We find $y = -2$ and $y = 3$ and since $x = y^2 - 4$, we get that the graphs intersect at $(0, -2)$ and $(5, 3)$. Putting all of this together, we get our final answer below.

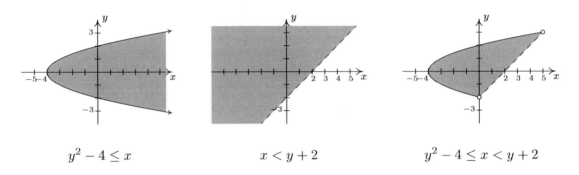

2. To solve this system of inequalities, we need to find all of the points (x, y) which satisfy both inequalities. To do this, we solve each inequality separately and take the set theoretic intersection of the solution sets. We begin with the inequality $x^2 + y^2 \geq 4$ which we rewrite as $x^2 + y^2 - 4 \geq 0$. The points which satisfy $x^2 + y^2 - 4 = 0$ form our friendly circle $x^2 + y^2 = 4$. Using test points $(0, 0)$ and $(0, 3)$ we find that our solution comprises the region outside the circle. As far as the circle itself, the point $(0, 2)$ satisfies the inequality, so the circle itself is part of the solution set. Moving to the inequality $x^2 - 2x + y^2 - 2y \leq 0$, we start with $x^2 - 2x + y^2 - 2y = 0$. Completing the squares, we obtain $(x - 1)^2 + (y - 1)^2 = 2$, which is a circle centered at $(1, 1)$ with a radius of $\sqrt{2}$. Choosing $(1, 1)$ to represent the inside of the circle, $(1, 3)$ as a point outside of the circle and $(0, 0)$ as a point on the circle, we find that the solution to the inequality is the inside of the circle, including the circle itself. Our final answer, then, consists of the points on or outside of the circle $x^2 + y^2 = 4$ which lie on or

inside the circle $(x-1)^2 + (y-1)^2 = 2$. To produce the most accurate graph, we need to find where these circles intersect. To that end, we solve the system

$$\begin{cases} (E1) & x^2 + y^2 = 4 \\ (E2) & x^2 - 2x + y^2 - 2y = 0 \end{cases}$$

We can eliminate both the x^2 and y^2 by replacing $E2$ with $-E1 + E2$. Doing so produces $-2x - 2y = -4$. Solving this for y, we get $y = 2 - x$. Substituting this into $E1$ gives $x^2 + (2-x)^2 = 4$ which simplifies to $x^2 + 4 - 4x + x^2 = 4$ or $2x^2 - 4x = 0$. Factoring yields $2x(x-2)$ which gives $x = 0$ or $x = 2$. Substituting these values into $y = 2 - x$ gives the points $(0, 2)$ and $(2, 0)$. The intermediate graphs and final solution are below.

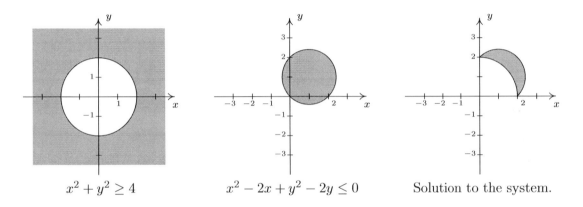

$x^2 + y^2 \geq 4$ \qquad $x^2 - 2x + y^2 - 2y \leq 0$ \qquad Solution to the system.

8.7.1 Exercises

In Exercises 1 - 6, solve the given system of nonlinear equations. Sketch the graph of both equations on the same set of axes to verify the solution set.

1. $\begin{cases} x^2 - y = 4 \\ x^2 + y^2 = 4 \end{cases}$
2. $\begin{cases} x^2 + y^2 = 4 \\ x^2 - y = 5 \end{cases}$
3. $\begin{cases} x^2 + y^2 = 16 \\ 16x^2 + 4y^2 = 64 \end{cases}$
4. $\begin{cases} x^2 + y^2 = 16 \\ 9x^2 - 16y^2 = 144 \end{cases}$
5. $\begin{cases} x^2 + y^2 = 16 \\ \frac{1}{9}y^2 - \frac{1}{16}x^2 = 1 \end{cases}$
6. $\begin{cases} x^2 + y^2 = 16 \\ x - y = 2 \end{cases}$

In Exercises 9 - 15, solve the given system of nonlinear equations. Use a graph to help you avoid any potential extraneous solutions.

7. $\begin{cases} x^2 - y^2 = 1 \\ x^2 + 4y^2 = 4 \end{cases}$
8. $\begin{cases} \sqrt{x+1} - y = 0 \\ x^2 + 4y^2 = 4 \end{cases}$
9. $\begin{cases} x + 2y^2 = 2 \\ x^2 + 4y^2 = 4 \end{cases}$
10. $\begin{cases} (x-2)^2 + y^2 = 1 \\ x^2 + 4y^2 = 4 \end{cases}$
11. $\begin{cases} x^2 + y^2 = 25 \\ y - x = 1 \end{cases}$
12. $\begin{cases} x^2 + y^2 = 25 \\ x^2 + (y-3)^2 = 10 \end{cases}$
13. $\begin{cases} y = x^3 + 8 \\ y = 10x - x^2 \end{cases}$
14. $\begin{cases} x^2 - xy = 8 \\ y^2 - xy = 8 \end{cases}$
15. $\begin{cases} x^2 + y^2 = 25 \\ 4x^2 - 9y = 0 \\ 3y^2 - 16x = 0 \end{cases}$

16. A certain bacteria culture follows the Law of Uninhibited Growth, Equation 6.4. After 10 minutes, there are 10,000 bacteria. Five minutes later, there are 14,000 bacteria. How many bacteria were present initially? How long before there are 50,000 bacteria?

Consider the system of nonlinear equations below

$$\begin{cases} \dfrac{4}{x} + \dfrac{3}{y} = 1 \\ \dfrac{3}{x} + \dfrac{2}{y} = -1 \end{cases}$$

If we let $u = \frac{1}{x}$ and $v = \frac{1}{y}$ then the system becomes

$$\begin{cases} 4u + 3v = 1 \\ 3u + 2v = -1 \end{cases}$$

This associated system of linear equations can then be solved using any of the techniques presented earlier in the chapter to find that $u = -5$ and $v = 7$. Thus $x = \frac{1}{u} = -\frac{1}{5}$ and $y = \frac{1}{v} = \frac{1}{7}$.

We say that the original system is **linear in form** because its equations are not linear but a few substitutions reveal a structure that we can treat like a system of linear equations. Each system in Exercises 17 - 19 is linear in form. Make the appropriate substitutions and solve for x and y.

8.7 Systems of Non-Linear Equations and Inequalities

17. $\begin{cases} 4x^3 + 3\sqrt{y} = 1 \\ 3x^3 + 2\sqrt{y} = -1 \end{cases}$
18. $\begin{cases} 4e^x + 3e^{-y} = 1 \\ 3e^x + 2e^{-y} = -1 \end{cases}$
19. $\begin{cases} 4\ln(x) + 3y^2 = 1 \\ 3\ln(x) + 2y^2 = -1 \end{cases}$

20. Solve the following system
$$\begin{cases} x^2 + \sqrt{y} + \log_2(z) = 6 \\ 3x^2 - 2\sqrt{y} + 2\log_2(z) = 5 \\ -5x^2 + 3\sqrt{y} + 4\log_2(z) = 13 \end{cases}$$

In Exercises 21 - 26, sketch the solution to each system of nonlinear inequalities in the plane.

21. $\begin{cases} x^2 - y^2 \leq 1 \\ x^2 + 4y^2 \geq 4 \end{cases}$
22. $\begin{cases} x^2 + y^2 < 25 \\ x^2 + (y-3)^2 \geq 10 \end{cases}$

23. $\begin{cases} (x-2)^2 + y^2 < 1 \\ x^2 + 4y^2 < 4 \end{cases}$
24. $\begin{cases} y > 10x - x^2 \\ y < x^3 + 8 \end{cases}$

25. $\begin{cases} x + 2y^2 > 2 \\ x^2 + 4y^2 \leq 4 \end{cases}$
26. $\begin{cases} x^2 + y^2 \geq 25 \\ y - x \leq 1 \end{cases}$

27. Systems of nonlinear equations show up in third semester Calculus in the midst of some really cool problems. The system below came from a problem in which we were asked to find the dimensions of a rectangular box with a volume of 1000 cubic inches that has minimal surface area. The variables x, y and z are the dimensions of the box and λ is called a Lagrange multiplier. With the help of your classmates, solve the system.[5]

$$\begin{cases} 2y + 2z = \lambda yz \\ 2x + 2z = \lambda xz \\ 2y + 2x = \lambda xy \\ xyz = 1000 \end{cases}$$

28. According to Theorem 3.16 in Section 3.4, the polynomial $p(x) = x^4 + 4$ can be factored into the product linear and irreducible quadratic factors. In this exercise, we present a method for obtaining that factorization.

    (a) Show that p has no real zeros.

    (b) Because p has no real zeros, its factorization must be of the form $(x^2 + ax + b)(x^2 + cx + d)$ where each factor is an irreducible quadratic. Expand this quantity and gather like terms together.

    (c) Create and solve the system of nonlinear equations which results from equating the coefficients of the expansion found above with those of $x^4 + 4$. You should get four equations in the four unknowns a, b, c and d. Write $p(x)$ in factored form.

29. Factor $q(x) = x^4 + 6x^2 - 5x + 6$.

---

[5]If using λ bothers you, change it to w when you solve the system.

8.7.2 Answers

1. $(\pm 2, 0)$, $(\pm\sqrt{3}, -1)$

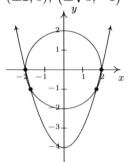

2. No solution

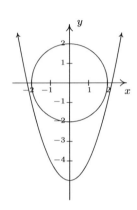

3. $(0, \pm 4)$

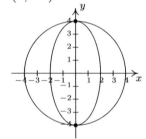

4. $(\pm 4, 0)$

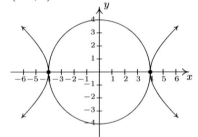

5. $\left(\pm\frac{4\sqrt{7}}{5}, \pm\frac{12\sqrt{2}}{5}\right)$

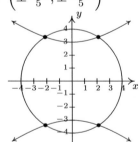

6. $\left(1+\sqrt{7}, -1+\sqrt{7}\right)$, $\left(1-\sqrt{7}, -1-\sqrt{7}\right)$

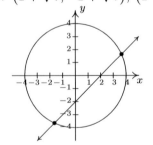

7. $\left(\pm\frac{2\sqrt{10}}{5}, \pm\frac{\sqrt{15}}{5}\right)$

8. $(0, 1)$

9. $(0, \pm 1)$, $(2, 0)$

10. $\left(\frac{4}{3}, \pm\frac{\sqrt{5}}{3}\right)$

11. $(3, 4)$, $(-4, -3)$

12. $(\pm 3, 4)$

13. $(-4, -56)$, $(1, 9)$, $(2, 16)$

14. $(-2, 2)$, $(2, -2)$

15. $(3, 4)$

16. Initially, there are $\frac{250000}{49} \approx 5102$ bacteria. It will take $\frac{5\ln(49/5)}{\ln(7/5)} \approx 33.92$ minutes for the colony to grow to 50,000 bacteria.

8.7 SYSTEMS OF NON-LINEAR EQUATIONS AND INEQUALITIES 649

17. $\left(-\sqrt[3]{5}, 49\right)$ 18. No solution   19. $\left(e^{-5}, \pm\sqrt{7}\right)$

20. $(1, 4, 8)$, $(-1, 4, 8)$

21. $\begin{cases} x^2 - y^2 \le 1 \\ x^2 + 4y^2 \ge 4 \end{cases}$

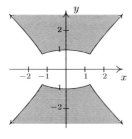

22. $\begin{cases} x^2 + y^2 < 25 \\ x^2 + (y-3)^2 \ge 10 \end{cases}$

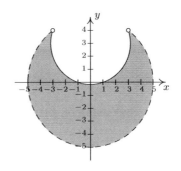

23. $\begin{cases} (x-2)^2 + y^2 < 1 \\ x^2 + 4y^2 < 4 \end{cases}$

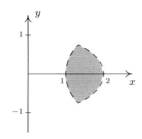

24. $\begin{cases} y > 10x - x^2 \\ y < x^3 + 8 \end{cases}$

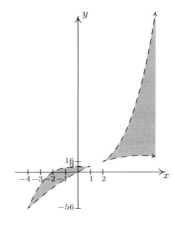

25. $\begin{cases} x + 2y^2 > 2 \\ x^2 + 4y^2 \le 4 \end{cases}$

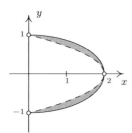

26. $\begin{cases} x^2 + y^2 \ge 25 \\ y - x \le 1 \end{cases}$

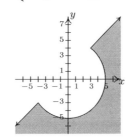

27. $x = 10$, $y = 10$, $z = 10$, $\lambda = \frac{2}{5}$

28. (c) $x^4 + 4 = (x^2 - 2x + 2)(x^2 + 2x + 2)$

29. $x^4 + 6x^2 - 5x + 6 = (x^2 - x + 1)(x^2 + x + 6)$

Chapter 9

Sequences and the Binomial Theorem

9.1 Sequences

When we first introduced a function as a special type of relation in Section 1.3, we did not put any restrictions on the domain of the function. All we said was that the set of x-coordinates of the points in the function F is called the domain, and it turns out that any subset of the real numbers, regardless of how weird that subset may be, can be the domain of a function. As our exploration of functions continued beyond Section 1.3, we saw fewer and fewer functions with 'weird' domains. It is worth your time to go back through the text to see that the domains of the polynomial, rational, exponential, logarithmic and algebraic functions discussed thus far have fairly predictable domains which almost always consist of just a collection of intervals on the real line. This may lead some readers to believe that the only important functions in a College Algebra text have domains which consist of intervals and everything else was just introductory nonsense. In this section, we introduce **sequences** which are an important class of functions whose domains are the set of natural numbers.[1] Before we get to far ahead of ourselves, let's look at what the term 'sequence' means mathematically. Informally, we can think of a sequence as an infinite list of numbers. For example, consider the sequence

$$\frac{1}{2}, -\frac{3}{4}, \frac{9}{8}, -\frac{27}{16}, \ldots \tag{1}$$

As usual, the periods of ellipsis, ..., indicate that the proposed pattern continues forever. Each of the numbers in the list is called a **term**, and we call $\frac{1}{2}$ the 'first term', $-\frac{3}{4}$ the 'second term', $\frac{9}{8}$ the 'third term' and so forth. In numbering them this way, we are setting up a function, which we'll call a per tradition, between the natural numbers and the terms in the sequence.

---

[1] Recall that this is the set $\{1, 2, 3, \ldots\}$.

n	$a(n)$
1	$\frac{1}{2}$
2	$-\frac{3}{4}$
3	$\frac{9}{8}$
4	$-\frac{27}{16}$
$\vdots$	$\vdots$

In other words, $a(n)$ is the nth term in the sequence. We formalize these ideas in our definition of a sequence and introduce some accompanying notation.

> **Definition 9.1.** A **sequence** is a function a whose domain is the natural numbers. The value $a(n)$ is often written as a_n and is called the nth **term** of the sequence. The sequence itself is usually denoted using the notation: a_n, $n \geq 1$ or the notation: $\{a_n\}_{n=1}^{\infty}$.

Applying the notation provided in Definition 9.1 to the sequence given (1), we have $a_1 = \frac{1}{2}$, $a_2 = -\frac{3}{4}$, $a_3 = \frac{9}{8}$ and so forth. Now suppose we wanted to know a_{117}, that is, the 117th term in the sequence. While the pattern of the sequence is apparent, it would benefit us greatly to have an explicit formula for a_n. Unfortunately, there is no general algorithm that will produce a formula for every sequence, so any formulas we do develop will come from that greatest of teachers, experience. In other words, it is time for an example.

Example 9.1.1. Write the first four terms of the following sequences.

1. $a_n = \dfrac{5^{n-1}}{3^n}$, $n \geq 1$

2. $b_k = \dfrac{(-1)^k}{2k+1}$, $k \geq 0$

3. $\{2n - 1\}_{n=1}^{\infty}$

4. $\left\{\dfrac{1 + (-1)^i}{i}\right\}_{i=2}^{\infty}$

5. $a_1 = 7$, $a_{n+1} = 2 - a_n$, $n \geq 1$

6. $f_0 = 1$, $f_n = n \cdot f_{n-1}$, $n \geq 1$

Solution.

1. Since we are given $n \geq 1$, the first four terms of the sequence are a_1, a_2, a_3 and a_4. Since the notation a_1 means the same thing as $a(1)$, we obtain our first term by replacing every occurrence of n in the formula for a_n with $n = 1$ to get $a_1 = \frac{5^{1-1}}{3^1} = \frac{1}{3}$. Proceeding similarly, we get $a_2 = \frac{5^{2-1}}{3^2} = \frac{5}{9}$, $a_3 = \frac{5^{3-1}}{3^3} = \frac{25}{27}$ and $a_4 = \frac{5^{4-1}}{3^4} = \frac{125}{81}$.

2. For this sequence we have $k \geq 0$, so the first four terms are b_0, b_1, b_2 and b_3. Proceeding as before, replacing in this case the variable k with the appropriate whole number, beginning with 0, we get $b_0 = \frac{(-1)^0}{2(0)+1} = 1$, $b_1 = \frac{(-1)^1}{2(1)+1} = -\frac{1}{3}$, $b_2 = \frac{(-1)^2}{2(2)+1} = \frac{1}{5}$ and $b_3 = \frac{(-1)^3}{2(3)+1} = -\frac{1}{7}$. (This sequence is called an **alternating** sequence since the signs alternate between + and −. The reader is encouraged to think what component of the formula is producing this effect.)

9.1 SEQUENCES

3. From $\{2n-1\}_{n=1}^{\infty}$, we have that $a_n = 2n-1$, $n \geq 1$. We get $a_1 = 1$, $a_2 = 3$, $a_3 = 5$ and $a_4 = 7$. (The first four terms are the first four odd natural numbers. The reader is encouraged to examine whether or not this pattern continues indefinitely.)

4. Here, we are using the letter i as a counter, not as the imaginary unit we saw in Section 3.4. Proceeding as before, we set $a_i = \frac{1+(-1)^i}{i}$, $i \geq 2$. We find $a_2 = 1$, $a_3 = 0$, $a_4 = \frac{1}{2}$ and $a_5 = 0$.

5. To obtain the terms of this sequence, we start with $a_1 = 7$ and use the equation $a_{n+1} = 2 - a_n$ for $n \geq 1$ to generate successive terms. When $n = 1$, this equation becomes $a_{1+1} = 2 - a_1$ which simplifies to $a_2 = 2 - a_1 = 2 - 7 = -5$. When $n = 2$, the equation becomes $a_{2+1} = 2 - a_2$ so we get $a_3 = 2 - a_2 = 2 - (-5) = 7$. Finally, when $n = 3$, we get $a_{3+1} = 2 - a_3$ so $a_4 = 2 - a_3 = 2 - 7 = -5$.

6. As with the problem above, we are given a place to start with $f_0 = 1$ and given a formula to build other terms of the sequence. Substituting $n = 1$ into the equation $f_n = n \cdot f_{n-1}$, we get $f_1 = 1 \cdot f_0 = 1 \cdot 1 = 1$. Advancing to $n = 2$, we get $f_2 = 2 \cdot f_1 = 2 \cdot 1 = 2$. Finally, $f_3 = 3 \cdot f_2 = 3 \cdot 2 = 6$. □

Some remarks about Example 9.1.1 are in order. We first note that since sequences are functions, we can graph them in the same way we graph functions. For example, if we wish to graph the sequence $\{b_k\}_{k=0}^{\infty}$ from Example 9.1.1, we graph the equation $y = b(k)$ for the values $k \geq 0$. That is, we plot the points $(k, b(k))$ for the values of k in the domain, $k = 0, 1, 2, \ldots$. The resulting collection of points is the graph of the sequence. Note that we do not connect the dots in a pleasing fashion as we are used to doing, because the domain is just the whole numbers in this case, not a collection of intervals of real numbers. If you feel a sense of nostalgia, you should see Section 1.2.

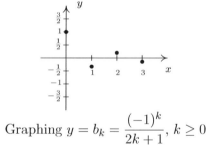

Graphing $y = b_k = \dfrac{(-1)^k}{2k+1}$, $k \geq 0$

Speaking of $\{b_k\}_{k=0}^{\infty}$, the astute and mathematically minded reader will correctly note that this technically isn't a sequence, since according to Definition 9.1, sequences are functions whose domains are the *natural* numbers, not the *whole* numbers, as is the case with $\{b_k\}_{k=0}^{\infty}$. In other words, to satisfy Definition 9.1, we need to shift the variable k so it starts at $k = 1$ instead of $k = 0$. To see how we can do this, it helps to think of the problem graphically. What we want is to shift the graph of $y = b(k)$ to the right one unit, and thinking back to Section 1.7, we can accomplish this by replacing k with $k - 1$ in the definition of $\{b_k\}_{k=0}^{\infty}$. Specifically, let $c_k = b_{k-1}$ where $k - 1 \geq 0$. We get $c_k = \frac{(-1)^{k-1}}{2(k-1)+1} = \frac{(-1)^{k-1}}{2k-1}$, where now $k \geq 1$. We leave to the reader to verify that $\{c_k\}_{k=1}^{\infty}$ generates the same list of numbers as does $\{b_k\}_{k=0}^{\infty}$, but the former satisfies Definition

9.1, while the latter does not. Like so many things in this text, we acknowledge that this point is pedantic and join the vast majority of authors who adopt a more relaxed view of Definition 9.1 to include any function which generates a list of numbers which can then be matched up with the natural numbers.[2] Finally, we wish to note the sequences in parts 5 and 6 are examples of sequences described **recursively**. In each instance, an initial value of the sequence is given which is then followed by a **recursion equation** – a formula which enables us to use known terms of the sequence to determine other terms. The terms of the sequence in part 6 are given a special name: $f_n = n!$ is called **n-factorial**. Using the '!' notation, we can describe the factorial sequence as: $0! = 1$ and $n! = n(n-1)!$ for $n \geq 1$. After $0! = 1$ the next four terms, written out in detail, are $1! = 1 \cdot 0! = 1 \cdot 1 = 1$, $2! = 2 \cdot 1! = 2 \cdot 1 = 2$, $3! = 3 \cdot 2! = 3 \cdot 2 \cdot 1 = 6$ and $4! = 4 \cdot 3! = 4 \cdot 3 \cdot 2 \cdot 1 = 24$. From this, we see a more informal way of computing $n!$, which is $n! = n \cdot (n-1) \cdot (n-2) \cdots 2 \cdot 1$ with $0! = 1$ as a special case. (We will study factorials in greater detail in Section 9.4.) The world famous Fibonacci Numbers are defined recursively and are explored in the exercises. While none of the sequences worked out to be the sequence in (1), they do give us some insight into what kinds of patterns to look for. Two patterns in particular are given in the next definition.

Definition 9.2. Arithmetic and Geometric Sequences: Suppose $\{a_n\}_{n=k}^{\infty}$ is a sequence[a]

- If there is a number d so that $a_{n+1} = a_n + d$ for all $n \geq k$, then $\{a_n\}_{n=k}^{\infty}$ is called an **arithmetic sequence**. The number d is called the **common difference**.

- If there is a number r so that $a_{n+1} = ra_n$ for all $n \geq k$, then $\{a_n\}_{n=k}^{\infty}$ is called a **geometric sequence**. The number r is called the **common ratio**.

[a]Note that we have adjusted for the fact that not all 'sequences' begin at $n = 1$.

Both arithmetic and geometric sequences are defined in terms of recursion equations. In English, an arithmetic sequence is one in which we proceed from one term to the next by always *adding* the fixed number d. The name 'common difference' comes from a slight rewrite of the recursion equation from $a_{n+1} = a_n + d$ to $a_{n+1} - a_n = d$. Analogously, a geometric sequence is one in which we proceed from one term to the next by always *multiplying* by the same fixed number r. If $r \neq 0$, we can rearrange the recursion equation to get $\frac{a_{n+1}}{a_n} = r$, hence the name 'common ratio.' Some sequences are arithmetic, some are geometric and some are neither as the next example illustrates.[3]

Example 9.1.2. Determine if the following sequences are arithmetic, geometric or neither. If arithmetic, find the common difference d; if geometric, find the common ratio r.

1. $a_n = \dfrac{5^{n-1}}{3^n}$, $n \geq 1$

2. $b_k = \dfrac{(-1)^k}{2k+1}$, $k \geq 0$

3. $\{2n - 1\}_{n=1}^{\infty}$

4. $\dfrac{1}{2}, -\dfrac{3}{4}, \dfrac{9}{8}, -\dfrac{27}{16}, \ldots$

---

[2]We're basically talking about the 'countably infinite' subsets of the real number line when we do this.
[3]Sequences which are both arithmetic and geometric are discussed in the Exercises.

9.1 SEQUENCES

Solution. A good rule of thumb to keep in mind when working with sequences is "When in doubt, write it out!" Writing out the first several terms can help you identify the pattern of the sequence should one exist.

1. From Example 9.1.1, we know that the first four terms of this sequence are $\frac{1}{3}, \frac{5}{9}, \frac{25}{27}$ and $\frac{125}{81}$. To see if this is an arithmetic sequence, we look at the successive differences of terms. We find that $a_2 - a_1 = \frac{5}{9} - \frac{1}{3} = \frac{2}{9}$ and $a_3 - a_2 = \frac{25}{27} - \frac{5}{9} = \frac{10}{27}$. Since we get different numbers, there is no 'common difference' and we have established that the sequence is *not* arithmetic. To investigate whether or not it is geometric, we compute the ratios of successive terms. The first three ratios

$$\frac{a_2}{a_1} = \frac{\frac{5}{9}}{\frac{1}{3}} = \frac{5}{3}, \quad \frac{a_3}{a_2} = \frac{\frac{25}{27}}{\frac{5}{9}} = \frac{5}{3} \quad \text{and} \quad \frac{a_4}{a_3} = \frac{\frac{125}{81}}{\frac{25}{27}} = \frac{5}{3}$$

suggest that the sequence is geometric. To prove it, we must show that $\frac{a_{n+1}}{a_n} = r$ for all n.

$$\frac{a_{n+1}}{a_n} = \frac{\frac{5^{(n+1)-1}}{3^{n+1}}}{\frac{5^{n-1}}{3^n}} = \frac{5^n}{3^{n+1}} \cdot \frac{3^n}{5^{n-1}} = \frac{5}{3}$$

This sequence is geometric with common ratio $r = \frac{5}{3}$.

2. Again, we have Example 9.1.1 to thank for providing the first four terms of this sequence: $1, -\frac{1}{3}, \frac{1}{5}$ and $-\frac{1}{7}$. We find $b_1 - b_0 = -\frac{4}{3}$ and $b_2 - b_1 = \frac{8}{15}$. Hence, the sequence is not arithmetic. To see if it is geometric, we compute $\frac{b_1}{b_0} = -\frac{1}{3}$ and $\frac{b_2}{b_1} = -\frac{3}{5}$. Since there is no 'common ratio,' we conclude the sequence is not geometric, either.

3. As we saw in Example 9.1.1, the sequence $\{2n - 1\}_{n=1}^{\infty}$ generates the odd numbers: $1, 3, 5, 7, \ldots$. Computing the first few differences, we find $a_2 - a_1 = 2$, $a_3 - a_2 = 2$, and $a_4 - a_3 = 2$. This suggests that the sequence is arithmetic. To verify this, we find

$$a_{n+1} - a_n = (2(n+1) - 1) - (2n - 1) = 2n + 2 - 1 - 2n + 1 = 2$$

This establishes that the sequence is arithmetic with common difference $d = 2$. To see if it is geometric, we compute $\frac{a_2}{a_1} = 3$ and $\frac{a_3}{a_2} = \frac{5}{3}$. Since these ratios are different, we conclude the sequence is not geometric.

4. We met our last sequence at the beginning of the section. Given that $a_2 - a_1 = -\frac{5}{4}$ and $a_3 - a_2 = \frac{15}{8}$, the sequence is not arithmetic. Computing the first few ratios, however, gives us $\frac{a_2}{a_1} = -\frac{3}{2}, \frac{a_3}{a_2} = -\frac{3}{2}$ and $\frac{a_4}{a_3} = -\frac{3}{2}$. Since these are the only terms given to us, we assume that the pattern of ratios continue in this fashion and conclude that the sequence is geometric. $\square$

We are now one step away from determining an explicit formula for the sequence given in (1). We know that it is a geometric sequence and our next result gives us the explicit formula we require.

> **Equation 9.1. Formulas for Arithmetic and Geometric Sequences:**
>
> - An arithmetic sequence with first term a and common difference d is given by
>
> $$a_n = a + (n-1)d, \quad n \geq 1$$
>
> - A geometric sequence with first term a and common ratio $r \neq 0$ is given by
>
> $$a_n = ar^{n-1}, \quad n \geq 1$$

While the formal proofs of the formulas in Equation 9.1 require the techniques set forth in Section 9.3, we attempt to motivate them here. According to Definition 9.2, given an arithmetic sequence with first term a and common difference d, the way we get from one term to the next is by adding d. Hence, the terms of the sequence are: $a, a+d, a+2d, a+3d, \ldots$. We see that to reach the nth term, we add d to a exactly $(n-1)$ times, which is what the formula says. The derivation of the formula for geometric series follows similarly. Here, we start with a and go from one term to the next by multiplying by r. We get a, ar, ar^2, ar^3 and so forth. The nth term results from multiplying a by r exactly $(n-1)$ times. We note here that the reason $r = 0$ is excluded from Equation 9.1 is to avoid an instance of 0^0 which is an indeterminant form.[4] With Equation 9.1 in place, we finally have the tools required to find an explicit formula for the nth term of the sequence given in (1). We know from Example 9.1.2 that it is geometric with common ratio $r = -\frac{3}{2}$. The first term is $a = \frac{1}{2}$ so by Equation 9.1 we get $a_n = ar^{n-1} = \frac{1}{2}\left(-\frac{3}{2}\right)^{n-1}$ for $n \geq 1$. After a touch of simplifying, we get $a_n = \frac{(-3)^{n-1}}{2^n}$ for $n \geq 1$. Note that we can easily check our answer by substituting in values of n and seeing that the formula generates the sequence given in (1). We leave this to the reader. Our next example gives us more practice finding patterns.

Example 9.1.3. Find an explicit formula for the nth term of the following sequences.

1. $0.9, 0.09, 0.009, 0.0009, \ldots$
2. $\frac{2}{5}, 2, -\frac{2}{3}, -\frac{2}{7}, \ldots$
3. $1, -\frac{2}{7}, \frac{4}{13}, -\frac{8}{19}, \ldots$

Solution.

1. Although this sequence may seem strange, the reader can verify it is actually a geometric sequence with common ratio $r = 0.1 = \frac{1}{10}$. With $a = 0.9 = \frac{9}{10}$, we get $a_n = \frac{9}{10}\left(\frac{1}{10}\right)^{n-1}$ for $n \geq 0$. Simplifying, we get $a_n = \frac{9}{10^n}$, $n \geq 1$. There is more to this sequence than meets the eye and we shall return to this example in the next section.

2. As the reader can verify, this sequence is neither arithmetic nor geometric. In an attempt to find a pattern, we rewrite the second term with a denominator to make all the terms appear as fractions. We have $\frac{2}{5}, \frac{2}{1}, -\frac{2}{3}, -\frac{2}{7}, \ldots$. If we associate the negative '−' of the last two terms with the denominators we get $\frac{2}{5}, \frac{2}{1}, \frac{2}{-3}, \frac{2}{-7}, \ldots$. This tells us that we can tentatively sketch out the formula for the sequence as $a_n = \frac{2}{d_n}$ where d_n is the sequence of denominators.

---

[4]See the footnotes on page 237 in Section 3.1 and page 418 of Section 6.1.

9.1 SEQUENCES

Looking at the denominators $5, 1, -3, -7, \ldots$, we find that they go from one term to the next by subtracting 4 which is the same as adding -4. This means we have an arithmetic sequence on our hands. Using Equation 9.1 with $a = 5$ and $d = -4$, we get the nth denominator by the formula $d_n = 5 + (n-1)(-4) = 9 - 4n$ for $n \geq 1$. Our final answer is $a_n = \frac{2}{9-4n}$, $n \geq 1$.

3. The sequence as given is neither arithmetic nor geometric, so we proceed as in the last problem to try to get patterns individually for the numerator and denominator. Letting c_n and d_n denote the sequence of numerators and denominators, respectively, we have $a_n = \frac{c_n}{d_n}$. After some experimentation,[5] we choose to write the first term as a fraction and associate the negatives '$-$' with the numerators. This yields $\frac{1}{1}, \frac{-2}{7}, \frac{4}{13}, \frac{-8}{19}, \ldots$. The numerators form the sequence $1, -2, 4, -8, \ldots$ which is geometric with $a = 1$ and $r = -2$, so we get $c_n = (-2)^{n-1}$, for $n \geq 1$. The denominators $1, 7, 13, 19, \ldots$ form an arithmetic sequence with $a = 1$ and $d = 6$. Hence, we get $d_n = 1 + 6(n-1) = 6n - 5$, for $n \geq 1$. We obtain our formula for $a_n = \frac{c_n}{d_n} = \frac{(-2)^{n-1}}{6n-5}$, for $n \geq 1$. We leave it to the reader to show that this checks out. $\square$

While the last problem in Example 9.1.3 was neither geometric nor arithmetic, it did resolve into a combination of these two kinds of sequences. If handed the sequence $2, 5, 10, 17, \ldots$, we would be hard-pressed to find a formula for a_n if we restrict our attention to these two archetypes. We said before that there is no general algorithm for finding the explicit formula for the nth term of a given sequence, and it is only through experience gained from evaluating sequences from explicit formulas that we learn to begin to recognize number patterns. The pattern $1, 4, 9, 16, \ldots$ is rather recognizable as the squares, so the formula $a_n = n^2$, $n \geq 1$ may not be too hard to determine. With this in mind, it's possible to see $2, 5, 10, 17, \ldots$ as the sequence $1 + 1, 4 + 1, 9 + 1, 16 + 1, \ldots$, so that $a_n = n^2 + 1$, $n \geq 1$. Of course, since we are given only a small *sample* of the sequence, we shouldn't be too disappointed to find out this isn't the *only* formula which generates this sequence. For example, consider the sequence defined by $b_n = -\frac{1}{4}n^4 + \frac{5}{2}n^3 - \frac{31}{4}n^2 + \frac{25}{2}n - 5$, $n \geq 1$. The reader is encouraged to verify that it also produces the terms $2, 5, 10, 17$. In fact, it can be shown that given any finite sample of a sequence, there are infinitely many explicit formulas all of which generate those same finite points. This means that there will be infinitely many correct answers to some of the exercises in this section.[6] Just because your answer doesn't match ours doesn't mean it's wrong. As always, when in doubt, write your answer out. As long as it produces the same terms in the same order as what the problem wants, your answer is correct.

Sequences play a major role in the Mathematics of Finance, as we have already seen with Equation 6.2 in Section 6.5. Recall that if we invest P dollars at an annual percentage rate r and compound the interest n times per year, the formula for A_k, the amount in the account after k compounding periods, is $A_k = P\left(1 + \frac{r}{n}\right)^k = \left[P\left(1 + \frac{r}{n}\right)\right]\left(1 + \frac{r}{n}\right)^{k-1}$, $k \geq 1$. We now spot this as a geometric sequence with first term $P\left(1 + \frac{r}{n}\right)$ and common ratio $\left(1 + \frac{r}{n}\right)$. In retirement planning, it is seldom the case that an investor deposits a set amount of money into an account and waits for it to grow. Usually, additional payments of principal are made at regular intervals and the value of the investment grows accordingly. This kind of investment is called an **annuity** and will be discussed in the next section once we have developed more mathematical machinery.

---
[5] Here we take 'experimentation' to mean a frustrating guess-and-check session.
[6] For more on this, see When Every Answer is Correct: Why Sequences and Number Patterns Fail the Test.

9.1.1 Exercises

In Exercises 1 - 13, write out the first four terms of the given sequence.

1. $a_n = 2^n - 1$, $n \geq 0$

2. $d_j = (-1)^{\frac{j(j+1)}{2}}$, $j \geq 1$

3. $\{5k - 2\}_{k=1}^{\infty}$

4. $\left\{\dfrac{n^2 + 1}{n + 1}\right\}_{n=0}^{\infty}$

5. $\left\{\dfrac{x^n}{n^2}\right\}_{n=1}^{\infty}$

6. $\left\{\dfrac{\ln(n)}{n}\right\}_{n=1}^{\infty}$

7. $a_1 = 3$, $a_{n+1} = a_n - 1$, $n \geq 1$

8. $d_0 = 12$, $d_m = \dfrac{d_{m-1}}{100}$, $m \geq 1$

9. $b_1 = 2$, $b_{k+1} = 3b_k + 1$, $k \geq 1$

10. $c_0 = -2$, $c_j = \dfrac{c_{j-1}}{(j+1)(j+2)}$, $j \geq 1$

11. $a_1 = 117$, $a_{n+1} = \dfrac{1}{a_n}$, $n \geq 1$

12. $s_0 = 1$, $s_{n+1} = x^{n+1} + s_n$, $n \geq 0$

13. $F_0 = 1$, $F_1 = 1$, $F_n = F_{n-1} + F_{n-2}$, $n \geq 2$ (This is the famous Fibonacci Sequence)

In Exercises 14 - 21 determine if the given sequence is arithmetic, geometric or neither. If it is arithmetic, find the common difference d; if it is geometric, find the common ratio r.

14. $\{3n - 5\}_{n=1}^{\infty}$

15. $a_n = n^2 + 3n + 2$, $n \geq 1$

16. $\dfrac{1}{3}, \dfrac{1}{6}, \dfrac{1}{12}, \dfrac{1}{24}, \ldots$

17. $\left\{3\left(\dfrac{1}{5}\right)^{n-1}\right\}_{n=1}^{\infty}$

18. $17, 5, -7, -19, \ldots$

19. $2, 22, 222, 2222, \ldots$

20. $0.9, 9, 90, 900, \ldots$

21. $a_n = \dfrac{n!}{2}$, $n \geq 0$.

In Exercises 22 - 30, find an explicit formula for the nth term of the given sequence. Use the formulas in Equation 9.1 as needed.

22. $3, 5, 7, 9, \ldots$

23. $1, -\dfrac{1}{2}, \dfrac{1}{4}, -\dfrac{1}{8}, \ldots$

24. $1, \dfrac{2}{3}, \dfrac{4}{5}, \dfrac{8}{7}, \ldots$

25. $1, \dfrac{2}{3}, \dfrac{1}{3}, \dfrac{4}{27}, \ldots$

26. $1, \dfrac{1}{4}, \dfrac{1}{9}, \dfrac{1}{16}, \ldots$

27. $x, -\dfrac{x^3}{3}, \dfrac{x^5}{5}, -\dfrac{x^7}{7}, \ldots$

9.1 SEQUENCES

28. $0.9, 0.99, 0.999, 0.9999, \ldots$ 29. $27, 64, 125, 216, \ldots$ 30. $1, 0, 1, 0, \ldots$

31. Find a sequence which is both arithmetic and geometric. (Hint: Start with $a_n = c$ for all n.)

32. Show that a geometric sequence can be transformed into an arithmetic sequence by taking the natural logarithm of the terms.

33. Thomas Robert Malthus is credited with saying, "The power of population is indefinitely greater than the power in the earth to produce subsistence for man. Population, when unchecked, increases in a geometrical ratio. Subsistence increases only in an arithmetical ratio. A slight acquaintance with numbers will show the immensity of the first power in comparison with the second." (See this webpage for more information.) Discuss this quote with your classmates from a sequences point of view.

34. This classic problem involving sequences shows the power of geometric sequences. Suppose that a wealthy benefactor agrees to give you one penny today and then double the amount she gives you each day for 30 days. So, for example, you get two pennies on the second day and four pennies on the third day. How many pennies do you get on the 30th day? What is the total dollar value of the gift you have received?

35. Research the terms 'arithmetic mean' and 'geometric mean.' With the help of your classmates, show that a given term of a arithmetic sequence a_k, $k \geq 2$ is the arithmetic mean of the term immediately preceding, a_{k-1} it and immediately following it, a_{k+1}. State and prove an analogous result for geometric sequences.

36. Discuss with your classmates how the results of this section might change if we were to examine sequences of other mathematical things like complex numbers or matrices. Find an explicit formula for the nth term of the sequence $i, -1, -i, 1, i, \ldots$. List out the first four terms of the matrix sequences we discussed in Exercise 8.3.1 in Section 8.3.

9.1.2 Answers

1. $0, 1, 3, 7$

2. $-1, -1, 1, 1$

3. $3, 8, 13, 18$

4. $1, 1, \frac{5}{3}, \frac{5}{2}$

5. $x, \frac{x^2}{4}, \frac{x^3}{9}, \frac{x^4}{16}$

6. $0, \frac{\ln(2)}{2}, \frac{\ln(3)}{3}, \frac{\ln(4)}{4}$

7. $3, 2, 1, 0$

8. $12, 0.12, 0.0012, 0.000012$

9. $2, 7, 22, 67$

10. $-2, -\frac{1}{3}, -\frac{1}{36}, -\frac{1}{720}$

11. $117, \frac{1}{117}, 117, \frac{1}{117}$

12. $1, x+1, x^2+x+1, x^3+x^2+x+1$

13. $1, 1, 2, 3$

14. arithmetic, $d = 3$

15. neither

16. geometric, $r = \frac{1}{2}$

17. geometric, $r = \frac{1}{5}$

18. arithmetic, $d = -12$

19. neither

20. geometric, $r = 10$

21. neither

22. $a_n = 1 + 2n$, $n \geq 1$

23. $a_n = \left(-\frac{1}{2}\right)^{n-1}$, $n \geq 1$

24. $a_n = \frac{2^{n-1}}{2n-1}$, $n \geq 1$

25. $a_n = \frac{n}{3^{n-1}}$, $n \geq 1$

26. $a_n = \frac{1}{n^2}$, $n \geq 1$

27. $\frac{(-1)^{n-1} x^{2n-1}}{2n-1}$, $n \geq 1$

28. $a_n = \frac{10^n - 1}{10^n}$, $n \geq 1$

29. $a_n = (n+2)^3$, $n \geq 1$

30. $a_n = \frac{1+(-1)^{n-1}}{2}$, $n \geq 1$

9.2 Summation Notation

In the previous section, we introduced sequences and now we shall present notation and theorems concerning the sum of terms of a sequence. We begin with a definition, which, while intimidating, is meant to make our lives easier.

> **Definition 9.3. Summation Notation:** Given a sequence $\{a_n\}_{n=k}^{\infty}$ and numbers m and p satisfying $k \leq m \leq p$, the summation from m to p of the sequence $\{a_n\}$ is written
> $$\sum_{n=m}^{p} a_n = a_m + a_{m+1} + \ldots + a_p$$
> The variable n is called the **index of summation**. The number m is called the **lower limit of summation** while the number p is called the **upper limit of summation**.

In English, Definition 9.3 is simply defining a short-hand notation for adding up the terms of the sequence $\{a_n\}_{n=k}^{\infty}$ from a_m through a_p. The symbol Σ is the capital Greek letter sigma and is shorthand for 'sum'. The lower and upper limits of the summation tells us which term to start with and which term to end with, respectively. For example, using the sequence $a_n = 2n - 1$ for $n \geq 1$, we can write the sum $a_3 + a_4 + a_5 + a_6$ as

$$\begin{aligned} \sum_{n=3}^{6}(2n-1) &= (2(3)-1) + (2(4)-1) + (2(5)-1) + (2(6)-1) \\ &= 5 + 7 + 9 + 11 \\ &= 32 \end{aligned}$$

The index variable is considered a 'dummy variable' in the sense that it may be changed to any letter without affecting the value of the summation. For instance,

$$\sum_{n=3}^{6}(2n-1) = \sum_{k=3}^{6}(2k-1) = \sum_{j=3}^{6}(2j-1)$$

One place you may encounter summation notation is in mathematical definitions. For example, summation notation allows us to define polynomials as functions of the form

$$f(x) = \sum_{k=0}^{n} a_k x^k$$

for real numbers a_k, $k = 0, 1, \ldots n$. The reader is invited to compare this with what is given in Definition 3.1. Summation notation is particularly useful when talking about matrix operations. For example, we can write the product of the ith row R_i of a matrix $A = [a_{ij}]_{m \times n}$ and the jth column C_j of a matrix $B = [b_{ij}]_{n \times r}$ as

$$Ri \cdot Cj = \sum_{k=1}^{n} a_{ik} b_{kj}$$

Again, the reader is encouraged to write out the sum and compare it to Definition 8.9. Our next example gives us practice with this new notation.

Example 9.2.1.

1. Find the following sums.

   (a) $\displaystyle\sum_{k=1}^{4} \frac{13}{100^k}$
   (b) $\displaystyle\sum_{n=0}^{4} \frac{n!}{2}$
   (c) $\displaystyle\sum_{n=1}^{5} \frac{(-1)^{n+1}}{n}(x-1)^n$

2. Write the following sums using summation notation.

   (a) $1 + 3 + 5 + \ldots + 117$
   (b) $1 - \dfrac{1}{2} + \dfrac{1}{3} - \dfrac{1}{4} + - \ldots + \dfrac{1}{117}$
   (c) $0.9 + 0.09 + 0.009 + \ldots 0.\underbrace{0 \cdots 0}_{n-1 \text{ zeros}}9$

Solution.

1. (a) We substitute $k = 1$ into the formula $\frac{13}{100^k}$ and add successive terms until we reach $k = 4$.

    $$\begin{aligned}\sum_{k=1}^{4} \frac{13}{100^k} &= \frac{13}{100^1} + \frac{13}{100^2} + \frac{13}{100^3} + \frac{13}{100^4} \\ &= 0.13 + 0.0013 + 0.000013 + 0.00000013 \\ &= 0.13131313\end{aligned}$$

   (b) Proceeding as in (a), we replace every occurrence of n with the values 0 through 4. We recall the factorials, $n!$ as defined in number Example 9.1.1, number 6 and get:

    $$\begin{aligned}\sum_{n=0}^{4} \frac{n!}{2} &= \frac{0!}{2} + \frac{1!}{2} + \frac{2!}{2} + \frac{3!}{2} = \frac{4!}{2} \\ &= \frac{1}{2} + \frac{1}{2} + \frac{2 \cdot 1}{2} + \frac{3 \cdot 2 \cdot 1}{2} + \frac{4 \cdot 3 \cdot 2 \cdot 1}{2} \\ &= \frac{1}{2} + \frac{1}{2} + 1 + 3 + 12 \\ &= 17\end{aligned}$$

   (c) We proceed as before, replacing the index n, but *not* the variable x, with the values 1 through 5 and adding the resulting terms.

9.2 Summation Notation

$$\sum_{n=1}^{5} \frac{(-1)^{n+1}}{n}(x-1)^n = \frac{(-1)^{1+1}}{1}(x-1)^1 + \frac{(-1)^{2+1}}{2}(x-1)^2 + \frac{(-1)^{3+1}}{3}(x-1)^3$$

$$+ \frac{(-1)^{1+4}}{4}(x-1)^4 + \frac{(-1)^{1+5}}{5}(x-1)^5$$

$$= (x-1) - \frac{(x-1)^2}{2} + \frac{(x-1)^3}{3} - \frac{(x-1)^4}{4} + \frac{(x-1)^5}{5}$$

2. The key to writing these sums with summation notation is to find the pattern of the terms. To that end, we make good use of the techniques presented in Section 9.1.

   (a) The terms of the sum 1, 3, 5, etc., form an arithmetic sequence with first term $a = 1$ and common difference $d = 2$. We get a formula for the nth term of the sequence using Equation 9.1 to get $a_n = 1 + (n-1)2 = 2n - 1$, $n \geq 1$. At this stage, we have the formula for the terms, namely $2n - 1$, and the lower limit of the summation, $n = 1$. To finish the problem, we need to determine the upper limit of the summation. In other words, we need to determine which value of n produces the term 117. Setting $a_n = 117$, we get $2n - 1 = 117$ or $n = 59$. Our final answer is

   $$1 + 3 + 5 + \ldots + 117 = \sum_{n=1}^{59}(2n - 1)$$

   (b) We rewrite all of the terms as fractions, the subtraction as addition, and associate the negatives '−' with the numerators to get

   $$\frac{1}{1} + \frac{-1}{2} + \frac{1}{3} + \frac{-1}{4} + \ldots + \frac{1}{117}$$

   The numerators, 1, −1, etc. can be described by the geometric sequence[1] $c_n = (-1)^{n-1}$ for $n \geq 1$, while the denominators are given by the arithmetic sequence[2] $d_n = n$ for $n \geq 1$. Hence, we get the formula $a_n = \frac{(-1)^{n-1}}{n}$ for our terms, and we find the lower and upper limits of summation to be $n = 1$ and $n = 117$, respectively. Thus

   $$1 - \frac{1}{2} + \frac{1}{3} - \frac{1}{4} + - \ldots + \frac{1}{117} = \sum_{n=1}^{117} \frac{(-1)^{n-1}}{n}$$

   (c) Thanks to Example 9.1.3, we know that one formula for the nth term is $a_n = \frac{9}{10^n}$ for $n \geq 1$. This gives us a formula for the summation as well as a lower limit of summation. To determine the upper limit of summation, we note that to produce the $n - 1$ zeros to the right of the decimal point before the 9, we need a denominator of 10^n. Hence, n is

---

[1] This is indeed a geometric sequence with first term $a = 1$ and common ratio $r = -1$.
[2] It is an arithmetic sequence with first term $a = 1$ and common difference $d = 1$.

the upper limit of summation. Since n is used in the limits of the summation, we need to choose a different letter for the index of summation.[3] We choose k and get

$$0.9 + 0.09 + 0.009 + \ldots 0.\underbrace{0\cdots0}_{n-1 \text{ zeros}}9 = \sum_{k=1}^{n} \frac{9}{10^k}$$

□

The following theorem presents some general properties of summation notation. While we shall not have much need of these properties in Algebra, they do play a great role in Calculus. Moreover, there is much to be learned by thinking about why the properties hold. We invite the reader to prove these results. To get started, remember, "When in doubt, write it out!"

Theorem 9.1. Properties of Summation Notation: Suppose $\{a_n\}$ and $\{b_n\}$ are sequences so that the following sums are defined.

- $\displaystyle\sum_{n=m}^{p} (a_n \pm b_n) = \sum_{n=m}^{p} a_n \pm \sum_{n=m}^{p} b_n$

- $\displaystyle\sum_{n=m}^{p} c\, a_n = c \sum_{n=m}^{p} a_n$, for any real number c.

- $\displaystyle\sum_{n=m}^{p} a_n = \sum_{n=m}^{j} a_n + \sum_{n=j+1}^{p} a_n$, for any natural number $m \leq j < j+1 \leq p$.

- $\displaystyle\sum_{n=m}^{p} a_n = \sum_{n=m+r}^{p+r} a_{n-r}$, for any whole number r.

We now turn our attention to the sums involving arithmetic and geometric sequences. Given an arithmetic sequence $a_k = a + (k-1)d$ for $k \geq 1$, we let S denote the sum of the first n terms. To derive a formula for S, we write it out in two different ways

$$\begin{array}{rccccccc}
S &=& a &+& (a+d) &+& \ldots &+& (a+(n-2)d) &+& (a+(n-1)d) \\
S &=& (a+(n-1)d) &+& (a+(n-2)d) &+& \ldots &+& (a+d) &+& a
\end{array}$$

If we add these two equations and combine the terms which are aligned vertically, we get

$$2S = (2a + (n-1)d) + (2a + (n-1)d) + \ldots + (2a + (n-1)d) + (2a + (n-1)d)$$

The right hand side of this equation contains n terms, all of which are equal to $(2a + (n-1)d)$ so we get $2S = n(2a + (n-1)d)$. Dividing both sides of this equation by 2, we obtain the formula

---

[3]To see why, try writing the summation using 'n' as the index.

9.2 Summation Notation

$$S = \frac{n}{2}(2a + (n-1)d)$$

If we rewrite the quantity $2a + (n-1)d$ as $a + (a + (n-1)d) = a_1 + a_n$, we get the formula

$$S = n\left(\frac{a_1 + a_n}{2}\right)$$

A helpful way to remember this last formula is to recognize that we have expressed the sum as the product of the number of terms n and the *average* of the first and n^{th} terms.

To derive the formula for the geometric sum, we start with a geometric sequence $a_k = ar^{k-1}$, $k \geq 1$, and let S once again denote the sum of the first n terms. Comparing S and rS, we get

$$\begin{array}{rcccccccccc} S & = & a & + & ar & + & ar^2 & + & \ldots & + & ar^{n-2} & + & ar^{n-1} \\ rS & = & & & ar & + & ar^2 & + & \ldots & + & ar^{n-2} & + & ar^{n-1} & + & ar^n \end{array}$$

Subtracting the second equation from the first forces all of the terms except a and ar^n to cancel out and we get $S - rS = a - ar^n$. Factoring, we get $S(1-r) = a(1 - r^n)$. Assuming $r \neq 1$, we can divide both sides by the quantity $(1-r)$ to obtain

$$S = a\left(\frac{1 - r^n}{1 - r}\right)$$

If we distribute a through the numerator, we get $a - ar^n = a_1 - a_{n+1}$ which yields the formula

$$S = \frac{a_1 - a_{n+1}}{1 - r}$$

In the case when $r = 1$, we get the formula

$$S = \underbrace{a + a + \ldots + a}_{n \text{ times}} = n\,a$$

Our results are summarized below.

> **Equation 9.2. Sums of Arithmetic and Geometric Sequences:**
>
> - The sum S of the first n terms of an arithmetic sequence $a_k = a + (k-1)d$ for $k \geq 1$ is
>
> $$S = \sum_{k=1}^{n} a_k = n\left(\frac{a_1 + a_n}{2}\right) = \frac{n}{2}(2a + (n-1)d)$$
>
> - The sum S of the first n terms of a geometric sequence $a_k = ar^{k-1}$ for $k \geq 1$ is
>
>   1. $S = \sum_{k=1}^{n} a_k = \dfrac{a_1 - a_{n+1}}{1-r} = a\left(\dfrac{1-r^n}{1-r}\right)$, if $r \neq 1$.
>
>   2. $S = \sum_{k=1}^{n} a_k = \sum_{k=1}^{n} a = na$, if $r = 1$.

While we have made an honest effort to derive the formulas in Equation 9.2, formal proofs require the machinery in Section 9.3. An application of the arithmetic sum formula which proves useful in Calculus results in formula for the sum of the first n natural numbers. The natural numbers themselves are a sequence[4] 1, 2, 3, ... which is arithmetic with $a = d = 1$. Applying Equation 9.2,

$$1 + 2 + 3 + \ldots + n = \frac{n(n+1)}{2}$$

So, for example, the sum of the first 100 natural numbers[5] is $\frac{100(101)}{2} = 5050$.

An important application of the geometric sum formula is the investment plan called an **annuity**. Annuities differ from the kind of investments we studied in Section 6.5 in that payments are deposited into the account on an on-going basis, and this complicates the mathematics a little.[6] Suppose you have an account with annual interest rate r which is compounded n times per year. We let $i = \frac{r}{n}$ denote the interest rate per period. Suppose we wish to make ongoing deposits of P dollars at the *end* of each compounding period. Let A_k denote the amount in the account after k compounding periods. Then $A_1 = P$, because we have made our first deposit at the *end* of the first compounding period and no interest has been earned. During the second compounding period, we earn interest on A_1 so that our initial investment has grown to $A_1(1+i) = P(1+i)$ in accordance with Equation 6.1. When we add our second payment at the end of the second period, we get

$$A_2 = A_1(1+i) + P = P(1+i) + P = P(1+i)\left(1 + \frac{1}{1+i}\right)$$

The reason for factoring out the $P(1+i)$ will become apparent in short order. During the third compounding period, we earn interest on A_2 which then grows to $A_2(1+i)$. We add our third

---

[4]This is the identity function on the natural numbers!

[5]There is an interesting anecdote which says that the famous mathematician Carl Friedrich Gauss was given this problem in primary school and devised a very clever solution.

[6]The reader may wish to re-read the discussion on compound interest in Section 6.5 before proceeding.

9.2 Summation Notation

payment at the end of the third compounding period to obtain

$$A_3 = A_2(1+i) + P = P(1+i)\left(1 + \frac{1}{1+i}\right)(1+i) + P = P(1+i)^2\left(1 + \frac{1}{1+i} + \frac{1}{(1+i)^2}\right)$$

During the fourth compounding period, A_3 grows to $A_3(1+i)$, and when we add the fourth payment, we factor out $P(1+i)^3$ to get

$$A_4 = P(1+i)^3\left(1 + \frac{1}{1+i} + \frac{1}{(1+i)^2} + \frac{1}{(1+i)^3}\right)$$

This pattern continues so that at the end of the kth compounding, we get

$$A_k = P(1+i)^{k-1}\left(1 + \frac{1}{1+i} + \frac{1}{(1+i)^2} + \ldots + \frac{1}{(1+i)^{k-1}}\right)$$

The sum in the parentheses above is the sum of the first k terms of a geometric sequence with $a = 1$ and $r = \frac{1}{1+i}$. Using Equation 9.2, we get

$$1 + \frac{1}{1+i} + \frac{1}{(1+i)^2} + \ldots + \frac{1}{(1+i)^{k-1}} = 1\left(\frac{1 - \frac{1}{(1+i)^k}}{1 - \frac{1}{1+i}}\right) = \frac{(1+i)\left(1 - (1+i)^{-k}\right)}{i}$$

Hence, we get

$$A_k = P(1+i)^{k-1}\left(\frac{(1+i)\left(1 - (1+i)^{-k}\right)}{i}\right) = \frac{P\left((1+i)^k - 1\right)}{i}$$

If we let t be the number of years this investment strategy is followed, then $k = nt$, and we get the formula for the future value of an **ordinary annuity**.

Equation 9.3. Future Value of an Ordinary Annuity: Suppose an annuity offers an annual interest rate r compounded n times per year. Let $i = \frac{r}{n}$ be the interest rate per compounding period. If a deposit P is made at the end of each compounding period, the amount A in the account after t years is given by

$$A = \frac{P\left((1+i)^{nt} - 1\right)}{i}$$

The reader is encouraged to substitute $i = \frac{r}{n}$ into Equation 9.3 and simplify. Some familiar equations arise which are cause for pause and meditation. One last note: if the deposit P is made a the *beginning* of the compounding period instead of at the end, the annuity is called an **annuity-due**. We leave the derivation of the formula for the future value of an annuity-due as an exercise for the reader.

Example 9.2.2. An ordinary annuity offers a 6% annual interest rate, compounded monthly.

1. If monthly payments of $50 are made, find the value of the annuity in 30 years.

2. How many years will it take for the annuity to grow to $100,000?

Solution.

1. We have $r = 0.06$ and $n = 12$ so that $i = \frac{r}{n} = \frac{0.06}{12} = 0.005$. With $P = 50$ and $t = 30$,

$$A = \frac{50\left((1+0.005)^{(12)(30)} - 1\right)}{0.005} \approx 50225.75$$

Our final answer is $50,225.75.

2. To find how long it will take for the annuity to grow to $100,000, we set $A = 100000$ and solve for t. We isolate the exponential and take natural logs of both sides of the equation.

$$\begin{aligned}
100000 &= \frac{50\left((1+0.005)^{12t} - 1\right)}{0.005} \\
10 &= (1.005)^{12t} - 1 \\
(1.005)^{12t} &= 11 \\
\ln\left((1.005)^{12t}\right) &= \ln(11) \\
12t \ln(1.005) &= \ln(11) \\
t &= \tfrac{\ln(11)}{12\ln(1.005)} \approx 40.06
\end{aligned}$$

This means that it takes just over 40 years for the investment to grow to $100,000. Comparing this with our answer to part 1, we see that in just 10 additional years, the value of the annuity nearly doubles. This is a lesson worth remembering. □

We close this section with a peek into Calculus by considering *infinite* sums, called **series**. Consider the number $0.\overline{9}$. We can write this number as

$$0.\overline{9} = 0.9999... = 0.9 + 0.09 + 0.009 + 0.0009 + ...$$

From Example 9.2.1, we know we can write the sum of the first n of these terms as

$$0.\underbrace{9\cdots9}_{n \text{ nines}} = .9 + 0.09 + 0.009 + \ldots 0.\underbrace{0\cdots0}_{n-1 \text{ zeros}}9 = \sum_{k=1}^{n} \frac{9}{10^k}$$

Using Equation 9.2, we have

9.2 Summation Notation

$$\sum_{k=1}^{n} \frac{9}{10^k} = \frac{9}{10}\left(\frac{1 - \frac{1}{10^{n+1}}}{1 - \frac{1}{10}}\right) = 1 - \frac{1}{10^{n+1}}$$

It stands to reason that $0.\overline{9}$ is the same value of $1 - \frac{1}{10^{n+1}}$ as $n \to \infty$. Our knowledge of exponential expressions from Section 6.1 tells us that $\frac{1}{10^{n+1}} \to 0$ as $n \to \infty$, so $1 - \frac{1}{10^{n+1}} \to 1$. We have just argued that $0.\overline{9} = 1$, which may cause some distress for some readers.[7] Any non-terminating decimal can be thought of as an infinite sum whose denominators are the powers of 10, so the phenomenon of adding up infinitely many terms and arriving at a finite number is not as foreign of a concept as it may appear. We end this section with a theorem concerning geometric series.

Theorem 9.2. Geometric Series: Given the sequence $a_k = ar^{k-1}$ for $k \geq 1$, where $|r| < 1$,

$$a + ar + ar^2 + \ldots = \sum_{k=1}^{\infty} ar^{k-1} = \frac{a}{1-r}$$

If $|r| \geq 1$, the sum $a + ar + ar^2 + \ldots$ is not defined.

The justification of the result in Theorem 9.2 comes from taking the formula in Equation 9.2 for the sum of the first n terms of a geometric sequence and examining the formula as $n \to \infty$. Assuming $|r| < 1$ means $-1 < r < 1$, so $r^n \to 0$ as $n \to \infty$. Hence as $n \to \infty$,

$$\sum_{k=1}^{n} ar^{k-1} = a\left(\frac{1-r^n}{1-r}\right) \to \frac{a}{1-r}$$

As to what goes wrong when $|r| \geq 1$, we leave that to Calculus as well, but will explore some cases in the exercises.

---

[7] To make this more palatable, it is usually accepted that $0.\overline{3} = \frac{1}{3}$ so that $0.\overline{9} = 3\left(0.\overline{3}\right) = 3\left(\frac{1}{3}\right) = 1$. Feel better?

9.2.1 Exercises

In Exercises 1 - 8, find the value of each sum using Definition 9.3.

1. $\sum_{g=4}^{9}(5g+3)$
2. $\sum_{k=3}^{8}\frac{1}{k}$
3. $\sum_{j=0}^{5}2^j$
4. $\sum_{k=0}^{2}(3k-5)x^k$

5. $\sum_{i=1}^{4}\frac{1}{4}(i^2+1)$
6. $\sum_{n=1}^{100}(-1)^n$
7. $\sum_{n=1}^{5}\frac{(n+1)!}{n!}$
8. $\sum_{j=1}^{3}\frac{5!}{j!(5-j)!}$

In Exercises 9 - 16, rewrite the sum using summation notation.

9. $8+11+14+17+20$
10. $1-2+3-4+5-6+7-8$
11. $x-\frac{x^3}{3}+\frac{x^5}{5}-\frac{x^7}{7}$
12. $1+2+4+\cdots+2^{29}$
13. $2+\frac{3}{2}+\frac{4}{3}+\frac{5}{4}+\frac{6}{5}$
14. $-\ln(3)+\ln(4)-\ln(5)+\cdots+\ln(20)$
15. $1-\frac{1}{4}+\frac{1}{9}-\frac{1}{16}+\frac{1}{25}-\frac{1}{36}$
16. $\frac{1}{2}(x-5)+\frac{1}{4}(x-5)^2+\frac{1}{6}(x-5)^3+\frac{1}{8}(x-5)^4$

In Exercises 17 - 28, use the formulas in Equation 9.2 to find the sum.

17. $\sum_{n=1}^{10} 5n+3$
18. $\sum_{n=1}^{20} 2n-1$
19. $\sum_{k=0}^{15} 3-k$

20. $\sum_{n=1}^{10}\left(\frac{1}{2}\right)^n$
21. $\sum_{n=1}^{5}\left(\frac{3}{2}\right)^n$
22. $\sum_{k=0}^{5} 2\left(\frac{1}{4}\right)^k$

23. $1+4+7+\ldots+295$
24. $4+2+0-2-\ldots-146$
25. $1+3+9+\ldots+2187$

26. $\frac{1}{2}+\frac{1}{4}+\frac{1}{8}+\ldots+\frac{1}{256}$
27. $3-\frac{3}{2}+\frac{3}{4}-\frac{3}{8}+-\cdots+\frac{3}{256}$
28. $\sum_{n=1}^{10} -2n+\left(\frac{5}{3}\right)^n$

In Exercises 29 - 32, use Theorem 9.2 to express each repeating decimal as a fraction of integers.

29. $0.\overline{7}$
30. $0.\overline{13}$
31. $10.\overline{159}$
32. $-5.8\overline{67}$

9.2 Summation Notation

In Exercises 33 - 38, use Equation 9.3 to compute the future value of the annuity with the given terms. In all cases, assume the payment is made monthly, the interest rate given is the annual rate, and interest is compounded monthly.

33. payments are $300, interest rate is 2.5%, term is 17 years.

34. payments are $50, interest rate is 1.0%, term is 30 years.

35. payments are $100, interest rate is 2.0%, term is 20 years

36. payments are $100, interest rate is 2.0%, term is 25 years

37. payments are $100, interest rate is 2.0%, term is 30 years

38. payments are $100, interest rate is 2.0%, term is 35 years

39. Suppose an ordinary annuity offers an annual interest rate of 2%, compounded monthly, for 30 years. What should the monthly payment be to have $100,000 at the end of the term?

40. Prove the properties listed in Theorem 9.1.

41. Show that the formula for the future value of an annuity due is

$$A = P(1+i)\left[\frac{(1+i)^{nt} - 1}{i}\right]$$

42. Discuss with your classmates what goes wrong when trying to find the following sums.[8]

   (a) $\sum_{k=1}^{\infty} 2^{k-1}$

   (b) $\sum_{k=1}^{\infty} (1.0001)^{k-1}$

   (c) $\sum_{k=1}^{\infty} (-1)^{k-1}$

---

[8] When in doubt, write them out!

9.2.2 Answers

1. 213
2. $\frac{341}{280}$
3. 63
4. $-5 - 2x + x^2$
5. $\frac{17}{2}$
6. 0
7. 20
8. 25
9. $\sum_{k=1}^{5}(3k+5)$
10. $\sum_{k=1}^{8}(-1)^{k-1}k$
11. $\sum_{k=1}^{4}(-1)^{k-1}\frac{x^{2k-1}}{2k-1}$
12. $\sum_{k=1}^{30}2^{k-1}$
13. $\sum_{k=1}^{5}\frac{k+1}{k}$
14. $\sum_{k=3}^{20}(-1)^k \ln(k)$
15. $\sum_{k=1}^{6}\frac{(-1)^{k-1}}{k^2}$
16. $\sum_{k=1}^{4}\frac{1}{2k}(x-5)^k$
17. 305
18. 400
19. -72
20. $\frac{1023}{1024}$
21. $\frac{633}{32}$
22. $\frac{1365}{512}$
23. 14652
24. -5396
25. 3280
26. $\frac{255}{256}$
27. $\frac{513}{256}$
28. $\frac{17771050}{59049}$
29. $\frac{7}{9}$
30. $\frac{13}{99}$
31. $\frac{3383}{333}$
32. $-\frac{5809}{990}$
33. $76,163.67
34. $20,981.40
35. $29,479.69
36. $38,882.12
37. 49,272.55
38. 60,754.80

39. For $100,000, the monthly payment is ≈ $202.95.

9.3 Mathematical Induction

The Chinese philosopher Confucius is credited with the saying, "A journey of a thousand miles begins with a single step." In many ways, this is the central theme of this section. Here we introduce a method of proof, Mathematical Induction, which allows us to *prove* many of the formulas we have merely *motivated* in Sections 9.1 and 9.2 by starting with just a single step. A good example is the formula for arithmetic sequences we touted in Equation 9.1. Arithmetic sequences are defined recursively, starting with $a_1 = a$ and then $a_{n+1} = a_n + d$ for $n \geq 1$. This tells us that we start the sequence with a and we go from one term to the next by successively adding d. In symbols,

$$a, a+d, a+2d, a+3d, a+4d + \ldots$$

The pattern *suggested* here is that to reach the nth term, we start with a and add d to it exactly $n-1$ times, which lead us to our formula $a_n = a + (n-1)d$ for $n \geq 1$. But how do we *prove* this to be the case? We have the following.

> **The Principle of Mathematical Induction (PMI):** Suppose $P(n)$ is a sentence involving the natural number n.
>
> **IF**
>
> 1. $P(1)$ is true **and**
>
> 2. whenever $P(k)$ is true, it follows that $P(k+1)$ is also true
>
> **THEN** the sentence $P(n)$ is true for all natural numbers n.

The Principle of Mathematical Induction, or PMI for short, is exactly that - a principle.[1] It is a property of the natural numbers we either choose to accept or reject. In English, it says that if we want to prove that a formula works for all natural numbers n, we start by showing it is true for $n = 1$ (the '**base step**') and then show that if it is true for a generic natural number k, it must be true for the next natural number, $k + 1$ (the '**inductive step**'). The notation $P(n)$ acts just like function notation. For example, if $P(n)$ is the sentence (formula) '$n^2 + 1 = 3$', then $P(1)$ would be '$1^2 + 1 = 3$', which is false. The construction $P(k+1)$ would be '$(k+1)^2 + 1 = 3$'. As usual, this new concept is best illustrated with an example. Returning to our quest to prove the formula for an arithmetic sequence, we first identify $P(n)$ as the formula $a_n = a + (n-1)d$. To prove this formula is valid for all natural numbers n, we need to do two things. First, we need to establish that $P(1)$ is true. In other words, is it true that $a_1 = a + (1-1)d$? The answer is yes, since this simplifies to $a_1 = a$, which is part of the definition of the arithmetic sequence. The second thing we need to show is that whenever $P(k)$ is true, it follows that $P(k+1)$ is true. In other words, we *assume* $P(k)$ is true (this is called the '**induction hypothesis**') and *deduce* that $P(k+1)$ is also true. Assuming $P(k)$ to be true seems to invite disaster - after all, isn't this essentially what we're trying to prove in the first place? To help explain this step a little better, we show how this works for specific values of n. We've already established $P(1)$ is true, and we now want to show that $P(2)$

---

[1] Another word for this you may have seen is 'axiom.'

is true. Thus we need to show that $a_2 = a + (2-1)d$. Since $P(1)$ is true, we have $a_1 = a$, and by the definition of an arithmetic sequence, $a_2 = a_1 + d = a + d = a + (2-1)d$. So $P(2)$ is true. We now use the fact that $P(2)$ is true to show that $P(3)$ is true. Using the fact that $a_2 = a + (2-1)d$, we show $a_3 = a + (3-1)d$. Since $a_3 = a_2 + d$, we get $a_3 = (a + (2-1)d) + d = a + 2d = a + (3-1)d$, so we have shown $P(3)$ is true. Similarly, we can use the fact that $P(3)$ is true to show that $P(4)$ is true, and so forth. In general, if $P(k)$ is true (i.e., $a_k = a + (k-1)d$) we set out to show that $P(k+1)$ is true (i.e., $a_{k+1} = a + ((k+1) - 1)d$). Assuming $a_k = a + (k-1)d$, we have by the definition of an arithmetic sequence that $a_{k+1} = a_k + d$ so we get $a_{k+1} = (a + (k-1)d) + d = a + kd = a + ((k+1) - 1)d$. Hence, $P(k+1)$ is true.

In essence, by showing that $P(k+1)$ must always be true when $P(k)$ is true, we are showing that the formula $P(1)$ can be used to get the formula $P(2)$, which in turn can be used to derive the formula $P(3)$, which in turn can be used to establish the formula $P(4)$, and so on. Thus as long as $P(k)$ is true for some natural number k, $P(n)$ is true for all of the natural numbers n which follow k. Coupling this with the fact $P(1)$ is true, we have established $P(k)$ is true for all natural numbers which follow $n = 1$, in other words, all natural numbers n. One might liken Mathematical Induction to a repetitive process like climbing stairs.[2] If you are sure that (1) you can get on the stairs (the base case) and (2) you can climb from any one step to the next step (the inductive step), then presumably you can climb the entire staircase.[3] We get some more practice with induction in the following example.

Example 9.3.1. Prove the following assertions using the Principle of Mathematical Induction.

1. The sum formula for arithmetic sequences: $\sum_{j=1}^{n}(a + (j-1)d) = \dfrac{n}{2}(2a + (n-1)d)$.

2. For a complex number z, $(\overline{z})^n = \overline{z^n}$ for $n \geq 1$.

3. $3^n > 100n$ for $n > 5$.

4. Let A be an $n \times n$ matrix and let A' be the matrix obtained by replacing a row R of A with cR for some real number c. Use the definition of determinant to show $\det(A') = c\det(A)$.

Solution.

1. We set $P(n)$ to be the equation we are asked to prove. For $n = 1$, we compare both sides of the equation given in $P(n)$

$$\sum_{j=1}^{1}(a + (j-1)d) \stackrel{?}{=} \dfrac{1}{2}(2a + (1-1)d)$$

$$a + (1-1)d \stackrel{?}{=} \dfrac{1}{2}(2a)$$

$$a = a \checkmark$$

---

[2] Falling dominoes is the most widely used metaphor in the mainstream College Algebra books.
[3] This is how Carl climbed the stairs in the Cologne Cathedral. Well, that, and encouragement from Kai.

9.3 Mathematical Induction

This shows the base case $P(1)$ is true. Next we assume $P(k)$ is true, that is, we assume

$$\sum_{j=1}^{k}(a+(j-1)d) = \frac{k}{2}(2a+(k-1)d)$$

and attempt to use this to show $P(k+1)$ is true. Namely, we must show

$$\sum_{j=1}^{k+1}(a+(j-1)d) = \frac{k+1}{2}(2a+(k+1-1)d)$$

To see how we can use $P(k)$ in this case to prove $P(k+1)$, we note that the sum in $P(k+1)$ is the sum of the first $k+1$ terms of the sequence $a_k = a+(k-1)d$ for $k \geq 1$ while the sum in $P(k)$ is the sum of the first k terms. We compare both side of the equation in $P(k+1)$.

$$\underbrace{\sum_{j=1}^{k+1}(a+(j-1)d)}_{\text{summing the first } k+1 \text{ terms}} \stackrel{?}{=} \frac{k+1}{2}(2a+(k+1-1)d)$$

$$\underbrace{\sum_{j=1}^{k}(a+(j-1)d)}_{\text{summing the first } k \text{ terms}} + \underbrace{(a+(k+1-1)d)}_{\text{adding the }(k+1)\text{st term}} \stackrel{?}{=} \frac{k+1}{2}(2a+kd)$$

$$\underbrace{\frac{k}{2}(2a+(k-1)d)}_{\text{Using } P(k)} + (a+kd) \stackrel{?}{=} \frac{(k+1)(2a+kd)}{2}$$

$$\frac{k(2a+(k-1)d) + 2(a+kd)}{2} \stackrel{?}{=} \frac{2ka+k^2d+2a+kd}{2}$$

$$\frac{2ka+2a+k^2d+kd}{2} = \frac{2ka+2a+k^2d+kd}{2} \checkmark$$

Since all of our steps on both sides of the string of equations are reversible, we conclude that the two sides of the equation are equivalent and hence, $P(k+1)$ is true. By the Principle of Mathematical Induction, we have that $P(n)$ is true for all natural numbers n.

2. We let $P(n)$ be the formula $(\overline{z})^n = \overline{z^n}$. The base case $P(1)$ is $(\overline{z})^1 = \overline{z^1}$, which reduces to $\overline{z} = \overline{z}$ which is true. We now assume $P(k)$ is true, that is, we assume $(\overline{z})^k = \overline{z^k}$ and attempt to show that $P(k+1)$ is true. Since $(\overline{z})^{k+1} = (\overline{z})^k \, \overline{z}$, we can use the induction hypothesis and

write $(\overline{z})^k = \overline{z^k}$. Hence, $(\overline{z})^{k+1} = (\overline{z})^k \, \overline{z} = \overline{z^k} \, \overline{z}$. We now use the product rule for conjugates[4] to write $\overline{z^k} \, \overline{z} = \overline{z^k z} = \overline{z^{k+1}}$. This establishes $(\overline{z})^{k+1} = \overline{z^{k+1}}$, so that $P(k+1)$ is true. Hence, by the Principle of Mathematical Induction, $(\overline{z})^n = \overline{z^n}$ for all $n \geq 1$.

3. The first wrinkle we encounter in this problem is that we are asked to prove this formula for $n > 5$ instead of $n \geq 1$. Since n is a natural number, this means our base step occurs at $n = 6$. We can still use the PMI in this case, but our conclusion will be that the formula is valid for all $n \geq 6$. We let $P(n)$ be the inequality $3^n > 100n$, and check that $P(6)$ is true. Comparing $3^6 = 729$ and $100(6) = 600$, we see $3^6 > 100(6)$ as required. Next, we assume that $P(k)$ is true, that is we assume $3^k > 100k$. We need to show that $P(k+1)$ is true, that is, we need to show $3^{k+1} > 100(k+1)$. Since $3^{k+1} = 3 \cdot 3^k$, the induction hypothesis gives $3^{k+1} = 3 \cdot 3^k > 3(100k) = 300k$. We are done if we can show $300k > 100(k+1)$ for $k \geq 6$. Solving $300k > 100(k+1)$ we get $k > \frac{1}{2}$. Since $k \geq 6$, we know this is true. Putting all of this together, we have $3^{k+1} = 3 \cdot 3^k > 3(100k) = 300k > 100(k+1)$, and hence $P(k+1)$ is true. By induction, $3^n > 100n$ for all $n \geq 6$.

4. To prove this determinant property, we use induction on n, where we take $P(n)$ to be that the property we wish to prove is true for all $n \times n$ matrices. For the base case, we note that if A is a 1×1 matrix, then $A = [a]$ so $A' = [ca]$. By definition, $\det(A) = a$ and $\det(A') = ca$ so we have $\det(A') = c \det(A)$ as required. Now suppose that the property we wish to prove is true for all $k \times k$ matrices. Let A be a $(k+1) \times (k+1)$ matrix. We have two cases, depending on whether or not the row R being replaced is the first row of A.

CASE 1: The row R being replaced is the first row of A. By definition,

$$\det(A') = \sum_{p=1}^{n} a'_{1p} C'_{1p}$$

where the $1p$ cofactor of A' is $C'_{1p} = (-1)^{(1+p)} \det\left(A'_{1p}\right)$ and A'_{1p} is the $k \times k$ matrix obtained by deleting the 1st row and pth column of A'.[5] Since the first row of A' is c times the first row of A, we have $a'_{1p} = c a_{1p}$. In addition, since the remaining rows of A' are identical to those of A, $A'_{1p} = A_{1p}$. (To obtain these matrices, the first row of A' is removed.) Hence $\det\left(A'_{1p}\right) = \det\left(A_{1p}\right)$, so that $C'_{1p} = C_{1p}$. As a result, we get

$$\det(A') = \sum_{p=1}^{n} a'_{1p} C'_{1p} = \sum_{p=1}^{n} c \, a_{1p} C_{1p} = c \sum_{p=1}^{n} a_{1p} C_{1p} = c \det(A),$$

as required. Hence, $P(k+1)$ is true in this case, which means the result is true in this case for all natural numbers $n \geq 1$. (You'll note that we did not use the induction hypothesis at all in this case. It is possible to restructure the proof so that induction is only used where

---
[4]See Exercise 54 in Section 3.4.
[5]See Section 8.5 for a review of this notation.

9.3 Mathematical Induction

it is needed. While mathematically more elegant, it is less intuitive, and we stand by our approach because of its pedagogical value.)

CASE 2: The row R being replaced is the not the first row of A. By definition,

$$\det(A') = \sum_{p=1}^{n} a'_{1p} C'_{1p},$$

where in this case, $a'_{1p} = a_{1p}$, since the first rows of A and A' are the same. The matrices A'_{1p} and A_{1p}, on the other hand, are different but in a very predictable way – the row in A'_{1p} which corresponds to the row cR in A' is exactly c times the row in A_{1p} which corresponds to the row R in A. In other words, A'_{1p} and A_{1p} are $k \times k$ matrices which satisfy the induction hypothesis. Hence, we know $\det\left(A'_{1p}\right) = c \det\left(A_{1p}\right)$ and $C'_{1p} = c\, C_{1p}$. We get

$$\det(A') = \sum_{p=1}^{n} a'_{1p} C'_{1p} = \sum_{p=1}^{n} a_{1p} c\, C_{1p} = c \sum_{p=1}^{n} a_{1p} C_{1p} = c \det(A),$$

which establishes $P(k+1)$ to be true. Hence by induction, we have shown that the result holds in this case for $n \geq 1$ and we are done. $\square$

While we have used the Principle of Mathematical Induction to prove some of the formulas we have merely motivated in the text, our main use of this result comes in Section 9.4 to prove the celebrated Binomial Theorem. The ardent Mathematics student will no doubt see the PMI in many courses yet to come. Sometimes it is explicitly stated and sometimes it remains hidden in the background. If ever you see a property stated as being true 'for all natural numbers n', it's a solid bet that the formal proof requires the Principle of Mathematical Induction.

9.3.1 Exercises

In Exercises 1 - 7, prove each assertion using the Principle of Mathematical Induction.

1. $\sum_{j=1}^{n} j^2 = \dfrac{n(n+1)(2n+1)}{6}$

2. $\sum_{j=1}^{n} j^3 = \dfrac{n^2(n+1)^2}{4}$

3. $2^n > 500n$ for $n > 12$

4. $3^n \geq n^3$ for $n \geq 4$

5. Use the Product Rule for Absolute Value to show $|x^n| = |x|^n$ for all real numbers x and all natural numbers $n \geq 1$

6. Use the Product Rule for Logarithms to show $\log(x^n) = n\log(x)$ for all real numbers $x > 0$ and all natural numbers $n \geq 1$.

7. $\begin{bmatrix} a & 0 \\ 0 & b \end{bmatrix}^n = \begin{bmatrix} a^n & 0 \\ 0 & b^n \end{bmatrix}$ for $n \geq 1$.

8. Prove Equations 9.1 and 9.2 for the case of geometric sequences. That is:

    (a) For the sequence $a_1 = a$, $a_{n+1} = ra_n$, $n \geq 1$, prove $a_n = ar^{n-1}$, $n \geq 1$.

    (b) $\sum_{j=1}^{n} ar^{n-1} = a\left(\dfrac{1-r^n}{1-r}\right)$, if $r \neq 1$, $\sum_{j=1}^{n} ar^{n-1} = na$, if $r = 1$.

9. Prove that the determinant of a lower triangular matrix is the product of the entries on the main diagonal. (See Exercise 8.3.1 in Section 8.3.) Use this result to then show $\det(I_n) = 1$ where I_n is the $n \times n$ identity matrix.

10. Discuss the classic 'paradox' All Horses are the Same Color problem with your classmates.

9.3 Mathematical Induction

9.3.2 Selected Answers

1. Let $P(n)$ be the sentence $\sum_{j=1}^{n} j^2 = \dfrac{n(n+1)(2n+1)}{6}$. For the base case, $n=1$, we get

$$\sum_{j=1}^{1} j^2 \stackrel{?}{=} \dfrac{(1)(1+1)(2(1)+1)}{6}$$

$$1^2 = 1 \checkmark$$

We now assume $P(k)$ is true and use it to show $P(k+1)$ is true. We have

$$\sum_{j=1}^{k+1} j^2 \stackrel{?}{=} \dfrac{(k+1)((k+1)+1)(2(k+1)+1)}{6}$$

$$\sum_{j=1}^{k} j^2 + (k+1)^2 \stackrel{?}{=} \dfrac{(k+1)(k+2)(2k+3)}{6}$$

$$\underbrace{\dfrac{k(k+1)(2k+1)}{6}}_{\text{Using } P(k)} + (k+1)^2 \stackrel{?}{=} \dfrac{(k+1)(k+2)(2k+3)}{6}$$

$$\dfrac{k(k+1)(2k+1)}{6} + \dfrac{6(k+1)^2}{6} \stackrel{?}{=} \dfrac{(k+1)(k+2)(2k+3)}{6}$$

$$\dfrac{k(k+1)(2k+1) + 6(k+1)^2}{6} \stackrel{?}{=} \dfrac{(k+1)(k+2)(2k+3)}{6}$$

$$\dfrac{(k+1)(k(2k+1) + 6(k+1))}{6} \stackrel{?}{=} \dfrac{(k+1)(k+2)(2k+3)}{6}$$

$$\dfrac{(k+1)\left(2k^2 + 7k + 6\right)}{6} \stackrel{?}{=} \dfrac{(k+1)(k+2)(2k+3)}{6}$$

$$\dfrac{(k+1)(k+2)(2k+3)}{6} = \dfrac{(k+1)(k+2)(2k+3)}{6} \checkmark$$

By induction, $\sum_{j=1}^{n} j^2 = \dfrac{n(n+1)(2n+1)}{6}$ is true for all natural numbers $n \geq 1$.

4. Let $P(n)$ be the sentence $3^n > n^3$. Our base case is $n=4$ and we check $3^4 = 81$ and $4^3 = 64$ so that $3^4 > 4^3$ as required. We now assume $P(k)$ is true, that is $3^k > k^3$, and try to show $P(k+1)$ is true. We note that $3^{k+1} = 3 \cdot 3^k > 3k^3$ and so we are done if we can show $3k^3 > (k+1)^3$ for $k \geq 4$. We can solve the inequality $3x^3 > (x+1)^3$ using the techniques of Section 5.3, and doing so gives us $x > \dfrac{1}{\sqrt[3]{3}-1} \approx 2.26$. Hence, for $k \geq 4$, $3^{k+1} = 3 \cdot 3^k > 3k^3 > (k+1)^3$ so that $3^{k+1} > (k+1)^3$. By induction, $3^n > n^3$ is true for all natural numbers $n \geq 4$.

6. Let $P(n)$ be the sentence $\log(x^n) = n \log(x)$. For the duration of this argument, we assume $x > 0$. The base case $P(1)$ amounts checking that $\log(x^1) = 1\log(x)$ which is clearly true. Next we assume $P(k)$ is true, that is $\log(x^k) = k\log(x)$ and try to show $P(k+1)$ is true. Using the Product Rule for Logarithms along with the induction hypothesis, we get

$$\log\left(x^{k+1}\right) = \log\left(x^k \cdot x\right) = \log\left(x^k\right) + \log(x) = k\log(x) + \log(x) = (k+1)\log(x)$$

Hence, $\log\left(x^{k+1}\right) = (k+1)\log(x)$. By induction $\log(x^n) = n\log(x)$ is true for all $x > 0$ and all natural numbers $n \geq 1$.

9. Let A be an $n \times n$ lower triangular matrix. We proceed to prove the $\det(A)$ is the product of the entries along the main diagonal by inducting on n. For $n = 1$, $A = [a]$ and $\det(A) = a$, so the result is (trivially) true. Next suppose the result is true for $k \times k$ lower triangular matrices. Let A be a $(k+1) \times (k+1)$ lower triangular matrix. Expanding $\det(A)$ along the first row, we have

$$\det(A) = \sum_{p=1}^{n} a_{1p} C_{1p}$$

Since $a_{1p} = 0$ for $2 \leq p \leq k+1$, this simplifies $\det(A) = a_{11} C_{11}$. By definition, we know that $C_{11} = (-1)^{1+1} \det(A_{11}) = \det(A_{11})$ where A_{11} is $k \times k$ matrix obtained by deleting the first row and first column of A. Since A is lower triangular, so is A_{11} and, as such, the induction hypothesis applies to A_{11}. In other words, $\det(A_{11})$ is the product of the entries along A_{11}'s main diagonal. Now, the entries on the main diagonal of A_{11} are the entries $a_{22}, a_{33}, \ldots, a_{(k+1)(k+1)}$ from the main diagonal of A. Hence,

$$\det(A) = a_{11} \det(A_{11}) = a_{11} \left(a_{22} a_{33} \cdots a_{(k+1)(k+1)}\right) = a_{11} a_{22} a_{33} \cdots a_{(k+1)(k+1)}$$

We have $\det(A)$ is the product of the entries along its main diagonal. This shows $P(k+1)$ is true, and, hence, by induction, the result holds for all $n \times n$ upper triangular matrices. The $n \times n$ identity matrix I_n is a lower triangular matrix whose main diagonal consists of all 1's. Hence, $\det(I_n) = 1$, as required.

9.4 The Binomial Theorem

In this section, we aim to prove the celebrated **Binomial Theorem**. Simply stated, the Binomial Theorem is a formula for the expansion of quantities $(a+b)^n$ for natural numbers n. In Elementary and Intermediate Algebra, you should have seen specific instances of the formula, namely

$$\begin{aligned} (a+b)^1 &= a+b \\ (a+b)^2 &= a^2 + 2ab + b^2 \\ (a+b)^3 &= a^3 + 3a^2b + 3ab^2 + b^3 \end{aligned}$$

If we wanted the expansion for $(a+b)^4$ we would write $(a+b)^4 = (a+b)(a+b)^3$ and use the formula that we have for $(a+b)^3$ to get $(a+b)^4 = (a+b)\left(a^3 + 3a^2b + 3ab^2 + b^3\right) = a^4 + 4a^3b + 6a^2b^2 + 4ab^3 + b^4$. Generalizing this a bit, we see that if we have a formula for $(a+b)^k$, we can obtain a formula for $(a+b)^{k+1}$ by rewriting the latter as $(a+b)^{k+1} = (a+b)(a+b)^k$. Clearly this means Mathematical Induction plays a major role in the proof of the Binomial Theorem.[1] Before we can state the theorem we need to revisit the sequence of factorials which were introduced in Example 9.1.1 number 6 in Section 9.1.

> **Definition 9.4. Factorials:** For a whole number n, $\boldsymbol{n}$ **factorial**, denoted $n!$, is the term f_n of the sequence $f_0 = 1$, $f_n = n \cdot f_{n-1}$, $n \geq 1$.

Recall this means $0! = 1$ and $n! = n(n-1)!$ for $n \geq 1$. Using the recursive definition, we get: $1! = 1 \cdot 0! = 1 \cdot 1 = 1$, $2! = 2 \cdot 1! = 2 \cdot 1 = 2$, $3! = 3 \cdot 2! = 3 \cdot 2 \cdot 1 = 6$ and $4! = 4 \cdot 3! = 4 \cdot 3 \cdot 2 \cdot 1 = 24$. Informally, $n! = n \cdot (n-1) \cdot (n-2) \cdots 2 \cdot 1$ with $0! = 1$ as our 'base case.' Our first example familiarizes us with some of the basic computations involving factorials.

Example 9.4.1.

1. Simplify the following expressions.

   (a) $\dfrac{3!\,2!}{0!}$
   (b) $\dfrac{7!}{5!}$
   (c) $\dfrac{1000!}{998!\,2!}$
   (d) $\dfrac{(k+2)!}{(k-1)!}$, $k \geq 1$

2. Prove $n! > 3^n$ for all $n \geq 7$.

Solution.

1. We keep in mind the mantra, "When in doubt, write it out!" as we simplify the following.

   (a) We have been programmed to react with alarm to the presence of a 0 in the denominator, but in this case $0! = 1$, so the fraction is defined after all. As for the numerator, $3! = 3 \cdot 2 \cdot 1 = 6$ and $2! = 2 \cdot 1 = 2$, so we have $\dfrac{3!\,2!}{0!} = \dfrac{(6)(2)}{1} = 12$.

---

[1] It's pretty much the reason Section 9.3 is in the book.

(b) We have $7! = 7 \cdot 6 \cdot 5 \cdot 4 \cdot 3 \cdot 2 \cdot 1 = 5040$ while $5! = 5 \cdot 4 \cdot 3 \cdot 2 \cdot 1 = 120$. Dividing, we get $\frac{7!}{5!} = \frac{5040}{120} = 42$. While this is correct, we note that we could have saved ourselves some of time had we proceeded as follows

$$\frac{7!}{5!} = \frac{7 \cdot 6 \cdot 5 \cdot 4 \cdot 3 \cdot 2 \cdot 1}{5 \cdot 4 \cdot 3 \cdot 2 \cdot 1} = \frac{7 \cdot 6 \cdot \cancel{5} \cdot \cancel{4} \cdot \cancel{3} \cdot \cancel{2} \cdot \cancel{1}}{\cancel{5} \cdot \cancel{4} \cdot \cancel{3} \cdot \cancel{2} \cdot \cancel{1}} = 7 \cdot 6 = 42$$

In fact, should we want to fully exploit the recursive nature of the factorial, we can write

$$\frac{7!}{5!} = \frac{7 \cdot 6 \cdot 5!}{5!} = \frac{7 \cdot 6 \cdot \cancel{5!}}{\cancel{5!}} = 42$$

(c) Keeping in mind the lesson we learned from the previous problem, we have

$$\frac{1000!}{998!\,2!} = \frac{1000 \cdot 999 \cdot 998!}{998! \cdot 2!} = \frac{1000 \cdot 999 \cdot \cancel{998!}}{\cancel{998!} \cdot 2!} = \frac{999000}{2} = 499500$$

(d) This problem continues the theme which we have seen in the previous two problems. We first note that since $k + 2$ is larger than $k - 1$, $(k+2)!$ contains all of the factors of $(k-1)!$ and as a result we can get the $(k-1)!$ to cancel from the denominator. To see this, we begin by writing out $(k+2)!$ starting with $(k+2)$ and multiplying it by the numbers which precede it until we reach $(k-1)$: $(k+2)! = (k+2)(k+1)(k)(k-1)!$. As a result, we have

$$\frac{(k+2)!}{(k-1)!} = \frac{(k+2)(k+1)(k)(k-1)!}{(k-1)!} = \frac{(k+2)(k+1)(k)\cancel{(k-1)!}}{\cancel{(k-1)!}} = k(k+1)(k+2)$$

The stipulation $k \geq 1$ is there to ensure that all of the factorials involved are defined.

2. We proceed by induction and let $P(n)$ be the inequality $n! > 3^n$. The base case here is $n = 7$ and we see that $7! = 5040$ is larger than $3^7 = 2187$, so $P(7)$ is true. Next, we assume that $P(k)$ is true, that is, we assume $k! > 3^k$ and attempt to show $P(k+1)$ follows. Using the properties of the factorial, we have $(k+1)! = (k+1)k!$ and since $k! > 3^k$, we have $(k+1)! > (k+1)3^k$. Since $k \geq 7$, $k + 1 \geq 8$, so $(k+1)3^k \geq 8 \cdot 3^k > 3 \cdot 3^k = 3^{k+1}$. Putting all of this together, we have $(k+1)! = (k+1)k! > (k+1)3^k > 3^{k+1}$ which shows $P(k+1)$ is true. By the Principle of Mathematical Induction, we have $n! > 3^n$ for all $n \geq 7$. □

Of all of the mathematical animals we have discussed in the text, factorials grow most quickly. In problem 2 of Example 9.4.1, we proved that $n!$ overtakes 3^n at $n = 7$. 'Overtakes' may be too polite a word, since $n!$ thoroughly trounces 3^n for $n \geq 7$, as any reasonable set of data will show. It can be shown that for any real number $x > 0$, not only does $n!$ eventually overtake x^n, but the ratio $\frac{x^n}{n!} \to 0$ as $n \to \infty$.[2]

Applications of factorials in the wild often involve counting arrangements. For example, if you have fifty songs on your mp3 player and wish arrange these songs in a playlist in which the order of the

---

[2]This fact is far more important than you could ever possibly imagine.

9.4 THE BINOMIAL THEOREM   683

songs matters, it turns out that there are 50! different possible playlists. If you wish to select only ten of the songs to create a playlist, then there are $\frac{50!}{40!}$ such playlists. If, on the other hand, you just want to select ten song files out of the fifty to put on a flash memory card so that now the order no longer matters, there are $\frac{50!}{40!10!}$ ways to achieve this.[3] While some of these ideas are explored in the Exercises, the authors encourage you to take courses such as Finite Mathematics, Discrete Mathematics and Statistics. We introduce these concepts here because this is how the factorials make their way into the Binomial Theorem, as our next definition indicates.

Definition 9.5. Binomial Coefficients: Given two whole numbers n and j with $n \geq j$, the binomial coefficient $\binom{n}{j}$ (read, n choose j) is the whole number given by

$$\binom{n}{j} = \frac{n!}{j!(n-j)!}$$

The name 'binomial coefficient' will be justified shortly. For now, we can physically interpret $\binom{n}{j}$ as the number of ways to select j items from n items where the order of the items selected is unimportant. For example, suppose you won two free tickets to a special screening of the latest Hollywood blockbuster and have five good friends each of whom would love to accompany you to the movies. There are $\binom{5}{2}$ ways to choose who goes with you. Applying Definition 9.5, we get

$$\binom{5}{2} = \frac{5!}{2!(5-2)!} = \frac{5!}{2!3!} = \frac{5 \cdot 4}{2} = 10$$

So there are 10 different ways to distribute those two tickets among five friends. (Some will see it as 10 ways to decide which three friends have to stay home.) The reader is encouraged to verify this by actually taking the time to list all of the possibilities.

We now state anf prove a theorem which is crucial to the proof of the Binomial Theorem.

Theorem 9.3. For natural numbers n and j with $n \geq j$,

$$\binom{n}{j-1} + \binom{n}{j} = \binom{n+1}{j}$$

The proof of Theorem 9.3 is purely computational and uses the definition of binomial coefficients, the recursive property of factorials and common denominators.

---

[3] For reference,

$$50! = 30414093201713378043612608166064768844377641568960512000000000000,$$
$$\frac{50!}{40!} = 37276043023296000, \text{ and}$$
$$\frac{50!}{40!10!} = 10272278170$$

$$\begin{aligned}
\binom{n}{j-1} + \binom{n}{j} &= \frac{n!}{(j-1)!(n-(j-1))!} + \frac{n!}{j!(n-j)!} \\
&= \frac{n!}{(j-1)!(n-j+1)!} + \frac{n!}{j!(n-j)!} \\
&= \frac{n!}{(j-1)!(n-j+1)(n-j)!} + \frac{n!}{j(j-1)!(n-j)!} \\
&= \frac{n!\,j}{j(j-1)!(n-j+1)(n-j)!} + \frac{n!(n-j+1)}{j(j-1)!(n-j+1)(n-j)!} \\
&= \frac{n!\,j}{j!(n-j+1)!} + \frac{n!(n-j+1)}{j!(n-j+1)!} \\
&= \frac{n!\,j + n!(n-j+1)}{j!(n-j+1)!} \\
&= \frac{n!\,(j + (n-j+1))}{j!(n-j+1)!} \\
&= \frac{(n+1)n!}{j!(n+1-j))!} \\
&= \frac{(n+1)!}{j!((n+1)-j))!} \\
&= \binom{n+1}{j} \checkmark
\end{aligned}$$

We are now in position to state and prove the Binomial Theorem where we see that binomial coefficients are just that - coefficients in the binomial expansion.

> **Theorem 9.4. Binomial Theorem:** For nonzero real numbers a and b,
> $$(a+b)^n = \sum_{j=0}^{n} \binom{n}{j} a^{n-j} b^j$$
> for all natural numbers n.

To get a feel of what this theorem is saying and how it really isn't as hard to remember as it may first appear, let's consider the specific case of $n = 4$. According to the theorem, we have

9.4 THE BINOMIAL THEOREM

$$\begin{aligned}(a+b)^4 &= \sum_{j=0}^{4}\binom{4}{j}a^{4-j}b^j \\ &= \binom{4}{0}a^{4-0}b^0 + \binom{4}{1}a^{4-1}b^1 + \binom{4}{2}a^{4-2}b^2 + \binom{4}{3}a^{4-3}b^3 + \binom{4}{4}a^{4-4}b^4 \\ &= \binom{4}{0}a^4 + \binom{4}{1}a^3b + \binom{4}{2}a^2b^2 + \binom{4}{3}ab^3 + \binom{4}{4}b^4\end{aligned}$$

We forgo the simplification of the coefficients in order to note the pattern in the expansion. First note that in each term, the total of the exponents is 4 which matched the exponent of the binomial $(a+b)^4$. The exponent on a begins at 4 and decreases by one as we move from one term to the next while the exponent on b starts at 0 and increases by one each time. Also note that the binomial coefficients themselves have a pattern. The upper number, 4, matches the exponent on the binomial $(a+b)^4$ whereas the lower number changes from term to term and matches the exponent of b in that term. This is no coincidence and corresponds to the kind of counting we discussed earlier. If we think of obtaining $(a+b)^4$ by multiplying $(a+b)(a+b)(a+b)(a+b)$, our answer is the sum of all possible products with exactly four factors - some a, some b. If we wish to count, for instance, the number of ways we obtain 1 factor of b out of a total of 4 possible factors, thereby forcing the remaining 3 factors to be a, the answer is $\binom{4}{1}$. Hence, the term $\binom{4}{1}a^3b$ is in the expansion. The other terms which appear cover the remaining cases. While this discussion gives an indication as to *why* the theorem is true, a formal proof requires Mathematical Induction.[4]

To prove the Binomial Theorem, we let $P(n)$ be the expansion formula given in the statement of the theorem and we note that $P(1)$ is true since

$$\begin{aligned}(a+b)^1 &\stackrel{?}{=} \sum_{j=0}^{1}\binom{1}{j}a^{1-j}b^j \\ a+b &\stackrel{?}{=} \binom{1}{0}a^{1-0}b^0 + \binom{1}{1}a^{1-1}b^1 \\ a+b &= a+b\checkmark\end{aligned}$$

Now we assume that $P(k)$ is true. That is, we assume that we can expand $(a+b)^k$ using the formula given in Theorem 9.4 and attempt to show that $P(k+1)$ is true.

---

[4]and a fair amount of tenacity and attention to detail.

$$\begin{aligned}(a+b)^{k+1} &= (a+b)(a+b)^k \\ &= (a+b)\sum_{j=0}^{k}\binom{k}{j}a^{k-j}b^j \\ &= a\sum_{j=0}^{k}\binom{k}{j}a^{k-j}b^j + b\sum_{j=0}^{k}\binom{k}{j}a^{k-j}b^j \\ &= \sum_{j=0}^{k}\binom{k}{j}a^{k+1-j}b^j + \sum_{j=0}^{k}\binom{k}{j}a^{k-j}b^{j+1}\end{aligned}$$

Our goal is to combine as many of the terms as possible within the two summations. As the counter j in the first summation runs from 0 through k, we get terms involving a^{k+1}, $a^k b$, $a^{k-1}b^2$, ..., ab^k. In the second summation, we get terms involving $a^k b$, $a^{k-1}b^2$, ..., ab^k, b^{k+1}. In other words, apart from the first term in the first summation and the last term in the second summation, we have terms common to both summations. Our next move is to 'kick out' the terms which we cannot combine and rewrite the summations so that we can combine them. To that end, we note

$$\sum_{j=0}^{k}\binom{k}{j}a^{k+1-j}b^j = a^{k+1} + \sum_{j=1}^{k}\binom{k}{j}a^{k+1-j}b^j$$

and

$$\sum_{j=0}^{k}\binom{k}{j}a^{k-j}b^{j+1} = \sum_{j=0}^{k-1}\binom{k}{j}a^{k-j}b^{j+1} + b^{k+1}$$

so that

$$(a+b)^{k+1} = a^{k+1} + \sum_{j=1}^{k}\binom{k}{j}a^{k+1-j}b^j + \sum_{j=0}^{k-1}\binom{k}{j}a^{k-j}b^{j+1} + b^{k+1}$$

We now wish to write

$$\sum_{j=1}^{k}\binom{k}{j}a^{k+1-j}b^j + \sum_{j=0}^{k-1}\binom{k}{j}a^{k-j}b^{j+1}$$

as a single summation. The wrinkle is that the first summation starts with $j=1$, while the second starts with $j=0$. Even though the sums produce terms with the same powers of a and b, they do so for different values of j. To resolve this, we need to shift the index on the second summation so that the index j starts at $j=1$ instead of $j=0$ and we make use of Theorem 9.1 in the process.

9.4 The Binomial Theorem

$$\sum_{j=0}^{k-1} \binom{k}{j} a^{k-j} b^{j+1} = \sum_{j=0+1}^{k-1+1} \binom{k}{j-1} a^{k-(j-1)} b^{(j-1)+1}$$

$$= \sum_{j=1}^{k} \binom{k}{j-1} a^{k+1-j} b^{j}$$

We can now combine our two sums using Theorem 9.1 and simplify using Theorem 9.3

$$\sum_{j=1}^{k} \binom{k}{j} a^{k+1-j} b^{j} + \sum_{j=0}^{k-1} \binom{k}{j} a^{k-j} b^{j+1} = \sum_{j=1}^{k} \binom{k}{j} a^{k+1-j} b^{j} + \sum_{j=1}^{k} \binom{k}{j-1} a^{k+1-j} b^{j}$$

$$= \sum_{j=1}^{k} \left[\binom{k}{j} + \binom{k}{j-1} \right] a^{k+1-j} b^{j}$$

$$= \sum_{j=1}^{k} \binom{k+1}{j} a^{k+1-j} b^{j}$$

Using this and the fact that $\binom{k+1}{0} = 1$ and $\binom{k+1}{k+1} = 1$, we get

$$(a+b)^{k+1} = a^{k+1} + \sum_{j=1}^{k} \binom{k+1}{j} a^{k+1-j} b^{j} + b^{k+1}$$

$$= \binom{k+1}{0} a^{k+1} b^{0} + \sum_{j=1}^{k} \binom{k+1}{j} a^{k+1-j} b^{j} + \binom{k+1}{k+1} a^{0} b^{k+1}$$

$$= \sum_{j=0}^{k+1} \binom{k+1}{j} a^{(k+1)-j} b^{j}$$

which shows that $P(k+1)$ is true. Hence, by induction, we have established that the Binomial Theorem holds for all natural numbers n.

Example 9.4.2. Use the Binomial Theorem to find the following.

1. $(x-2)^4$

2. 2.1^3

3. The term containing x^3 in the expansion $(2x+y)^5$

Solution.

1. Since $(x-2)^4 = (x+(-2))^4$, we identify $a = x$, $b = -2$ and $n = 4$ and obtain

$$(x-2)^4 = \sum_{j=0}^{4} \binom{4}{j} x^{4-j}(-2)^j$$
$$= \binom{4}{0}x^{4-0}(-2)^0 + \binom{4}{1}x^{4-1}(-2)^1 + \binom{4}{2}x^{4-2}(-2)^2 + \binom{4}{3}x^{4-3}(-2)^3 + \binom{4}{4}x^{4-4}(-2)^4$$
$$= x^4 - 8x^3 + 24x^2 - 32x + 16$$

2. At first this problem seem misplaced, but we can write $2.1^3 = (2 + 0.1)^3$. Identifying $a = 2$, $b = 0.1 = \frac{1}{10}$ and $n = 3$, we get

$$\begin{aligned}\left(2 + \frac{1}{10}\right)^3 &= \sum_{j=0}^{3} \binom{3}{j} 2^{3-j} \left(\frac{1}{10}\right)^j \\ &= \binom{3}{0} 2^{3-0}\left(\frac{1}{10}\right)^0 + \binom{3}{1} 2^{3-1}\left(\frac{1}{10}\right)^1 + \binom{3}{2} 2^{3-2}\left(\frac{1}{10}\right)^2 + \binom{3}{3} 2^{3-3}\left(\frac{1}{10}\right)^3 \\ &= 8 + \frac{12}{10} + \frac{6}{100} + \frac{1}{1000} \\ &= 8 + 1.2 + 0.06 + 0.001 \\ &= 9.261 \end{aligned}$$

3. Identifying $a = 2x$, $b = y$ and $n = 5$, the Binomial Theorem gives

$$(2x + y)^5 = \sum_{j=0}^{5} \binom{5}{j}(2x)^{5-j} y^j$$

Since we are concerned with only the term containing x^3, there is no need to expand the entire sum. The exponents on each term must add to 5 and if the exponent on x is 3, the exponent on y must be 2. Plucking out the term $j = 2$, we get

$$\binom{5}{2}(2x)^{5-2} y^2 = 10(2x)^3 y^2 = 80x^3 y^2$$

$\square$

We close this section with Pascal's Triangle, named in honor of the mathematician Blaise Pascal. Pascal's Triangle is obtained by arranging the binomial coefficients in the triangular fashion below.

9.4 The Binomial Theorem

$$\binom{0}{0}$$

$$\binom{1}{0} \quad \binom{1}{1}$$

$$\binom{2}{0} \quad \binom{2}{1} \quad \binom{2}{2}$$

$$\binom{3}{0} \quad \binom{3}{1} \quad \binom{3}{2} \quad \binom{3}{3}$$

$$\binom{4}{0} \quad \binom{4}{1} \quad \binom{4}{2} \quad \binom{4}{3} \quad \binom{4}{4}$$

$$\vdots$$

Since $\binom{n}{0} = 1$ and $\binom{n}{n} = 1$ for all whole numbers n, we get that each row of Pascal's Triangle begins and ends with 1. To generate the numbers in the middle of the rows (from the third row onwards), we take advantage of the additive relationship expressed in Theorem 9.3. For instance, $\binom{1}{0} + \binom{1}{1} = \binom{2}{1}$, $\binom{2}{0} + \binom{2}{1} = \binom{3}{1}$ and so forth. This relationship is indicated by the arrows in the array above. With these two facts in hand, we can quickly generate Pascal's Triangle. We start with the first two rows, 1 and 1 1. From that point on, each successive row begins and ends with 1 and the middle numbers are generated using Theorem 9.3. Below we attempt to demonstrate this building process to generate the first five rows of Pascal's Triangle.

$$
\begin{array}{c}
1 \\
1 \quad 1 \\
\boxed{1} \quad \underline{1+1} \quad \boxed{1}
\end{array}
\longrightarrow
\begin{array}{c}
1 \\
1 \quad 1 \\
1 \quad 2 \quad 1
\end{array}
$$

$$
\begin{array}{c}
1 \\
1 \quad 1 \\
1 \quad 2 \quad 1 \\
\boxed{1} \quad \underline{1+2} \quad \underline{2+1} \quad \boxed{1}
\end{array}
\longrightarrow
\begin{array}{c}
1 \\
1 \quad 1 \\
1 \quad 2 \quad 1 \\
1 \quad 3 \quad 3 \quad 1
\end{array}
$$

$$
\begin{array}{c}
1 \\
1 \quad 1 \\
1 \quad 2 \quad 1 \\
1 \quad 3 \quad 3 \quad 1 \\
\boxed{1} \quad \underline{1+3} \quad \underline{3+3} \quad \underline{3+1} \quad \boxed{1}
\end{array}
\longrightarrow
\begin{array}{c}
1 \\
1 \quad 1 \\
1 \quad 2 \quad 1 \\
1 \quad 3 \quad 3 \quad 1 \\
1 \quad 4 \quad 6 \quad 4 \quad 1
\end{array}
$$

To see how we can use Pascal's Triangle to expedite the Binomial Theorem, suppose we wish to expand $(3x - y)^4$. The coefficients we need are $\binom{4}{j}$ for $j = 0, 1, 2, 3, 4$ and are the numbers which form the fifth row of Pascal's Triangle. Since we know that the exponent of $3x$ in the first term is 4 and then decreases by one as we go from left to right while the exponent of $-y$ starts at 0 in the first term and then increases by one as we move from left to right, we quickly obtain

$$\begin{aligned}(3x - y)^4 &= (1)(3x)^4 + (4)(3x)^3(-y) + (6)(3x)^2(-y)^2 + 4(3x)(-y)^3 + 1(-y)^4 \\ &= 81x^4 - 108x^3y + 54x^2y^2 - 12xy^3 + y^4\end{aligned}$$

We would like to stress that Pascal's Triangle is a very quick method to expand an *entire* binomial. If only a term (or two or three) is required, then the Binomial Theorem is definitely the way to go.

9.4 The Binomial Theorem

9.4.1 Exercises

In Exercises 1 - 9, simplify the given expression.

1. $(3!)^2$
2. $\dfrac{10!}{7!}$
3. $\dfrac{7!}{2^3 3!}$
4. $\dfrac{9!}{4!3!2!}$
5. $\dfrac{(n+1)!}{n!}$, $n \geq 0$.
6. $\dfrac{(k-1)!}{(k+2)!}$, $k \geq 1$.
7. $\binom{8}{3}$
8. $\binom{117}{0}$
9. $\binom{n}{n-2}$, $n \geq 2$

In Exercises 10 - 13, use Pascal's Triangle to expand the given binomial.

10. $(x+2)^5$
11. $(2x-1)^4$
12. $\left(\frac{1}{3}x + y^2\right)^3$
13. $\left(x - x^{-1}\right)^4$

In Exercises 14 - 17, use Pascal's Triangle to simplify the given power of a complex number.

14. $(1 + 2i)^4$
15. $\left(-1 + i\sqrt{3}\right)^3$
16. $\left(\dfrac{\sqrt{3}}{2} + \dfrac{1}{2}i\right)^3$
17. $\left(\dfrac{\sqrt{2}}{2} - \dfrac{\sqrt{2}}{2}i\right)^4$

In Exercises 18 - 22, use the Binomial Theorem to find the indicated term.

18. The term containing x^3 in the expansion $(2x - y)^5$

19. The term containing x^{117} in the expansion $(x+2)^{118}$

20. The term containing $x^{\frac{7}{2}}$ in the expansion $(\sqrt{x} - 3)^8$

21. The term containing x^{-7} in the expansion $\left(2x - x^{-3}\right)^5$

22. The constant term in the expansion $\left(x + x^{-1}\right)^8$

23. Use the Prinicple of Mathematical Induction to prove $n! > 2^n$ for $n \geq 4$.

24. Prove $\displaystyle\sum_{j=0}^{n} \binom{n}{j} = 2^n$ for all natural numbers n. (HINT: Use the Binomial Theorem!)

25. With the help of your classmates, research <u>Patterns and Properties of Pascal's Triangle.</u>

26. You've just won three tickets to see the new film, '$8.\overline{9}$.' Five of your friends, Albert, Beth, Chuck, Dan, and Eugene, are interested in seeing it with you. With the help of your classmates, list all the possible ways to distribute your two extra tickets among your five friends. Now suppose you've come down with the flu. List all the different ways you can distribute the three tickets among these five friends. How does this compare with the first list you made? What does this have to do with the fact that $\binom{5}{2} = \binom{5}{3}$?

9.4.2 Answers

1. 36
2. 720
3. 105
4. 1260
5. $n+1$
6. $\frac{1}{k(k+1)(k+2)}$
7. 56
8. 1
9. $\frac{n(n-1)}{2}$
10. $(x+2)^5 = x^5 + 10x^4 + 40x^3 + 80x^2 + 80x + 32$
11. $(2x-1)^4 = 16x^4 - 32x^3 + 24x^2 - 8x + 1$
12. $\left(\frac{1}{3}x + y^2\right)^3 = \frac{1}{27}x^3 + \frac{1}{3}x^2y^2 + xy^4 + y^6$
13. $\left(x - x^{-1}\right)^4 = x^4 - 4x^2 + 6 - 4x^{-2} + x^{-4}$
14. $-7 - 24i$
15. 8
16. i
17. -1
18. $80x^3y^2$
19. $236x^{117}$
20. $-24x^{\frac{7}{2}}$
21. $-40x^{-7}$
22. 70

Index

n^{th} root
    of a complex number, 1000, 1001
    principal, 397
n^{th} Roots of Unity, 1006
u-substitution, 273
x-axis, 6
x-coordinate, 6
x-intercept, 25
y-axis, 6
y-coordinate, 6
y-intercept, 25

abscissa, 6
absolute value
    definition of, 173
    inequality, 211
    properties of, 173
acidity of a solution
    pH, 432
acute angle, 694
adjoint of a matrix, 622
alkalinity of a solution
    pH, 432
amplitude, 794, 881
angle
    acute, 694
    between two vectors, 1035, 1036
    central angle, 701
    complementary, 696
    coterminal, 698
    decimal degrees, 695
    definition, 693
    degree, 694
    DMS, 695
    initial side, 698
    measurement, 693
    negative, 698
    obtuse, 694
    of declination, 761
    of depression, 761
    of elevation, 753
    of inclination, 753
    oriented, 697
    positive, 698
    quadrantal, 698
    radian measure, 701
    reference, 721
    right, 694
    standard position, 698
    straight, 693
    supplementary, 696
    terminal side, 698
    vertex, 693
angle side opposite pairs, 896
angular frequency, 708
annuity
    annuity-due, 667
    ordinary
        definition of, 666
        future value, 667
applied domain of a function, 60
arccosecant
    calculus friendly
        definition of, 831
        graph of, 830
        properties of, 831

trigonometry friendly
    definition of, 828
    graph of, 827
    properties of, 828
arccosine
    definition of, 820
    graph of, 819
    properties of, 820
arccotangent
    definition of, 824
    graph of, 824
    properties of, 824
arcsecant
    calculus friendly
        definition of, 831
        graph of, 830
        properties of, 831
    trigonometry friendly
        definition of, 828
        graph of, 827
        properties of, 828
arcsine
    definition of, 820
    graph of, 820
    properties of, 820
arctangent
    definition of, 824
    graph of, 823
    properties of, 824
argument
    of a complex number
        definition of, 991
        properties of, 995
    of a function, 55
    of a logarithm, 425
    of a trigonometric function, 793
arithmetic sequence, 654
associative property
    for function composition, 366
    matrix
        addition, 579
        matrix multiplication, 585
        scalar multiplication, 581
    vector
        addition, 1015
        scalar multiplication, 1018
asymptote
    horizontal
        formal definition of, 304
        intuitive definition of, 304
        location of, 308
    of a hyperbola, 531
    slant
        determination of, 312
        formal definition of, 311
    slant (oblique), 311
    vertical
        formal definition of, 304
        intuitive definition of, 304
        location of, 306
augmented matrix, 568
average angular velocity, 707
average cost, 346
average cost function, 82
average rate of change, 160
average velocity, 706
axis of symmetry, 191

back substitution, 560
bearings, 905
binomial coefficient, 683
Binomial Theorem, 684
Bisection Method, 277
BMI, body mass index, 355
Boyle's Law, 350
buffer solution, 478

cardioid, 951
Cartesian coordinate plane, 6
Cartesian coordinates, 6
Cauchy's Bound, 269
center
    of a circle, 498
    of a hyperbola, 531
    of an ellipse, 516

central angle, 701
change of base formulas, 442
characteristic polynomial, 626
Charles's Law, 355
circle
    center of, 498
    definition of, 498
    from slicing a cone, 495
    radius of, 498
    standard equation, 498
    standard equation, alternate, 519
circular function, 744
$\text{cis}(\theta)$, 995
coefficient of determination, 226
cofactor, 616
Cofunction Identities, 773
common base, 420
common logarithm, 422
commutative property
    function composition does not have, 366
    matrix
        addition, 579
    vector
        addition, 1015
        dot product, 1034
complementary angles, 696
Complex Factorization Theorem, 290
complex number
    n^{th} root, 1000, 1001
    n^{th} Roots of Unity, 1006
    argument
        definition of, 991
        properties of, 995
    conjugate
        definition of, 288
        properties of, 289
    definition of, 2, 287, 991
    imaginary part, 991
    imaginary unit, i, 287
    modulus
        definition of, 991
        properties of, 993
    polar form
        cis-notation, 995
        principal argument, 991
        real part, 991
        rectangular form, 991
        set of, 2
complex plane, 991
component form of a vector, 1013
composite function
    definition of, 360
    properties of, 367
compound interest, 470
conic sections
    definition, 495
conjugate axis of a hyperbola, 532
conjugate of a complex number
    definition of, 288
    properties of, 289
Conjugate Pairs Theorem, 291
consistent system, 553
constant function
    as a horizontal line, 156
    formal definition of, 101
    intuitive definition of, 100
constant of proportionality, 350
constant term of a polynomial, 236
continuous, 241
continuously compounded interest, 472
contradiction, 549
coordinates
    Cartesian, 6
    polar, 919
    rectangular, 919
correlation coefficient, 226
cosecant
    graph of, 801
    of an angle, 744, 752
    properties of, 802
cosine
    graph of, 791
    of an angle, 717, 730, 744
    properties of, 791

cost
- average, 82, 346
- fixed, start-up, 82
- variable, 159

cost function, 82

cotangent
- graph of, 805
- of an angle, 744, 752
- properties of, 806

coterminal angle, 698

Coulomb's Law, 355

Cramer's Rule, 619

curve
- orientated, 1048

cycloid, 1056

decibel, 431

decimal degrees, 695

decreasing function
- formal definition of, 101
- intuitive definition of, 100

degree measure, 694

degree of a polynomial, 236

DeMoivre's Theorem, 997

dependent system, 554

dependent variable, 55

depreciation, 420

Descartes' Rule of Signs, 273

determinant of a matrix
- definition of, 614
- properties of, 616

Difference Identity
- for cosine, 771, 775
- for sine, 773, 775
- for tangent, 775

difference quotient, 79

dimension
- of a matrix, 567

direct variation, 350

directrix
- of a conic section in polar form, 981
- of a parabola, 505

discriminant
- of a conic, 979
- of a quadratic equation, 195
- trichotomy, 195

distance
- definition, 10
- distance formula, 11

distributive property
- matrix
  - matrix multiplication, 585
  - scalar multiplication, 581
- vector
  - dot product, 1034
  - scalar multiplication, 1018

DMS, 695

domain
- applied, 60
- definition of, 45
- implied, 58

dot product
- commutative property of, 1034
- definition of, 1034
- distributive property of, 1034
- geometric interpretation, 1035
- properties of, 1034
- relation to orthogonality, 1037
- relation to vector magnitude, 1034
- work, 1042

Double Angle Identities, 776

earthquake
- Richter Scale, 431

eccentricity, 522, 981

eigenvalue, 626

eigenvector, 626

ellipse
- center, 516
- definition of, 516
- eccentricity, 522
- foci, 516
- from slicing a cone, 496
- guide rectangle, 519
- major axis, 516
- minor axis, 516

INDEX 1073

reflective property, 523
standard equation, 519
vertices, 516
ellipsis (...), 31, 651
empty set, 2
end behavior
of $f(x) = ax^n, n$ even, 240
of $f(x) = ax^n, n$ odd, 240
of a function graph, 239
polynomial, 243
entry
in a matrix, 567
equation
contradiction, 549
graph of, 23
identity, 549
linear of n variables, 554
linear of two variables, 549
even function, 95
Even/Odd Identities, 770
exponential function
algebraic properties of, 437
change of base formula, 442
common base, 420
definition of, 418
graphical properties of, 419
inverse properties of, 437
natural base, 420
one-to-one properties of, 437
solving equations with, 448
extended interval notation, 756

Factor Theorem, 258
factorial, 654, 681
fixed cost, 82
focal diameter of a parabola, 507
focal length of a parabola, 506
focus
of a conic section in polar form, 981
focus (foci)
of a hyperbola, 531
of a parabola, 505
of an ellipse, 516

free variable, 552
frequency
angular, 708, 881
of a sinusoid, 795
ordinary, 708, 881
function
(absolute) maximum, 101
(absolute, global) minimum, 101
absolute value, 173
algebraic, 399
argument, 55
arithmetic, 76
as a process, 55, 378
average cost, 82
circular, 744
composite
definition of, 360
properties of, 367
constant, 100, 156
continuous, 241
cost, 82
decreasing, 100
definition as a relation, 43
dependent variable of, 55
difference, 76
difference quotient, 79
domain, 45
even, 95
exponential, 418
Fundamental Graphing Principle, 93
identity, 168
increasing, 100
independent variable of, 55
inverse
definition of, 379
properties of, 379
solving for, 384
uniqueness of, 380
linear, 156
local (relative) maximum, 101
local (relative) minimum, 101
logarithmic, 422

notation, 55
odd, 95
one-to-one, 381
periodic, 790
piecewise-defined, 62
polynomial, 235
price-demand, 82
product, 76
profit, 82
quadratic, 188
quotient, 76
range, 45
rational, 301
revenue, 82
smooth, 241
sum, 76
transformation of graphs, 120, 135
zero, 95
fundamental cycle
of $y = \cos(x)$, 791
Fundamental Graphing Principle
for equations, 23
for functions, 93
for polar equations, 938
Fundamental Theorem of Algebra, 290

Gauss-Jordan Elimination, 571
Gaussian Elimination, 557
geometric sequence, 654
geometric series, 669
graph
hole in, 305
horizontal scaling, 132
horizontal shift, 123
of a function, 93
of a relation, 20
of an equation, 23
rational function, 321
reflection about an axis, 126
transformations, 135
vertical scaling, 130
vertical shift, 121
greatest integer function, 67

growth model
limited, 475
logistic, 475
uninhibited, 472
guide rectangle
for a hyperbola, 532
for an ellipse, 519

Half-Angle Formulas, 779
harmonic motion, 885
Henderson-Hasselbalch Equation, 446
Heron's Formula, 914
hole
in a graph, 305
location of, 306
Hooke's Law, 350
horizontal asymptote
formal definition of, 304
intuitive definition of, 304
location of, 308
horizontal line, 23
Horizontal Line Test (HLT), 381
hyperbola
asymptotes, 531
branch, 531
center, 531
conjugate axis, 532
definition of, 531
foci, 531
from slicing a cone, 496
guide rectangle, 532
standard equation
horizontal, 534
vertical, 534
transverse axis, 531
vertices, 531
hyperbolic cosine, 1062
hyperbolic sine, 1062
hyperboloid, 542

identity
function, 367
matrix, additive, 579

INDEX

matrix, multiplicative, 585
  statement which is always true, 549
imaginary axis, 991
imaginary part of a complex number, 991
imaginary unit, i, 287
implied domain of a function, 58
inconsistent system, 553
increasing function
  formal definition of, 101
  intuitive definition of, 100
independent system, 554
independent variable, 55
index of a root, 397
induction
  base step, 673
  induction hypothesis, 673
  inductive step, 673
inequality
  absolute value, 211
  graphical interpretation, 209
  non-linear, 643
  quadratic, 215
  sign diagram, 214
inflection point, 477
information entropy, 477
initial side of an angle, 698
instantaneous rate of change, 161, 472, 707
integer
  definition of, 2
  greatest integer function, 67
  set of, 2
intercept
  definition of, 25
  location of, 25
interest
  compound, 470
  compounded continuously, 472
  simple, 469
Intermediate Value Theorem
  polynomial zero version, 241
interrobang, 321
intersection of two sets, 4

interval
  definition of, 3
  notation for, 3
  notation, extended, 756
inverse
  matrix, additive, 579, 581
  matrix, multiplicative, 602
  of a function
    definition of, 379
    properties of, 379
    solving for, 384
    uniqueness of, 380
inverse variation, 350
invertibility
  function, 382
invertible
  function, 379
  matrix, 602
irrational number
  definition of, 2
  set of, 2
irreducible quadratic, 291

joint variation, 350

Kepler's Third Law of Planetary Motion, 355
Kirchhoff's Voltage Law, 605

latus rectum of a parabola, 507
Law of Cosines, 910
Law of Sines, 897
leading coefficient of a polynomial, 236
leading term of a polynomial, 236
Learning Curve Equation, 315
least squares regression line, 225
lemniscate, 950
limaçon, 950
line
  horizontal, 23
  least squares regression, 225
  linear function, 156
  of best fit, 225
  parallel, 166

perpendicular, 167
point-slope form, 155
slope of, 151
slope-intercept form, 155
vertical, 23
linear equation
    n variables, 554
    two variables, 549
linear function, 156
local maximum
    formal definition of, 102
    intuitive definition of, 101
local minimum
    formal definition of, 102
    intuitive definition of, 101
logarithm
    algebraic properties of, 438
    change of base formula, 442
    common, 422
    general, "base b", 422
    graphical properties of, 423
    inverse properties of, 437
    natural, 422
    one-to-one properties of, 437
    solving equations with, 459
logarithmic scales, 431
logistic growth, 475
LORAN, 538
lower triangular matrix, 593

main diagonal, 585
major axis of an ellipse, 516
Markov Chain, 592
mathematical model, 60
matrix
    addition
        associative property, 579
        commutative property, 579
        definition of, 578
        properties of, 579
    additive identity, 579
    additive inverse, 579
    adjoint, 622

augmented, 568
characteristic polynomial, 626
cofactor, 616
definition, 567
determinant
    definition of, 614
    properties of, 616
dimension, 567
entry, 567
equality, 578
invertible, 602
leading entry, 569
lower triangular, 593
main diagonal, 585
matrix multiplication
    associative property of, 585
    definition of, 584
    distributive property, 585
    identity for, 585
    properties of, 585
minor, 616
multiplicative inverse, 602
product of row and column, 584
reduced row echelon form, 570
rotation, 986
row echelon form, 569
row operations, 568
scalar multiplication
    associative property of, 581
    definition of, 580
    distributive properties, 581
    identity for, 581
    properties of, 581
    zero product property, 581
size, 567
square matrix, 586
sum, 578
upper triangular, 593
maximum
    formal definition of, 102
    intuitive definition of, 101
measure of an angle, 693

midpoint
  definition of, 12
  midpoint formula, 13
minimum
  formal definition of, 102
  intuitive definition of, 101
minor, 616
minor axis of an ellipse, 516
model
  mathematical, 60
modulus of a complex number
  definition of, 991
  properties of, 993
multiplicity
  effect on the graph of a polynomial, 245, 249
  of a zero, 244

natural base, 420
natural logarithm, 422
natural number
  definition of, 2
  set of, 2
negative angle, 698
Newton's Law of Cooling, 421, 474
Newton's Law of Universal Gravitation, 351

oblique asymptote, 311
obtuse angle, 694
odd function, 95
Ohm's Law, 350, 605
one-to-one function, 381
ordered pair, 6
ordinary frequency, 708
ordinate, 6
orientation, 1048
oriented angle, 697
oriented arc, 704
origin, 7
orthogonal projection, 1038
orthogonal vectors, 1037
overdetermined system, 554

parabola
  axis of symmetry, 191
  definition of, 505
  directrix, 505
  focal diameter, 507
  focal length, 506
  focus, 505
  from slicing a cone, 496
  graph of a quadratic function, 188
  latus rectum, 507
  reflective property, 510
  standard equation
    horizontal, 508
    vertical, 506
  vertex, 188, 505
  vertex formulas, 194
paraboloid, 510
parallel vectors, 1030
parameter, 1048
parametric equations, 1048
parametric solution, 552
parametrization, 1048
partial fractions, 628
Pascal's Triangle, 688
password strength, 477
period
  circular motion, 708
  of a function, 790
  of a sinusoid, 881
periodic function, 790
pH, 432
phase, 795, 881
phase shift, 795, 881
pi, π, 700
piecewise-defined function, 62
point of diminishing returns, 477
point-slope form of a line, 155
polar coordinates
  conversion into rectangular, 924
  definition of, 919
  equivalent representations of, 923
  polar axis, 919
  pole, 919

polar form of a complex number, 995
polar rose, 950
polynomial division
    dividend, 258
    divisor, 258
    factor, 258
    quotient, 258
    remainder, 258
    synthetic division, 260
polynomial function
    completely factored
        over the complex numbers, 291
        over the real numbers, 291
    constant term, 236
    definition of, 235
    degree, 236
    end behavior, 239
    leading coefficient, 236
    leading term, 236
    variations in sign, 273
    zero
        lower bound, 274
        multiplicity, 244
        upper bound, 274
positive angle, 698
Power Reduction Formulas, 778
power rule
    for absolute value, 173
    for complex numbers, 997
    for exponential functions, 437
    for logarithms, 438
    for radicals, 398
    for the modulus of a complex number, 993
price-demand function, 82
principal, 469
principal n^{th} root, 397
principal argument of a complex number, 991
principal unit vectors, $\hat{\imath}, \hat{\jmath}$, 1024
Principle of Mathematical Induction, 673
product rule
    for absolute value, 173
    for complex numbers, 997
    for exponential functions, 437
    for logarithms, 438
    for radicals, 398
    for the modulus of a complex number, 993
Product to Sum Formulas, 780
profit function, 82
projection
    x−axis, 45
    y−axis, 46
    orthogonal, 1038
Pythagorean Conjugates, 751
Pythagorean Identities, 749

quadrantal angle, 698
quadrants, 8
quadratic formula, 194
quadratic function
    definition of, 188
    general form, 190
    inequality, 215
    irreducible quadratic, 291
    standard form, 190
quadratic regression, 228
Quotient Identities, 745
quotient rule
    for absolute value, 173
    for complex numbers, 997
    for exponential functions, 437
    for logarithms, 438
    for radicals, 398
    for the modulus of a complex number, 993

radian measure, 701
radical
    properties of, 398
radicand, 397
radioactive decay, 473
radius
    of a circle, 498
range
    definition of, 45
rate of change
    average, 160

INDEX   1079

  instantaneous, 161, 472
  slope of a line, 154
rational exponent, 398
rational functions, 301
rational number
  definition of, 2
  set of, 2
Rational Zeros Theorem, 269
ray
  definition of, 693
  initial point, 693
real axis, 991
Real Factorization Theorem, 292
real number
  definition of, 2
  set of, 2
real part of a complex number, 991
Reciprocal Identities, 745
rectangular coordinates
  also known as Cartesian coordinates, 919
  conversion into polar, 924
rectangular form of a complex number, 991
recursion equation, 654
reduced row echelon form, 570
reference angle, 721
Reference Angle Theorem
  for cosine and sine, 722
  for the circular functions, 747
reflection
  of a function graph, 126
  of a point, 10
regression
  coefficient of determination, 226
  correlation coefficient, 226
  least squares line, 225
  quadratic, 228
  total squared error, 225
relation
  algebraic description, 23
  definition, 20
  Fundamental Graphing Principle, 23
Remainder Theorem, 258

revenue function, 82
Richter Scale, 431
right angle, 694
root
  index, 397
  radicand, 397
Roots of Unity, 1006
rotation matrix, 986
rotation of axes, 974
row echelon form, 569
row operations for a matrix, 568

scalar multiplication
  matrix
    associative property of, 581
    definition of, 580
    distributive properties of, 581
    properties of, 581
  vector
    associative property of, 1018
    definition of, 1017
    distributive properties of, 1018
    properties of, 1018
scalar projection, 1039
secant
  graph of, 800
  of an angle, 744, 752
  properties of, 802
secant line, 160
sequence
  n^{th} term, 652
  alternating, 652
  arithmetic
    common difference, 654
    definition of, 654
    formula for n^{th} term, 656
    sum of first n terms, 666
  definition of, 652
  geometric
    common ratio, 654
    definition of, 654
    formula for n^{th} term, 656
    sum of first n terms, 666

recursive, 654
series, 668
set
    definition of, 1
    empty, 2
    intersection, 4
    roster method, 1
    set-builder notation, 1
    sets of numbers, 2
    union, 4
    verbal description, 1
set-builder notation, 1
Side-Angle-Side triangle, 910
Side-Side-Side triangle, 910
sign diagram
    algebraic function, 399
    for quadratic inequality, 214
    polynomial function, 242
    rational function, 321
simple interest, 469
sine
    graph of, 792
    of an angle, 717, 730, 744
    properties of, 791
sinusoid
    amplitude, 794, 881
    baseline, 881
    frequency
        angular, 881
        ordinary, 881
    graph of, 795, 882
    period, 881
    phase, 881
    phase shift, 795, 881
    properties of, 881
    vertical shift, 881
slant asymptote, 311
slant asymptote
    determination of, 312
    formal definition of, 311
slope
    definition, 151

of a line, 151
    rate of change, 154
    slope-intercept form of a line, 155
    smooth, 241
sound intensity level
    decibel, 431
square matrix, 586
standard position of a vector, 1019
standard position of an angle, 698
start-up cost, 82
steady state, 592
stochastic process, 592
straight angle, 693
Sum Identity
    for cosine, 771, 775
    for sine, 773, 775
    for tangent, 775
Sum to Product Formulas, 781
summation notation
    definition of, 661
    index of summation, 661
    lower limit of summation, 661
    properties of, 664
    upper limit of summation, 661
supplementary angles, 696
symmetry
    about the x-axis, 9
    about the y-axis, 9
    about the origin, 9
    testing a function graph for, 95
    testing an equation for, 26
synthetic division tableau, 260
system of equations
    back-substitution, 560
    coefficient matrix, 590
    consistent, 553
    constant matrix, 590
    definition, 549
    dependent, 554
    free variable, 552
    Gauss-Jordan Elimination, 571
    Gaussian Elimination, 557

inconsistent, 553
independent, 554
leading variable, 556
linear
   n variables, 554
   two variables, 550
linear in form, 646
non-linear, 637
overdetermined, 554
parametric solution, 552
triangular form, 556
underdetermined, 554
unknowns matrix, 590

tangent
   graph of, 804
   of an angle, 744, 752
   properties of, 806
terminal side of an angle, 698
Thurstone, Louis Leon, 315
total squared error, 225
transformation
   non-rigid, 129
   rigid, 129
transformations of function graphs, 120, 135
transverse axis of a hyperbola, 531
Triangle Inequality, 183
triangular form, 556

underdetermined system, 554
uninhibited growth, 472
union of two sets, 4
Unit Circle
   definition of, 501
   important points, 724
unit vector, 1023
Upper and Lower Bounds Theorem, 274
upper triangular matrix, 593

variable
   dependent, 55
   independent, 55
variable cost, 159

variation
   constant of proportionality, 350
   direct, 350
   inverse, 350
   joint, 350
variations in sign, 273
vector
   x-component, 1012
   y-component, 1012
   addition
      associative property, 1015
      commutative property, 1015
      definition of, 1014
      properties of, 1015
   additive identity, 1015
   additive inverse, 1015, 1018
   angle between two, 1035, 1036
   component form, 1012
   Decomposition Theorem
      Generalized, 1040
      Principal, 1024
   definition of, 1012
   direction
      definition of, 1020
      properties of, 1020
   dot product
      commutative property of, 1034
      definition of, 1034
      distributive property of, 1034
      geometric interpretation, 1035
      properties of, 1034
      relation to magnitude, 1034
      relation to orthogonality, 1037
      work, 1042
   head, 1012
   initial point, 1012
   magnitude
      definition of, 1020
      properties of, 1020
      relation to dot product, 1034
   normalization, 1024
   orthogonal projection, 1038

orthogonal vectors, 1037
parallel, 1030
principal unit vectors, $\hat{\imath}$, $\hat{\jmath}$, 1024
resultant, 1013
scalar multiplication
  associative property of, 1018
  definition of, 1017
  distributive properties, 1018
  identity for, 1018
  properties of, 1018
  zero product property, 1018
scalar product
  definition of, 1034
  properties of, 1034
scalar projection, 1039
standard position, 1019
tail, 1012
terminal point, 1012
triangle inequality, 1044
unit vector, 1023

velocity
  average angular, 707
  instantaneous, 707
  instantaneous angular, 707

vertex
  of a hyperbola, 531
  of a parabola, 188, 505
  of an angle, 693
  of an ellipse, 516

vertical asymptote
  formal definition of, 304
  intuitive definition of, 304
  location of, 306

vertical line, 23
Vertical Line Test (VLT), 43

whole number
  definition of, 2
  set of, 2
work, 1041
wrapping function, 704

zero

multiplicity of, 244
of a function, 95
upper and lower bounds, 274

Printed in Poland
by Amazon Fulfillment
Poland Sp. z o.o., Wrocław